Integration on locally compact spaces

Integration on locally compact spaces

N. DINCULEANU
Professor of Mathematics
University of Bucarest
Romania

NOORDHOFF INTERNATIONAL PUBLISHING
LEYDEN

A division of A. W. Sijthoff
International Publishing Company B.V.,
Leyden, The Netherlands

ISBN: 90 286 0453 7

Library of Congress Catalog Card Number: 70-119881

Printed in The Netherlands

CONTENTS

PART II

PREFACE

This book is a continuation of my book 'Vector measures' in which general measure and integration theory in abstract spaces is studied.

The present book treats an important special case in the theory of integration, namely, integration in locally compact spaces. The importance of this case is justified by the following reasons:

1. Locally compact spaces enjoy the property that measures defined on such spaces possess properties similar to the classical model, namely, Lebesgue measure on the real line. One of the features of locally compact spaces is the existence of a rich class of subsets having remarkable properties, namely the compact sets. Another important feature is the existence of a locally countable family of disjoint compact sets, whose union is almost equal to the whole space. This enables one to establish classical theorems, such as the Lebesgue–Nikodym theorem and the Lebesgue-Fubini theorem without further restrictions on the measure.

2. Locally compact spaces are, at the same time, general enough, so that an integration theory developed on such spaces can be applied successfully to many areas in analysis, for example harmonic analysis on locally compact groups and potential theory.

3. We mention, finally, that integration on abstract spaces can always be reduced to integration on locally compact spaces.

Usually, starting from a measure defined as a set function, we construct an integral, which is a linear functional on the space of integrable functions. The classical Riesz theorem, extended by Kakutani to compact spaces, shows that, conversely, every continuous linear functional on the space of continuous functions on a compact space, is, in fact, an integral with respect to some regular measure. This theorem suggests the identification of regular measures with continuous linear functionals. Using this identification, N. Bourbarki has constructed a theory of integration on locally compact spaces, in which a measure is defined as a functional rather than a set function.

In this book we have followed closely the Bourbaki point of view, which has been adapted to the case of vector measures defined as linear operations on the space $\mathscr{K}_E(T)$ of continuous vector functions with compact support taking their values in a Banach space E.

Dominated measures can be regarded as being defined on the space $\mathcal{K}(T)$ of real functions, since if the measure takes on values in a Banach space X and if $X \subset \mathcal{L}(E, F)$ then we can 'extend' the measure to the space $\mathcal{K}_E(T)$.

The book is divided into three parts. In the first part we shall study positive and vector measures, the upper integral, $\mathcal{L}^p$ spaces, integrable functions, measurable functions and the lifting property of $\mathcal{L}^\infty$.

The second part deals specifically with vector measures. The subject matter is the following. Measures defined by densities, absolutely continuous measures and Lebesgue–Nikodym type theorems with applications to integral representation of linear operations on $\mathcal{L}^p$ spaces; sums of measures, images of measures, induced measures and measures on product spaces. Using the existence of a lifting of $\mathcal{L}^\infty$, Lebesgue–Nikodym theorem and integral representations of linear operations on $\mathcal{L}^p$ spaces are proved, without requiring separability conditions on the Banach spaces.

In the third part we study measures on locally compact groups; Haar measure, convolution (for vector functions), group algebras (of vector functions and measures), representations of groups and group algebras, and harmonic analysis on abelian groups.

Lastly we treat measures defined on spaces of vector fields, Lebesgue and Orlicz spaces of vector fields and integral representation of linear operations on these spaces.

We remark that most results proved by Bourbaki for the upper integral remain valid for the essential upper integral. On the other hand, some important results, for example those pertaining to measures defined by densities, images of measures and induced measures, are valid, in general, only for the essential upper integral. For a unified treatment of the above results we have made some changes in presentation. In this book, the terms *upper integral, negligible function, integrable function*, correspond respectively to Bourbaki's *essential upper integral, locally negligible function, essentially integrable function.* With this alteration, all L^p-spaces can be defined as quotient spaces $\mathcal{L}^p/\mathcal{N}$ for $1 \leqslant p < \infty$, as well as for $p = \infty$, where $\mathcal{N}$ is the set of negligible functions. Also the Hölder inequality $\int^* fg\mathrm{d}\mu \leqslant N_p(f)N_q(g)$ is valid for positive functions f and g even when $p=1$ and $q=\infty$.

The Lebesgue–Fubini theorem is not valid for Bourbaki's essentially integrable functions. Therefore, with the above mentioned alterations, the Lebesgue–Fubini theorem has to be stated with the additional hypothesis that the function vanish outside a sequence of integrable open sets. Bourbaki's treatment of product measures also include similar restrictions.

We mention, finally, that most of the results stated in this book for Banach-valued measures can be extended to measures taking their values in locally convex spaces, by using the defining family of seminorms.

This book is a translation in English of a Romanian version published in 1965.

N. Dinculeanu

PART I

Chapter I

MEASURES ON LOCALLY COMPACT SPACES

§1 Definition of measure

1 *The space of continuous functions with compact support*

Throughout this book we shall denote by T a locally compact space. Let E be a Banach space. The norm of an element $x \in E$ will be denoted by $|x|$. We shall denote by $\mathscr{K}_E(T)$ the vector space of all continuous functions $f: T \to E$ with compact support. The support of the function f is the closure of the set $\{t \mid t \in T, f(t) \neq 0\}$.

For each subset $A \subset T$, $\mathscr{K}_E(T, A)$ is the subspace of $\mathscr{K}_E(T)$ of functions with the support contained in A.

$\mathscr{C}_E(T)$ or $\mathscr{C}_E$ is the space of *continuous* functions $f: T \to E$.

In case T is compact, we have $\mathscr{K}_E(T) = \mathscr{C}_E(T)$. If $E = R$, we shall write $\mathscr{K}(T)$, $\mathscr{K}(T, A)$, $\mathscr{C}(T)$ instead of $\mathscr{K}_R(T)$, $\mathscr{K}_R(T, A)$, $\mathscr{C}_R(T)$ respectively. $\mathscr{K}_+(T)$ or $\mathscr{K}_+$ is the set of continuous *positive* functions with compact support defined on T.

The space $\mathscr{K}(T)$ is ordered by the relation $f \leqslant g$:

$$f \leqslant g \text{ means } f(t) \leqslant g(t) \text{ for every } t \in T .$$

If $f \in \mathscr{K}_E(T)$, then $|f| \in \mathscr{K}_+(T)$, where $|f|$ denotes the function $t \to |f(t)|$.

For every function $f \in \mathscr{K}_E(T)$ we write

$$\|f\| = \sup_{t \in T} |f(t)| .$$

The mapping $f \to \|f\|$ is a norm on each space $\mathscr{K}_E(T, A)$ and defines on this space the topology of uniform convergence on T.

We can also consider on the space $\mathscr{K}_E(T)$ the topology of the compact convergence. This is the locally convex topology defined by the family of seminorms

$$\|f\|_K = \sup_{t\in K}|f(t)|$$

where K runs over the set of the compact subsets of T.

Still another topology which can be considered on the space $\mathscr{K}_E(T)$ is the *inductive limit of the topologies of uniform convergence on each of the subspaces* $\mathscr{K}_E(T, K)$, when K runs over the set of compact subsets of T.

If G is an open set and if $f: G \to E$ is a continuous function with compact support, then f can be extended to a continuous function on the whole space T, with compact support contained in G, by assigning to f the value 0 outside G.

In this way the space $\mathscr{K}_E(G)$ can be identified with the subspace $\mathscr{K}_E(T, G)$ of $\mathscr{K}_E(T)$ and we can consider $\mathscr{K}_E(G)$ itself as a subspace of $\mathscr{K}_E(T)$.

We remind the reader of the following two propositions:

1.1 PROPOSITION. *For every compact set $K \subset T$ and for every open set $G \supset K$, there exists a continuous function φ on T, with values in* $[0, 1]$, *such that $\varphi(t)=1$ for $t \in K$ and $\varphi(t)=0$ for $t \notin G$.*

1.2 PROPOSITION. *Let $K \subset T$ be a compact set. Every continuous function f defined on K, with values in a Banach space E, can be extended to a continuous function with compact support f_1 defined on T with values in E.*

Other properties of continuous functions with compact support, which will be used further, are given in the following propositions.

1.3 PROPOSITION. *For every compact set $K \subset T$ and every finite cover $(G_i)_{1\leqslant i\leqslant n}$ of K, consisting of open sets, there exists a finite family $(f)_{1\leqslant i\leqslant n}$ of continuous mappings of T into* $[0, 1]$, *such that*

$$\Sigma_{1\leqslant i\leqslant n} f_i(t) = 1 \text{ for } t \in K\,,$$

$$\Sigma_{1\leqslant i\leqslant n} f_i(t) \leqslant 1 \text{ for every } t \in T$$

and such that for every i, the support of f_i is contained in G_i.

Let us remark first that there exists a finite family $(V_i)_{1\leqslant i\leqslant n}$ of relatively compact open sets covering K, such that $\overline{V}_i \subset G_i$ for $i=1, 2, \ldots, n$.

In fact since each point of T has a fundamental system of compact neighbourhoods, for every point $x \in K$ there exists a relatively compact open neighbourhood U_x of x such that $\overline{U}_x$ is contained in every open set G_i which contains x. The open neighbourhoods U_x form a cover of K; it is then possible to extract a finite cover $(U_{x_j})_{1\leqslant j\leqslant m}$ of K. For each i, $1\leqslant i\leqslant n$, we take the union V_i of those sets U_{x_j} which are contained in G_i. Since $\bar{V}_i$ is compact, there exists a continuous function $g_i : T\to[0, 1]$ equal to 1 on $\bar{V}_i$ and to zero on $T\setminus G_i$. Therefore we have $\Sigma_{1\leqslant i\leqslant n}g_i(t)>0$, for $t\in\bar{V}$, where $V=\cup_{1\leqslant i\leqslant n}V_i$. The restriction of the function $[\Sigma_{1\leqslant i\leqslant n}g_i(t)]^{-1}$ to the compact set $\bar{V}$ is continuous, therefore it can be extended to a continuous real function h on T. We have

$$h(t)=[\Sigma_{1\leqslant i\leqslant n}g_i(t)]^{-1} \text{ for } t\in\bar{V} .$$

There exists also a continuous function $\varphi : T\to[0, 1]$, equal to 1 on K and to zero on $T\setminus V$. For each i, $1\leqslant i\leqslant n$, put

$$f_i(t)=g_i(t)\varphi(t)h(t) .$$

The function f_i is defined on T, has its values in $[0, 1]$, is continuous on T and has its support contained in $\bar{V}_i\subset G_i$. For $t\in K$ we have $\varphi(t)=1$; therefore

$$\Sigma_{1\leqslant i\leqslant n}f_i(t)=\Sigma_{1\leqslant i\leqslant n}g_i(t)h(t)=1 .$$

For $t\in V$ we have $\varphi(t)\leqslant 1$; therefore

$$\Sigma_{1\leqslant i\leqslant n}f_i(t)\leqslant\Sigma_{1\leqslant i\leqslant n}g_i(t)h(t)=1 .$$

For $t\notin V$ we have $\varphi(t)=0$; therefore, $\Sigma_{1\leqslant i\leqslant n}f_i(t)=0\leqslant 1$.

The functions $(f_i)_{1\leqslant i\leqslant n}$ fulfill the required conditions and the proposition is proved.

The family of functions $(f_i)_{1\leqslant i\leqslant n}$ is called a *continuous partition of unity on K, subordinated to the open cover* $(G_i)_{1\leqslant i\leqslant n}$ of K.

1.4 Proposition (*Dini*). *Let* $H\subset\mathscr{K}_+$ *be a set directed for the relation* $\leqslant$. *If* $\sup_{g\in H}g\in\mathscr{K}_+$, *then the filter of sections of H converges uniformly to the upper envelope* $\sup_{g\in H}g$, *that is*

$$\lim_{g\in H}g=\sup_{g\in H}g .$$

Let us write $f=\sup_{g\in H}g$. Let K be the support of f. Each function $g\in H$ has the support contained in K because $0\leqslant g\leqslant f$.

Let $\varepsilon>0$ be given; for each point $s\in K$, there exists a function g_s such that $g_s(s)>f(s)-\varepsilon$. Since the functions g_s and f are continuous, there exists an open neighbourhood V_s of s such that

$$g_s(t) > f(t) - \varepsilon \text{ for } t \in V_s .$$

K being a compact set, there exists a finite number of open sets $V_1, \ldots, V_n$ which cover K and a finite number of functions $g_1, \ldots, g_n$ from H such that for each i we have

$$g_i(t) > f(t) - \varepsilon \text{ for } t \in V_i .$$

But H is directed; therefore there exists a function $g \in H$ such that $g \geqslant g_i$ for $i = 1, 2, \ldots, n$. Then

$$g(t) > f(t) - \varepsilon \text{ for every } t \in T ,$$

hence $g \geqslant f - \varepsilon$. As $g \leqslant f$, it follows that f is the uniform limit of the directed set H, $f = \lim_{g \in H} g$ and the proposition is proved.

1.5 PROPOSITION. *Let $K \subset T$ be a compact set, $\varphi : T \to [0, 1]$ a continuous function with compact support, equal to 1 on K, and $\boldsymbol{f}$ a function of $\mathscr{K}_E(T, K)$. For every number $\varepsilon > 0$, there exists a finite family $(\varphi_i)_{1 \leqslant i \leqslant n}$ of continuous mappings of T into $[0, 1]$, with support contained in the support of φ, and a finite family $(a_i)_{1 \leqslant i \leqslant n}$ of elements of E, such that we have*

$$|\boldsymbol{f}(t) - \Sigma_{1 \leqslant i \leqslant n} a_i \varphi_i(t)| \leqslant \varepsilon \varphi(t) \text{ for every } t \in T .$$

The set $U = \{t \in T \mid \varphi(t) > 0\}$ is open, relatively compact and $K \subset U$.

Let $t \in \overline{U}$ be given. As $\boldsymbol{f}$ is continuous, there exists an open neighbourhood V_t of t such that $|\boldsymbol{f}(t) - \boldsymbol{f}(t')| < \varepsilon/2$ for every $t' \in V_t$. Then for $t', t'' \in V_t$ we have $|\boldsymbol{f}(t') - \boldsymbol{f}(t'')| < \varepsilon$. Since $\overline{U}$ is compact, there exists a finite family $(V_i)_{1 \leqslant i \leqslant n}$ of open sets covering K, such that for each i we have

$$|\boldsymbol{f}(t') - \boldsymbol{f}(t'')| < \varepsilon \text{ for every } t', t'' \in V_i.$$

Let $(h_i)_{1 \leqslant i \leqslant n}$ be a continuous partition of unity on $\overline{U}$, subordinated to the covering $(V_i)_{1 \leqslant i \leqslant n}$ of $\overline{U}$. We have

$$\Sigma_{1 \leqslant i \leqslant n} h_i(t) = 1 \text{ for } t \in \overline{U} .$$

Put $\varphi_i = h_i \varphi$, $1 \leqslant i \leqslant n$; the functions φ_i are continuous mappings of T into $[0, 1]$, with support contained in $\overline{U} \cap V_i$ and $\Sigma_{1 \leqslant i \leqslant n} \varphi_i = \varphi$.

In each set V_i choose a point t_i and take $a_i = \boldsymbol{f}(t_i) \in E$. If $t \in V_i$ we have

$$|\boldsymbol{f}(t) \varphi_i(t) - a_i \varphi_i(t)| = \varphi_i(t) |\boldsymbol{f}(t) - \boldsymbol{f}(t_i)| \leqslant \varepsilon \varphi_i(t) .$$

The inequality

$$|f(t)\varphi_i(t) - a_i\varphi_i(t)| \leqslant \varepsilon\varphi_i(t)$$

is valid also for $t \notin V_i$, because in this case $\varphi_i(t) = 0$. Taking into account the fact that $f = f\Sigma_{1 \leqslant i \leqslant n}\varphi_i$, we deduce

$$|f(t) - \Sigma_{1 \leqslant i \leqslant n} a_i\varphi_i(t)| = |\Sigma_{1 \leqslant i \leqslant n}\varphi_i(t)f(t) - \Sigma_{1 \leqslant i \leqslant n} a_i\varphi_i(t)| \leqslant$$
$$\leqslant \Sigma_{1 \leqslant i \leqslant n}|\varphi_i(t)f(t) - a_i\varphi_i(t)| \leqslant$$
$$\leqslant \Sigma_{1 \leqslant i \leqslant n}\varphi_i(t)\varepsilon = \varepsilon\varphi(t)$$

for every $t \in T$. The proposition is proved.

1.6 COROLLARY. *For every compact set $K \subset T$ and every compact neighbourhood U of K, the space $\mathscr{H}_E(T, U)$ of linear combinations $\Sigma_{1 \leqslant i \leqslant n} a_i\varphi_i$, with $a_i \in E$ and $\varphi_i \in \mathscr{K}_+(T, U)$, is dense in the space $\mathscr{K}_E(T, K)$ for the topology of the uniform convergence.*

1.7 DEFINITION. *We say that a subspace $\mathscr{H}$ of the space $\mathscr{K}_E(T)$ is rich, if for every compact set $K \subset T$ there exists a relatively compact neighbourhood U of K, such that every function f of $\mathscr{K}_E(T, K)$ can be approximated uniformly with functions of $\mathscr{H}$, with support contained in U.*

If T is compact, to say that $\mathscr{H}$ is rich, means that it is dense in $\mathscr{C}_E(T)$. If T is not compact, then every rich space is dense in $\mathscr{K}_E(T)$, but the converse is not true.

Using this definition, corollary 1.6 can be stated in the following way:

The space $\mathscr{H}_E(T)$ of the linear combinations $\Sigma_{1 \leqslant i \leqslant n} a_i\varphi_i$ with $a_i \in E$ and $\varphi_i \in \mathscr{K}_+(T)$ is rich.

1.8 PROPOSITION. *The space $\mathscr{K}_E(T)$ is dense in the space $\mathscr{C}_E(T)$ for the topology of the compact convergence.*

In fact, let $f \in \mathscr{C}_E(T)$. For every compact set $K \subset T$ there exists a function $\varphi \in \mathscr{K}_+(T)$, such that $\varphi(t) = 1$ for $t \in K$; then $g = f\varphi \in \mathscr{K}_E(T)$ and $f(t) = g(t)$ for $t \in K$. Therefore the function f can be approximated uniformly, on each compact set, by functions g of $\mathscr{K}_E(T)$. It follows that the space $\mathscr{K}_E(T)$ is dense in the space $\mathscr{C}_E(T)$ for the topology of the uniform convergence on compact sets.

2 *Definition of measure*

Let E and F be two Banach spaces (both of them being either real or complex Banach spaces). In particular, one or both of E and F can be equal to R or C, and C can be considered as a real or complex Banach space.

1.9 DEFINITION. *We shall call a vector measure on T with respect to the spaces E and F, or an (E, F)-measure on T, any linear mapping $\boldsymbol{m} : \mathscr{K}_E(T) \to F$ having the property that for each compact set $K \subset T$, the restriction of $\boldsymbol{m}$ to the subspace $\mathscr{K}_E(T, K)$ is continuous for the topology of uniform convergence.*

The value $\boldsymbol{m}(\boldsymbol{f})$ of the measure $\boldsymbol{m}$ for a function $\boldsymbol{f} \in \mathscr{K}_E(T)$ is called *the integral of $\boldsymbol{f}$ with respect to $\boldsymbol{m}$* and is also denoted by $\int \boldsymbol{f} \mathrm{d}\boldsymbol{m}$ or $\int \boldsymbol{f}(t) \mathrm{d}\boldsymbol{m}(t)$.

To say that the operation $\boldsymbol{m} : \mathscr{K}_E(T) \to F$ is a measure on T, means that $\boldsymbol{m}$ satisfies the following conditions:

(1) $\boldsymbol{m}(\boldsymbol{f}+\boldsymbol{g}) = \boldsymbol{m}(\boldsymbol{f}) + \boldsymbol{m}(\boldsymbol{g})$ for $\boldsymbol{f}, \boldsymbol{g} \in \mathscr{K}_E(T)$

(2) $\boldsymbol{m}(\alpha \boldsymbol{f}) = \alpha \boldsymbol{m}(\boldsymbol{f})$ for $\boldsymbol{f} \in \mathscr{K}_E(T)$ and α scalar

(3) for each compact set $K \subset T$ there exists a number $a_K > 0$ such that we have

$$|\boldsymbol{m}(\boldsymbol{f})| \leqslant a_K \|\boldsymbol{f}\| \text{ for every function } \boldsymbol{f} \in \mathscr{K}_E(T, K).$$

If no confusion is possible concerning the spaces E and F we shall say, simply, measure, instead of (E, F)-measure.

One can verify immediately that the set of (E, F)-measures on T is a vector space with respect to real or complex scalars, (according to whether E and F are considered real or complex vector spaces), for the usual operations $\boldsymbol{m}+\boldsymbol{n}$ and $\alpha\boldsymbol{m}$.

If T is *compact*, to say that $\boldsymbol{m}$ is a measure, means that the linear mapping $\boldsymbol{m} : \mathscr{C}_E(T) \to F$ is continuous for the topology of the uniform convergence. The space of (E, F)-measures on T coincides in this case with the space $\mathscr{L}(\mathscr{C}_E, F)$ of the linear continuous mappings of $\mathscr{C}_E$ into F. In particular, if $F=C$ (or $F=R$), the space of (E, C)-measures on T (respectively the space of (E, R)-measures on T) is the dual of the space $\mathscr{C}_E(T)$.

If T is not compact, a measure $\boldsymbol{m}$ is not continuous, in general, on the whole $\mathscr{K}_E(T)$, for the topology of the uniform convergence.

Remark. If we consider on the space $\mathscr{K}_E(T)$ the *inductive limit* of the topologies of uniform convergence, on each subspace $\mathscr{K}_E(T, K)$ when K runs over the set of the compact subsets of T, then to say that $\boldsymbol{m}$ is an (E, F)-measure means that the linear mappings $\boldsymbol{m} : \mathscr{K}_E(T) \to F$ is continuous for this topology.

For each (E, F)-measure $\boldsymbol{m}$ on T we denote

$$\|\boldsymbol{m}\| = \sup_{\|\boldsymbol{f}\| \leqslant 1} |\boldsymbol{m}(\boldsymbol{f})|, \quad \boldsymbol{f} \in \mathscr{K}_E(T).$$

The following properties are immediate:

(1) $0 \leqslant \|\boldsymbol{m}\| \leqslant +\infty$; $\|\boldsymbol{m}\| = 0$ *if and only if* $\boldsymbol{m} = 0$
(2) $\|\alpha \boldsymbol{m}\| = |\alpha| \, \|\boldsymbol{m}\|$, α *scalar*
(3) $\|\boldsymbol{m} + \boldsymbol{n}\| \leqslant \|\boldsymbol{m}\| + \|\boldsymbol{n}\|$.

(We adopt the convention $0 \cdot \infty = 0$. According to this convention, the property (2) is true also if $\alpha = 0$ and $\|\boldsymbol{m}\| = +\infty$).

If T is compact, to say that $\boldsymbol{m}$ is a measure, means $\|\boldsymbol{m}\| < +\infty$. The mapping $\boldsymbol{m} \to \|\boldsymbol{m}\|$ is, in this case, a norm on the space of (E, F)-measures on T.

Examples

(1) For every point $t \in T$, the mapping $f \to f(t)$ of $\mathscr{K}(T)$ into R is a measure on T. This measure is denoted by ε_t and is called the measure defined by the unit mass placed in the point t.

Then we have $\varepsilon_t(f) = f(t)$ for $f \in \mathscr{K}(T)$.

If E is a Banach space, the mapping $\boldsymbol{f} \to \boldsymbol{f}(t)$ of $\mathscr{K}_E(T)$ into E is also a measure and it is still denoted by ε_t.

(2) Let $N \subset T$ be a set with the property that $N \cap K$ is a *finite* set, for every compact set $K \subset T$, and let $\alpha : T \to R$ be a function such that $N = \{t \in T \mid \alpha(t) \neq 0\}$. For every function $f \in \mathscr{K}(T)$ put

$$\mu(f) = \Sigma_{t \in T} \, \alpha(t) f(t).$$

The sum in the right-hand side is well defined because the number of the non-zero terms is finite. It is easy to prove that the mapping $f \to \mu(f)$ is linear. For each compact set $K \subset T$ put

$$a_K = \Sigma_{t \in K} \, \alpha(t).$$

Then we have

$$|\mu(f)| \leqslant a_K \|f\|, \text{ for } f \in \mathscr{K}(T, K)$$

therefore μ is a measure on T.

We say that μ is a *discrete* measure *defined by the masses* $\alpha(t)$ *placed in the points* $t \in T$.

If T is a discrete space, every measure $\mu : \mathscr{K}(T) \to R$ is discrete.

In fact, for each $t \in T$ let us denote by e_t the function defined on T by $e_t(t)=1$ and $e_t(s)=0$ if $s \neq t$. Then for every function $f \in \mathscr{K}(T)$ we have

$$f = \Sigma_{t \in T} f(t)\, e_t\, .$$

If $\mu : \mathscr{K}(T) \to R$ is a measure, and if we put $\alpha(t) = \mu(e_t)$, then we have

$$\mu(f) = \Sigma_{t \in T} f(t)\, \mu(e_t) = \Sigma_{t \in T} \alpha(t) f(t) \quad \text{for } f \in \mathscr{K}(T)\, .$$

Therefore μ is a discrete measure.

(3) More generally, let $\alpha : T \to R$ be a function such that for every compact set $K \subset T$ we have $a_K = \Sigma_{t \in K} |\alpha(t)| < +\infty$. It follows that the set $\{t \in K \mid \alpha(t) \neq 0\}$ is at most countable. If for each function $f \in \mathscr{K}(T)$ we denote

$$\mu(f) = \Sigma_{t \in T} \alpha(t) f(t)\, ,$$

then μ is a measure on T.

(4) Let $\mu : \mathscr{K}(T) \to R$ be a measure and $g : T \to R$ a continuous function. For every function $f \in \mathscr{K}(T)$, the function fg is continuous and with compact support. The mapping $f \to \mu(fg)$ of $\mathscr{K}(T)$ into R is a measure. In fact, it is linear; if for each compact set $K \subset T$, a_K satisfies the inequality

$$|\mu(f)| \leqslant a_K \, \|f\| \quad \text{for } f \in \mathscr{K}(T, K)$$

then, denoting $b_K = a_K \|g\|$, we have

$$|\mu(fg)| \leqslant a_K \, \|f\| \, \|g\| = b_K \|f\| \quad \text{for } f \in \mathscr{K}(T, K)$$

therefore, the mapping $f \to \mu(fg)$ is a measure. This measure is denoted by $g\mu$ and is called the product of the measure μ by the function g. Then we have

$$(g\mu)(f) = \mu(fg) \quad \text{for } f \in \mathscr{K}(T)\, .$$

We shall study later the product of a measure by functions which are not necessarily continuous.

1.10 Proposition. *Let H, E, F, G be Banach spaces. If $u : H \to E$ and $v : F \to G$ are continuous linear mappings and $\boldsymbol{m} : \mathscr{K}_E(T) \to F$ is a measure, then the mapping $\boldsymbol{n} : \mathscr{K}_H(T) \to G$ defined by the equality*

$$\boldsymbol{n}(\boldsymbol{f}) = v \circ \boldsymbol{m}\,(u \circ \boldsymbol{f}) \; \textit{for } \boldsymbol{f} \in \mathscr{K}_H(T)$$

is a measure and $\|\boldsymbol{n}\| \leqslant |v| \, \|\boldsymbol{m}\| \, |u|$.

The linearity of the mapping $\boldsymbol{n}$ follows from the linearity of the mappings u, $\boldsymbol{m}$ and v.

Let $K \subset T$ be a compact set and let $a_K > 0$ be such that

$$|\boldsymbol{m}(\boldsymbol{g})| \leqslant a_K \|\boldsymbol{g}\| \text{ for every } \boldsymbol{g} \in \mathscr{K}_E(T, K).$$

For every function $\boldsymbol{f} \in \mathscr{K}_H(T, K)$, we have $u \circ \boldsymbol{f} \in \mathscr{K}_E(T, K)$, and

$$\|u \circ \boldsymbol{f}\| = \sup_{t \in T} |u(\boldsymbol{f}(t))| \leqslant |u| \sup_{t \in T} |\boldsymbol{f}(t)| = |u| \, \|\boldsymbol{f}\|$$

therefore

$$|v \circ \boldsymbol{m}(u \circ \boldsymbol{f})| \leqslant |v| \, |\boldsymbol{m}(u \circ \boldsymbol{f})| \leqslant |v| \, a_K \|u \circ \boldsymbol{f}\| \leqslant |v| \, |u| \, a_K \|\boldsymbol{f}\|$$

hence writing $b_K = |v| \, |u| \, a_K$,

$$|\boldsymbol{n}(\boldsymbol{f})| \leqslant b_K \|\boldsymbol{f}\| \text{ for } \boldsymbol{f} \in \mathscr{K}_H(T, K).$$

This means that $\boldsymbol{n}$ is a measure. Then we have

$$|\boldsymbol{n}(\boldsymbol{f})| = |v \circ \boldsymbol{m}(u \circ \boldsymbol{f})| \leqslant |v| \, |\boldsymbol{m}(u \circ \boldsymbol{f})| \leqslant |v| \, \|\boldsymbol{m}\| \, \|u \circ \boldsymbol{f}\| \leqslant |v| \, \|\boldsymbol{m}\| \, |u| \, \|\boldsymbol{f}\|,$$

Consequently, $\|\boldsymbol{n}\| \leqslant |v| \, \|\boldsymbol{m}\| \, |u|$. Thus the proposition is proved.

1.11 COROLLARY. *Let H, E, F, be Banach spaces. If $u : H \to E$ is a continuous linear mapping and $\boldsymbol{m} : \mathscr{K}_E(T) \to F$ is a measure, then the mapping $\boldsymbol{n} : \mathscr{K}_H(T) \to F$ defined by the equality*

$$\boldsymbol{n}(\boldsymbol{f}) = \boldsymbol{m}(u \circ \boldsymbol{f}) \textit{ for } \boldsymbol{f} \in \mathscr{K}_H(T)$$

is a measure and $\|\boldsymbol{n}\| \leqslant |u| \, \|\boldsymbol{m}\| \leqslant +\infty$.

In fact, in proposition 1.10 we take $G = F$ and $v(y) = y$ for $y \in F$.

1.12 COROLLARY. *If $\boldsymbol{m} : \mathscr{K}_E(T) \to F$ is a measure and E_1 is a Banach subspace of E, then the restriction $\boldsymbol{m}_1 : \mathscr{K}_{E_1}(T) \to F$ of $\boldsymbol{m}$ to the subspace $\mathscr{K}_{E_1}(T)$ is a measure and $\|\boldsymbol{m}_1\| \leqslant \|\boldsymbol{m}\| \leqslant +\infty$.*

In the previous corollary we take $H = E_1$, and u the canonical injection of E_1 into E, $u(x) = x$ for $x \in E_1$, and we remark that

$$\boldsymbol{m}_1(\boldsymbol{f}) = \boldsymbol{m}(u \circ \boldsymbol{f}) \text{ for } \boldsymbol{f} \in \mathscr{K}_{E_1}(T)$$

and $|u| = 1$.

1.13 COROLLARY. *If $\boldsymbol{m} : \mathscr{K}_E(T) \to F$ is a measure, then, for any $x \in E$, the mapping $\boldsymbol{m}_x : \mathscr{K}(T) \to F$ defined by the equality*

$$\boldsymbol{m}_x(\varphi) = \boldsymbol{m}(\varphi x) \text{ for } \varphi \in \mathscr{K}(T)$$

is a measure and $\|\boldsymbol{m}_x\| \leqslant \|\boldsymbol{m}\| \, |x|$. Furthermore, we have

$$\boldsymbol{m}_{x+y} = \boldsymbol{m}_x + \boldsymbol{m}_y \text{ and } \boldsymbol{m}_{\alpha x} = \alpha \boldsymbol{m}_x, \text{ for } x, y \in E \text{ and } \alpha \text{ scalar}.$$

Indeed, the mapping $u(\alpha) = \alpha x$ of R into E is linear and continuous, $|u| = |x|$, and $\boldsymbol{m}_x(\varphi) = \boldsymbol{m}(u \circ \varphi)$. Therefore we can apply corollary 1.11. The equalities $\boldsymbol{m}_{x+y} = \boldsymbol{m}_x + \boldsymbol{m}_y$ and $\boldsymbol{m}_{\alpha x} = \alpha \boldsymbol{m}_x$ can be verified without difficulty.

1.14 COROLLARY. *Let E, F, G be Banach spaces. If $v : F \to G$ is a continuous linear mapping and $\boldsymbol{m} : \mathscr{K}_E(T) \to F$ is a measure, then the mapping $\boldsymbol{n} : \mathscr{K}_E(T) \to G$ defined by the equality*

$$\boldsymbol{n}(\boldsymbol{f}) = v \circ \boldsymbol{m}(\boldsymbol{f}) \text{ for } \boldsymbol{f} \in \mathscr{K}_E(T)$$

is a measure and $\|\boldsymbol{n}\| \leqslant |v| \, \|\boldsymbol{m}\|$.

In fact, in proposition 1.10 we take $H = E$ and $u(x) = x$ for $x \in H$.

1.15 COROLLARY. *If $\boldsymbol{m} : \mathscr{K}_E(T) \to F$ is a measure, then for every $z' \in F'$, the mapping $z' \circ \boldsymbol{m} : \mathscr{K}_E(T) \to C$ defined by the equality*

$$(z' \circ \boldsymbol{m})(\boldsymbol{f}) = \langle \boldsymbol{m}(\boldsymbol{f}), z' \rangle \text{ for } \boldsymbol{f} \in \mathscr{K}_E(T)$$

is a measure. We have $\|z' \circ \boldsymbol{m}\| \leqslant |z'| \, \|\boldsymbol{m}\|$ and $\|\boldsymbol{m}\| = \sup_{|z'| \leqslant 1} \|z' \circ \boldsymbol{m}\|$.

In corollary 1.14 we take $G = C$ and $v = z'$. Then we have

$$\|\boldsymbol{m}\| = \sup_{\|\boldsymbol{f}\| \leqslant 1} |\boldsymbol{m}(\boldsymbol{f})| = \sup_{\|\boldsymbol{f}\| \leqslant 1} \sup_{|z'| \leqslant 1} |\langle \boldsymbol{m}(\boldsymbol{f}), z' \rangle| =$$

$$= \sup_{|z'| \leqslant 1} \sup_{\|\boldsymbol{f}\| \leqslant 1} |(z' \circ \boldsymbol{m})(\boldsymbol{f})| =$$

$$= \sup_{|z'| \leqslant 1} \|z' \circ \boldsymbol{m}\|.$$

1.16 COROLLARY. *If $\boldsymbol{m} : \mathscr{K}_E(T) \to F$ is a measure, then for every $x \in E$ and every $z' \in F'$ the mapping $z' \circ \boldsymbol{m}_x : \mathscr{K}(T) \to C$ defined by the equality*

$$(z' \circ \boldsymbol{m}_x)(\varphi) = \langle \boldsymbol{m}(\varphi x), z' \rangle \text{ for } \varphi \in \mathscr{K}(T)$$

is a measure. Furthermore, we have $\|z' \circ \boldsymbol{m}_x\| < |z'| \, \|\boldsymbol{m}\| \, |x|$ and $z' \circ \boldsymbol{m}_x = (z' \circ \boldsymbol{m})_x$.

In fact

$$(z' \circ \boldsymbol{m})_x(\varphi) = (z' \circ \boldsymbol{m})(\varphi x) = \langle \boldsymbol{m}(\varphi x), z' \rangle = \langle \boldsymbol{m}_x(\varphi), z' \rangle = (z' \circ \boldsymbol{m}_x)(\varphi)$$

whence $(z' \circ \boldsymbol{m})_x = z' \circ \boldsymbol{m}_x$. The other assertions follow from proposition 1.10 or from corollaries 1.13 and 1.15.

We shall say that a mapping $x \to \bar{x}$ from a vector space X into X is an *involution* if it satisfies the following conditions:

(1) $\overline{x+y} = \bar{x} + \bar{y}$

(2) $\overline{\alpha x} = \bar{\alpha}\bar{x}$, α scalar ($\bar{\alpha}$ is the complex-conjugate of α)

(3) $\bar{\bar{x}} = x$

If X is an algebra we require in addition

(4) $\overline{xy} = \bar{y}\bar{x}$.

If X is a normed space, we shall suppose, without specification that the involution is isometric:

(5) $|\bar{x}| = |x|$.

If X is a real vector space, property (2) is written $\overline{\alpha x} = \alpha\bar{x}$, therefore, in this case, the involution is a linear mapping.

Let E be a Banach space with involution. For every function $\boldsymbol{f}: T \to E$ we shall denote by $\bar{\boldsymbol{f}}: T \to E$ the function defined by the equality

$$\bar{\boldsymbol{f}}(t) = \overline{\boldsymbol{f}(t)} \text{ for } t \in T .$$

The mapping $\boldsymbol{f} \to \bar{\boldsymbol{f}}$ is an involution on the space of all functions from T into E. We have $|\bar{\boldsymbol{f}}| = |\boldsymbol{f}|$ and $\|\bar{\boldsymbol{f}}\| = \|\boldsymbol{f}\|$. If $\boldsymbol{f}$ is continuous then, $\bar{\boldsymbol{f}}$ is continuous, and $\boldsymbol{f}$ and $\bar{\boldsymbol{f}}$ have the same support.

1.17 Lemma. *For every continuous linear mapping $U : E \to F$, the mapping $\bar{U} : E \to F$ defined by the equality*

$$\bar{U}x = U\bar{x} \text{ for } x \in E ,$$

is linear, continuous and $|\bar{U}| = |U|$. The mapping $U \to \bar{U}$ is an involution on the space $\mathscr{L}(E, F)$.

One can verify immediately that $\bar{U}$ is linear. Then we have

$$|\bar{U}x| = |\overline{U\bar{x}}| = |U\bar{x}| \leqslant |U|\,|\bar{x}| = |U|\,|x| ,$$

therefore $|\bar{U}| \leqslant |U|$ and $\bar{U} \in \mathscr{L}(E, F)$.

Since $\bar{\bar{U}} = U$, we deduce that $|U| = |\bar{\bar{U}}| \leqslant |\bar{U}|$ and consequently $|\bar{U}| = |U|$.

The other properties of the involution $U \to \bar{U}$ can be verified without difficulty.

Remark. The equality $\overline{U}x = \overline{U\overline{x}}$ is equivalent to the equality

$$\overline{Ux} = \overline{U}\overline{x}$$

1.18 PROPOSITION. *Let E and F be two Banach spaces with involution. If $\boldsymbol{m} : \mathscr{K}_E(T) \to F$ is a measure, then the mapping $\overline{\boldsymbol{m}} : \mathscr{K}_E(T) \to F$ defined by the equality*

$$\overline{\boldsymbol{m}}(f) = \overline{\boldsymbol{m}(\bar{f})}\,, \text{ for } f \in \mathscr{K}_E(T)$$

is a measure and $\|\overline{\boldsymbol{m}}\| = \|\boldsymbol{m}\| \leqslant +\infty$. The mapping $\boldsymbol{m} \to \overline{\boldsymbol{m}}$ is an involution on the space of all measures $\boldsymbol{m} : \mathscr{K}_E(T) \to F$.

It is easy to verify that $\overline{\boldsymbol{m}}$ is linear. Let $K \subset T$ be a compact set and $a_K > 0$ be such that

$$|\boldsymbol{m}(\boldsymbol{f})| < a_K \|\boldsymbol{f}\|\,, \text{ for } \boldsymbol{f} \in \mathscr{K}_E(T, K)\,.$$

Then for every function $\boldsymbol{f} \in \mathscr{K}_E(T, K)$ we have

$$|\overline{\boldsymbol{m}}(\boldsymbol{f})| = |\overline{\boldsymbol{m}(\bar{\boldsymbol{f}})}| = |\boldsymbol{m}(\bar{\boldsymbol{f}})| < a_K \|\bar{\boldsymbol{f}}\| = a_K \|\boldsymbol{f}\|\,,$$

therefore $\overline{\boldsymbol{m}}$ is a measure.

The involution properties of the mapping $\boldsymbol{m} \to \overline{\boldsymbol{m}}$ are verified without difficulty and the same is true for the equality $\|\overline{\boldsymbol{m}}\| = \|\boldsymbol{m}\|$.

Remark. For every measure $\boldsymbol{m} : \mathscr{K}_E(T) \to F$ and function $\boldsymbol{f} \in \mathscr{K}_E(T)$ we have

$$\overline{\boldsymbol{m}(\boldsymbol{f})} = \overline{\boldsymbol{m}}(\bar{\boldsymbol{f}})\,.$$

3 *Two methods to define a measure*

1.19 PROPOSITION. *Let $(G_\alpha)_{\alpha \in A}$ be a family of open sets which cover T and, for each $\alpha \in A$, let $\boldsymbol{m}_\alpha : \mathscr{K}_E(G_\alpha) \to F$ be a vector measure on G_α. If for each pair (α, β) of indices, the restrictions of the measures $\boldsymbol{m}_\alpha$ and $\boldsymbol{m}_\beta$ to the subspace $\mathscr{K}_E(G_\alpha \cap G_\beta)$ are equal, then there exists a unique measure $\boldsymbol{m} : \mathscr{K}_E(T) \to F$ such that its restriction to the space $\mathscr{K}_E(G_\alpha)$ is equal to $\boldsymbol{m}_\alpha$, for every $\alpha \in A$.*

(a) Let $f \in \mathscr{K}_E(T)$ be an arbitrary function and K its support. There exists a finite family $(G_{\alpha_i})_{1 \leqslant i \leqslant n}$ which covers K. Let $(\varphi_i)_{1 \leqslant i \leqslant n}$ be a continuous partition of unity on K, subordinated to the covering $(G_{\alpha_i})_{1 \leqslant i \leqslant n}$; each function φ_i has values in $[0, 1]$, and compact support contained in G_{α_i}, and $\Sigma_{1 \leqslant i \leqslant n} \varphi_i(t) = 1$ for $t \in K$. Then

$$\boldsymbol{f} = \Sigma_{1 \leqslant i \leqslant n} \boldsymbol{f} \varphi_i$$

and $\boldsymbol{f} \varphi_i \in \mathscr{K}_E(G_{\alpha_i})$. Put

$$\boldsymbol{m}(\boldsymbol{f}) = \Sigma_{1 \leqslant i \leqslant n} \boldsymbol{m}_{\alpha_i}(\boldsymbol{f} \varphi_i) .$$

We shall show that $\boldsymbol{m}(\boldsymbol{f})$ depends on $\boldsymbol{f}$ only. Indeed, let $(G_{\beta_j})_{1 \leqslant j \leqslant m}$ be another open finite covering of K and let $(\psi_j)_{1 \leqslant j \leqslant m}$ be a continuous partition of unity on K, subordinated to the covering $(G_{\beta_j})_{1 \leqslant j \leqslant m}$; each function ψ_j has values in $[0, 1]$, and compact support contained in G_{β_j}, and $\Sigma_{1 \leqslant j \leqslant m} \psi_j(t)$ equals 1 for $t \in K$. Then

$$\boldsymbol{f} = \Sigma_{1 \leqslant j \leqslant m} \boldsymbol{f} \psi_j$$

and $\boldsymbol{f} \psi_j \in \mathscr{K}_E(G_{\beta_j})$.

For each i we have

$$\boldsymbol{f} \varphi_i = \Sigma_{1 \leqslant j \leqslant m} \boldsymbol{f} \varphi_i \psi_j \in \mathscr{K}_E(G_{\alpha_i})$$

therefore

$$\boldsymbol{m}_{\alpha_i}(\boldsymbol{f} \varphi_i) = \Sigma_{1 \leqslant j \leqslant m} \boldsymbol{m}_{\alpha_i}(\boldsymbol{f} \varphi_i \psi_j)$$

whence

$$\Sigma_{1 \leqslant i \leqslant n} \boldsymbol{m}_{\alpha_i}(\boldsymbol{f} \varphi_i) = \Sigma_{1 \leqslant i \leqslant n} \Sigma_{1 \leqslant j \leqslant m} \boldsymbol{m}_{\alpha_i}(\boldsymbol{f} \varphi_i \psi_j) .$$

Also, for each j we have

$$\boldsymbol{f} \psi_j = \Sigma_{1 \leqslant i \leqslant n} \boldsymbol{f} \varphi_i \psi_j \in \mathscr{K}_E(G_{\beta_j})$$

therefore

$$\boldsymbol{m}_{\beta_j}(\boldsymbol{f} \psi_j) = \Sigma_{1 \leqslant i \leqslant n} \boldsymbol{m}_{\beta_j}(\boldsymbol{f} \varphi_i \psi_j)$$

whence

$$\Sigma_{1 \leqslant j \leqslant m} \boldsymbol{m}_{\beta_j}(\boldsymbol{f} \psi_j) = \Sigma_{1 \leqslant j \leqslant m} \Sigma_{1 \leqslant i \leqslant n} \boldsymbol{m}_{\beta_j}(\boldsymbol{f} \varphi_i \psi_j) .$$

But for each pair of indices (i, j), the function $\boldsymbol{f} \varphi_i \psi_j$ belongs to the space $\mathscr{K}_E(G_{\alpha_i} \cap G_{\beta_j})$ and the restrictions of the measures $\boldsymbol{m}_{\alpha_i}$ and $\boldsymbol{m}_{\beta_j}$ to this space coincide:

$$\boldsymbol{m}_{\alpha_i}(\boldsymbol{f} \varphi_i \psi_j) = \boldsymbol{m}_{\beta_j}(\boldsymbol{f} \varphi_i \psi_j) .$$

It follows that

$$\Sigma_{1 \leqslant i \leqslant n} \boldsymbol{m}_{\alpha_i}(\boldsymbol{f}\varphi_i) = \Sigma_{1 \leqslant j \leqslant m} \boldsymbol{m}_{\beta_j}(\boldsymbol{f}\psi_j) .$$

Therefore, the equality

$$\boldsymbol{m}(\boldsymbol{f}) = \Sigma_{1 \leqslant i \leqslant n} \boldsymbol{m}_{\alpha_i}(\boldsymbol{f}\varphi_i)$$

defines $\boldsymbol{m}(\boldsymbol{f})$ without ambiguity.

We have thus defined a mapping $\boldsymbol{m} : \mathscr{K}_E(T) \to F$.

(b) The restriction of $\boldsymbol{m}$ to $\mathscr{K}_E(G_\alpha)$ coincides with $\boldsymbol{m}_\alpha$. In fact, if $\boldsymbol{f} \in \mathscr{K}_E(G_\alpha)$, the support of $\boldsymbol{f}$ is covered by the family formed by G_α only; if $\varphi : T \to [0, 1]$ has compact support contained in G_α and $\varphi(t)=1$ for $t \in K$, we have $\boldsymbol{f} = \boldsymbol{f}\varphi$ and

$$\boldsymbol{m}(\boldsymbol{f}) = \boldsymbol{m}_\alpha(\boldsymbol{f}\varphi) = \boldsymbol{m}_\alpha(\boldsymbol{f}) .$$

(c) The mapping $\boldsymbol{m}$ is linear. Let $\boldsymbol{f}, \boldsymbol{g} \in \mathscr{K}_E(T)$. There exists a compact set K which contains the support of both $\boldsymbol{f}$ and $\boldsymbol{g}$. If $(G_{\alpha_i})_{1 \leqslant i \leqslant n}$ is a finite covering of K and $(\varphi_i)_{1 \leqslant i \leqslant n}$ is a corresponding partition of unity on K we have

$$\boldsymbol{m}(\boldsymbol{f}) = \Sigma\, \boldsymbol{m}_{\alpha_i}(\boldsymbol{f}\varphi_i)$$

$$\boldsymbol{m}(\boldsymbol{g}) = \Sigma\, \boldsymbol{m}_{\alpha_i}(\boldsymbol{g}\varphi_i) .$$

But

$$\boldsymbol{f} + \boldsymbol{g} = \Sigma(\boldsymbol{f} + \boldsymbol{g})\varphi_i$$

therefore

$$\boldsymbol{m}(\boldsymbol{f}+\boldsymbol{g}) = \Sigma\, \boldsymbol{m}_{\alpha_i}[(\boldsymbol{f}+\boldsymbol{g})\varphi_i] = \Sigma\, \boldsymbol{m}_{\alpha_i}(\boldsymbol{f}\varphi_i + \boldsymbol{g}\varphi_i) =$$

$$= \Sigma\, \boldsymbol{m}_{\alpha_i}(\boldsymbol{f}\varphi_i) + \Sigma\, \boldsymbol{m}_{\alpha_i}(\boldsymbol{g}\varphi_i) = \boldsymbol{m}(\boldsymbol{f}) + \boldsymbol{m}(\boldsymbol{g})$$

The equality $\boldsymbol{m}(c\boldsymbol{f}) = c\boldsymbol{m}(\boldsymbol{f})$ (c scalar), is proved in a similar way.

(d) Let us show now that $\boldsymbol{m}$ is a measure. Let $K \subset T$ be a compact set, $(G_{\alpha_i})_{1 \leqslant i \leqslant n}$ a finite covering of K and $(\varphi_i)_{1 \leqslant i \leqslant n}$ a continuous partition of unity on K subordinated to this covering. For each i, let K_i be the support of the function φ_i. Since $\boldsymbol{m}_{\alpha_i}$ is a measure, there exists a number $a_i > 0$ such that

$$|\boldsymbol{m}_{\alpha_i}(\boldsymbol{g})| \leqslant a_i \, \|\boldsymbol{g}\| \quad \text{for} \quad \boldsymbol{g} \in \mathscr{K}(T, K_i) .$$

Therefore, for every function $\boldsymbol{f} \in \mathscr{K}_E(T, K)$ we have

$$|\boldsymbol{m}_{\alpha_i}(\boldsymbol{f}\varphi_i)| \leqslant a_i \, \|\boldsymbol{f}\varphi_i\| \leqslant a_i \|\boldsymbol{f}\|$$

and consequently

$$|\boldsymbol{m}(\boldsymbol{f})| \leqslant (\Sigma_{1 \leqslant i \leqslant n} a_i) \, \|\boldsymbol{f}\| \, .$$

It follows that the restriction of $\boldsymbol{m}$ to the space $\mathcal{K}_E(T, K)$ is continuous; that is, $\boldsymbol{m}$ is a measure.

(e) It remains to prove the uniqueness of $\boldsymbol{m}$. Let $\boldsymbol{m}' : \mathcal{K}_E(T) \to F$ be a measure such that for each $\alpha \in A$, its restriction to the space $\mathcal{K}_E(G_\alpha)$ is equal to $\boldsymbol{m}_\alpha$. Let us show that $\boldsymbol{m}' = \boldsymbol{m}$. Let $\boldsymbol{f} \in \mathcal{K}_E(T)$, and K the support of $\boldsymbol{f}$, $(G_{\alpha_i})_{1 \leqslant i \leqslant n}$ a finite covering of K and $(\varphi_i)_{1 \leqslant i \leqslant n}$ a continuous partition of unity on K, subordinated to this covering. We have

$$\boldsymbol{f} = \Sigma_{1 \leqslant i \leqslant n} \boldsymbol{f} \varphi_i$$

and $\boldsymbol{f} \varphi_i \in \mathcal{K}_E(G_{\alpha_i})$ for each i. Then

$$\boldsymbol{m}'(\boldsymbol{f}) = \Sigma_{1 \leqslant i \leqslant n} \boldsymbol{m}'(\boldsymbol{f} \varphi_i) = \Sigma_{1 \leqslant i \leqslant n} \boldsymbol{m}_{\alpha_i}(\boldsymbol{f} \varphi_i) = \boldsymbol{m}(\boldsymbol{f})$$

therefore $\boldsymbol{m}' = \boldsymbol{m}$.

Thus the proposition is proved.

1.20 COROLLARY. (*The principle of localization*). *Let $(G_\alpha)_{\alpha \in A}$ be a family of open sets and $\boldsymbol{m}, \boldsymbol{n} : \mathcal{K}_E(T) \to F$ two measures on T. If for each $\alpha \in A$, the restrictions of the measures $\boldsymbol{m}$ and $\boldsymbol{n}$ to the space $\mathcal{K}_E(G_\alpha)$ are equal, then the restrictions of $\boldsymbol{m}$ and $\boldsymbol{n}$ to the space $\mathcal{K}_E(\cup_{\alpha \in A} G_\alpha)$ are equal also. If, in addition, the family $(G_\alpha)_{\alpha \in A}$ covers T, then $\boldsymbol{m}$ and $\boldsymbol{n}$ are equal.*

1.21 PROPOSITION. *Let $\mathcal{H}_E$ be a rich subspace of the space $\mathcal{K}_E(T)$. A linear mapping $\boldsymbol{n} : \mathcal{H}_E \to F$ can be extended to a measure $\boldsymbol{m} : \mathcal{K}_E(T) \to F$, if and only if for every compact set $K \subset T$, the restriction of $\boldsymbol{n}$ to the subspace of functions of $\mathcal{H}_E$ with support in K is continuous for the topology of the uniform convergence. Then the extension is unique.*

Let us suppose first that $\boldsymbol{m} : \mathcal{K}_E(T) \to F$ is a measure whose restriction to the space $\mathcal{H}_E$ is equal to $\boldsymbol{n}$. For each compact set $K \subset T$ there exists a number a_K such that

$$|\boldsymbol{m}(\boldsymbol{f})| \leqslant a_K \, \|\boldsymbol{f}\| \quad \text{for } \boldsymbol{f} \in \mathcal{K}_E(T, K) \, .$$

In particular, the preceding relation is true for the functions of $\mathcal{H}_E$ with support contained in K. This means that the restriction $\boldsymbol{n}$ of $\boldsymbol{m}$ to $\mathcal{H}_E$ is continuous on the subspace $\mathcal{H}_E \cap \mathcal{K}_E(T, K)$, for the topology of the uniform convergence.

Conversely, let $\boldsymbol{n} : \mathcal{H}_E \to F$ be a linear mapping which fulfills the condition required in the statement. Let $K \subset T$ be a compact set and let U be a compact neighbourhood of K such that the space $\mathscr{H}_E(U) = \mathscr{H}_E \cap \mathscr{K}_E(T, U)$ is dense in $\mathscr{K}_E(T, K)$ for the topology of the uniform convergence. Since $\boldsymbol{n}$ is continuous on $\mathscr{H}_E(U)$ for this topology, it can be extended uniquely to a continuous linear mapping $\boldsymbol{n}_U : \overline{\mathscr{H}_E(U)} \to F$. Since $\mathscr{K}_E(T, K) \subset \overline{\mathscr{H}_E(U)}$, let us denote by $\boldsymbol{m}_{K,U}$ the restriction of $\boldsymbol{n}_U$ to $\mathscr{K}_E(T, K)$.

The mapping $\boldsymbol{m}_{K,U} : \mathscr{K}_E(T, K) \to F$ does not depend on U, but depends on K only. If U' and U'' are two compact neighbourhoods of K such that $\mathscr{H}_E(U')$ and $\mathscr{H}_E(U'')$ are dense in $\mathscr{K}_E(T, K)$, then $U = U' \cup U''$ is a compact neighbourhood of K and $\mathscr{H}_E(U)$ is dense in $\mathscr{K}_E(T, K)$ too. Because $\overline{\mathscr{H}_E(U')} \subset \overline{\mathscr{H}_E(U)}$ and $\overline{\mathscr{H}_E(U'')} \subset \overline{\mathscr{H}_E(U)}$, it follows that $\boldsymbol{n}_{U'}$ is the restriction of $\boldsymbol{n}_U$ to $\overline{\mathscr{H}_E(U')}$ and $\boldsymbol{n}_{U''}$ is the restriction of $\boldsymbol{n}_U$ to $\overline{\mathscr{H}_E(U'')}$. Then it follows that $\boldsymbol{m}_{K,U'} = \boldsymbol{m}_{K,U}$ and $\boldsymbol{m}_{K,U''} = \boldsymbol{m}_{K,U}$ whence $\boldsymbol{m}_{K,U'} = \boldsymbol{m}_{K,U''}$ and therefore $\boldsymbol{m}_{K,U}$ is independent on U.

Let us denote by $\boldsymbol{m}_K$ the linear mapping of $\mathscr{K}_E(T, K)$ into F, defined above.

For two compact sets K' and K'' such that $K' \subset K''$, the restriction of $\boldsymbol{m}_{K''}$ to the subspace $\mathscr{K}_E(T, K')$ coincides with $\boldsymbol{m}_{K'}$; in fact, if U is a compact neighbourhood of K'' such that $\mathscr{K}_E(T, K'') \subset \overline{\mathscr{H}_E(U)}$, then $\mathscr{K}_E(T, K') \subset \overline{\mathscr{H}_E(U)}$ and therefore the restriction of $\boldsymbol{m}_{K'',U}$ to the space $\mathscr{K}_E(T, K')$ coincides with the restriction $\boldsymbol{m}_{K',U}$ of $\boldsymbol{n}_U$ to $\mathscr{K}_E(T, K')$.

For every function $\boldsymbol{f} \in \mathscr{K}_E(T)$ put

$$\boldsymbol{m}(\boldsymbol{f}) = \boldsymbol{m}_K(\boldsymbol{f})$$

where K is an arbitrary compact set such that $\boldsymbol{f}$ vanishes outside K. It follows from the preceding considerations that $\boldsymbol{m}(\boldsymbol{f})$ does not depend on K. Since $\boldsymbol{m}_K$ is linear on $\mathscr{K}_E(T, K)$, we deduce that $\boldsymbol{m}(\alpha\boldsymbol{f}) = \alpha\boldsymbol{m}(\boldsymbol{f})$ for every scalar α. If $\boldsymbol{f}, \boldsymbol{g} \in \mathscr{K}_E(T)$, there exists a compact set K outside which both functions vanish. As $\boldsymbol{m}$ is linear on $\mathscr{K}_E(T, K)$, we deduce that $\boldsymbol{m}(\boldsymbol{f} + \boldsymbol{g}) = \boldsymbol{m}(\boldsymbol{f}) + \boldsymbol{m}(\boldsymbol{g})$. Thus the mapping $\boldsymbol{m} : \mathscr{K}_E(T) \to K$, defined above, is linear.

Since for every compact set $K \subset T$, the restriction of $\boldsymbol{m}$ to $\mathscr{K}_E(T, K)$ coincides with $\boldsymbol{m}_K$, and $\boldsymbol{m}_K$ is continuous on $\mathscr{K}_E(T, K)$ for the topology of the uniform convergence, it follows that $\boldsymbol{m}$ is a measure.

The restriction of $\boldsymbol{m}$ to the space $\mathscr{H}_E$ coincides with $\boldsymbol{n}$: if $\boldsymbol{f} \in \mathscr{H}_E \cap \mathscr{K}_E(T, K)$, then $\boldsymbol{m}(\boldsymbol{f}) = \boldsymbol{m}_K(\boldsymbol{f})$ and $\boldsymbol{m}_K(\boldsymbol{f}) = \boldsymbol{n}(\boldsymbol{f})$, therefore $\boldsymbol{m}(\boldsymbol{f}) = \boldsymbol{n}(\boldsymbol{f})$.

If $\boldsymbol{m}'$ is another measure whose restriction to $\mathscr{H}_E$ is equal to $\boldsymbol{n}$, it follows that $\boldsymbol{m}$ and $\boldsymbol{m}'$ coincide on each space $\overline{\mathscr{H}_E(U)}$ and therefore $\boldsymbol{m} = \boldsymbol{m}'$. Thus the proposition is completely proved.

1.22 COROLLARY. *Let $\boldsymbol{m}_1, \boldsymbol{m}_2 : \mathscr{K}_E(T) \to F$ be two measures. If*

$$\boldsymbol{m}_1(\varphi x) = \boldsymbol{m}_2(\varphi x)$$

for every function $\varphi \in \mathscr{K}_+(T)$ and every $x \in E$, then $\boldsymbol{m}_1 = \boldsymbol{m}_2$.

In fact, we have

$$\boldsymbol{m}_1(\Sigma_{1 \leqslant i \leqslant n} \varphi_i x_i) = \boldsymbol{m}_2(\Sigma_{1 \leqslant i \leqslant n} \varphi_i x_i)$$

for every function $\Sigma \varphi_i x_i$ with $x_i \in E$ and $\varphi_i \in \mathscr{K}_+(T)$. As the space $\mathscr{H}_E$ of these functions is rich (corollary 1.6) it follows that $\boldsymbol{m}_1(\boldsymbol{f}) = \boldsymbol{m}_2(\boldsymbol{f})$ for every function $\boldsymbol{f} \in \mathscr{K}_E(T)$, therefore $\boldsymbol{m}_1 = \boldsymbol{m}_2$.

In particular, if $\boldsymbol{m}_1$, $\boldsymbol{m}_2 : \mathscr{K}(T) \to F$ are two measures satisfying the equality

$$\boldsymbol{m}_1(\varphi) = \boldsymbol{m}_2(\varphi) \text{ for } \varphi \in \mathscr{K}_+(T)$$

then $\boldsymbol{m}_1 = \boldsymbol{m}_2$.

§2 Properties of measures

1 *Integration of scalar functions with respect to a vector measure*

Let E and F be two Banach spaces, both of them over the same field of scalars.

2.1 PROPOSITION. *For every measure $\boldsymbol{m} : \mathscr{K}_E(T) \to F$ there exists a measure $\boldsymbol{\mu} : \mathscr{K}(T) \to \mathscr{L}(E, F)$ and only one, satisfying the equality*

$$\boldsymbol{\mu}(\varphi)x = \boldsymbol{m}(\varphi x) \textit{ for every } \varphi \in \mathscr{K}(T) \textit{ and } x \in E.$$

The mapping $\boldsymbol{m} \to \boldsymbol{\mu}$ is injective, linear and $\|\boldsymbol{\mu}\| \leqslant \|\boldsymbol{m}\| \leqslant +\infty$.

(a) For each compact set $K \subset T$ choose a number $a_K > 0$ such that

$$|\boldsymbol{m}(\boldsymbol{f})| \leqslant a_K \|\boldsymbol{f}\| \text{ for every function } \boldsymbol{f} \in \mathscr{K}_E(T, K).$$

For every function $\varphi \in \mathscr{K}(T)$ and every $x \in E$ we have $\varphi x \in \mathscr{K}_E(T)$. Put

$$\boldsymbol{m}_x(\varphi) = \boldsymbol{m}(\varphi x).$$

For every fixed function $\varphi \in \mathscr{K}(T)$, the mapping $\boldsymbol{\mu}(\varphi) : E \to F$ defined by the equality

$$\boldsymbol{\mu}(\varphi)x = \boldsymbol{m}_x(\varphi) = \boldsymbol{m}(\varphi x), \text{ for } x \in E$$

is linear. The linear mapping $\mu(\varphi)$ is also continuous because, denoting by K the support of φ, we have

$$|\boldsymbol{\mu}(\varphi)x| = |\boldsymbol{m}(\varphi x)| \leqslant a_K \|\varphi x\| = a_K \|\varphi\| \, |x|$$

for every $x \in E$. Hence $\boldsymbol{\mu}(\varphi) \in \mathscr{L}(E, F)$. The mapping $\boldsymbol{\mu} : \mathscr{K}(T) \to \mathscr{L}(E, F)$ is linear and for every compact set $K \subset T$ we have

$$|\boldsymbol{\mu}(\varphi)| = \sup_{|x| \leqslant 1} |\boldsymbol{\mu}(\varphi)x| \leqslant a_K \|\varphi\| \text{ for every } \varphi \in \mathscr{K}(T, K).$$

Hence $\boldsymbol{\mu}$ is a measure and from its definition it follows that we have

$$\boldsymbol{\mu}(\varphi)x = \boldsymbol{m}(\varphi x) \text{ for } \varphi \in \mathscr{K}(T) \text{ and } x \in E.$$

(b) The correspondence $\boldsymbol{m} \to \boldsymbol{\mu}$ established above is injective. In fact, if for two measures $\boldsymbol{m}$, $\boldsymbol{m}' : \mathscr{K}_E(T) \to F$ there corresponds the same measure $\boldsymbol{\mu} : \mathscr{K}(T) \to \mathscr{L}(E, F)$, then for every function $\varphi \in \mathscr{K}(T)$ and every $x \in E$ we have

$$\boldsymbol{\mu}(\varphi)x = \boldsymbol{m}'(\varphi x) = \boldsymbol{m}(\varphi x).$$

From the corollary 1.22 it follows then that $\boldsymbol{m}' = \boldsymbol{m}$. Therefore the correspondence $\boldsymbol{m} \to \boldsymbol{\mu}$ is injective.

(c) Let $\boldsymbol{m}$, $\boldsymbol{m}' : \mathscr{K}_E(T) \to F$ be two measures and $\boldsymbol{\mu}, \boldsymbol{\mu}' : \mathscr{K}(T) \to \mathscr{L}(E, F)$ the corresponding measures satisfying the equalities

$$\boldsymbol{\mu}(\varphi)x = \boldsymbol{m}(\varphi x) \text{ and } \boldsymbol{\mu}'(\varphi)x = \boldsymbol{m}'(\varphi x)$$

for $\varphi \in \mathscr{K}(T)$ and $x \in E$. Then

$$\begin{aligned}(\boldsymbol{\mu} + \boldsymbol{\mu}')(\varphi)x &= [\boldsymbol{\mu}(\varphi) + \boldsymbol{\mu}'(\varphi)]x = \boldsymbol{\mu}(\varphi)x + \boldsymbol{\mu}'(\varphi)x = \\ &= \boldsymbol{m}(\varphi x) + \boldsymbol{m}'(\varphi x) = (\boldsymbol{m} + \boldsymbol{m}')(\varphi x).\end{aligned}$$

Let $\boldsymbol{\nu} : \mathscr{K}(T) \to \mathscr{L}(E, F)$ be the measure corresponding to the measure $\boldsymbol{m} + \boldsymbol{m}'$:

$$\boldsymbol{\nu}(\varphi)x = (\boldsymbol{m} + \boldsymbol{m}')(\varphi)x \text{ for } \varphi \in \mathscr{K}(T) \text{ and } x \in E.$$

It follows that

$$(\boldsymbol{\mu} + \boldsymbol{\mu}')(\varphi)x = \boldsymbol{\nu}(\varphi)x \text{ for } \varphi \in \mathscr{K}(T) \text{ and } x \in E$$

therefore

$$(\boldsymbol{\mu} + \boldsymbol{\mu}')(\varphi) = \boldsymbol{\nu}(\varphi) \text{ for } \varphi \in \mathscr{K}(T)$$

whence

$$\boldsymbol{\mu} + \boldsymbol{\mu}' = \boldsymbol{\nu}\,.$$

Consequently the correspondence $\boldsymbol{m} \to \boldsymbol{\mu}$ is additive. The homogeneity is proved in the same way.

(d) It remains to verify the relation

$$\|\boldsymbol{\mu}\| \leqslant \|\boldsymbol{m}\|\,.$$

For every function $\varphi \in \mathscr{K}(T)$ and every $\varepsilon > 0$, there exists $x \in E$ with $|x| = 1$ and

$$|\boldsymbol{\mu}(\varphi)| < |\boldsymbol{\mu}(\varphi)x| + \varepsilon\,.$$

Then

$$|\boldsymbol{\mu}(\varphi)| - \varepsilon < |\boldsymbol{\mu}(\varphi)x| = |\boldsymbol{m}(\varphi x)| \leqslant \|\boldsymbol{m}\|\ \|\varphi x\| = \|\boldsymbol{m}\|\ \|\varphi\|\,.$$

ε being arbitrary, we deduce

$$|\boldsymbol{\mu}(\varphi)| \leqslant \|\boldsymbol{m}\|\ \|\varphi\|$$

whence

$$\|\boldsymbol{\mu}\| \leqslant \|\boldsymbol{m}\| \leqslant +\infty\,.$$

Thus the proposition is proved.

2.2 COROLLARY. *Let E and F be two Banach spaces with involution and* $\boldsymbol{m} : \mathscr{K}_E(T) \to F$ *a measure. If* $\mu : \mathscr{K}(T) \to \mathscr{L}(E, F)$ *is the measure corresponding to* $\boldsymbol{m}$*, then* $\bar{\mu}$ *is the measure corresponding to* $\bar{\boldsymbol{m}}$.

Indeed, we have

$$\bar{\boldsymbol{m}}(f) = \overline{\boldsymbol{m}(\bar{f})},\ \text{for } f \in \mathscr{K}_E(T)$$

$$\boldsymbol{m}(\varphi x) = \mu(\varphi)x,\ \text{for } \varphi \in \mathscr{K}(T) \text{ and } x \in E$$

and

$$\bar{\mu}(\varphi) = \overline{\mu(\varphi)},\ \text{for } \varphi \in \mathscr{K}(T)\,.$$

Let $\nu : \mathscr{K}(T) \to \mathscr{L}(E, F)$ be the unique measure satisfying the equality

$$\bar{\boldsymbol{m}}(\varphi x) = \nu(\varphi)x,\ \text{for } \varphi \in \mathscr{K}(T) \text{ and } x \in E\,.$$

Then, for every $\varphi \in \mathscr{K}(T)$ and every $x \in E$, we have

$$\nu(\varphi)x = \bar{\boldsymbol{m}}(\varphi x) = \overline{\boldsymbol{m}(\varphi\bar{x})} = \overline{\mu(\varphi)\bar{x}} = \bar{\mu}(\varphi)\bar{\bar{x}} = \bar{\mu}(\varphi)x\ ;$$

therefore $\bar{\mu}(\varphi) = \nu(\varphi)$ for every $\varphi \in \mathscr{K}(T)$, that is $\bar{\mu} = \nu$. But then, $\bar{\boldsymbol{m}}(\varphi x) = \bar{\mu}(\varphi)x$, for $\varphi \in \mathscr{K}(T)$ and $x \in E$.

Remarks

1. There are measures $\mu : \mathscr{K}(T) \to \mathscr{L}(E, F)$ which cannot be obtained as in proposition 2.1 from measures $\boldsymbol{m} : \mathscr{K}_E(T) \to F$. We shall see further that if the measure $\mu : \mathscr{K}(T) \to \mathscr{L}(E, F)$ has an additional property (to be dominated), then it can be "extended" to a measure $\boldsymbol{m} : \mathscr{K}_E(T) \to F$ which satisfies the equality $\boldsymbol{m}(\varphi x) = \mu(\varphi)x$, $\varphi \in \mathscr{K}(T)$, $x \in E$.

2. If E is a complex Banach space, then for every function $\varphi \in \mathscr{K}_C(T)$ and every $x \in E$ we have $\varphi x \in \mathscr{K}_E(T)$. We can prove in the same way that for every measure $\boldsymbol{m} : \mathscr{K}_E(T) \to F$ there exists a unique measure $\mu' : \mathscr{K}_C(T) \to \mathscr{L}(E, F)$ satisfying the equality

$$\mu'(\varphi)x = \boldsymbol{m}(\varphi x) \text{ for every } \varphi \in \mathscr{K}_C(T) \text{ and } x \in E\,.$$

As a matter of fact the measure μ' can be obtained by extending the measure μ from $\mathscr{K}(T)$ to $\mathscr{K}_C(T)$, as it will be shown later.

Because of the injective correspondence $\boldsymbol{m} \to \mu$, we shall identify, later, the measure $\boldsymbol{m}$ with the corresponding measure μ. For functions $\varphi \in \mathscr{K}(T)$, the integral $\int \varphi\, d\mu$ will be also denoted by $\int \varphi\, d\boldsymbol{m}$ and will be called the *integral of the scalar function φ with respect to the vector measure $\boldsymbol{m}$*.

The equality in the statement of proposition 2.1 is then written

$$\boldsymbol{m}(\varphi)x = \boldsymbol{m}(\varphi x) \text{ or } \int \varphi x\, d\boldsymbol{m} = x \int \varphi\, d\boldsymbol{m}, \text{ for } \varphi \in \mathscr{K}(T) \text{ and } x \in E\,.$$

We notice that if $\varphi \in \mathscr{K}(T)$, then the integral $\int \varphi\, d\boldsymbol{m}$ is an element of $\mathscr{L}(E, F)$, while, if $\boldsymbol{f} \in \mathscr{K}_E(T)$, the integral $\int \boldsymbol{f}\, d\boldsymbol{m}$ is an element of F.

2. *Measures defined for scalar functions*

Let X be a Banach space.

2.3 PROPOSITION. *Let $\mu : \mathscr{K}_+ \to X$ be a mapping satisfying the following three conditions*

(1) $\mu(\varphi_1 + \varphi_2) = \mu(\varphi_1) + \mu(\varphi_2)$, *for* $\varphi_1, \varphi_2 \in \mathscr{K}_+$;

(2) $\mu(\alpha\varphi) = \alpha\mu(\varphi)$, *for* $\alpha \geqslant 0$ *and* $\varphi \in \mathscr{K}_+$;

(3) *for every compact set $K \subset T$, there exists a number $a_K > 0$ such that*

$$|\mu(\varphi)| \leqslant a_K \|\varphi\| \textit{ for } \varphi \in \mathscr{K}_+(T, K)\,.$$

Then μ can be extended uniquely to a measure $\boldsymbol{m} : \mathscr{K}(T) \to X$. The correspondence $\mu \to \boldsymbol{m}$ is linear.

Let f be an arbitrary function from $\mathscr{K}(T)$. The function f can be written as a difference of positive functions from $\mathscr{K}_+(T)$. If

$$f = \varphi_1 - \psi_1 = \varphi_2 - \psi_2, \text{ with } \varphi_1, \varphi_2, \psi_1, \psi_2 \in \mathscr{K}_+(T)$$

then $\varphi_1 + \psi_2 = \varphi_2 + \psi_1$. From property (1) we deduce

$$\mu(\varphi_1) + \mu(\psi_2) = \mu(\varphi_2) + \mu(\psi_1)$$

whence

$$\mu(\varphi_1) - \mu(\psi_1) = \mu(\varphi_2) - \mu(\psi_2) .$$

If $\boldsymbol{f} = \varphi - \psi$ with φ and ψ from $\mathscr{K}_+$, we put, by definition,

$$\boldsymbol{m}(f) = \mu(\varphi) - \mu(\psi) .$$

It follows that $\boldsymbol{m}(f)$ does not depend on the particular way in which f is written as a difference of positive functions from $\mathscr{K}_+$.

For a function φ from $\mathscr{K}_+$ we have $\varphi = \varphi - 0$, therefore

$$\boldsymbol{m}(\varphi) = \mu(\varphi) - \mu(0) = \mu(\varphi) .$$

Hence, $\boldsymbol{m}$ is an extension of μ from $\mathscr{K}_+$ to $\mathscr{K}(T)$. Let us show that $\boldsymbol{m}$ is a measure.

Let $f, g \in \mathscr{K}(T)$. Write

$$f = f^+ - f^- \text{ and } g = g^+ - g^- .$$

The functions f^+, f^-, g^+, g^- are from $\mathscr{K}_+$ and

$$f + g = (f^+ + g^+) - (f^- + g^-) .$$

Then

$$\begin{aligned} \boldsymbol{m}(f+g) &= \mu(f^+ + g^+) - \mu(f^- + g^-) = \\ &= \mu(f^+) + \mu(g^+) - \mu(f^-) - \mu(g^-) = \\ &= (f^+) - \mu(f^-) + \mu(g^+) - \mu(g^-) = \boldsymbol{m}(f) + \boldsymbol{m}(g). \end{aligned}$$

If $\alpha \geqslant 0$, then $\alpha f = \alpha f^+ - \alpha f^-$ and

$$\begin{aligned} \boldsymbol{m}(\alpha f) &= \mu(\alpha f^+) - \mu(\alpha f^-) = \alpha\mu(f^+) - \alpha\mu(f^-) = \\ &= \alpha(\mu(f^+) - \mu(f^-)) = \alpha \boldsymbol{m}(f) . \end{aligned}$$

If $\alpha < 0$, then $\alpha f = \alpha f^+ - \alpha f^- = (-\alpha) f^- - (-\alpha) f^+$, therefore

$$\boldsymbol{m}(\alpha f) = \mu[(-\alpha) f^-] - \mu[(-\alpha) f^+] = \\ = -\alpha(\mu(f^-) - \mu(f^+)) = \alpha(\mu(f^+) - \mu(f^-)) = \alpha \boldsymbol{m}(f).$$

Hence $\boldsymbol{m}$ is linear.

Let $K \subset T$ be a compact set and $f \in \mathscr{K}(T, K)$. We have

$$f = f^+ - f^-, \quad |f| = f^+ + f^- \text{ and } f^+ \leqslant |f|, f^- \leqslant |f|.$$

Then

$$|\boldsymbol{m}(f)| = |\mu(f^+) - \mu(f^-)| \leqslant |\mu(f^+)| + |\mu(f^-)| \leqslant \\ \leqslant a_K \|f^+\| + a_K \|f^-\| \leqslant 2a_K \|f\|$$

therefore $\boldsymbol{m}$ is a measure on T.

If $\boldsymbol{n} : \mathscr{K}(T) \to X$ is a measure which satisfies the equality

$$\boldsymbol{n}(\varphi) = \mu(\varphi) \text{ for } \varphi \in \mathscr{K}_+(T),$$

then

$$\boldsymbol{n}(f) = \boldsymbol{n}(f^+ - f^-) = \boldsymbol{n}(f^+) - \boldsymbol{n}(f^-) = \mu(f^+) - \mu(f^-) = \boldsymbol{m}(f)$$

for every $f \in \mathscr{K}(T)$, therefore $\boldsymbol{n} = \boldsymbol{m}$. It follows that the extension of μ to $\mathscr{K}(T)$, as a measure, is unique.

The linearity of the correspondence $\mu \to \boldsymbol{m}$ follows from the fact that if $\boldsymbol{m}, \boldsymbol{n} : \mathscr{K}(T) \to X$ are two measures whose restrictions to $\mathscr{K}_+(T)$ are μ and ν, then the restriction of $\boldsymbol{m} + \boldsymbol{n}$ to $\mathscr{K}_+(T)$ is $\mu + \nu$ and the restriction of $\alpha \boldsymbol{m}$ to $\mathscr{K}_+(T)$ is $\alpha\mu$. Thus the proposition is proved.

Remark. It will be shown further that for a mapping $\mu : \mathscr{K}_+(T) \to R$ to be extended to $\mathscr{K}(T)$ as a measure it is enough to be additive and positive.

2.4 PROPOSITION. *If X is a complex Banach space and $\mu : \mathscr{K}(T) \to X$ is a measure, then μ can be extended uniquely to the whole space $\mathscr{K}_C(T)$ as a measure $\boldsymbol{m} : \mathscr{K}_C(T) \to X$. The correspondence $\mu \to \boldsymbol{m}$ is linear and $\|\mu\| \leqslant \|\boldsymbol{m}\| \leqslant 2\|\mu\| < +\infty$. If, in addition, X is a Banach space with involution, the correspondence $\mu \to \boldsymbol{m}$ preserves the involution on the corresponding spaces of measure.*

Let $f \in \mathscr{K}_C(T)$ be an arbitrary function. The function f can be written uniquely in the form

$$f = f_1 + i f_2 \text{ with } f_1, f_2 \in \mathscr{K}(T).$$

We put by definition

$$m(f) = \mu(f_1) + i\mu(f_2).$$

If $f \in \mathscr{K}(T)$, then $f = g + i0$, therefore

$$m(f) = \mu(f) + i\mu(0) = \mu(f).$$

Hence m is an extension of μ from $\mathscr{K}(T)$ to $\mathscr{K}_C(T)$. If $f = f_1 + if_2$ and $g = g_1 + ig_2$ with f_1, f_2, g_1, g_2 from $\mathscr{K}(T)$, then

$$f + g = (f_1 + g_1) + i(f_2 + g_2)$$

therefore

$$m(f+g) = \mu(f_1 + g_1) + i\mu(f_2 + g_2) = \mu(f_1) + \mu(g_1) + i\mu(f_2) + i\mu(g_2) =$$
$$= \mu(f_1) + i\mu(f_2) + \mu(g_1) + i\mu(g_2) = m(f) + m(g).$$

If $z = \alpha + i\beta$ with α and β real, then

$$zf = (\alpha + i\beta)(f_1 + if_2) = (\alpha f_1 - \beta f_2) + i(\beta f_1 + \alpha f_2)$$

therefore

$$m(zf) = \mu(\alpha f_1 - \beta f_2) + i\mu(\beta f_1 + \alpha f_2) =$$
$$= \alpha\mu(f_1) - \beta\mu(f_2) + i(\beta\mu(f_1) + \alpha\mu(f_2)) =$$
$$= (\alpha + i\beta)(\mu(f_1) + i\mu(f_2)) = zm(f).$$

Hence m is a (complex) linear mapping from $\mathscr{K}_C(T)$ into X.

Let $K \subset T$ be a compact set and let $a_K > 0$ be such that

$$|\mu(\varphi)| \leqslant a_K \|\varphi\| \quad \text{for } \varphi \in \mathscr{K}(T, K).$$

For every function $f = f_1 + if_2$ from $\mathscr{K}_C(T, K)$ with f_1 and f_2 from $\mathscr{K}(T)$, we have $|f_1| \leqslant |f|$ and $|f_2| \leqslant |f|$, therefore $f_1, f_2 \in \mathscr{K}(T, K)$. Then

$$|m(f)| = |\mu(f_1) + i\mu(f_2)| \leqslant |\mu(f_1)| + |\mu(f_2)| \leqslant$$
$$\leqslant a_K \|f_1\| + a_K \|f_2\| \leqslant 2a_K \|f\|.$$

Therefore m is a measure.

For $f \in \mathscr{K}(T)$ we have

$$|\mu(f)| = |m(f)| \leqslant \|m\| \, \|f\|$$

therefore

$$\|\mu\| \leqslant \|m\|.$$

For $f=f_1+if_2$ we have

$$|\boldsymbol{m}(f)|=|\mu(f_1)+i\mu(f_2)|\leqslant|\mu(f_1)|+|\mu(f_2)|\leqslant$$
$$\leqslant\|\mu\|\;\|f_1\|+\|\mu\|\;\|f_2\|\leqslant 2\|\mu\|\;\|f\|$$

therefore $\|\boldsymbol{m}\|\leqslant 2\|\mu\|$.

If $\boldsymbol{m}':\mathscr{K}_C(T)\to X$ is a measure extending μ, for every function $f=f_1+if_2$ with f_1 and f_2 from $\mathscr{K}(T)$ we have

$$\boldsymbol{m}'(f)=\boldsymbol{m}'(f_1)+i\boldsymbol{m}'(f_2)==\mu(f_1)+i\mu(f_2)=\boldsymbol{m}(f)$$

therefore $\boldsymbol{m}'=\boldsymbol{m}$.

If $\boldsymbol{m}, \boldsymbol{n}:\mathscr{K}_C(T)\to X$ are two measures whose restrictions to $\mathscr{K}(T)$ are μ and ν, and $\alpha\in C$, then the restrictions to $\mathscr{K}(T)$ of the measures $\boldsymbol{m}+\boldsymbol{n}$ and $\alpha\boldsymbol{m}$ are $\mu+\nu$ and $\alpha\mu$ respectively. Therefore the correspondence $\boldsymbol{m}\to\mu$ is linear. If $x\to\bar{x}$ is an involution on X and if $\boldsymbol{m}$ corresponds to μ, then the measure $\bar{\boldsymbol{m}}$ corresponds to $\bar{\mu}$, that is, if $\bar{\mu}(\varphi)=\overline{\mu(\varphi)}$ for $\varphi\in\mathscr{K}(T)$, then $\bar{\boldsymbol{m}}(f)=\overline{\boldsymbol{m}(f)}$ for $f\in\mathscr{K}_C(T)$. Thus the proposition is proved.

Remarks

1. If X is not a complex space, then the measure $\mu:\mathscr{K}(T)\to X$ can be extended to a measure $\boldsymbol{m}:\mathscr{K}_C(T)\to X_C$, where X_C is the Banach space $\{(x+iy)\,|\,x, y\in X\}$, with the addition

$$(x+iy)+(x'+iy')=(x+x')+i(y+y'),$$

the multiplication by scalars

$$(\alpha+i\beta)(x+iy)=(\alpha x-\beta y)+i(\alpha y+\beta x)$$

and the norm

$$|x+iy|=(|x|^2+|y|^2)^{\frac{1}{2}}\,.$$

2. If $X=C$, it will be shown further that $\|\boldsymbol{m}\|=\|\mu\|$ (see corollary 2.12).

For functions $f\in\mathscr{K}_C(T)$, the integral $\int f\mathrm{d}\boldsymbol{m}$ will be also denoted by $\int f\mathrm{d}\mu$ and it will be called *the integral* of the complex function f with respect to the measure μ.

2.5 Proposition. *Let $H\subset\mathscr{K}_+$ be a set directed for the relation $\leqslant$ and $\boldsymbol{m}:\mathscr{K}(T)\to X$ a measure defined for scalar functions. If $\sup_{g\in H} g\in\mathscr{K}_+$, then the mapping $g\to\boldsymbol{m}(g)$ has a limit in X following the filter of sections of H and*

$$\lim_{g\in H}\boldsymbol{m}(f)=\boldsymbol{m}(\sup_{g\in H} g)=\boldsymbol{m}(\lim_{g\in H} g)\,.$$

Let us denote $f = \sup_{g \in H} g = \lim_{g \in H} g$ (proposition 1.4), the limit being understood in the sense of the uniform convergence. Let K be the support of f. Since $0 \leqslant g \leqslant f$ for every $g \in H$, it follows that $H \subset \mathscr{K}(T, K)$. The restriction of the measure $\boldsymbol{m}$ to the subspace $\mathscr{K}(T, K)$ being continuous for the topology of the uniform convergence, we deduce that

$$\lim_{g \in H} \boldsymbol{m}(g) = \boldsymbol{m}(f) = \boldsymbol{m}(\sup_{g \in H} g) = \boldsymbol{m}(\lim_{g \in H} g)$$

and the proposition is proved.

2.6 COROLLARY *Let $\boldsymbol{m} : \mathscr{K}(T) \to X$ be a measure defined for scalar functions. For every function $f \in \mathscr{K}_+$, we have then*

$$\boldsymbol{m}(f) = \lim_{g \leqslant f} \boldsymbol{m}(g), \quad g \in \mathscr{K}_+ .$$

In fact, the set $H = \{g \mid g \in \mathscr{K}_+,\ g \leqslant f\}$ is directed for the relation $\leqslant$ and $\sup_{g \in H} g = f \in \mathscr{K}_+$.

2.7 PROPOSITION. *Let $\mu : \mathscr{K}(T) \to \mathscr{L}(E, F)$ be a measure, $\varphi_i \in \mathscr{K}(T)$ and $x_i \in E$, $i = 1, 2, \ldots, n$. If $\Sigma_{1 \leqslant i \leqslant n} \varphi_i(t)\, x_i \equiv 0$, then*

$$\Sigma_{1 \leqslant i \leqslant n} \mu(\varphi_i) x_i = 0 .$$

If $x_1 = x_2 = \ldots = x_n = 0$, the equality $\Sigma_{1 \leqslant i \leqslant n} \mu(\varphi_i) x_i = 0$ is evidently true. Let us suppose then that the real space generated by the elements $x_1, x_2, \ldots, x_n$ is different from $\{0\}$. Let $y_1, y_2, \ldots, y_m$ be a basis of this space. Then each element x_i is written $x_i = \Sigma_{1 \leqslant j \leqslant m} \alpha_{ij} y_j$, $\alpha_{ij} \in R$, therefore

$$\Sigma_{1 \leqslant i \leqslant n} \varphi_i x_i = \Sigma_{1 \leqslant i \leqslant n} \varphi_i \Sigma_{1 \leqslant j \leqslant m} \alpha_{ij} y_j =$$
$$= \Sigma_{1 \leqslant j \leqslant m} (\Sigma_{1 \leqslant i \leqslant n} \alpha_{ij} \varphi_i) y_i = \Sigma_{1 \leqslant j \leqslant m} \psi_j y_j ,$$

where the functions $\psi_j = \Sigma \alpha_{ij} \varphi_i$ belong to the space $\mathscr{K}(T)$. From the equality

$$\Sigma_{1 \leqslant i \leqslant n} \varphi_i(t)\, x_j \equiv 0$$

it follows

$$\Sigma_{1 \leqslant j \leqslant m} \psi_j(t)\, y_j \equiv 0.$$

Since the elements $y_1, \ldots, y_m$ are linearly independent it follows that for each $t \in T$ we have $\psi_1(t) = 0, \ldots, \psi_m(t) = 0$, that is $\psi_1 = \psi_2 = \ldots = \psi_m = 0$. Then $\mu(\psi_j) = 0$, $j = 1, 2, \ldots, m$, therefore

$$\Sigma_{1\leqslant i\leqslant n}\mu(\varphi_i)x_i=\Sigma_{1\leqslant i\leqslant n}\mu(\varphi_i)\Sigma_{1\leqslant j\leqslant m}\alpha_{ij}y_j=$$
$$=\Sigma_{1\leqslant j\leqslant m}\mu(\Sigma_{1\leqslant i\leqslant n}\alpha_{ij}\varphi_i)y_j=$$
$$=\Sigma_{1\leqslant j\leqslant m}\mu(\psi_j)y_j=0$$

and the proposition is proved.

2.8 Corollary. *Let $\mu:\mathscr{K}(T)\to\mathscr{L}(E,F)$ be a measure. If*

$$\Sigma_{1\leqslant i\leqslant n}\varphi_i x_i=\Sigma_{1\leqslant j\leqslant m}\psi_j y_j$$

where φ_i, $\psi_j\in\mathscr{K}(T)$ and x_i, $y_j\in E$, then

$$\Sigma_{1\leqslant i\leqslant n}\mu(\varphi_i)x_i=\Sigma_{1\leqslant j\leqslant m}\mu(\psi_j)y_j\,.$$

In fact

$$\Sigma_{1\leqslant i\leqslant n}\varphi_i(t)x_i-\Sigma_{1\leqslant j\leqslant m}\psi_j(t)y_j\equiv 0,$$

therefore

$$\Sigma_{1\leqslant i\leqslant n}\mu(\varphi_i)x_i-\Sigma_{1\leqslant j\leqslant m}\mu(\psi_j)y_j=0\,,$$

which yields the desired equality.

3 *Scalar measures*

The complex valued measures $\mu:\mathscr{K}(T)\to C$, defined for real functions, are called *complex* measures.

The real valued measures $\mu:\mathscr{K}(T)\to R$, defined for real functions, are called *real* measures.

The real and complex measures are called *scalar* measures.

Every real measure is at the same time a complex measure, since $R\subset C$. Therefore, the scalar measures can be always considered with values in C.

For every real measure μ we have $\bar{\mu}=\mu$.

A positive measure is a scalar measure μ which has positive values for positive functions:

$$\mu(\varphi)\geqslant 0 \text{ for } \varphi\geqslant 0,\ \varphi\in\mathscr{K}(T)\,.$$

Every positive measure μ is a real measure.

In fact, if $\varphi\in\mathscr{K}(T)$, then φ can be written $\varphi=\varphi_1-\varphi_2$ with φ_1 and φ_2 from $\mathscr{K}_+(T)$. Then $\mu(\varphi_1)\geqslant 0$ and $\mu(\varphi_2)\geqslant 0$, therefore

$$\mu(\varphi)=\mu(\varphi_1)-\mu(\varphi_2)\in R.$$

From proposition 2.4 it follows that every measure $\mu : \mathcal{K}(T) \to C$ can be extended in a unique way to a measure $m : \mathcal{K}_C(T) \to C$. For this reason, if we identify a measure $\mu : \mathcal{K}(T) \to C$ with its extension $m : \mathcal{K}_C(T) \to C$, the measures $m : \mathcal{K}_C(T) \to C$ defined for complex functions, will be called *scalar measures* too. Among the complex measures $m : \mathcal{K}_C(T) \to C$, the *real measures* are characterized by the property *of taking on real values for real functions* and the *positive measures* are characterized by the property of taking on *positive values* for *positive functions*. If μ is a complex measure, we have $\overline{\mu(f)} = \bar{\mu}(\bar{f})$, for $f \in \mathcal{K}_C(T)$. If μ is real, we have $\overline{\mu(f)} = \mu(\bar{f})$ for $f \in \mathcal{K}_C(T)$.

We shall denote by $\mathcal{M}_C$ the space of all complex measures, by $\mathcal{M}(T)$ the space of all real measures and by $\mathcal{M}_+(T)$ the set of all positive measures on T.

It is easy to verify that if μ and ν are two positive measures on T and $\alpha \geqslant 0$, then $\mu + \nu$ and $\alpha\mu$ are also positive measures on T. It follows that $\mathcal{M}_+(T)$ is a *convex cone* in the vector space $\mathcal{M}(T)$.

We define on $\mathcal{M}(T)$ the relation $\mu \leqslant \nu$ by the inequality:

$$\mu(\varphi) \leqslant \nu(\varphi) \text{ for every function } \varphi \geqslant 0 \text{ from } \mathcal{K}_+(T).$$

It follows that a measure μ is positive if and only if $\mu \geqslant 0$, and that $\mu \leqslant \nu$ if and only if $\nu - \mu$ is a positive measure.

It is easy to verify that the relation $\mu \leqslant \nu$ is an *order relation* on $\mathcal{M}(T)$, consistent with the structure of vector space of $\mathcal{M}(T)$:

(1) $\mu \leqslant \mu$ for every $\mu \in \mathcal{M}(T)$;

(2) $\mu \leqslant \nu$ and $\nu \leqslant \mu \Rightarrow \mu = \nu$ (according to corollary 1.22);

(3) $\mu \leqslant \nu$ and $\nu \leqslant \lambda \Rightarrow \mu \leqslant \lambda$;

(4) $\mu \leqslant \nu \Rightarrow \mu + \lambda \leqslant \nu + \lambda$;

(5) $\mu \geqslant 0$ and $\alpha \geqslant 0 \Rightarrow \alpha\mu \geqslant 0$.

Thus, $\mathcal{M}(T)$ is an ordered vector space for the relation $\mu \leqslant \nu$.

2.9 PROPOSITION. *Every complex measure m on T can be written uniquely in the form*

$$m = \mu + i\nu,$$

where μ and ν are real measures on T.

For every real function $\varphi \in \mathcal{K}(T)$ put

$$\mu(\varphi) = \operatorname{Re}[m(\varphi)], \quad \nu(\varphi) = \operatorname{Im}[m(\varphi)].$$

The real numbers $\mu(\varphi)$ and $\nu(\varphi)$ are uniquely determined by the equality

$$m(\varphi)=\mu(\varphi)+i\nu(\varphi).$$

It is easy to verify that μ and ν are real linear functionals on $\mathscr{K}(T)$, since the functions $z\to\mathrm{Re}(z)$ and $z\to\mathrm{Im}(z)$ are linear.

Let $K\subset T$ be a compact set and $a_K>0$ such that

$$|m(\varphi)|\leqslant a_K\|\varphi\| \text{ for every } \varphi\in\mathscr{K}(T,K).$$

Since

$$|\mu(\varphi)|\leqslant|m(\varphi)| \text{ and } |\nu(\varphi)|\leqslant|m(\varphi)|$$

we have

$$|\mu(\varphi)|\leqslant a_K\|\varphi\| \text{ and } |\nu(\varphi)|\leqslant a_K\|\varphi\|$$

for every function $\varphi\in\mathscr{K}(T,K)$. Therefore μ and ν are real measures on T.

The equality

$$m(\varphi)=\mu(\varphi)+i\nu(\varphi)$$

valid for *real* functions, remains valid also for complex functions because the extension from $\mathscr{K}(T)$ to $\mathscr{K}_C(T)$ is a linear operation.

2.10 COROLLARY. *If* $m=\mu+i\nu$, *then* $\overline{m}=\mu-i\nu$.

4 *Integration of vector functions with respect to a scalar measure*

2.11 THEOREM. *For every real (respectively complex) measure* $\mu:\mathscr{K}(T)\to C$ *and for every real (respectively complex) Banach space* E, *there exists a unique measure* $\boldsymbol{m}:\mathscr{K}_E(T)\to E$ *satisfying the equality*

$$\boldsymbol{m}(\varphi x)=\mu(\varphi)x \textit{ for every } \varphi\in\mathscr{K}(T) \textit{ and } x\in E.$$

If E *is a Banach space with involution, then*

$$\overline{\boldsymbol{m}}(\varphi x)=\bar{\mu}(\varphi)x, \textit{ for } \varphi\in\mathscr{K}(T) \textit{ and } x\in E.$$

The correspondence $\mu\to\boldsymbol{m}$ *is injective, linear and* $\|\boldsymbol{m}\|=\|\mu\|$.

Denote by $\mathscr{H}_E$ the set of the linear combinations of the form $g=\Sigma_{1\leqslant i\leqslant n}\varphi_i x_i$ of functions φ_i from $\mathscr{K}(T)$ with coefficients x_i from E. For each such function $g=\Sigma_{1\leqslant i\leqslant n}\varphi_i x_i$ put

$$\boldsymbol{n}(g)=\Sigma_{1\leqslant i\leqslant n}\mu(\varphi_i)\,x_i\in E.$$

From the corollary 2.8 it follows that $\boldsymbol{n}(\boldsymbol{g})$ does not depend on the particular form in which $\boldsymbol{g}$ is written as a linear combination of functions from $\mathscr{K}(T)$ with coefficients from E, but it depends on $\boldsymbol{g}$ only.

We have defined in this way a mapping $\boldsymbol{n} : \mathscr{H}_E \to E$ which, in particular, satisfies the equality

$$\boldsymbol{n}(\varphi x) = \mu(\varphi)x \text{ for } \varphi \in \mathscr{K}(T) \text{ and } x \in E .$$

It is easily verified that the application $\boldsymbol{n}$ is linear.

Let us show now that the linear mapping $\boldsymbol{n}$ fulfills the conditions of proposition 1.21.

Let $K \subset T$ be a compact set and $\boldsymbol{g} = \Sigma_{1 \leqslant i \leqslant n} \varphi_i x_i$ a function from $\mathscr{H}_E$ with support in K. Let $a_K > 0$ be such that

$$|\mu(\varphi)| \leqslant a_K \|\varphi\| \text{ for } \varphi \in \mathscr{K}(T, K) .$$

There exists an element $z' \in E'$ with $|z'| = 1$ such that

$$|\boldsymbol{n}(\boldsymbol{g})| = \langle \boldsymbol{n}(\boldsymbol{g}), z' \rangle.$$

We have

$$\begin{aligned} \langle \boldsymbol{n}(\boldsymbol{g}), z' \rangle &= \langle \Sigma \mu(\varphi_i) x_i, z' \rangle = \Sigma \mu(\varphi_i) \langle x_i, z' \rangle = \\ &= \Sigma \mu(\varphi_i \langle x_i, z' \rangle) = \mu(\Sigma \langle \varphi_i x_i, z' \rangle) = \\ &= \mu(\langle \Sigma \varphi_i x_i, z' \rangle) = \mu(\langle \boldsymbol{g}, z' \rangle) \end{aligned}$$

therefore

$$\begin{aligned} |\boldsymbol{n}(\boldsymbol{g})| &= \langle \boldsymbol{n}(\boldsymbol{g}), z' \rangle = \mu(\langle \boldsymbol{g}, z' \rangle) \leqslant \\ &\leqslant a_K \| \langle \boldsymbol{g}, z' \rangle \| \leqslant a_K \|\boldsymbol{g}\| \, |z'| = a_K \|\boldsymbol{g}\| . \end{aligned}$$

Hence

$$\|\boldsymbol{n}(\boldsymbol{g})\| \leqslant a_K \|\boldsymbol{g}\| \text{ for every } \boldsymbol{g} \in \mathscr{H}_E \cap \mathscr{K}_E(T, K)$$

that is, $\boldsymbol{n}$ is continuous on $\mathscr{H}_E \cap \mathscr{K}_E(T, K)$ for the topology of the uniform convergence.

By proposition 1.21, there exists then a unique measure $\boldsymbol{m} : \mathscr{K}_E(T) \to E$ whose restriction to the space $\mathscr{H}_E$ is equal to $\boldsymbol{n}$. For $\varphi \in \mathscr{K}(T)$ and $x \in E$ we have

$$\boldsymbol{m}(\varphi x) = \boldsymbol{n}(\varphi x) = \mu(\varphi) x .$$

If another measure $\boldsymbol{m}' : \mathscr{K}_E(T) \to E$ satisfies the equality

$$\boldsymbol{m}'(\varphi x) = \mu(\varphi) x \text{ for } \varphi \in \mathscr{K}(T) \text{ and } x \in E ,$$

then we have $\boldsymbol{m}'(\boldsymbol{g})=\boldsymbol{n}(\boldsymbol{g})$ for every $\boldsymbol{g}\in\mathscr{H}_E$ and therefore $\boldsymbol{m}'=\boldsymbol{m}$.

If $\boldsymbol{m}$ and $\boldsymbol{n}$ are the measures corresponding to the scalar measures μ and ν, then the measures $\boldsymbol{m}+\boldsymbol{n}$, and $\alpha\boldsymbol{m}$, α scalar, correspond to the measures $\mu+\nu$ and $\alpha\mu$ respectively.

In fact

$$\boldsymbol{m}(\varphi x)=\mu(\varphi)x \text{ and } \boldsymbol{n}(\varphi x)=\nu(\varphi)x$$

for $x\in E$ and $\varphi\in\mathscr{K}(T)$. Then

$$(\boldsymbol{m}+\boldsymbol{n})(\varphi x)=\boldsymbol{m}(\varphi)x+\boldsymbol{n}(\varphi x)=\mu(\varphi)x+\nu(\varphi)x=(\mu+\nu)(\varphi)x$$

and

$$(\alpha\boldsymbol{m})(\varphi x)=\alpha\boldsymbol{m}(\varphi x)+\alpha\mu(\varphi)x=(\alpha\mu)(\varphi)x\,.$$

Therefore, the mapping $\mu\to\boldsymbol{m}$ is linear. If E is with involution, we have

$$\bar{m}(\varphi x)=\overline{m(\varphi\bar{x})}=\overline{\mu(\varphi)\bar{x}}=\bar{\mu}(\varphi)x \text{ for } \varphi\in\mathscr{K}(T) \text{ and } x\in E.$$

It remains to verify the equality $||\mu||=||\boldsymbol{m}||$.

For every function $\varphi\in\mathscr{K}(T)$ and $x\in E$ with $|x|=1$ we have

$$|\mu(\varphi)|=|\mu(\varphi)|\,|x|=|\mu(\varphi)x|=|\boldsymbol{m}(\varphi x)|\leqslant\|\boldsymbol{m}\|\;\|\varphi x\|=\|\boldsymbol{m}\|\;\|\varphi\|$$

whence

$$\|\mu\|\leqslant\|m\|\leqslant+\infty\,.$$

Conversely, let $\boldsymbol{f}\in\mathscr{K}_E(T)$ be an arbitrary function and let K be the support of $\boldsymbol{f}$. Let U be a compact neighbourhood of K. Let $\varepsilon>0$ be given. There exists then a function $\boldsymbol{g}=\Sigma_{1\leqslant i\leqslant n}\varphi_i x_i$ with $\varphi_i\in\mathscr{K}(T,U)$ and $x_i\in E$, such that

$$|\boldsymbol{f}(t)-\boldsymbol{g}(t)|\leqslant\min(\varepsilon/a_U,\varepsilon)\,.$$

Then, $\|\boldsymbol{g}\|\leqslant\|\boldsymbol{f}\|+\varepsilon$; on the other hand, $\boldsymbol{f}-\boldsymbol{g}\in\mathscr{K}(T,U)$, therefore

$$|\boldsymbol{m}(\boldsymbol{f}-\boldsymbol{g})|<a_U\|\boldsymbol{f}-\boldsymbol{g}\|<\varepsilon$$

whence

$$|\boldsymbol{m}(\boldsymbol{f})|\leqslant|\boldsymbol{m}(\boldsymbol{g})|+\varepsilon\,.$$

Let $z'\in F'$ with $|z'|=1$ be such that

$$\langle\boldsymbol{m}(\boldsymbol{g}),z'\rangle=|\boldsymbol{m}(\boldsymbol{g})|\,.$$

We have

$$|\boldsymbol{m}(\boldsymbol{f})| - \varepsilon < |\boldsymbol{m}(\boldsymbol{g})| = \langle \boldsymbol{m}(\boldsymbol{g}), z' \rangle = \mu(\langle \boldsymbol{g}, z' \rangle) \leqslant$$
$$\leqslant \|\mu\| \, \|\langle \boldsymbol{g}, z' \rangle\| \leqslant \|\mu\| . \|\boldsymbol{g}\| < \|\mu\| (\|\boldsymbol{f}\| + \varepsilon).$$

ε being arbitrary, it follows that

$$|\boldsymbol{m}(\boldsymbol{f})| \leqslant \|\mu\| \, \|\boldsymbol{f}\|$$

and then $\|\boldsymbol{m}\| \leqslant \|\mu\|$. Taking into account the converse inequality proved above, it follows that $\|\mu\| = \|\boldsymbol{m}\|$. Thus the proposition is proved.

2.12 Corollary. *If $\mu : \mathcal{K}(T) \to C$ is a scalar measure and if $m : \mathcal{K}_C(T) \to C$ is the measure which extends μ from $\mathcal{K}(T)$ to $\mathcal{K}_C(T)$, then $\|\boldsymbol{m}\| = \|\mu\|$.*

In fact, the measure m satisfies the equality

$$m(\varphi z) = \mu(\varphi) z \text{ for } \varphi \in \mathcal{K}(T) \text{ and } z \in C .$$

Because of the correspondence $\mu \to m$, we identify the measure $\mu : \mathcal{K}(T) \to C$ with the measure $\boldsymbol{m} : \mathcal{K}_E(T) \to E$. For functions $\boldsymbol{f} \in \mathcal{K}_E(T)$, the integral $\int \boldsymbol{f} \mathrm{d}\boldsymbol{m}$ is also denoted by $\int \boldsymbol{f} \mathrm{d}\mu$ and it is called *the integral of the vector function* $\boldsymbol{f}$ *with respect to the scalar measure* μ. The equality in the statement of theorem 2.11 is written then

$$\mu(\varphi x) = \mu(\varphi) x \text{ or } \int \varphi x \mathrm{d}\mu = x \int \varphi \mathrm{d}\mu$$

for $\varphi \in \mathcal{K}(T)$ and $x \in E$.

2.13 Theorem. *Let $\mu : \mathcal{K}(T) \to C$ be a real (respectively complex) measure, E and F two real (respectively complex) Banach spaces. For every continuous linear mapping $U : E \to F$ we have*

$$U\left(\int \boldsymbol{f} \mathrm{d}\mu\right) = \int U(\boldsymbol{f}(t)) \, \mathrm{d}\mu(t), \; \boldsymbol{f} \in \mathcal{K}_E(T) .$$

For every function $\boldsymbol{g} = \Sigma \varphi_i x_i$ from $\mathcal{H}_E(T)$ with $\varphi_i \in \mathcal{K}(T)$ and $x_i \in E$ we have

$$U\left(\int g \, \mathrm{d}\mu\right) = U\left(\Sigma_{1 \leqslant i \leqslant n} \mu(\varphi_i) x_i\right) = \Sigma_{1 \leqslant i \leqslant n} \mu(\varphi_i) U x_i =$$
$$= \mu\left(\Sigma_{1 \leqslant i \leqslant n} \varphi_i U x_i\right) = \mu\left(U \, \Sigma_{1 \leqslant i \leqslant n} \varphi_i x_i\right) =$$
$$= \int U(g(t)) \, \mathrm{d}\mu(t)$$

and

$$U \circ g = \Sigma_{1 \leqslant i \leqslant n} \varphi_i U x_i \in \mathcal{H}_F(T) .$$

Let $\boldsymbol{f} \in \mathcal{K}_E(T)$ be arbitrary and let K be the support of $\boldsymbol{f}$. Let V be a compact neighbourhood of K and let $(\boldsymbol{g}_n)$ be a sequence of functions from $\mathcal{H}_E$ with support in V, uniformly convergent to $\boldsymbol{f}$. Then the functions $U \circ \boldsymbol{g}_n$ are from $\mathcal{H}_E(T)$, have the support contained in V and are uniformly convergent to $U \circ \boldsymbol{f}$. Since μ is continuous on $\mathcal{K}_E(T, V)$ and $\mathcal{K}_F(T, V)$ for the topology of the uniform convergence, passing to the limit in the equalities

$$U(\int \boldsymbol{g}_n \mathrm{d}\mu) = \int U(\boldsymbol{g}_n(t))\mathrm{d}\mu(t)$$

we obtain

$$U(\int \boldsymbol{f} \mathrm{d}\mu) = \int U(\boldsymbol{f}(t))\mathrm{d}\mu(t)\,.$$

2.14 Corollary. *Let $\mu : \mathcal{K}(T) \to C$ be a real (respectively complex) measure and E a real (respectively complex) Banach space. For every $z' \in F'$ we have*

$$\langle \int \boldsymbol{f} \mathrm{d}\mu, z' \rangle = \int \langle \boldsymbol{f}, z' \rangle \mathrm{d}\mu, \ \boldsymbol{f} \in \mathcal{K}(T)\,.$$

5 *Positive measures*

We shall prove first an auxiliary proposition.

2.15 Proposition. *An additive functional $\mu : \mathcal{K}(T) \to R$ is positive if and only if it is increasing.*

Suppose first that μ is positive:

$$\mu(\varphi) \geqslant 0 \text{ for } \varphi \geqslant 0 \text{ from } \mathcal{K}(T)$$

and let us show that μ is increasing,

$$\mu(\varphi) \leqslant \mu(\psi) \text{ if } \varphi \leqslant \psi\,.$$

In fact, if $\varphi \leqslant \psi$, $(\varphi, \psi \in \mathcal{K}(T))$, then $\psi - \varphi \geqslant 0$ therefore $\mu(\psi - \varphi) \geqslant 0$. But

$$\mu(\psi) = \mu(\psi - \varphi + \varphi) = \mu(\psi - \varphi) + \mu(\varphi)$$

therefore

$$\mu(\psi - \varphi) = \mu(\psi) - \mu(\varphi)$$

and then

$$\mu(\psi)-\mu(\varphi)\geqslant 0$$

whence

$$\mu(\varphi)\leqslant\mu(\psi)\,.$$

Conversely, supposing that μ is increasing, it follows that if $\varphi\geqslant 0$, $\varphi\in\mathscr{K}(T)$, then $\mu(\varphi)\geqslant\mu(0)=0$, thus μ is positive.

2.16 COROLLARY. *Every positive measure μ on T is increasing*

2.17 COROLLARY. *If μ is a positive measure on T, then*

$$\|\mu\|=\sup\nolimits_{0\leqslant f\leqslant 1}\mu(f)\leqslant +\infty,\quad f\in\mathscr{K}(T)\,.$$

Evidently

$$\sup\nolimits_{0\leqslant f\leqslant 1}\mu(f)\leqslant \sup\nolimits_{|g|\leqslant 1}|\mu(g)|=\|\mu\|,\quad f,g\in\mathscr{K}(T)\,.$$

Conversely, let $g\in\mathscr{K}$ with $|g|\leqslant 1$. Since μ is increasing, from the relation $-|g|\leqslant g\leqslant|g|$ we deduce that

$$-\mu(|g|)\leqslant\mu(g)\leqslant\mu(|g|),$$

therefore

$$|\mu(g)|\leqslant\mu(|g|)\leqslant\sup\nolimits_{f\in\mathscr{K},\,0\leqslant f\leqslant 1}\mu(f),\qquad f\in\mathscr{K}(T)\,,$$

and then

$$\|\mu\|\leqslant\sup\nolimits_{0\leqslant f\leqslant 1}\mu(f),\quad f\in\mathscr{K}(T)\,.$$

2.18 COROLLARY. *If μ and v are two positive measures on T and if $\mu\leqslant v$, then* $\|\mu\|\leqslant\|v\|$.

In fact, for every function $f\geqslant 0$ from $\mathscr{K}$ we have $\mu(f)\leqslant v(f)$. Taking in this inequality the supremum for $f\in\mathscr{K}$, $0\leqslant f\leqslant 1$, we get $\|\mu\|\leqslant\|v\|$.

The next proposition shows that for positive measures, the continuity condition in definition 1.9 is superfluous.

2.19 PROPOSITION. *Every linear positive functional $\mu:\mathscr{K}(T)\to R$, is a positive measure on T.*

Let $K\subset T$ be an arbitrary compact set and let $\varphi_0: T\to[0,1]$ be a continuous function with compact support, such that $\varphi_0(t)=1$ for every $t\in K$. For every function $\varphi\in\mathscr{K}(T, K)$ we have

$$\|\varphi\|\varphi_0\leqslant\varphi\leqslant\|\varphi\|\varphi_0\,.$$

Since the functional μ is positive, it is also increasing, hence

$$-\|\varphi\|\mu(\varphi_0)\leqslant\mu(\varphi)\leqslant\|\varphi\|\mu(\varphi_0)$$

whence

$$|\mu(\varphi)|\leqslant\mu(\varphi_0)\|\varphi\|\,.$$

Taking $a_K=\mu(\varphi_0)$, it follows that

$$|\mu(\varphi)|\leqslant a_K\|\varphi\| \text{ for every } \varphi\in\mathscr{K}(T,K)$$

therefore μ is a positive measure on T and the proposition is proved.

As any measure defined on $\mathscr{K}(T)$, the positive measures are uniquely determined by the values taken on $\mathscr{K}_+(T)$; however, positive measures have an additional property.

2.20 Proposition. *Every additive and positive functional $\mu:\mathscr{K}_+(T)\to R$ can be extended uniquely to a positive measure on T.*

For every function $f\in\mathscr{K}(T)$, we write $f=\varphi-\psi$ with $\varphi, \psi\in\mathscr{K}_+(T)$ and we define

$$m(f)=\mu(\varphi)-\mu(\psi)\,.$$

It can be shown, as in the proof of proposition 2.3, that m depends on f only, extends μ and is additive on $\mathscr{K}(T)$. It follows that m is a *positive* functional:

$$m(f)\geqslant 0, \text{ for } f\in\mathscr{K}_+(T),$$

therefore it is increasing (proposition 2.15). It remains to show that m is *homogeneous.* Let us remark first that if $f=\varphi-\psi$ with $\varphi, \psi\in\mathscr{K}_+$, then $-f=\psi-\varphi$, therefore.

$$m(-f)=\mu(\psi)-\mu(\varphi)=-m(f)\,.$$

Let $f\in\mathscr{K}(T)$. For every natural number n we have

$$m(nf)=m(f+f+\ldots+f)=m(f)+m(f)+\ldots+m(f)=n\cdot m(f)$$

and

$$m(-nf)=m(n(-f))=n\cdot m(-f)=n(-m(f))=-n\cdot m(f)\,.$$

If p and q are two integers and $q \neq 0$, then

$$qm(pq^{-1}f) = m(q \cdot pq^{-1}f) = m(pf) = pm(f)$$

therefore

$$m(pq^{-1}f) = pq^{-1}m(f) .$$

Hence, for every rational number r we have

$$m(rf) = rm(f) .$$

Let α be an arbitrary real number. There exist two sequences (r_n) and (s_n) of rational numbers, convergent to α, such that for every n,

$$r_n \leqslant \alpha \leqslant s_n .$$

If f is positive, we have

$$r_n f \leqslant \alpha f \leqslant s_n f ;$$

since m is increasing, we deduce

$$r_n m(f) = m(r_n f) \leqslant m(\alpha f) \leqslant m(s_n f) = s_n m(f)$$

and

$$\lim_{n\to\infty} r_n m(f) = \lim_{n\to\infty} s_n m(f) = \alpha m(f) .$$

Taking the limit in the previous inequalities we deduce that

$$\alpha m(f) = m(\alpha, f) .$$

If $f = f_1 - f_2$ with f_1 and f_2 from $\mathscr{K}_+(T)$, then

$$\alpha f = \alpha f_1 - \alpha f_2$$

and αf_1 and αf_2 are either both positive or both negative. Then

$$m(\alpha f) = m(\alpha f_1) - m(\alpha f_2) = \alpha m(f_1) - \alpha m(f_2) = \\ = \alpha(m(f_1) - m(f_2)) = \alpha m(f).$$

Thus the proposition is proved.

2.21 Proposition. *Let μ be a positive measure on T. For every set $H \subset \mathscr{K}_+$ directed for the relation $\leqslant$, the positive functional $g \to \mu(g)$ has a (finite or $+\infty$) limit following the filter of sections of H and*

$$\lim_{g \in H} \mu(g) = \sup_{g \in H} \mu(g).$$

If $\sup_{g\in H} g \in \mathscr{K}_+$, *then*

$$\lim_{g\in H} \mu(g) = \mu(\lim_{g\in H} g) = \mu(\sup_{g\in H} g) .$$

Let us denote $a=\sup_{g\in H} \mu(g)$. We have $0\leqslant a\leqslant \infty$. Let $\alpha < a$. There exists a function $g_\alpha \in H$ such that $\alpha < \mu(g_\alpha) \leqslant a$. Then for every function $g \geqslant g_\alpha$ from H we have $\mu(g) \geqslant \mu(g_\alpha)$, therefore $\alpha < \mu(g) \leqslant a$. We deduce that every neighbourhood of a contains the image under μ of a section $S(g_\alpha) = \{g \mid g \in H, g \geqslant g_\alpha\}$ of H. This means that the function μ has the limit a, when g runs over the filter of sections of H:

$$a = \lim_{g\in H} \mu(g) = \sup_{g\in H} \mu(g) .$$

2.22 COROLLARY. *Let μ be a positive measure on T. For every function $f\geqslant 0$ defined on T we have*

$$0\leqslant \lim_{0\leqslant g\leqslant f} \mu(g) = \sup_{0\leqslant g\leqslant f} \mu(g) \leqslant \infty, \quad g \in \mathscr{K} .$$

In fact, the set $H=\{g \mid g\in \mathscr{K}_+, g\leqslant f\}$ is directed for the relation $\leqslant$.

2.23 COROLLARY. *If μ is a positive measure on T, then*

$$\|\mu\| = \lim_{0\leqslant g\leqslant 1} \mu(g) \leqslant +\infty, \quad g \in \mathscr{K} .$$

In fact, taking $f(t)\equiv 1$ in corollary 2.22 we deduce,

$$\lim_{0\leqslant g\leqslant 1} \mu(g) = \sup_{0\leqslant g\leqslant 1} \mu(g) \leqslant +\infty, \quad g\in \mathscr{K} .$$

The statement follows then from corollary 2.17.

2.24 COROLLARY. *If μ and ν are two positive measures on T, then*

$$\|\mu+\nu\| = \|\mu\| + \|\nu\| .$$

In fact, $\mu+\nu$ is a positive measure on T, therefore

$$\|\mu+\nu\| = \lim_{0\leqslant g\leqslant 1} (\mu+\nu)(g) = \lim_{0\leqslant g\leqslant 1} (\mu(g)+\nu(g)) =$$
$$= \lim_{0\leqslant g\leqslant 1} \mu(g) + \lim_{0\leqslant g\leqslant 1} \nu(g) = \|\mu\| + \|\nu\| .$$

6 *Measures with scalar values*

Let E be a (real or complex) Banach space. We shall prove first the following lemma:

2.25 LEMMA. *Let* $\boldsymbol{h} \in \mathscr{K}_E(T)$ *and* $g \in \mathscr{K}_+(T)$ *be such that* $0 \leqslant g \leqslant |\boldsymbol{h}|$. *Put* $\sigma_{\boldsymbol{h}}(t) = \boldsymbol{h}(t)/|\boldsymbol{h}(t)|$ *if* $\boldsymbol{h}(t) \neq 0$ *and* $\sigma_{\boldsymbol{h}}(t) = 0$ *if* $\boldsymbol{h}(t) = 0$. *Then* $\sigma_{\boldsymbol{h}}|\boldsymbol{h}| = \boldsymbol{h}$ *and* $\sigma_{\boldsymbol{h}} g \in \mathscr{K}_E(T)$.

The equality $\sigma_{\boldsymbol{h}}|\boldsymbol{h}| = \boldsymbol{h}$ is immediate. The function $\sigma_{\boldsymbol{h}} g$ has a compact support (contained in the support of g). Let us prove that $\sigma_{\boldsymbol{h}} g$ is continuous.

Let $t_0 \in T$. If $\boldsymbol{h}(t_0) \neq 0$, then there exists an open neighbourhood V of t_0 such that $\boldsymbol{h}(t) \neq 0$ for every $t \in V$. Then the function $|\boldsymbol{h}|$ is continuous and nonzero on V, therefore the function $\sigma_{\boldsymbol{h}} = \boldsymbol{h}/|\boldsymbol{h}|$ is continuous on V, in particular it is continuous at t_0; consequently the function $\sigma_{\boldsymbol{h}} g$ is also continuous at t_0.

If $\boldsymbol{h}(t_0) = 0$, then $g(t_0) = 0$. Let $\varepsilon > 0$; there exists a neighbourhood V of t_0 such that $g(t) \leqslant \varepsilon$ for every $t \in V$. Then

$$|\sigma_{\boldsymbol{h}}(t) g(t)| = |\sigma_{\boldsymbol{h}}(t)| g(t) \leqslant g(t) \leqslant \varepsilon$$

for $t \in V$. Since $\sigma_{\boldsymbol{h}}(t_0) g(t_0) = 0$, it follows that $\sigma_{\boldsymbol{h}} g$ is continuous at t_0.

Now let $\boldsymbol{m} : \mathscr{K}_E(T) \to C$ be a measure with scalar values.

For each function $\varphi \geqslant 0$ from $\mathscr{K}_+(T)$ put

$$\mu(\varphi) = \sup_{|\boldsymbol{f}| \leqslant \varphi} |\boldsymbol{m}(\boldsymbol{f})|, \quad \boldsymbol{f} \in \mathscr{K}_E(T). \tag{1}$$

2.26 PROPOSITION. *The functional* μ *defined on* $\mathscr{K}_+(T)$ *by the equality* (1) *has the following properties*

(1) $0 \leqslant \mu(\varphi) < +\infty$ *for every* $\varphi \in \mathscr{K}_+(T)$;

(2) $\mu(\varphi_1 + \varphi_2) = \mu(\varphi_1) + \mu(\varphi_2)$, $\varphi_1, \varphi_2 \in \mathscr{K}_+(T)$;

(3) $|\boldsymbol{m}(\boldsymbol{f})| \leqslant \mu(|\boldsymbol{f}|)$ *for every* $\boldsymbol{f} \in \mathscr{K}_E(T)$.

In fact, let $\varphi \in \mathscr{K}_+(T)$ and let K be its support; let $a_K > 0$ be such that

$$|\boldsymbol{m}(\boldsymbol{f})| \leqslant a_K \|\boldsymbol{f}\| \quad \text{for every } \boldsymbol{f} \in \mathscr{K}_E(T, K).$$

Every function $\boldsymbol{f} \in \mathscr{K}_E(T)$, such that $|\boldsymbol{f}| \leqslant \varphi$, has its support contained in K, therefore

$$\mu(\varphi) = \sup_{|\boldsymbol{f}| \leqslant \varphi} |\boldsymbol{m}(\boldsymbol{f})| \leqslant \sup_{|\boldsymbol{f}| \leqslant \varphi} a_K \|\boldsymbol{f}\| \leqslant a_K \|\varphi\| < +\infty .$$

The inequality $\mu(\varphi) \geqslant 0$ is evident and thus property (1) is proved.

Property (3) follows from equality (1) taking $\varphi = |\boldsymbol{f}|$.

It remains to show that μ is additive on $\mathscr{K}_+(T)$.

Let $\varphi, \varphi' \in \mathscr{K}_+(T)$ and $\varepsilon > 0$. There exist two functions $f, f' \in \mathscr{K}_E(T)$ with $|\boldsymbol{f}| \leqslant \varphi$, $|\boldsymbol{f}'| \leqslant \varphi'$ and

$$\mu(\varphi) \leqslant |\boldsymbol{m}(\boldsymbol{f})| + \varepsilon/2, \quad \mu(\varphi') \leqslant |\boldsymbol{m}(\boldsymbol{f}')| + \varepsilon/2 .$$

There exist two complex numbers θ and θ' with $|\theta| = |\theta'| = 1$, such that

$$|\boldsymbol{m}(\boldsymbol{f})| = \theta \boldsymbol{m}(\boldsymbol{f}) = \boldsymbol{m}(\theta \boldsymbol{f}),$$
$$|\boldsymbol{m}(\boldsymbol{f}')| = \theta' \boldsymbol{m}(\boldsymbol{f}') = \boldsymbol{m}(\theta' \boldsymbol{f}') .$$

Then

$$|\theta \boldsymbol{f} + \theta' \boldsymbol{f}'| \leqslant |\theta \boldsymbol{f}| + |\theta' \boldsymbol{f}'| \leqslant |\boldsymbol{f}| + |\boldsymbol{f}'| \leqslant \varphi + \varphi'$$

and

$$\mu(\varphi) + \mu(\varphi') \leqslant |\boldsymbol{m}(\boldsymbol{f})| + |\boldsymbol{m}(\boldsymbol{f}')| + \varepsilon = \boldsymbol{m}(\theta \boldsymbol{f} + \theta' \boldsymbol{f}') + \varepsilon \leqslant \mu(\varphi + \varphi') + \varepsilon .$$

ε being arbitrary, it follows

$$\mu(\varphi) + \mu(\varphi') \leqslant \mu(\varphi + \varphi') .$$

In order to prove the converse inequality, let $\boldsymbol{h} \in \mathscr{K}_E(T)$ be such that $|\boldsymbol{h}| \leqslant \varphi + \varphi'$ and

$$\mu(\varphi + \varphi') \leqslant |\boldsymbol{m}(\boldsymbol{h})| + \varepsilon .$$

The function $g = \inf(|\boldsymbol{h}|, \varphi)$ is positive, belongs to $\mathscr{K}_+(T)$ and we have $g \leqslant |\boldsymbol{h}|$ and $g \leqslant \varphi$. The function $g' = |\boldsymbol{h}| - g$ belongs also to $\mathscr{K}_+(T)$ and we have $g' \leqslant |\boldsymbol{h}|$ and $g' \leqslant \varphi'$. In fact, if $|\boldsymbol{h}(t_0)| \leqslant \varphi(t_0)$, then $g(t_0) = |\boldsymbol{h}(t_0)|$ therefore $g'(t_0) = 0$; if $|\boldsymbol{h}(t_0)| > \varphi(t_0)$, then $g(t_0) = \varphi(t_0)$ and

$$g'(t_0) = |\boldsymbol{h}(t_0)| - g(t_0) = |\boldsymbol{h}(t_0)| - \varphi(t_0) \leqslant \varphi(t_0) + \varphi'(t_0) - \varphi(t_0) = \varphi'(t_0) .$$

Consider now the function $\sigma_{\boldsymbol{h}}$ defined in the previous lemma. We have

$$\boldsymbol{h} = \sigma_{\boldsymbol{h}} |\boldsymbol{h}| = \sigma_{\boldsymbol{h}} g + \sigma_{\boldsymbol{h}} g'$$

and the functions $\sigma_{\boldsymbol{h}} g$ and $\sigma_{\boldsymbol{h}} g'$ belong to $\mathscr{K}_E(T)$. We have also

$$|\sigma_{\boldsymbol{h}} g| \leqslant g \leqslant \varphi \quad \text{and} \quad |\sigma_{\boldsymbol{h}} g'| \leqslant g' \leqslant \varphi'$$

therefore

$$\mu(\varphi + \varphi') \leqslant \boldsymbol{m}(\sigma_{\boldsymbol{h}} g + \sigma_{\boldsymbol{h}} g') + \varepsilon \leqslant$$
$$\leqslant |\boldsymbol{m}(\sigma_{\boldsymbol{h}} g)| + |\boldsymbol{m}(\sigma_{\boldsymbol{h}} g')| + \varepsilon \leqslant \mu(\varphi) + \mu(\varphi') + \varepsilon .$$

ε being arbitrary it follows

$$\mu(\varphi + \varphi') \leqslant \mu(\varphi) + \mu(\varphi')$$

and then

$$\mu(\varphi+\varphi')=\mu(\varphi)+\mu(\varphi')$$

and the proposition is proved.

From proposition 2.20 it follows that there exists a unique positive measure on T, denoted also by μ, which satisfies the equality (1) for every function $\varphi \geqslant 0$ from $\mathscr{K}(T)$.

2.27 PROPOSITION. *The positive measure μ on T defined by the equality* (1), *is the smallest positive measure on T which satisfies the inequality*

$$|\boldsymbol{m}(\boldsymbol{f})| \leqslant \mu(|\boldsymbol{f}|) \text{ for every } \boldsymbol{f} \in \mathscr{K}_E(T). \tag{2}$$

In fact, let ν be a positive measure on T such that

$$|\boldsymbol{m}(\boldsymbol{f})| \leqslant \nu(|\boldsymbol{f}|) \text{ for every } \boldsymbol{f} \in \mathscr{K}_E(T).$$

For every function $\varphi \in \mathscr{K}_+(T)$ we have

$$\mu(\varphi)=\sup_{|\boldsymbol{f}| \leqslant \varphi}|\boldsymbol{m}(\boldsymbol{f})| \leqslant \sup_{|\boldsymbol{f}| \leqslant \varphi} \nu(|\boldsymbol{f}|) \leqslant \nu(\varphi)$$

that is $\mu \leqslant \nu$.

The positive measure defined by the equality (1) will be denoted by $|\boldsymbol{m}|$ and called the *modulus* (or the variation) of $\boldsymbol{m}$.

If μ is a *positive* measure on T, then $|\mu|=\mu$.

Using this notation, the inequality (2) can be written

$$|\int \boldsymbol{f} \mathrm{d}\boldsymbol{m}| \leqslant \int |\boldsymbol{f}| \mathrm{d}|\boldsymbol{m}| \text{ for every } \boldsymbol{f} \in \mathscr{K}_E(T) ;$$

if μ is a positive measure, then

$$|\int \varphi \mathrm{d}\mu| \leqslant \int |\varphi| \mathrm{d}\mu \text{ for every } \varphi \in \mathscr{K}(T).$$

2.28 PROPOSITION. *We have $\|\boldsymbol{m}\| = \|\mu\| \leqslant +\infty$ if $\mu=|\boldsymbol{m}|$.*

In fact

$$\|\boldsymbol{m}\| = \sup_{\|\boldsymbol{f}\| \leqslant 1} |\boldsymbol{m}(\boldsymbol{f})| \leqslant \sup_{\|\boldsymbol{f}\| \leqslant 1} \mu(|\boldsymbol{f}|) \leqslant$$

$$\leqslant \sup_{\|\varphi\| \leqslant 1} |\mu(\varphi)| = \|\mu\| , \quad \boldsymbol{f} \in \mathscr{K}_E, \varphi \in \mathscr{K} .$$

In order to prove the converse inequality, take $\varphi \in \mathscr{K}(T)$ with $\|\varphi\| \leqslant 1$. Since $\mu \geqslant 0$, we have $|\mu| = \mu$, therefore

$$|\mu(\varphi)| \leqslant \mu(|\varphi|) = \sup_{|f| \leqslant |\varphi|} |\boldsymbol{m}(f)| \leqslant \sup_{\|f\| \leqslant 1} |\boldsymbol{m}(f)| = \|\boldsymbol{m}\| , f \in \mathscr{K}_E .$$

whence

$$\|\mu\| = \sup_{\|\varphi\| \leqslant 1} |\mu(\varphi)| \leqslant \|\boldsymbol{m}\| , \quad \varphi \in \mathscr{K} .$$

Consequently $\|\boldsymbol{m}\| = \|\mu\|$.

2.29 Corollary. *We have*

$$\|\boldsymbol{m}\| = \sup_{0 \leqslant \varphi \leqslant 1} |\boldsymbol{m}|(\varphi) \leqslant \infty, \quad \varphi \in \mathscr{K} .$$

In fact, for the positive measure $\mu = |\boldsymbol{m}|$ we have

$$\|\mu\| = \sup_{0 \leqslant \varphi \leqslant 1} \mu(\varphi) , \quad \varphi \in \mathscr{K} .$$

2.30 Corollary. *The measure* **m** *is continuous on* $\mathscr{K}_E(T)$ *for the topology of the uniform convergence if and only if the measure* $|\boldsymbol{m}|$ *is continuous on* $\mathscr{K}(T)$ *for this topology.*

2.31 Proposition. *Every real measure* μ *is the difference of two positive measures.*

Let $|\mu|$ be the modulus of μ. For every function $\varphi \geqslant 0$ from $\mathscr{K}(T)$ we have

$$|\mu(\varphi)| \leqslant |\mu|(\varphi)$$

therefore

$$-|\mu|(\varphi) \leqslant \mu(\varphi) \leqslant |\mu|(\varphi) .$$

It follows that

$$-|\mu| \leqslant \mu \leqslant |\mu| \text{ and } -|\mu| \leqslant -\mu \leqslant |\mu|$$

therefore $|\mu| - \mu \geqslant 0$ and $|\mu| + \mu \geqslant 0$. Let us denote

$$\mu_+ = \tfrac{1}{2}(|\mu| + \mu) \text{ and } \mu_- = \tfrac{1}{2}(|\mu| - \mu).$$

Then μ_+ and μ_- are positive measures and we have

$$\mu = \mu_+ - \mu_- .$$

Remark. We have also

$$|\mu| = \mu_+ + \mu_- .$$

The measures μ_+ and μ_- are called respectively the positive and negative part of the real measure μ.

A real measure μ can be written in different ways as a difference of two positive measure. For example, if $\nu \geqslant 0$, then

$$\mu_+ + \nu \geqslant 0, \ \mu_- + \nu \geqslant 0 \text{ and } \mu = (\mu_+ + \nu) - (\mu_- - \nu) .$$

The measures μ_+ and μ_- are the only positive measures which satisfy at the same time the equalities

$$\mu = \mu_+ - \mu_- \text{ and } |\mu| = \mu_+ + \mu_- .$$

2.32 COROLLARY. *If μ is a real measure, then*

$$\|\mu\| = \|\mu_+\| + \|\mu_-\| \leqslant +\infty.$$

In fact, $|\mu| = \mu_+ + \mu_-$; therefore (corollary 2.24) we have

$$\| \, |\mu| \, \| = \|\mu_+\| + \|\mu_-\| .$$

On the other hand $\|\mu\| = \| \, |\mu| \, \|$, therefore

$$\|\mu\| = \|\mu_+\| + \|\mu_-\| .$$

§3 Dominated measures

1 *Dominated families of real measures*

We shall prove that in the ordered vector space $\mathscr{M}(T)$ of the real measures on T, every dominated (majorized) set has a supremum (least upper bound).

We prove first

3.1 PROPOSITION. *For every real measure μ we have*

$$|\mu| = \sup(\mu, -\mu), \ \mu_+ = \sup(\mu, 0), \ \mu_- = \sup(-\mu, 0) .$$

From the inequality

$$-|\mu| \leqslant \mu \leqslant |\mu|$$

we deduce

$$|\mu| \geqslant \mu \text{ and } |\mu| \geqslant -\mu$$

therefore

$$|\mu| \geqslant \sup(-\mu, \mu).$$

Let us show that if $\nu \geqslant \mu$ and $\nu \geqslant -\mu$, then $\nu \geqslant |\mu|$; it will then follow that $|\mu| = \sup(-\mu, \mu)$. For every function $\varphi \in \mathscr{K}(T)$ we have $\mu(\varphi) \leqslant \nu(\varphi)$ and $-\mu(\varphi) \leqslant \nu(\varphi)$ therefore $|\mu(\varphi)| \leqslant \nu(\varphi)$. In particular, it follows that $\nu(\varphi) \geqslant 0$ for $\varphi \in \mathscr{K}_+(T)$, that is $\nu \geqslant 0$.

Now let $\varphi \in \mathscr{K}(T)$ be arbitrary. Denote

$$\varphi_+ = \tfrac{1}{2}(|\varphi| + \varphi) \text{ and } \varphi_- = \tfrac{1}{2}(|\varphi| - \varphi).$$

Then $\varphi_+, \varphi_- \in \mathscr{K}(T)$ and

$$\varphi = \varphi_+ - \varphi_-; \ |\varphi| = \varphi_+ + \varphi_-.$$

We have then

$$|\mu(\varphi)| = |\mu(\varphi_+) - \mu(\varphi_-)| \leqslant |\mu(\varphi_+)| + |\mu(\varphi_-)| \leqslant \nu(\varphi_+) - \nu(\varphi_-) = \nu(|\varphi|).$$

As $|\mu|$ is the smallest measure satisfying the inequality $|\mu(\varphi)| \leqslant |\mu|(\varphi)$ for every $\varphi \in \mathscr{K}(T)$, it follows that $|\mu| \leqslant \nu$ and then

$$|\mu| = \sup(\mu, -\mu).$$

The other equalities follow from the relations (which have to be understood in the following way: if the supremum in one side of the equality exists, then the supremum in the other side exists also and the equality itself is true).

$$\sup(\mu, \nu) + \lambda = \sup(\mu + \lambda, \nu + \lambda),$$

$$\sup(\alpha\mu, \alpha\nu) = \alpha \sup(\mu, \nu), \ \alpha \geqslant 0.$$

Namely

$$|\mu| + \mu = \sup(\mu, -\mu) + \mu = \sup(2\mu, 0)$$

$$|\mu| - \mu = \sup(\mu, -\mu) - \mu = \sup(0, -2\mu)$$

whence

$$\mu_+ = \tfrac{1}{2}(|\mu| + \mu) = \sup(\mu, 0),$$

$$\mu_- = \tfrac{1}{2}(|\mu| - \mu) = \sup(0, -\mu) = \sup(-\mu, 0).$$

3.2 PROPOSITION. *If μ is a real measure then*

$$\inf(\mu_+, \mu_-) = 0 \text{ and } \sup(\mu_+, \mu_-) = |\mu| .$$

If μ_1 and μ_2 are two positive measures with $\mu = \mu_1 - \mu_2$ and if $\inf(\mu_1, \mu_2) = 0$, then $\mu_1 = \mu_+$ and $\mu_2 = \mu_-$.

Denote $\lambda = \inf(\mu_+, \mu_-)$ and consider the measures $\nu_1 = \mu_+ - \lambda$ and $\nu_2 = \mu_- - \lambda$. We have, evidently, $\nu_1 \geqslant 0$ and $\nu_2 \geqslant 0$ and $\mu = \nu_1 - \nu_2$, whence $\mu \leqslant \nu_1$. Then

$$\mu_+ = \sup(\mu, 0) \leqslant \sup(\nu, 0) = \nu .$$

It follows that $\lambda = \mu_+ - \nu \leqslant 0$. But $\lambda \geqslant 0$, therefore $\lambda = 0$, that is

$$\inf(\mu_+, \mu_-) = 0 .$$

If $\mu = \mu_1 - \mu_2$ and $\mu_1 \geqslant 0$, $\mu_2 \geqslant 0$, then $\mu \leqslant \mu_1$ and

$$\mu_+ = \sup(\mu, 0) \leqslant \sup(\mu_1, 0) = \mu_1 .$$

Denote $\lambda = \mu_1 - \mu_+$. We have

$$\mu_2 = \mu_1 - \mu = \mu_1 - \mu_+ + \mu_- = \lambda + \mu_- .$$

From the equalities $\mu_1 = \lambda + \mu_+$ and $\mu_2 = \lambda + \mu_-$ it follows that $\lambda \leqslant \mu_1$ and $\lambda \leqslant \mu_2$; therefore $\lambda \leqslant \inf(\mu_1, \mu_2) = 0$. But $\lambda \geqslant 0$, therefore $\lambda = 0$, whence $\mu_1 = \mu_+$ and $\mu_2 = \mu_-$.

The equality $\sup(\mu_+, \mu_-) = |\mu|$ is obtained in the following way:

$$\sup(\mu_+, \mu) = (\sup(\mu_+, \mu_-) - \mu_-) + \mu_- =$$
$$= \sup(\mu_+ - \mu_-, 0) + \mu_- = \sup(\mu, 0) + \mu_- = \mu_+ + \mu_- = |\mu| .$$

3.3 PROPOSITION. *Every finite set of real measures has a supremum and an infimum.*

It is sufficient to prove the proposition for two measures μ and ν.

In proposition 3.1 it was proved that there exists $\sup(\mu - \nu, 0) = (\mu - \nu_+)$. Then there exists also $\sup(\mu, \nu)$ and

$$\sup(\mu, \nu) = \sup(\mu - \nu, 0) + \nu = (\mu - \nu)_+ + \nu .$$

Let us remark further that $\sup(\mu, \nu) = \sup(-\mu + \nu, 0) + \mu = (\mu - \nu)_- + \mu$.

Then it follows that

$$\sup(\mu, \nu) = \tfrac{1}{2}(\mu + \nu + (\mu - \nu)_+ + (\mu - \nu)_-) ,$$

that is

$$\sup(\mu, \nu) = \tfrac{1}{2}(\mu + \nu + |\mu - \nu|).$$

We deduce then that there exists $\inf(\mu, \nu) = -\sup(-\mu, -\nu)$ and the proposition is proved.

Remark. Proposition 3.3 asserts that the ordered space $\mathscr{M}(T)$ of real measures on T is *reticulated.* A reticulated vector space is called a Riesz space.

3.4 Proposition. *Every nonempty set A of positive measures, dominated and directed for the relation $\leqslant$, has a supremum.*

Since A is dominated, there exists a measure $\mu_0 \geqslant 0$ such that $\mu \leqslant \mu_0$ for every $\mu \in A$.

For every positive function $\varphi \in \mathscr{K}_+(T)$ put

$$\nu(\varphi) = \sup_{\mu \in A} \mu(\varphi).$$

Then we have

$$0 \leqslant \nu(\varphi) \leqslant \mu_0(\varphi) < +\infty \quad \text{for } \varphi \in \mathscr{K}_+(T).$$

Since A is directed we have

$$\nu(\varphi) = \lim_{\mu \in A} \mu(\varphi) \quad \text{for } \varphi \in \mathscr{K}_+(T).$$

From this it follows that ν is additive on $\mathscr{K}_+(T)$; therefore, it can be extended to a measure on T, still denoted by ν. From the definition of ν it follows that $\mu \leqslant \nu$ for every $\mu \in A$, and also that $\nu \leqslant \mu_0$ for every measure μ_0 dominating A, therefore ν is the smallest measure dominating A:

$$\nu = \sup A.$$

3.5 Proposition. *Every nonempty dominated set H of real measures has a supremum.*

Let A be the set of the least upper bounds of all finite families of measures from H. The set A is directed for the relation $\leqslant$.

In fact, let $\mu_1, \mu_2 \in A$; there exist two finite sets C_1 and C_2 of H such that $\mu_1 = \sup C_1$ and $\mu_2 = \sup C_2$. Then $\mu_3 = \sup(C_1 \cup C_2) \in A$ and $\mu_1 \leqslant \mu_3$, $\mu_2 \leqslant \mu_3$.

On the other hand, every measure dominating H is also a measure dominating A, therefore A is dominated. Let $\mu_0 \in A$ and $A_0 = \{\mu \mid \mu \in A, \mu \geqslant \mu_0\}$. The set $A_0 - \mu_0 = \{\mu - \mu_0 \mid \mu \in A, \mu \geqslant \mu_0\}$ is directed for the relation $\leqslant$ is dominated and consists of positive measures, therefore there exists $\sup(A_0 - \mu_0)$. Then there exists also

$$\sup A_0 = \sup(A_0 - \mu_0) + \mu_0$$

and therefore there exists

$$\sup A = \sup A_0 = \nu .$$

It follows that ν is the least upper bound of the set H.

In fact, every majorant λ of H is also a majorant of A; therefore $\lambda \geqslant \nu$. On the other hand, ν is a majorant of H, since for every $\mu \in H$ there exists $\mu_1 \in A$ such that $\mu \leqslant \mu_1$; therefore $\mu \leqslant \nu$. It follows, therefore, that $\nu = \sup H$.

3.6 COROLLARY. *Every nonempty minorized set H of real measures has an infimum.*

In fact, $-H = \{-\mu \mid \mu \in H\}$ is nonempty and dominated, therefore it has a supremum. Then

$$\inf H = -\sup(-H) .$$

Remark. A reticulated vector space in which every nonempty majorized subset has a supremum is called a *completely reticulated* space. The space $\mathscr{M}(T)$ of real measures is therefore completely reticulated.

2 *Dominated measures*

Let E and F be two (real or complex) Banach spaces.

3.7 DEFINITION. *We say that a measure $\boldsymbol{m} : \mathscr{K}_E(T) \to F$ is dominated, or majorized, by a positive measure ν, if*

$$|\boldsymbol{m}(f)| \leqslant \nu(|f|) \textit{ for every function } f \in \mathscr{K}_E(T) .$$

We say that the measure $\boldsymbol{m} : \mathscr{K}_E(T) \to F$ is dominated, or majorized, if there exists a positive measure dominating $\boldsymbol{m}$.

The continuity property of a measure on the spaces $\mathscr{K}_E(T, K)$ follows from the fact that it is dominated.

3.8 PROPOSITION. *Every linear dominated mapping* $\boldsymbol{m}: \mathscr{K}_E(T) \to E$ *is a dominated measure.*

Let ν be a measure $\geqslant 0$ on T which dominates $\boldsymbol{m}$:

$$|\boldsymbol{m}(\boldsymbol{f})| \leqslant \nu(|\boldsymbol{f}|), \text{ for } \boldsymbol{f} \in \mathscr{K}_E(T).$$

Let $K \subset T$ be a compact set and let $a_K > 0$ be such that

$$|\nu(\varphi)| \leqslant a_K \| \varphi \| \text{ for every } \varphi \in \mathscr{K}(T, K).$$

For every function $\boldsymbol{f} \in \mathscr{K}_E(T, K)$ we have $|\boldsymbol{f}| \in \mathscr{K}(T, K)$ and $\|\boldsymbol{f}\| = \| \, |\boldsymbol{f}| \, \|$, therefore

$$|\boldsymbol{m}(\boldsymbol{f})| \leqslant \nu(|\boldsymbol{f}|) \leqslant a_K \|\boldsymbol{f}\| .$$

This means that $\boldsymbol{m}$ is a measure.

There are measures which are not dominated, as we shall see in the following

Example. Take $T = [0, 1]$, $E = R$, $F = \mathscr{C}(T)$ and $\boldsymbol{m}$ the identity mapping of $\mathscr{C}(T)$ into F. Evidently, $\boldsymbol{m}$ is continuous on $\mathscr{C}(T)$, therefore it is a measure; but $\boldsymbol{m}$ is not majorized. In fact, let us suppose the contrary, that $\boldsymbol{m}$ is dominated by the positive measure ν:

$$\|\varphi\| = |\boldsymbol{m}(\varphi)| \leqslant \nu(|\varphi|) \text{ for every } \varphi \in \mathscr{C}(T).$$

Let us show, first, that for every compact set $K \subset T$ with nonempty interior, and for every positive function $\psi \in \mathscr{C}(T)$, equal to 1 on K, we have $\nu(\psi) \geqslant 1$. Since Int $K \neq \varnothing$, there exists a function $\varphi \in \mathscr{C}(T, K)$ different from 0. Then $|\varphi| \leqslant \|\varphi\| \psi$ and $\nu(|\varphi|) \leqslant \|\varphi\| \, \nu(\psi)$; therefore $0 < \|\varphi\| = |\boldsymbol{m}(\varphi)| \leqslant \nu(|\varphi|) \leqslant \|\varphi\| \, \nu(\psi)$ whence $\nu(\psi) \geqslant 1$.

For every natural number n, put $K_n = [(4n-1)^{-1}, (4n-2)^{-1}]$ and take $\psi_n \in \mathscr{C}_+(T)$ with support contained in $[(4n)^{-1}, (4n-3)^{-1}]$ such that $\psi_n(t) = 1$ for $t \in K_n$. Then for every n we have $\nu(\psi_n) \geqslant 1$ and $\Sigma_{1 \leqslant i \leqslant n} \psi_i \leqslant 1$, therefore $n \leqslant \Sigma_{1 \leqslant i \leqslant n} \nu(\psi_n) = \nu(\Sigma_{1 \leqslant i \leqslant n} \psi_n) \leqslant \nu(1)$, whence $\nu(1) = +\infty$; that is absurd, since the function identically equal to 1 belongs to $\mathscr{C}(T)$.

For measures with scalar values we have

3.9 PROPOSITION. *Every measure* $\boldsymbol{m}: \mathscr{K}_E(T) \to C$ *with scalar values is dominated by the measure* $\mu = |\boldsymbol{m}|$.

In fact, the measure $|\boldsymbol{m}|$ majorizes $\boldsymbol{m}$, is the smallest positive measure with this property (proposition 2.27) and $\|\boldsymbol{m}\| = \|\mu\|$.

In particular, the scalar measures are dominated. A positive measure μ is dominated by itself, since $|\mu| = \mu$.

We shall show now that every dominated measure $\boldsymbol{m}$ possesses a smallest dominating positive measure which will be denoted by $|\boldsymbol{m}|$:

3.10 PROPOSITION. *If* $\boldsymbol{m}: \mathscr{K}_E(T) \to F$ *is a dominated measure, then there exists a smallest positive measure* $|\boldsymbol{m}|$ *which dominates* $\boldsymbol{m}$ *and* $\|\boldsymbol{m}\| \leqslant \| \, |\boldsymbol{m}| \, \| \leqslant +\infty$.

If E and F are two Banach spaces with involution, then $\bar{\boldsymbol{m}}$ *is dominated and* $|\bar{\boldsymbol{m}}| = |\boldsymbol{m}|$.

Let ν be a positive measure which dominates $\boldsymbol{m}$:

$$|\boldsymbol{m}(\boldsymbol{f})| \leqslant \nu(|\boldsymbol{f}|) \text{ for } \boldsymbol{f} \in \mathscr{K}_E(T).$$

For every $z' \in F'$ consider the measure $z' \circ \boldsymbol{m}: \mathscr{K}_E(T) \to C$ defined by the equality

$$(z' \circ \boldsymbol{m})(\boldsymbol{f}) = \langle \boldsymbol{m}(\boldsymbol{f}), z' \rangle \text{ for } \boldsymbol{f} \in \mathscr{K}_E(T).$$

For every function $\boldsymbol{f} \in \mathscr{K}_E(T)$ we have

$$|(z' \circ \boldsymbol{m})(\boldsymbol{f})| \leqslant |z'| \, |\boldsymbol{m}(\boldsymbol{f})| \leqslant |z'| \, \nu(|\boldsymbol{f}|)$$

therefore the measure $z' \circ \boldsymbol{m}$ is majorized by the positive measure $|z'| \nu$. But $z' \circ \boldsymbol{m}$ is a measure with scalar values, and $|z' \circ \boldsymbol{m}|$ is the smallest positive measure dominating it. It follows that

$$|z' \circ \boldsymbol{m}| \leqslant |z'| \nu \text{ for every } z' \in F'.$$

The family $(|z' \circ \boldsymbol{m}|)$ of positive measures is majorized by ν in the completely reticulated space $\mathscr{M}(T)$. It follows that this family has a supremum μ:

$$|z' \circ \boldsymbol{m}| \leqslant \mu \leqslant \nu \text{ for every } z' \in F' \text{ with } |z'| \leqslant 1.$$

For every function $\boldsymbol{f} \in \mathscr{K}_E(T)$ we have

$$|\boldsymbol{m}(\boldsymbol{f})| = \sup_{|z'| \leqslant 1} |\langle \boldsymbol{m}(\boldsymbol{f}), z' \rangle| =$$
$$= \sup_{|z'| \leqslant 1} |(z' \circ \boldsymbol{m})(\boldsymbol{f})| \leqslant \sup_{|z'| \leqslant 1} |z' \circ \boldsymbol{m}|(|\boldsymbol{f}|) \leqslant \mu(|\boldsymbol{f}|)$$

therefore μ dominates $\boldsymbol{m}$. As $\mu \leqslant \nu$ and ν is an arbitrary measure dominating $\boldsymbol{m}$, it follows that $|\boldsymbol{m}| = \mu$ is the smallest measure dominating $\boldsymbol{m}$.

Furthermore,

$$\|\boldsymbol{m}\| = \sup_{\|f\| \leqslant 1} |\boldsymbol{m}(f)| \leqslant \sup_{\|f\| \leqslant 1} \mu(|f|) \leqslant \sup_{\|\varphi\| \leqslant 1} |\mu(\varphi)| =$$
$$= \|\mu\| = \| \, |\boldsymbol{m}| \, \|, \quad f \in \mathcal{K}_E, \ \varphi \in \mathcal{K} .$$

Suppose that E and F are with involution. For every function $f \in \mathcal{K}_E(T)$ we have

$$|\bar{\boldsymbol{m}}(f)| = |\overline{\boldsymbol{m}(\bar{f})}| = |\boldsymbol{m}(\bar{f})| \leqslant |\boldsymbol{m}|(|\bar{f}|) = |\boldsymbol{m}|(|f|),$$

therefore $\bar{\boldsymbol{m}}$ is dominated and $|\bar{\boldsymbol{m}}| \leqslant |\boldsymbol{m}|$. Then we have $|\boldsymbol{m}| = |\bar{\bar{\boldsymbol{m}}}| \leqslant |\bar{\boldsymbol{m}}|$, therefore $|\bar{\boldsymbol{m}}| = |\boldsymbol{m}|$ and the proposition is proved.

As with measures with scalar values, the smallest positive measure $|\boldsymbol{m}|$ which dominates a dominated measure $\boldsymbol{m}$, is called the *modulus* (or the variation) of $\boldsymbol{m}$.

For every function $\varphi \in \mathcal{K}_+(T)$ we have

$$|\boldsymbol{m}|(\varphi) \geqslant \sup_{|f| \leqslant \varphi} |\boldsymbol{m}(f)|, \ \|\boldsymbol{m}\| \leqslant \| \, |\boldsymbol{m}| \, \| , \quad f \in \mathcal{K}_E$$

and, as it was shown (propositions 2.27 and 2.28), if the measure $\boldsymbol{m}$ has scalar values, then

$$|\boldsymbol{m}|(\varphi) = \sup_{|f| \leqslant \varphi} |\boldsymbol{m}(f)|, \quad \|\boldsymbol{m}\| = \| \, |\boldsymbol{m}| \, \|, \quad f \in \mathcal{K}_E .$$

In case $\boldsymbol{m}$ has not scalar values, the inequalities can be strict.

We shall denote by $\mathcal{M}_{E,F}(T)$ or $\mathcal{M}_{E,F}$ the set of dominated measures $\boldsymbol{m} : \mathcal{K}_E(T) \to F$.

For the set $\mathcal{M}_{R,C}$ of complex measures and the set $\mathcal{M}_{R,R}$ of real measures we shall use the previous notations, $\mathcal{M}_C(T)$ and $\mathcal{M}(T)$ respectively.

3.11 PROPOSITION. *The set $\mathcal{M}_{E,F}(T)$ of all dominated measures $\boldsymbol{m} : \mathcal{K}_E(T) \to F$ is a vector space. If $\boldsymbol{m}, \boldsymbol{n} \in \mathcal{M}_{E,F}$ and α is a scalar, we have*

$$|\boldsymbol{m} + \boldsymbol{n}| \leqslant |\boldsymbol{m}| + |\boldsymbol{n}| \ \textit{and} \ |\alpha \boldsymbol{m}| = |\alpha| \, |\boldsymbol{m}| .$$

Let $\boldsymbol{m}, \boldsymbol{n} \in \mathcal{M}_{E,F}$ and α a scalar. Denote $\mu = |\boldsymbol{m}|$ and $\nu = |\boldsymbol{n}|$. For every $f \in \mathcal{K}_E(T)$ we have

$$|(\boldsymbol{m} + \boldsymbol{n})(f)| = |\boldsymbol{m}(f) + \boldsymbol{n}(f)| \leqslant |\boldsymbol{m}(f)| + |\boldsymbol{n}(f)| \leqslant$$
$$\leqslant \mu(|f|) + \nu(|f|) = (\mu + \nu)(|f|)$$

therefore $\boldsymbol{m}+\boldsymbol{n}$ is dominated and $|\boldsymbol{m}+\boldsymbol{n}| \leqslant \mu+\nu$, that is,

$$|\boldsymbol{m}+\boldsymbol{n}| \leqslant |\boldsymbol{m}|+|\boldsymbol{n}| .$$

Further, for every $\boldsymbol{f} \in \mathscr{K}_E(T)$ we have

$$|\alpha \boldsymbol{m}(\boldsymbol{f})| = |\alpha|\,|\boldsymbol{m}(\boldsymbol{f})| \leqslant |\alpha|\,\mu(|\boldsymbol{f}|)$$

therefore $\alpha \boldsymbol{m}$ is dominated and $|\alpha \boldsymbol{m}| \leqslant |\alpha| \mu$, i.e.

$$|\alpha \boldsymbol{m}| \leqslant |\alpha|\,|\boldsymbol{m}|.$$

If $\alpha=0$, the equality $|\alpha \boldsymbol{m}| = |\alpha|\,|\boldsymbol{m}|$ is satisfied. Suppose $\alpha \neq 0$. If λ majorizes $\alpha \boldsymbol{m}$, then $\lambda/|\alpha|$ majorizes $\boldsymbol{m}$, therefore $\lambda/|\alpha| \geqslant |\boldsymbol{m}|$, whence $\lambda \geqslant |\alpha|\,|\boldsymbol{m}|$. It follows that $|\alpha|\,|\boldsymbol{m}|$ is the smallest measure which majorizes $\alpha \boldsymbol{m}$, and therefore

$$|\alpha \boldsymbol{m}| = |\alpha|\,|\boldsymbol{m}|.$$

Thus the proposition is proved.

3.12 Corollary. *If m is a complex measure and if $m=\mu+i\nu$, where μ and ν are real measures, then*

$$|\mu| \leqslant |m|,\ |\nu| \leqslant |m| \ \text{ and } \ |m| \leqslant |\mu|+|\nu| .$$

For every real function $f \in \mathscr{K}(T)$ we have $|\mu(f)| \leqslant |m(f)|$, therefore for $\varphi \in \mathscr{K}_+(T)$ we have

$$|\mu|(\varphi) = \sup_{|f| \leqslant \varphi} |\mu(f)| \leqslant \sup_{|f| \leqslant \varphi} |m(f)| \leqslant |m|(\varphi) .$$

3.13 Proposition. *Let $\boldsymbol{m} : \mathscr{K}_E(T) \to E$ be a measure and $\mu : \mathscr{K}(T) \to \mathscr{L}(E, F)$ be the measure satisfying the equality*

$$\mu(\varphi)x = \boldsymbol{m}(\varphi x) \ \textit{for} \ \varphi \in \mathscr{K}(T) \ \textit{and} \ x \in E .$$

If $\boldsymbol{m}$ is dominated, then μ is dominated and $|\mu| \leqslant |\boldsymbol{m}|$.

In fact, let $\varphi \in \mathscr{K}(T)$ and $\varepsilon > 0$. There exists $x \in E$ with $|x|=1$ and $|\mu(\varphi)| < |\mu(\varphi)x| + \varepsilon$. Then

$$|\mu(\varphi)| - \varepsilon < |\mu(\varphi)x| = |\boldsymbol{m}(\varphi x)| \leqslant |\boldsymbol{m}|(|\varphi x|) = $$
$$= |\boldsymbol{m}|(|\varphi|\,|x|) = |\boldsymbol{m}|(|\varphi|) .$$

ε being arbitrary, we deduce $|\mu(\varphi)| \leqslant |\boldsymbol{m}|(|\varphi|)$, therefore μ is dominated by $|\boldsymbol{m}|$. It follows then that $|\mu| \leqslant |\boldsymbol{m}|$.

Remark. It will be proved (theorem 8.61) that, conversely, if μ is dominated, then $\boldsymbol{m}$ is dominated and we have the equality $|\mu| = |\boldsymbol{m}|$.

Moreover, it will be proved that every dominated measure $\mu : \mathscr{K}(T) \to \mathscr{L}(E, F)$ can be "extended" to a measure $\boldsymbol{m} : \mathscr{K}_E(T) \to F$ which satisfies the equality

$$\mu(\varphi)x = \boldsymbol{m}(\varphi x), \qquad \varphi \in \mathscr{K}(T), \quad x \in E.$$

If we identify the measures $\boldsymbol{m}$ and μ, the inequality

$$|\textstyle\int \varphi \, \mathrm{d}\mu| \leqslant \int |\varphi| \mathrm{d}|\boldsymbol{m}|$$

is written

$$|\textstyle\int \varphi \, \mathrm{d}\boldsymbol{m}| \leqslant \int |\varphi| \mathrm{d}|\boldsymbol{m}|, \text{ for } \varphi \in \mathscr{K}(T).$$

We also have the inequality

$$|\textstyle\int \boldsymbol{f} \mathrm{d}\boldsymbol{m}| \leqslant \int |\boldsymbol{f}| \mathrm{d}|\boldsymbol{m}|, \text{ for } \boldsymbol{f} \in \mathscr{K}_E(T).$$

3.14 Proposition. *Let $\mu : \mathscr{K}(T) \to C$ be a real (respectively complex) measure, E a real (respectively complex) Banach space and $\boldsymbol{m} : \mathscr{K}_E(T) \to E$ the unique measure which satisfies the equality*

$$\boldsymbol{m}(\varphi x) = \mu(\varphi)x \text{ for } \varphi \in \mathscr{K}(T) \text{ and } x \in E.$$

The measure $\boldsymbol{m}$ is dominated if and only if μ is dominated. In this case $|\boldsymbol{m}| = |\mu|$.

If $\boldsymbol{m}$ is dominated, for every function $\varphi \in \mathscr{K}(T)$ and $x \in E$ with $|x| = 1$ we have

$$|\mu(\varphi)| = |\mu(\varphi)| \, |x| = |\mu(\varphi)x| = |\boldsymbol{m}(\varphi x)| \leqslant |\boldsymbol{m}|(|\varphi x|) = |\boldsymbol{m}|(|\varphi|),$$

therefore μ is dominated and $|\mu| \leqslant |\boldsymbol{m}|$.

Conversely, assume that μ is dominated. Let $\boldsymbol{f} \in \mathscr{K}_E(T)$; let K be the support of $\boldsymbol{f}$ and U a compact neighbourhood of K. Let $\varepsilon > 0$; there exists a function $\boldsymbol{g} = \Sigma \varphi_i x_i$, with $\varphi_i \in \mathscr{K}(T)$ and $x_i \in E$, with support contained in U, such that

$$|\boldsymbol{m}(\boldsymbol{f})| < |\boldsymbol{m}(\boldsymbol{g})| + \varepsilon$$

and

$$|g| \leqslant |f| + \varepsilon\varphi_0 ,$$

where $\varphi_0 \in \mathscr{K}(T)$ and $\varphi_0(t) = 1$ for $t \in U$. Let $z' \in F'$ with $|z'| = 1$ and $\langle m(g), z' \rangle = |m(g)|$. Then

$$|m(f)| - \varepsilon < |m(g)| = \langle m(g), z' \rangle = \mu(\langle g, z' \rangle) \leqslant$$
$$\leqslant |\mu|(|\langle g, z' \rangle|) \leqslant |\mu|(|g|) \leqslant |\mu|(|f| + \varepsilon\varphi_0) =$$
$$= |\mu|(|f| + \varepsilon|\mu|(\varphi_0) .$$

Since ε is arbitrary, we deduce

$$|m(f)| \leqslant |\mu|(|f|) .$$

It follows that m is dominated and $|m| \leqslant |\mu|$.

It follows that if m and μ are dominated, we have $|m| \leqslant |\mu|$ and $|\mu| \leqslant |m|$, therefore $|m| = |\mu|$. The proposition is proved.

If we identify the measures m and μ, the inequality

$$|\int f \mathrm{d}m| \leqslant \int |f| \mathrm{d}|m| \quad \text{for } f \in \mathscr{K}_E(T)$$

is written

$$|\int f \mathrm{d}\mu| \leqslant \int |f| \mathrm{d}|\mu| \quad \text{for } f \in \mathscr{K}_E(T) .$$

Further, we have the equality

$$|\int \varphi \mathrm{d}\mu| \leqslant \int |\varphi| \mathrm{d}|\mu| \quad \text{for } \varphi \in \mathscr{K}(T) .$$

3.15 COROLLARY. *If $\mu : \mathscr{K}(T) \to C$ is a scalar measure and $m : \mathscr{K}_C(T) \to C$ is its extension to $\mathscr{K}_C(T)$, then $|m| = |\mu|$.*

3. *Bounded measures*

3.16 DEFINITION. *We say that a positive measure μ on T is bounded if the linear mapping $\mu : \mathscr{K}(T) \to R$ is continuous for the topology of the uniform convergence.*

We say that a vector measure $m : \mathscr{K}_E(T)F$ is bounded if it is dominated by a bounded positive measure.

To say that a positive measure μ is bounded means that μ belongs to the dual of the normed space $\mathscr{K}(T)$ (with the norm $\|f\| = \sup_{t \in T} |f(t)|$), and this means that $\|\mu\| < +\infty$, where

$$\|\mu\| = \sup_{\|f\| \leqslant 1} |\mu(f)|, \quad f \in \mathscr{K}(T).$$

If μ and ν are two bounded positive measures and $\alpha \geqslant 0$, then $\mu+\nu$ and $\alpha\mu$ are bounded positive measures, since

$$\|\mu+\nu\| = \|\mu\| + \|\nu\| < +\infty \text{ and } \|\alpha\mu\| = |\alpha| \, \|\mu\| < +\infty .$$

Also, if μ and ν are two positive measures such that $\mu \leqslant \nu$ and if ν is bounded, then μ is bounded too, since

$$\|\mu\| \leqslant \|\nu\| < +\infty .$$

We deduce that:

A *dominated* vector measure $\boldsymbol{m} : \mathscr{K}_E(T) \to F$ is bounded if and only if its modulus $|\boldsymbol{m}|$ is a bounded positive measure.

Denote by $\mathscr{M}^1_{E,F}(T)$ the set of bounded dominated vector measures $m : \mathscr{K}_E(T) \to F$. Then $\mathscr{M}^1_{E,F}(T)$ is a subset of the space $\mathscr{M}_{E,F}(T)$ of all dominated measures. The set of bounded scalar measures will be denoted by $\mathscr{M}^1_C(T)$, and the set of bounded real measures by $\mathscr{M}^1(T)$. The set of bounded positive measures is denoted by $\mathscr{M}^1_+(T)$.

3.17 PROPOSITION. *The set $\mathscr{M}^1_{E,F}(T)$ is a vector space and the mapping $\boldsymbol{m} \to \|\boldsymbol{m}\|$ is a norm on this space.*

Let $\boldsymbol{m}$ and $\boldsymbol{n}$ be two measures from $\mathscr{M}^1_{E,F}(T)$ and α a scalar.

The measures $\boldsymbol{m}$ and $\boldsymbol{n}$ are dominated and the positive measures $|\boldsymbol{m}|$ and $|\boldsymbol{n}|$ are bounded.

It follows first that $\boldsymbol{m}+\boldsymbol{n}$ and $\alpha\boldsymbol{m}$ are bounded. From the inequalities:

$$|\boldsymbol{m}+\boldsymbol{n}| \leqslant |\boldsymbol{m}| + |\boldsymbol{n}| \text{ and } |\alpha\boldsymbol{m}| = |\alpha| \, |\boldsymbol{m}|$$

and from the fact that $|\boldsymbol{m}| + |\boldsymbol{n}|$ and $|\alpha| \, |\boldsymbol{m}|$ are bounded, it follows that $|\boldsymbol{m}+\boldsymbol{n}|$ and $|\alpha\boldsymbol{m}|$ are bounded measures and the proposition is proved.

From the inequality

$$\|\boldsymbol{m}\| \leqslant \| \, |\boldsymbol{m}| \, \|$$

we deduce that every *bounded* measure $\boldsymbol{m} : \mathscr{K}_E(T) \to F$ is a linear *continuous* mapping on $\mathscr{K}_E(T)$ for the topology of the uniform convergence.

The converse is not always true: there exist continuous vector measures which are not dominated (see example following proposition 3.8).

3.18 PROPOSITION. *A measure with scalar values $\boldsymbol{m} : \mathscr{K}_E(T) \to C$ is bounded* *if and only if it is a continuous linear mapping on $\mathscr{K}_E(T)$ for the topology of the uniform convergence.*

In fact, such a measure $\boldsymbol{m}$ is always dominated and if $\mu = |\boldsymbol{m}|$, then $\|\boldsymbol{m}\| = \|\mu\|$. If $\boldsymbol{m}$ is continuous, then $\|\mu\| = \|\boldsymbol{m}\| < +\infty$; therefore μ is bounded, hence $\boldsymbol{m}$ is bounded.

Conversely, if $\boldsymbol{m}$ is bounded, then $\|\boldsymbol{m}\| = \|\mu\| < +\infty$, therefore $\boldsymbol{m}$ is continuous.

From this proposition we deduce that the space $\mathscr{M}^1_{E,C}(T)$ of the bounded measures $\boldsymbol{m} : \mathscr{K}_E(T) \to C$ coincides with the dual of the normed space $\mathscr{K}_E(T)$; therefore $\mathscr{M}^1_{E,C}(T)$ is a Banach space for the norm $\|\boldsymbol{m}\|$.

In particular, $\mathscr{M}^1_C(T)$ and $\mathscr{M}^1(T)$ are Banach spaces.

3.19 Proposition. *Let $\mu : \mathscr{K}(T) \to C$ be a scalar measure, E a Banach space and $\boldsymbol{m} : \mathscr{K}_E(T) \to E$ the unique measure which satisfies the equality*

$$\boldsymbol{m}(\varphi x) = \mu(\varphi)x\,, \text{ for } \varphi \in \mathscr{K}(T),\ x \in E\,.$$

The measure $\boldsymbol{m}$ is bounded if and only if μ is bounded.

In fact, $\|\boldsymbol{m}\| = \|\mu\|$ (theorem 2.11).

It follows that the set of all bounded scalar measures $\mathscr{M}(T)$ can be isometrically embedded in the set $\mathscr{L}(\mathscr{K}_E(T), E)$ of linear continuous mappings of $\mathscr{K}_E(T)$ into E.

3.20 Proposition. *Let $\boldsymbol{m} : \mathscr{K}(T) \to C$ be a complex measure and μ and ν the real measures satisfying $m = \mu + i\nu$. The measure m is bounded if and only if the real measures μ and ν are bounded.*

The assertion follows from the inequalities

$$|\mu| \leqslant |m|,\ |\nu| \leqslant |m|,\ |m| \leqslant |\mu| + |\nu|\,.$$

3.21 Proposition. *A real measure $\mu : \mathscr{K}(T) \to R$ is bounded if and only if μ_+ and μ_- are bounded positive measures.*

The assertion follows from the equality

$$|\mu| = \mu_+ + \mu_-$$

and from the inequalities $\mu_+ \leqslant |\mu|$ and $\mu_- \leqslant |\mu|$.

§4 The support of a measure

Let E and F be two Banach spaces and $\boldsymbol{m} : \mathscr{K}_E(T) \to F$ a measure.

4.1 DEFINITION. *The support of the measure* $\boldsymbol{m}$ *is the set of all points* $t \in T$ *having the property that for every neighbourhood* V *of* t, *there exists a function* $\boldsymbol{f} \in \mathscr{K}_E(T, V)$ *such that* $\boldsymbol{m}(\boldsymbol{f}) \neq 0$.

Denote the support of $\boldsymbol{m}$ by $S(\boldsymbol{m})$. To say that a point $t \in T$ does not belong to the support of $\boldsymbol{m}$ means that there exists a neighbourhood V of t, such that the restriction of $\boldsymbol{m}$ to the space $\mathscr{K}_E(T, V)$ vanishes identically.

If G is an open set, and the restriction of the measures $\boldsymbol{m}$ to the space $\mathscr{K}_E(G)$ is zero, we say that $\boldsymbol{m}$ *possesses no mass in the set* G. Let G_0 be the union of all open sets in which $\boldsymbol{m}$ possesses no mass. From the principle of localization we deduce that $\boldsymbol{m}$ possesses no mass in G_0; therefore G_0 is the largest open set in which $\boldsymbol{m}$ possesses no mass.

4.2 PROPOSITION. *The support of the measure* $\boldsymbol{m}$ *is the complement of the largest open set in which* $\boldsymbol{m}$ *possesses no mass.*

Let G_0 be the largest open set in which $\boldsymbol{m}$ possesses no mass. This means that $\boldsymbol{m}(\boldsymbol{f})=0$ for every function $\boldsymbol{f} \in \mathscr{K}_E(T, G_0)$. For every point $t \in G_0$ we take $V = G_0$ and deduce that $t \notin S(\boldsymbol{m})$; therefore, $G_0 \cap S(\boldsymbol{m}) = \varnothing$, whence

$$S(\boldsymbol{m}) \subset T \setminus G_0 .$$

Conversely, if $t \notin G_0$, there exists no open neighbourhood G of t, in which $\boldsymbol{m}$ possesses no mass, since every open set in which $\boldsymbol{m}$ possesses no mass is contained in G_0. Therefore, if $t \notin G_0$, then for an arbitrary neighbourhood V of t, the restriction of $\boldsymbol{m}$ to $\mathscr{K}_E(T, V)$ is not identically zero and therefore $t \in S(\boldsymbol{m})$. It follows that $T \setminus G_0 \subset S(\boldsymbol{m})$. The opposite inclusion being already proved we deduce that $S(\boldsymbol{m}) = T \setminus G_0$.

4.3 COROLLARY. *The support of a measure is a closed set.*

4.4 PROPOSITION. *If* $\boldsymbol{m}, \boldsymbol{n} : \mathscr{K}_E(T) \to F$ *are measures and* α *is scalar* $\neq 0$, *then*

$$S(\boldsymbol{m}+\boldsymbol{n}) \subset S(\boldsymbol{m}) \cup S(\boldsymbol{n}) \text{ and } S(\alpha \boldsymbol{m}) = S(\boldsymbol{m}) .$$

In fact, none of these measures possesses mass in the open set

$$T\setminus(S(\boldsymbol{m})\cup S(\boldsymbol{n})).$$

It follows that the measure $\boldsymbol{m}+\boldsymbol{n}$ possesses no mass in this set, therefore

$$T\setminus(S(\boldsymbol{m})\cup S(\boldsymbol{n}))\subset T\setminus S(\boldsymbol{m}+\boldsymbol{n})$$

whence

$$S(\boldsymbol{m}+\boldsymbol{n})\subset S(\boldsymbol{m})\cup S(\boldsymbol{n}).$$

The equality $S(\alpha\boldsymbol{m})=S(\boldsymbol{m})$ is straightforward since the measures $\boldsymbol{m}$ and $\alpha\boldsymbol{m}$ are zero simultaneously.

For positive measures, we obtain additional properties.

4.5 Proposition. *If $\mu\geqslant 0$ and $\nu\geqslant 0$ are positive measures and if $\mu\leqslant\nu$, then $S(\mu)\subset S(\nu)$.*

In fact, if ν possesses no mass in an open set G, then μ also possesses no mass in G. It follows that $T\setminus S(\nu)\subset T\setminus S(\mu)$ whence

$$S(\mu)\subset S(\nu).$$

4.6 Proposition. *If μ and ν are two positive measures, then*

$$S(\mu+\nu)=S(\mu)\cup S(\nu).$$

The inclusion $S(\mu+\nu)\subset S(\mu)\cup S(\nu)$ is true for every measure.

Then we have $\mu\leqslant\mu+\nu$ and $\nu\leqslant\mu+\nu$, therefore

$$S(\mu)\subset S(\mu+\nu)\ \text{ and }\ S(\nu)\subset S(\mu+\nu)$$

whence

$$S(\mu)\cup S(\nu)\subset S(\mu+\nu).$$

From these inclusions the required equality follows.

4.7 Proposition. *If $(\mu_i)_{1\leqslant i\leqslant n}$ is a finite family of positive measures then*

$$S(\sup_{1\leqslant i\leqslant n}\mu_i)=\cup_{1\leqslant i\leqslant n}S(\mu_i).$$

Write $\lambda=\sup_{1\leqslant i\leqslant n}\mu_i$. We have $\mu_i\leqslant\lambda$, therefore $S(\mu_i)\subset S(\lambda)$ for $i=1, 2, \dots, n$, whence

$$\cup_{1\leqslant i\leqslant n}S(\mu_i)\subset S(\lambda).$$

On the other hand, $\lambda \leqslant \mu_1 + \mu_2 + \dots + \mu_n$, therefore

$$S(\lambda) \subset S(\mu_1 + \dots + \mu_n) = S(\mu_1) \cup \dots \cup S(\mu_n) = \cup_{1 \leqslant i \leqslant n} S(\mu_i),$$

Consequently

$$S(\lambda) = \cup_{1 \leqslant i \leqslant n} S(\mu_i).$$

4.8 PROPOSITION. *If H is a majorized family of positive measures, then*

$$S(\sup_{\mu \in H} \mu) = \overline{\cup_{\mu \in H} S(\mu)}$$

Denote $\lambda = \sup_{\mu \in H} \mu$. We have $\mu \leqslant \lambda$, therefore $S(\mu) \subset S(\lambda)$ for every $\mu \in H$, whence

$$\cup_{\mu \in H} S(\mu) \subset S(\lambda).$$

Since $S(\lambda)$ is a closed set, we deduce

$$\overline{\cup_{\mu \in H} S(\mu)} \subset S(\lambda).$$

In order to prove the converse inclusion, consider the set A of least upper bounds of the finite subsets of H. The set A is directed, majorized and

$$\lambda = \sup_{\nu \in A} \nu = \lim_{\nu \in A} \nu.$$

For every function $\varphi \in \mathscr{K}(T)$ we have therefore

$$\lambda(\varphi) = \lim_{\nu \in A} \nu(\varphi).$$

If follows that if no measure $\nu \in A$ has mass in an open set G, then λ also has no mass in G. This means that if $G \subset \cap_{\nu \in A} (T \setminus S(\nu))$, then $G \subset T \setminus S(\lambda)$, that is

$$\operatorname{Int} \cap_{\nu \in A} (T \setminus S(\nu)) \subset T \setminus S(\lambda),$$

or equivalently

$$T \setminus \overline{\cup_{\nu \in A} S(\nu)} \subset T \setminus S(\lambda),$$

whence

$$\overline{\cup_{\nu \in A} S(\nu)} \supset S(\lambda).$$

Each measure $\nu \in A$ is the supremum of a finite family $\mu_1, \dots, \mu_n$ of measures from H, therefore

$$S(\nu) = \cup_{1 \leqslant i \leqslant n} S(\mu_i) \subset \cup_{\mu \in H} S(\mu) .$$

It follows that

$$\cup_{\nu \in A} S(\nu) \subset \cup_{\mu \in H} S(\mu)$$

and then

$$\overline{\cup_{\mu \in H} S(\mu)} \supset S(\lambda) .$$

We deduce then that

$$\overline{\cup_{\mu \in H} S(\mu)} = S(\lambda)$$

and the proposition is proved.

4.9 PROPOSITION. *If* $\boldsymbol{m} : \mathscr{K}_E(T) \to F$ *is a dominated measure, then* $S(\boldsymbol{m}) = S(|\boldsymbol{m}|)$.

Consider first the case of measures with scalar values $\boldsymbol{m} : \mathscr{K}_E(T) \to C$. For every function $\varphi \in \mathscr{K}_+(T)$ we have

$$|\boldsymbol{m}|(\varphi) = \sup_{|f| \leqslant \varphi} |\boldsymbol{m}(\boldsymbol{f})| .$$

It follows that if the restriction of $\boldsymbol{m}$ to a subspace $\mathscr{K}_E(T, G)$ is zero, then the restriction of $|\boldsymbol{m}|$ to the space $\mathscr{K}(T, G)$ is zero. We deduce that if $\boldsymbol{m}$ possesses no mass in an open set G, then $|\boldsymbol{m}|$ also possesses no mass in G. Therefore

$$T \setminus S(\boldsymbol{m}) \subset T \setminus S(|\boldsymbol{m}|)$$

whence

$$S(\boldsymbol{m}) \supset S(|\boldsymbol{m}|) .$$

Consider now a measure $\boldsymbol{m} : \mathscr{K}_E(T) \to F$ with values in F. For every $z' \in F'$, consider the measure $z' \circ \boldsymbol{m} : \mathscr{K}_E(T) \to C$ defined by the equality

$$(z' \circ \boldsymbol{m})(\boldsymbol{f}) = \langle \boldsymbol{m}(\boldsymbol{f}), z' \rangle \quad \text{for } \boldsymbol{f} \in \mathscr{K}_E(T) .$$

If $\boldsymbol{m}$ possesses no mass in an open set G, then $z' \circ \boldsymbol{m}$ also possesses no mass in G, therefore

$$S(\boldsymbol{m}) \supset S(z' \circ \boldsymbol{m}) .$$

From the first part of the proof it follows that

$$S(z' \circ \boldsymbol{m}) \supset S(|z' \circ \boldsymbol{m}|)$$

therefore

$$S(\boldsymbol{m}) \supset S(|z' \circ \boldsymbol{m}|)$$

But

$$|\boldsymbol{m}| = \sup_{|z'| \leqslant 1} |z' \circ \boldsymbol{m}|,$$

therefore

$$S(|\boldsymbol{m}|) = \overline{\cup_{|z'| \leqslant 1} S(|z' \circ \boldsymbol{m}|)} \subset S(\boldsymbol{m}).$$

The opposite inclusion follows from the inequality

$$|\boldsymbol{m}(\boldsymbol{f})| \leqslant |\boldsymbol{m}| (|\boldsymbol{f}|) \text{ for } \boldsymbol{f} \in \mathscr{K}_F(T).$$

If $|\boldsymbol{m}|$ possess no mass in G, then $\boldsymbol{m}$ also possess no mass in G, therefore

$$T \setminus S(|\boldsymbol{m}|) \subset T \setminus S(\boldsymbol{m})$$

that is

$$S(|\boldsymbol{m}|) \supset S(\boldsymbol{m}).$$

It follows then that

$$S(\boldsymbol{m}) = S(|\boldsymbol{m}|)$$

and the proposition is proved.

4.10 Corollary. *For every real measure μ we have*

$$S(\mu) = S(\mu_+) \cup S(\mu_-).$$

In fact, $|\mu| = \mu_+ + \mu_-$ and μ_+ and μ_- are positive measures, therefore

$$S(\mu) = S(|\mu|) = S(\mu_+ + \mu_-) = S(\mu_+) \cup S(\mu_-).$$

4.11 Corollary. *If $\boldsymbol{m}$ and $\boldsymbol{n}$ are two dominated vector measures and if $|\boldsymbol{m}| \leqslant |\boldsymbol{n}|$, then*

$$S(\boldsymbol{m}) \subset S(\boldsymbol{n}).$$

4.12 Corollary. *If m is a complex measure and if $m = \mu + i\nu$, where μ and ν are real measures, then*

$$S(m) = S(\mu) \cup S(\nu).$$

We have $|\mu| \leqslant |m|$, $|\nu| \leqslant |m|$ and $|m| \leqslant |\mu| + |\nu|$, therefore

$$S(\mu) \subset S(m) \text{ and } S(\nu) \subset S(m)$$

whence

$$S(\mu) \cup S(\nu) \subset S(m) .$$

On the other hand

$$S(m) \subset S(|\mu| + |\nu|) = S(|\mu|) \cup S(|\nu|) = S(\mu) \cup S(\nu) .$$

Therefore

$$S(m) = S(\mu) \cup S(\nu) .$$

4.13 PROPOSITION. *If $\boldsymbol{m} : \mathscr{K}_E(T) \to F$ is a measure and if $\mu : \mathscr{K}(T) \to \mathscr{L}(E, F)$ is a measure satisfying the equality*

$$\boldsymbol{m}(\varphi x) = \mu(\varphi)x, \text{ for } \varphi \in \mathscr{K}(T) \text{ and } x \in E ,$$

then $S(\boldsymbol{m}) = S(\mu)$.

If the restriction of μ to the subspace $\mathscr{K}(T, G)$ is zero, G being open, then for every function $\boldsymbol{g} = \Sigma_{1 \leqslant i \leqslant n} \varphi_i x_i$, with $\varphi_i \in \mathscr{K}(T, G)$, and $x_i \in E$ we have $\boldsymbol{m}(\boldsymbol{g}) = 0$. Since every function $\boldsymbol{f} \in \mathscr{K}_E(T, G)$ can be uniformly approximated by functions of the form $\Sigma_{1 \leqslant i \leqslant n} \varphi_i x_i$, with $\varphi_i \in \mathscr{K}(T, G)$ and $x_i \in E$, and $\boldsymbol{m}$ is continuous on $\mathscr{K}_E(T, G)$, it follows that the restriction of $\boldsymbol{m}$ to $\mathscr{K}_E(T, G)$ is zero. Therefore

$$T \setminus S(\mu) \subset T \setminus S(\boldsymbol{m}) ,$$

consequently

$$S(\mu) \supset S(\boldsymbol{m}) .$$

On the other hand, if the restriction of $\boldsymbol{m}$ to $\mathscr{K}_E(T, G)$ is zero, we have $\boldsymbol{m}(\varphi x) = 0$ for every $\varphi \in \mathscr{K}(T, G)$ and every $x \in E$, that is $\mu(\varphi) = 0$ for every $\varphi \in \mathscr{K}(T, G)$. Therefore $T \setminus S(\boldsymbol{m}) \subset T \setminus S(\mu)$, whence

$$S(\boldsymbol{m}) \supset S(\mu) ,$$

consequently

$$S(\boldsymbol{m}) = S(\mu) .$$

4.14 COROLLARY. *If μ is a scalar measure, then its extension $\boldsymbol{m} : \mathscr{K}_E(T) \to E$ (or $\boldsymbol{m} : \mathscr{K}_C(T) \to C$) has support equal to the support of μ.*

4.15 PROPOSITION. *Let $\boldsymbol{m} : \mathscr{K}_E(T) \to F$ be a measure and $\boldsymbol{f} \in \mathscr{K}_E(T)$. If the function $\boldsymbol{f}$ is zero on the support of $\boldsymbol{m}$, then $\boldsymbol{m}(\boldsymbol{f}) = 0$.*

Let K be the support of $\boldsymbol{f}$ and let $a_K > 0$ be such that

$$|\boldsymbol{m}|(\boldsymbol{g})| \leqslant a_K \|\boldsymbol{g}\| \text{ for } \boldsymbol{g} \in \mathscr{K}_E(T, K) .$$

Let $\varepsilon > 0$ and $V = \{t \in T | |\boldsymbol{f}(t)| < \varepsilon\}$; V is an open set and $S(\boldsymbol{m}) \subset V$. The set $T \setminus V$ is compact, $T \setminus S(\boldsymbol{m})$ is open and $K \setminus V \subset T \setminus S(\boldsymbol{m})$. There exists then a continuous function $\varphi : T \to [0, 1]$ with compact support such that $\varphi(t) = 1$ for $t \in K \setminus V$ and with support contained in $T \setminus S(\boldsymbol{m})$. Then $\boldsymbol{f}\varphi \in \mathscr{K}_E(T)$, and the support of the function $\boldsymbol{f}\varphi$ is contained in $T \setminus S(\boldsymbol{m})$. Since the measure $\boldsymbol{m}$ possesses no mass in $T \setminus S(\boldsymbol{m})$ it follows that $\boldsymbol{m}(\boldsymbol{f}\varphi) = 0$. Then we have $|\boldsymbol{f}\varphi| \leqslant |\boldsymbol{f}|$ and $\boldsymbol{f}(t) = \boldsymbol{f}(t)\varphi(t)$ for $t \in K \setminus V$; therefore $|\boldsymbol{f}(t) - \boldsymbol{f}(t)\varphi(t)| \leqslant 2\varepsilon$ for every $t \in T$. The support of the function $\boldsymbol{f} - \boldsymbol{f}\varphi$ is contained in K, therefore

$$|\boldsymbol{m}(\boldsymbol{f}) - \boldsymbol{f}\varphi)| \leqslant a_K \|\boldsymbol{f} - \boldsymbol{f}\varphi\| \leqslant 2a_K\varepsilon .$$

But

$$\boldsymbol{m}(\boldsymbol{f} - \boldsymbol{f}\varphi) = \boldsymbol{m}(\boldsymbol{f}) - \boldsymbol{m}(\boldsymbol{f}\varphi) = \boldsymbol{m}(\boldsymbol{f}) ,$$

therefore

$$|\boldsymbol{m}(\boldsymbol{f})| \leqslant 2a_K\varepsilon .$$

ε being arbitrary, it follows that $|\boldsymbol{m}(\boldsymbol{f})| = 0$, that is $\boldsymbol{m}(\boldsymbol{f}) = 0$.

4.16 COROLLARY. *Let $\boldsymbol{m}: \mathscr{K}_E(T) \to F$ be a measure, $\boldsymbol{f}$ and $\boldsymbol{g}$ two functions from $\mathscr{K}_E(T)$. If $\boldsymbol{f}$ and $\boldsymbol{g}$ are equal on the support of the measure $\boldsymbol{m}$, then $\boldsymbol{m}(\boldsymbol{f}) = \boldsymbol{m}(\boldsymbol{g})$.*

The positive measures have a converse property.

4.17 PROPOSITION. *Let μ be a positive measure and φ a positive function from $\mathscr{K}(T)$. If $\mu(\varphi) = 0$, then the function φ is zero on the support of the measure μ.*

Let $t_0 \in T$ be such that $\varphi(t_0) > 0$. Let us prove that $t_0 \notin S(\mu)$. Let α be a number such that $0<\alpha<\varphi(t_0)$. There exists a neighbourhood V of t_0 such that $\varphi(t)>\alpha$ for $t\in V$.

For every function $\psi\geqslant 0$ from $\mathscr{K}(T, V)$ put $\beta=\|\psi\|$; we have $\psi\leqslant\beta\alpha^{-1}\varphi$, therefore $\mu(\psi)\leqslant\beta\alpha^{-1}\mu(\varphi)=0$. If $\psi\in\mathscr{K}(T, V)$, then $\psi=\psi_1-\psi_2$ with $\psi_1, \psi_2\in\mathscr{K}_+(T, V)$; therefore $\mu(\psi)=\mu(\psi_1)-\mu(\psi_2)=0$. Hence $t_0\notin S(\mu)$. It follows that if $t\in S(\mu)$, then $\varphi(t)=0$ and the proposition is proved.

4.18 PROPOSITION. *A measure $\boldsymbol{m}:\mathscr{K}_E(T)\to F$ can be extended to a linear mapping of $\mathscr{C}_E(T)$ into F, continuous for the topology of the compact convergence, if and only if $\boldsymbol{m}$ has compact support. In this case the extension is unique.*

If $\boldsymbol{m}$ can be extended to a linear mapping of $\mathscr{C}_E(T)$ into F, continuous for the topology of the compact convergence, then $\boldsymbol{m}$ is continuous on $\mathscr{K}_E(T)$ for this topology. The topology of the compact convergence is a locally convex topology generated by the seminorms

$$\|\boldsymbol{f}\|_K=\sup_{t\in K}|\boldsymbol{f}(t)|$$

when K runs over the set of the compact subsets of T. Since $\boldsymbol{m}$ is continuous on $\mathscr{K}_E(T)$ for the topology defined by the family of seminorms $\|\boldsymbol{f}\|_K$, $\boldsymbol{m}$ is continuous for a certain seminorm from this family. Therefore there exists a number $a>0$ and a compact set $K\subset T$, such that

$$|\boldsymbol{m}(\boldsymbol{f})|<a\|\boldsymbol{f}\|_K \text{ for every function } \boldsymbol{f}\in\mathscr{K}_E(T)\,.$$

If the support of a function $\boldsymbol{g}\in\mathscr{K}_E(T)$ does not meet K, we have $\|\boldsymbol{g}\|_K=0$ and then $\boldsymbol{m}(\boldsymbol{g})=0$. It follows that the support of the measure $\boldsymbol{m}$ is contained in K and therefore it is compact.

Conversely, suppose that the support $S(\boldsymbol{m})$ of the measure $\boldsymbol{m}$ is compact. Let U be a compact neighbourhood of $S(\boldsymbol{m})$ and $\varphi: T\to[0, 1]$ a continuous function with compact support H, equal to 1 on U. For every function $\boldsymbol{f}\in\mathscr{K}_E(T)$ we have $\boldsymbol{g}=\boldsymbol{f}\varphi\in\mathscr{K}_E(T, H)$ and $\boldsymbol{f}(t)=\boldsymbol{g}(t)$ for $t\in U$, therefore also for $t\in S(\boldsymbol{m})$. It follows that $\boldsymbol{m}(\boldsymbol{f})=\boldsymbol{m}(\boldsymbol{g})$. If $a_H>0$ is such that

$$|\boldsymbol{m}(\boldsymbol{h})|\leqslant a_H\|\boldsymbol{h}\| \text{ for } \boldsymbol{h}\in\mathscr{K}_E(T, H)\,,$$

then

$$|\boldsymbol{m}(\boldsymbol{f})|=|\boldsymbol{m}(\boldsymbol{g})|\leqslant a_H\|\boldsymbol{g}\|\leqslant a_H\sup_{t\in H}|\boldsymbol{f}(t)|=a_H\|\boldsymbol{f}\|_H\cdot$$

It follows that $\boldsymbol{m}$ is continuous on $\mathscr{K}_E(T)$ for the topology of the compact convergence. Since $\mathscr{K}_E(T)$ is dense in $\mathscr{C}_E(T)$ for this topology, it follows that

$\boldsymbol{m}$ can be extended uniquely to a linear mapping of $\mathscr{C}_E(T)$ into F, continuous for this topology. Thus the proposition is proved.

4.19 COROLLARY. *If $\boldsymbol{m} : \mathscr{K}_E(T) \to F$ is a measure with compact support, then $\|\boldsymbol{m}\| < +\infty$.*

Indeed, $\boldsymbol{m}$ is continuous on $\mathscr{K}_E(T)$ for the topology of the compact convergence; there exists a number $a > 0$ and a compact set K such that

$$|\boldsymbol{m}(\boldsymbol{f})| \leqslant a \|\boldsymbol{f}\|_K \text{ for every } \boldsymbol{f} \in \mathscr{K}_E(T).$$

Then

$$|\boldsymbol{m}(\boldsymbol{f})| \leqslant a \|\boldsymbol{f}\| \text{ for every } \boldsymbol{f} \in \mathscr{K}_E(T).$$

Therefore it follows that $\|\boldsymbol{m}\| \leqslant a < \infty$.

4.20 PROPOSITION. *If $\mu : \mathscr{K}(T) \to C$ is a measure with finite support, $S(\mu) = \{a_1, a_2, \dots, a_n\}$, then μ is a linear combination of the pointwise measures ε_{a_i}.*

In fact, if $f \in \mathscr{K}(T)$ and if $f(a_i) = 0$, $i = 1, 2, \dots, n$, then $\mu(f) = 0$. But the equalities $f(a_i) = 0$ can be written $\varepsilon_{a_i}(f) = 0$. Then from linear algebra it is known that μ is a linear combination of the measures ε_{a_i}.

Chapter II

$\mathscr{L}^p$ SPACES. INTEGRABLE FUNCTIONS

§5 The upper integral

1 *Lower semicontinuous functions*

A numerical function $f: T \to \overline{R}$ (with finite or infinite values) is called *lower semicontinuous*, if for every point $t_0 \in T$ and every number $\varepsilon > 0$, there exists a neighbourhood V of t_0 such that

$$f(t) \geqslant f(t_0) - \varepsilon \text{ for every } t \in V .$$

Every continuous function $f: T \to R$ is, evidently, lower semicontinuous.

The constant function $f(t) \equiv +\infty$ is lower semicontinuous.

We shall denote by $\mathscr{I}_+(T)$ or $\mathscr{I}_+$ the set of *lower semicontinuous positive* functions defined on T, with finite or infinite values. We have $\mathscr{K}_+(T) \subset \mathscr{I}_+(T)$.

The functions of $\mathscr{I}_+$ have the following properties:

(1) If $f \in \mathscr{I}_+$ and $\alpha \geqslant 0$, then $\alpha f \in \mathscr{I}_+$; (we use the convention $0 \cdot \infty = 0$).

(2) If (f_α) is an arbitrary family of functions of $\mathscr{I}_+$, then

$$\sup f_\alpha \in \mathscr{I}_+ \text{ and } \Sigma f_\alpha \in \mathscr{I}_+ ,$$

(3) If (f_α) is a *finite* family of functions of $\mathscr{I}_+$ then

$$\inf_\alpha f_\alpha \in \mathscr{I}_+ .$$

In fact:

(1) If $\alpha = 0$, we have $\alpha f(t) \equiv 0$, hence $\alpha f \in \mathscr{I}_+$; if $\alpha > 0$, the property results from the fact that the inequality $f(t) \geqslant f(t_0) - \varepsilon/\alpha$ for $t \in V$ implies the inequality $\alpha f(t) \geqslant \alpha f(t_0) - \varepsilon$ for $t \in V$.

(2) Let $f = \sup_\alpha f_\alpha$, $t_0 \in T$ and $\varepsilon > 0$. There exists an index α' such that $f_{\alpha'}(t_0) > f(t_0) - \varepsilon/2$ and a neighbourhood V of t_0 such that, for $t \in V$, we

have $f_{\alpha'}(t) \geqslant f_{\alpha'}(t_0) - (\varepsilon/2) > f(t_0) - \varepsilon$. Then for $t \in V$, we have $f(t) = \sup_\alpha f_\alpha(t) \geqslant f(t_0) - \varepsilon$, therefore $f \in \mathscr{I}_+$.

The sum of a finite family of functions from $\mathscr{I}_+$ belongs to $\mathscr{I}_+$. If (f_α) is an arbitrary family, we have

$$\Sigma_\alpha f_\alpha = \sup_J \Sigma_{\alpha \in J} f_\alpha ,$$

where J runs over the set of finite subsets of indices. Since $\Sigma_{\alpha \in J} f_\alpha \in \mathscr{I}_+$, from the first part of the proof it follows

$$\Sigma f_\alpha \in \mathscr{I}_+ .$$

(3) Let $f_1, \ldots, f_n$ be a finite family of functions from $\mathscr{I}_+$, $t_0 \in T$ and $\varepsilon > 0$. There exist n neighbourhoods $V_1, \ldots, V_n$ of t_0, such that

$$f_i(t) \geqslant f_i(t_0) - \varepsilon \text{ for } t \in V_i, \quad i = 1, 2, \ldots, n .$$

Then

$$\inf_i f_i(t) \geqslant \inf_i f_i(t_0) - \varepsilon \text{ for } t \in V = \cap_{1 \leqslant i \leqslant n} V_i ,$$

and V is a neighbourhood of t_0. Therefore $\inf_i f_i \in \mathscr{I}_+$.

5.1 PROPOSITION. *Every function $f \in \mathscr{I}_+$ is the upper envelope of the directed set (for the relation $\leqslant$) $\{g \mid g \in \mathscr{K}_+, g \leqslant f\}$.*

In fact, if $f(t_0) = 0$, then $g \in \mathscr{K}_+$ and $g \leqslant f$ implies $g(t_0) = 0$, therefore

$$f(t_0) = \sup_{g \leqslant f} g(t_0), \; g \in \mathscr{K}_+ .$$

Suppose $f(t_0) > 0$; let $\varepsilon > 0$ be an arbitrary number such that $f(t_0) - \varepsilon > 0$. There exists a neighbourhood V of t_0 such that

$$f(t) \geqslant f(t_0) - \varepsilon \text{ for } t \in V .$$

There exists, also, a function $g_\varepsilon \in \mathscr{K}_+$ with support contained in V, such that

$$g_\varepsilon(t_0) = f(t_0) - \varepsilon \text{ and } g_\varepsilon(t) \leqslant f(t_0) - \varepsilon \text{ for } t \in V .$$

It follows that $g_\varepsilon(t) \leqslant f(t)$ for every $t \in T$, that is $g_\varepsilon \leqslant f$ and

$$f(t_0) = \sup_{g \leqslant f} g(t_0), \; g \in \mathscr{K}_+ .$$

Therefore, for every $t \in T$ we have

$$f(t) = \sup_{g \leqslant f} g(t), \; g \in \mathscr{K}_+ .$$

A function $f: T \to \bar{R}$ is called *upper semicontinuous* if for every point $t_0 \in T$ and every number $\varepsilon > 0$, there exists a neighbourhood V of t_0 such that

$$f(t) \leqslant f(t_0) + \varepsilon \text{ for every } t \in V.$$

The function f is upper semicontinuous if and only if $-f$ is lower semicontinuous.

2 *The upper integral of lower semicontinuous functions*

Let μ be a positive measure on T.

5.2 Definition. *The number $\mu^*(f)$ defined by the equality*

$$\mu^*(f) = \sup_{g \in \mathscr{K}_+,\, g \leqslant f} \mu(g)$$

is called the upper integral (with respect to μ) of the function $f \in \mathscr{I}_+$.

If $f \in \mathscr{K}_+$, it can be immediately proved that $\mu^*(f) = \mu(f)$. Therefore μ^* is an extension of the measure μ from $\mathscr{K}_+$ to $\mathscr{I}_+$. Since the set $A = \{g \mid g \in \mathscr{K},\, g \leqslant f\}$ is directed for the relation $\leqslant$ and the measure μ is increasing, it follows that the function $g \to \mu(g)$ has a (finite or infinite) limit when g runs over the filter of sections of A, and the limit is equal to the supremum of the function μ on this set:

$$\lim_{g \leqslant f} \mu(g) = \sup_{g \leqslant f} \mu(g),\ g \in \mathscr{K}.$$

Then it follows that the upper integral $\mu^*(f)$ can also be defined by the equality:

$$\mu^*(f) = \lim_{g \leqslant f} \mu(g),\ g \in \mathscr{K}_+.$$

From the definition 5.2 we deduce immediately the following properties:

(1) $0 \leqslant \mu^*(f) \leqslant +\infty$ for $f \in \mathscr{I}_+$; $\mu^*(0) = 0$;
(2) $\mu^*(f_1) \leqslant \mu^*(f_2)$ if $f_1 \leqslant f_2, f_1, f_2 \in \mathscr{I}_+$;
(3) $\mu^*(\alpha f) = \alpha \mu^*(f)$ for $\alpha \geqslant 0, f \in \mathscr{I}_+$.

Example. Let α be a finite, positive, numerical function on T, such that for every compact set $K \subset T$, we have $\Sigma_{t \in K} \alpha(t) < +\infty$. Let μ be the positive measure on T defined by the equality

$$\mu(f) = \Sigma_{t \in T} \alpha(t) f(t) \text{ for } f \in \mathscr{K}(T).$$

Let us prove that for every function $f \in \mathscr{I}_+$ we have

$$\mu^*(f) = \Sigma_{t\in T}\, \alpha(t) f(t)$$

(putting $\alpha(t)f(t)=0$ if $\alpha(t)=0$ and $f(t)=+\infty$).

For every function $g \in \mathscr{K}_+$ with $g \leqslant f$ we have

$$\mu(g) = \Sigma_{t\in T}\, \alpha(t) g(t) \leqslant \Sigma_{t\in T}\, \alpha(t) f(t)\,,$$

therefore

$$\mu^*(f) \leqslant \Sigma_{t\in T}\, \alpha(t) f(t)\,.$$

Now take a number $a < \Sigma_{t\in T}\,\alpha(t) f(t)$. There exists a finite subset $M \subset T$ with $a < \Sigma_{t\in M}\,\alpha(t) f(t)$. For very $t \in M$ and every number $b_t < f(t)$, there exists a function $g_t \in \mathscr{K}_+$ with $g_t \leqslant f$ and $g_t(t) > b_t$. If we denote $g = \sup_{t\in M} g_t$, we have $g \in \mathscr{K}_+$, $g \leqslant f$ and $\mu(g) \geqslant \Sigma_{t\in M}\,\alpha(t) b_t$; therefore

$$\mu^*(f) \geqslant \Sigma_{t\in M}\,\alpha(t) b_t\,.$$

The last sum can be made as close to $\Sigma_{t\in M}\,\alpha(t) f(t)$ as we like, by taking suitable numbers b_t. It follows that $\mu^*(f) > a$. Since a is arbitrary we deduce

$$\mu^*(f) \geqslant \Sigma_{t\in T}\,\alpha(t) f(t)\,,$$

therefore

$$\mu^*(f) = \Sigma_{t\in T}\,\alpha(t) f(t)\,.$$

5.3 THEOREM. *For every set $H \subset \mathscr{I}_+$ directed for the relation $\leqslant$ we have*

$$\mu^*(\sup_{g\in H} g) = \sup_{g\in H} \mu^*(g) = \lim_{g\in H} \mu^*(g)\,.$$

Write $f = \sup g$. If $H \subset \mathscr{K}_+$ and $f \in \mathscr{K}_+$, then according to proposition 2.21, we have

$$\mu(f) = \lim_{g\in H} \mu(g)\,.$$

Since μ is increasing, $(\mu(g))_{g\in H}$ is a directed family of numbers, therefore

$$\lim_{g\in H} \mu(g) = \sup_{g\in H} \mu(g)$$

and the theorem is proved in this case.

Suppose, now, that H is arbitrary. We have $g \leqslant f$, therefore $\mu^*(g) \leqslant \mu^*(f)$ for every $g \in H$ and hence

$$\sup_{g \in H} \mu^*(g) \leqslant \mu^*(f) = \mu^*(\sup_{g \in H} g).$$

It remains to prove the converse inequality. In order to do this it is sufficient to show that for every function $\psi \in \mathscr{K}_+$ such that $\psi \leqslant f$, we have

$$\mu(\psi) \leqslant \sup_H \mu^*(g).$$

For every function $g \in H$ denote $A_g = \{\varphi \in \mathscr{K}_+, \varphi \leqslant g\}$ and $A = \cup_{g \in H} A_g$. Since H is directed, A is directed too and we have $f = \sup_{\varphi \in A} \varphi$. Let ψ be an arbitrary function from $\mathscr{K}_+$ such that $\psi \leqslant f$. Observe that

$$\psi = \sup_{\varphi \in A} [\inf(\psi, \varphi)].$$

But $\psi \in \mathscr{K}_+$ and $\inf(\psi, \varphi) \in \mathscr{K}_+$. From the first part of the proof it follows that

$$\mu(\psi) = \sup_{\varphi \in A} \mu[\inf(\psi, \varphi)].$$

For every $\varphi \in A$ there exists $g \in H$ such that $\varphi \in A_g$, therefore $\varphi \leqslant g$ and hence

$$\mu(\inf(\psi, \varphi)) \leqslant \mu(\varphi) \leqslant \mu^*(g) \leqslant \sup \mu^*(g)$$

whence

$$\mu(\psi) \leqslant \sup_{g \in H} \mu^*(g).$$

Then it follows that

$$\mu^*(f) = \sup_{\psi \leqslant f} \mu(\psi) \leqslant \sup \mu^*(g), \quad \psi \in \mathscr{K}_+.$$

The equality $\sup \mu^*(g) = \mathit{lim}\ \mu^*(g)$ follows from the fact that H is directed.

5.4 THEOREM. *If* $f_1, f_2 \in \mathscr{I}_+$, then

$$\mu^*(f_1 + f_2) = \mu^*(f_1) + \mu^*(f_2).$$

The set $H = \{\varphi_1 + \varphi_2 \mid \varphi_1 \leqslant f_1, \varphi_2 \leqslant f_2, \varphi_1, \varphi_2 \in \mathscr{K}_+\}$ is directed for the relation $\leqslant$, and $\sup H = f_1 + f_2$. By the previous theorem we have

$$\begin{aligned} \mu^*(f_1 + f_2) &= \sup \mu(\varphi_1 + \varphi_2) = \sup[\mu(\varphi_1) + \mu(\varphi_2)] = \\ &= \sup \mu(\varphi_1) + \sup \mu(\varphi_2) = \mu^*(f_1) + \mu^*(f_2) \end{aligned}$$

the least upper bounds being considered for $\varphi_1 \leqslant f_1$, $\varphi_2 \leqslant f_2$, $\varphi_1, \varphi_2 \in \mathscr{K}_+$, and the theorem is proved.

Theorem 5.4 asserts the fact that the upper integral is finitely additive on $\mathscr{I}_+$. Moreover:

5.5 PROPOSITION. *For every family $(f_i)_{i\in I}$ of functions of $\mathscr{I}_+$ we have*

$$\mu^*(\Sigma_{i\in I} f_i) = \Sigma_{i\in I}\, \mu^*(f_i)\,.$$

For every finite subset $J \subset I$, we deduce by recurrence from the preceding theorem that

$$\mu^*(\Sigma_{i\in J} f_i) = \Sigma_{i\in J}\, \mu^*(f_i)\,.$$

Denote $g_J = \Sigma_{i\in J} f_i$; we have $g_J \in \mathscr{I}_+$. When J runs over the set of finite subsets of I, the functions g_J form a directed set for the relation $\leqslant$.

From theorem 5.3 it follows that

$$\mu^*(\Sigma_{i\in I} f_i) = \mu^*(\sup_J g_J) = \sup_J \mu^*(g_J) =$$
$$= \sup_J \mu^*(\Sigma_{i\in J} f_i) = \sup_J \Sigma_{i\in J}\, \mu^*(f_i) = \Sigma_{i\in I}\, \mu^*(f_i)\,.$$

3 *The outer measure of open sets*

The characteristic function φ_G of an open set $G \subset T$ is lower semicontinuous.

For every open set $G \subset T$, the upper integral $\mu^*(\varphi_G)$ is called *the exterior or outer measure of the set* G and is denoted by $\mu^*(G)$.

We have $\|\mu\| = \mu^*(T)$.

In fact, since $\varphi_T(t) \equiv 1$, we have

$$\mu^*(T) = \mu^*(\varphi_T) = \sup_{0\leqslant g\leqslant 1} \mu(g) =$$
$$= \sup_{0\leqslant g\leqslant 1} (\sup_{|h|\leqslant g} |\mu(h)|) = \sup_{\|h\|\leqslant 1} |\mu(h)| = \|\mu\|, \ g, h \in \mathscr{K}\,.$$

From the properties of the upper integral we deduce the following properties of the outer measure of open sets:

(1) $0 \leqslant \mu^*(G) \leqslant +\infty$ *for every open set* $G \subset T$; $\mu^*(\varnothing) = 0$.

(2) *If* $G_1 \subset G_2$, *then* $\mu^*(G_1) \leqslant \mu^*(G_2)$.

In fact, in this case $\varphi_{G_1} \leqslant \varphi_{G_2}$.

(3) *If* $\mathscr{G}$ *is a set of open subsets of* T, *directed for the relation* $\subset$, *then*

$$\mu^*(\cup_{G\in\mathscr{G}}\, G) = \sup_{G\in\mathscr{G}}\, \mu^*(G)\,.$$

In fact, the family $(\varphi_G)_{G\in\mathscr{G}}$ is directed for $\leqslant$ and consists of functions from $\mathscr{I}_+$ and its upper envelope is the characteristic function of the union $\cup_{G\in\mathscr{G}}\, G$.

(4) *If* $(G_i)_{i\in I}$ *is an arbitrary family of open sets, then*

$$\mu^*(\cup_{i\in I} G_i) \leqslant \Sigma_{i\in I}\,\mu^*(G_i)\,.$$

If the sets G_i are mutually disjoint, then

$$\mu^*(\cup_{i\in I} G_i) = \Sigma_{i\in I}\,\mu^*(G_i)\,.$$

Put $G = \cup_{i\in I} G_i$. Then we have

$$\varphi_G = \sup_{i\in I} \varphi_{G_i} \leqslant \Sigma_{i\in I}\,\varphi_{G_i}\,,$$

therefore

$$\mu^*(G) \leqslant \mu^*(\Sigma_{i\in I}\,\varphi_{G_i}) = \Sigma_{i\in I}\,\mu^*(\varphi_{G_i}) = \Sigma_{i\in I}\,\mu^*(G_i)\,.$$

If, in addition, the sets G_i are mutually disjoint, then

$$\varphi_G = \Sigma_{i\in I}\,\varphi_{G_i}\,,$$

therefore

$$\mu^*(G) = \mu^*(\Sigma_{i\in I}\,\varphi_{G_i}) = \Sigma_{i\in I}\,\mu^*(\varphi_{G_i}) = \Sigma_{i\in I}\,\mu^*(G_i)$$

(5) *If the open set G is relatively compact, then $\mu^*(G) < +\infty$.*

In fact, in this case, there exists a function $f \in \mathscr{K}_+$ such that $\varphi_G \leqslant f$ and therefore

$$\mu^*(G) = \mu^*(\varphi_G) \leqslant \mu^*(f) < +\infty\,.$$

(6) *If $f \geqslant 0$ is lower semicontinuous and if $\mu^*(f) < +\infty$, then the set $A = \{t \mid f(t) \neq 0\}$ is equal to the union of a sequence (G_n) of open sets with $\mu^*(G_n) < +\infty$.*

In fact, the sets

$$G_n = \{t \mid f(t) > n^{-1}\}$$

are open and we have

$$A = \cup_{1 \leqslant n < \infty} G_n\,.$$

From the inequality $f \geqslant n^{-1}\varphi_{G_n}$ we deduce $\varphi_{G_n} \leqslant nf$, therefore

$$\mu^*(G_n) \leqslant n\,\mu^*(f) < +\infty\,.$$

4. *The upper integral of positive functions*

We define first the upper integral of functions $f \geqslant 0$ with compact support.

5.6 DEFINITION. *Let $f \geqslant 0$ be a (finite or infinite) numerical function with compact support. The number $\mu^*(f)$ defined by the equality*

$$\mu^*(f) = \inf_{h \geqslant f} \mu^*(h), \quad h \in \mathscr{I}_+ .$$

is called the upper integral (with respect to μ) of the function f.

If $f \geqslant 0$ is a lower semicontinuous function, *with compact support*, then the upper integral $\mu^*(f)$ defined above is equal to the upper integral previously defined; in fact, in the above definition we can take $h = f$.

Since the set $B = \{h \in \mathscr{I}_+ \mid h \geqslant f\}$ is directed for the relation $\geqslant$, and the upper integral $\mu^*(h)$ is increasing on B, it follows that the function $h \to \mu^*(h)$ has a (finite or infinite) limit when h runs over the filter of sections of B, and the limit is equal to the greatest lower bound of the function μ^* on the set B:

$$\lim_{h \geqslant f} \mu^*(h) = \inf_{h \geqslant f} \mu^*(h), \quad h \in \mathscr{I}_+ .$$

It follows that the upper integral $\mu^*(f)$ of a function $f \geqslant 0$ with *compact support* can also be defined by the equality

$$\mu^*(f) = \lim_{h \geqslant f} \mu^*(h), \quad h \in \mathscr{I}_+ .$$

If f and g are two functions $\geqslant 0$ with compact support, and if $f \leqslant g$, then $\mu^*(f) \leqslant \mu^*(g)$.

5.7 PROPOSITION. *For every lower semicontinuous function $f \geqslant 0$ we have*

$$\mu^*(f) = \sup_K \mu^*(f\varphi_K)$$

when K runs over the set of compact subsets of T.

In fact, for every compact set $K \subset T$, the function $f\varphi_K$ has compact support and we have $f\varphi_K \leqslant f$, and $f \in \mathscr{I}_+$. By the preceding remark we have

$$\mu^*(f\varphi_K) \leqslant \mu^*(f),$$

therefore

$$\sup_K \mu^*(f\varphi_K) \leqslant \mu^*(f).$$

Conversely, let $g \in \mathscr{K}_+(T)$, with $g \leqslant f$ and let K be the support of g. Then $g \leqslant f\varphi_K$, therefore

$$\mu(g) = \mu^*(g) \leqslant \mu^*(f\varphi_K) \leqslant \sup_K \mu^*(f\varphi_K)$$

whence

$$\mu^*(f) = \sup_{g \leqslant f} \mu^*(g) \leqslant \sup_K \mu^*(f\varphi_K), \ g \in \mathscr{K}_+ ,$$

consequently

$$\mu^*(f) = \sup_K \mu^*(f\varphi_K) , \ K \text{ compact} .$$

Remark. For every function $f \geqslant 0$ with compact support we have also

$$\mu^*(f) = \sup_K \mu^*(f\varphi_K), \ K \text{ compact}.$$

In fact, let K_0 be the support of f. For every compact set $K \subset T$ we have $f\varphi_K \leqslant f$, therefore $\mu^*(f\varphi_K) \leqslant \mu^*(f)$ and thus

$$\mu^*(f) = \mu^*(f\varphi_{K_0}) \leqslant \sup_K \mu^*(f\varphi_K) \leqslant \mu^*(f) .$$

5.8 DEFINITION. *The number $\mu^*(f)$ defined by the equality*

$$\mu^*(f) = \sup_K \mu^*(f\varphi_K)$$

when K runs over the set of compact subsets of T, is called the upper integral (with respect to μ) of a (finite or infinite) function $f \geqslant 0$.

The upper integral $\mu^*(f)$ of a function $f \geqslant 0$ will also be denoted by $\int^* f \, d\mu$ or $\int^* f(t) d\mu(t)$.

From the preceding proposition and remark, it follows that for *lower semicontinuous* functions $f \geqslant 0$ and for functions $f \geqslant 0$ with *compact support*, the upper integral defined here coincides with that previously defined. Therefore, we extended the definition of the upper integral μ^* from $\mathscr{I}_+$ to the set of all function $f \geqslant 0$ defined on T.

Example. Let α be a finite positive function defined on T such that for every compact set $K \subset T$ we have $\Sigma_{t \in K} \alpha(t) < +\infty$ and let μ be the positive measure defined by the equality

$$\mu(f) = \Sigma_{t \in T} \alpha(t) f(t) \text{ for } f \in \mathscr{K}(T) .$$

We saw that for every lower semicontinuous function $f \geqslant 0$ we have

$$\mu^*(f) = \Sigma_{t \in T} \alpha(t) f(t) .$$

Let $f \geqslant 0$ be arbitary. We have

$$\mu^*(f) \geqslant \Sigma_{t\in T}\,\alpha(t)f(t)\,.$$

Let us prove that if

$$\mu^*(f) < +\infty\,,$$

then

$$\mu^*(f) = \Sigma_{t\in T}\,\alpha(t)f(t)\,.$$

Suppose first that f has compact support. Then there exists a function $h\in\mathscr{I}_+$ with $h\geqslant f$ and $\mu^*(h) < +\infty$. Put $u(t)=h(t)-f(t)$ if the difference is defined and $u(t)=0$ if $f(t)=h(t)=+\infty$.

The sum $\Sigma_{t\in T}\,\alpha(t)\,u(t)$ is finite. For every $\varepsilon>0$ there exists a finite subset $M\subset T$ with $\Sigma_{t\in M}\,\alpha(t)\,u(t) > \Sigma_{t\in T}\,\alpha(t)\,u(t)-\varepsilon$. Let h_1 be the function equal to $f(t)$ for $t\in M$ and to $h(t)$ for $t\notin M$. The function h_1 is lower semicontinuous and $h_1\geqslant f$.

Then we have

$$\Sigma_{t\in T}\,\alpha(t)\,h_1(t) = \Sigma_{t\in T}\,\alpha(t)\,h(t) - \Sigma_{t\in M}\,\alpha(t)\,u(t) \leqslant \Sigma_{t\in T}\,\alpha(t)f(t)+\varepsilon\,,$$

therefore

$$\mu^*(f) \leqslant \Sigma_{t\in T}\,\alpha(t)f(t)+\varepsilon\,.$$

Since ε is arbitrary, we deduce

$$\mu^*(f) \leqslant \Sigma_{t\in T}\,\alpha(t)f(t)$$

and therefore

$$\mu^*(f) = \Sigma_{t\in T}\,\alpha(t)f(t)\,.$$

If $f\geqslant 0$ is an arbitrary function with $\mu^*(f) < +\infty$, then for every compact set $K\subset T$ we have

$$\mu^*(f\varphi_K) = \Sigma_{t\in K}\,\alpha(t)f(t)$$

and passing to the limit when K runs over the directed family of compact sets we obtain

$$\mu^*(f) = \Sigma_{t\in T}\,\alpha(t)f(t)\,.$$

5.9 Proposition. *For every function $f\geqslant 0$ we have the inequality*

$$\mu^*(f) \leqslant \inf_{h\geqslant f}\mu^*(h)\,,\quad h\in\mathscr{I}_+\,.$$

In fact, for every function $h \geqslant f$ from $\mathscr{I}_+$ and every compact set $K \subset T$ we have $h \geqslant f\varphi_K$ and $f\varphi_K$ has compact support, therefore

$$\mu^*(f\varphi_K) \leqslant \mu^*(h) .$$

Then

$$\mu^*(f\varphi_K) \leqslant \inf_{h \geqslant f} \mu^*(h) , \quad h \in \mathscr{I}_+$$

whence

$$\mu^*(f) = \sup_K \mu^*(f\varphi_K) \leqslant \inf_{h \geqslant f} \mu^*(h) , \quad h \in \mathscr{I}_+ .$$

Remark. For functions $f \geqslant 0$ with compact support and for lower semi-continuous functions $f \geqslant 0$, we even have the equality

$$\mu^*(f) = \inf_{h \geqslant f} \mu^*(h), \quad h \in \mathscr{I}_+ .$$

From the preceding proposition it follows that the equality is valid for every function $f \geqslant 0$ with $\mu^*(f) = +\infty$.

It will be shown further that the equality is true for every function $f \geqslant 0$ for which there exists a lower semicontinuous function $h \geqslant f$ with $\mu^*(h) < +\infty$.

But the inequality in proposition 5.9 may be strict: there are functions $f \geqslant 0$ with $\mu^*(f) < +\infty$, such that $\mu^*(h) = +\infty$ for every function $h \in \mathscr{I}_+$ with $h \geqslant f$.

We shall show further that sometimes (for instance, if T is countable at infinity, or, more generally, if it is the union of a sequence (A_n) of open sets with $\mu^*(\varphi_{A_n}) < +\infty$ for each n) the equality is true for every function $f \geqslant 0$.

Remark. N. Bourbaki calls *upper integral* of a function $f \geqslant 0$ the number $\inf_{h \geqslant f} \mu^*(h)$, $h \in \mathscr{I}_+$ (which is denoted $\mu^*(f)$ by him) and *essential upper integral* of f, the number $\sup_K \mu^*(f\varphi_K)$ (which is denoted $\overline{\mu^*}(f)$ by him).

The upper integral of the positive functions has the following properties:

(1) $0 \leqslant \mu^*(f) \leqslant +\infty$,

(2) $\mu^*(f_1) \leqslant \mu^*(f_2)$ if $0 \leqslant f_1 \leqslant f_2$;

(3) $\mu^*(\alpha f) = \alpha\mu^*(f)$ if $f \geqslant 0$ and $\alpha \geqslant 0$.

In fact, these properties, valid for functions from $\mathscr{I}_+$, are preserved by passing to the supremum, for positive functions with compact support, and then, by passing to the infimum, for arbitrary positive functions.

The upper integral, which was additive for functions from $\mathscr{I}_+$ is, in general, only subadditive for arbitrary positive functions:

5.10 PROPOSITION. *If* $f_1 \geqslant 0$ *and* $f_2 \geqslant 0$ *then*

$$\mu^*(f_1+f_2) \leqslant \mu^*(f_1)+\mu^*(f_2).$$

In fact, let $K \subset T$ be a compact set, and let the functions $h_1, h_2 \in \mathcal{I}_+$ be such that $f_1\varphi_K \leqslant h_1$ and $f_2\varphi_K \leqslant h_2$. Then $(f_1+f_2)\varphi_K \leqslant h_1+h_2$ and $h_1+h_2 \in \mathcal{I}_+$, therefore

$$\mu^*((f_1+f_2)\varphi_K) \leqslant \mu^*(h_1+h_2) = \mu^*(h_1)+\mu^*(h_2)$$

whence

$$\begin{aligned}\mu^*((f_1+f_2)\varphi_K) &\leqslant \inf_{h_1 \geqslant f_1\varphi_K,\, h_2 \geqslant f_2\varphi_K}[\mu^*(h_1)+\mu^*(h_2)] = \\ &= \inf_{h_1 \geqslant f_1\varphi_K} \mu^*(h_1) + \inf_{h_2 \geqslant f_2\varphi_K} \mu^*(h_2) = \\ &= \mu^*(f_1\varphi_K)+\mu^*(f_2\varphi_K).\end{aligned}$$

Then it follows that

$$\begin{aligned}\mu^*(f_1+f_2) = \sup_K \mu^*((f_1+f_2)\varphi_K) &\leqslant \sup_K \mu^*(f_1\varphi_K)+\sup_K \mu^*(f_2\varphi_K) = \\ &= \mu^*(f_1)+\mu^*(f_2).\end{aligned}$$

Remark. The inequality may be strict for certain functions. It will be shown further that for certain classes of functions (for example, for integrable functions) the upper integral is additive.

Since the set $C = \{f\varphi_K | K \text{ compact}\}$ is directed for the relation $\leqslant$, and μ^* is increasing, it follows that the function $f \to \mu^*(f)$ has a limit when f runs over the filter of sections of C, and this limit is equal to the supremum of the function μ^* on the set C:

$$\lim_K \mu^*(f\varphi_K) = \sup_K \mu^*(f\varphi_K), \; K \text{ compact}.$$

It follows that the upper integral $\mu^*(f)$ of an arbitrary function $f \geqslant 0$ can be also defined by the equality

$$\mu^*(f) = \lim_K \mu^*(f\varphi_K).$$

The permutability property of the upper integral with the supremum, valid for *arbitrary* directed families of function from $\mathcal{I}_+$, is valid in general only for *countable* families of positive functions.

5.11 PROPOSITION. *Let (f_n) be an increasing sequence of functions $\geqslant 0$ and $f = \sup_n f_n$. If for each n we have*

$$\mu^*(f_n) = \inf_{h \geqslant f_n} \mu^*(h), \quad h \in \mathscr{I}_+ ,$$

then

$$\mu^*(f) = \inf_{h \geqslant f} \mu^*(h) \quad \textit{and} \quad \mu^*(f) = \sup_n \mu^*(f_n), \quad h \in \mathscr{I}_+ .$$

Since $f_n \leqslant f$ for each n, we have

$$\mu^*(f_n) \leqslant \mu^*(f) ,$$

therefore

$$\sup_n \mu^*(f_n) \leqslant \mu^*(f) .$$

If $\sup \mu^*(f_n) = +\infty$, it follows that $\mu^*(f) = +\infty$ and $\mu^*(h) = +\infty$ for every function $h \geqslant f$ from $\mathscr{I}_+$ and the proposition is proved in this case (without any condition on the functions f_n).

Suppose now that $\sup \mu^*(f_n) < +\infty$, therefore $\mu^*(f_n) < +\infty$ for each n.

Let $\varepsilon > 0$. For each n there exists a function $h_n \in \mathscr{I}_+$ such that $f_n \leqslant h_n$ and $\mu^*(f_n) \leqslant \mu^*(h_n) \leqslant \mu^*(f_n) + \varepsilon/2^n$. For each n put

$$g_n = \sup(h_1, \ldots, h_n) .$$

The function g_n belongs to $\mathscr{I}_+$, the sequence (g_n) is increasing and $f_n \leqslant g_n$ for each n. In addition,

$$\mu^*(g_n) \leqslant \mu^*(f_n) + \varepsilon(1 - 2^{-n})$$

for each n. In fact, for $n = 1$, we have $g_1 = h_1$ and the inequality is written $\mu^*(h_1) \leqslant \mu^*(f_1) + \varepsilon/2$; this inequality is indeed satisfied. Suppose the inequality is true for an arbitrary n and prove it for $n+1$.

Observe that

$$g_{n+1} = \sup(g_n, h_{n+1}), \quad h_{n+1} \geqslant f_{n+1} \geqslant f_n \quad \text{and} \quad g_n \geqslant f_n$$

therefore

$$\inf(g_n, h_{n+1}) \geqslant f_n .$$

Then

$$\inf(g_n, h_{n+1}) + \sup(g_n, h_{n+1}) = g_n + h_{n+1} ,$$

that is

$$\inf(g_n, h_{n+1}) + g_{n+1} = g_n + h_{n+1}\,.$$

Since these functions are lower semicontinuous, we deduce

$$\mu^*(\inf(g_n, h_{n+1}) + \mu^*(g_{n+1}) = \mu^*(g_n) + \mu^*(h_{n+1})$$

therefore

$$\begin{aligned}\mu^*(g_{n+1}) &= \mu^*(g_n) + \mu^*(h_{n+1}) - \mu^*(\inf(g_n, h_{n+1})) \leqslant \\ &\leqslant \mu^*(g_n) + \mu^*(h_{n+1}) - \mu^*(f_n) \leqslant \\ &\leqslant \mu^*(f_{n+1}) + \varepsilon 2^{-n-1} + \varepsilon(1-2^{-n}) \\ &= \mu^*(f_{n+1}) + \varepsilon(1-2^{-n-1})\,.\end{aligned}$$

Therefore, the inequality

$$\mu^*(g_n) \leqslant \mu^*(f_n) + \varepsilon(1-2^{-n})$$

is true for every n.

Now denote $g = \sup g_n$. The functions g_n and g are lower semicontinuous, therefore $\mu^*(g) = \sup \mu^*(g_n)$, whence

$$\mu^*(g) \leqslant \sup_n \mu^*(f_n) + \varepsilon\,.$$

But $f = \sup_n f_n \leqslant g$, therefore

$$\mu^*(f) \leqslant \mu^*(g) \leqslant \sup_n \mu^*(f_n) + \varepsilon\,,$$

ε being arbitrary, it follows

$$\mu^*(f) \leqslant \sup_n \mu^*(f_n)\ ;$$

consequently,

$$\mu^*(f) = \sup \mu^*(f_n)\,.$$

The inequality $\mu^*(g) \leqslant \sup_n \mu^*(f_n) + \varepsilon$ can now be written $\mu^*(g) \leqslant \mu^*(f) + \varepsilon$. Since $g \in \mathscr{I}_+$ and $g \geqslant f$, we deduce

$$\mu^*(f) = \inf_{h \geqslant f} \mu^*(h),\ h \in \mathscr{I}_+\,.$$

Thus the proposition is proved.

5.12 Corollary. *If the function $f \geqslant 0$ has support contained in the union of a sequence (K_n) of compact sets, then*

$$\mu^*(f) = \inf_{h \geqslant f} \mu^*(h),\ h \in \mathscr{I}_+\,.$$

If for every n we take $A_n = \cup_{1 \leqslant i \leqslant n} K_i$, the sets A_n are compact and the sequence (A_n) is increasing; therefore the sequence $(f\varphi_{A_n})$ is increasing and $f = \sup_n f\varphi_{A_n}$. Since the functions $f\varphi_{A_n}$ have compact support, we have

$$\mu^*(f\varphi_{A_n}) = \inf_{h \geqslant f\varphi_{A_n}} \mu^*(h) \text{ for every } n, \ h \in \mathscr{I}_+ .$$

By the preceding proposition we have then

$$\mu^*(f) = \inf_{h \geqslant f} \mu^*(h), \quad h \in \mathscr{I}_+ .$$

5.13 COROLLARY. *If the space T is countable at infinity, in particular if T is compact, then we have*

$$\mu^*(f) = \inf_{h \geqslant f} \mu^*(h), \ h \in \mathscr{I}_+ ,$$

for every function $f \geqslant 0$.

Remark. It will be proved further (proposition 11.19) that the equality in corollary 5.13 still remains valid for functions vanishing on the complement of the union of a sequence (G_n) of open sets with $\mu^*(\varphi_{G_n}) < +\infty$ for each n.

It will then follow that if T is a union of a sequence (G_n) of open sets with $\mu^*(\varphi_{G_n}) < +\infty$, in particular if μ^* is bounded, the equality is true for every function $f \geqslant 0$.

5.14 THEOREM. *For every increasing sequence (f_n) of functions $\geqslant 0$ we have*

$$\mu^*(\sup_n f_n) = \sup_n \mu^*(f_n) .$$

For every compact set $K \subset T$ and every n we have

$$\mu^*(f_n \varphi_K) = \inf_{h \geqslant f_n \varphi_K} \mu^*(h) , \ h \in \mathscr{I}_+ ,$$

the sequence $(f_n \varphi_K)$ is increasing and $\sup_n (f_n \varphi_K) = (\sup_n f_n)\varphi_K$. By the preceding proposition we have

$$\mu^*((\sup_n f_n)\varphi_K) = \sup_n \mu^*(f_n \varphi_K) .$$

Taking the supremum for all the compact sets $K \subset T$ we obtain

$$\mu^*(\sup_n f_n) = \sup_K \mu^*((\sup_n f_n)\varphi_K) = \sup_K \sup_n \mu^*(f_n \varphi_K) =$$
$$= \sup_n \sup_K \mu^*(f_n \varphi_K) = \sup_n \mu^* f_n .$$

and the theorem is proved.

Remark. If (f_n) is a *decreasing* sequence of functions $\geqslant 0$, we have, in general, only the inequality

$$\mu^*(\inf_n f_n) \leqslant \inf_n \mu^*(f_n).$$

5.15 Corollary. *For every countable set H of positive functions, directed for the relation $\leqslant$, we have*

$$\mu^*(\sup_{f\in H} f) = \sup_{f\in H} \mu^*(f).$$

In fact, $\sup_{f\in H} f \geqslant f$, therefore $\mu^*(\sup_{f\in H} f) \geqslant \mu^*(f)$ for every $f\in H$, whence

$$\mu^*(\sup_{f\in H} f) \geqslant \sup_{f\in H} \mu^*(f).$$

In order to prove the converse inequality, let (f_n) be the sequence of the functions of H, arranged in an arbitrary order. We extract from this sequence an increasing subsequence (f_{n_k}) in the following way: $f_{n_1} = f_1$ and for every k we choose $f_{n_{k+1}} \geqslant \sup(f_{n_k}, f_k)$. Then $\sup_k f_{n_k} = \sup_{f\in H} f$ therefore

$$\mu^*(\sup_{f\in H} f) = \mu^*(\sup_k f_{n_k}) = \sup_k \mu^*(f_{n_k}) \leqslant \sup_{f\in H} \mu^*(f).$$

Remark. If H is not countable, the corollary is no longer true.

For instance, for the Lebesgue measure on the real line and for the directed set H of the characteristic functions φ_A of all finite sets $A \subset R$ we have $\mu^*(\varphi_A) = 0$; therefore $\sup_{\varphi_A\in H} \mu^*(\varphi_A) = 0$, but $\sup_{\varphi_A\in H} \varphi_A = 1$ and $\mu(\sup_{\varphi_A\in H} \varphi_A) = +\infty$.

5.16 Proposition. *For every sequence (f_n) of positive functions we have*

$$\mu^*(\Sigma_{1\leqslant n<\infty} f_n) \leqslant \Sigma_{1\leqslant n<\infty} \mu^*(f_n).$$

Consider the increasing sequence of functions $g_n = \Sigma_{1\leqslant k\leqslant n} f_k$ and apply theorem 5.14 taking into account that μ^* is finitely subadditive:

$$\mu^*(\Sigma_{1\leqslant n<\infty} f_n) = \mu^*(\sup_n \Sigma_{1\leqslant k\leqslant n} f_k) = \sup_n \mu^*(\Sigma_{1\leqslant k\leqslant n} f_k) \leqslant$$
$$\leqslant \sup_n \Sigma_{1\leqslant k\leqslant n} \mu^*(f_k) = \Sigma_{1\leqslant k<\infty} \mu^*(f_k).$$

5.17 Lemma (*Fatou*). *For every sequence (f_n) of positive functions we have*

$$\mu^*(\liminf f_n) \leqslant \liminf \mu^*(f_n).$$

For every n put $g_n = \inf_{p \geqslant 0} f_{n+p}$. The sequence (g_n) is increasing; therefore

$$\mu^*(\sup_n g_n) = \sup_n \mu^*(g_n) .$$

But

$$\sup_n g_n = \sup_n (\inf_{p \geqslant 0} f_{n+p}) = \liminf f_n ,$$

therefore

$$\mu^*(\liminf f_n) = \sup \mu^*(g_n) .$$

On the other hand, $g_n \leqslant f_{n+p}$ for $p \geqslant 0$. Consequently

$$\mu^*(g_n) \leqslant \mu^*(f_{n+p}) .$$

Then

$$\mu^*(g) \leqslant \inf_{p \geqslant 0} (f_{n+p})$$

hence

$$\mu^*(\liminf f_n) \leqslant \sup_n \mu^*(g_n) \leqslant \sup_n (\inf_{p \geqslant 0} \mu^*(f_{n+p})) =$$
$$= \liminf \mu^*(f_n) .$$

5.18 Corollary. *Let (f_n) be a sequence of functions $\geqslant 0$. If $\mu \neq 0$ and if $\lim_n f_n(t) = +\infty$ for every $t \in T$, then $\lim_n \mu^*(f_n) = +\infty$.*

In fact, taking $f_0(t) \equiv +\infty$, we have $f_0(t) = \lim_n f_n(t) = \liminf_n f_n(t)$, therefore, by Fatou's lemma,

$$\mu^*(f_0) \leqslant \liminf \mu^*(f_n) .$$

It remains to show that $\mu^*(f_0) = +\infty$. Since $\mu \neq 0$, there exists a function $g \in \mathscr{K}_+$ such that $\mu(g) > 0$. Since $f_0 \geqslant g$, we deduce $\mu^*(f_0) \geqslant \mu(g) > 0$. For every $k \in N$ we have $f_0 = k f_0$, therefore $\mu^*(f_0) = k\mu^*(f_0) \geqslant k\ \mu(g)$, whence $\mu^*(f_0) = +\infty$. It follows that $\liminf \mu^*(f_n) = +\infty$, therefore the sequence $(\mu^*(f_n))$ has a limit and

$$\lim \mu^*(f_n) = \liminf \mu^*(f_n) = +\infty .$$

5.19 Proposition. *If μ and ν are two positive measures and $\alpha > 0$, then*

(1) $(\mu + \nu)^* = \mu^* + \nu^*$;

(2) $(\alpha\mu)^* = \alpha\mu^*$;
(3) $\mu^* \leqslant \nu^*$ *if* $\mu \leqslant \nu$,

For $f \in \mathscr{I}_+$ we have

$$(\mu+\nu)^*(f) = \lim_{g \leqslant f}(\mu+\nu)(g) = \lim_{g \leqslant f}(\mu(g)+\nu(g)) =$$

$$= \lim_{g \leqslant f}\mu(g) + \lim_{g \leqslant f}\nu(g) = \mu^*(f)+\nu^*(f), \quad g \in \mathscr{K}_+ .$$

If $f \geqslant 0$ has compact support, the equality (1) can be proved in the same way, taking the limit when g runs over the set of functions $g \geqslant f$, $g \in \mathscr{I}_+$, directed for the relation $\geqslant$. If $f \geqslant 0$ is arbitrary, we proceed in the same way:

$$(\mu+\nu)^*(f) = \lim_K (\mu+\nu)^*(f\varphi_K) = \lim_K (\mu^*(f\varphi_K)+\nu^*(f\varphi_K)) =$$
$$= \lim_K \mu^*(f\varphi_K) + \lim_K \nu^*(f\varphi_K) = \mu^*(f)+\nu^*(f) .$$

The equality (2) is proved in a similar way.

If $\mu \leqslant \nu$, then $\nu = \mu + (\nu - \mu)$ and $\nu - \mu \geqslant 0$, therefore, for every function $f \geqslant 0$, we have

$$\nu^*(f) = \mu^*(f) + (\nu-\mu)^*(f) .$$

Since $(\nu-\mu)^*(f) \geqslant 0$, if follows that $\mu^*(f) \leqslant \nu^*(f)$, therefore $\mu^* \leqslant \nu^*$.

Thus the proposition is proved.

5 *The outer measure of sets*

For every set $A \subset T$, the upper integral $\mu^*(\varphi_A)$ is called the *exterior* or *outer measure* of the set A (for the measure μ) and is denoted by $\mu^*(A)$.

The properties of the outer measure follow from those of the upper integral:

(1) $0 \leqslant \mu^*(A) \leqslant +\infty$.
(2) $\mu^*(A) \leqslant \mu^*(B)$ *if* $A \subset B$.
(3) *If* (A_n) *is an increasing sequence of subsets of* T, *then*

$$\mu^*(\cup_{1 \leqslant n < \infty} A_n) = \sup_n \mu^*(A_n) .$$

In fact

$$\sup_n \varphi_A = \varphi_{\cup_{1 \leqslant n < \infty} A_n}$$

(4) *For every sequence* (A_n) *of subsets of* T *we have*

$\mu^*(\cup_{1 \leqslant n < \infty} A_n) \leqslant \Sigma_{1 \leqslant n < \infty} \mu^*(A_n)$.

The inequality may be strict even if the sets are mutually disjoint.

(5) *For every set $A \subset T$ contained in the union of a sequence of compact sets, we have*

$\mu^*(A) = \inf_{G \supset A} \mu^*(G)$, *G open* .

Let $0 < \varepsilon < 1$. Since the function φ_A has the support contained in the union of a sequence of compact sets, we have

$\mu^*(\varphi_A) = \inf \mu^*(h)$, $h \geqslant \varphi_A$, $h \in \mathscr{I}_+$.

Therefore, there exists a function $f \in \mathscr{I}_+$ with $\varphi_A \leqslant f$ and $\mu^*(A) \leqslant \mu^*(f) \leqslant \mu^*(A) + \varepsilon$.

The set $G = \{t \mid f(t) > 1 - \varepsilon\}$ is open, because f is lower semicontinuous. We have $G \supset A$ and $f \geqslant (1-\varepsilon)\varphi_G$, therefore, $(1-\varepsilon)^{-1} f \geqslant \varphi_G$, whence

$\mu^*(G) \leqslant (1-\varepsilon)^{-1} \mu^*(f) \leqslant (1-\varepsilon)^{-1}(\mu^*(A) + \varepsilon)$.

Then

$\inf_{G \supset A} \mu^*(G) \leqslant (1-\varepsilon)^{-1}(\mu^*(A) + \varepsilon)$;

ε being arbitrary we obtain

$\mu^*(A) \geqslant \inf_{G \supset A} \mu^*(G)$, G open ,

whence

$\mu^*(A) = \inf_{G \supset A} \mu^*(G)$, G open .

Property (5) is, in particular, true for every relatively compact set $A \subset T$.

From property (5) it follows that if T is *countable at infinity*, in particular if T is compact, the equality

$\mu^*(A) = \inf_{G \supset A} \mu^*(G)$, G open ,

is true for *every* set $A \subset T$.

Remark. The equality is evidently true also if $\mu^*(A) = +\infty$. It will be shown further that the equality remains true for every set A contained in an open set G with $\mu^*(G) < +\infty$.

However there are sets $A \subset T$ with $\mu^*(A) < +\infty$ and $\mu^*(G) = +\infty$ for every open set $G \supset A$.

(6) *For every relatively compact set* $A \subset T$ *we have* $\mu^*(A) < +\infty$. In fact, there exists a relatively compact open set $G \supset A$. Then

$$\mu^*(A) \leqslant \mu^*(G) < +\infty .$$

(7) *For every set* $A \subset T$ *we have*

$$\mu^*(A) = \sup_K \mu^*(A \cap K), \quad K \text{ compact} .$$

In fact,

$$\varphi_A \cdot \varphi_K = \varphi_{A \cap K} \cdot$$

Remark. For every set $A \subset T$, the number $\mu_*(A)$ defined by the equality

$$\mu_*(A) = \sup_{K \subset A} \mu^*(K), \quad K \text{ compact} ,$$

is called the *interior* or *inner measure* of the set A.

We have $\mu_*(A) \leqslant \mu^*(A)$ and the inequality may be strict.

§6 Negligible functions and sets

1 *Positive negligible functions*

Let μ be a positive measure on T.

6.1 Definition. *We say that a function* $f \geqslant 0$ *is negligible for the measure* μ, *or* μ*-negligible, if* $\mu^*(f) = 0$. *We say that a set* $A \subset T$ *is* μ*-negligible if its characteristic function* φ_A *is* μ*-negligible, that is if* $\mu^*(A) = 0$.

Remark. N. Bourbaki calls *negligible functions* the functions $f \geqslant 0$ for which $\inf_{h \geqslant f} \mu^*(h) = 0$, $h + \mathscr{I}_+$, and *locally negligible functions*, the functions $f \geqslant 0$ for which $\sup_K \mu^*(f\varphi_K) = 0$, K compact.

If there is no danger of confusion about the measure μ we shall say simply, negligible function (set) instead of μ-negligible function (set).

We list some properties of negligible functions:

(1) *If* f *is negligible,* $\alpha \geqslant 0$ *and* $0 \leqslant g \leqslant \alpha f$, *then* g *is negligible.*

In fact,

$$0 \leqslant \mu^*(g) \leqslant \alpha\mu^*(f) = 0 .$$

In particular, if f is negligible and $\alpha \geqslant 0$, then αf is negligible.

(2) *If* (f_n) *is a sequence of negligible positive functions, then the sum* $\Sigma_{1 \leqslant n < \infty} f_n$ *and the upper envelope* $\sup_n f_n$ *are negligible.*

In fact,

$$\mu^*(\Sigma_{1 \leqslant n < \infty} f_n) \leqslant \Sigma_{1 \leqslant n < \infty} \mu^*(f_n) = 0$$

and $\sup_n f_n \leqslant \Sigma_{1 \leqslant n < \infty} f_n$, therefore

$$\mu^*(\sup_n f_n) \leqslant \mu^*(\Sigma_{1 \leqslant n < \infty} f_n) = 0 .$$

(3) *A function* $f \geqslant 0$ *is negligible if and only if* $f\varphi_K$ *is negligible for every compact set* $K \subset T$.

In fact,

$$\mu^*(f) = \sup_K \mu^*(f\varphi_K) .$$

6.2 Proposition. *A function* $f \in \mathscr{I}_+$ *is* μ*-negligible if and only if* f *vanishes identically on the support of* μ.

Let S be the support of μ. If $\mu^*(f) = 0$, we have $\mu(g) = 0$ for every function $g \leqslant f$ from $\mathscr{K}_+$; and for the functions $g \in \mathscr{K}_+$ with $\mu(g) = 0$ we have $g(t) = 0$ for $t \in S$. Then it follows that

$$f(t) = \sup_{g \leqslant f} g(t) = 0, \; g \in \mathscr{K}_+, \text{ for } \; t \in S .$$

Conversely, if $f(t) = 0$ for every $t \in S$, then $g(t) = 0$, for every $t \in S$ and every function $g \leqslant f$ from $\mathscr{K}_+$, and for the functions $g \in \mathscr{K}_+$ which are zero on S we have $\mu(g) = 0$. Then it follows that $\mu^*(f) = 0$.

The properties of negligible sets are deduced from those of negligible functions:

(1) *Every subset* $B \subset A$ *of a negligible set* A *is negligible.*

(2) *The union of a sequence* (A_n) *of negligible sets is a negligible set.*

(3) *A set* A *is negligible if and only if* $A \cap K$ *is negligible, for every compact set* $K \subset T$.

6.3 Proposition. *The complement of the support of the measure* μ *is the biggest* μ *-negligible open set.*

For every open set G we have $\varphi_G \in \mathscr{I}_+$. From proposition 6.2 it follows that G is μ-negligible if and only if G is disjoint from the support S of the measure μ, that is if and only if $G \subset T \setminus S$. As $T \setminus S$ is open and disjoint from S, it follows that $T \setminus S$ is μ-negligible, and namely, the biggest μ-negligible open set.

6.4 PROPOSITION. *Let μ and ν be two positive measures. A function $f \geqslant 0$ is $\mu+\nu$-negligible if and only if it is μ-negligible and ν-negligible.*

In fact

$$(\mu+\nu)^*(f) = \mu^*(f) + \nu^*(f).$$

6.5 COROLLARY. *A set $A \subset T$ is $\mu+\nu$-negligible if and only if A is μ-negligible and ν-negligible.*

6.6 PROPOSITION. *Let μ and ν be two positive measures. If $\mu \leqslant \nu$, then every ν-negligible function $f \geqslant 0$ is μ-negligible and every ν-negligible set $A \subset T$ is μ-negligible.*

In fact, if $\mu \leqslant \nu$, then $\mu^*(f) \leqslant \nu^*(f)$ and $\mu^*(A) \leqslant \nu^*(A)$.

2 *Properties true almost everywhere*

Let $P(t)$ be a property defined for every point $t \in T$, possibly false for some points and true for others.

6.7 DEFINITION. *We say that the property $P(t)$ is true almost everywhere with respect to a positive measure μ, or μ-almost everywhere, if the set N of points t of T for which the property $P(t)$ is not true is μ-negligible.*

If there is no danger of confusion about the measure μ, we shall simply say "almost everywhere" instead of μ-almost everywhere.

6.8 THEOREM. *A function $f \geqslant 0$ is μ-negligible if and only if $f(t)=0$, μ-almost everywhere.*

Assume first that f is μ-negligible, that is $\mu^*(f)=0$. Denote $N=\{t \mid t \in T, f(t) \neq 0\}$. We have $\varphi_N \leqslant \sup_n (nf)$ and nf is μ-negligible, for every n. It follows that φ_N, therefore N too, is μ-negligible, that is $f(t)=0$, μ-almost everywhere.

Conversely assume that $f(t)=0$, μ-almost everywhere, that is $\mu^*(N)=0$. Then $f \leqslant \sup n\varphi_N$, and $n\varphi_N$ is μ-negligible for every n, whence f is μ-negligible.

This theorem suggests to call *μ-negligible every mapping of T into a vector space, vanishing μ-almost everywhere.*

6.9 Proposition. *If two functions $f \geqslant 0$ and $g \geqslant 0$ are equal μ-almost everywhere, then $\mu^*(f)=\mu^*(g)$.*

Assume first that $f \leqslant g$ and denote $N=\{t \mid t \in T, f(t) \neq g(t)\}$. The set N is μ-negligible. The function h defined by $h(t)=0$ if $t \notin N$ and $h(t)=+\infty$ if $t \in N$ is μ-negligible, since it vanishes μ-almost everywhere, and we have $f \leqslant g \leqslant f+h$. Then

$$\mu^*(f) \leqslant \mu^*(g) \leqslant \mu^*(f+h) \leqslant \mu^*(f)+\mu^*(h) \leqslant \mu^*(f)$$

whence

$$\mu^*(f)=\mu^*(g).$$

If f and g are arbitrary (equal μ-almost everywhere), we have $f \leqslant \sup(f, g)$, and f and $\sup(f, g)$ are equal μ-almost everywhere. From the first part of the proof it follows that

$$\mu^*(f)=\mu^*[\sup(f, g)].$$

Similarly, it follows that

$$\mu^*(g)=\mu^*[\sup(f, g)],$$

therefore

$$\mu^*(f)=\mu^*(g).$$

6.10 Proposition. *Let $f \geqslant 0$. If $\mu^*(f)<+\infty$, then $f(t)<+\infty$, μ-almost everywhere.*

Let $N=\{t \mid t \in T, f(t)=+\infty\}$. For every n we have $n\varphi_N \leqslant f$, therefore $\mu^*(n\varphi_N) \leqslant \mu^*(f)<+\infty$ whence

$$n\mu^*(\varphi_N)=\mu^*(n\varphi_N)<+\infty.$$

It follows that $\mu^*(\varphi_N)=0$, consequently $f(t)<+\infty$, μ-almost everywhere.

3 *Classes of equivalent functions*

Let μ be a positive measure on T and E an arbitrary set.

We say that two mappings f and g of T into E are *equivalent with respect* to μ or μ-equivalent, if $f(t)=g(t)$, μ-almost everywhere; we write $f \sim g$.

The relation $f \sim g$ has the following properties:

(1) $f \sim f$ for any $f: T \to E$;

(2) $f \sim g \Rightarrow g \sim f$;

(3) $f \sim g$ and $g \sim h \Rightarrow f \sim h$.

Therefore $f \sim g$ is an equivalence relation in the set E^T of all mappings of T into E.

The equivalence class of a function f will be denoted by $\tilde{f}$. Suppose, now, that E is a *vector space*. We say that a function $f: T \to E$ is μ-negligible if it vanishes μ-almost everywhere.

If f is μ-negligible and if $g \sim f$, then g is also μ-negligible and $\tilde{f}=\tilde{0}$.

The set of equivalence classes of functions $f: T \to E$ can be organized as a vector space, defining the sum $\tilde{f}+\tilde{g}$ and the product by scalars $\alpha\tilde{f}$ as the equivalence classes of the functions $f+g$ and αf respectively. The classes $\tilde{f}+\tilde{g}$ and $\alpha\tilde{f}$ depend only on the classes $\tilde{f}$ and $\tilde{g}$ and are independent on the functions f and g. In fact, it is easily verified that if $f \sim f'$ and $g \sim g'$, then

$$f+g \sim f'+g' \text{ and } \alpha f \sim af' .$$

If $\boldsymbol{f}: T \to E$ and $\varphi: T \to R$, we define the product $\tilde{\varphi} \cdot \tilde{\boldsymbol{f}}$ to be the equivalence class of the function $\varphi \boldsymbol{f}$.

If E is an *algebra* and $\boldsymbol{f}, \boldsymbol{g}: T \to E$, the product $\tilde{\boldsymbol{f}} \cdot \tilde{\boldsymbol{g}}$ is defined to be the equivalence class of the function $\boldsymbol{f} \cdot \boldsymbol{g}$.

The set of the equivalence classes of (*finite*)*numerical* functions defined on T is a *ring*.

The set of equivalence classes of the functions $\boldsymbol{f}: T \to E$ is a *module* over the ring of equivalence classes of numerical functions defined on T.

If E is an algebra, the set of equivalence classes of functions $\boldsymbol{f}: T \to E$ is an *algebra*.

If E is a normed space, two mappings $\boldsymbol{f}, \boldsymbol{g}: T \to E$ are equivalent if and only if the numerical function $|\boldsymbol{f}(t)-\boldsymbol{g}(t)|$ is μ-negligible.

In fact, the relation $\boldsymbol{f}(t)=\boldsymbol{g}(t)$ is equivalent to the relation $\boldsymbol{f}(t)-\boldsymbol{g}(t)=0$ which is equivalent to $|\boldsymbol{f}(t)-\boldsymbol{g}(t)|=0$.

6.11 PROPOSITION. *Two continuous mappings f and g of T into a separated topological space E are μ-equivalent if and only if they are equal on the support of μ.*

Since the set $G=\{t|t\in T, f(t)\neq g(t)\}$ is open, it is μ-negligible if and only if it is disjoint from the support of μ. It follows that $f\sim g$ if and only if $f(t)=g(t)$ for every point t in the support of μ.

4 *Functions defined almost everywhere*

Let $A\subset T$ and let f be a function defined on A with values in an arbitrary set E. If $T\setminus A$ is μ-negligible, we say that the function f is defined μ-almost everywhere.

Let f and g be E-valued functions defined μ-almost everywhere. We say that f and g are μ-equivalent and we write $f\sim g$ if the set of points in which both functions are defined and equal, differs from T by a μ-negligible set.

In particular, a function f defined μ-almost everywhere is equivalent to every function f defined on the whole space T equal to f on the domain of f and taking on arbitrary values in the rest of the points. Any two such extensions of f to the whole space T are, evidently, equivalent.

If f is a function defined μ-almost everywhere, then the equivalence class of every extension g of f to the whole space T will be called the equivalence class of f and will be denoted by $\tilde{f}$. The equivalence class $\tilde{f}$ does not depend on g, but on f only. With this notation two functions f and g defined μ-almost everywhere are equivalent if and only if $\tilde{f}=\tilde{g}$.

If E is a vector space, an E-valued function $\boldsymbol{f}$ defined μ-almost everywhere will be still called μ-negligible if $\tilde{\boldsymbol{f}}=\tilde{0}$.

The set of E-valued functions defined almost everywhere is not a vector space or even an additive group, because the function identically zero is a neutral element for the addition, but if $\boldsymbol{f}$ is not defined everywhere, there exists no function $\boldsymbol{g}$ such that $\boldsymbol{f}+\boldsymbol{g}=0$. However the set of equivalence classes is a vector space (and even an algebra if E is an algebra).

5 *Functions with values in $\overline{R}$.*

We say that a real function f defined μ-almost everywhere on T, taking on finite or infinite values, is *finite μ-almost everywhere*, if the set of points $t\in T$ in which t is *defined* and *finite* has a μ-negligible complement.

A function f which is *finite μ-almost everywhere* is equivalent to a function which is defined and finite everywhere. That is why $\tilde{f}$ is defined as the equivalence class of a function equivalent to f, defined and finite everywhere. In this way we can define the sum $\tilde{f}+\tilde{g}$ and the product $\alpha\tilde{f}$.

If f and g are two numerical functions (with finite or infinite values) defined almost everywhere, and if $f(t) \leqslant g(t)$ almost everywhere, then for every function $f_1 \sim f$ and every function $g_1 \sim g$ we have still $f_1(t) \leqslant g(t)$ almost everywhere. Therefore, we can write $\tilde{f} \leqslant \tilde{g}$ and this relation is a property of the classes $\tilde{f}$ and $\tilde{g}$.

The relation $\tilde{f} \leqslant \tilde{g}$ is an order relation on the set of equivalences classes of the numerical functions defined on T.

A μ-almost everywhere finite function is still called μ-negligible if it is equivalent to 0.

§7 $\mathscr{L}^p$-spaces

1 *Hölder and Minkowski inequalities*

In order to prove these important inequalities we need the following lemma:

7.1 Lemma *If* $p>1$, $q>1$ *and* $p^{-1}+q^{-1}=1$ *and if* $a \geqslant 0$ *and* $b \geqslant 0$ *then*

$$ab \leqslant a^p p^{-1} + b^q q^{-1} .$$

In fact, if either $a=0$ or $b=0$, the inequality is evidently true. Suppose then that $a>0$ and $b>0$. We start with the function $\varphi(x)=x^p p^{-1}+x^{-q}q^{-1}$ defined for $x>0$. This function is differentiable and

$$\varphi'(x)=x^{p-1}-x^{-q-1} .$$

φ' vanishes only at $x=1$; if $x<1$, then $\varphi'(x)>0$ and if $x>1$, then $\varphi'(x)<0$; consequently, the function φ has a minimum at $x=1$:

$$\varphi(x)=x^p p^{-1}+x^{-q}q^{-1} \geqslant \varphi(1)=p^{-1}+q^{-1}=1 .$$

Now writing the obtained inequality, $1 \leqslant x^p p^{-1}+x^{-q} q^{-1}$, for $x=a^{1/q} b^{-1/p}$ it follows

$$1 \leqslant a^{p-1} (bp)^{-1} + b^{q-1} (aq)^{-1} ,$$

since $p/q=p-1$ and $q/p=q-1$. Multiplying by ab we obtain

$$ab \leqslant a^p p^{-1} + b^q q^{-1}$$

and the lemma is proved.

For each number $p \geqslant 1$ we consider the number q such that:

$$q = +\infty \quad \text{if } p = 1,$$
$$q = p/(p-1) \text{ if } p > 1.$$

The numbers p and q are called *conjugate* to each other. We have $p^{-1} + q^{-1} = 1$.

If $1 \leqslant p \leqslant 2$, then $2 \leqslant q \leqslant +\infty$. If $p = 2$, then $q = 2$.

7.2 HÖLDER'S INEQUALITY. *Let μ be a positive measure, f and g two positive functions defined on T with finite or $+\infty$ values. If $p > 1$ and $p^{-1} + q^{-1} = 1$, then:*

$$\int^* fg \, d\mu \leqslant \left(\int^* f^p \, d\mu\right)^{1/p} \left(\int^* g^q \, d\mu\right)^{1/q}.$$

In fact, if the right-hand side is zero, then we have, for instance, $\int^* f^p \, d\mu = 0$ (even if $\int^* g^q \, d\mu = +\infty$), that is f^p is μ-negligible. It follows that f^p vanishes μ-almost everywhere, therefore f and fg are zero μ-almost everywhere, whence $\int^* fg \, d\mu = 0$, and thus the inequality is satisfied in this case. Suppose now that the integrals from the right-hand side are different from zero. If one of these integrals is $+\infty$, the inequality is satisfied. It remains to consider the case where

$$0 < \int^* f^p \, d\mu < +\infty \text{ and } 0 < \int^* g^q \, d\mu < +\infty.$$

We shall suppose first that f and g are *finite* everywhere. For each $t \in T$, take in the inequality $ab \leqslant a^p p^{-1} + b^q q^{-1}$,

$$a = f(t)/\left(\int^* f^p \, d\mu\right)^{1/p} \text{ and } b = g(t)/\left(\int^* g^q \, d\mu\right)^{1/q}.$$

Then

$$f(t) g(t)/\left(\int^* f^p \, d\mu\right)^{1/p} \left(\int^* g^q \, d\mu\right)^{1/q} \leqslant$$
$$\leqslant f(t)^p / p \int^* f^p \, d\mu + g(t)^q / q \int^* g^q \, d\mu.$$

Taking the upper integrals in these inequalities we get

$$\int^* fg \, d\mu / \left(\int^* f^p \, d\mu\right)^{1/p} \left(\int^* g^q \, d\mu\right)^{1/q}$$
$$\leqslant \int^* f^p \, d\mu / p \int^* f^p \, d\mu + \int^* g^q \, d\mu / q \int^* g^q \, d\mu =$$
$$= 1/p + 1/q = 1$$

whence

$$\int^* fg\,d\mu \leqslant \left(\int^* f^p\,d\mu\right)^{1/p}\left(\int^* g^q\,d\mu\right)^{1/q}.$$

Finally, if f and g are not finite everywhere, then from the inequalities $\int^* f^p\,d\mu < +\infty$ and $\int^* g^q\,d\mu < +\infty$ it follows that f^p and g^q, therefore f and g, are finite μ-almost everywhere, and therefore each of them is equivalent to a function f_1 respectively g_1 which is *finite* everywhere. Then

$$\int^* f_1 g_1\,d\mu \leqslant \left(\int^* f_1^p\,d\mu\right)^{1/p}\left(\int^* g_1^q\,d\mu\right)^{1/q}.$$

But $f_1^p \sim f^p$, $g_1^q \sim g^q$ and $f_1 g_1 \sim fg$, therefore $\int^* f_1^p\,d\mu = \int^* f^p\,d\mu$, $\int^* g_1^q\,d\mu = \int^* g^q\,d\mu$ and $\int^* f_1 g_1\,d\mu = \int^* fg\,d\mu$; substituting in the above inequality, the required inequality follows also for f and g.

7.3 Minkowski's inequality. *Let μ be a positive measure, f and g two positive functions on T with finite or $+\infty$ values, and $1 \leqslant p < +\infty$.*

Then

$$\left(\int^* (f+g)^p\,d\mu\right)^{1/p} \leqslant \left(\int^* f^p\,d\mu\right)^{1/p} + \left(\int^* g^p\,d\mu\right)^{1/p}.$$

In fact, if $p=1$ we have $\int^* (f+g)\,d\mu \leqslant \int^* f\,d\mu + \int^* g\,d\mu$, since μ^* is subadditive. Let $p>1$. If $\int^* (f+g)^p\,d\mu = 0$, the inequality is satisfied. Suppose then that $0 < \int^* (f+g)^p\,d\mu < +\infty$. Using Hölder's inequality, we obtain successively:

$$\begin{aligned}
\int^* (f+g)^p\,d\mu &= \int^* (f+g)(f+g)^{p-1}\,d\mu \leqslant \\
&\leqslant \int^* f(f+g)^{p-1}\,d\mu + \int^* g(f+g)^{p-1}\,d\mu \leqslant \\
&\leqslant \left(\int^* f^p\,d\mu\right)^{1/p}\left(\int^* (f+g)^{q(p-1)}\,d\mu\right)^{1/q} + \\
&\quad + \left(\int^* g^p\,d\mu\right)^{1/p}\left(\int^* (f+g)^{q(p-1)}\,d\mu\right)^{1/q} = \\
&= \left[\left(\int^* f^p\,d\mu\right)^{1/p} + \left(\int^* g^p\,d\mu\right)^{1/p}\right]\left(\int^* (f+g)^p\,d\mu\right)^{1/q}
\end{aligned}$$

since $q(p-1)=p$. Dividing by $\left(\int^* (f+g)^p\,d\mu\right)^{1/q}$ and observing that $1-1/q = 1/p$, we obtain the desired inequality.

It remains to consider the case where

$$\int^* (f+g)^p\,d\mu = +\infty.$$

We shall prove that in this case at least one of the integrals $\int^* f^p\,d\mu$, $\int^* g^p\,d\mu$ is equal to $+\infty$. In order to do so, observe that if $0 \leqslant a \leqslant +\infty$, and $0 \leqslant b \leqslant +\infty$, then

$$(a+b)^p \leqslant 2^p(a^p+b^p)\,.$$

In fact, assuming, for instance, that $a \leqslant b$, we have

$$(a+b)^p \leqslant (b+b)^p = 2^p b^p \leqslant 2^p(a^p+b^p)\,.$$

For each point $t \in T$ we have then

$$(f(t)+g(t))^p \leqslant 2^p(f(t)^p+g(t)^p)\,,$$

whence

$$\int^* (f+g)^p \mathrm{d}\mu \leqslant 2^p \int^* f^p \mathrm{d}\mu + 2^p \int^* g^p \mathrm{d}\mu\,.$$

Since the left-hand side is equal to $+\infty$, it follows that at least one of the integrals from the right-hand side is equal to $+\infty$.

If, for example, $\int^* f^p \mathrm{d}\mu = +\infty$, then $(\int^* f^p \mathrm{d}\mu)^{1/p} = +\infty$ and the required inequality is satisfied with equality.

Thus the Minkowski inequality is proved.

2 N_p *seminorms*

Let μ be a positive measure on T, and E a (real or complex) Banach space.

For every function $\boldsymbol{f}: T \to E$ and every number p such that $1 \leqslant p < +\infty$, we denote

$$N_p(\boldsymbol{f}, \mu) = (\int^* |\boldsymbol{f}|^p \mathrm{d}\mu)^{1/p}\,.$$

When there is no danger of confusion, we shall write $N_p(\boldsymbol{f})$ instead of $N_p(\boldsymbol{f}, \mu)$.

We list some properties of the functional $N_p(\boldsymbol{f})$:

(1) $0 \leqslant N_p(\boldsymbol{f}) \leqslant +\infty$; $N_p(\boldsymbol{f}) = 0$ if and only if $\boldsymbol{f}$ is μ-negligible;

(2) $N_p(\alpha\boldsymbol{f}) = |\alpha| N_p(\boldsymbol{f})$, α scalar;

(3) $N_p(\boldsymbol{f}+\boldsymbol{g}) \leqslant N_p(\boldsymbol{f}) + N_p(\boldsymbol{g})$;

(4) $N_p(\boldsymbol{f}) \leqslant N_p(\boldsymbol{g})$ if $|\boldsymbol{f}| \leqslant |\boldsymbol{g}|$ almost everywhere.

In fact: property (1) follows from the fact that the upper integral is *positive* (finite or $+\infty$), and $|\boldsymbol{f}(t)|^p = 0$ if and only if $\boldsymbol{f}(t) = 0$; property (2) follows from the fact that the upper integral μ^* is positively homogeneous; property (3) is deduced from Minkowski's inequality, observing that $|\boldsymbol{f}+\boldsymbol{g}| \leqslant |\boldsymbol{f}| + |\boldsymbol{g}|$ and that μ^* is increasing; property (4) follows from the fact that μ^* is increasing.

From the above properties there follows the inequality

(5) $|N_p(\boldsymbol{f}) - N_p(\boldsymbol{g})| \leqslant N_p(\boldsymbol{f}-\boldsymbol{g})\,.$

In fact,

$$|f| = |f-g+g| \leqslant |f-g| + |g|,$$

therefore

$$N_p(f) \leqslant N_p(f-g) + N_p(g)$$

whence

$$N_p(f) - N_p(g) \leqslant N_p(f-g).$$

Interchanging f with g we obtain

$$N_p(g) - N_p(f) \leqslant N_p(f-g).$$

From the two inequalities we deduce

$$|N_p(f) - N_p(f) \leqslant N_p(f-g).$$

We have also the equality

(6) $N_p(f) = \sup_K N_p(f\varphi_K)$, *K compact*.

In fact, we have $|f\varphi_K|^p = |f|^p \varphi_K$ therefore

$$\int^* |f|^p \varphi_K \mathrm{d}\mu = \int^* |f\varphi_K|^p \mathrm{d}\mu = N_p(f\varphi_K)^p.$$

Then

$$N_p(f)^p = \int^* |f|^p \mathrm{d}\mu = \sup_K \int^* |f|^p \varphi_K \mathrm{d}\mu =$$
$$= \sup_K N_p(f\varphi_K)^p = (\sup_K N_p(f\varphi_K))^p,$$

whence

$$N_p(f) = \sup_K N_p(f\varphi_K), \ K \text{ compact}.$$

For every numerical function $f: T \to \overline{R}$, (with finite of infinite values), we still denote

$$N_p(f) = (\int^* |f|^p \mathrm{d}\mu)^{1/p}.$$

The above properties still remain valid for functions with values in $\overline{R}$ provided that the sum $f+g$ in property (3) or the difference $f-g$ in property (5) is *defined on T*.

For functions $f: T \to \overline{R}$ we have the following property:

(7) *If* $N_p(f) < +\infty$ *for one single* p, $1 \leqslant p < +\infty$, *then* $f(t)$ *is finite, almost everywhere.*

In fact, $\int^* f(t)^p \mathrm{d}\mu < +\infty$, therefore $|f(t)|^p < +\infty$ almost everywhere; then $|f(t)|$ and therefore $f(t)$ is also finite almost everywhere.

Hölder's inequality can now be written:

$$\int^* fg\,\mathrm{d}\mu \leqslant N_p(f)\,N_q(g)$$

for $1 < p$, $q < +\infty$, $p^{-1} + q^{-1} = 1$, $f \geqslant 0$, $g \leqslant 0$. We shall define later $N_\infty(g)$. Then it will be shown that Hölder's inequality still remains valid in case $p = 1$ and $q = +\infty$.

7.4 Lemma. If $0 \leqslant a \leqslant +\infty$, $0 \leqslant b \leqslant +\infty$ and $1 \leqslant p < +\infty$ then we have

$$a^p + b^p \leqslant (a+b)^p .$$

If $a+b=0$, then $a=0$ and $b=0$, therefore the inequality is satisfied with equality.

The inequality is also satisfied if $a+b = +\infty$. Suppose $0 < a+b < +\infty$. Then

$$a/(a+b) \leqslant 1 \text{ and } b/(a+b) \leqslant 1,$$

therefore

$$a^p/(a+b)^p \leqslant a/(a+b) \text{ and } b^p/(a+b)^p \leqslant b/(a+b) .$$

Adding these inequalities, we obtain

$$a^p/(a+b)^p + b^p/(a+b)^p \leqslant a/(a+b) + b/(a+b) = 1 .$$

Multiplying by $(a+b)^p$ we obtain the desired inequality.

From this lemma we deduce

7.5 Proposition. *If μ and ν are two positive measures, then for every function $f \geqslant 0$ and for $1 \leqslant p < +\infty$ we have*

$$N_p(f, \mu+\nu) \leqslant N_p(f, \mu) + N_p(f, \nu) \leqslant 2N_p(f, \mu+\nu).$$

In fact,

$$(\mu+\nu)^* = \mu^* + \nu^* ,$$

therefore,

$$\int^* |f|^p \,\mathrm{d}(\mu+\nu) = \int^* |f|^p \mathrm{d}\mu + \int^* |f|^p \mathrm{d}\nu$$

whence

$$N_p(f, \mu+\nu)^p = N_p(f, \mu)^p + N_p(f, \nu)^p \leqslant (N_p(f, \mu) + N_p(f, \nu))^p .$$

Consequently

$$N_p(f, \mu+\nu) \leqslant N_p(f, \mu) + N_p(f, \nu) .$$

Because we have $\mu \leqslant \mu+\nu$ and $\nu \leqslant \mu+\nu$, we deduce

$$\int^* |f|^p \mathrm{d}\mu \leqslant \int^* |f|^p \mathrm{d}(\mu+\nu) \text{ and } \int^* |f|^p \mathrm{d}\nu \leqslant \int^* |f|^p \mathrm{d}(\mu+\nu),$$

that is,

$$N_p(f, \mu) \leqslant N_p(f, \mu+\nu) \text{ and } N_p(f, \nu) \leqslant N_p(f, \mu+\nu)$$

whence

$$N_p(f, \mu) + N_p(f, \nu) \leqslant 2N_p(f, \mu+\nu) .$$

7.6 THEOREM (*of the countable convexity*). *For every sequence* (f_n) *of positive* (*finite of infinite*) *functions defined on* T, *we have*

$$N_p(\Sigma_{1 \leqslant n < \infty} f_n) \leqslant \Sigma_{1 \leqslant n < \infty} N_p(f_n) .$$

From property (3) we deduce, by recurrence,

$$N_p(\Sigma_{1 \leqslant k \leqslant n} f_k) \leqslant \Sigma_{1 \leqslant k \leqslant n} N_p(f_k)$$

for every natural n. Denoting $f = \Sigma_{1 \leqslant n < \infty} f_n$, we have

$$f = \sup_n \Sigma_{1 \leqslant k \leqslant n} f_k \text{ and } f^p = \sup_n (\Sigma_{1 \leqslant k \leqslant n} f_k)^p .$$

The sequence $(\Sigma_{1 \leqslant k \leqslant n} f_k)^p$ is increasing, and therefore by theorem (5.14) we can write

$$\int^* f^p \mathrm{d}\mu = \int^* \sup_n (\Sigma_{1 \leqslant k \leqslant n} f_k)^p \mathrm{d}\mu = \sup_n \int^* (\Sigma_{1 \leqslant k \leqslant n} f_k)^p \mathrm{d}\mu$$

whence

$$\begin{aligned}(\textstyle\int^* f^p \mathrm{d}\mu)^{1/p} &= [\sup_n \textstyle\int^* (\Sigma_{1 \leqslant k \leqslant n} f_k)^p \mathrm{d}\mu]^{1/p} \\ &= \sup_n [\textstyle\int^* (\Sigma_{1 \leqslant k \leqslant n} f_k)^p \mathrm{d}\mu]^{1/p} ,\end{aligned}$$

consequently,

$$N_p(f) = \sup_n N_p(\Sigma_{1 \leqslant k \leqslant n} f_k) \leqslant \sup_n \Sigma_{1 \leqslant k \leqslant n} N_p(f_k) = \Sigma_{1 \leqslant k < \infty} N_p(f_k) .$$

7.7 PROPOSITION. *Let f and g be two functions defined on T with values in a Banach space E. If f and g are equivalent, then $N_p(f-g)=0$ for $1 \leqslant p < +\infty$. Conversely, if $N_p(f-g)=0$ for one single $p \geqslant 1$, then f and g are equivalent.*

Indeed, if f and g are equivalent then $f(t)-g(t)=0$, μ-almost everywhere and thus $|f(t)-g(t)|^p=0$, μ-almost everywhere. Then it follows that for every p such that $1 \leqslant p < +\infty$, we have

$$N_p(f-g) = (\int^* |f-g|^p \mathrm{d}\mu)^{1/p} = 0 .$$

Conversely, if $N_p(f-g)=0$ for some $p \geqslant 1$, then

$$\int^* |f-g|^p \mathrm{d}\mu = 0 .$$

Therefore $|f-g|$ is μ-negligible, consequently $f-g$ is μ-negligible too. It follows then that $f(t)=g(t)$, μ-almost everywhere. This means that f and g are equivalent, and the proposition is proved.

Let f and g be two functions defined on T with values in E or $\overline{R}$. If f and g are equivalent, then the numerical functions $|f|^p$ and $|g|^p$ are equivalent and

$$\int^* |f|^p \mathrm{d}\mu = \int^* |g|^p \mathrm{d}\mu$$

whence $N_p(f)=N_p(g)$. Therefore, $N_p(f)$ depends only on the class $\tilde{f}$ of f. This gives rise to the definition

$$N_p(\tilde{f}) = N_p(f) .$$

We have the following properties:

(1) $0 \leqslant N_p(\tilde{f}) \leqslant +\infty$, $N_p(\tilde{f})=0$ if and only if $\tilde{f}=\tilde{0}$;

(2) $N_p(\alpha\tilde{f}) = |\alpha|\, N_p(\tilde{f})$, α scalar;

(3) $N_p(\tilde{f}+\tilde{g}) \leqslant N_p(\tilde{f}) + N_p(\tilde{g})$, if $f+g$ is defined on T.

In addition, there follows the inequality

(4) $|N_p(\tilde{f}) - N_p(\tilde{g})| \leqslant N_p(\tilde{f}-\tilde{g})$.

If, now, f is a function defined *μ-almost everywhere* with values in E or $\overline{R}$, we put $N_p(f)=N_p(\tilde{f})$; in this case, $N_p(f)$ has the same properties as the functions defined everywhere.

Remark. If $0 < p < 1$, then the function $N_p(f)=(\int^* |f|^p \mathrm{d}\mu)^{1/p}$ is no longer subadditive, i.e. the inequality $N_p(f+g) \leqslant N_p(f)+N_p(g)$ is no longer valid in general.

3 $\mathscr{F}^p$ spaces

Let μ be a positive measure on T, and E a (real or complex) Banach space. We shall denote by $\mathscr{F}_E^p(T, \mu)$, or $\mathscr{F}_E^p(\mu)$, or $\mathscr{F}_E^p$ the set of functions $f: T \to E$ with $N_p(f) < +\infty$.

If $E=R$ we shall write $\mathscr{F}^p$ instead of $\mathscr{F}_R^p$.

We emphasize that the set $\mathscr{F}_E^p$ (respectively $\mathscr{F}^p$) consists of functions *defined on the whole space* T (respectively *defined and finite* on the whole space T).

From the properties of the function $N_p(f)$ it follows that $\mathscr{F}_E^p$ is a vector space (real or complex, according to whether E is real or complex) and that $N_p(f)$ is a *seminorm* on this space.

We shall consider on this space the topology defined by the seminorm N_p, called the *topology of the convergence in the mean of order p*.

The topology defined by the seminorm N_1 on the space $\mathscr{F}_E^1$ is called, simply, the *topology of the convergence in the mean*.

The adherence of 0 in the space $\mathscr{F}_E^p$ is the subspace $\mathscr{N}_E$ of μ-negligible functions (for which $N_p(f)=0$).

7.8 PROPOSITION. *If (f_n) is a sequence of functions of $\mathscr{F}_E^p$ convergent in the mean of order p to a function $f \in \mathscr{F}_E^p$, then*

$$\lim_n N_p(f_n) = N_p(f).$$

In fact, to say that (f_n) converges to f in the mean of order p means that $\lim_n N_p(f_n - f) = 0$. Then the assertion follows from the inequality

$$|N_p(f_n) - N_p(f)| \leqslant N_p(f_n - f).$$

7.9 PROPOSITION. *If (f_n) is a sequence of functions of $\mathscr{F}_E^p$ such that $\Sigma_{1 \leqslant n < \infty} N_p(f_n) < +\infty$, then:*

(1) *the series $\Sigma_{1 \leqslant n < \infty} f_n(t)$ is absolutely convergent (in E), almost everywhere;*

(2) *the function $f: T \to E$ defined by*

$f(t) = \Sigma_{1 \leqslant n < \infty} f_n(t)$, *if the series is convergent in t,*
$f(t) = 0$, *if the series is not convergent in t,*

belongs to the space $\mathscr{F}_E^p$;

(3) *the series* $\Sigma_{1 \leqslant n < \infty} f_n$ *converges in the mean of order* p *to* f; *more precisely for every natural number* n *we have*

$$N_p(f - \Sigma_{1 \leqslant k \leqslant n} f_k) \leqslant \Sigma_{k > n} N_p(f_k).$$

In fact,

$$N_p(\Sigma_{1 \leqslant n < \infty} |f_n|) \leqslant \Sigma_{1 \leqslant n < \infty} N_p(f_n) < +\infty,$$

therefore, $\Sigma_{1 \leqslant n < \infty} |f_n(t)| < +\infty$, μ-almost everywhere, which proves (1). Since E is complete, it follows that the series $\Sigma_{1 \leqslant n < \infty} f_n(t)$ is convergent (in E), μ-almost everywhere. For each $t \in T$ we have

$$|f(t)| \leqslant \Sigma_{1 \leqslant n < \infty} |f_n(t)|$$

whence

$$N_p(f) \leqslant N_p|\Sigma_{1 \leqslant n < \infty} |f|) < +\infty$$

therefore $f \in \mathscr{F}_E^p$ and thus (2) is proved. For each natural n we have

$$|f(t) - \Sigma_{k \leqslant n} f_k(t)| \leqslant \Sigma_{k > n} |f_k(t)|, \ \mu\text{-almost everywhere.}$$

Then

$$N_p(f - \Sigma_{k \leqslant n} f_k) \leqslant N_p(\Sigma_{k > n} |f_k|) \leqslant \Sigma_{k > n} N_p(f_k).$$

Since the series $\Sigma_{1 \leqslant n < \infty} N_p(f_n)$ is convergent, for every number $\varepsilon > 0$ there exists a number $N(\varepsilon)$ such that for every $n \geqslant N(\varepsilon)$ we have $\Sigma_{k > n} N_p(f_k) < \varepsilon$, therefore $N_p(f - \Sigma_{k \leqslant n} f_k) < \varepsilon$. This means that the series $\Sigma_{1 \leqslant n < \infty} f_n$ is convergent in the mean of order p to f, and the proposition is completely proved.

7.10 PROPOSITION. *From every Cauchy sequence* (f_n) *of functions of* $\mathscr{F}_E^p$ *we can extract a subsequence* (f_{n_k}) *with the following properties*:

(1) $\Sigma_{1 \leqslant k < \infty} N_p(f_{n_{k+1}} - f_{n_k}) < +\infty$;

(2) *there exists a function* $g \geqslant 0$ *such that* $N_p(g) < +\infty$ *and* $|f_{n_k}(t)| \leqslant g(t)$ *for every* $k \in N$ *and every* $t \in T$.

Since (f_n) is a Cauchy sequence in $\mathscr{F}_E^p$, for every number $\varepsilon > 0$ there exists a number $n(\varepsilon)$ such that, for every $n \geqslant n(\varepsilon)$ and $m \geqslant n(\varepsilon)$, we have $N_p(f_n - f_m) < \varepsilon$.

Taking $\varepsilon = 1/2$ and $n_1 \geqslant n(1/2)$, we have

$$N_p(f_n - f_{n_1}) < 1/2 \text{ for every } n \geqslant n(1/2).$$

Taking $\varepsilon = 1/2^2$ and $n_2 > n\ (1/2^2)$ such that $n_2 > n_1$, we have

$$N_p(f_n - f_{n_2}) < 1/2^2 \text{ for every } n \geqslant n(1/2^2)$$

and

$$N_p(f_{n_2} - f_{n_1}) < 1/2\,.$$

Suppose that we have chosen $n_k \geqslant n(1/2^k)$ such that

$$N_p(f_n - f_{n_k}) < 1/2^k \text{ for every } n \geqslant n(1/2^k)\,.$$

Taking $\varepsilon = 1/2^{k+1}$ and $n_{k+1} \geqslant n(1/2^{k+1})$ such that $n_{k+1} > n_k$, we have

$$N_p(f_n - f_{n_{k+1}}) < 1/2^{k+1}, \text{ for every } n \geqslant n(1/2^{k+1})$$

and

$$N_p(f_{n_{k+1}} - f_{n_k}) < 1/2^k\,.$$

Thus we have proved, by recurrence, that we can extract a subsequence f_{n_k} with the following property:

$$N_p(f_{n_{k+1}} - f_{n_k}) < 1/2^k, \text{ for every } k \in N\,.$$

Then

$$\Sigma_{1 \leqslant k < \infty} N_p(f_{n_{k+1}} - f_{n_k}) < \Sigma_{1 \leqslant k < \infty} 1/2^k < +\infty\,.$$

For each $t \in T$, denote

$$h(t) = \Sigma_{1 \leqslant k < \infty} |f_{n_{k+1}}(t) - f_{n_k}(t)| \leqslant +\infty\,.$$

We have

$$N_p(h) = N_p(\Sigma_{1 \leqslant k < \infty} |f_{n_{k+1}} - f_{n_k}|) \leqslant \Sigma_{1 \leqslant k < \infty} N_p(f_{n_{k+1}} - f_{n_k}) < +\infty\,.$$

Writing $g = h + |f_{n_1}|$, we have $g \geqslant 0$ and

$$N_p(g) = N_p(h + |f_{n_1}|) \leqslant N_p(h) + N_p(f_{n_1}) < +\infty\,.$$

On the other hand,

$$g = h + |f_{n_1}| = \Sigma_{1 \leqslant k < \infty} |f_{n_{k+1}} - f_{n_k}| + |f_{n_1}| \geqslant \Sigma_{k<i} |f_{n_{k+1}} - f_{n_k}| + |f_{n_1}| \geqslant$$
$$\geqslant \Sigma_{k<i} [|f_{n_{k+1}}| - |f_{n_k}|] + |f_{n_1}| = |f_{n_i}|$$

whence it follows that $|f_{n_i}| \leqslant g$ for every $i \in N$, that is $|f_{n_i}(t)| \leqslant g(t)$ for every $i \in N$ and every $t \in T$, and the proposition is proved.

Remark. If the functions (f_n) are *continuous*, then the function g can be chosen to be *lower semicontinuous*.

In fact, with the notation from the proof of proposition 7.10, the function h is lower semicontinuous and, therefore, the function $g = h + |f_{n_1}|$ is lower semicontinuous too.

7.11 THEOREM. *The space $\mathscr{F}_E^p$ is complete.*

Let (f_n) be a Cauchy sequence of functions of $\mathscr{F}_E^p$. By proposition 7.10, there exists a subsequence (f_{n_k}) such that

$$\Sigma_{1 \leqslant k < \infty} N_p(f_{n_{k+1}} - f_{n_k}) < +\infty .$$

From proposition 7.9 it follows that there exists a function $f \in \mathscr{F}_E^p$ such that the series

$$\Sigma_{1 \leqslant k < \infty} [f_{n_{k+1}}(t) - f_{n_k}(t)]$$

converges to $f(t)$, μ-almost everywhere and in the mean of order p:

$$\lim_i N_p(f - \Sigma_{k<i} [f_{n_{k+1}} - f_{n_k}] = 0 .$$

But

$$f - \Sigma_{k<i} [f_{n_{k+1}} - f_{n_k}] = f - (f_{n_2} - f_{n_1}) - (f_{n_3} - f_{n_2}) - \ldots - (f_{n_i} - f_{n_{i+1}}) =$$
$$= f + f_{n_1} - f_{n_i}$$

therefore

$$\lim_i N_p(f + f_{n_1} - f_{n_i}) = 0 .$$

This means that the subsequence (f_{n_i}) of the Cauchy sequence (f_n) converges μ-almost everywhere and in the mean of order p to the function $f + f_{n_1} \in \mathscr{F}_E^p$. Then it follows that the Cauchy sequence (f_n) also converges in the mean of order p to $f + f_{n_1}$ and the theorem is proved.

From the proof of this theorem and from proposition 7.10 it follows:

7.12 PROPOSITION. *For every Cauchy sequence (f_n) of functions of $\mathscr{F}_E^p$ there exists a subsequence (f_{n_k}), converging μ-almost everywhere and in the mean of order p to a function of $\mathscr{F}_E^p$, and there is a function $g \geqslant 0$ such that $N_p(g) < +\infty$ and $|f_{n_k}(t)| \leqslant g(t)$ for every k and every $t \in T$.*

7.13 PROPOSITION. *Let (f_n) be a sequence of functions of $\mathscr{F}_E^p$. If all the functions (f_n) have their supports contained in the same compact set $K \subset T$ and if* *the sequence* (f_n) *converges uniformly to a function* f_0, *then* $f_0 \in \mathscr{F}_E^p$ *and* (f_n) *converges to* f_0 *in the mean of order p.*

In fact, let $\varepsilon > 0$; there exists then an index $n(\varepsilon)$, such that for every $n > n(\varepsilon)$ we have

$$|f_n(t) - f_0(t)| < \varepsilon, \text{ for every } t \in T .$$

Since f_n have their supports contained in K, it follows, that the limit f_0 also has its support contained in K. Therefore for $t \notin K$ we have

$$|f_n(t) - f_0(t)| = 0 .$$

Let φ be a *continuous* function *with compact support* defined on T with values in $[0, 1]$ such that $\varphi(t) = 1$ for $t \in K$. Then

$$|f_n(t) - f_0(t)| \leqslant \varepsilon\varphi(t), \text{ for } t \in T ,$$

hence

$$N_p(f_n - f_0) \leqslant \varepsilon N_p(\varphi) < +\infty .$$

It follows, on the one hand, that

$$N_p(f_0) \leqslant N_p(f_n - f_0) + N_p(f_n) < +\infty$$

therefore $f_0 \in \mathscr{F}_E^p$, and on the other hand

$$\lim_n N_p(f_n - f_0) = 0 .$$

Thus the proposition is proved.

Remark. The proposition asserts that on the set of functions $f \in \mathscr{F}_E^p$ which have support contained in the same compact set $K \subset T$, the topology of the uniform convergence is finer than the topology of the convergence in the mean of order p.

7.14 PROPOSITION. *If* μ *and* ν *are two positive measures and if* $\mu \geqslant \nu$ *then* $\mathscr{F}_E^p(\mu) \subset \mathscr{F}_E^p(\nu)$ *and the topology of the space* $\mathscr{F}_E^p(\mu)$ *is finer than the topology induced by the space* $\mathscr{F}_E^p(\nu)$.

In fact, $\mu^* \geqslant \nu^*$; therefore for every function $f: T \to E$ we have

$$\int^* |f|^p \mathrm{d}\mu \geqslant \int^* |f|^p \mathrm{d}\nu$$

whence

$$N_p(f, \mu) \geqslant N_p(f, \nu).$$

From this inequality it follows that $\mathscr{F}_E^p(\mu) \subset \mathscr{F}_E^p(\nu)$ and that on $\mathscr{F}_E^p(\mu)$, the topology of the convergence in the mean of order p, for the measure μ, is finer than the topology of the convergence in the mean of order p, for ν.

7.15 PROPOSITION. *If μ and ν are two positive measures and $\alpha > 0$ then:*

(1) $\mathscr{F}_E^p(\alpha, \mu) = \mathscr{F}_E^p(\mu)$;

(2) $\mathscr{F}_E^p(\mu + \nu) = \mathscr{F}_E^p(\mu) \cap \mathscr{F}_E^p(\nu)$.

The proposition follows from the relations

$$N_p(\alpha f) = \alpha N_p(f)$$

and

$$N_p(f, \mu + \nu) \leqslant N_p(f, \mu) + N_p(f, \nu) \leqslant 2N_p(f, \mu + \nu).$$

4 $\mathscr{L}^p$ spaces

Let μ be a positive measure on T, and E a (real or complex) Banach space. The space $\mathscr{K}_E(T)$ of the continuous mappings $f: T \to E$ with compact support is contained in each space $\mathscr{F}_E^p$, $(1 \leqslant p < +\infty)$.

In fact, if $f \in \mathscr{K}_E(T)$, then $|f| \in \mathscr{K}(T)$; therefore $|f|^p \in \mathscr{K}(T)$, consequently,

$$N_p(f) = (\int |f|^p \mathrm{d}\mu)^{1/p} < +\infty .$$

We shall denote by $\mathscr{L}_E^p(T, \mu)$ or $\mathscr{L}_E^p(\mu)$ the closure of the space $\mathscr{K}_E(T)$ in the space $\mathscr{F}_E^p$ and by $L_E^p(T, \mu)$ or $L_E^p(\mu)$ the quotient space $\mathscr{L}_E^p/\mathscr{N}_E$ of the space $\mathscr{L}_E^p$ by the space $\mathscr{N}_E$ of the μ-negligible functions.

Instead of $\mathscr{L}_R^p$ and L_R^p we shall write $\mathscr{L}^p$ respectively L^p. The space $\mathscr{L}_E^p$ is a real or complex *vector space* according to whether E is real or complex.

Since the constant zero-function is continuous with compact support (the support is empty) and since the adherence of this function in the space $\mathscr{F}_E^p$ is the set $\mathscr{N}_E$ of μ-negligible functions, it follows that every μ-negligible function $f: T \to E$ belongs to the space $\mathscr{L}_E^p$, for every p with $1 \leqslant p < +\infty$.

We remark that the functions of $\mathscr{L}_E^p$ (respectively $\mathscr{L}^p$) are defined (respectively defined and *finite*) on the *whole space* T.

The functions of $\mathscr{L}_E^1$ are said to be *integrable* with respect to μ, or μ-integrable.

The functions f of $\mathscr{L}_E^p$ are called *functions with integrable p power*, or p-integrable functions. This name will be justified further by the fact that if $f \in \mathscr{L}_E^p$, then the numerical function $|f|^p$ is integrable (corollary 7.45).

By extension, we shall call p-integrable functions, every E- (respectively $\bar{R}$) valued function defined *almost everywhere* on T (respectively defined and *finite almost everywhere* on T) equivalent to a function $f \in \mathscr{L}_E^p$ (respectively $f \in \mathscr{L}^p$).

Most of the following results, stated for functions of $\mathscr{L}_E^p$ (respectively $\mathscr{L}^p$) still remain valid for p-integrable functions defined (respectively defined and finite) almost everywhere on T.

Every function $f: T \to E$, equivalent to a function from $\mathscr{L}_E^p$ belongs to $\mathscr{L}_E^p$ too.

7.16 Proposition. *The space* $\mathscr{K}_E(T)$ *is dense in each space* $\mathscr{L}_E^p$, $(1 \leqslant p < +\infty)$: *for every function* $f \in \mathscr{L}_E^p$ *and every* $\varepsilon > 0$ *there exists a function* $g \in \mathscr{K}_E(T)$ *such that* $N_p(f-g) < \varepsilon$.

7.17 Theorem. *The space* $\mathscr{L}_E^p$ *is complete. The space* L_E^p *is a Banach space.*

In fact, $\mathscr{L}_E^p$ is closed in the complete space $\mathscr{F}_E^p$, therefore $\mathscr{L}_E^p$ also is complete. But then the quotient space $\mathscr{L}_E^p/\mathscr{N}_E$ is complete too. Since $N_p(\tilde{f})$ is a norm on L_E^p, it follows that L_E^p is a Banach space.

Later, the norm $N_p(\tilde{f})$ will be also denoted by $\|\tilde{f}\|_p$.

7.18 Proposition. *For every sequence* (f_n) *of functions in* $\mathscr{L}_E^p$ *converging in the mean of order p to a function* $f \in \mathscr{L}_E^p$, *there exists a subsequence* (f_{n_k}) *converging* μ*-almost everywhere to* f, *and a function* $g \geqslant 0$ *such that* $N_p(g) < +\infty$ *and* $|f_{n_k}(t)| \leqslant g(t)$ *for every k and every* $t \in T$.

Since the sequence (f_n) is convergent (in $\mathscr{L}_E^p$), it is a Cauchy sequence of functions in $\mathscr{F}_E^p$. By proposition 7.12, there exists a subsequence (f_{n_k}), converging in the mean of order p, and almost everywhere, to a function $h \in \mathscr{F}_E^p$, and a function $g \geqslant 0$ such that $N_p(g) < +\infty$ and $|f_{n_k}(t)| \geqslant g(t)$ for every k and every $t \in T$.

But (f_{n_k}) converges in the mean of order p to f, as a subsequence of the sequence (f_n). It follows that $N_p(f-h) = 0$, that is the functions f and h are equivalent and hence (f_{n_k}) converges μ-almost everywhere also to f.

7.19 COROLLARY. *If (f_n) is a convergent sequence in the space $\mathscr{L}_E^p$ and if $(f_n(t))$ has a limit $f(t) \in E$ almost everywhere, then the function f (defined almost everywhere) is p-integrable and (f_n) converges also in the mean of order p to f.*

Indeed, there exists a subsequence (f_{n_k}) converging almost everywhere and in the mean of order p to a function $h \in \mathscr{L}_E^p$. It follows that the Cauchy sequence (f_n) converges in the mean of order p to h.

Since (f_n) converges almost everywhere to f, it follows that the subsequence (f_{n_k}) converges to f almost everywhere, hence $f(t) = h(t)$ almost everywhere.

It follows that f is p-integrable and (f_n) converges to f in the mean of order p.

7.20 PROPOSITION. *Let $\mathscr{A}$ be a dense subset of $\mathscr{L}_E^p$. For every function $f \in \mathscr{L}_E^p$ there exists a sequence (f_n) of functions from $\mathscr{A}$, converging to f in the mean of order p and almost everywhere, and a function $g \geqslant 0$ such that $N_p(g) < +\infty$ and $|f_n(t)| \leqslant g(t)$ for every n and every $t \in T$.*

Indeed, for every natural n there exists a function $h_n \in \mathscr{A}$ such that $N_p(h_n - f) < n^{-1}$. Then the sequence (h_n) consists of functions from $\mathscr{A}$ and converges in the mean of order p to f. By proposition 7.18, the sequence (h_n) contains a subsequence (f_n) converging almost everywhere and in the mean of order p to f, and there exists a function $g \geqslant 0$ such that $N_p(g) < +\infty$ and $|f_n(t)| \leqslant g(t)$ for every n and every $t \in T$.

7.21 COROLLARY. *For every function $f \in \mathscr{L}_E^p$ there exists a sequence (f_n) of functions from $\mathscr{K}_E(T)$, converging to f in the mean of order p and almost everywhere, and a lower semicontinuous function $g \geqslant 0$, such that $N_p(g) < +\infty$ and $|f_n(t)| \leqslant g(t)$ for every n and every $t \in T$.*

In fact, $\mathscr{K}_E(T)$ is a dense subspace of $\mathscr{L}_E^p$. The fact that the function g can be chosen to be lower semicontinuous follows from the remark after proposition 7.10.

Remark. Later (proposition 8.53) we shall prove that the Borel step functions are p-integrable and form a dense subset in $\mathscr{L}_E^p$. It will then follow that every function of $\mathscr{L}_E^p$ can be approximated almost everywhere and in the mean of order p by Borel step functions.

5 Relationships between $\mathscr{L}_E^p(\mu)$ and $\mathscr{L}_F^p(\mu)$

7.22 PROPOSITION. *The linear combinations $\Sigma_{1\leqslant i\leqslant n} a_i\varphi_i$ of positive continuous functions φ_i with compact support, with coefficients a_i from E, form a dense subspace in $\mathscr{L}_E^p$.*

In fact, every function $f\in\mathscr{K}_E(T)$ can be uniformly approximated by functions of the form $\Sigma\, a_i\varphi_i$ where $a_i\in E$ and φ_i are functions from $\mathscr{K}_+(T)$, with support contained in a fixed compact neighbourhood of the support of f. This means that the functions of the form $\Sigma\, a_i\varphi_i$ with $\varphi_i\in\mathscr{K}(T)$ and $a_i\in E$ form a dense subset in $\mathscr{K}_E(T)$ for the topology of the uniform convergence, hence also for the coarser topology of the convergence in the mean of order p. Since $\mathscr{K}_E(T)$ is dense in $\mathscr{L}_E^p$ for this latter topology, it follows that the set of the functions of the form $\Sigma\, a_i\varphi_i$ is dense in $\mathscr{L}_E^p$ for this topology.

7.23 COROLLARY. *The linear combinations $\Sigma_{1\leqslant i\leqslant n} a_i\varphi_i$ of positive functions of $\mathscr{L}^p$ with coefficients a_i from E, form a dense subspace in $\mathscr{L}_E^p$.*

Let $f=\Sigma_{1\leqslant i\leqslant n} a_i\varphi_i$ be a function with $\varphi_i\in\mathscr{L}^p$ and $a_i\in E$ and let $\varepsilon>0$. For each i there exists a function $\psi_i\in\mathscr{K}(T)$ such that $N_p(\varphi_i-\psi_i)<\varepsilon/A$, where $A>0$ is chosen such that $\Sigma_{1\leqslant i\leqslant n}|a_i|<A$. Then the function $g=\Sigma_{1\leqslant i\leqslant n} a_i\psi_i$ belongs to the space $\mathscr{K}_E(T)$, and

$$N_p(f-g)=N_p(\Sigma_{1\leqslant i\leqslant n} a_i(\varphi_i-\psi_i))\leqslant$$
$$\leqslant\Sigma_{1\leqslant i\leqslant n}|a_i|\, N_p(\varphi_i-\psi_i)\leqslant\Sigma_{1\leqslant i\leqslant n}|a_i|\,\varepsilon/A\leqslant\varepsilon$$

hence $f\in\mathscr{L}_E^p$.

Since the functions $\Sigma_{1\leqslant i\leqslant n} a_i\varphi_i$ with $a_i\in E$, and $\varphi_i\in\mathscr{L}^p$, contain the functions of the same form, with $\varphi_i\in\mathscr{K}(T)$ and since these are dense in $\mathscr{L}_E^p$ it follows that the former ones also are dense in $\mathscr{L}_E^p$.

7.24 THEOREM. *Let E and F be two Banach spaces and $U: E\to F$ a linear continuous mapping. For every function $f\in\mathscr{L}_E^p$ the function $U\circ f$ belongs to the space $\mathscr{L}_F^p$.*

In fact, let $f\in\mathscr{L}_E^p$ and $\varepsilon>0$; there exists a function $g\in\mathscr{K}_E(T)$ such that $N_p(g-f)<\varepsilon$. But for every $t\in T$ we have

$$|(U\circ f)(t)-(U\circ g)(t)=|U(f(t))-U(g(t))|=|U[f(t)-g(t)]|\leqslant$$
$$\leqslant\|U\|\;|f(t)-g(t)|$$

hence

$$N_p(U \circ \boldsymbol{f} - U \circ \boldsymbol{g}) \leqslant \|U\| \, N_p(\boldsymbol{f} - \boldsymbol{g}) \leqslant \varepsilon \|U\| .$$

As $U \circ \boldsymbol{g} \in \mathscr{K}_F(T)$, it follows that $U \circ \boldsymbol{f} \in \mathscr{L}^p_E$ and the theorem is proved.

7.25 COROLLARY. *For every function $\boldsymbol{f} \in \mathscr{L}^p_E$ and every element $a' \in E'$, the scalar function $t \to \langle \boldsymbol{f}(t), a' \rangle$ belongs to the space $\mathscr{L}^p$ or $\mathscr{L}^p_C$ according to whether E is a real or complex Banach space.*

7.26 PROPOSITION. *If $\boldsymbol{f} \in \mathscr{L}^p_E$, then $|\boldsymbol{f}| \in \mathscr{L}^p$ and the mapping $\boldsymbol{f} \to |\boldsymbol{f}|$ is uniformly continuous (for the topology of the convergence in the mean of order p).*

In fact, let $\boldsymbol{f} \in \mathscr{L}^p_E$ and $\varepsilon > 0$; there exists a function $\boldsymbol{g} \in \mathscr{K}_E(T)$ such that $N_p(\boldsymbol{f} - \boldsymbol{g}) < \varepsilon$. Then $|\boldsymbol{f}| \in \mathscr{K}_+(T)$ and since $||\boldsymbol{f}| - |\boldsymbol{g}|| \leqslant |\boldsymbol{f} - \boldsymbol{g}|$ we have $N_p(|\boldsymbol{f}| - |\boldsymbol{g}|) < \varepsilon$, whence $|\boldsymbol{f}| \in \mathscr{L}^p$. The fact that the mapping $\boldsymbol{f} \to |\boldsymbol{f}|$ is uniformly continuous follows from the inequality

$$N_p(|\boldsymbol{f}_1| - |\boldsymbol{f}_2|) \leqslant N_p(\boldsymbol{f}_1 - \boldsymbol{f}_2), \ \boldsymbol{f}_1, \boldsymbol{f}_2 \in \mathscr{L}^p_E .$$

6 *Relationships between $\mathscr{L}^p_E(\mu)$ and $\mathscr{L}^p_E(\nu)$*

7.27 PROPOSITION. *Let μ and ν be two positive measures. If $\mu \geqslant \nu$, then $\mathscr{L}^p_E(\mu) \subset \mathscr{L}^p_E(\nu)$, $1 \leqslant p < +\infty$, and the topology of the space $\mathscr{L}^p_E(\mu)$ is finer than the topology induced on $\mathscr{L}^p_E(\mu)$ by the topology of $\mathscr{L}^p_E(\nu)$.*

From the inequality $\mu^* \geqslant \nu^*$ we deduce

$$N_p(\boldsymbol{f}, \mu) \geqslant N_p(\boldsymbol{f}, \nu) .$$

If $\boldsymbol{f} \in \mathscr{L}^p_E(\nu)$, there exists a sequence $(\boldsymbol{g}_n)$ of functions from $\mathscr{K}_E(T)$ such that

$$\lim N_p(\boldsymbol{f} - \boldsymbol{g}_n, \mu) = 0 .$$

Then

$$\lim N_p(\boldsymbol{f} - \boldsymbol{g}_n, \nu) = 0 ,$$

therefore $\boldsymbol{f} \in \mathscr{L}^p_E(\nu)$. Hence $\mathscr{L}^p_E(\mu) \subset \mathscr{L}^p_E(\nu)$.

The assertion concerning the topologies follows from the inequality

$$N_p(\boldsymbol{f}, \mu) \geqslant N_p(\boldsymbol{f}, \nu) .$$

7.28 PROPOSITION. *If μ is a positive measure and $\alpha > 0$, then $\mathscr{L}_E^p(\alpha, \mu) = \mathscr{L}_E^p(\mu)$, $1 \leq p < +\infty$ and the topologies of the two spaces are equivalent.*

In fact, we have

$$N_p(f, \alpha\mu) = \alpha N_p(f, \mu).$$

Remark. We shall show (proposition 11.15) that if μ and v are positive measures, then

$$\mathscr{L}_E^p(\mu + v) = \mathscr{L}_E^p(\mu) \cap \mathscr{L}_E^p(v).$$

From proposition 7.27 we can deduce only the inclusion

$$\mathscr{L}_E^p(\mu + v) \subset \mathscr{L}_E^p(\mu) \cap \mathscr{L}_E^p(v)$$

(since $\mu + v \geq \mu$ and $\mu + v \geq v$).

7 *p-integrable real functions*

7.29 PROPOSITION. *A real function f defined on T belongs to the space $\mathscr{L}^p$ if and only if the functions f^+ and f^- belong to the space $\mathscr{L}^p$.*

In fact, if $f \in \mathscr{L}^p$, then $|f| \in \mathscr{L}^p$ and hence the function $f^+ = 2^{-1}(|f| + f)$ and the function $f^- = 2^{-1}(|f| - f)$ belong to $\mathscr{L}^p$. Conversely, if f^+ and f^- belong to the space $\mathscr{L}^p$, then $f = f^+ - f^- \in \mathscr{L}^p$.

7.30 COROLLARY. *The upper envelope and the lower envelope of a finite family of functions of $\mathscr{L}^p$, belong to $\mathscr{L}^p$.*

In fact,

$$\sup(f, g) = 2^{-1}(f + g + |f - g|),$$
$$\inf(f, g) = 2^{-1}(f + g - |f - g|).$$

If $f, g \in \mathscr{L}^p$, then $f + g \in \mathscr{L}^p$ and $|f - g| \in \mathscr{L}^p$; it follows that $\sup(f, g)$ and $\inf(f, g)$ belong also to $\mathscr{L}^p$. It can then be proved by induction that the assertion remains true for every finite family of functions from $\mathscr{L}^p$.

7.31 PROPOSITION. *Let f be a finite or infinite numerical function, defined almost everywhere on T. If for each number $\varepsilon > 0$, there exist two p-integrable* *functions g and h such that $g \leqslant f \leqslant h$, μ-almost everywhere and $N_p(h-g) < \varepsilon$, then f is p-integrable.*

In fact, since g and h are p-integrable, they are finite almost everywhere; from the inequalities $g \leqslant f \leqslant h$ it follows that f is finite almost everywhere.

Let f_1, g_1, h_1 be three functions defined and finite everywhere equivalent respectively to f, g, h.

It follows that g_1 and h_1 are p-integrable, hence $g_1, h_1 \in \mathscr{L}^p$. We have $N_p(h_1 - g_1) < \varepsilon$ and $g_1 \leqslant f_1 \leqslant h_1$ almost everywhere; therefore, $0 \leqslant f_1 - g_1 \leqslant h_1 - g_1$ almost everywhere. It follows that

$$N_p(f_1 - g_1) \leqslant N_p(h_1 - g_1) < \varepsilon \text{ and } N_p(f_1) \leqslant N_p(h_1) < +\infty .$$

Hence $f_1 \in \mathscr{F}^p$ and it can be approximated in the mean of order p by functions from $\mathscr{L}^p$. Since $\mathscr{L}^p$ is closed, we deduce that $f_1 \in \mathscr{L}^p$. From the relation $f \sim f_1$ it follows that f is p-integrable and the proposition is proved.

7.32 Lemma. *If $0 \leqslant \beta \leqslant \alpha \leqslant +\infty$ and if β is finite, then*

$$(\alpha - \beta)^p \leqslant \alpha^p - \beta^p \textit{ for } 1 \leqslant p < \infty .$$

In fact, $\alpha - \beta \geqslant 0$ and denoting $a = \alpha - \beta$ and $b = \beta$, from lemma 7.4 we deduce

$$(\alpha - \beta)^p + \beta^p \leqslant (\alpha - \beta + \beta)^p = \alpha^p ,$$

whence the required inequality follows.

7.33 Proposition. *If f and g are two positive functions from $\mathscr{L}^p$ and if $f \geqslant g$, then*

$$(N_p(f-g))^p \leqslant (N_p(f))^p - (N_p(g))^p .$$

From the above lemma we deduce $(f-g)^p \leqslant f^p - g^p$. If f and g are continuous with compact support, then the functions $(f-g)^p$, f^p and g^p are also continuous with compact support. From the fact that the integral of the continuous functions with compact support is increasing, we deduce

$$\int (f-g)^p \mathrm{d}\mu \leqslant \int f^p \mathrm{d}\mu - \int g^p \mathrm{d}\mu$$

that is

$$(N_p(f-g))^p \leqslant (N_p(f))^p - N_p(g))^p .$$

Assume, now, that f and g are arbitrary functions from $\mathscr{L}^p$ such that $f \geqslant g \geqslant 0$. We have $f-g \geqslant 0$ and $f-g \in \mathscr{L}^p$. There exist then two sequences (g_n) and (h_n) of continuous positive functions with compact support, converging in the mean of order p, respectively to g and $f-g$. The functions $f_n = g_n + h_n$ are positive, continuous, with compact support and $\lim f_n = \lim g_n + \lim h_n = g+f-g=f$ the limit being in the sense of the convergence in the mean of order p. We have also $f_n \geqslant g_n$ for every n. From the first part of the proof it follows that

$$(N_p(f_n - g_n))^p \leqslant (N_p(f_n))^p - (N_p g_n))^p .$$

Noticing that the sequence $(f_n - g_n)$ converges in the mean of order p to $f-g$ and passing to the limit in the last inequality, we obtain

$$(N_p(f-g))^p \leqslant (N_p(f))^p - (N_p(g))^p .$$

7.34 PROPOSITION. *Every function $f \geqslant 0$ in $\mathscr{L}^p(\mu)$ vanishes on the complement of the union of a sequence of compact sets and of a negligible set.*

In fact, let $f \geqslant 0$ be a function from $\mathscr{L}^p$. We have

$$N_p(f) = \sup_K N_p(f\varphi_K), \quad K \text{ compact} .$$

Hence, there exists an increasing sequence (K_n) of compact sets such that

$$N_p(f) = \sup_n N_p(f\varphi_{K_n}) = \lim_n N_p(f\varphi_{K_n}) .$$

Denoting $A = \cup_{1 \leqslant n < \infty} K_n$ and $f' = f\varphi_A$ we have

$$f-f' = f - f\varphi_A \leqslant f - f\varphi_{K_n}, \text{ for each } n ,$$

therefore

$$N_p(f-f') \leqslant N_p(f - f\varphi_{K_n}) .$$

From the preceding proposition we deduce

$$N_p(f - f\varphi_{K_n})^p \leqslant N_p(f)^p - N_p(f\varphi_{K_n})^p$$

whence

$$\lim_n N_p(f - f\varphi_{K_n}) = 0 .$$

We deduce then that $N_p(f-f') = 0$, hence $f-f'$ is μ-negligible, that is f and f' are equivalent. It follows that f is zero on the complementary set of the union of A and a negligible set.

7.35 THEOREM. *Let (f_n) be an increasing sequence of functions of $\mathscr{L}^p$ and $f=\sup_n f_n$. The function f is p-integrable if and only if* $\sup_n N_p(f_n) < +\infty$. *In this case, the sequence (f_n) converges to f in the mean of order p. If the functions f_n are $\geqslant 0$, then*

$$N_p(\sup_n f_n) = \sup_n N_p(f) = \lim_n N_p(f_n)\,.$$

Assume first, that $f_n \geqslant 0$ for every $n \in N$. If f is p-integrable, we have $N_p(f) < +\infty$. For every $n \in N$ we have $f_n \leqslant f$, therefore $N_p(f_n) \leqslant N_p(f)$, whence $\sup_n N_p(f_n) \leqslant N_p(f) < +\infty$.

Conversely, assume that $\sup_n N_p(f_n) < +\infty$. Since the sequence (f_n) is increasing and the seminorm N_p is increasing too, it follows that the numerical sequence $(N_p(f_n))$ is increasing; since it is also bounded, it is convergent. Then the sequence of the powers $(N_p(f_n))^p$ is also convergent, hence it is a numerical Cauchy sequence. For every $\varepsilon > 0$, there exists then a number $n(\varepsilon)$ such that, for every $n \geqslant n(\varepsilon)$ and $k \geqslant 1$, we have

$$N_p(f_{n+k})^p - N_p(f_n)^p < \varepsilon^p\,.$$

From the inequality proved in lemma 7.32,

$$N_p(f_{n+k} - f_n)^p \leqslant N_p(f_{n+k})^p - N_p(f_n)^p,$$

we deduce that

$$N_p(f_{n+k} - f_n) < \varepsilon\,,$$

that is, (f_n) is a Cauchy sequence of functions of $\mathscr{L}^p$. Since $\mathscr{L}^p$ is complete, the sequence (f_n) converges in the mean of order p to a function of $\mathscr{L}^p$. But the sequence (f_n) is increasing, therefore we have

$$\lim_n f_n(t) = \sup_n f_n(t) = f(t) \text{ for every } t \in T\,.$$

Hence the Cauchy sequence (f_n) converges pointwise to f. It follows that f is p-integrable and (f_n) converges to f in the mean of order p:

$$\lim_n N_p(f - f_n) = 0\,.$$

It follows that

$$\lim_n N_p(f_n) = N_p(f) = N_p(\sup_n f_n)\,.$$

The equality

$$\lim_n N_p(f_n) = \sup_n N_p(f_n)$$

follows from the fact that the numerical sequence $(N_p(f_n))$ is increasing.

Assume now that (f_n) is an arbitrary increasing sequence of functions from $\mathscr{L}^p$. For each $n \in N$ put $g_n = f_n + f_1^-$. The functions g_n are $\geqslant 0$, belong to the space $\mathscr{L}^p$, and the sequence (g_n) is increasing. From the first part of the proof we deduce that the function $g = \sup g_n$ is p-integrable if and only if $\sup_n N_p(g_n) < +\infty$. We have

$$g = \sup g_n = \sup f_n + f_1^- = f + f_1^- ;$$

therefore, g is p-integrable if and only if f is p-integrable. Then we have

$$N_p(g_n) = N_p(f_n + f_1^-) \leqslant N_p(f_n) + N_p(f_1^-)$$

and

$$N_p(f_n) = N_p(g_n - f_1) \leqslant N_p(g_n) + N_p(f_1^-),$$

hence, $\sup_n N_p(f_n) < +\infty$ if and only if $\sup_n N_p(g_n) < +\infty$.

It follows that $f = \sup f_n$ is p-integrable if and only if

$$\sup_n N_p(f_n) < +\infty .$$

In this case, $g = \sup g_n$ is p-integrable, hence the sequence (g_n) converges in the mean of order p to g:

$$\lim_n N_p(g_n - g) = 0 .$$

But

$$f_n - f = g_n - f_1^- - (g - f_1^-) = g_n - g ,$$

hence

$$\lim_n N_p(f_n - f) = 0 ,$$

that is, the sequence (f_n) converges in the mean of order p to f.

Thus the theorem is proved.

7.36 Corollary. *If (f_n) is a decreasing sequence of positive functions of $\mathscr{L}^p$, then the function $f = \inf f_n$ belongs to the space $\mathscr{L}^p$, the sequence (f_n) converges in the mean of order p to f and we have*

$$N_p(\inf f_n) = \inf_n N_p(f_n) = \lim_n N_p(f_n) .$$

In fact, the sequence $g_n = f_1 - f_n$ is increasing, consists of positive functions and

$$N_p(g_n) = N_p(f_1 - f_n) \leqslant N_p(f_1) + N_p(f_n) \leqslant N_p(f_1) + N_p(f_1),$$

therefore

$$\sup_n N_p(g_n) \leqslant 2N_p(f_1) < +\infty .$$

From the preceding theorem it follows that the function $g = \sup g_n = f_1 - \inf f_n$ has p-integrable power and that the sequence (g_n) converges in the mean of order p to g.

Let us remark that the function $\inf f_n$ is defined and finite everywhere on T, hence the function g is defined and finite everywhere; being p-integrable, it follows that $g \in \mathscr{L}^p$, hence

$$f = \inf f_n = f_1 - g \in \mathscr{L}^p .$$

Since the sequence (g_n) converges in the mean of order p to $g = f_1 - f$, it follows that the sequence $f_n = f_1 - g_n$ converges in the mean of order p to $f_1 - g = f$:

$$\lim_n N_p(f_n - f) = 0 .$$

It follows then that

$$\inf_n N_p(f_n) = \lim_n N_p(f_n) = N_p(f) = N_p(\inf f_n) .$$

7.37 COROLLARY. *Let (f_n) be a decreasing sequence of functions of $\mathscr{L}^p$. The function $f = \inf f_n$ is p-integrable if and only if $\sup_n N_p(f_n) < +\infty$. In this case the sequence (f_n) converges in the mean of order p to f.*

The sequence $g_n = -f_n$ is increasing and $N_p(g_n) = N_p(f_n)$. Then, $\sup g_n = -\inf f_n$. It follows that $\inf f_n$ is p-integrable if and only if $\sup g_n$ is p-integrable. But $\sup g_n$ is p-integrable, if and only if $\sup_n N_p(g_n) = \sup_n N_p(f_n) < +\infty$. It follows that $\inf f_n$ is p-integrable if and only if $\sup N_p(f_n) < +\infty$.

If $f = \inf f_n$ is p-integrable, then $g = \sup g_n$ is p-integrable, therefore (g_n) converges in the mean of order p to g. It follows that $f_n = -g_n$ converges in the mean of order p to $f = -g$.

7.38 COROLLARY. *The upper envelope $f = \sup f_n$ of a sequence (f_n) of functions of $\mathscr{L}^p$ is p-integrable if and only if there exists a function $g \geqslant 0$ (not necessarily from $\mathscr{L}^p$) such that $\int^* g^p \, d\mu < +\infty$ and $f_n \leqslant g$ for every n.*

If f is p-integrable, then f^+ is p-integrable and taking $g = f^+$ we have $g \geqslant 0$, $\int^* g^p \, d\mu < +\infty$ and $f_n \leqslant f \leqslant f^+$, hence $f_n \leqslant g$ for every n.

Conversely, assume that there exists $g \geqslant 0$ with $\int^* g^p \, d\mu < +\infty$ such that $f_n \leqslant g$ for every n. For every n put $g_n = \sup_{k \leqslant n} f_k \leqslant g$. The sequence (g_n) is increasing and consists of functions from $\mathscr{L}^p$. The functions $h_n = g_n + g_1^-$ are $\geqslant 0$ and belong to the space $\mathscr{L}^p$; the sequence (h_n) is also increasing and $g_n + g_1^- \leqslant g + g_1^-$, hence

$$N_p(h_n) = N_p(g_n + g_1^-) \leqslant N_p(g + g_1^-) < +\infty .$$

From theorem 7.34 it follows that $\sup h_n$ is p-integrable; therefore, the function $f = \sup f_n = \sup g_n = \sup h_n - g_1^-$ is p-integrable.

7.39 COROLLARY. *The lower envelope $f = \inf f_n$ of a sequence (f_n) of functions of $\mathscr{L}^p$ is p-integrable if and only if there exists a function $g \leqslant 0$ such that $f_n \geqslant g$ for every n and $\int^* |g|^p \, d\mu < +\infty$.*

For every n, denote $h_n = -f_n$. We have $\sup h_n = -\inf f_n$. If $\inf f_n$ is p-integrable, then $\sup h_n$ is p-integrable and hence, according to corollary 7.38, there exists a function $h \geqslant 0$ such that $\int^* h^p \, d\mu < +\infty$ and $h_n \leqslant h$ for every n. Taking $g = -h$ we have $g \leqslant 0$, $f_n \geqslant g$ and $\int^* |g|^p \, d\mu < +\infty$.

Conversely, if there exists $g \leqslant 0$ with $\int^* |g|^p \, d\mu < +\infty$ such that $f_n \geqslant g$, then $h = -g \geqslant 0$, $\int^* h^p \, d\mu < +\infty$ and $h_n \leqslant h$; hence $\sup h_n$ is p-integrable. It follows then that the function $\inf f_n = -\sup h_n$ is also p-integrable.

7.40 COROLLARY. *Let A be a countable set of indices, ordered and directed for an order relation $\leqslant$, and let $(f_\alpha)_{\alpha \in A}$ be a set of positive functions of $\mathscr{L}^p$. If there exists a function $g \geqslant 0$ such that $N_p(g) < +\infty$ and $f_\alpha \leqslant g$ for every $\alpha \in A$, then the function* $\limsup_{\alpha \in A} f_\alpha$ *is p-integrable and*

$$\limsup_{\alpha \in A} N_p(f_\alpha) \leqslant N_p(\limsup_{\alpha \in A} f_\alpha) .$$

For each index $\alpha \in A$ consider the section

$$S_\alpha = \{\beta \mid \beta \in A, \beta \geqslant \alpha\} .$$

We can construct a *decreasing* sequence (A_n) of sections, such that each section S_α contains a section from this sequence.

To this aim, arrange in a sequence

$$\beta_1, \beta_2, \ldots, \beta_n, \ldots$$

in an arbitrary order, all the elements of the set A. Take $A_1 = S_{\beta_1}$. There exists an index α_2 such that $\alpha_2 \geqslant \beta_1$ and $\alpha_2 \geqslant \beta_2$. If we take $A_2 = S_{\alpha_2}$, we have $A_1 \supset A_2$, $S_{\beta_2} \supset A_2$. Assuming that we have chosen $A_n = S_{\alpha_n}$ we can find and index α_{n+1} such that $\alpha_{n+1} \geqslant \alpha_n$ and $\alpha_{n+1} \geqslant \beta_{n+1}$. We take $A_{n+1} = S_{\alpha_{n+1}}$ and we have then $A_n \supset A_{n+1}$ and $S_{\beta_{n+1}} \supset A_{n+1}$. The sequence (A_n) constructed in this way corresponds to the above requirement.

For every index n put $g_n = \sup_{\alpha \in A_n} f_\alpha$. From the relation $A_n \supset A_{n+1}$ for every n, it follows that the sequence (g_n) is *decreasing*. Since the family $(f_\alpha)_{\alpha \in A}$ is countable and since $f_\alpha \leqslant g$ and $N_p(g) < +\infty$, from corollary 7.38 we deduce that g_n is p-integrable. On the other hand, $g_n \geqslant f_\alpha$ for $\alpha \in A_n$ hence $N_p(g_n) \geqslant N_p(f_\alpha)$ for $\alpha \in A$, therefore,

$$N_p(g_n) \geqslant \sup_{\alpha \in A_n} N_p(f_\alpha) .$$

Since the sequence (A_n) is decreasing, it follows that putting $\varphi_n = \sup_{\alpha \in A_n} N_p(f_\alpha)$, the sequence (φ_n) is *decreasing* and

$$\inf_n (\sup_{\alpha \in A_n} N_p(f_\alpha)) = \lim_n (\sup_{\alpha \in A_n} N_p(f_\alpha)) = \lim \inf_{\alpha \in A} N_p(f_\alpha) .$$

Since the sequence (g_n) is decreasing and consists of p-integrable functions, from corollary 7.36, it follows that the function $\inf g_n$ is p-integrable and

$$N_p(\inf g_n) = \inf_n N_p(g_n) \geqslant \inf_n (\sup_{\alpha \in A_n} N_p(f_\alpha)) = \lim \inf_{\alpha \in A} N_p(f_\alpha) .$$

But

$$\inf g_n = \inf_n (\sup_{\alpha \in A_n} f_\alpha) = \lim \inf_{\alpha \in A} f_\alpha$$

and hence

$$N_p (\lim \inf_{\alpha \in A} f_\alpha) \geqslant \lim \inf_{\alpha \in A} N_p(f_\alpha) .$$

8 *Lebesgue's theorem*

7.41 LEBESGUE'S THEOREM. *If (f_n) is a sequence of functions of $\mathscr{L}^p_E$ such that:*

(1) *the sequence $(f_n(t))$ converges almost everywhere to a limit $f(t) \in E$;*

(2) *there exists a numerical function $g \geqslant 0$ such that $\int^* g^p \mathrm{d}\mu < +\infty$ and $|f_n(t)| \leqslant g(t)$, almost everywhere, for each n;*

then the function f (defined μ-almost everywhere) is p-integrable and the sequence (f_n) converges in the mean of order p to f.

For the proof, we apply corollary 7.40 to the family $g_{m,n} = |f_m - f_n|$ of positive functions, taking the set of indices $A = \{(m, n) | m, n \in N\}$, with the order relation $(m, n) \leqslant (m', n')$ equivalent with $m \leqslant m'$ and $n \leqslant n'$; A is directed for this order relation, hence the family $(g_{m,n})_{(m,n)\in A}$ is directed. We have $\lim_{m,n} g_{m,n}(t) = \lim_{(m,n)\in A} g_{m,n}(t) = 0$ almost everywhere, $|g_{m,n}(t)| \leqslant 2g(t)$, almost everywhere, and $N_p(2g) < +\infty$. According to corollary 7.40, we have

$$\limsup_{(m,n)\in A} N_p(g_{m,n}) \leqslant N_p(\limsup_{(m,n)\in A} g_{m,n}) = N_p(0) = 0\,,$$

and since $N_p(f_m - f_n) \geqslant 0$, it follows that

$$\lim_{m,n} N_p(f_m - f_n) = 0\,.$$

This means that (f_n) is a Cauchy sequence in $\mathscr{L}^p_E$. Since (f_n) converges almost everywhere to f, it follows that f is p-integrable and that

$$\lim_n N_p(f_n - f) = 0\,,$$

that is, the sequence (f_n) converges to f in the mean of order p.

7.42 COROLLARY. *Let A be a countable set of indices, ordered and directed for an order relation $\leqslant$ and let $(f_\alpha)_{\alpha\in A}$ be a family of functions of the space $\mathscr{L}^p_E$.*
If:

(1) *the family $(f_\alpha(t))_{\alpha\in A}$ has a limit $f(t)\in E$, almost everywhere;*
(2) *there exists a numerical function $g \geqslant 0$ such that $\int^* g^p \,\mathrm{d}\mu < +\infty$ and such that for each $\alpha\in A$ we have $|f_\alpha(t)| \leqslant g(t)$ almost everywhere;*

then the function f is p-integrable and the family $(f_\alpha)_{\alpha\in A}$ converges in the mean of order p to f.

Let (A_n) be a decreasing sequence of sections such that each section $S_\alpha = \{\beta \mid \beta\in A, \beta \geqslant \alpha\}$ contains a section from the sequence (see the proof of corollary 7.40). For each n, choose an index $\alpha_n \in A$. The sequence (f_{α_n}) is pointwise convergent almost everywhere to f; by hypothesis (2) and Lebesgue's theorem we deduce that the function f is p-integrable and that $\lim_n N_p(f_{\alpha_n} - f) = 0$. It remains to show that $\lim_{\alpha\in A} N_p(f_\alpha - f) = 0$. This means that for every $\varepsilon > 0$, there exists a section A_n such that for every $\alpha\in A_n$ we have $N_p(f_\alpha - f) < \varepsilon$. Assume, on the contrary, that $\lim_\alpha N_p(f_\alpha - f) \neq 0$. This means that there exists a number $\varepsilon_0 > 0$ with the property that in every section A_n there exists an index α_n such that $N_p(f_{\alpha_n} - f) > \varepsilon_0$. Thus we get a sequence (f_{α_n}) with $\alpha_n \in A_n$ for each n. From the first part of the proof it follows that $\lim_n N_p(f_{\alpha_n} - f) = 0$, which contradicts the inequalities $N_p(f_{\alpha_n} - f) > \varepsilon_0 > 0$, for each n. Hence it follows that

$$\lim_{\alpha \in A} N_p(f_\alpha - f) = 0 ,$$

and the corollary is proved.

Remark. In the proof we have made no use of the fact that the directed set A is countable, but only of the fact that the filter of sections has a countable basis. The corollary remains valid for every family A of indices on which there exists a filter $\mathscr{F}$ with countable basis. In this case the convergence almost everywhere or in the mean of order p must be understood following the filter $\mathscr{F}$.

If the filter $\mathscr{F}$ does not have a countable basis, the corollary is no longer valid in general.

9 *Relationships between $\mathscr{L}_E^p(\mu)$ and $\mathscr{L}_E^q(\mu)$*

7.43 THEOREM. *Let $1 \leqslant p < +\infty$ and $1 \leqslant q < +\infty$. We have $f \in \mathscr{L}_E^p$ if and only if $|f|^{p/q-1} f \in \mathscr{L}_E^q$.*

Let $f \in \mathscr{L}_E^p$. There exists a sequence h_n of functions of $\mathscr{K}_E(T)$ converging in the mean of order p to f. This sequence contains a subsequence (h_{n_k}) such that

$$\Sigma_{1 \leqslant k < \infty} N_p(h_{n_{k+1}} - h_{n_k}) < +\infty$$

and such that the series

$$\Sigma_{1 \leqslant k < \infty} (h_{n_{k+1}} - h_{n_k})$$

is convergent almost everywhere. If for each k we put $f_{k+1} = h_{n_{k+1}} - h_{n_k}$ and $f_1 = h_{n_1}$, then the functions (f_n) belong to $\mathscr{K}_E(T)$ and we have

$$\Sigma_{1 \leqslant n < \infty} N_p(f_n) < +\infty \text{ and } \Sigma_{1 \leqslant n < \infty} f_n(t) = f(t), \text{ almost everywhere.}$$

Put

$$g_n = |f_1 + f_2 + \ldots + f_n|^{p/q-1} (f_1 + f_2 + \ldots + f_n) .$$

The functions g_n belong to $\mathscr{K}_E(T)$ and

$$|g_n|^q = |f_1 + f_2 + \ldots + f_n|^p \leqslant (|f_1| + |f_2| + \ldots + |f_n|)^p \leqslant (\Sigma_{1 \leqslant n < \infty} |f_n|)^p .$$

If we denote $h = (\Sigma_{1 \leqslant n < \infty} |f_n|)^{p/q}$, we have $h \geqslant 0$, $|g_n| \leqslant h$ for each n consequently,

$$N_p(h)^q = \int^* h^q \, d\mu = \int^* (\Sigma_{1 \leqslant n < \infty} |f_n|)^p \, d\mu = [N_p(\Sigma_{1 \leqslant n < \infty} |f_n|)]^p \leqslant$$
$$\leqslant [\Sigma_{1 \leqslant n < \infty} N_p(f_n)]^p < +\infty .$$

Since

$$\lim_n \boldsymbol{g}_n(t) = \lim_n |\Sigma_{k\leqslant n} \boldsymbol{f}_k(t)|^{p/q-1} \Sigma_{k\leqslant n} \boldsymbol{f}_k(t) =$$
$$= |\Sigma_{1\leqslant k<\infty} \boldsymbol{f}_k(t)|^{p/q-1} \Sigma_{1\leqslant k<\infty} \boldsymbol{f}_k(t) =$$
$$= |\boldsymbol{f}(t)|^{p/q-1} \boldsymbol{f}(t), \text{ almost everywhere },$$

by Lebesgue's theorem it follows that the limit function $|\boldsymbol{f}|^{p/q-1}\boldsymbol{f}$ is q-integrable.

Conversely, if the function $\boldsymbol{g} = |\boldsymbol{f}|^{p/q-1}\boldsymbol{f}$ is q-integrable, then $\boldsymbol{f} = |\boldsymbol{g}|^{q/p-1}\boldsymbol{g}$, and from the first part of the proof it follows that $\boldsymbol{f}$ is p-integrable.

Remark. It can be shown that $\boldsymbol{f} \to |\boldsymbol{f}|^{p/q-1}\boldsymbol{f}$ is a homeomorphism of the space $\mathscr{L}^p_E$ onto the space $\mathscr{L}^q_E$.

7.44 COROLLARY. *We have $\boldsymbol{f} \in \mathscr{L}^p_E$ if and only if $|\boldsymbol{f}|^{p-1}\boldsymbol{f} \in \mathscr{L}^1_E$.*

7.45 COROLLARY. *A function $f \geqslant 0$ belongs to the space $\mathscr{L}^p$ if and only if $f^p \in \mathscr{L}^1$.*

10 *Directed sets in L^p*

In §6, section 5, we defined an order relation $\tilde{f} \leqslant \tilde{g}$ in the set F of equivalence classes of numerical functions defined and finite μ-almost everywhere on T.

If $\tilde{f}_1, \tilde{f}_2 \in F$, then the supremum $g = \sup(f_1, f_2)$ and the infimum $h = \inf(f_1, f_2)$ are defined and finite almost everywhere, and we have

$$\tilde{g} = \sup(\tilde{f}_1, \tilde{f}_2), \ \tilde{h} = \inf(\tilde{f}_1, \tilde{f}_2),$$

hence F is a reticulated vector space (or a vector lattice, or a Riesz space).

By corollary 7.30, the subspace L^p of F is also reticulated for the order relation $\tilde{f} \leqslant \tilde{g}$ and the supremum in L^p of two elements $\tilde{f}, \tilde{g} \in L^p$ coincides with their supremum in F.

7.46 PROPOSITION. *In the Banach lattice L^p, the mapping $\tilde{f} \to |\tilde{f}|$ is uniformly continuous and the set P of elements $\tilde{f} \geqslant 0$ is closed.*

The uniform continuity of the mapping $\tilde{f} \to |\tilde{f}|$ follows from the inequality

$$\| \, |\tilde{f}| - |\tilde{g}| \, \|_p \leqslant \| \tilde{f} - \tilde{g} \|_p \, .$$

Let $\tilde{f} \in \bar{P}$; there exists a sequence $(\tilde{g}_n)$ of elements from P such that $\tilde{g}_n \to \tilde{f}$ in L^p.

Since the mapping $\tilde{f} \to |\tilde{f}|$ is continuous, it follows that $|\tilde{g}_n| \to |\tilde{f}|$ in L^p. But $\tilde{g}_n \geqslant 0$, hence $\tilde{g}_n = |\tilde{g}_n|$, whence $|\tilde{f}| = \tilde{f}$. It follows that $\tilde{f} \geqslant 0$, therefore $\tilde{f} \in P$ and hence P is closed.

7.47 Theorem. *Let $H \subset L^p$ be a set directed for the relation $\leqslant$, and consisting of classes $\geqslant 0$. The set H has a supremum in L^p if and only if $\sup_{\tilde{f} \in H} \|\tilde{f}\|_p < \infty$.*

In this case sup H is the limit (in L^p) of the filter of sections of H.

If there exists the supremum $\tilde{g} = \sup H$ in L^p, then $\tilde{f} \leqslant \tilde{g}$, hence $\|\tilde{f}\|_p \leqslant \|\tilde{g}\|_p$ for every $\tilde{f} \in H$; consequently,

$$\sup_{\tilde{f} \in H} \|\tilde{f}\|_p \leqslant \|\tilde{g}\|_p < +\infty .$$

Conversely, assume that $\sup_{\tilde{f} \in H} \|\tilde{f}\|_p < +\infty$. Since the mapping $\tilde{f} \to \|\tilde{f}\|_p$ is increasing, it follows that this mapping has a limit when $\tilde{f}$ runs over the filter of sections of H and, moreover,

$$\lim_{\tilde{f} \in H} \|\tilde{f}\|_p = \sup_{\tilde{f} \in H} \|\tilde{f}\|_p .$$

From the inequality (proposition 7.33)

$$(\|\tilde{f} - \tilde{h}\|_p)^p \leqslant (\|\tilde{f}\|_p)^p - (\|\tilde{h}\|_p)^p ,$$

we deduce that the filter of sections of H is a Cauchy filter in the Banach space L^p, hence it has a limit $\tilde{g} \in L^p$. It follows then that $\tilde{g}$ is the supremum of the directed set H and the theorem is proved.

7.48 Corollary. *If $\tilde{g}$ is the supremum of H in L^p, then*

$$\|\tilde{g}\|_p = \lim_{\tilde{f} \in H} \|\tilde{f}\|_p = \sup_{\tilde{f} \in H} \|\tilde{f}\|_p .$$

In fact, the mapping $\tilde{f} \to \|\tilde{f}\|_p$ is continuous and increasing.

7.49 Corollary. *Every non-empty dominated set $A \subset L^p$ has a supremum in L^p.*

The set B of suprema of finite subsets of A is directed for the relation $\leqslant$.

Let $\tilde{h} \in B$ and the section $B_{\tilde{h}} = \{\tilde{f} \in A ; \tilde{f} \geqslant \tilde{h}\}$.

The set $H = B_{\tilde{h}} - \tilde{h}$ consists of elements $\geqslant 0$, is dominated and directed for the relation $\leqslant$, hence it has a supremum $\tilde{g}$ in L^p. It follows that $\tilde{g} + \tilde{h}$ is the supremum of the section $B_{\tilde{h}}$. Then we deduce that $\tilde{g} + \tilde{h}$ is the supremum of the set A.

Remark. The conclusion of theorem 7.47 is no longer true for functions from $\mathscr{L}^p$. If $M \subset \mathscr{L}^p$ is a set, directed for the relation $\leqslant$ and consisting of functions $\geqslant 0$ such that $\sup_{f \in M} N_p(f) < +\infty$, the equivalence class of the upper envelope $g = \sup_{f \in M} f$ is not, in general, identical to the supremum in L^p of the set $\{\tilde{f}; f \in M\}$. Moreover, g does not necessarily belong to the space $\mathscr{L}^p$, and even if $g \in \mathscr{L}^p$, the equality $N_p(g) = \sup_{f \in M} N_p(f)$ is not true in general.

§8 Integrable functions

1 *Definition of integrable functions with respect to a dominated measure*

Let E and F be two Banach spaces, both of them over the same field. In particular, one or both spaces E and F can be equal to R or to C.

Let $\boldsymbol{m} : \mathscr{K}_E(T) \to F$ be a *dominated* measure and $|\boldsymbol{m}|$ the modulus of $\boldsymbol{m}$. In particular, we can have $\boldsymbol{m} = |\boldsymbol{m}|$ if $\boldsymbol{m}$ is positive.

Consider the space $\mathscr{L}^1_E(|\boldsymbol{m}|)$, endowed with the topology of the convergence in the mean defined by the seminorm

$$N_1(\boldsymbol{f}, |\boldsymbol{m}|) = \int^* |\boldsymbol{f}| \, \mathrm{d}|\boldsymbol{m}| \, .$$

The functions of the space $\mathscr{L}^1_E(|\boldsymbol{m}|)$ are said to be *integrable with respect to the positive measure* $|\boldsymbol{m}|$, or $|\boldsymbol{m}|$-integrable.

The functions $\boldsymbol{g}$ of $\mathscr{K}_E(T)$ satisfy the inequality

$$\left|\int \boldsymbol{g} \, \mathrm{d}\boldsymbol{m}\right| \leqslant \int |\boldsymbol{g}| \, \mathrm{d}|\boldsymbol{m}| \, .$$

For the functions $\boldsymbol{g}$ of $\mathscr{K}_E(T)$ we have $|\boldsymbol{g}| \in \mathscr{K}_+(T)$, therefore

$$\int^* |\boldsymbol{g}| \, \mathrm{d}|\boldsymbol{m}| = \int |\boldsymbol{g}| \, \mathrm{d}|\boldsymbol{m}| \, ,$$

hence

$$N_1(\boldsymbol{g}, |\boldsymbol{m}|) = \int |\boldsymbol{g}| \, \mathrm{d}|\boldsymbol{m}| \, ;$$

therefore the above inequality can be written

$$|\int g\,\mathrm{d}\boldsymbol{m}| \leqslant N_1(g, |\boldsymbol{m}|).$$

This inequality shows that the linear mapping $g \to \int g\,\mathrm{d}\boldsymbol{m}$ of $\mathscr{K}_E(T)$ into F is *continuous on* $\mathscr{K}_E(T)$ *for the topology of the convergence in the mean.* Since the space $\mathscr{K}_E(T)$ is dense in the space $\mathscr{L}_E^1(|\boldsymbol{m}|)$ for this topology, the linear application $g \to \int g\,\mathrm{d}\boldsymbol{m}$ of $\mathscr{K}_E(T)$ into F can be extended uniquely to a continuous linear mapping of the whole space $\mathscr{L}_E^1(|\boldsymbol{m}|)$ into F.

8.1 DEFINITION. *A function $f: T \to E$ is said to be integrable with respect to the dominated measure $\boldsymbol{m}$, or $\boldsymbol{m}$-integrable, if it is integrable with respect to $|\boldsymbol{m}|$. For every $\boldsymbol{m}$-integrable function $f \in \mathscr{L}_E^1(|\boldsymbol{m}|)$, the value at f of the extension to $\mathscr{L}_E^1(|\boldsymbol{m}|)$ of the mapping $g \to \int g\,\mathrm{d}\boldsymbol{m}$ of $\mathscr{K}_E(T)$ into F is called the integral of f with respect to $\boldsymbol{m}$ and it is denoted by $\boldsymbol{m}(f)$, or $\int f\mathrm{d}\boldsymbol{m}$, or $\int f(t)\mathrm{d}\boldsymbol{m}(t)$.*

The space of $\boldsymbol{m}$-integrable functions $f: T \to E$ will be denoted by $\mathscr{L}_E^1(\boldsymbol{m})$. Hence $\mathscr{L}_E^1(\boldsymbol{m}) = \mathscr{L}_E^1(|\boldsymbol{m}|)$. The space $\mathscr{L}_E^p(|\boldsymbol{m}|)$ of p-summable functions $f: T \to E$ with respect to the measure $|\boldsymbol{m}|$ will also be denoted, occasionally, by $\mathscr{L}_E^p(\boldsymbol{m})$.

8.2 PROPOSITION. *For every function $f \in \mathscr{L}_E^p(\boldsymbol{m})$, we have*

$$|\int f\mathrm{d}\boldsymbol{m}| \leqslant N_1(f), |\boldsymbol{m}|).$$

In fact, let (f_n) be a sequence of functions from $\mathscr{K}_E(T)$, converging in the mean to f. Since the integral and the seminorm are continuous in this topology as functions of f, we have

$$\int f_n\mathrm{d}\boldsymbol{m} \to \int f\mathrm{d}\boldsymbol{m} \quad \text{and} \quad N_1(f_n, |\boldsymbol{m}|) \to N_1(f, |\boldsymbol{m}|).$$

But for the functions f_n from $\mathscr{K}_E(T)$ we have the inequality

$$|\int f_n\mathrm{d}\boldsymbol{m}| \leqslant N_1(f_n, |\boldsymbol{m}|).$$

Passing to the limit in this inequality, we obtain the required inequality in the statement.

The seminorm $N_1(f, |\boldsymbol{m}|)$ will also be denoted by $N_1(f, \boldsymbol{m})$:

$$N_1(f, \boldsymbol{m}) = \int^* |f|\,\mathrm{d}|\boldsymbol{m}|.$$

With this notation, the above inequality can be written

$$|\int f\mathrm{d}\boldsymbol{m}| \leqslant N_1(f, \boldsymbol{m}) \quad \text{for} \quad f \in \mathscr{L}_E^1(\boldsymbol{m}).$$

2 *Integration with respect to a scalar measure*

Let $\mu : \mathscr{K}(T) \to C$ be a *real* (*complex*) *measure* and E a real (complex) Banach space. In particular, μ can be a positive measure, and E *can be the space* R *of the real numbers or the space* C *of complex numbers* (considered as real or complex vector spaces, according to whether the measure μ is real or complex).

As we have shown (theorem 2.11 and proposition 3.13), there exists a unique measure $\boldsymbol{m} : \mathscr{K}_E(T) \to E$ satisfying the equality

$$\boldsymbol{m}(\varphi x) = \mu(\varphi) x \text{ for } \varphi \in \mathscr{K}(T) \text{ and } x \in E .$$

The measure $\boldsymbol{m}$ is dominated and $|\boldsymbol{m}| = |\mu|$.

We can apply definition 8.1 for the measure $\boldsymbol{m}$. Since $|\boldsymbol{m}| = |\mu|$, we have

$$\mathscr{L}_E^1(\boldsymbol{m}) = \mathscr{L}_E^1(|\mu|) = \mathscr{L}_E^1(\mu) .$$

We have already agreed to identify the measures $\boldsymbol{m}$ and μ and to write $\int \boldsymbol{f} \mathrm{d}\mu$ instead of $\int \boldsymbol{f} \mathrm{d}\boldsymbol{m}$ for the functions $\boldsymbol{f}$ from $\mathscr{K}_E(T)$. With this identification, we have

$$\int \varphi x \, \mathrm{d}\mu = x \int \varphi \, \mathrm{d}\mu \text{ for } \varphi \in \mathscr{K}(T) \text{ and } x \in E .$$

The $\boldsymbol{m}$-integrable functions $\boldsymbol{f} : T \to E$ will be called μ-integrable functions and the space $\mathscr{L}_E^1(\boldsymbol{m})$ will be denoted by $\mathscr{L}_E^1(\mu)$. Instead of $\int \boldsymbol{f} \mathrm{d}\boldsymbol{m}$ we shall write $\int \boldsymbol{f} \mathrm{d}\mu$ and we shall say that this is the integral of the function $\boldsymbol{f} : T \to E$ with respect to the scalar measure μ.

The seminorm $N_1(\boldsymbol{f}, \boldsymbol{m})$ will be denoted by $N_1(\boldsymbol{f}, \mu)$:

$$N_1(\boldsymbol{f}, \mu) = \int^* |\boldsymbol{f}| \, \mathrm{d}|\mu| \; ;$$

with these notations we have

$$\left|\int \boldsymbol{f} \mathrm{d}\mu\right| \leqslant N_1(\boldsymbol{f}, \mu), \text{ for } \boldsymbol{f} \in \mathscr{L}_E^1(\mu) .$$

8.3 Theorem. *Let μ be a real (respectively complex) measure, E and F two real (complex) Banach spaces and $U : E \to F$ a linear continuous mapping. For every μ-integrable function $\boldsymbol{f} : T \to E$, the function $U \circ \boldsymbol{f} : T \to F$ is μ-integrable and*

$$\int U(\boldsymbol{f}(t)) \mathrm{d}\mu(t) = U\left(\int \boldsymbol{f}(t) \mathrm{d}\mu(t)\right) .$$

We have already proved (theorem 7.24) that if $\boldsymbol{f}$ is a $|\mu|$-integrable function from $\mathscr{L}^1_E(\mu)$, then $U \circ \boldsymbol{f} \in \mathscr{L}^1_F(\mu)$; hence, $U \circ \boldsymbol{f}$ is $|\mu|$-integrable. It has also been proved that the equality of the statement is true for functions $\boldsymbol{f} \in \mathscr{K}_E(T)$ (theorem 2.13).

Now let $\boldsymbol{f}$ be an arbitrary function from $\mathscr{L}_E(\mu)$ and let $(\boldsymbol{f}_n)$ be a sequence of functions from $\mathscr{K}_E(T)$, converging in the mean to $\boldsymbol{f}$. Then:

$$\lim \int \boldsymbol{f}_n \mathrm{d}\mu = \int \boldsymbol{f} \mathrm{d}\mu, \text{ hence } \lim U\left(\int \boldsymbol{f}_n \mathrm{d}\mu\right) = U\left(\int \boldsymbol{f} \mathrm{d}\mu\right).$$

From the inequality

$$N_1(U \circ \boldsymbol{f}_n - U \circ \boldsymbol{f}, |\mu|) = \int^* |U \circ (\boldsymbol{f}_n - \boldsymbol{f})| \mathrm{d}|\mu| \leqslant$$
$$\leqslant |U| \int |\boldsymbol{f}_n - \boldsymbol{f}| \mathrm{d}|\mu| = |U| N_1(\boldsymbol{f}_n - \boldsymbol{f}, |\mu|),$$

it follows that $U \circ \boldsymbol{f}_n$ converges in the mean to $U \circ \boldsymbol{f}$, hence

$$\lim \int U(\boldsymbol{f}_n(t)) \mathrm{d}\mu(t) = \int U(\boldsymbol{f}(t)) \mathrm{d}\mu(t).$$

Passing to the limit in the equality

$$\int U(\boldsymbol{f}_n(t)) \mathrm{d}\mu(t) = U\left(\int \boldsymbol{f}_n(t) \mathrm{d}\mu(t)\right),$$

we get the required equality.

8.4 Corollary. *Let μ be a scalar measure, E and F two Banach spaces. If $\boldsymbol{f}: T \to \mathscr{L}(E, F)$ is a μ-integrable function and $x \in E$, then function $t \to \boldsymbol{f}(t)x$ (with values in F) is μ-integrable and*

$$\int \boldsymbol{f}(t) x \mathrm{d}\mu(t) = \left(\int \boldsymbol{f}(t) \mathrm{d}\mu(t)\right) x.$$

We apply the preceding theorem, for the linear continuous mapping $Uz = xz$ of $\mathscr{L}(E, F)$ into F.

8.5 Corollary. *Let μ be a scalar measure and E a Banach space. If φ is a μ-integrable scalar function and $x \in E$, then $\varphi x : T \to E$ is integrable and*

$$\int \varphi x \,\mathrm{d}\mu = x \int \varphi \,\mathrm{d}\mu.$$

We apply either corollary 8.4 taking $F = E$ and considering $\varphi : T \to \mathscr{L}(E, F)$, or theorem 8.3 for the linear continuous mapping $U\alpha = \alpha x$ of C (or R) into E.

Remark. Corollary 8.5 shows that we can get the same result if, starting with a scalar measure $\mu : \mathscr{K}(T) \to C$, we first define the measure $\mu : \mathscr{K}_E(T) \to E$ by extending the equality

$$\mu(\varphi x) = \mu(\varphi)x, \text{ for } \varphi \in \mathscr{K}(T) \text{ and } x \in E,$$

and then we extend μ to the space $\mathscr{L}_E^1(\mu)$, or if we extend first the measure μ to the space $\mathscr{L}^1(\mu)$ and then we define the integral of the functions from $\mathscr{L}_E^1(\mu)$ by extending the equality

$$\int \varphi x \, d\mu = x \int \varphi \, d\mu, \text{ for } \varphi \in \mathscr{L}^1(\mu) \text{ and } x \in E.$$

8.6 COROLLARY. *Let μ be a scalar measure and E a Banach space. If $\boldsymbol{f} : T \to E$ is a μ-integrable function and $z' \in E'$, then the scalar function $t \to \langle \boldsymbol{f}(t), z' \rangle$ is μ-integrable and*

$$\int \langle \boldsymbol{f}, z' \rangle \, d\mu = \langle \int \boldsymbol{f} \, d\mu, z' \rangle.$$

3 *Integration of scalar functions with respect to a dominated measure*

Let E and F be two Banach spaces (over the same field of scalars), $\boldsymbol{m} : \mathscr{K}_E(T) \to F$ a dominated measure and $|\boldsymbol{m}|$ its modulus.

As we have seen (proposition 2.1) there exists a unique measure $\boldsymbol{\mu} : \mathscr{K}(T) \to \mathscr{L}(E, F)$ which satisfies the equality

$$\boldsymbol{\mu}(\varphi)x = \boldsymbol{m}(\varphi x) \text{ for } \varphi \in \mathscr{K}(T) \text{ and } x \in E.$$

The measure $\boldsymbol{\mu}$ is also dominated and $|\boldsymbol{\mu}| \leqslant |\boldsymbol{m}|$, hence

$$|\int \varphi \, d\boldsymbol{\mu}| \leqslant \int |\varphi| \, d|\boldsymbol{\mu}| \leqslant \int |\varphi| \, d|\boldsymbol{m}|, \text{ for } \varphi \in \mathscr{K}(T).$$

It will be shown later that we even have the equality $|\boldsymbol{\mu}| = |\boldsymbol{m}|$.

From the inequality $|\boldsymbol{\mu}| \leqslant |\boldsymbol{m}|$, we deduce

$$\int^* |\varphi| \, d|\boldsymbol{\mu}| \leqslant \int^* |\varphi| \, d|\boldsymbol{m}|, \text{ hence } N_1(\varphi, |\boldsymbol{\mu}|) \leqslant N_1(\varphi, |\boldsymbol{m}|).$$

It follows that

$$\mathscr{L}^1(|\boldsymbol{m}|) \subset \mathscr{L}^1(|\boldsymbol{\mu}|) = \mathscr{L}^1(\boldsymbol{\mu}).$$

The $|\boldsymbol{m}|$-integrable functions $\varphi \in \mathscr{L}^1(|\boldsymbol{m}|)$ are at the same time $\boldsymbol{\mu}$-integrable.

We have already agreed to identify the measures $\boldsymbol{m}$ and $\boldsymbol{\mu}$ and to write $\int \varphi \, d\boldsymbol{m}$ instead of $\int \varphi \, d\boldsymbol{\mu}$ for functions φ from $\mathscr{K}(T)$. With this notation, we have

$$\left(\int \varphi \, d\boldsymbol{m}\right)x = \int \varphi x \, d\boldsymbol{m}, \text{ for } \varphi \in \mathscr{K}(T) \text{ and } x \in E$$

and

$$|\int \varphi \, d\boldsymbol{m}| \leqslant \int |\varphi| \, d|\boldsymbol{m}| = N_1(\varphi, |\boldsymbol{m}|), \text{ for } \varphi \in \mathscr{K}(T).$$

By this identification the μ-integrable numerical functions φ from $\mathscr{L}^1(|\boldsymbol{m}|)$ will be called $\boldsymbol{m}$-integrable functions and the space $\mathscr{L}^1(|\boldsymbol{m}|)$ will be denoted by $\mathscr{L}^1(\boldsymbol{m})$. For the functions φ from $\mathscr{L}^1(\boldsymbol{m})$, we write $\int \varphi \, d\boldsymbol{m}$ instead of $\int \varphi \, d\boldsymbol{\mu}$ and we shall say that $\int \varphi \, d\boldsymbol{m}$ is the integral of the numerical function φ with respect to the vector measure $\boldsymbol{m}$.

The seminorm $N_1(\varphi, |\boldsymbol{m}|)$ will be denoted by $N_1(\varphi, \boldsymbol{m})$:

$$N_1(\varphi, \boldsymbol{m}) = \int^* |\varphi| \, d(|\boldsymbol{m}|) .$$

With this notation we have

$$|\int \varphi \, d\boldsymbol{m}| \leqslant N_1(\varphi, \boldsymbol{m}) \text{ for } \varphi \in \mathscr{L}^1(\boldsymbol{m}) .$$

Remark. Further on (theorem 8.61), after the equality $|\boldsymbol{\mu}| = |\boldsymbol{m}|$ will be proved, it will follow that $\mathscr{L}^1(\boldsymbol{m}) = \mathscr{L}^1(\boldsymbol{\mu})$ and the identification of the measures $\boldsymbol{m}$ and $\boldsymbol{\mu}$ will be completely justified

8.7 PROPOSITION. *If* $\boldsymbol{m} : \mathscr{K}_E \to F$ *is a dominated measure, then for every* $\boldsymbol{m}$*-integrable numerical function* $\varphi \in \mathscr{L}^1(\boldsymbol{m})$ *and every* $x \in E$, *we have*

$$\int \varphi x \, d\boldsymbol{m} = (\int \varphi \, d\boldsymbol{m}) x .$$

Let $\varphi \in \mathscr{L}^1(\boldsymbol{m})$ and let (φ_n) be a sequence of functions from $\mathscr{K}(T)$ converging in $\mathscr{L}^1(\boldsymbol{m})$ to φ. For the functions φ_n of $\mathscr{K}(T)$ we have

$$\boldsymbol{m}(\varphi_n) x = \boldsymbol{m}(\varphi_n x), \quad x \in E .$$

Since $\varphi_n \to \varphi$ in $\mathscr{L}^1(\boldsymbol{m})$, we have $\boldsymbol{m}(\varphi_n) \to \boldsymbol{m}(\varphi)$ in $\mathscr{L}(E, F)$; therefore, $\boldsymbol{m}(\varphi_n) x \to \boldsymbol{m}(\varphi) x$ in F, for every $x \in E$. On the other hand, since $\varphi_n \to \varphi$ in $\mathscr{L}^1(\boldsymbol{m})$, we have $\varphi_n x \to \varphi x$ in $\mathscr{L}^1_E(\boldsymbol{m})$ for each $x \in E$ and hence $\boldsymbol{m}(\varphi_n x) \to \boldsymbol{m}(\varphi x)$. Passing to the limit in the above equality, we obtain

$$\boldsymbol{m}(\varphi) x = \boldsymbol{m}(\varphi x) ,$$

that is,

$$\int \varphi x \, d\boldsymbol{m} = (\int \varphi \, d\boldsymbol{m}) x .$$

Remark. This proposition shows that we get the same result if, starting with a dominated measure $\boldsymbol{m} : \mathscr{K}_E(T) \to F$, either we define first the measure

$\boldsymbol{m}: \mathscr{K}(T) \to \mathscr{L}(E, F)$ by the equality

$$\boldsymbol{m}(\varphi)x = \boldsymbol{m}(\varphi x), \text{ for } \varphi \in \mathscr{K}(T) \text{ and } x \in E$$

and then we extend it to $\mathscr{L}^1(\boldsymbol{m})$, or we extend first the measure $\boldsymbol{m}$ to $\mathscr{L}_E^1(\boldsymbol{m})$ and then define the integral of the functions from $\mathscr{L}^1(|\boldsymbol{m}|)$ by the equality

$$(\int \varphi \mathrm{d}\boldsymbol{m})x = \int \varphi x \mathrm{d}\boldsymbol{m}.$$

4 *Integrable functions defined almost everywhere*

From the preceding sections it follows that for every (vector or scalar) dominated measure $\boldsymbol{m}$, we can integrate with respect to $\boldsymbol{m}$ both vector and scalar functions. The integrability with respect to $\boldsymbol{m}$ is reduced to the integrability with respect to the positive measure $|\boldsymbol{m}|$.

From the inequality

$$|\int \boldsymbol{f} \mathrm{d}\boldsymbol{m}| \leqslant N_1(\boldsymbol{f}, \boldsymbol{m}) = \int^* |\boldsymbol{f}| \mathrm{d}|\boldsymbol{m}|,$$

valid for every $\boldsymbol{m}$-integrable function $\boldsymbol{f}$, we deduce that, if $\boldsymbol{f}$ is $|\boldsymbol{m}|$-negligible, we have $\int \boldsymbol{f} \mathrm{d}\boldsymbol{m} = 0$. The converse implication is not true in general; we may have $\int \boldsymbol{f} \mathrm{d}\boldsymbol{m} = 0$ for a function $\boldsymbol{f}$ vanishing at no points.

The $|\boldsymbol{m}|$-negligible functions will be also called $\boldsymbol{m}$-negligible functions.

For properties which are valid $|\boldsymbol{m}|$-almost everywhere, we shall say also that they are valid $\boldsymbol{m}$-almost everywhere.

If $\boldsymbol{f}$ and $\boldsymbol{g}$ are two equivalent functions from $\mathscr{L}_E^1(\boldsymbol{m})$, then $N_1(\boldsymbol{f} - \boldsymbol{g}, \boldsymbol{m}) = 0$, hence

$$\int \boldsymbol{f} \mathrm{d}\boldsymbol{m} = \int \boldsymbol{g} \mathrm{d}\boldsymbol{m}.$$

It follows that the integral $\int \boldsymbol{f} \mathrm{d}\boldsymbol{m}$ depends only on the equivalence class $\tilde{\boldsymbol{f}}$ of $\boldsymbol{f}$. That is why we put, by definition,

$$\boldsymbol{m}(\tilde{\boldsymbol{f}}) = \int \boldsymbol{f} \mathrm{d}\boldsymbol{m}, \text{ for } \tilde{\boldsymbol{f}} \in L_E^1.$$

The mapping $\tilde{\boldsymbol{f}} \to \boldsymbol{m}(\tilde{\boldsymbol{f}})$ defined on the space L_E^1 is continuous and

$$|\boldsymbol{m}(\tilde{\boldsymbol{f}})| \leqslant N_1(\boldsymbol{f}, \boldsymbol{m}).$$

If now $\boldsymbol{f}$ is an E-valued function defined $\boldsymbol{m}$-almost everywhere on T and if $\boldsymbol{f}$ is equivalent to an $\boldsymbol{m}$-integrable function from $\mathscr{L}_E^1(\boldsymbol{m})$ (defined everywhere), that is if $\tilde{\boldsymbol{f}} \in L_E^1(\boldsymbol{m})$, then we say also that $\boldsymbol{f}$ is $\boldsymbol{m}$-integrable and we put, by definition,

$$\int \boldsymbol{f}\,\mathrm{d}\boldsymbol{m} = \boldsymbol{m}(\tilde{\boldsymbol{f}})\,.$$

$\int \boldsymbol{f}\,\mathrm{d}\boldsymbol{m}$ is called also the integral of $\boldsymbol{f}$ with respect to $\boldsymbol{m}$ and it satisfies the inequality

$$|\int \boldsymbol{f}\,\mathrm{d}\boldsymbol{m}| \leqslant N_1(\boldsymbol{f}, \boldsymbol{m})\,.$$

If f is a numerical function, defined and finite $\boldsymbol{m}$-almost everywhere on T, and if f is equivalent to an $\boldsymbol{m}$-integrable function from $\mathscr{L}^1(\boldsymbol{m})$ (defined and finite everywhere), that is if $\tilde{\boldsymbol{f}} \in L^1(\boldsymbol{m})$, we say also that f is $\boldsymbol{m}$-integrable, and the integral $\int f\,\mathrm{d}\boldsymbol{m}$ is defined by the equality

$$\int f\,\mathrm{d}\boldsymbol{m} = \boldsymbol{m}(\tilde{f})\,,$$

and it satisfies the inequality

$$|\int f\,\mathrm{d}\boldsymbol{m}| \leqslant N_1(f, \boldsymbol{m})\,.$$

Most of the following results are stated for integrable functions from $\mathscr{L}_E^1$ or from $\mathscr{L}^1$, but they remain valid for integrable functions which are defined almost everywhere, or are defined and finite almost everywhere.

5 *Properties of integrable functions*

Let μ be a *positive* measure. In particular, μ can be the modulus $|\boldsymbol{m}|$ of a dominated measure $\boldsymbol{m}$.

8.8 Proposition. *For every μ-integrable function $f \geqslant 0$, we have*

$$\int f\,\mathrm{d}\mu = \int^* f\,\mathrm{d}\mu = N_1(f, \mu)\,.$$

Let (g_n) be a sequence of functions of $\mathscr{K}(T)$ converging in the mean to f:

$$\lim N_1(g_n - f, \mu) = 0\,.$$

Since $f \geqslant 0$, we have

$$|\,|g_n| - f\,| = |\,|g_n| - |f|\,| \leqslant |g_n - f|\,,$$

whence

$$\lim N_1(|g_n| - f, \mu) = 0\,.$$

Hence, the sequence $(|g_n|)$ of positive functions from $\mathscr{K}(T)$ converges in the mean to f; the integral being continuous, we deduce

$$\lim \int |g_n| \, \mathrm{d}\mu = \int f \mathrm{d}\mu .$$

The seminorm N_1 being also continuous on $\mathscr{L}^1(\mu)$, we have

$$\lim N_1(|g_n|, \mu) = N_1(f, \mu) .$$

But for the functions $|g_n|$ from $\mathscr{K}_+(T)$ we have

$$\int |g_n| \, \mathrm{d}\mu = \int^* |g_n| \, \mathrm{d}\mu = N_1(|g_n|, \mu) .$$

Passing to the limit in these equalities, we get

$$\int f \mathrm{d}\mu = \int^* f \mathrm{d}\mu = N_1(f, \mu) .$$

8.9 Corollary. *For every μ-integrable function $\boldsymbol{f}$ of $\mathscr{L}_E^1(\mu)$, $|\boldsymbol{f}|$ is a μ-integrable function of $\mathscr{L}^1(\mu)$ and*

$$\int |\boldsymbol{f}| \, \mathrm{d}\mu = \int^* |\boldsymbol{f}| \, \mathrm{d}\mu = N_1(\boldsymbol{f}, \mu) .$$

The fact that $\boldsymbol{f} \in \mathscr{L}_E^1(\mu)$, implies $|\boldsymbol{f}| \in \mathscr{L}^1(\mu)$, has already been proved (proposition 7.26). The equality follows from the above proposition.

8.10 Corollary. *A function $\boldsymbol{f}$ is μ-negligible if and only if it is μ-integrable and $\int |\boldsymbol{f}| \mathrm{d}\mu = 0$.*

In fact, if $\boldsymbol{f}$ is μ-negligible, then $\boldsymbol{f}$ is μ-integrable and $\boldsymbol{f}(t) = 0$, μ-almost everywhere, hence $|\boldsymbol{f}(t)| = 0$, μ-almost everywhere. Therefore $\int^* |\boldsymbol{f}| \mathrm{d}\mu = 0$ and by corollary 8.9 we have $\int |\boldsymbol{f}| \mathrm{d}\mu = 0$.

Conversely, if $\boldsymbol{f}$ is μ-integrable and if $\int |\boldsymbol{f}| \mathrm{d}\mu = 0$, then $\int^* |\boldsymbol{f}| \mathrm{d}\mu = \int |\boldsymbol{f}| \mathrm{d}\mu$ therefore $\int^* |\boldsymbol{f}| \mathrm{d}\mu = 0$, that is $|\boldsymbol{f}|$ is μ-negligible, hence $\boldsymbol{f}$ is also μ-negligible.

8.11 Corollary. *Two μ-integrable functions $\boldsymbol{f}$ and $\boldsymbol{g}$ from $\mathscr{L}_E^1(\mu)$ are equivalent if and only if*

$$\int |\boldsymbol{f} - \boldsymbol{g}| \, \mathrm{d}\mu = 0 .$$

In fact, the functions $\boldsymbol{f}$ and $\boldsymbol{g}$ from $\mathscr{L}_E^1(\mu)$ are equivalent if and only if $\boldsymbol{f} - \boldsymbol{g}$ is μ-negligible.

8.12 PROPOSITION. *If f is a function from $\mathscr{L}_E^p(\mu)$, $1 \leqslant p < +\infty$, then $|f|^p$ is a μ-integrable function from $\mathscr{L}^1(\mu)$ and*

$$N_p(f, \mu) = (\int |f|^p \mathrm{d}\mu)^{1/p} .$$

In fact, if $f \in \mathscr{L}_E^p(\mu)$, then (corollary 7.44) the function $|f|^{p-1} f$ belongs to the space $\mathscr{L}_E^1(\mu)$, that is, it is μ-integrable.

From corollary 8.9, we deduce that the function

$$|f|^p = | \, |f|^{p-1} f |$$

is μ-integrable and

$$\int |f|^p \mathrm{d}\mu = \int^* |f|^p \mathrm{d}\mu .$$

Then

$$N_p(f, \mu) = (\int^* |f|^p \mathrm{d}\mu)^{1/p} = (\int |f|^p \mathrm{d}\mu)^{1/p} .$$

8.13 COROLLARY. *For every dominated measure $\boldsymbol{m}$ defined on $\mathscr{K}_E(T)$ and for every $\boldsymbol{m}$-integrable function f from $\mathscr{L}_E^1(\boldsymbol{m})$ or from $\mathscr{L}^1(\boldsymbol{m})$, we have*

$$|\int f \mathrm{d}\boldsymbol{m}| \leqslant \int |f| \mathrm{d}|\boldsymbol{m}|, \ |\int |f| \mathrm{d}\boldsymbol{m}| \leqslant \int |f| \mathrm{d}|\boldsymbol{m}| \ \textit{and} \ |\int f \mathrm{d}|\boldsymbol{m}|| \leqslant \int |f| \mathrm{d}|\boldsymbol{m}| .$$

In fact, for every function f from $\mathscr{L}_E^1(\boldsymbol{m})$ or from $\mathscr{L}^1(\boldsymbol{m})$ we have

$$|\int f \mathrm{d}\boldsymbol{m}| \leqslant N_1(f, \boldsymbol{m}) = N_1(f, |\boldsymbol{m}|)$$

$$|\int |f| \mathrm{d}\boldsymbol{m}| \leqslant N_1(|f|, \boldsymbol{m}) = N_1(f, |\boldsymbol{m}|)$$

$$|\int f \mathrm{d}|\boldsymbol{m}|| \leqslant N_1(f, |\boldsymbol{m}|)$$

and

$$N_1(f, |\boldsymbol{m}|) = \int |f| \mathrm{d}|\boldsymbol{m}| .$$

8.14 PROPOSITION. *Let μ and ν be two positive measures. If $\mu \leqslant \nu$ and if the function $f: T \to E$ is ν-integrable, then f is μ-integrable and*

$$\int |f| \, \mathrm{d}\mu \leqslant \int |f| \mathrm{d}\nu .$$

In fact, since $\nu \geqslant \mu$, we have $\mathscr{L}_E^1(\nu) \subset \mathscr{L}_E^1(\mu)$; hence if f is ν-integrable, then f is μ-integrable. In this case we have

$$\int |f| \mathrm{d}\mu = \int^* |f| \mathrm{d}\mu \leqslant \int^* |f| \mathrm{d}\nu = \int |f| \mathrm{d}\nu .$$

Let E and F be two (not necessarily distinct) Banach spaces and $\boldsymbol{m} : \mathscr{K}_E \to F$ a *dominated* measure. In particular $\boldsymbol{m}$ can be a scalar measure and $\boldsymbol{m} = |\boldsymbol{m}|$.

8.15 Proposition. *Let $(\boldsymbol{f}_n)$ be a sequence of $\boldsymbol{m}$-integrable functions from $\mathscr{L}_E^1 \boldsymbol{m}$ (or from $\mathscr{L}^1(\boldsymbol{m})$). If all the functions $\boldsymbol{f}_n$ have their supports contained in the same compact set $K \subset T$ and if the sequence $(\boldsymbol{f}_n)$ converges uniformly on T to a function $\boldsymbol{f}_0$, then $\boldsymbol{f}_0$ is $\boldsymbol{m}$-integrable and*

$$\int \boldsymbol{f}_0 \, \mathrm{d}\boldsymbol{m} = \lim \int \boldsymbol{f}_n \, \mathrm{d}\boldsymbol{m} \, .$$

In fact, from proposition 7.13, it follows that $\boldsymbol{f}_0 \in \mathscr{L}_E^1(|\boldsymbol{m}|)$ or $\boldsymbol{f}_0 \in \mathscr{L}^1(|\boldsymbol{m}|)$ and that the sequence $(\boldsymbol{f}_n)$ converges in the mean to $\boldsymbol{f}_0$. Then it follows that

$$\lim \int \boldsymbol{f}_n \, \mathrm{d}\boldsymbol{m} = \int \boldsymbol{f}_0 \, \mathrm{d}\boldsymbol{m} \, .$$

8.16 Theorem. *Let $(\boldsymbol{f}_n)$ be a sequence of $\boldsymbol{m}$-integrable functions from $\mathscr{L}_E^1(\boldsymbol{m})$ (or from $\mathscr{L}^1(\boldsymbol{m})$). If the sequence $(\boldsymbol{f}_n)$ converges $\boldsymbol{m}$-almost everywhere to a function $\boldsymbol{f}_0$, and if there exists a function $\varphi \geqslant 0$ defined on T such that $\int^* \varphi \, \mathrm{d}|\boldsymbol{m}| < +\infty$ and $|\boldsymbol{f}_n(t)| \leqslant \varphi(t)$ almost everywhere, for each $n \in N$, then the function $\boldsymbol{f}_0$ (defined $\boldsymbol{m}$-almost everywhere on T) is $\boldsymbol{m}$-integrable and*

$$\int \boldsymbol{f}_0 \, \mathrm{d}\boldsymbol{m} = \lim \int \boldsymbol{f}_n \, \mathrm{d}\boldsymbol{m} \, .$$

By Lebesgue's theorem the function $\boldsymbol{f}_0$ is equivalent to a function from $\mathscr{L}_E^1(|\boldsymbol{m}|)$, therefore $\boldsymbol{f}_0$ is $\boldsymbol{m}$-integrable, and the sequence $(\boldsymbol{f}_n)$ converges in the mean to $\boldsymbol{f}_0$. Then it follows that

$$\lim \int \boldsymbol{f}_n \, \mathrm{d}\boldsymbol{m} = \int \boldsymbol{f}_0 \, \mathrm{d}\boldsymbol{m} \, .$$

8.17 Corollary. *Let $(\boldsymbol{f}_n)$ be a sequence of $\boldsymbol{m}$-integrable functions from $\mathscr{L}_E^1(\boldsymbol{m})$ (or from $\mathscr{L}^1(\boldsymbol{m})$). If the series $\Sigma_{1 \leqslant n < \infty} \boldsymbol{f}_n(t)$ converges $\boldsymbol{m}$-almost everywhere to a function $\boldsymbol{f}_0$ and if there exists a function $\varphi \geqslant 0$ defined on T such that $\int^* \varphi \, \mathrm{d}|\boldsymbol{m}| < +\infty$ and $|\Sigma_{k \leqslant n} \boldsymbol{f}_k(t)| \leqslant \varphi(t)$, $|\boldsymbol{m}|$-almost everywhere, for each $n \in N$, then the function $\boldsymbol{f}_0$ (defined $\boldsymbol{m}$-almost everywhere on T) is $\boldsymbol{m}$-integrable and*

$$\int \boldsymbol{f}_0 \, \mathrm{d}\boldsymbol{m} = \Sigma_{1 \leqslant n < \infty} \int \boldsymbol{f}_n \, \mathrm{d}\boldsymbol{m} \, .$$

The preceding theorem is applied to the sequence $\boldsymbol{g}_n = \Sigma_{k \leqslant n} \boldsymbol{f}_k$.

8.18 PROPOSITION. *Let* (f_n) *be an increasing (respectively decreasing) sequence of* $\boldsymbol{m}$*-integrable numerical functions from* $\mathscr{L}^1(\boldsymbol{m})$. *The upper envelope* $f=\sup f_n$ *(respectively the lower envelope* $f=\inf f_n$*) is* $\boldsymbol{m}$*-integrable if and only if* $\sup \int f_n \mathrm{d}|\boldsymbol{m}| < +\infty$ *(respectively* $\inf \int f_n \mathrm{d}|\boldsymbol{m}| > -\infty$*).*

In this case we have

$$\int f \mathrm{d}\boldsymbol{m} = \lim \int f_n \mathrm{d}\boldsymbol{m} .$$

Assume that f is $\boldsymbol{m}$-integrable. Then we have $\sup N_1(f, |\boldsymbol{m}|) < +\infty$, irrespective to whether the sequence (f_n) is increasing or decreasing (theorem 7.35 and corollary 7.37). From the inequality

$$\left|\int f_n \mathrm{d}|\boldsymbol{m}|\right| \leqslant \int |f_n| \mathrm{d}|\boldsymbol{m}| \leqslant N_1(f_n, |\boldsymbol{m}|),$$

it follows that

$$\sup \int f_n \mathrm{d}|\boldsymbol{m}| \leqslant \sup \left|\int f_n \mathrm{d}|\boldsymbol{m}|\right| \leqslant \sup N_1(f_n, |\boldsymbol{m}|) < +\infty$$

and also

$$\inf \int f_n \mathrm{d}|\boldsymbol{m}| \geqslant -\sup \left|\int f_n \mathrm{d}|\boldsymbol{m}|\right| > -\infty .$$

From theorem 7.35 and corollary 7.37, it follows that, in this case, the sequence (f_n) converges in the mean to f, hence

$$\lim \int f_n \mathrm{d}\boldsymbol{m} = \int f \mathrm{d}\boldsymbol{m} .$$

Conversely, assume that the sequence (f_n) is increasing and that $\sup \int f_n \mathrm{d}|\boldsymbol{m}| < +\infty$. Then the functions $g_n = f_n + f_1^-$ are *positive*, $|\boldsymbol{m}|$-integrable and form an increasing sequence, and

$$N_1(g_n, |\boldsymbol{m}|) = \int g_n \mathrm{d}|\boldsymbol{m}| = \int f_n \mathrm{d}|\boldsymbol{m}| + \int f_1^- \mathrm{d}|\boldsymbol{m}| .$$

It follows that

$$\sup N_1(g_n, |\boldsymbol{m}|) = \sup \int f_n \mathrm{d}|\boldsymbol{m}| + \int f_1^- \mathrm{d}|\boldsymbol{m}| < +\infty ,$$

therefore the function $g = \sup g_n = \sup f_n + f_1^- = f + f_1^-$ is $|\boldsymbol{m}|$-integrable. From this we deduce that the function $f = g - f_1^-$ is also $|\boldsymbol{m}|$-integrable.

If (f_n) is decreasing and $\inf \int f_n \mathrm{d}|\boldsymbol{m}| > -\infty$, then the functions $g_n = -f_n$ from an increasing sequence and $\sup \int g_n \mathrm{d}|\boldsymbol{m}| = -\inf \int f_n \mathrm{d}|\boldsymbol{m}| < +\infty$.

From the first part of the proof it follows that the function $g = \sup g_n = -\inf f_n = -f$ is $|\boldsymbol{m}|$-integrable, hence $f = \inf f_n = -g$ is also $|\boldsymbol{m}|$-integrable.

Thus the proposition is proved.

8.19 PROPOSITION. *Let (f_n) be a sequence of $\boldsymbol{m}$-integrable numerical functions from $\mathscr{L}^1(\boldsymbol{m})$. The upper envelope $f = \sup f_n$ is $\boldsymbol{m}$-integrable, if and only if there exists a function $g \geqslant 0$ such that $\int^* g \mathrm{d}|\boldsymbol{m}| < +\infty$ and $f_n \leqslant g$ for every n.*

In fact, we take $p=1$ and $\mu = |\boldsymbol{m}|$ in corollary 7.38.

8.20 PROPOSITION. *Let (f_n) be a sequence of $\boldsymbol{m}$-integrable numerical functions from $\mathscr{L}^1(\boldsymbol{m})$. The lower envelope $f = \inf f_n$ is $\boldsymbol{m}$-integrable if and only if there exists a function $g \leqslant 0$ such that $\int^* |g| \mathrm{d}|\boldsymbol{m}| < +\infty$ and $f_n \geqslant g$ for every n.*

In corollary 7.39 we take $p=1$ and $\mu = |\boldsymbol{m}|$.

8.21 COROLLARY. *If (f_n) is a sequence of $\boldsymbol{m}$-integrable positive functions from $\mathscr{L}^1(\boldsymbol{m})$, then the lower envelope $f = \inf f_n$ is an $\boldsymbol{m}$-integrable function from $\mathscr{L}^1(\boldsymbol{m})$.*

We take $g(t) = 0$ in the preceding proposition.

7 *Characterization of integrable numerical functions*

Let μ be a positive measure on T. In particular, μ can be the modulus $|\boldsymbol{m}|$ of a scalar or vector measure.

8.22 PROPOSITION. *A numerical positive (finite or infinite) lower semicontinuous function $f \geqslant 0$ is μ-integrable if and only if $\int^* f \mathrm{d}\mu < +\infty$.*

The condition $\int^* f \mathrm{d}\mu < +\infty$ is necessary for every μ-integrable function $f \geqslant 0$.

Conversely, assume that $\int^* f \mathrm{d}\mu < +\infty$. From the definition of the upper integral $\int^* f \mathrm{d}\mu = \mu^*(f)$ for functions $f \in \mathscr{I}_+$, it follows that for every $\varepsilon \geqslant 0$, there exists a continuous function $g \geqslant 0$ with compact support, such that

$$g \leqslant f \text{ and } \mu^*(f) \leqslant \mu(g) + \varepsilon .$$

But $f - g$ is $\geqslant 0$ and lower semicontinuous, hence

$$\mu^*(f) = \mu^*(g + f - g) = \mu^*(g) + \mu^*(f - g) ,$$

whence

$$N_1(f-g, \mu) = \mu^*(f-g) = \mu^*(f) - \mu(g) < \varepsilon .$$

It follows that f is μ-integrable.

8.23 COROLLARY. *A positive, finite, upper semicontinuous function $f \geqslant 0$ with compact support is μ-integrable if (and only if) $\int^* f \mathrm{d}\mu < +\infty$.*

If $\int^* f \mathrm{d}\mu < +\infty$, then there exists a function $h \in \mathscr{I}_+$ such that $f \leqslant h$ and $\int^* h \mathrm{d}\mu < +\infty$. The function h is finite μ-almost everywhere. The function $h-f$ is defined everywhere, is lower semicontinuous and we have

$$\int^* (h-f) \mathrm{d}\mu \leqslant \int^* h \mathrm{d}\mu < +\infty ,$$

hence $h-f$ is μ-integrable. Then we have $f(t) = h(t) - (h(t) - f(t))$, μ-almost everywhere; since h and $h-f$ are μ-integrable, it follows that f is also μ-integrable.

Remark. It will be shown later (corollary 10.18 and theorem 11.10) that every upper semicontinuous function $f \geqslant 0$ with $\int^* f \mathrm{d}\mu < +\infty$, is μ-integrable even if its support is not compact.

8.24 THEOREM. *A function $f \geqslant 0$ is μ-integrable if and only if for each number $\varepsilon > 0$, there exist a positive, finite, upper semicontinuous function $g \leqslant f$, with compact support, and a lower semicontinuous μ-integrable function $h \geqslant g$, such that $f(t) \leqslant h(t)$, μ-almost everywhere, and $\int (h-g) \mathrm{d}\mu < \varepsilon$.*

The condition is sufficient: h and g are μ-integrable and $h-f \leqslant h-g$, hence $N_1(h-f) \leqslant N_1(h-g) = \int (h-g) \mathrm{d}\mu < \varepsilon$. It follows that $N_1(f) \leqslant N_1(f) + N_1(h-f) < +\infty$, therefore $f \in \mathscr{F}^1$. It follows also that f is adherent to the closed space $\mathscr{L}^1$, hence f is μ-integrable.

Conversely, assume that f is μ-integrable. There exists a sequence (K_n) of compact sets such that, denoting $A = \cup_{1 \leqslant n < \infty} K_n$ and $f' = f\varphi_A$, we have $f' = f$, μ-almost everywhere. The function f' is also μ-integrable and $\int f' \mathrm{d}\mu = \int f \mathrm{d}\mu$.

Let $\varepsilon > 0$. There exists a function $\varphi \in \mathscr{K}_+(T)$ such that $N_1(f' - \varphi) < \varepsilon/4$, that is $\mu^*(|f' - \varphi|) < \varepsilon/4$. Since $|f' - \varphi|$ vanishes outside the union of a sequence of compact sets, there exists a lower semicontinuous function $\psi \geqslant |f' - \varphi|$ such that $\mu^*(\psi) < \varepsilon/2$. The function ψ is therefore μ-integrable.

We have

$$-\psi(t) \leqslant f'(t) - \varphi(t) \leqslant \psi(t), \text{ for } t \in T .$$

Since $\varphi(t)$ is finite, we deduce

$$\varphi(t)-\psi(t)\leqslant f'(t)\leqslant \varphi(t)+\psi(t), \text{ for } t\in T .$$

Since $f'(t)\geqslant 0$, we even have $(\varphi(t)-\psi(t))^+ \leqslant f(t)\leqslant \varphi(t)+\psi(t)$, for $t\in T$.

The function $g=(\varphi-\psi)^+$ is $\geqslant 0$, upper semicontinuous, finite (since $\psi\geqslant 0$, hence $\varphi-\psi$ does not take on the value $+\infty$) and with compact support (since if $\varphi(t)=0$ and $\psi(t)>0$, then $g(t)=0$); the function $h=\varphi+\psi$ is lower semicontinuous and we have $g\leqslant f'\leqslant h$ and

$$\int (h-g)\,d\mu \leqslant \int [(\varphi+\psi)-(\varphi-\psi)]\,d\mu = 2\int \psi\,d\mu < \varepsilon .$$

Since $f(t)=f'(t)$ μ-almost everywhere, we deduce $f(t)\leqslant h(t)$, μ-almost everywhere.

Thus the theorem is proved.

Remark. From the preceding proof it follows that if f is dominated by an integrable lower semicontinuous function, then h can be chosen such that $h(t)\geqslant f(t)$ everywhere.

8.25 COROLLARY. *Let* $\boldsymbol{m}:\mathscr{K}(T)\to F$ *be a dominated measure.*

For every $|\boldsymbol{m}|$*-integrable numerical function* $f\geqslant 0$*, there exists an increasing sequence* (g_n) *of positive finite upper semicontinuous functions with compact support, and a decreasing sequence* (h_n) *of* μ*-integrable lower semicontinuous positive functions such that*

(1) $g_n(t)\leqslant f(t)$ *and* $g_n(t)\leqslant h_n(t)$*, for every* $t\in T$ *and every* $n\in N$*,* $f(t)\leqslant h_n(t)$*,* μ*-almost everywhere, for each* $n\in N$*;*

(2) $f(t)=\sup g_n(t)$*,* $|\boldsymbol{m}|$*-almost everywhere, and* $f(t)=\inf h_n(t)$*,* $|\boldsymbol{m}|$*-almost everywhere;*

(3) $\int f\,d\boldsymbol{m}=\lim \int g_n\,d\boldsymbol{m}=\lim \int h_n\,d\boldsymbol{m}$.

For every n there exist a finite upper semicontinuous function $\psi_n\geqslant 0$ with compact support and a lower semicontinuous μ-integrable function $\varphi_n\geqslant 0$ such that

$$\psi_n\leqslant f,\quad \psi_n\leqslant \varphi_n,\quad f(t)\leqslant \varphi_n(t),\quad |\boldsymbol{m}|\text{-almost everywhere}$$

and

$$\int (\varphi_n-\psi_n)\,d|\boldsymbol{m}| < n^{-1}.$$

For each n we take

$$g_n=\sup(\psi_1,\ldots,\psi_n) \text{ and } h_n=\inf(\varphi_1,\ldots,\varphi_n).$$

The functions g_n and h_n have the required properties and fulfill condition (1).

Let $g = \sup g_n$ and $h = \inf h_n$. Since $\sup \int^* g_n \mathrm{d}|\boldsymbol{m}| \leqslant \int^* f \mathrm{d}|\boldsymbol{m}| < +\infty$, it follows that g is $|\boldsymbol{m}|$-integrable and

$$\int (f - g_n) \mathrm{d}|\boldsymbol{m}| \leqslant \int (\varphi_n - \psi_n) \mathrm{d}|\boldsymbol{m}| \leqslant n^{-1},$$

hence

$$\int (f - g) \mathrm{d}|\boldsymbol{m}| = \lim \int (f - g_n) \mathrm{d}|\boldsymbol{m}| = 0 .$$

It follows that f and g are equivalent, that is, equal $|\boldsymbol{m}|$-almost everywhere. It can be shown, in the same way, that f and h are equal $|\boldsymbol{m}|$-almost everywhere and thus condition (2) is fulfilled. Condition (3) follows from proposition 8.18:

$$\int f \mathrm{d}\boldsymbol{m} = \int g \, \mathrm{d}\boldsymbol{m} = \lim \int g_n \, \mathrm{d}\boldsymbol{m} ,$$

$$\int f \mathrm{d}\boldsymbol{m} = \int h \, \mathrm{d}\boldsymbol{m} = \lim \int h_n \, \mathrm{d}\boldsymbol{m} .$$

Remark. If the function f is dominated by an integrable lower semicontinuous function, then the function h_n can be chosen such that $f(t) \leqslant h_n(t)$, for every $t \in T$ and every n.

8 *Integrable sets*

Let E and F be two, not necessarily distinct, Banach spaces, both of them over the same field of scalars. In particular one or both E and F can be equal to R or C.

Let $\boldsymbol{m} : \mathscr{K}_E(T) \to F$ be a dominated measure and $\mu = |\boldsymbol{m}|$. In particular $\boldsymbol{m}$ can be a positive measure and in this case $\mu = \boldsymbol{m}$.

8.26 Definition. *We say that a set $A \subset T$ is integrable with respect to $\boldsymbol{m}$, or $\boldsymbol{m}$-integrable, if its characteristic function φ_A is an $\boldsymbol{m}$-integrable function, from $\mathscr{L}^1(\boldsymbol{m})$.*

The integral $\int \varphi_A \mathrm{d}\boldsymbol{m}$ of the characteristic function φ_A with respect to $\boldsymbol{m}$ is called the measure of the set A and is denoted by $\boldsymbol{m}(A)$:

$$\boldsymbol{m}(A) = \int \varphi_A \mathrm{d}\boldsymbol{m} .$$

We remark that the measure $\boldsymbol{m}(A)$ of an $\boldsymbol{m}$-integrable set A is an element of the space $\mathscr{L}(E, F)$.

If m is a (real or complex) scalar measure, then $m(A)$ is a (real respectively complex) scalar. If m is a *positive* measure, then $0 \leqslant m(A) < +\infty$.

We remark also that a scalar measure m can always be extended to the space $\mathscr{K}_E(T)$ with values in E and then the measure $m(A)$ can be considered as an element of $\mathscr{L}(E, E)$, for every Banach space E.

For every $\boldsymbol{m}$-integrable set $A \subset T$ we have

$$|\boldsymbol{m}(A)| \leqslant \mu(A).$$

In fact $|\int \varphi_A \mathrm{d}\boldsymbol{m}| \leqslant \int \varphi_A \mathrm{d}\mu$ and $\boldsymbol{m}(A) = \int \varphi_A \mathrm{d}\boldsymbol{m}$ and $\mu(A) = \int \varphi_A \mathrm{d}\mu$.

If E and F are two Banach spaces with involution and if $A \subset T$ is $\boldsymbol{m}$-integrable, then A is $\bar{\boldsymbol{m}}$-integrable and we have

$$\bar{\boldsymbol{m}}(A) = \overline{\boldsymbol{m}(A)}.$$

8.27 PROPOSITION. *If A is a μ-integrable set, then $\mu(A) = \mu^*(A)$*

In fact, $\mu(A) = \int \varphi_A \mathrm{d}\mu$ and $\mu^*(A) = \int^* \varphi_A \mathrm{d}\mu$.

On the other hand, if A is μ-integrable, then the positive function φ_A is integrable with respect to the positive measure μ, hence $\int \varphi_A \mathrm{d}\mu = \int^* \varphi_A \mathrm{d}\mu$. Then it follows that $\mu(A) = \mu^*(A)$.

8.28 COROLLARY. *A set A is μ-negligible if and only if A is μ-integrable and $\mu(A) = 0$.*

In particular, the empty set $\varnothing$ is μ-integrable and $\mu(\varnothing) = 0$.

Denote by $\mathscr{B}(\boldsymbol{m})$ the set of $\boldsymbol{m}$-integrable sets $A \subset T$. We have $\mathscr{B}(\boldsymbol{m}) = \mathscr{B}(\mu)$. We shall show that $\mathscr{B}(\boldsymbol{m})$ is a *semitribe* (that is a non-empty class of parts, which is closed under differences, finite unions and countable intersections) which contains the semitribe $\mathscr{B}$ of all relatively compact Borel subsets of T.

8.29 PROPOSITION. *If (A_n) is a sequence of μ-integrable sets, then the intersection $\cap_{1 \leqslant n < \infty} A_n$ is μ-integrable. If there exists a set C such that $\mu^*(C) < +\infty$ (in particular, if C is μ-integrable) and such that $A_n \subset C$ for every $n \in N$, then the union $\cup_{1 \leqslant n < \infty} A_n$ is μ-integrable.*

Denote first $A = \cap_{1 \leqslant n < \infty} A_n$. We have $\varphi_A = \inf \varphi_{A_n}$ and the functions φ_{A_n} are positive and μ-integrable.

From corollary 8.21 it follows that φ_A is μ-integrable, hence A is μ-integrable.

Denote now $B = \cup_{1 \leqslant n < \infty} A_n$. We have $\varphi_B = \sup \varphi_{A_n}$, φ_{A_n} are μ-integrable, $\int^* \varphi_C \mathrm{d}\mu = \mu^*(C) < +\infty$ and $\varphi_{A_n} \leqslant \varphi_C$ for every $n \in N$. Then it follows that B is μ-integrable.

8.30 Corollary. *If* (A_n) *is a sequence of* μ*-integrable sets and if* $\Sigma_{1 \leqslant n < \infty} \mu(A_n) < +\infty$, *then the union* $\cup_{1 \leqslant n < \infty} A_n$ *is* μ*-integrable.*

Indeed, taking $C = \cup_{1 \leqslant n < \infty} A_n$ we have $A_n \subset C$ for every $n \in N$ and

$$\mu^*(C) = \mu^*(\cup_{1 \leqslant n < \infty} A_n) \leqslant \Sigma_{1 \leqslant n < \infty} \mu^*(A_n) < +\infty\ .$$

From the preceding proposition it follows that $\cup_{1 \leqslant n < \infty} A_n$ is μ-integrable.

8.31 Corollary. *The intersection and the union of a finite family* $(A_i)_{1 \leqslant i \leqslant n}$ *of* μ*-integrable sets, are* μ*-integrable.*

In fact, taking $A_{n+p} = A_n$ for every $p \in N$, the sequence $(A_i)_{1 \leqslant i < \infty}$ consists of μ-integrable functions, hence the intersection

$$\cap_{1 \leqslant i \leqslant n} A_i = \cap_{1 \leqslant i < \infty} A_i$$

is a μ-integrable set.

On the other hand, taking $A_{n+p} = \varnothing$ for every $p \in N$, the sequence $(A_i)_{1 \leqslant i < \infty}$ consists of μ-integrable functions and

$$\Sigma_{1 \leqslant i < \infty} \mu(A_i) = \Sigma_{1 \leqslant i \leqslant n} \mu(A_i) < +\infty\ .$$

It follows that the set

$$\cup_{1 \leqslant i \leqslant n} A_i = \cup_{1 \leqslant i < \infty} A_i$$

is μ-integrable.

8.32 Proposition. *If* A *and* B *are two* μ*-integrable sets, then* $A \setminus B$ *is integrable.*

Indeed

$$A \setminus B = A \setminus (A \cap B)$$

and $A \cap B \subset A$, therefore $\varphi_{A \setminus B} = \varphi_A - \varphi_{A \cap B}$. The sets A and $A \cap B$ are

μ-integrable, hence the functions φ_A and $\varphi_{A \cap B}$ are μ-integrable. It follows then that $\varphi_{A \setminus B}$ is integrable, hence $A \setminus B$ is μ-integrable.

This proposition and corollary 8.31 show that $\mathscr{B}(\boldsymbol{m})$ is a semitribe.

8.33 PROPOSITION. *The relatively compact Borel sets are integrable for every measure* μ.

In fact, the compact sets $K \subset T$ are μ-integrable, since their characteristic functions φ_K are finite, upper semicontinuous, with compact support and $\mu^*(\varphi_K) < +\infty$. Since the semitribe of the μ-integrable sets contains all the compact sets, it contains all the relatively compact Borel sets.

8.34 PROPOSITION. *An open set* $G \supset T$ *is* μ*-integrable if and only if* $\mu^*(G) < +\infty$.

In fact, the function φ_G is lower semicontinuous, therefore it is μ-integrable if and only if $\mu^*(\varphi_G) < +\infty$.

Remark. We shall prove later (corollary 11.13) that a closed set $A \subset T$ is μ-integrable if and only if $\mu^*(A) < +\infty$.

Now we shall prove that the mapping $A \to \boldsymbol{m}(A)$ is a countably additive set function.

8.35 PROPOSITION. *If* (A_n) *is a sequence of mutually disjoint,* $\boldsymbol{m}$*-integrable sets, and if the union* $\cup_{1 \leqslant n < \infty} A_n$ *is* $\boldsymbol{m}$*-integrable, then*

$$\boldsymbol{m}(\cup_{1 \leqslant n < \infty} A_n) = \Sigma_{1 \leqslant n < \infty} \boldsymbol{m}(A_n) .$$

In fact, if $A = \cup_{1 \leqslant n < \infty} A_n$, then $\varphi_A = \Sigma_{1 \leqslant n < \infty} \varphi_{A_n}$, $\Sigma_{1 \leqslant i \leqslant n} \varphi_{A_i} \leqslant \varphi_A$ for each $n \in N$ and $\int^* \varphi_A \mathrm{d}\mu = \mu(A) < +\infty$. It follows then from corollary 8.17 that

$$\boldsymbol{m}(A) = \Sigma_{1 \leqslant n < \infty} \boldsymbol{m}(A_n) .$$

8.36 COROLLARY. *If* $(A_i)_{1 \leqslant i \leqslant n}$ *is a finite family of mutually disjoint integrable sets, then*

$$\boldsymbol{m}(\cup_{1 \leqslant i \leqslant n} A_i) = \Sigma_{1 \leqslant i \leqslant n} \boldsymbol{m}(A_i) .$$

In fact, the union $\cup_{1 \leqslant i \leqslant n} A_i$ is $\boldsymbol{m}$-integrable.

Taking $A_{n+p}=\varnothing$ for any $p\in N$, $(A_i)_{1\leqslant i<+\infty}$ is a sequence of $\boldsymbol{m}$-integrable sets whose union

$$\cup_{1\leqslant i<\infty}A_i=\cup_{1\leqslant i\leqslant n}A_i$$

is $\boldsymbol{m}$-integrable. It follows then that

$$\boldsymbol{m}(\cup_{1\leqslant i\leqslant n}A_i=\boldsymbol{m}(\cup_{1\leqslant i<\infty}A_i)=$$
$$=\Sigma_{1\leqslant i<\infty}\boldsymbol{m}(A_i)=\Sigma_{1\leqslant i\leqslant n}\boldsymbol{m}(A_i)\,.$$

8.37 PROPOSITION. *If A and B are two $\boldsymbol{m}$-integrable sets and if $A\supset B$, then $\boldsymbol{m}(A\setminus B)=\boldsymbol{m}(A)-\boldsymbol{m}(B)$.*

Indeed, $\varphi_{A\setminus B}=\varphi_A-\varphi_B$, therefore $\int\varphi_{A\setminus B}\mathrm{d}\boldsymbol{m}=\int\varphi_A\mathrm{d}\boldsymbol{m}-\int\varphi_B\mathrm{d}\boldsymbol{m}$ whence

$$\boldsymbol{m}(A\setminus B)=\boldsymbol{m}(A)-\boldsymbol{m}(B)\,.$$

8.38 PROPOSITION. *Let (A_n) be an increasing sequence of $\boldsymbol{m}$-integrable sets. The union $A=\cup_{1\leqslant n<\infty}A_n$ is $\boldsymbol{m}$-integrable if and only if* $\sup\mu(A_n)<+\infty$. *In this case we have*

$$\boldsymbol{m}(\cup_{1\leqslant n<\infty}A_n)=\lim\boldsymbol{m}(A_n)\,.$$

In fact, $\varphi_A=\sup\varphi_{A_n}$ and (φ_{A_n}) is an increasing sequence of $\boldsymbol{m}$-integrable functions. Then the function φ_A is $\boldsymbol{m}$-integrable if and only if $\sup\int\varphi_{A_n}\mathrm{d}\mu<+\infty$ and in this case we have

$$\int\varphi_A\mathrm{d}\boldsymbol{m}=\lim\int\varphi_{A_n}\mathrm{d}\boldsymbol{m}\,.$$

It follows that A is $\boldsymbol{m}$-integrable if and only if $\sup\mu(A_n)<+\infty$ and then we have

$$\boldsymbol{m}(A)=\lim\boldsymbol{m}(A_n)\,.$$

8.39 PROPOSITION. *For every decreasing sequence (A_n) of $\boldsymbol{m}$-integrable sets we have*

$$\boldsymbol{m}(\cap_{1\leqslant n<\infty}A_n)=\lim\boldsymbol{m}(A_n)\,.$$

In fact, if $A=\cap_{1\leqslant n<\infty}A_n$, then $\varphi_A=\inf\varphi_{A_n}$ and (φ_{A_n}) is a decreasing sequence of $\boldsymbol{m}$-integrable positive functions. If follows then that

$$\boldsymbol{m}(A) = \lim \boldsymbol{m}(A_n) .$$

The positive measures have additional properties.

8.40 PROPOSITION. *For every finite or countable family* $(A_i)_{i \in I}$ *of* μ-*integrable sets with* μ-*integrable union, we have*

$$\mu(\cup A_i) \leqslant \Sigma \mu(A_i) .$$

Indeed, the sets A_i and $\cup A_i$ being μ-integrable, we have

$$\mu(A_i) = \mu^*(A_i) \text{ and } \mu(\cup A_i) = \mu^*(\cup A_i) .$$

The proposition follows then from the inequality

$$\mu^*(\cup A_i) \leqslant \Sigma \mu^*(A_i) .$$

8.41 PROPOSITION. *If* A *and* B *are two integrable sets and if* $A \subset B$, *then* $\mu(A) \leqslant \mu(B)$.

Indeed, $\varphi_A \leqslant \varphi_B$ hence $\int \varphi_A \mathrm{d}\mu \leqslant \int \varphi_B \mathrm{d}\mu$ whence $\mu(A) \leqslant \mu(B)$.

8.42 PROPOSITION. *If* (A_n) *is an increasing (respectively decreasing) sequence of* μ-*integrable sets with* μ-*integrable union, then*

$$\mu(\cup A_n) = \lim \mu(A_n) = \sup \mu(A_n)$$

(respectively $\mu(\cap A_n) = \lim \mu(A_n) = \inf \mu(A_n)$.

Indeed, the sequence $(\mu(A_n))$ is increasing (respectively decreasing) and

$$\sup \mu(A_n) = \lim \mu(A_n)$$

(respectively $\inf \mu(A_n) = \lim \mu(A_n)$) .

8.43 THEOREM. *Let* $A \subset T$ *be a set such that*

$$\mu^*(A) = \inf_{G \supset A} \mu^*(G), \ G \text{ open} .$$

The set A *is* μ-*integrable if and only if for every number* $\varepsilon > 0$ *there exists a compact set* K *and a* μ-*integrable open set* G, *such that* $K \subset A \subset G$ *and*

$$\mu(G \setminus K) = \mu(G) - \mu(K) < \varepsilon .$$

Assume first that for every $\varepsilon>0$, there exists a compact set $K_\varepsilon \subset A$ and a μ-integrable open set $G_\varepsilon \supset A$ such that $\mu(G_\varepsilon \setminus A_\varepsilon)<\varepsilon$. Then $\varphi_{K_\varepsilon} \leqslant \varphi_A \leqslant \varphi_{G_\varepsilon}$ therefore

$$\mu^*(\varphi_A-\varphi_{K_\varepsilon}) \leqslant \mu^*(\varphi_{G_\varepsilon}-\varphi_{K_\varepsilon})<\varepsilon$$

whence

$$\mu^*(\varphi_A) \leqslant \mu^*(\varphi_A-\varphi_{K_\varepsilon})+\mu(K_\varepsilon)< +\infty \,.$$

It follows first that $\varphi_A \in \mathscr{F}^1$ and then that φ_A can be approximated in the mean by μ-integrable functions φ_{K_ε}; therefore, φ_A is μ-integrable, and hence A is μ-integrable.

Conversely, assume that A is μ-integrable and let $\varepsilon>0$. There exists an open set $G \supset A$ such that $\mu^*(G)<\mu(A)+\varepsilon/2< +\infty$. The set G is μ-integrable and $\mu(G)-\mu(A)<\varepsilon/2$.

Since φ_A is μ-integrable, there exists an upper semicontinuous function $f \geqslant 0$ with compact support H, such that $f \leqslant \varphi_A$ and $\int (\varphi_A - f)\mathrm{d}\mu<\varepsilon/4$.
It follows that

$$\mu(A) \leqslant \int f \mathrm{d}\mu+\varepsilon/4.$$

Let $\delta>0$ be such that $\delta\mu(A)<\varepsilon/4$ and let $K=\{t \mid t \in H, f(t) \geqslant \delta\}$. The set K is closed and contained in H, hence K is compact. Since $f \leqslant \varphi_A$ we have $K \subset A$. The set $B=A\setminus K$ is μ-integrable and $f \leqslant \varphi_K+\delta\varphi_B$, hence

$$\int f\mathrm{d}\mu \leqslant \mu(K)+\delta\mu(B) \leqslant \mu(K)+\delta\mu(A) \leqslant \mu(K)+\varepsilon/4\,,$$

whence

$$\mu(A) \leqslant \int f\mathrm{d}\mu+\varepsilon/2 \leqslant \mu(K)+\varepsilon/2.$$

It follows then

$$\mu(G\setminus K)=\mu(G)-\mu(K)=\mu(G)-\mu(A)+\mu(A)-\mu(K)<\varepsilon$$

and the theorem is proved.

8.44 Corollary. *Let A be a set contained in the union of a sequence of compact sets. The set A is μ-integrable if and only if for every $\varepsilon>0$ there exists a compact set K and an open μ-integrable set G such that $K \subset A \subset G$ and $\mu(G\setminus K)<\varepsilon$.*

Indeed, in this case

$$\mu^*(A) = \inf_{G \supset A} \mu^*(G), \quad G \text{ open} .$$

In particular, the theorem is valid for relatively compact sets.

8.45 COROLLARY. *For every relatively compact Borel set A and every number* $\varepsilon > 0$, *there exists a compact set* $K \subset A$ *and a relatively compact open set* $G \supset A$ *such that, for every relatively compact Borel set* A' *with* $K \subset A' \subset G$, *we have* $|\boldsymbol{m}(A) - \boldsymbol{m}(A')| < \varepsilon$.

Indeed, the relatively compact Borel set A is μ-integrable. For $\varepsilon > 0$, there exists a compact set $K \subset A$ and an open μ-integrable set $G \supset A$, which can be chosen relatively compact, such that $\mu(G \setminus K) < \varepsilon/2$. If A' is a relatively compact Borel set and $K \subset A' \subset G$ we have

$$\begin{aligned} |\boldsymbol{m}(A) - \boldsymbol{m}(A')| &= |\boldsymbol{m}(A) - \boldsymbol{m}(K) - (\boldsymbol{m}(A') - \boldsymbol{m}(K))| = \\ &= |\boldsymbol{m}(A \setminus K) - \boldsymbol{m}(A' \setminus K)| \leqslant |\boldsymbol{m}(A \setminus K)| + |\boldsymbol{m}(A' \setminus K)| \leqslant \\ &\leqslant \mu(A \setminus K) + \mu(A' \setminus K) \leqslant 2\mu(G \setminus K) < \varepsilon . \end{aligned}$$

Corollary 8.45 states the fact that the measure $\boldsymbol{m}$ is *regular.*

8.46 THEOREM. *A set* $A \subset T$ *is* μ*-integrable if and only if for every number* $\varepsilon > 0$, *there exists a compact set* $K \subset A$ *such that* $\mu^*(A \setminus K) < \varepsilon$.

If for each $\varepsilon > 0$ there exists a compact set $K \subset A$ with $\mu^*(A \setminus K) < \varepsilon$, it can be shown as in theorem 8.43, that A is μ-integrable.

Conversely, assume that A is μ-integrable and let be $\varepsilon > 0$.

Since

$$\mu^*(A) = \sup_K \mu^*(A \cap K), \quad K \text{ compact} ,$$

there exists a compact set L such that

$$\mu^*(A) \leqslant \mu^*(A \cap L) + \varepsilon/2 .$$

The set $A \cap L$ is μ-integrable and relatively compact, hence

$$\mu^*(A \cap L) = \inf_G \mu^*(G), \quad G \text{ open}, \quad G \supset A \cap L .$$

By theorem 8.24, there exists a compact set $K \subset A \cap L$ and an open μ-integrable set $G \supset A \cap L$ such that $\mu(G \setminus K) < \varepsilon/2$. Then $K \subset A$ and

$$\begin{aligned} \mu(A \setminus K) &= \mu(A) - \mu(K) \leqslant \mu(A \cap L) - \mu(K) + \varepsilon/2 \leqslant \\ &\leqslant \mu(G) - \mu(K) + \varepsilon/2 < \varepsilon. \end{aligned}$$

8.47 COROLLARY. *If $A \subset T$ is an integrable set, then there exists a sequence (K_n) of mutually disjoint compact sets contained in A, such that the set $A \setminus \cup K_n$ is μ-negligible. Then we have*

$$\mu(A) = \Sigma_{1 \leqslant n < \infty} \mu(K_n) .$$

We define the sequence (K_n) by recurrence in the following way: we choose $K_1 \subset A$ such that $\mu(A \setminus K_1) < 1$; for every $n > 1$, we choose $K_n \subset A \setminus \cap_{i<n} K_i$ such that

$$\mu[(A - \cup_{i<n} K_i) - K_n] < n^{-1} .$$

This means that

$$\mu(A - \cup_{i \leqslant n} K_i) < n^{-1} \text{ for every } n \in N .$$

The sets K_n are mutually disjoint and

$$\mu^*(A \setminus \cup_{1 \leqslant i < \infty} K_i) \leqslant \mu(A \setminus \cup_{i \leqslant n} K_i) < n^{-1} \text{ for every } n \in N ,$$

therefore

$$\mu^*(A \setminus \cup_{1 \leqslant i < \infty} K_i) = 0 ,$$

that is, the set $N = A - \cup_{1 \leqslant i < \infty} K_i$ is μ-negligible. Then N is μ-integrable, hence the set $A \setminus N = \cup_{1 \leqslant i < \infty} K_i$ is μ-integrable and

$$\mu(A) = \mu(A \setminus N) = \mu(\cup_{1 \leqslant i < \infty} K_i) = \Sigma_{1 \leqslant i < \infty} \mu(K_i) .$$

8.48 COROLLARY. *If $A \subset T$ is a μ-integrable set, then there exists an increasing sequence (K_n) of compact set contained in A such that the set $A \setminus \cup_{1 \leqslant n < \infty} K_n$ is μ-negligible and*

$$\mu(A) = \sup \mu(K_n) = \lim \mu(K_n) .$$

If A is μ-integrable and if $\mu(A) = \inf_{G \supset A} \mu^*(G)$, G open, then there exists a decreasing sequence (G_n) of μ-integrable open sets which contain A, such that the set $\cap_{1 \leqslant n < \infty} G_n \setminus A$ is μ-negligible and

$$\mu(A) = \inf \mu(G_n) = \lim \mu(G_n) .$$

We apply theorems 8.43 and 8.46 for $\varepsilon = n^{-1}$; we obtain a sequence (H_n) of compact sets contained in A, such that $\mu(A) - \mu(H_n) < n^{-1}$, respectively a sequence $U_n \supset A$ of μ-integrable open sets such that $\mu(U_n) - \mu(A) < n^{-1}$. We deduce that

$$\lim \mu(H_n) = \mu(A) \text{ and } \lim \mu(U_n) = \mu(A) .$$

For each n denote $K_n = \cup_{1 \leqslant i \leqslant n} H_i$ respectively $G_n = \cap_{1 \leqslant i \leqslant n} U_i$. The sets (K_n) are compact, contained in A and form an increasing sequence; the sets (G_n) are μ-integrable and open, they contain A and form a decreasing sequence, and we have

$$\mu(H_n) \leqslant \mu(K_n) \leqslant \mu(A), \text{ respectively } \mu(A) \leqslant \mu(G_n) < \mu(U_n);$$

therefore

$$\lim \mu(K_n) = \mu(A) \text{ respectively } \mu(A) = \lim \mu(G_n) .$$

The equalities $\lim \mu(K_n) = \sup \mu(K_n)$ and $\lim \mu(G_n) = \inf \mu(G_n)$ follow from the fact that the sequence $(\mu(K_n))$ is increasing, and the sequence $(\mu(G_n))$ is decreasing. Then we have

$$\mu^*(A \setminus \cup_{1 \leqslant n < \infty} K_n) \leqslant \mu(A \setminus K_n) < n^{-1} \text{ for every } n ,$$

respectively

$$\mu(\cap_{1 \leqslant n < \infty} G_n \setminus A) \leqslant \mu(G_n \setminus A) \leqslant n^{-1}, \text{ for every n } ;$$

therefore

$$\mu^*(A \setminus \cup_{1 \leqslant n < \infty} K_n) = 0 \text{ respectively } \mu^*(\cap_{1 \leqslant n < \infty} G_n \setminus A) = 0$$

whence it follows that the set $A \setminus \cup_{1 \leqslant n < \infty} K_n$, respectively the set $\cap_{1 \leqslant n < \infty} G_n \setminus A$, is μ-negligible.

8.49 Corollary. *For every μ-integrable set $A \subset T$ we have*

$$\mu(A) = \sup_{K \subset A} \mu(K), \ K \text{ compact} .$$

If A is μ-integrable and

$$\mu(A) = \inf_{G \supset A} \mu^*(G), \ G \text{ open}$$

we have

$$\mu(A) = \inf_{G \supset A} \mu(G), \ G \text{ open, } \mu\text{-integrable} .$$

9 *Integrable step functions*

8.50 PROPOSITION. *Let μ be a positive measure. If A is a relatively compact μ-integrable set, then the function φ_A is p-summable for every p such that $1 \leqslant p < +\infty$.*

Let $\varepsilon > 0$; there exists a compact set $K \subset A$ and a μ-integrable open set $G \supset A$ such that $\mu(G) - \mu(K) < \varepsilon^p$. Since A is relatively compact, we can choose the set G to be relatively compact, replacing it, if necessary, by its intersection with a relatively compact open neighbourhood of $\bar{A}$.

There exists a continuous function $\varphi : T \to [0, 1]$ with compact support, equal to 1 on K and equal to zero on $T \setminus G$. Then $|\varphi_A(t) - \varphi(t)| \leqslant 1$ for every $t \in T$ and $\varphi(t) = \varphi_A(t)$ for $t \in K$ and for $t \in T \setminus G$. We deduce that $|\varphi_A(t) - \varphi(t)| \leqslant \varphi_{G-K}(t)$, hence

$$N_p(\varphi_A - \varphi) = (\textstyle\int^* |\varphi_A(t) - \varphi(t)|^p \mathrm{d}\mu)^{1/p} \leqslant$$
$$\leqslant (\textstyle\int \varphi_{G-K}(t) \mathrm{d}\mu)^{1/p} = [\mu(G \setminus K)]^{1/p} < \varepsilon .$$

Therefore φ_A can be approximated in the mean of order p by functions from $\mathscr{K}(T)$. It follows that $\varphi_A \in \mathscr{L}^p(\mu)$.

8.51 COROLLARY. *For every measure μ and every p with $1 \leqslant p < \infty$, the space $\mathscr{L}_E^p(\mu)$ contains the space $\mathscr{E}_E(\mathscr{B})$ of Borel step functions of the form $\Sigma_{1 \leqslant i \leqslant n} \varphi_{A_i} x_i$ with $A_i \in \mathscr{B}$ and $x_i \in E$.*

Indeed, every relatively compact Borel set A is μ-integrable with respect to every positive measure μ. It follows that for $A_i \in \mathscr{B}$ we have $\varphi_{A_i} \in \mathscr{L}^p(\mu)$, therefore for $x_i \in E$ we have $\varphi_{A_i} x \in \mathscr{L}_E^p(\mu)$, whence $\Sigma_{1 \leqslant i \leqslant n} \varphi_{A_i} x_i \in \mathscr{L}_E^p(\mu)$.

8.52 LEMMA. *Let Φ be a clan ($=$ring) of sets of T such that for every point $t \in T$, Φ contains a fundamental system of neighbourhoods of t. Then for every continuous function $f : T \to E$ with compact support K and for every neighbourhood U of K, there exists a sequence of Φ-step functions of the form $\Sigma \varphi_{A_i} x_i$ with $A_i \in \Phi$ and $x_i \in E$ with supports contained in U, converging uniformly to f.*

Let $\varepsilon > 0$. Since f is continuous, for every point $t_0 \in K$, there exists a neighbourhood $V_0 \in \Phi$ of t_0 such that $V_0 \subset U$ and

$$|f(t) - f(t_0)| < \varepsilon/2 \text{ for every } t \in V_0 .$$

If $t, t' \in V_0$, then $|f(t) - f(t')| < \varepsilon$.

Since K is compact, we can choose a finite family $V_1, \ldots, V_n$ of sets from Φ, contained in U, whose interiors cover K, such that for each i

$$|f(t) - f(t')| < \varepsilon \text{ if } t, t' \in V_i .$$

Denote $A_1 = V_1$ and for $1 < k \leqslant n$ define

$$A_k = V_k - \cup_{i<k} A_i .$$

The sets A_k belong to Φ, are contained in U, are mutually disjoint,

$$\cup_{1 \leqslant i \leqslant n} A_i = \cup_{1 \leqslant i \leqslant n} V_i \supset K$$

and $A_i \subset V_i$ for each i, hence

$$|f(t) - f(t')| < \varepsilon \text{ if } t, t' \in A_i .$$

For each i, choose a point $t_i \in A_i$ and take $x_i = f(t_i)$. Define the Φ-step function

$$f_\varepsilon = \Sigma_{1 \leqslant i \leqslant n} \varphi_{A_i} x_i .$$

The function f_ε has the support contained in U and for every $t \in T$ we have $|f(t) - f_\varepsilon(t)| < \varepsilon$. Indeed, if t does not belong to any set A_i, we have $f(t) = 0$ and $f_\varepsilon(t) = 0$; if t belongs to some set A_k, then

$$f_\varepsilon(t) = \Sigma_{1 \leqslant i \leqslant n} \varphi_{A_i}(t) x_i = x_k = f(t_k) ,$$

therefore

$$|f(t) - f_\varepsilon(t)| = |f(t) - f(t_k)| < \varepsilon .$$

Taking $\varepsilon = 1, \frac{1}{2}, \ldots, n^{-1}, \ldots$ we obtain a sequence (f_n) of step functions with supports contained in U, uniformly convergent to f.

Remark. We can choose Φ to be a clan which contains all the compact sets. In this case, the function f can be uniformly approximated by Φ-step functions with supports contained in K as well.

In fact, in the above proof we can choose $A_1 = V_1 \cap K$ and for $1 < k < n$,

$$A_k = (V_k \cap K) - \cup_{i<k} A_i .$$

The sets A_k can be chosen to be differences of compact sets.

In particular, Φ can be the semitribe $\mathscr{B}$ of relatively compact Borel sets.

8.53 PROPOSITION. *Let $\mathscr{C}$ be the clan generated by the compact sets. For every positive measure μ and every $1 \leqslant p < \infty$, the space $\mathscr{E}_E(\mathscr{C})$ of $\mathscr{C}$-step functions is dense in $\mathscr{L}_E^p(\mu)$.*

We remark first that the sets of $\mathscr{C}$ are relatively compact Borel sets; therefore, the functions of $\mathscr{E}_E(\mathscr{C})$ belong to the space $\mathscr{L}_E^p(\mu)$.

Let $\boldsymbol{f} \in \mathscr{K}_E(T)$; there exists a sequence $(\boldsymbol{g}_n)$ of $\mathscr{C}$-step functions, with supports contained in the support K of $\boldsymbol{f}$, uniformly converging to $\boldsymbol{f}$. It follows that the sequence $\boldsymbol{g}_n$ converges in the mean of order p to $\boldsymbol{f}$.

Hence, the space $\mathscr{E}_E(\mathscr{C})$ is dense in $\mathscr{K}_E(T)$ for the topology of the convergence in the mean of order p. Since $\mathscr{K}_E(T)$ is dense in $\mathscr{L}_E^p$ for this topology, we deduce that $\mathscr{E}_E(\mathscr{C})$ is dense in $\mathscr{L}_E^p$.

8.54 COROLARY. *The space $\mathscr{E}_E(\mathscr{B})$ of Borel step functions is dense in $\mathscr{L}_E^p(\mu)$ for every positive measure μ and every p with $1 \leqslant p < \infty$.*

8.55 COROLLARY. *Let $\boldsymbol{m}$ and $\boldsymbol{n}$ be two dominated measures. If we have $\boldsymbol{m}(K) = \boldsymbol{n}(K)$ for every compact set $K \subset T$, then $\boldsymbol{m} = \boldsymbol{n}$.*

We can easily verify that the set of all sets $A \subset T$ which are integrable with respect to $\boldsymbol{m}$ and $\boldsymbol{n}$, such that $\boldsymbol{m}(A) = \boldsymbol{n}(A)$, is a clan which contains the compact sets, hence contains also the clan $\mathscr{C}$ generated by them.

It follows then that for every $\mathscr{C}$-step function $\boldsymbol{g} = \Sigma_{1 \leqslant i \leqslant n} \varphi_{A_i} x_i$ we have

$$\int \boldsymbol{g} \,\mathrm{d}\boldsymbol{m} = \Sigma_{1 \leqslant i \leqslant n} \int \varphi_{A_i} x_i \mathrm{d}\boldsymbol{m} = \Sigma_{1 \leqslant i \leqslant n} \boldsymbol{m}(A_i) x_i = \\ = \Sigma_{1 \leqslant i \leqslant n} \boldsymbol{n}(A_i) x_i = \int \boldsymbol{g} \,\mathrm{d}\boldsymbol{n} \,.$$

The space $\mathscr{E}_E(\mathscr{C})$ is dense in the space $\mathscr{L}_E^1(|\boldsymbol{m}| + |\boldsymbol{n}|)$, for the seminorm $N_1(\boldsymbol{f}, |\boldsymbol{m}| + |\boldsymbol{n}|)$. Then we have

$$|\int \boldsymbol{g} \,\mathrm{d}\boldsymbol{m}| \leqslant \int |\boldsymbol{g}| \mathrm{d}|\boldsymbol{m}| \leqslant \int |\boldsymbol{g}| \mathrm{d}(|\boldsymbol{m}| + |\boldsymbol{n}|) = N_1(\boldsymbol{f}, |\boldsymbol{m}| + |\boldsymbol{n}|)$$

and also

$$|\int \boldsymbol{g} \,\mathrm{d}\boldsymbol{n}| \leqslant N_1(\boldsymbol{g}, |\boldsymbol{m}| + |\boldsymbol{n}|) \,,$$

hence the mappings $\int \boldsymbol{g} \,\mathrm{d}\boldsymbol{m}$ and $\int \boldsymbol{g} \,\mathrm{d}\boldsymbol{n}$ are continuous for the topology of the space $\mathscr{L}_E^1(|\boldsymbol{m}| + |\boldsymbol{n}|)$. As these mappings coincide on the dense subspace $\mathscr{E}_E(\mathscr{C})$ it follows that

$$\int \boldsymbol{f} \mathrm{d}\boldsymbol{m} = \int \boldsymbol{f} \mathrm{d}\boldsymbol{n}$$

for every function $f \in \mathscr{L}_E^1(|\boldsymbol{m}|+|\boldsymbol{n}|)$, in particular for the functions $f \in \mathscr{K}_E(T)$, hence $\boldsymbol{m}=\boldsymbol{n}$.

8.56 LEMMA. *Let $\mathscr{C}$ be a clan. For every finite family $(A_i)_{i\in I}$ of sets of $\mathscr{C}$ there exists a finite family $(B_j)_{j\in J}$ of disjoint subsets of $\mathscr{C}$ such that each set A_i is the union of a certain number of sets B_j.*

We take the family $(B_j)_{j\in J}$ to be the family of all sets of the form $\cap_{i\in I} C_i$, where $C_i = A_i$ at least for one i and if $C_i \neq A_i$, then $C_i = T\setminus A_i$. Each set A_k is the union of those sets B_j for which $C_k = A_k$.

8.57 LEMMA. *Let $\mathscr{C}$ be a clan. Every $\mathscr{C}$-step function $f=\Sigma_i \varphi_{A_i} x_i$ can be written in the form $f=\Sigma_j \varphi_{B_j} y_j$ where B_j are disjoint sets of $\mathscr{C}$.*

Let (B_j) be a finite family of disjoint sets of $\mathscr{C}$ such that each A_i, is the union of a certain number of sets B_j.

For each set B_k, let I_k be the set of those indices i for which $B_k \subset A_i$, and let $y_k = \Sigma_{i\in I_k} x_i$. Then $f=\Sigma\, \varphi_{B_j} y_j$. Indeed, let $t \in \cup A_i$. There exists a unique set $B_k \ni t$. Then $t \in A_i$ for each $i \in I_k$ and $t \notin A_i$ for $i \notin I_k$, hence

$$f(t) = \Sigma_i \varphi_{A_i}(t) x_i = \Sigma_{i\in I_k} \varphi_{A_i}(t) x_i = \Sigma_{i\in I_k} x_i = y_k = \Sigma_j \varphi_{B_j}(t) y_j .$$

8.58 LEMMA. *Let $\mathscr{C}$ be a clan and $\boldsymbol{m}: \mathscr{C} \to \mathscr{L}(E, F)$ an additive function. If $\Sigma_{1\leqslant i\leqslant n} \varphi_{A_i} x_i = 0$, where $A_i \in \mathscr{C}$ and $x_i \in E$, then*

$$\Sigma_{1\leqslant i\leqslant n}\, \boldsymbol{m}(A_i) x_i = 0 .$$

If $A_i = \varnothing$, we have $\boldsymbol{m}(A_i) = \varnothing$, therefore, without loss of generality, we may assume that the sets A_i are nonempty. By lemma 8.56, there exists a family $(B_j)_{j\in J}$ of disjoint sets of $\mathscr{C}$ such that each A_i is the union of a certain number of sets B_j. Again we can assume that the sets B_j are non-empty, by removing the empty ones.

For each i, denote by J_i the set of those indices j for which $B_j \subset A_i$. Hence

$$A_i = \cup_{j\in J_i} B_j ,$$

therefore

$$\varphi_{A_i} = \Sigma_{j\in J_i} \varphi_{B_j} = \Sigma_{j\in J} \alpha_{ij} \varphi_{B_j} ,$$

where $\alpha_{ij}=1$ for $j\in J_i$ and $\alpha_{ij}=0$ for $j\notin J_i$. Hence,

$$0=\Sigma_i \varphi_{A_i} x_i = \Sigma_i (\Sigma_j \alpha_{ij} \varphi_{B_j}) x_i = \Sigma_j (\Sigma_i x_i \alpha_{ij}) \varphi_{B_j},$$

consequently $\Sigma_i x_i \alpha_{ij}=0$ for each j. Then

$$\Sigma_i \boldsymbol{m}(A_i) x_i = \Sigma_i (\Sigma_j \alpha_{ij} \boldsymbol{m}(B_j)) x_i = \Sigma_j (\Sigma_i x_i \alpha_{ij}) \boldsymbol{m}(B_j) = 0 .$$

8.59 COROLLARY. *If* $\Sigma_i \varphi_{A_i} x_i = \Sigma_j \varphi_{B_j} y_j$, *where* $A_i, B_j \in \mathscr{C}$ *and* $x_i, y_j \in E$, *then*

$$\Sigma_i \boldsymbol{m}(A_i) x_i = \Sigma_j \boldsymbol{m}(B_j) y_j .$$

8.60 PROPOSITION. *Let* $\boldsymbol{m} : \mathscr{K}_E(T) \to F$ *be a dominated measure and* $\mu = |\boldsymbol{m}|$. *For every set* $A \in \mathscr{B}$ *we have*

$$\mu(A) = \sup \Sigma_i |\boldsymbol{m}(A_i)| ,$$

where the supremum is taken for all finite families (A_i) *of mutually disjoint relatively compact Borel sets contained in* A.

For every $\boldsymbol{m}$-integrable set $B \subset T$ we have

$$|\boldsymbol{m}(B)| \leqslant \mu(B) .$$

If (A_i) is a finite family of disjoint sets of $\mathscr{B}$ such that $\cup_i A_i \subset A$, then

$$\Sigma |\boldsymbol{m}(A_i)| \leqslant \Sigma \mu(A_i) = \mu(\cup A_i) \leqslant \mu(A)$$

whence follows the inequality

$$\nu(A) = \sup \Sigma |\boldsymbol{m}(A_i)| \leqslant \mu(A) .$$

The set function $\nu(A)$ defined for $A \in \mathscr{B}$ is *the variation* of the set function $\boldsymbol{m}(A)$. It is known (see, for example, N. Dinculeanu, *Vector measures*, Pergamon Press 1967), that ν is countably additive on $\mathscr{B}$. For every scalar step function $\varphi = \Sigma_{1 \leqslant i \leqslant n} \varphi_{A_i} \alpha_i$ of $\mathscr{E}(\mathscr{B})$, put

$$\nu(\varphi) = \Sigma_{1 \leqslant i \leqslant n} \nu(A_i) \alpha_i .$$

$\nu(\varphi)$ does not depend on the particular form in which φ is written as a step function (corollary 8.59). If the sets A_i are mutually disjoint we have

$$|\nu(\varphi)| \leqslant \Sigma_{1 \leqslant i \leqslant n} \nu(A_i) |\alpha_i| \leqslant \Sigma_{1 \leqslant i \leqslant n} \mu(A_i) |\alpha_i| = \mu(|\varphi|) = N_1(\varphi, \mu) .$$

The mapping $\varphi \to \nu(\varphi)$ can be extended to a linear continuous mapping denoted also by ν, of $\mathscr{L}^1(\mu)$ into R. For $\varphi \in \mathscr{L}^1(\mu)$ we have also

$$|\nu(\varphi)| \leqslant N_1(\varphi_1, \mu) = \mu(|\varphi|).$$

In particular, this inequality is valid for functions $\varphi \in \mathscr{K}(T)$, and it proves that ν is a positive measure and $\nu \leqslant \mu$.

At the same time, the inequality

$$|\boldsymbol{m}(A)| \leqslant \nu(A), \text{ for } A \in \mathscr{B}$$

can be extended to the inequality

$$|\int \boldsymbol{f} \mathrm{d}\boldsymbol{m}| \leqslant \int |\boldsymbol{f}| \mathrm{d}\nu$$

for the step functions $\boldsymbol{f} = \Sigma \varphi_{A_i} x_i$ of $\mathscr{E}_E(\mathscr{B})$ with the sets A_i mutually disjoint.

As both sides of the inequality are continuous functions of $\boldsymbol{f}$ (for the seminorm $N_1(\boldsymbol{f}, \mu)$), it follows that the inequality remains valid for functions $\boldsymbol{f} \in \mathscr{L}^1_E(\mu)$, in particular for functions $\boldsymbol{f} \in \mathscr{K}_E(T)$. This means that $\boldsymbol{m}$ is dominated by ν. As $\mu = |\boldsymbol{m}|$, μ is the smallest positive measure which dominates $\boldsymbol{m}$. It follows that $\mu \leqslant \nu$, therefore $\mu = \nu$, that is,

$$\mu(A) = \sup \Sigma |\boldsymbol{m}(A_i)|$$

10 *Integration of vector functions with respect to vector measures*

8.61 THEOREM. *Let X, E and F be three Banach spaces and $(x, y) \to x \cdot y$ a continuous bilinear mapping of $X \times E$ into F.*

If $\boldsymbol{\mu} : \mathscr{K}(T) \to X$ is a dominated measure, then there exists one and only one dominated measure $\boldsymbol{m} : \mathscr{K}_E(T) \to F$ which satisfies the equality

$$\boldsymbol{m}(\varphi y) = \boldsymbol{\mu}(\varphi) y, \text{ for } \varphi \in \mathscr{K}(T) \text{ and } y \in E.$$

The correspondence $\boldsymbol{\mu} \to \boldsymbol{m}$ is linear

If $|x \cdot y| \leqslant |x|\,|y|$ for every $x \in X$ and $y \in E$ then

$$|\boldsymbol{m}| \leqslant |\boldsymbol{\mu}|.$$

If $|x| = \sup_{|y| \leqslant 1} |xy|$ for every $x \in X$, in particular if $X \subset \mathscr{L}(E, F)$, then

$$|\boldsymbol{m}| = |\boldsymbol{\mu}|$$

For every step function $\boldsymbol{g} = \Sigma_{1 \leqslant i \leqslant n} \varphi_{A_i} y_i$ of $\mathscr{E}_E(\mathscr{B})$ put

$$\boldsymbol{m}(\boldsymbol{g}) = \Sigma_{1 \leqslant i \leqslant n} \boldsymbol{\mu}(A_i) y_i.$$

From corollary 8.59 it follows that $\boldsymbol{m}(\boldsymbol{g})$ does not depend on the particular form in which $\boldsymbol{g}$ is written as a step function. It is easily verified that the mapping $\boldsymbol{g} \to \boldsymbol{m}(\boldsymbol{g})$ of $\mathscr{E}_E(\mathscr{B})$ into F is linear.

Every step function $\boldsymbol{g} \in \mathscr{E}_E(\mathscr{B})$ can be written in the form $\boldsymbol{g} = \Sigma_{1 \leqslant i \leqslant n} \varphi_{A_i} y_i$ such that the sets A_i are mutually disjoint. Then

$$|\boldsymbol{g}| = \Sigma_{1 \leqslant i \leqslant n} \varphi_{A_i} |y_i|$$

is a step function of $\mathscr{E}(\mathscr{B})$, and if $a > 0$ is a number such that $|xy| \leqslant a|x|\,|y|$ for every $x \in X$ and $y \in E$, then we have

$$|\boldsymbol{m}(\boldsymbol{g})| = |\Sigma_{1 \leqslant i \leqslant n} \boldsymbol{\mu}(A_i) x_i| \leqslant a \Sigma_{i \leqslant 1 \leqslant n} |\boldsymbol{\mu}(A_i)|\,|x_i| \leqslant$$

$$\leqslant a \Sigma_{1 \leqslant i \leqslant n} |\boldsymbol{\mu}|(A_i)|x_i| = a \int \Sigma_{1 \leqslant i \leqslant n} \varphi_{A_i} |x_i| \mathrm{d}|\boldsymbol{\mu}| =$$

$$= a \int |\boldsymbol{g}| \mathrm{d}|\boldsymbol{\mu}| = a N_1(\boldsymbol{g}, |\boldsymbol{\mu}|) .$$

It follows that the mapping $\boldsymbol{g} \to \boldsymbol{m}(\boldsymbol{g})$ of $\mathscr{E}_E(\mathscr{B})$ into F is continuous for the topology defined by the seminorm $N_1(\boldsymbol{g}), |\boldsymbol{\mu}|)$. As $\mathscr{E}_E(\mathscr{B})$ is dense in the space $\mathscr{L}_E^1(|\boldsymbol{\mu}|)$ for this topology, the mapping $\boldsymbol{m} : \mathscr{E}_E(\mathscr{B}) \to F$ can be extended uniquely to a continuous linear mapping, denoted still by $\boldsymbol{m}$, of the space $\mathscr{L}_E^1(|\boldsymbol{\mu}|)$ into F, which still satisfies the inequality

$$|\boldsymbol{m}(\boldsymbol{f})| \leqslant a N_1(\boldsymbol{f}, |\boldsymbol{\mu}|) \text{ for } \boldsymbol{f} \in \mathscr{L}_E^1(|\boldsymbol{\mu}|) .$$

In particular, for the functions $\boldsymbol{f} \in \mathscr{K}_E(T)$ we have

$$|\boldsymbol{m}(\boldsymbol{f})| \leqslant a \int |\boldsymbol{f}| \mathrm{d}|\boldsymbol{\mu}| = \int |\boldsymbol{f}| \mathrm{d}(a|\boldsymbol{\mu}|) ;$$

therefore, the restriction of $\boldsymbol{m}$ to $\mathscr{K}_E(T)$ is a linear mapping dominated by the positive measure $a|\boldsymbol{\mu}|$, hence $\boldsymbol{m}$ is a *dominated measure*, and $|\boldsymbol{m}| \leqslant a|\boldsymbol{\mu}|$.

Now let $\varphi = \Sigma_{1 \leqslant i \leqslant n} \varphi_{A_i} \alpha_i$ be a scalar Borel step function and $y \in E$. The function

$$\varphi y = \Sigma \varphi_{A_i} \alpha_i y$$

is a Borel step function of $\mathscr{E}_E(\mathscr{B})$ and we have

$$\boldsymbol{m}(\varphi y) = \Sigma \boldsymbol{\mu}(A_i) \alpha_i y = (\Sigma \boldsymbol{\mu}(A_i) \alpha_i) y =$$

$$= (\int \Sigma \varphi_{A_i} \alpha_i \mathrm{d}\boldsymbol{\mu}) y = (\int \varphi \mathrm{d}\boldsymbol{\mu}) y = \boldsymbol{\mu}(\varphi) y .$$

If $\varphi \in \mathscr{L}^1(|\boldsymbol{\mu}|)$, then there exists a sequence (φ_n) of functions of $\mathscr{E}(\mathscr{B})$ converging to φ for the seminorm $N_1(\varphi, |\boldsymbol{\mu}|)$. Then the sequence $(\varphi_n y)$ converges to φy for the seminorm $N_1(\varphi), |\boldsymbol{\mu}|)$. Since $\boldsymbol{m}$ and $\boldsymbol{\mu}$ are continuous for the topologies defined by this seminorm on $\mathscr{L}_E^1(|\boldsymbol{\mu}|)$ and $\mathscr{L}^1(\boldsymbol{\mu})$ respectively we have

$$\lim \boldsymbol{m}(\varphi_n y) = \boldsymbol{m}(\varphi) y, \text{ and } \lim \boldsymbol{\mu}(\varphi_n) y = \boldsymbol{\mu}(\varphi) y .$$

Since for the functions φ_n we have the equality $\boldsymbol{m}(\varphi_n y) = \boldsymbol{\mu}(\varphi_n) y$, passing to the limit we obtain

$$\boldsymbol{m}(\varphi y) = \boldsymbol{\mu}(\varphi) y .$$

In particular, this equality is valid for every function $\varphi \in \mathscr{K}(T)$ and every $y \in E$.

If $\boldsymbol{m}' : \mathscr{K}_E(T) \to F$ is a dominated measure which also satisfies the equality

$$\boldsymbol{m}'(\varphi y) = \boldsymbol{\mu}(\varphi) y, \text{ for } \varphi \in \mathscr{K}(T) \text{ and } y \in E ,$$

then $\boldsymbol{m}' = \boldsymbol{m}$ by corollary 1.22.

The linearity of the correspondence $\boldsymbol{\mu} \to \boldsymbol{m}$ is easily verified.

If $|x \cdot y| \leqslant |x|\,|y|$ for every $x \in X$ and $y \in E$, then taking $a = 1$ in the above proof, we deduce that $|\boldsymbol{m}| \leqslant |\boldsymbol{\mu}|$.

Finally, if for each $x \in X$ we have $|x| = \sup_{|y| \leqslant 1} |xy|$, then X can be isometrically embedded in $\mathscr{L}(E, F)$. In this case, to the measure $\boldsymbol{m} : \mathscr{K}_E(T) \to F$ there corresponds a unique dominated measure $\boldsymbol{\mu}' : \mathscr{K}(T) \to \mathscr{L}(E, F)$, which satisfies the equality

$$\boldsymbol{m}(\varphi y) = \boldsymbol{\mu}'(\varphi) y, \text{ for } \varphi \in \mathscr{K}(T) \text{ and } y \in E ,$$

and $|\boldsymbol{\mu}'| \leqslant |\boldsymbol{m}|$. It follows that $\boldsymbol{\mu}(\varphi) y = \boldsymbol{\mu}'(\varphi)\, y$ for every $y \in E$ and $\varphi \in \mathscr{K}(T)$, therefore $\boldsymbol{\mu}(\varphi) = \boldsymbol{\mu}'(\varphi)$ for every $\varphi \in \mathscr{K}(T)$, that is $\boldsymbol{\mu} = \boldsymbol{\mu}'$. We deduce that in this case the correspondence $\boldsymbol{\mu} \to \boldsymbol{m}$ is injective and $|\boldsymbol{\mu}| \leqslant |\boldsymbol{m}|$, hence $|\boldsymbol{m}| = |\boldsymbol{\mu}|$.

Remark. By theorem 8.61, we can consider that the dominated measures are defined only on $\mathscr{K}(T)$ and then they are "extended" to $\mathscr{K}_E(T)$.

8.62 THEOREM. *Assume that E and F are two Banach spaces with involution and let* $\boldsymbol{m} : \mathscr{K}(T) \to \mathscr{L}(E, F)$ *be a dominated measure. We have* $\mathscr{L}_E^1(\overline{\boldsymbol{m}}) = \mathscr{L}_E^1(\boldsymbol{m})$ *and*

$$\overline{\int \boldsymbol{f} \,\mathrm{d}\boldsymbol{m}} = \int \overline{\boldsymbol{f}} \,\mathrm{d}\overline{\boldsymbol{m}}, \textit{ for } \boldsymbol{f} \in \mathscr{L}_E^1(\boldsymbol{m}) .$$

Indeed, $|\overline{\boldsymbol{m}}| = |\boldsymbol{m}|$, therefore $\mathscr{L}_E^1(\overline{\boldsymbol{m}}) = \mathscr{L}_E^1(\boldsymbol{m})$. Let now $\boldsymbol{f} \in \mathscr{L}_E^1(\boldsymbol{m})$ and let

$(\boldsymbol{f}_n)$ be a sequence of functions of $\mathscr{K}_E(T)$ converging in the mean to $\boldsymbol{f}$. Then we have

$$\lim \int \boldsymbol{f}_n \mathrm{d}\boldsymbol{m} = \int \boldsymbol{f} \mathrm{d}\boldsymbol{m} \quad \text{and} \quad \lim \int \boldsymbol{f}_n \mathrm{d}\overline{\boldsymbol{m}} = \int \boldsymbol{f} \mathrm{d}\overline{\boldsymbol{m}} .$$

But, for functions $\boldsymbol{f}_n$ from $\mathscr{K}_E(T)$ we have (see remark after proposition 1.18)

$$\overline{\int \boldsymbol{f}_n \mathrm{d}\boldsymbol{m}} = \int \overline{\boldsymbol{f}}_n \mathrm{d}\overline{\boldsymbol{m}}$$

whence, passing to the limit, we deduce

$$\overline{\int \boldsymbol{f} \mathrm{d}\boldsymbol{m}} = \int \overline{\boldsymbol{f}} \mathrm{d}\overline{\boldsymbol{m}} .$$

8.63 COROLLARY. *Let E be a Banach space with involution and μ a real measure on T. For every function $\boldsymbol{f} \in \mathscr{L}_E^1(\mu)$ we have*

$$\overline{\int \boldsymbol{f} \mathrm{d}\mu} = \int \overline{\boldsymbol{f}} \mathrm{d}\mu .$$

Chapter III

MEASURABLE FUNCTIONS. THE SPACE $\mathscr{L}^\infty$

§9 Measurable functions and sets

1 *Definition of measurable functions*

Let μ be a *positive* measure on T and let X be an arbitrary topological space.

9.1 DEFINITION. *A function $f: T \to X$ is said to be measurable with respect to the measure μ, or μ-measurable, if for every compact set $K \subset T$ and every $\varepsilon > 0$, there exists a compact set $K_1 \subset K$ such that the restriction of f to K_1 is continuous and $\mu(K \setminus K_1) < \varepsilon$.*

From the definition, it follows first that every continuous function $f: T \to X$ is μ-measurable for every positive measure μ. We shall prove later (theorem 11.10) that every p-summable (for the measure μ) function is μ-measurable.

If $\boldsymbol{m}$ is a dominated measure, we say that a function $f: T \to X$ is $\boldsymbol{m}$-measurable if f is $|\boldsymbol{m}|$-measurable.

9.2 PROPOSITION. *If the function $f: T \to X$ is μ-measurable, then for every μ-integrable set $A \subset T$ and every $\varepsilon > 0$, there exists a compact set $K_1 \subset A$ such that the restriction of f to K_1 is continuous and $\mu(A \setminus K_1) < \varepsilon$.*

Let A be a μ-integrable set and $\varepsilon > 0$. Since A is μ-integrable, there exists a compact set $K \subset A$ such that $\mu(A \setminus K) < \varepsilon/2$. Since f is μ-measurable, there exists a compact set $K_1 \subset K$ such that the restriction of f to K_1 is continuous and $\mu(K \setminus K_1) < \varepsilon/2$. Then:

$$A \setminus K_1 = (A \setminus K) \cup (K \setminus K_1) \text{ and } (A \setminus K) \cap (K \setminus K_1) = \varnothing ,$$

therefore

$$\mu(A \setminus K_1) = \mu(A \setminus K) + \mu(K \setminus K_1) < \varepsilon$$

and the proposition is proved.

9.3 Proposition. *Let $f: T \to X$ be a function. If for every compact set $K \subset T$ and every $\varepsilon > 0$, there exists a μ-integrable set $A_1 \subset K$ such that the restriction of f to A_1 is continuous and $\mu(K \setminus A_1) < \varepsilon$, then f is μ-measurable.*

Let $K \subset T$ be a compact set and $\varepsilon > 0$; let $A_1 \subset K$ be a μ-integrable set such that the restriction of f to A_1 is continuous and $\mu(K \setminus A_1) < \varepsilon/2$. Since A_1 is μ-integrable, there exists a compact set $K_1 \subset A_1$ such that $\mu(A_1 \setminus K_1) < \varepsilon/2$.

Then $K_1 \subset K$, the restriction of f to K_1 is continuous and

$$\mu(K \setminus K_1) = \mu(K \setminus A_1) + \mu(A_1 \setminus K_1) < \varepsilon,$$

therefore f is μ-measurable.

9.4 Proposition. *A function $f: T \to X$ is μ-measurable if and only if for every μ-integrable set $A \subset T$ and every $\varepsilon > 0$, there exists a μ-integrable set $A_1 \subset A$ such that the restriction of f to A_1 is continuous and $\mu(A \setminus A_1) < \varepsilon$.*

Assume first that f is μ-measurable. Let $A \subset T$ be a μ-integrable set and $\varepsilon > 0$. By proposition 9.2, there exists then a compact set $K_1 \subset A$ such that the restriction of f to K_1 is continuous and $\mu(A \setminus K_1) < \varepsilon$. Taking $A_1 = K_1$, the condition of the statement is satisfied.

Conversely, assume that this condition is satisfied for every μ-integrable set $A \subset T$. In particular, it is satisfied for every compact set K. By proposition 9.3, the function f is μ-measurable and the proposition is proved.

9.5 Proposition. *A function $f: T \to X$ is μ-measurable if and only if for every compact set $K \subset T$ there exists a μ-negligible set $N \subset K$ and a sequence (K_n) of mutually disjoint compact sets, with union equal to $K \setminus N$, such that the restriction of f to each set K_n is continuous.*

Assume first that f is μ-measurable. Let $K \subset T$ be a compact set. We can construct, by induction, a sequence (K_n) of mutually disjoint compact sets, such that the restriction of f to each set K_n is continuous and

$$\mu(K \setminus \cup_{1 \leqslant i \leqslant n} K_i) < n^{-1}, \text{ for each } n.$$

Indeed, since f is μ-measurable, there exists a compact set $K_1 \subset K$ such that the restriction of f to K_1 is continuous and $\mu(K \setminus K_1) < 1$. Assuming that we have found the compact sets $K_1, K_2, \ldots, K_{n-1}$ contained in K, the set $K \setminus \cup_{1 \leq i \leq n-1} K_i$ is μ-integrable; therefore, there exists a compact set $K_n \subset K \setminus \cup_{1 \leq i \leq n-1} K_i$ such that the restriction of f to K_n is continuous and

$$\mu((K \setminus \cup_{1 \leq i \leq n-1} K_i) - K_n) < n^{-1}.$$

The set K_n is disjoint from $K_1, \ldots, K_{n-1}$ and

$$(K \setminus \cup_{1 \leq i \leq n-1} K_i) \setminus K_n = K \setminus \cup_{1 \leq i \leq n} K_i,$$

therefore

$$\mu(K \setminus \cup_{1 \leq i \leq n} K_i) < n^{-1}.$$

If we denote $N = K \setminus \cup_{1 \leq i < \infty} K_i$, we have

$$\mu^*(N) = \mu^*(K \setminus \cup_{1 \leq i < \infty} K_i) \leq \mu^*(K \setminus \cup_{1 \leq i \leq n} K_i) < n^{-1}, \text{ for each } n,$$

therefore $\mu(N) = 0$, that is, N is μ-negligible, and the first part of the proposition is proved.

Conversely, assume that the condition of the statement is satisfied.

Let $K \subset T$ be a compact set, let $N \subset K$ be a negligible set and (K_n) a sequence of mutually disjoint compact sets such that the restriction of f to each set K_n is continuous and $\cup_{1 \leq n < \infty} K_n = K \setminus N$. We have

$$\mu(K) = \mu(K \setminus N) = \mu(\cup_{1 \leq n < \infty} K_n) = \Sigma_{1 \leq n < \infty} \mu(K_n) < +\infty.$$

Let $\varepsilon > 0$; there exists a number n such that $\Sigma_{i > n} \mu(K_i) < \varepsilon$. The set $H = K_1 \cup K_2 \cup \ldots \cup K_n$ is compact, the restriction of f to H is continuous (since the sets $K_1, \ldots, K_n$ are disjoint and compact and the restrictions of f to $K_1, \ldots, K_n$ are continuous). We have then

$$\mu(K \setminus H) = \mu(K \setminus \cup_{1 \leq i \leq n} K_i = \mu(N \cup_{i > n} K_i) = \Sigma_{i > n} \mu(K_i) < \varepsilon.$$

It follows that f is μ-measurable and the proposition is proved.

Reasoning as above we can prove the following three propositions.

9.6 PROPOSITION. *If the function $f: T \to X$ is μ-measurable, then for every μ-integrable set $A \subset T$, there exists a μ-negligible set $N \subset A$ and a sequence (K_n) of mutually disjoint compact sets with union equal to $A \setminus N$ such that the restriction of f to each set K_n is continuous.*

The proof uses Proposition 9.2.

9.7 PROPOSITION. *If the function $f: T \to X$ has the property that for every compact set $K \subset T$ there exists a μ-negligible set $N \subset K$ and a sequence (A_n) of mutually disjoint μ-integrable sets, with union equal to $A \setminus N$ and such that the restriction of f to each A_n is continuous, then f is μ-measurable.*

We use Proposition 9.3.

9.8 PROPOSITION. *A function $f: T \to X$ is μ-measurable if and only if for every μ-integrable set $A \subset T$, there exists a μ-negligible set $N \subset A$ and a sequence (A_n) of mutually disjoint μ-integrable sets with union equal to $A \setminus N$, such that the restriction of f to each set A_n is continuous.*

We use Proposition 9.4.

9.9 PROPOSITION. *Let μ and ν be two positive measures on T. A function $f: T \to X$ is $\mu + \nu$-measurable if and only if it is μ-measurable and ν-measurable.*

This proposition follows from the fact that a set N is $\mu + \nu$-negligible if and only if it is μ-negligible and ν-negligible.

9.10 PROPOSITION. *Let μ and ν be two positive measures. If $\nu \leqslant \mu$ then every μ-measurable function $f: T \to X$, is ν-measurable.*

Indeed, every μ-negligible set is ν-negligible.

2 *Composition of measurable functions*

9.11 PROPOSITION. *Let $(X_\alpha)_{\alpha \in I}$ be a finite or countable family of topological spaces. If for each $\alpha \in I$, $f_\alpha : T \to X_\alpha$ is a μ-measurable function, then the function $f(t) = (f_\alpha(t))_{\alpha \in I}$ defined on T with values in the product $X = \Pi_{\alpha \in I} X_\alpha$ is μ-measurable.*

Let $K \subset T$ be a compact set and $\varepsilon > 0$. For each $\alpha \in I$ let n_α be a number >0 such that

$$\Sigma_{\alpha \in I} n_\alpha < 1 .$$

For each α there exists a compact set $K_\alpha \subset K$ such that the restriction of the function f_α to K_α is continuous and $\mu(K \setminus K_\alpha) < \varepsilon n_\alpha$. The set $K_0 =$

$\cap_{\alpha\in I} K_\alpha$ is compact, contained in K, the restrictions of all functions f_α to K_0 are continuous; therefore, the restriction of the function f to K_0 is continuous, and

$$\mu(K\setminus K_0)\leqslant \mu(\cup_{\alpha\in I}(K\setminus K_\alpha)\leqslant \Sigma_{\alpha\in I}\,\mu(K\setminus K_\alpha)\leqslant \Sigma_{\alpha\in I}\,\varepsilon n_\alpha<\varepsilon\,.$$

It follows that the function f is μ-measurable and the proposition is proved.

9.12 PROPOSITION. *Let X and Y be two topological spaces. If $f: T\to X$ is a μ-measurable function and if $u: X\to Y$ is a continuous function, then the composition $u\circ f: T\to Y$ is μ-measurable.*

Let $K\subset T$ be a compact set and $\varepsilon>0$. There exists a compact set $K_1\subset K$ such such that the restriction of f to K_1 is continuous and $\mu(K\setminus K_1)<\varepsilon$. Then the restriction of $u\circ f$ to K_1 is continuous. It follows that $u\circ f$ is μ-measurable.

9.13 COROLLARY. *Let $(X_\alpha)_{\alpha\in I}$ be a finite or countable family of topological spaces, $X=\Pi_{\alpha\in I}X_\alpha$ and Y a topological space. If for each $\alpha\in I, f_\alpha: T\to X_\alpha$ is a μ-measurable function and if $u: X\to Y$ is a continuous function, then the function $t\to u[(f_\alpha(t))_{\alpha\in I}]$ defined on T with values in Y is μ-measurable.*

In fact, the function $t\to f(t)=(f_\alpha(t))_{\alpha\in I}$ defined on T with values in X is μ-measurable, hence the function $u\circ f$ is μ-measurable.

3 *Measurable vector functions*

9.14 PROPOSITION. *Let F, G, E be three topological vector spaces. If $(u, v)\to u\cdot v$ is a continuous mapping of $F\times G$ into E and if the functions $\boldsymbol{f}: T\to F$ and $\boldsymbol{g}: T\to G$ are μ-measurable, then the function $\boldsymbol{f}\cdot\boldsymbol{g}: T\to E$ is μ-measurable.*

It follows immediately from proposition 9.12 as a particular case.

9.15 COROLLARY. *Let E be a topological vector space. If $\boldsymbol{f}, \boldsymbol{g}: T\to E$ are μ-measurable functions and φ is a μ-measurable scalar function, then the function $\boldsymbol{f}+\boldsymbol{g}$ and $\varphi\boldsymbol{f}$ are μ-measurable.*

If E is a normed algebra, then the function $\boldsymbol{fg}$ is μ-measurable.

In fact, the mappings $(u, v)\to u+v$ and $(\alpha, u)\to\alpha u$ *of* $E\times E$ into E respectively

$R \times E$ (or $C \times E$) into E are continuous. If E is a normed algebra, the mapping $(u, v) \to u \cdot v$ is continuous.

In particular, taking $\varphi(t) = \alpha$, it follows that if f is μ-measurable the function αf is μ-measurable.

The set of the μ-measurable functions $\boldsymbol{f}: T \to E$ is a vector space and a module on the space of the μ-measurable scalar functions. If E is a normed algebra, the set of the μ-measurable functions $\boldsymbol{f}: T \to E$ is an algebra.

9.16 COROLLARY. *Let E be a topological vector space and E′ its dual. If the functions* $\boldsymbol{f}: T \to E$ *and* $\boldsymbol{g}: T \to E'$ *are μ-measurable, then the scalar function* $t \to \langle \boldsymbol{f}(t), \boldsymbol{g}(t) \rangle$ *is μ-measurable.*

In particular, the functions $t \to \langle \boldsymbol{f}(t), z' \rangle$ and $t \to \langle z, \boldsymbol{g}(t) \rangle$ are μ-measurable for every $z' \in E'$ and $z \in E$.

In fact, the functions $\boldsymbol{f}'(t) = z$ and $\boldsymbol{g}'(t) = z'$ are continuous, being constant, consequently they are μ-measurable.

9.17 PROPOSITION. *Let E be a normed space. If* $\boldsymbol{f}: T \to E$ *is a μ-measurable function, then the positive function* $|\boldsymbol{f}|$ *is μ-measurable.*

In fact, the function $\mu: E \to R$ defined by $u(z) = |z|$ for $z \in E$, is continuous.

9.18 PROPOSITION. *A complex function* $f: T \to C$ *is μ-measurable if and only if the real part* Re f *and the imaginary part* Im f *are μ-measurable real functions.*

If f is μ-measurable, then $f_1 = \mathrm{Re}\, f$ and $f_2 = \mathrm{Im}\, f$ are μ-measurable since the functions $z \to \mathrm{Re}\, z$ and $z \to \mathrm{Im}\, z$ are continuous.

Conversely, if f_1 and f_2 are μ-measurable, then the function $f = f_1 + if_2$ is μ-measurable.

4 *Measurable numerical functions*

9.19 PROPOSITION. *The upper envelope and the lower envelope of a finite family of μ-measurable numerical (not necessarily finite) functions, are μ-measurable functions.*

In fact, the mappings $(u, v) \to \sup(u, v)$ and $(u, v) \to \inf(u, v)$ of $\overline{R} \times \overline{R}$ into $\overline{R}$ are continuous.

9.20 PROPOSITION. *If f is a μ-measurable numerical (not necessarily finite) function, then the function $|f|$ is μ-measurable.*

In fact, the mapping $x \to |x|$ of $\bar{R}$ into $\bar{R}$ is continuous.

9.21 PROPOSITION. *Let f and g be two μ-measurable numerical functions. If the function $f+g$ (respectively $f-g, fg, f/g, f^g$) is defined at every point $t \in T$, then it is μ-measurable.*

If $f+g$ is defined on T, then the image A of T by the mapping $t \to (f(t), g(t))$ does not contain the points $(+\infty, -\infty)$ and $(-\infty, +\infty)$ hence the restriction of the mapping $(u, v) \to u+v$ to A is continuous, consequently $f+g$ is μ-measurable. Similar proof for the other functions.

9.22 COROLLARY. *A numerical function f is μ-measurable if and only if f^+ and f^- are μ-measurable.*

Note that the positive functions f^+ and f^- do not take on simultaneously the value $+\infty$, hence $f^+ - f^-$ is defined at every point of T. If f^+ and f^- are μ-measurable, then the function $f = f^+ - f^-$ is μ-measurable.

Conversely, if f is μ-measurable, then the functions $f^+ = \sup(f, 0)$ and $f^- = \sup(-f, 0)$ are μ-measurable

5 *Measurable sets*

Let μ be a positive measure on T.

9.23 DEFINITION. *We say that a set $A \subset T$ is μ-measurable if its characteristic function φ_A is μ-measurable.*

9.24 PROPOSITION. *A set $A \subset T$ is μ-measurable if and only if for every compact set $K \subset T$ and every $\varepsilon > 0$, there exists a compact set $K_1 \subset K$ such that $\mu(K \setminus K_1) < \varepsilon$ and such that the sets $K_1 \cap A$ and $K_1 \setminus A$ are compact.*

Assume first that A is μ-measurable, hence that φ_A is μ-measurable. Let $K \subset T$ be a compact set and $\varepsilon > 0$. There exists a compact set $K_1 \subset K$ such that $\mu(K \setminus K_1) < \varepsilon$ and the restriction of φ_A to K_1 is continuous. Then the set $K_1' = \varphi_A^{-1}(1) \cap K_1 = A \cap K_1$ is closed, hence compact, and also the set $K_1'' =$

$\varphi_A^{-1}(0) \cap K_1 = K_1 \setminus A$ is compact. Conversely, assume that the stated condition is fulfilled. We have $\varphi_A(t) = 1$ for $t \in K_1 \cap A$ and $\varphi_A(t) = 0$ for $t \in K_1 \setminus A$, consequently the restriction of φ_A to each of the sets $K_1 \cap A$ and $K_1 \setminus A$ is continuous. As these sets are compact and disjoint, it follows that the restriction of φ_A to the union K_1 of these sets is also continuous, hence φ_A is μ-measurable, consequently A is μ-measurable.

9.25 PROPOSITION. *A set $A \subset T$ is μ-measurable if and only if for every compact set $K \subset T$, there exists a μ-negligible set $N \subset K$ and a sequence (K_n) of mutually disjoint compact sets, with union $K \setminus N$, such that each set K_n is contained either in $K \cap A$ or in $K \setminus A$.*

To prove this we use proposition 9.5 and the fact that the restriction of the function φ_A to a compact set H is continuous, if and only if the sets $H \cap A$ and $H \setminus A$ are compact.

9.26 COROLLARY. *Every μ-integrable set $A \subset T$ is μ-measurable.*

In fact, let $K \subset T$ be a compact set. The sets $K \cap A$ and $K \setminus A$ are μ-integrable. There exists then two μ-negligible sets $N' \subset K \cap A$ and $N'' \subset K \setminus A$ and two sequences (K_n') and (K_n'') of mutually disjoint compact sets, such that

$$N' \cup (\cup_{1 \leqslant n < \infty} K_n') = K \cap A \quad \text{and} \quad N'' \cup (\cup_{1 \leqslant n < \infty} K_n'') = K \setminus A .$$

Then the set $N = N' \cup N'' \subset K$ is μ-negligible and $K \setminus N$ is the union of the sequences (K_n') and (K_n'') of mutually disjoint compact sets such that each set of these sequences is contained either in $K \cap A$ or in $K \setminus A$. It follows that A is μ-measurable.

In particular, every μ-negligible set is μ-measurable.

9.27 PROPOSITION. *A set $A \subset T$ is μ-measurable if and only if $A \cap K$ is μ-integrable, for every compact set $K \subset T$.*

Assume first that A is μ-measurable. Let $K \subset T$ be a compact set. There exists a sequence of mutually disjoint compact sets (K_n), and a μ-negligible set $N \subset K \cap A$ such that $(K \cap A) \setminus N = \cup_{1 \leqslant n < \infty} K_n$. We have

$$\Sigma_{1 \leqslant n < \infty} \mu(K_n) = \sup_n \Sigma_{1 \leqslant i \leqslant n} \mu(K_i) = \sup_n \mu(\cup_{1 \leqslant i \leqslant n} K_i) \leqslant \mu(K) < +\infty ,$$

hence $\cup_{1 \leqslant n < \infty} K_n$ is μ-integrable. As N is also μ-integrable and $K \cap A = N \cup (\cup_{1 \leqslant n < \infty} K_n)$, it follows that the set $K \cap A$ is also μ-integrable.

Conversely, assume that $A \cap K$ is μ-integrable, for every compact set $K \subset T$.

Let $K \subset T$ be a compact set. The set $K \cap A$ being μ-integrable, the set $K \setminus A$ is also μ-integrable. Then each of the sets $K \cap A$ and $K \setminus A$ is the union of a μ-negligible set and of a sequence of mutually disjoint compact sets. It follows that the set K is the union of a μ-negligible set and of a sequence of mutually disjoint compact sets, each of them contained either in $K \cap A$ or in $K \setminus A$. Thus A is μ-measurable and the proposition is proved.

9.28 COROLLARY. *Every μ-measurable set contained in a μ-integrable set is μ-integrable.*

Let A be a μ-measurable set and let $B \supset A$ be a μ-integrable set. There exists a sequence (K_n) of mutually disjoint compact sets and a μ-negligible set N, with union B. Since A is μ-measurable, the sets $A \cap K_n$ are μ-integrable and mutually disjoint, the set $A \cap N$ is μ-negligible, consequently μ-integrable, and their union is A.

The set $A \setminus N = \cup_{1 \leqslant n < \infty} K_n$ is integrable, hence

$$\Sigma_{1 \leqslant n < \infty} \mu(K_n) = \mu(\cup_{1 \leqslant n < \infty} K_n) = \mu(A \setminus N) < +\infty .$$

Then

$$\Sigma_{1 \leqslant n < \infty} \mu(A \cap K_n) + \mu(A \cap N) \leqslant \Sigma_{1 \leqslant n < \infty} \mu(K_n) < +\infty .$$

It follows that the set $(A \cap N) \cup (\cup_{1 \leqslant n < \infty} (A \cap K_n)) = A$ is μ-integrable and the proposition is proved.

9.29 COROLLARY. *The open sets and the closed sets are measurable for every measure μ.*

In fact, if A is closed or open and K is compact, then the set $A \cap K$ is a relatively compact Borel set, hence it is integrable for every measure μ. Then A is μ-measurable for every measure μ.

In particular, the whole space T is measurable for every measure μ.

We shall denote by $\mathscr{B}_{\lambda}$ the class of the Borel sets, that is, sets $A \subset T$ which have the property that $A \cap B \in \mathscr{B}$ for every $A \in \mathscr{B}$.

9.30 COROLLARY. *The Borel sets of $\mathscr{B}_{\lambda}$ are measurable for every measure μ.*

In fact, if $A \in \mathscr{B}_\lambda$ and K is compact, the set $A \cap K$ is relatively compact Borel set, hence it is integrable for every measure μ.

9.31 PROPOSITION. *The union and the intersection of a sequence (A_n) of μ-measurable sets are μ-measurable sets.*

Let K be a compact set. The sets $K \cap A_n$ are μ-integrable and they are contained in the integrable set K. It follows that the union and the intersection of the sets $K \cap A_n$ are integrable. From the relation

$$K \cap (\cap A_n) = \cap (K \cap A_n),$$

it follows that the set $K \cap (\cap A_n)$ is μ-integrable, hence $\cap A_n$ is μ-measurable.

From the relation

$$K \cap (\cup A_n) = \cup (K \cap A_n)$$

it follows that the set $K \cap (\cup A_n)$ is μ-integrable, hence $\cup A_n$ is μ-measurable and the proposition is proved.

9.32 COROLLARY. *The set of the μ-measurable subsets is a tribe ($=\sigma$-ring) which contains the whole space T.*

In fact, if A and B are measurable, then $A \setminus B$ is measurable, since $\varphi_{A \setminus B} = \varphi_A - \varphi_A \varphi_B$.

9.33 COROLLARY. *Let X be a topological space and the function $f: T \to X$. If f is μ-measurable, then the inverse image by f of an open or closed set of X is a μ-measurable subset of T.*

Let $A \subset X$ be a closed set. Let $K \subset T$ be a compact set. Since f is μ-measurable, there exists a μ-negligible set $N \subset K$ and a partition (K_n) of $K \setminus N$, consisting of compact sets, such that the restriction of f to each K_n is continuous. The intersection $K_n \cap f^{-1}(A)$ is the inverse image of A by the restriction of f to K_n. As this restriction is continuous and A is closed, it follows that the set $K_n \cap f^{-1}(A)$ is closed in K_n, hence it is compact.

Thus, $K \cap f^{-1}(A)$ is the union of the μ-negligible set $N \cap f^{-1}(A)$ and of the mutually disjoint compact sets $K_n \cap f^{-1}(A)$. It follows that $f^{-1}(A)$ is μ-measurable.

If A is open, then $B = T \setminus A$ is closed and $f^{-1}(A) = T \setminus f^{-1}(B)$; since $f^{-1}(B)$ is μ-measurable, it follows that $f^{-1}(A)$ is also μ-measurable.

9.34 PROPOSITION. *Let X be a separated topological space and $f\colon T\to X$ a function taking on a finite set of values, $f(T)=\{a_1, a_2, \dots, a_n\}$. The function f is μ-measurable if and only if the sets $A_i=f^{-1}(a_i)$ are μ-measurable.*

Assume first that the set A_i are μ-measurable. Let $K\subset T$ be a compact set. For each i the set $K\cap A_i$ is μ-integrable; there exists then a μ-negligible set $N_i\subset K\cap A_i$ and a partition $(K_{in})_{n\in N}$ of $(K\cap A_i)\setminus N_i$ consisting of compact sets. The restriction of f to each K_i is constant, hence it is continuous. As the sets $K\cap A_i$ are mutually disjoint and their union is K, it follows that the set K is the union of the μ-negligible set $N=\cup_{1\leqslant i\leqslant m}N_i$ and of the countable family $(K_{in})_{n\in N, 1\leqslant i\leqslant m}$ of mutually disjoint sets, such that the restriction of f to each K_{in} is continuous. Thus, f is μ-measurable.

Conversely, if f is μ-measurable, the sets $A_i=f^{-1}(a_i)$ are measurable since, X being separated, the sets $\{a_i\}$ are closed.

The functions $f\colon T\to X$ which take on only a finite number of values are called *step-functions*, as in the case when X is a vector space.

If the sets $A_i=f^{-1}(a_i)$ are Borel sets, f will be called a *Borel-step-function*. Evidently, the Borel step-functions are measurable for every measure.

6 Measurable functions defined on measurable sets

9.35 PROPOSITION. *Let $A\subset T$ be a μ-measurable set and $g: T\to X$ a μ-measurable function. If the function $f: T\to X$ is equal to g on A and is constant on $T\setminus A$, then f is μ-measurable.*

Let $K\subset T$ be a compact set. Since A is μ-measurable, the sets $K\cap A$ and $K\setminus A$ are μ-integrable. As g is μ-measurable, the μ-integrable set $K\cap A$ is the union of a μ-negligible set N and of a sequence $(A_n)_{1\leqslant n<+\infty}$ of mutually disjoint μ-integrable sets such that the restriction of g to each A_n is continuous. Since on $K\cap A$ f and g coincide, it follows that the restriction of f to each A_n is continuous. The set $A_0=K\setminus A$ is μ-integrable, disjoint from the sets A_n and the restriction of f to A_0 is constant, consequently it is continuous. Thus, K is the union of the μ-negligible set N and of the sequence $(A_n)_{0\leqslant n<+\infty}$ of mutually disjoint μ-integrable sets such that the restriction of f to each A_n is continuous. It follows that f is μ-integrable and the proposition is proved.

9.36 DEFINITION. *Let $A \subset T$ be a μ-measurable set. We say that a function $f: A \to X$ is μ-measurable, if extending it to a function $f': T \to X$ which is constant on $T \setminus A$, the function f' is μ-measurable.*

From the preceding proposition it follows that the definition is independent of the constant value taken by the extention of f on $T \setminus A$.

The properties of the measurable functions defined on T are preserved for the measurable functions defined on a subset $A \subset T$.

9.37 PROPOSITION. *If $f: T \to X$ is a μ-measurable function and $A \subset T$ is a μ-measurable set, then the restriction of f to A is μ-measurable.*

Denote by g an extension to T, of the restriction of f to A, such that g is constant on $T \setminus A$. For $t \in A$ we have $g(t) = f(t)$. It follows that g is μ-measurable (proposition 9.35), hence the restriction of g to A is μ-measurable.

9.38 PROPOSITION. *If f and g are two μ-measurable numerical functions, then the functions $f+g, f-g, fg, f/g, f^g$ are μ-measurable on the sets on which they are defined.*

7 *The localization principle*

9.39 PROPOSITION (*Localization principle*). *If the function $f: T \to X$ has the property that for every point $s \in T$ there exists a neighbourhood V_s of s and a μ-measurable mapping $g_s: V_s \to X$ such that $f(t) = g_s(t)$, μ-almost everywhere on V_s, then f is μ-measurable.*

Note first that for each point $s \in T$ there exists an open, relatively compact neighbourhood U_s of s, contained in V_s; we have $f(t) = g_s(t)$, μ-almost everywhere on U_s.

Let K be a compact set. There exists a finite number of relatively compact open sets $U_1, U_2, \ldots, U_n$, which cover K, and a finite number of μ-measurable functions $g_1, g_2, \ldots, g_n$, such that for each i, $1 \leqslant i \leqslant n$, we have $f(t) = g_i(t)$, μ-almost everywhere on U_i.

Put $A_i = K \cap U_i$. We have $K_n = \cup_{1 \leqslant i \leqslant n} A_i$; the sets A_i are Borel sets and $f(t) = g_i(t)$, μ-almost everywhere on A_i. There exists a partition of K consisting of a finite number B_j, $1 \leqslant j \leqslant m$, of relatively compact Borel sets, such that each A_i is the union of a certain number of sets B_j. We attach to each set B_j one of the functions g_i for which $B_j \subset A_i$, and we denote it by h_j. The functions h_j are μ-measurable and for each j we have $f(t)=h_j(t)$, μ-almost everywhere on B_j. Let N'_j be the μ-negligible set of B_j such that for $t \in B_j \setminus N'_j$ we have $f(t)=h_j(t)$. Each set $B_j \setminus N'_j$ is μ-integrable; since h_j is measurable, there exists a μ-negligible set $N''_j \subset B_j \setminus N'_j$ and a sequence $(K_{nj})_{n \in N}$ of compact sets with union $B_j \setminus (N'_j \cup N''_j)$ such that the restriction of h_j to each set K_{nj} is continuous.

Since on each set K_{nj} the function f is equal to the function h_j, it follows that the restriction of f to K_{nj} is continuous. Denoting $N = \cup_{1 \leqslant j \leqslant m}(N'_j \cup N''_j)$, the set N is μ-negligible, and the countable family $(K_{nj})_{n \in N, 1 \leqslant j \leqslant m}$ of mutually disjoint compact sets is a partition of $K \setminus N$ and the restriction of f to each K_{nj} is continuous. It follows then that f is μ-measurable. Thus the proposition is proved.

9.40 COROLLARY. *If the function $g: T \to X$ is μ-measurable then every function $f: T \to X$ which is equal to g, μ-almost everywhere, is μ-measurable.*

In fact, the set $M = \{t \mid t \in T, f(t) = g(t)\}$ is negligible; choosing for each point $s \in T$ a relatively compact open neighbourhood V_s, the set $V_s \cap M$ is μ-negligible, hence $f(t)=g(t)$, μ-almost everywhere on V_s. The sets V_s are μ-integrable, and taking $g_s = g$ for each $s \in T$, the functions g_s are μ-measurable. From the preceding proposition it follows then that f is μ-measurable.

9.41 COROLLARY. *If the function $f: T \to X$ has the property that $f\varphi_K$ is μ-measurable for every compact set $K \subset T$, then f is μ-measurable.*

For each point $s \in T$, let V_s be a compact neighbourhood of s and $g_s = f\varphi_{V_s}$. By hypothesis, g_s is μ-measurable. Since $g_s(t) = f(t)$ for $t \in V_s$, it follows that f is μ-measurable.

§10 Sequences of measurable functions

1 *Functions with values in metric spaces*

Let μ be a positive measure on T.

10.1 THEOREM (*Egoroff*). *Let E be a metric space and (f_n) a sequence of*

μ-measurable functions defined on T with values in E. If the $\lim f_n(t) = f(t)$ *exists, μ-almost everywhere on T, then for every compact set $K \subset T$ and every $\varepsilon > 0$, there exists a compact set $K_1 \subset K$ such that $\mu(K \setminus K_1) < \varepsilon$ and such that the restrictions of the functions f_n to K_1 are continuous and converge uniformly on K_1 to f.*

Let $K \subset T$ be a compact set and $\varepsilon > 0$. Since the mapping $t \to (f_n(t))$ of T into the product E^N is μ-measurable, there exists a compact set $K_0 \subset K$ such that $\mu(K \setminus K_0) < \varepsilon/2$ and the restriction of the function $t \to (f_n(t))$ to K_0 is continuous. This means that the restriction of all functions to K_0 are continuous. Denote by d the distance on E.

For each pair (p, q) of natural numbers and every natural number r denote

$$A_{p,q}(r) = \{t \mid t \in K_0,\ d(f_p(t), f_q(t)) \geqslant r^{-1}\}\ .$$

Since the restrictions of f_p and f_q to K_0 are continuous, and the distance $(x, y) \to d(x, y)$ is continuous on $E \times E$, it follows that the set $A_{p,q}(r)$ is closed in K_0, hence it is compact. For each pair (n, r) of natural numbers denote

$$B_n(r) = \cup_{p \geqslant n,\, q \geqslant n} A_{p,q}(r)\ .$$

The set $B_n(r)$ is μ-integrable, as it is the union of a countable family of μ-integrable sets contained in the μ-integrable set K_0.

A point $t \in K_0$ belongs to $B_n(r)$ if and only if there exists $p \geqslant n$ and $q \geqslant n$ such that $d(f_p(t), f_q(t)) \geqslant r^{-1}$.

Let $t \in K_0$ be a point at which $f_n(t) \to f(t)$. There exists a number $n \in N$ such that for every $p \geqslant n$ and $q \geqslant n$ we have $d(f_p(t), f_q(t)) < r^{-1}$, hence $t \notin B_n(r)$. Since the sequence $(f_n(t))$ converges to $f(t)$ almost everywhere on K_0, it follows that the intersection $\cap_{1 \leqslant n < \infty} B_n(r)$ is μ-negligible, since it does not contain any point of convergence of the sequence (f_n). The sequence $(B_n(r))$ being decreasing, we deduce

$$\lim_n \mu(B_n(r)) = 0.$$

There exists an integer $n(r)$ such that $\mu(B_{n(r)}(r)) < \varepsilon/2^{r+2}$. Let $B = \cup_{1 \leqslant r < \infty} B_{n(r)}(r)$. The set B is μ-integrable, as it is the union of the μ-integrable sets $B_{n(r)}(r)$ contained in the compact set K_0, and we have

$$\mu(B) \leqslant \Sigma_{1 \leqslant r < \infty} \mu(B_{n(r)}(r)) \leqslant \varepsilon/4.$$

The set $C=K_0\setminus B$ is μ-integrable and the sequence (f_n) converges uniformly to f on C. There exists a compact set $K_1\subset C$ such that $\mu(C\setminus K_1)<\varepsilon/4$. Thus $K_1\subset K$, $\mu(K\setminus K_1)=\mu(K\setminus K_0)+\mu(K_0\setminus C)+\mu(C\setminus K_1)<\varepsilon$, and the sequence (f_n) also converges uniformly on K_1. Thus the theorem is proved.

10.2 COROLLARY. *Let E be a metric space and (f_n) a sequence of μ-measurable mappings of T into E. If the sequence $(f_n(t))$ has a limit $f(t)$, μ-almost everywhere on E, then the function f (defined μ-almost everywhere) is μ-measurable.*

In fact, for every compact set $K\subset T$ and every $\varepsilon>0$, there exists a compact set $K_1\subset K$ such that the restriction of the functions f_n to K_1 are continuous, f_n is uniformly convergent on K_1 to f and $\mu(K\setminus K_1)<\varepsilon$. It follows that the limit f is continuous on K_1, hence f is μ-measurable.

10.3 THEOREM. *Let E be a metric space. A function $f: T\to E$ is μ-measurable if and only if for every compact set $K\subset T$, there exists a sequence (g_n) of Borel step functions with values in E, converging to f, μ-almost everywhere on K.*

If for each compact set $K\subset T$ there exists a sequence (g_n) of μ-measurable step function converging to f, μ-almost everywhere on K, then the restriction of f to K is μ-measurable. As K is arbitrary, it follows that f is μ-measurable.

Conversely, assume that f is μ-measurable. Let $K\subset T$ be a compact set. There exists a μ-negligible set $N\subset K$ and a partition $(K_m)_{m\in N}$ of $K\setminus N$, consisting of compact sets, such that the restriction of f to each K_m is continuous. For each n we shall define the Borel step function g_n in the following way:

Consider the sets K_i with $i\leqslant n$. For each set K_i there exists a partition (A_{ij}) of K_i, consisting of Borel sets, such that for $t', t''\in A_{ij}$ we have $d(f(t'), f(t''))\leqslant n^{-1}$. In fact, for $t_0\in K_i$ there exists an open neighbourhood V of t_0 such that for $t\in V$ we have $d(f(t)), f(t_0))<2^{-1}n^{-1}$, since the restriction of f to K_i is continuous. Then for $t', t''\in V$ we have

$$d(f(t'),f(t''))\leqslant d(f(t'), f(t_0))+d(f(t_0),f(t''))<n^{-1}.$$

There exists a finite number of open sets V_j which cover K_i, such that for $t', t''\in V_j$ we have $d(f(t'),f(t''))\leqslant n^{-1}$. There exists then a finite family of mutually disjoint Borel sets whose union is K_i such that each set A_{ij} is contained in a set V_j. It follows that for $t', t''\in A_{ij}$ we have also $d(f(t'), f(t''))\leqslant n^{-1}$. In each set A_{ij} take a point t_{ij} and denote $a_{ij}=f(t_{ij})\in E$. Let also $a\in E$. We define the function g_n by

$$g_n(t) = a_{ij}, \text{ if } t \in A_{ij} \quad (i \leqslant n),$$
$$g_n(t) = a, \text{ if } t \notin \cup_{i \leqslant n} K_i .$$

We have then $d(g_n(t), f(t)) \leqslant n^{-1}$ for $t \in \cup_{i \leqslant n} K_i$. If follows that the sequence (g_n) converges to f on the set $\cup_{1 \leqslant i < \infty} K_i = K \setminus N$, that is g_n tends to f, μ-almost everywhere on K. Thus the theorem is proved.

Remark. If F is a normed space, taking $a=0$ we can get the functions g_n with compact support.

10.4 Corollary. *Let E be a metric space. If the locally compact space T is countable at infinity, every μ-measurable function $f: T \to E$ is the limit μ-almost everywhere on T of a sequence (g_n) of Borel step functions.*

The space T is the union of a sequence (B_n) of compact sets. The sets $C_n = \cup_{1 \leqslant i \leqslant n} B_i$ are compact and form an increasing sequence, and their union is T. The sets

$$D_1 = C_1, D_2 = C_2 \setminus C_1, \ldots, D_n = C_n \setminus C_{n-1}, \ldots$$

are μ-integrable and mutually disjoint.

Since f is μ-measurable, each set D_n is the union of a negligible set and of a sequence of mutually disjoint compact sets such that the restriction of f to each of them is continuous.

Thus, the whole space T can be written as a union of a μ-negligible set N and of a sequence (K_n) of mutually disjoint compact sets such that the restriction of f to each set K_n is continuous. The reasoning goes on as in the preceding proposition.

Remark. If E is a normed space, the functions g_n can be chosen with compact support.

10.5 Theorem. *Let E be a metric space. A function $f: T \to E$ is μ-measurable if and only if it satisfies the following conditions:*

(a) *for every closed sphere $S \subset E$, the set $f^{-1}(S)$ is μ-measurable,*

(b) *for every compact set $K \subset T$, there exists a μ-negligible set $N \subset K$ and a countable set $H \subset E$ such that $f(K \setminus N) \subset \bar{H}$.*

Assume first that f is μ-measurable. Then the inverse image by the function f of every closed set of E is a μ-measurable set, hence condition (a) is fulfilled.

Let $K \subset T$ be a compact set. There exists a μ-negligible set $N \subset K$ and a sequence (g_n) of Borel functions converging to f on $K \setminus N$. If we take $H = \cup g_n(T)$, then H is countable and $f(t) = \lim g_n(t) \in \bar{H}$, for every $t \in K \setminus N$; that is $f(K \setminus N) \subset \bar{H}$ and thus condition (b) is also fulfilled.

Conversely, assume that f fulfils conditions (a) and (b).

Let $K \subset T$ be a compact set. By condition (b), there exists a μ-negligible set $N \subset K$ and a countable set $H = \{a_1, a_2, \ldots, a_n \ldots\} \subset E$ such that $f(K \setminus N) \subset \bar{H}$. Instead of the function f we take a function g, equal to f on $K \setminus N$, such that $g(K) \subset \bar{H}$. For example, we can take the function g constant on N, with $g(N) \in H$. The functions f and g are equal μ-almost everywhere on K.

For each pair (n, p) of natural numbers, denote $A_{n,p} = \{t \mid t \in K, d(g(t), a_n) \leqslant p^{-1}\}$. Thus, $A_{n,p}$ is the inverse image by g of the closed sphere S with center a_n and radius p^{-1}. The set $A_{n,p}$ differs from the set $f^{-1}(S)$ by a set of measure zero. But by condition (a), $f^{-1}(S)$ is μ-measurable. It follows that $A_{n,p}$ is also measurable. We have $\cup_{1 \leqslant n < \infty} A_{n,p} = K$. For p fixed, we define the sequence $(B_{n,p})_{n \in N}$ of mutually disjoint μ-measurable sets, with union K, in the following way: $B_{1,p} = A_{1,p}$ and for $n > 1$,

$$B_{n,p} = A_{n,p} \setminus \cup_{i<n} A_{i,p}\,.$$

We define the function $g_{k,p}$ by

$$g_{k,p}(t) = a_i \text{ if } t \in B_{i,p}, \quad 1 \leqslant i \leqslant k,$$
$$g_{k,p}(t) = b \in H \text{ if } t \notin \cup_{i<k} B_{i,p}.$$

The functions $g_{k,p}$ are measurable step functions and

$$\lim_k g_{k,p}(t) = a_n \text{ if } t \in B_{n,p}, \quad n \geqslant 1$$
$$\lim_k g_{k,p}(t) = b \text{ if } t \notin K\,.$$

The function $g_p(t) = \lim_k g_{k,p}(t)$ is μ-measurable and

$$\lim_p g_p(t) = g(t) \text{ for } t \in K,$$
$$\lim_p g_p(t) = b \quad \text{for } t \notin K\,.$$

It follows that the restriction of g to K is μ-measurable. As $f(t) = g(t)$, μ-almost everywhere on K, the restriction of the function f to K is also μ-measurable. By the localization principle we deduce that f is μ-measurable. Thus the theorem is proved.

10.6 Corollary. *Let E be a metric space of countable type (that is, there exists* *a countable everywhere dense set). A function $f: T \to E$ is μ-measurable if and only if the inverse image by f of every closed sphere $S \subset E$ is a μ-measurable set.*

Remarks

1. In theorem 10.5 and in corollary 10.6, condition (a) can be replaced by the condition that the inverse image by f of every open sphere of E is μ-measurable.

2. Condition (a) may be weakened by asking that the inverse image by f of the spheres of rational radius to be μ-measurable.

10.7 PROPOSITION. *If E is a compact metric space, every μ-measurable function $f: T \to E$ is the uniform limit on T of a sequence of μ-measurable step (not necessarily Borel) functions.*

Since E is compact, for each natural n, there exists a finite number of closed spheres S_i of radius n^{-1} which cover E. Since f is μ-measurable, the sets $A_i = f^{-1}(S_i)$ are μ-measurable and cover T. There exists then a finite number of μ-measurable sets B_i which cover E such that $B_i \subset A_i$ for each i. In fact, we take $B_1 = A_1$ and for $i > 1$, $B_i = A_i \setminus \cup_{j<i} B_j$. In each set B_i we take a point t_i and denote $b_i = f(t) \in E$. Define the function g_n by the equality $g_n(t) = b_i$ if $t \in B_i$. The function g_n is a μ-measurable step function and $d(f(t), g_n(t)) \leqslant 2n^{-1}$ for every $t \in T$.

It follows that the sequence (g_n) converges uniformly on T to f and the proposition is proved.

2 *Functions with values in Banach spaces*

10.8 THEOREM. *Let E be a Banach space. If $f: T \to E$ is a μ-measurable function, for every compact set $K \subset T$ there exists a μ-negligible set $N \subset K$ and a sequence of Borel step functions $\boldsymbol{g}_n : T \to E$, with the support contained in K, such that $|\boldsymbol{g}_n(t)| \leqslant |\boldsymbol{f}(t)|$ for every $t \in T$ and $\boldsymbol{g}_n(t) \to \boldsymbol{f}(t)$ for $t \in K \setminus N$.*

Let $K \subset T$ be a compact set. Since $\boldsymbol{f}$ is μ-measurable, there exists a μ-negligible set $N \subset K$ and a sequence (K_n) of mutually disjoint compact sets with union $K \setminus N$ and such that the restriction of f to each K_n is continuous.

For each n consider the sets K_i with $i \leqslant n$. For each K_i there exists a partition (A_{ij}) consisting of Borel sets such that for $t', t'' \in A_{ij}$ we have

$$|\boldsymbol{f}(t') - \boldsymbol{f}(t'')| < n^{-1}.$$

Let $t_{ij} \in A_{ij}$ and $a_{ij} = \boldsymbol{f}(t_{ij})$. Define now the function $\boldsymbol{g}_n$ by:

$\boldsymbol{g}_n(t) = 0$, if $|a_{ij}| \leqslant n^{-1}$ and $t \in A_{ij}$,

$\boldsymbol{g}_n(t) = a_{ij}[1-(n|a_{ij}|)^{-1}]$, if $|a_{ij}| \geqslant n^{-1}$ and $t \in A_{ij}$,

$\boldsymbol{g}_n(t) = 0$, if $t \notin \cup_{i \leqslant n} A_{ij}$.

Then $\boldsymbol{g}_n$ is a Borel step function with the support contained in K, $|\boldsymbol{g}_n(t)| \leqslant |\boldsymbol{f}(t)|$ for $t \in T$ and $\lim \boldsymbol{g}_n(t) = \boldsymbol{f}(t)$ for $t \in K \setminus N$. Thus the theorem is proved.

10.9 PROPOSITION. *If E is finite-dimensional, every μ-measurable bounded function $f: T \to E$ is the uniform limit on T of a sequence $(\boldsymbol{g}_n)$ of μ-measurable step functions.*

In fact, $\boldsymbol{f}$ may be considered with values in the compact space $\overline{\boldsymbol{f}(T)}$ (see proposition 10.7).

In particular, the proposition is true for real or complex functions.

10.10 THEOREM. *Let E and F be two Banach spaces. A function $\boldsymbol{f}: T \to \mathscr{L}(E, F)$ is μ-measurable if and only if the following conditions are satisfied:*

(a) *for every element $x \in E$, the function $t \to \boldsymbol{f}(t)x$ (with values in F) is μ-measurable;*

(b) *for every compact set $K \subset T$ there exists a μ-negligible set $N \subset K$ and a countable set $H \subset \mathscr{L}(E, F)$ such that $\boldsymbol{f}(K \setminus N) \subset \bar{H}$.*

If $\boldsymbol{f}$ is μ-measurable and $x \in E$, then the function $t \to \boldsymbol{f}(t)x$ is μ-measurable, since the mapping $u \to ux$ of $\mathscr{L}(E, F)$ into F is continuous; also, condition (b) is satisfied, by theorem 10.5.

Conversely, assume that conditions (a) and (b) are fulfilled. Since $\mathscr{L}(E, F)$ is a metric space, by theorem 10.5, it is sufficient to show that the inverse image by f of every closed sphere $S \subset \mathscr{L}(E, F)$ is μ-measurable. Let S be a closed sphere with center a and positive radius $r > 0$. Let $K \subset T$ be a compact set. By condition (b) there exists a μ-negligible set $N \subset K$ and a countable set $H \subset \mathscr{L}(E, F)$ such that $\boldsymbol{f}(K \setminus N) \subset \bar{H}$. Define the function $\boldsymbol{g}: T \to \mathscr{L}(E, F)$ equal to $\boldsymbol{f}$ on $T \setminus N$ and such that $\boldsymbol{g}(K) \cap H$ (for example taking $\boldsymbol{g}(t) = b \in H$ for $t \in N$). The set $H \cup \{a\}$ is countable. Arrange in a sequence $\{a_1, a_2, \ldots, a_n, \ldots\}$ the elements of the set $H \cup \{a\}$. For each element a_n and each natural number m there exists an element $x_{n,m} \in E$ with $|x_{n,m}| = 1$ and

$$|a_n x_{n,m}| > |a_n| + m^{-1}.$$

Then

$$|a_n| = \sup_m |a_n x_{n,m}|.$$

Let V be the closed vector space generated by the set $H \cup \{a\}$. For every $u \in V$ we have $|u| = \sup_{n,m} |u x_{n,m}|$.

For each n and m, the function $t \to \boldsymbol{g}(t) x_{n,m} - a x_{n,m}$ is μ-measurable, by hypothesis (a), since $\boldsymbol{f} = \boldsymbol{g}$, μ-almost everywhere. Then the function $t \to |\boldsymbol{g}(t) x_{n,m} - a x_{n,m}|$ is μ-measurable, consequently the set

$$A_{n,m} = \{t \mid |\boldsymbol{g}(t) x_{n,m} - a x_{n,m}| \leqslant r |x_{n,n}|\}$$

is μ-measurable, being the inverse image by the numerical function $t \to |\boldsymbol{g}(t) x_{n,m} - a x_{n,m}|$ of the closed set $\{s; |s| \leqslant r |x_{n,m}|\}$ on the line.

Noticing that we have

$$|\boldsymbol{g}(t) - a| = \sup_{n,m} |(\boldsymbol{g}(t) - a) x_{n,m}| \quad \text{for } t \in K,$$

it follows that

$$K \cap \boldsymbol{g}^{-1}(S) = \cap_{n,m} (A_{n,m} \cap K),$$

hence the set $K \cap \boldsymbol{g}^{-1}(S)$ is μ-measurable. Then the set $K \cap \boldsymbol{f}^{-1}(S)$ is μ-measurable too, since it differs from $K \cap \boldsymbol{g}^{-1}(S)$ by a μ-negligible set. By the localization principle, $\boldsymbol{f}^{-1}(S)$ is μ-measurable and thus condition (a) of theorem 10.5 is fulfilled. By the same theorem, $\boldsymbol{f}$ is μ-measurable, thus the theorem is proved.

10.11 Corollary. *If there exists a countable set $H \subset \mathscr{L}(E, F)$ such that $\boldsymbol{f}(T) \subset \bar{H}$ (in particular if $\mathscr{L}(E, F)$ is of countable type), then $\boldsymbol{f}$ is μ-measurable if and only if the function $t \to \boldsymbol{f}(t)x$ is μ-measurable for every $x \in E$.*

10.12 Corollary. *A function $\boldsymbol{f}: T \to E$ is μ-measurable if and only if the following conditions are fulfilled:*

(a) *for every functional $z' \in E'$, the numerical function $t \to \langle \boldsymbol{f}(t), z' \rangle$ is μ-measurable;*

(b) *for every compact set $K \subset T$, there exists a μ-negligible set $N \subset K$ and a countable set $H \subset E$ such that $\boldsymbol{f}(K \setminus N) \in \bar{H}$.*

10.13 Corollary. *Let E be a Banach space of countable type. A function $\boldsymbol{f}: T \to E$ is μ-measurable if and only if the numerical function $t \to \langle \boldsymbol{f}(t), z' \rangle$ is μ-measurable for every $z' \in E'$.*

10.14 PROPOSITION. *If (f_n) is a sequence of μ-measurable numerical functions (taking on finite or infinite values), then the functions* $\sup f_n$, $\inf f_n$, $\limsup f_n$, $\liminf f_n$, *are μ-measurable.*

The extended real line is homeomorphic, for example, with the segment $[-1, 1]$. Hence, we can define on $\bar{R}$ a distance $d(x, y)$ compatible with its topology. $\bar{R}$ is then a metric space.

The function $\sup f_n$ is the limit of the increasing sequence of μ-measurable functions $g_n = \sup(f_1, \ldots, f_n)$, hence $\sup f_n$ is μ-measurable too.

The function $\limsup f_n$ is the limit of the decreasing sequence of functions $h_n = \sup_{p \geqslant 0} f_{n+p}$, which are μ-measurable by the first part of the proof. It follows that $\limsup f_n$ is μ-measurable too.

Furthermore, we have $\inf f_n = -\sup(-f_n)$ and $\liminf f_n = -\limsup(-f_n)$; hence, by the above proof, these functions are also μ-measurable.

10.15 PROPOSITION. *If the numerical function (taking on finite or infinite values) $f: T \to \bar{R}$ is μ-measurable, then the sets.*

(1) $f^{-1}[a, b] = \{t \mid a \leqslant f(t) \leqslant b\}$

(2) $f^{-1}(a, b] = \{t \mid a < f(t) \leqslant b\}$

(3) $f^{-1}[a, b) = \{t \mid a \leqslant f(t) < b\}$

(4) $f^{-1}(a, b) = \{t \mid a < f(t) < b\}$

are μ-measurable for every numbers a and b such that $-\infty \leqslant a \leqslant b \leqslant +\infty$.

In fact, if f is μ-measurable, the sets from (1) and (4) are μ-measurable, since they are inverse images by f of the sets $[a, b]$ and (a, b) which are closed, respectively open in $\bar{R}$.

In particular, for every point $a \in \bar{R}$, the set $f^{-1}(a)$ is μ-measurable.

The sets from (2) and (3) are μ-measurable too, since they are differences of μ-measurable sets:

$$f^{-1}(a, b] = f^{-1}[a, b] \setminus f^{-1}(a),$$

$$f^{-1}[a, b) = f^{-1}[a, b] \setminus f^{-1}(b).$$

Thus the proposition is proved.

In particular, if f is μ-measurable, the following sets are μ-measurable:

(5) $f^{-1}[a, +\infty] = \{t | f(t) \geqslant a\}$,

(6) $f^{-1}(a, +\infty] = \{t | f(t) > a\}$,

(7) $f^{-1}[-\infty, b] = \{t | f(t) \leqslant b\}$,

(8) $f^{-1}[-\infty, b) = \{t | f(t) < b\}$

(9) $f^{-1}(a) = \{t | f(t) = a\}$,

for every numbers $a, b \in \overline{R}$.

For instance, the sets

$$\{t | f(t) = -\infty\}, \{t | f(t) \geqslant -\infty\}, \{t | f(t) > -\infty\},$$
$$\{t | f(t) = +\infty\}, \{t | f(t) \leqslant +\infty\}, \{t | f(t) < +\infty\}$$

are μ*-measurable.*

10.16 Corollary. *If* f *and* g *are two* μ*-measurable numerical function (taking on finite or infinite values), then the sets*

$$\{t | f(t) < g(t)\}, \{t | f(t) \leqslant g(t)\}, \{t | f(t) = g(t)\}$$

are μ*-measurable.*

In fact, for every rational number r, the sets $\{t | f(t) < r\}$, $\{t | r < g(t)\}$ are μ-measurable, hence the set

$$\{t | f(t) < g(t)\} = \cup_{r \in Q}(\{t | f(t) < r\} \cap \{t | r < g(t)\})$$

is μ-measurable too, since it is the union of a sequence of μ-measurable sets.

It follows then that the set

$$\{t | f(t) > g(t)\} = \{t | g(t) < f(t)\}$$

is also μ-measurable, hence the sets

$$\{t | f(t) \leqslant g(t)\} = T \setminus \{t | f(t) > g(t)\}$$
$$\{t | f(t) = g(t)\} = \{t | f(t) \leqslant g(t)\} \setminus \{t | f(t) < g(t)\}$$

are μ-measurable too.

10.17 Proposition. *Let* D *be a countable set, everywhere dense in* R. *A numerical function (taking on finite or infinite values)* $f: T \to \overline{R}$ *is* μ*-measurable if and only if one of the following conditions is fulfilled.*

(1) $\{t | a \leqslant f(t) \leqslant b\}$ and $\{t | f(t) = +\infty)$ are μ-measurable, for every $a, b \in D$;

(2) $\{t \mid a < f(t) \leqslant b\}$ and $\{t \mid f(t) = +\infty\}$ are μ-measurable, for every $a, b \in D$;

(3) $\{t \mid a \leqslant f(t) < b\}$ and $\{t \mid f(t) = +\infty\}$ are μ-measurable for every $a, b \in D$;

(4) $\{t \mid a < f(t) < b\}$ and $\{t \mid f(t) = +\infty\}$ are μ-measurable for every $a, b \in D$;

(5) $\{t \mid f(t) \geqslant a\}$ is μ-measurable for every $a \in D$;

(6) $\{t \mid f(t) > a\}$ is μ-measurable for every $a \in D$;

(7) $\{t \mid f(t) \leqslant b\}$ is μ-measurable for every $b \in D$;

(8) $\{t \mid f(t) < b\}$ is μ-measurable for every $b \in D$.

Assume that condition (1) is satisfied. For every numbers $\alpha, \beta \in R$, we have

$$\{t \mid \alpha \leqslant f(t) \leqslant \beta\} = \cap_{a \leqslant \alpha, \beta \leqslant b} \{t \mid a \leqslant f(t) \leqslant b\}.$$

Since D is countable, in the right-hand side we have the intersection of a countable family of μ-measurable sets, hence the set $\{t \mid \alpha \leqslant f(t) \leqslant \beta\}$ is μ-measurable for every $\alpha, \beta \in R$. It follows then successively that the following sets are μ-measurable:

$$\{t \mid \alpha \leqslant f(t) < +\infty\} = \cup_{1 \leqslant n < \infty} \{t \mid \alpha \leqslant f(t) \leqslant \alpha + n\},$$

$$\{t \mid \alpha < f(t) < +\infty\} = \cup_{1 \leqslant n < \infty} \{t \mid \alpha + n^{-1} \leqslant f(t) < +\infty\},$$

$$\{t \mid \alpha \leqslant f(t) \leqslant +\infty\} = \{t \mid \alpha \leqslant f(t) < +\infty\} \cup \{t \mid f(t) = +\infty\},$$

$$\{t \mid \alpha < f(t) \leqslant +\infty\} = \{t \mid \alpha < f(t) < +\infty\} \cup \{t \mid f(t) = +\infty\},$$

$$\{t \mid -\infty \leqslant f(t) \leqslant \beta\} = T \setminus \{t \mid \beta < f(t) \leqslant +\infty\}.$$

Consequently, for every closed interval $[\alpha, \beta] \cap \bar{R}$, the set

$$f^{-1}[\alpha, \beta] = \{t \mid \alpha \leqslant f(t) \leqslant \beta\}$$

is μ-measurable. On the space $\bar{R}$, the function

$$d(x, y) = x(|x|+1)^{-1} - y(|y|+1)^{-1}$$

is a distance compatible with the topology of $\bar{R}$, and the closed spheres $S \subset \bar{R}$ for this distance are closed intervals $[\alpha, \beta] \subset \bar{R}$. Since the space $\bar{R}$ contains a countable everywhere dense set, we do deduce from theorem 10.5 that f is μ-measurable.

The cases (2), (3) and (4) can be reduced to the preceding one, by observing that for $\alpha, \beta \in \overline{R}$, we have

$$\{t \mid \alpha \leqslant t \leqslant \beta\} = \cap_{a \leqslant \alpha, \beta \geqslant b} \{t \mid a < f(t) \leqslant b\} = \cap_{a \leqslant \alpha, \beta \geqslant b} \{t \mid a \leqslant f(t) < b\} = \\ = \cap_{a \leqslant \alpha, \beta \geqslant b} \{t \mid a < f(t) < b\} .$$

The cases (5) and (6) can be reduced to the case (1) by observing that

$$\{t \mid f(t) = +\infty\} = \cap_{a \in D} \{t \mid f(t) \geqslant a\} = \cap_{a \in D} \{t \mid f(t) > a\}$$

and then that for $\alpha, \beta \in R$, we have

$$\{t \mid f(t) \geqslant \alpha\} = \cap_{a \leqslant \alpha} \{t \mid f(t) \geqslant a\} = \cap_{a \leqslant \alpha} \{t \mid f(t) > a\} ,$$

$$\{t \mid f(t) > \beta\} = \cup_{n \in N} \{t \mid f(t) \geqslant \beta + n^{-1}\} ,$$

$$\{t \mid \alpha \leqslant f(t) \leqslant \beta\} = \{t \mid f(t) \geqslant \alpha\} \setminus \{t \mid f(t) > \beta\} .$$

The cases (7) and (8) can be reduced to the case (1) by observing that

$$\{t \mid f(t) \leqslant \beta\} = \cap_{b \geqslant \beta} \{t \mid f(t) \leqslant b\} = \cap_{b \geqslant \beta} \{t \mid f(t) > b\} ,$$

$$\{t \mid f(t) < \alpha\} = \cup_{n \in N} \{t \mid f(t) \leqslant \alpha - n^{-1}\} ,$$

$$\{t \mid \alpha \leqslant f(t) \leqslant \beta\} = \{t \mid f(t) \leqslant \beta\} \setminus \{t \mid f(t) < \alpha\}$$

and then

$$\{t \mid f(t) < -\infty\} = \cup_{b \in D} \{t \mid f(t) \leqslant b\} \setminus \cup_{b \in D} \{t \mid f(t) < b\} ,$$

$$\{t \mid f(t) = +\infty\} = T \setminus \{t \mid f(t) < +\infty\} .$$

Thus the proposition is proved.

Remark. The set $\{t \mid f(t) = +\infty\}$ in conditions (1)–(4) can be replaced by the set $\{t \mid f(t) = -\infty\}$. If either $+\infty \in D$ or $-\infty \in D$, the condition that the set $\{t \mid f(t) = +\infty\}$ should be μ-measurable is superfluous. This condition is superfluous too if the function f is everywhere *finite*.

The set D may be the set Q of the rational numbers.

10.18 Corollary. *Every lower semicontinuous or upper semicontinuous function is measurable for every measure μ.*

In fact, if f is lower semicontinuous, the set $\{t \mid f(t) \leqslant a\}$ is closed for every $a \in \overline{R}$; if f is upper semicontinuous, the set $\{t \mid f(t) \geqslant a\}$ is closed for every $a \in \overline{R}$; and the closed sets are measurable for every measure μ.

10.19 PROPOSITION. *If the numerical function (taking on finite or infinite values)* $f: T \to \overline{R}$ *is positive and* μ*-measurable, then for every compact set* $K \subset T$*, there exists a sequence* (f_n) *of positive Borel step functions with the supports contained in* K *such that* $f_n(t) \leqslant f(t)$ *for every* $t \in T$ *and* $f_n(t) \to f(t)$*,* μ*-almost everywhere on* K.

If f is everywhere finite, f takes on values in the Banach space R. By theorem 10.8, for every compact set K, there exists a sequence (g_n) of Borel step functions with the supports contained in K such that $|g_n(t)| \leqslant f(t)$ for $t \in T$ and $g_n(t) \to f(t)$, μ-almost everywhere on K. The positive functions $f_n(t) = |g_n(t)|$ fulfill the conditions of the proposition.

Assume now that f takes on also the value $+\infty$. The mapping $u(x) = x(|x|+1)^{-1}$ of the space $\overline{R}$ onto the segment $[-1, 1]$ is strictly increasing and continuous. Its inverse mapping $v(y) = y(1-|y|)^{-1}$ defined on $[-1, 1]$ with values in $\overline{R}$ is also strictly increasing and continuous.

The function

$$h(t) = u(f(t)) = f(t)(f(t)+1)^{-1}$$

is μ-measurable and positive with values in $[0, 1]$. By the first part of the proof, there exists a sequence (h_n) of positive Borel step functions with the supports in K such that $h_n(t) \to h(t)$, μ-almost everywhere on K. The positive step functions

$$f_n(t) = v(h_n(t)) = h_n(t)(1-|h_n(t)|)^{-1}$$

fulfill the conditions of the proposition.

10.20 PROPOSITION. *If* $f: T \to R$ *is a bounded* μ*-measurable function, there exists a sequence* (f_n) *of* μ*-measurable step functions converging uniformly to* f.

In fact, the function f can be considered with values in the space $\overline{f(T)}$, which is compact. Then we apply proposition 10.7.

§11 Integrability criteria

1 *The additivity of the upper integral*

Let μ be a positive measure on T.

We have seen that for lower semicontinuous functions, the upper integral is additive.

The upper integral is also additive for positive integrable functions, since for these functions we have

$$\int^* f\mathrm{d}\mu = \int f\mathrm{d}\mu .$$

We shall prove that the upper integral is still additive in some other cases.

11.1 PROPOSITION. *If f and g are two μ-measurable positive functions (taking on finite or infinite values), then*

$$\int^* (f+g)\mathrm{d}\mu = \int^* f\mathrm{d}\mu + \int^* g\mathrm{d}\mu .$$

Assume first that f and g vanish outside a compact set K.

If, for instance, $\int^* f\mathrm{d}\mu = +\infty$, then from the inequality $f \leqslant f+g$ we deduce

$$\int^* f\mathrm{d}\mu \leqslant \int^* (f+g)\mathrm{d}\mu$$

hence

$$\int^* (f+g)\mathrm{d}\mu = +\infty$$

and the equality is proved in this case. Now assume that $\int^* f\mathrm{d}\mu < \infty$ and $\int^* g\mathrm{d}\mu < \infty$. We shall prove that f and g are μ-integrable.

There exists a sequence (f_n) of positive Borel step functions, converging to f almost everywhere on K, such that $f_n \leqslant f$ for every n (proposition 10.9). Since the functions f_n are μ-integrable, we deduce from Lebesgue's theorem that f is μ-integrable. One can show in the same way that g is integrable. Then it follows that

$$\int^* (f+g)\mathrm{d}\mu = \int (f+g)\mathrm{d}\mu = \int f\mathrm{d}\mu + \int g\mathrm{d}\mu = \int^* f\mathrm{d}\mu + \int^* g\mathrm{d}\mu .$$

Now assume that f and g are positive and measurable. By the first part of the proof, for every compact set $K \subset T$, we have

$$\int^* (f+g)\varphi_K\mathrm{d}\mu = \int^* f\varphi_K\mathrm{d}\mu + \int^* g\varphi_K\mathrm{d}\mu.$$

Passing to the limit along the directed set of the compact sets $K \cap T$ we get (remark after proposition 5.10).

$$\int^* (f+g)\mathrm{d}\mu = \int^* f\mathrm{d}\mu + \int^* g\mathrm{d}\mu .$$

11.2 COROLLARY. *If A and B are two disjoint* μ*-measurable sets, then*

$$\mu^*(A \cup B) = \mu^*(A) + \mu^*(B).$$

In fact

$$\varphi_{A \cup B} = \varphi_A + \varphi_B .$$

We shall prove that the upper integral is also additive for some functions which are not measurable. First we prove the following

11.3 LEMMA. *Let f and g be two positive functions (taking on finite or infinite values). If g is* μ*-measurable and with compact support, then*

$$\int^* fg \, d\mu = \inf \int^* hg \, d\mu$$

when h runs over the set of μ*-measurable functions with* $h \geqslant f$.

The inequality

$$\int^* fg \, d\mu \leqslant \inf \int^* hg \, d\mu$$

follows from the fact that $fg \leqslant hg$, whence $\int^* fg \, d\mu \leqslant \int^* hg \, d\mu$ for every function $h \geqslant f$.

Conversely, let $\varphi \geqslant fg$ be a lower semicontinuous function. We consider the μ-measurable function h_0 defined as follows:

$$h_0(t) = +\infty \text{ if either } g(t) = 0 \text{ or } g(t) = +\infty ,$$
$$h_0(t) = \varphi(t)/g(t) \text{ if } 0 < g(t) < +\infty .$$

We have $h_0 g \leqslant \varphi$. If $0 < g(t) < +\infty$, then we have

$$h_0(t) g(t) = \varphi(t) \geqslant f(t) g(t)$$

hence $h_0(t) \geqslant f(t)$. This inequality is still satisfied by the other values of t, since if $g(t) = 0$ or if $g(t) = +\infty$, then we have $h_0(t) = +\infty$. Consequently, $h_0 \geqslant f$. Then

$$\int^* \varphi \, d\mu \geqslant \int^* h_0 g \, d\mu \geqslant \inf_{h \geqslant f} \int^* hg \, d\mu, \; h \text{ measurable} .$$

Since the lower semicontinuous function $\varphi \geqslant fg$ is arbitrary and fg has compact support, we have

$$\int^* fg \, d\mu \geqslant \inf_{h \geqslant f} \int^* hg \, d\mu$$

and thus the lemma is proved.

11.4 PROPOSITION. *Let f, g_1, g_2 be three positive functions (taking on finite or infinite values). If g_1 and g_2 are μ-measurable, then*

$$\int^* f(g_1+g_2)\mathrm{d}\mu = \int^* fg_1\,\mathrm{d}\mu + \int^* fg_2\,\mathrm{d}\mu.$$

We have, first, $f(g_1+g_2)=fg_1+fg_2$, hence

$$\int^* f(g_1+g_2)\mathrm{d}\mu = \int^* (fg_1+fg_2)\mathrm{d}\mu \leqslant \int^* fg_1\,\mathrm{d}\mu + \int^* fg_2\,\mathrm{d}\mu\,.$$

Let now $h \geqslant f$ be a μ-measurable function and $K \subset T$ an arbitrary compact set. We have

$$\int^* h(g_1+g_2)\varphi_K\mathrm{d}\mu = \int^* hg_1\varphi_K\mathrm{d}\mu + \int^* hg_2\varphi_K\mathrm{d}\mu$$

since $hg_1\varphi_K$ and $hg_2\varphi_K$ are μ-measurable (proposition 11.1).

But

$$\int^* hg_1\varphi_K\mathrm{d}\mu \geqslant \int^* fg_1\varphi_K\mathrm{d}\mu \quad\text{and}\quad \int^* hg_2\varphi_K\mathrm{d}\mu \geqslant \int^* fg_2\varphi_K\mathrm{d}\mu\,,$$

hence

$$\int^* h(g_1+g_2)\varphi_K\mathrm{d}\mu \geqslant \int^* fg_1\varphi_K\mathrm{d}\mu + \int^* fg_2\varphi_K\mathrm{d}\mu.$$

Taking in the left-hand side the infimum for all μ-measurable function $h \geqslant f$, we get from the preceding lemma

$$\int^* f(g_1+g_2)\varphi_K\mathrm{d}\mu \geqslant \int^* fg_1\varphi_K\mathrm{d}\mu + \int^* fg_2\varphi_K\mathrm{d}\mu\,,$$

hence

$$\int^* f(g_1+g_2)\varphi_K\mathrm{d}\mu = \int^* fg_1\varphi_K\mathrm{d}\mu + \int^* fg_2\varphi_K\mathrm{d}\mu\,.$$

Then we have

$$\int^* f(g_1+g_2)\mathrm{d}\mu = \lim_K \int^* f(g_1+g_2)\varphi_K\mathrm{d}\mu = \lim_K \int^* fg_1\varphi_K\mathrm{d}\mu +$$
$$+\lim_K \int fg_2\varphi_K\mathrm{d}\mu = \int^* fg_1\,\mathrm{d}\mu + \int^* fg_2\,\mathrm{d}\mu$$

and the proposition is proved.

11.5 COROLLARY. *If A and B are two disjoint measurable sets, then for every positive function f (taking on finite or infinite values) we have*

$$\int^* f\varphi_{A\cup B}\mathrm{d}\mu = \int^* f\varphi_A\mathrm{d}\mu + \int^* f\varphi_B\mathrm{d}\mu\,.$$

In fact, we have then $\varphi_{B\cup A}=\varphi_A+\varphi_B$.

2 *The support of p-integrable functions*

11.6 PROPOSITION. *Every positive function f with $\int^* f d\mu < +\infty$ vanishes μ-almost everywhere outside the union of a sequence of mutually disjoint compact sets.*

Since

$$\int^* f d\mu = \sup_K \int^* f\varphi_K d\mu, \ K \text{ compact},$$

there exists a sequence (K_n) of compact sets such that

$$\int^* f d\mu = \sup_n \int^* f\varphi_{K_n} d\mu .$$

Put $A = \cup_{1 \leqslant n < \infty} K_n$. For every n, the sets $T \setminus A$ and K_n are disjoint and μ-measurable, hence

$$\int^* f d\mu \geqslant \int^* f\varphi_{(T\setminus A)\cup K_n} d\mu = \int^* f\varphi_{T\setminus A} d\mu + \int^* f\varphi_{K_n} d\mu$$

whence we deduce, by taking the supremum in the right-hand side,

$$\int^* f d\mu \geqslant \int^* f\varphi_{T\setminus A} d\mu + \int^* f d\mu .$$

Since $\int^* f d\mu < +\infty$, it follows that $\int^* f\varphi_{T\setminus A} d\mu = 0$, that is, $f\varphi_{T\setminus A}$ is μ-negligible, hence $f(t) = f(t)\varphi_A(t)$, μ-almost everywhere.

Denote $A_1 = K_1$ and for every n, $A_n = K_n \setminus \cup_{i<n} K_i$. The sets A_n are integrable, mutually disjoint and $A = \cup_{1 \leqslant n < \infty} A_n$. Every set A_n is the union of a sequence $(K_{nm})_{m \in N}$ of mutually disjoint compact sets and of a μ-negligible set N_n. Then the set $N = \cup_{1 \leqslant n < \infty} N_n$ is μ-negligible, hence the function f vanishes μ-almost everywhere outside the double sequence (K_{nm}) of compact mutually disjoint sets.

11.7 COROLLARY. *Let f be a function defined on T with values in a Banach space E or in $\overline{R}$. If there exists $1 \leqslant p < +\infty$ with $N_p(f) < +\infty$, then f vanishes μ-almost everywhere outside the union of a sequence of mutually disjoint compact sets.*

In fact

$$N_p(f) = \left(\int^* |f|^p d\mu\right)^{1/p} < +\infty ,$$

hence

$$\int^* |f|^p d\mu < +\infty .$$

There exists a sequence (K_n) of mutually disjoint compact sets and a μ-negligible set N such that for every $t \notin N \cup \bigcup_{1 \leqslant n < \infty} K_n$ we have $|f(t)|^p = 0$, hence $f(t) = 0$.

11.8 COROLLARY. *Every set $A \subset T$ with $\mu^*(A) < +\infty$ is contained in the union of a μ-negligible set and of a sequence of mutually disjoint compact set.*

11.9 PROPOSITION. *Let f be a function defined on T with values in a Banach space E or in $\overline{R}$. If f has summable p-th power $(1 \leqslant p < +\infty)$, then the set $A = \{t \mid f(t) \neq 0\}$ is equal to the union of a μ-negligible set and of a sequence of mutually disjoint compact sets.*

Since $N_p(f) < +\infty$, the set A is contained in the union of a μ-negligible set N and of a sequence (K_n) of compact sets. Since f is μ-measurable, the set A is μ-measurable, hence the sets $A \cap K_n$ are μ-integrable. Every set $A \cap K_n$ is the union of a μ-negligible set N_n and of a sequence $(K_{nm})_{m \in N}$ of mutually disjoint compact sets. Since $A = (A \cap N) \cup (\bigcup_{1 \leqslant n < \infty} (A \cap K_n))$, the set A is the union of the μ-negligible set $(A \cap N) \cup (\bigcup_{1 \leqslant n < \infty} N_n)$ and of the double sequence (K_{nm}) of mutually disjoint compact sets.

3 *Integrability criteria*

By means of measurability we can now give the following criterion of integrability.

11.10 THEOREM. *Let E be a Banach space. A function $f: T \to E$ is p-summable $(1 \leqslant p < +\infty)$ if and only if f is μ-measurable and $N_p(f) < +\infty$.*

If f is p-summable, then $N_p(f) < +\infty$ and there exists a sequence (f_n) of continuous functions with compact support converging μ-almost everywhere to f. Since the functions f_n are μ-measurable, it follows that f is μ-measurable too.

Conversely, assume that f is μ-measurable and that $N_p(f) < +\infty$. It follows first that the set $A = \{t \mid f(t) \neq 0\}$ is equal to the union of a μ-negligible set $N \subset A$ and of a sequence (C_n) of mutually disjoint compact sets. Since f is measurable, every compact C_n is the union of a negligible set N_n and of a sequence $(C_{nm})_{m \in N}$ of mutually disjoint compact sets such that f restricted to every C_{nm} is continuous. Changing the notation if necessary, we may assume from the beginning that f restricted to every C_n is continuous.

For every n, the set $K_n = \cup_{1 \leqslant i \leqslant n} C_i$ is compact, the sequence (K_n) is increasing, and f restricted to K_n is continuous (since the compact sets $C_1, \ldots, C_n$ are disjoint and f restricted to each of them is continuous). We denote $f_n = f\varphi_{K_n}$.

Every function f_n has p-summable power. Let $\varepsilon > 0$ and U_n be a compact neighbourhood of K_n such that

$$\mu(U_n \setminus K_n) < \varepsilon^p (\| f_n \| + \varepsilon)^{-p}$$

(which is possible since the measure of a compact set is the infimum of the measures of the open sets containing this set; and $\| f_n \| = \sup_{t \in T} | f_n(t) | < +\infty$ since f_n restricted to every K_n is continuous). Since the restriction of f_n to K_n is continuous, it can be extended to a function g_n which is continuous on the whole space such that $\| g_n \| < \| f_n \| + \varepsilon$. In addition, we can choose the function g_n with the support in U_n (multiplying it if necessary by a continuous function $h : T \to [0, 1]$ equal to 1 on K_n and vanishing on $T \setminus U_n$), hence $g_n \in \mathscr{K}_E(T)$. Then

$$N_p(f_n - g_n) = \left(\int_{U_n \setminus K_n} |g_n|^p \, \mathrm{d}\mu\right)^{1/p} \leqslant (\| f_n \| + \varepsilon) \mu(U_n \setminus K_n)^{1/p} < \varepsilon \, .$$

Consequently $f_n \in \mathscr{L}_E^p$. On the other hand, the sequence (f_n) converges almost everywhere to f and $|f_n(t)| \leqslant |f(t)|$ for every $t \in T$ and every $n \in N$, and $N_p(f) < +\infty$. By Lebesgue's theorem it follows that f is p-integrable, and the theorem is proved.

11.11 Corollary. *A lower semicontinuous or an upper-semicontinuous positive function f is μ-integrable if and only if $\int^* \mathrm{d}\mu < \infty$.*

In fact, f is measurable (corollary 10.11).

11.12 Corollary. *A set $A \subset T$ is μ-integrable if and only if A is μ-measurable and $\mu^*(A) < \infty$.*

11.13 Corollary. *A set A which is either closed or open is μ-integrable if and only if $\mu^*(A) < \infty$.*

In fact, A is measurable (corollary 9.29).

11.14 Corollary. *For every p-summable function $f : T \to E$ and every μ-measurable set $A \subset T$, the function $f\varphi_A$ is p-summable.*

In fact, the function $f\varphi_A$ is μ-measurable and $N_p(f\varphi_A) \leqslant N_p(f) < +\infty$.

If $\boldsymbol{m}$ is a dominated measure, for every $\boldsymbol{m}$-integrable function $f: T \to E$ and every $|\boldsymbol{m}|$-measurable set $A \subset T$, we shall use the notation:

$$\int_A f \mathrm{d}\boldsymbol{m} = \int f\varphi_A \mathrm{d}\boldsymbol{m} .$$

11.15 PROPOSITION. *If μ and ν are two positive measures, then*

$$\mathscr{L}_E^p(\mu+\nu) = \mathscr{L}_E^p(\mu) \cap \mathscr{L}_E^p(\nu), \quad 1 \leqslant p < +\infty ,$$

and the topology defined on the space $\mathscr{L}_E^p(\mu+\nu)$ by the seminorm $N_p(f, \mu+\nu)$ is equivalent to the topology defined on the space $\mathscr{L}_E^p(\mu) \cap \mathscr{L}_E^p(\nu)$ by the seminorm $N_p(f, \mu) + N_p(f, \nu)$.

From the inequalities

$$N_p(f, \mu+\nu) \leqslant N_p(f, \mu) + N_p(f, \nu) \leqslant 2N_p(f, \mu+\nu)$$

it follows that $N_p(f, \mu+\nu) < +\infty$ if and only if $N_p(f, \mu) < +\infty$ and $N_p(f, \nu) < +\infty$.

On the other hand, the function $f: T \to E$ is $\mu+\nu$-measurable if and only if f is μ-measurable and ν-measurable.

Consequently $f \in \mathscr{L}_E^p(\mu+\nu)$ if and only if $f \in \mathscr{L}_E^p(\mu)$ and $f \in \mathscr{L}_E^p(\nu)$, that is,

$$\mathscr{L}_E^p(\mu+\nu) = \mathscr{L}_E^p(\mu) \cap \mathscr{L}_E^p(\nu).$$

The assertion concerning the equivalence of the topologies follows from the inequalities between the seminorms.

11.16 COROLLARY. *Let μ and ν be two positive measures. A function $f: T \to E$ is $\mu+\nu$-integrable if and only if it is μ-integrable and ν-integrable. Then we have*

$$\int f \mathrm{d}(\mu+\nu) = \int f \mathrm{d}\mu + \int f \mathrm{d}\nu .$$

The assertion concerning the integrability of f follows from the preceding proposition. If f is $\mu+\nu$-integrable, there exists a sequence (f_n) of functions from $\mathscr{K}_E(T)$ such that $\int |f-f_n| \mathrm{d}(\mu+\nu) \to 0$. Then $\int |f-f_n| \mathrm{d}\mu \to 0$ and $\int |f-f_n| \mathrm{d}\nu \to 0$.

It follows that

$$\int f_n \mathrm{d}\mu \to \int f \mathrm{d}\mu, \int f_n \mathrm{d}\nu \to \int f \mathrm{d}\nu \text{ and } \int f_n \mathrm{d}(\mu+\nu) \to \int f \mathrm{d}(\mu+\nu).$$

For the functions f_n of $\mathscr{K}_E(T)$ we have

$$\int f_n \mathrm{d}(\mu+\nu) = \int f_n \mathrm{d}\mu + \int f_n \mathrm{d}\nu .$$

Passing to the limit, we get the equality

$$\int f \mathrm{d}(\mu+\nu) = \int f \mathrm{d}\mu + \int f \mathrm{d}\nu .$$

11.17 PROPOSITION. *Let $\boldsymbol{m}, \boldsymbol{n} : \mathscr{K}_E(T) \to F$ be two dominated measures and $f: T \to E$ a function. If f is $\boldsymbol{m}$-integrable and $\boldsymbol{n}$-integrable, then f is $\boldsymbol{m}+\boldsymbol{n}$ integrable and*

$$\int f \mathrm{d}(\boldsymbol{m}+\boldsymbol{n}) = \int f \mathrm{d}\boldsymbol{m} + \int f \mathrm{d}\boldsymbol{n} .$$

From the inequality $|\boldsymbol{m}+\boldsymbol{n}| \leqslant |\boldsymbol{m}| + |\boldsymbol{n}|$, we deduce

$$\mathscr{L}_E^1(|\boldsymbol{m}|) \cap \mathscr{L}_E^1(|\boldsymbol{n}|) = \mathscr{L}_E^1(|\boldsymbol{m}|+|\boldsymbol{n}|) \subset \mathscr{L}_E^1(|\boldsymbol{m}+\boldsymbol{n}|),$$

hence, if f is $\boldsymbol{m}$-integrable and $\boldsymbol{n}$-integrable, then f is $|\boldsymbol{m}|+|\boldsymbol{n}|$ integrable and $|\boldsymbol{m}+\boldsymbol{n}|$-integrable. Let (f_n) be a sequence of functions from $\mathscr{K}_E(T)$, converging to f in the mean, for the measure $|\boldsymbol{m}|+|\boldsymbol{n}|$:

$$\lim \int |f_n - f| \mathrm{d}(|\boldsymbol{m}|+|\boldsymbol{n}|) = 0 .$$

Then

$$\lim \int |f_n - f| \mathrm{d}|\boldsymbol{m}| = 0, \quad \lim \int |f_n - f| \mathrm{d}|\boldsymbol{n}| = 0$$

(since $|\boldsymbol{m}| \leqslant |\boldsymbol{m}|+|\boldsymbol{n}|$ and $|\boldsymbol{n}| < |\boldsymbol{m}|+|\boldsymbol{n}|$) and

$$\lim \int |f_n - f| \mathrm{d}|\boldsymbol{m}+\boldsymbol{n}| = 0 .$$

We deduce

$$\lim \int f_n \mathrm{d}\boldsymbol{m} = \int f \mathrm{d}\boldsymbol{m}, \quad \lim \int f_n \mathrm{d}\boldsymbol{n} = \int f \mathrm{d}\boldsymbol{n}$$

and

$$\lim \int f_n \mathrm{d}(\boldsymbol{m}+\boldsymbol{n}) = \int f \mathrm{d}(\boldsymbol{m}+\boldsymbol{n}) .$$

But for the functions f_n of $\mathscr{K}_E(T)$ we have

$$\int f_n \mathrm{d}(\boldsymbol{m}+\boldsymbol{n}) = \int f_n \mathrm{d}\boldsymbol{m} + \int f_n \mathrm{d}\boldsymbol{n} .$$

Passing to the limit we get

$$\int f \mathrm{d}(\boldsymbol{m}+\boldsymbol{n}) = \int f \mathrm{d}\boldsymbol{m} + \int f \mathrm{d}\boldsymbol{n} .$$

4 *Computation of the upper integral for some functions*

11.18 PROPOSITION. *Let $f \geqslant 0$ be a (not necessarily finite) function defined on T. If there exists a lower semicontinuous function $h_0 \geqslant f$ with $\mu^*(h_0) < +\infty$, then*

$$\mu^*(f) = \inf_{h \geqslant f} \mu^*(h), \quad h \in \mathscr{I}_+ .$$

(a) Assume first that $\mu^*(f)=0$, and denote $A=\{t \mid f(t)>0\}$ and $G_0=\{t \mid h_0(t)>0\}$. The set G_0 is open and the set A is μ-negligible. The function h_0 is μ-integrable, therefore the function $h_0 - h_0 \varphi_A$ is μ-integrable, hence the set

$$G_0 \setminus A = \{t \mid h_0(t) - h_0(t) \varphi_A(t) > 0\}$$

is the union of an increasing sequence (K_n) of compact sets and of a μ-negligible set N. The sets $G_n = G_0 \setminus K_n$ are open, they form a decreasing sequence and their intersection $\cap_{1 \leqslant n < \infty} G_n = A \cup N$ is μ-negligible. Then the functions $h\varphi_{G_n}$ are lower semicontinuous, they form a decreasing sequence,

$$\inf_n h_0 \varphi_{G_n} = h_0 \varphi_{A \cup N} \geqslant h_0 \varphi_A \geqslant f$$

and $\inf_n \mu^*(h_0 \varphi_{G_n}) \leqslant \mu^*(h_0) < +\infty$. It follows that

$$\inf_n \mu^*(h_0 \varphi_{G_n}) = \mu^*(h_0 \varphi_{A \cup N}) = 0 ,$$

hence

$$\inf_{h \geqslant f} \mu^*(h) = 0 = \mu^*(f), \quad h \in \mathscr{I}_+ .$$

(b) Now we prove that if $A \subset T$ is a μ-negligible set such that

$$\mu^*(f\varphi_A) = \inf \mu^*(h), \quad h \geqslant f\varphi_A, \ h \in \mathscr{I}_+ ,$$

then for every compact set $K \subset T$ we have

$$\mu^*(f\varphi_{A \cup K}) = \inf \mu^*(h), \quad h \geqslant f\varphi_{A \cup K}, \ h \in \mathscr{I}_+ .$$

In fact, let $\varepsilon > 0$. There exist two lower semicontinuous functions $h_1 \geqslant f\varphi_A$ and $h_2 \geqslant f\varphi_K$ such that

$$\mu^*(h_1) \leqslant \mu^*(f\varphi_A) + \varepsilon/2 = \varepsilon/2$$

and

$$\mu^*(h_2) \leqslant \mu^*(f\varphi_K) + \varepsilon/2 \leqslant \mu^*(f\varphi_{K \cup A}) + \varepsilon/2.$$

Then $h_1 + h_2$ is lower semicontinuous, $h_1 + h_2 \geqslant f\varphi_{K \cup A}$ and

$$\mu^*(h_1+h_2)=\mu^*(h_1)+\mu^*(h_2)\leqslant \mu^*(f\varphi_{K\cup A})+\varepsilon\,,$$

hence

$$\mu^*(f\varphi_{K\cup A})=\inf \mu^*(h), \quad h\geqslant f\varphi_{K\cup A},\ h\in\mathscr{I}_+\,.$$

(c) Consider now an arbitrary positive function f and a lower semicontinuous function $h_0\geqslant f$ with $\mu^*(h_0)<+\infty$. Since h_0 is μ-integrable, the set $G_0=\{t\,|\,h_0(t)>0\}$ is the union of a μ-negligible set N and of an increasing sequence (K_n) of compact sets. Since $f\varphi_N\leqslant h_0$, we have

$$\mu^*(f\varphi_N)=\inf \mu^*(h), \quad h\geqslant f\varphi_N,\ h\in\mathscr{I}_+\,,$$

hence, for each n,

$$\mu^*(f\varphi_{K_n\cup N})=\inf \mu^*(h), \quad h\geqslant f\varphi_{K_n\cup N},\ h\in\mathscr{I}_+\,,$$

The sequence $(f\varphi_{K_n\cup N})$ is increasing and $\sup f\varphi_{K_n\cup N}=f$. Then we deduce from proposition 5.11 that

$$\mu^*(f)=\inf \mu^*(h), \quad h\geqslant f,\ h\in\mathscr{I}_+\,.$$

11.19 PROPOSITION. *Let f be a positive function defined on T. If the set $A=\{t\,|\,f(t)\neq 0\}$ is contained in the union of a sequence (G_n) of open sets with $\mu^*(G_n)<+\infty$ for each n, then*

$$\mu^*(f)=\inf_{h\geqslant f}\mu^*(h)\,,\ h\in\mathscr{I}_+\,.$$

(a) Assume first that $\mu^*(f)=0$ and that A is contained in an open set G with $\mu^*(G)<+\infty$. Since $n\varphi_A\leqslant n\varphi_G$ and $n\varphi_G$ is lower semicontinuous and $\mu^*(n\varphi_G)=n\mu^*(\varphi_G)<+\infty$, we deduce that for each n we have

$$0=\mu^*(n\varphi_A)=\inf \mu^*(h), \quad h\geqslant n\varphi_A,\ h\in\mathscr{I}_+\,.$$

By proposition 5.11, we have then

$$0=\mu^*(\sup n\varphi_A)=\inf \mu^*(h)\,, \quad h\geqslant \sup n\varphi_A,\ h\in\mathscr{I}_+\,.$$

Consequently, for each $\varepsilon>0$, there exists a function $h\in\mathscr{I}_+$ with $h\geqslant \sup_n n\varphi_A\geqslant f$ such that $\mu^*(h)<\varepsilon$, hence

$$\mu^*(f)=\inf \mu^*(h)=0, \quad h\geqslant f,\ h\in\mathscr{I}_+\,.$$

(b) Assume now that $\mu^*(f)\geqslant 0$ and that A is contained in an open set G with $\mu^*(G)<+\infty$. Since G is μ-integrable, there exists an increasing sequence (K_n) of compact sets contained in G such that the set $N = G \setminus \cup_{1 \leqslant n < \infty} K_n$ is μ-negligible. The function $f\varphi_N$ is μ-negligible and the set of points at which it is different from zero is contained in G, hence

$$\mu^*(f\varphi_N) = \inf \mu^*(h), \quad h \geqslant f\varphi_N, \ h \in \mathscr{I}_+ .$$

Then, as in the proof of proposition 11.18, we deduce that

$$\mu^*(f\varphi_{K_n \cup N}) = \inf \mu^*(h), \quad h \geqslant f\varphi_{K_n \cup N}, \ h \in \mathscr{I}_+ ,$$

and since $f = \sup_n f\varphi_{K_n \cup N}$, it follows that

$$\mu^*(f) = \inf \mu^*(h), \quad h \geqslant f, \ h \in \mathscr{I}_+ .$$

(c) If A is contained in the union of a sequence (G_n) of open sets with $\mu^*(G_n) < +\infty$, we may assume that the sequence (G_n) is increasing and then $f = \sup f\varphi_{G_n}$. Since for each n we have, according to (b),

$$\mu^*(f\varphi_{G_n}) = \inf \mu^*(h), \ h \geqslant f\varphi_{G_n}, \ h \in \mathscr{I}_+ ,$$

it follows that

$$\mu^*(f) = \inf \mu^*(h), \quad h \geqslant f, \ h \in \mathscr{I}_+ .$$

11.20 COROLLARY. *For every set $A \subset T$ contained in the union of a sequence (G_n) of open sets with $\mu(G_n) < +\infty$ we have*

$$\mu^*(A) = \inf_{G \supset A} \mu^*(G), \ G \text{ open} .$$

If $\mu^*(A) = +\infty$, we have $\mu^*(G) = +\infty$ for every open set $G \supset A$ and the stated equality is satisfied.

Assume that $\mu^*(A) < +\infty$. By the preceding proposition we have then

$$\mu^*(A) = \mu^*(\varphi_A) = \inf \mu^*(h), \quad h \geqslant \varphi_A, \ h \in \mathscr{I}_+ .$$

Let $0 < \varepsilon < 1$. There exists a function $h \in \mathscr{I}_+$ such that $\varphi_A \leqslant h$ and $\mu^*(A) \leqslant \mu^*(h) \leqslant \mu^*(A) + \varepsilon$. The set $G = \{t \mid h(t) > 1 - \varepsilon\}$ is open and contains A. But $h \geqslant (1-\varepsilon)\varphi_G$, therefore

$$\mu^*(G) \leqslant (1-\varepsilon)^{-1} \mu^*(h) \leqslant (1-\varepsilon)^{-1} (\mu^*(A) + \varepsilon) ,$$

hence

$$\inf_{G \supset A} \mu^*(G) \leqslant (1-\varepsilon)^{-1} (\mu^*(A) + \varepsilon).$$

ε being arbitrary, we obtain

$$\inf_{G\supset A}\mu^*(G)\leqslant \mu^*(A)\,,\quad G \text{ open}\,,$$

whence

$$\mu^*(A)=\inf_{G\supset A}\mu^*(G),\ G \text{ open}\,.$$

5 *Measurable functions defined locally*

11.21 COROLLARY. *We say that a set* $\mathscr{L}$ *of subsets of* T *is locally countable if for every point* $t\in T$, *there exists a neighbourhood* V *of* t *with the property that the set of the subsets* $A\in\mathscr{L}$ *which intersect* V *is at most countable.*

From this definition if follows that if $\mathscr{L}$ is locally countable then the set of subsets $A\in\mathscr{L}$ which intersect a *compact* set $K\subset T$, is at most countable, since K may be covered by a finite number of neighbourhoods V occurring in the above definition.

11.22 PROPOSITION. *Let* μ *be a positive measure on* T. *There exists then a locally countable set* $\mathscr{K}$ *of nonempty disjoint compact subsets such that the set* $T\setminus\cup_{K\in\mathscr{K}}K$ *is* μ*-negligible.*

If $\mu=0$, we take $\mathscr{K}$ consisting of a single nonempty compact set $K\subset T$ and the proposition is proved. Assume now that $\mu\neq 0$.

Consider the sets $\mathscr{L}$ of nonempty compact mutually disjoint subsets $K\subset T$ with the property that for every set $K\in\mathscr{L}$ and every open set $U\subset T$ with $U\cap K\neq\varnothing$ we have $\mu(U\cap K)>0$. The sets $\mathscr{L}$ form a subset $\mathscr{F}$ of the set $\mathscr{P}(\mathscr{P}(T))$ which we consider ordered by inclusion.

The set $\mathscr{F}$ is not empty. In fact, since $\mu\neq 0$, there exists at least a compact set $H\subset T$ with $\mu(H)>0$. Let $\mathscr{U}$ be the set of all the open subsets $U\subset T$ which intersect H and for which $\mu(H\cap T)=0$ and let $U_0=\cup_{U\in\mathscr{U}}U$. The set U_0 is open and does not contain H, since if we had $H\subset U_0$, then H might be covered by a finite number of sets from $\mathscr{U}$ and it would follow that $\mu(H)=0$. Consider the compact set $K=H\setminus U_0$. We have $\mu(K)>0$.

In fact, if we had $\mu(K)=0$, there would exist an open set $G\supset K$ with $\mu(G)<\mu(H)$; then H might be covered by G and by a finite number of sets from $\mathscr{U}$ and it would result $\mu(H)=\mu(G\cap H)<\mu(H)$.

If U is an open set with $K\cap U\neq\varnothing$, then $\mu(K\cap U)>0$. In fact, if we had $\mu(K \cap U)=0$, then it would result $\mu(H \cap U)=0$ hence $U \subset G_0$, therefore $K \cap U=\varnothing$.

The set $\mathscr{L}$ consisting of the subset K only belongs to $\mathscr{F}$; therefore $\mathscr{F}$ is not empty.

We can prove immediately that if $(\mathscr{L}_\alpha)_{\alpha\in A}$ is a totally ordered family of sets of $\mathscr{F}$, then $\mathscr{L}=\cup_{\alpha\in A}\mathscr{L}_\alpha$ also belongs to $\mathscr{F}$. Consequently, the set $\mathscr{F}$ is inductively ordered. By Zorn's lemma, $\mathscr{F}$ has at least one maximal element. Let $\mathscr{K}$ be a maximal element of $\mathscr{F}$. The set $\mathscr{K}$ consists of mutually disjoint nonempty compact subsets and for every set $K\in\mathscr{K}$ and every open set $U\subset T$ with $U\cap K\neq\varnothing$ we have $\mu(U\cap K)>0$.

It remains to prove that the set $\mathscr{K}$ has the other stated properties.

Let $t\in T$ and V be a compact neighbourhood of t. For every finite family $(K_i)_{1\leqslant i\leqslant n}$ of subsets of $\mathscr{K}$, which intersect V we have

$$\Sigma_{1\leqslant i\leqslant n}\mu(K_i\cap V)=\mu(V\cap(\cup_{1\leqslant i\leqslant n}K_i))\leqslant\mu(V)$$

since the sets K_i are disjoint. If we denote by $\mathscr{K}_V$ the sets of the subsets $K\in\mathscr{K}$ which intersect V, we have

$$\Sigma_{K\in\mathscr{K}_V}\mu(K\cap V)\leqslant\mu(V)<\infty\,.$$

Since $\mu(K\cap V)>0$ for $K\in\mathscr{K}_V$, it follows that the set $\mathscr{K}_V$ is at most countable, therefore $\mathscr{K}$ is locally countable.

We prove now that the set $N=T\setminus\cup_{K\in\mathscr{K}}K$ is μ-negligible.

Assume the contrary that N is not negligible. There exists then a compact set K' such that $N\cap K'$ is not μ-negligible. There exists a sequence (K_n) of sets of $\mathscr{K}$ which intersect K', therefore

$$N\cap K'=K'\setminus\cup_{1\leqslant n<\infty}(K_n\cap K)\,.$$

Hence $N\cap K'$ is a relatively compact Borel set, therefore it is μ-integrable. There exists then at least one compact set $H\subset N\cap K'$ with $\mu(H)>0$. Similarly, we construct a compact set $K\subset H$ such that for every open set U with $U\cap K\neq\varnothing$ we have $\mu(U\cap K)>0$. Then the set $\mathscr{K}\cup\{K\}$ belongs to $\mathscr{F}$ which contradicts the fact that the set $\mathscr{K}$ is a maximal element in $\mathscr{F}$.

Hence the set N is μ-negligible and the proposition is proved.

11.23 PROPOSITION. *Let E be a set, μ a positive measure on T, and for each compact set $K\subset T$, let $g_K: K\to E$ be a function. Assume that for each pair (K, K') of compact sets the functions g_K and $g_{K'}$ are equal μ-almost everywhere* *on $K \cap K'$. There exists then a function $f: T \to E$ such that for every compact set $K \subset T$ we have $f(t) = g_K(t)$, μ-almost everywhere on K.*

Let $\mathscr{K}$ be a locally countable set of nonempty compact mutually disjoint sets $K \subset T$ such that the set $N = T \setminus \cup_{K \in \mathscr{K}} K$ is μ-negligible.

We define the function f on T by

$$f(t) = g_K(t), \text{ if } t \in K \text{, and } K \in \mathscr{K}$$
$$f(t) = a \in E, \text{ if } t \in N \,.$$

Now let $H \subset T$ be an arbitrary compact set. The set of the subsets K of $\mathscr{K}$ which intersect H is at most countable.

Let (K_n) be the family of sets which intersect H. Since $N \cup (\cup_{K \in \mathscr{K}} K) = T$, we have $(H \cap N) \cup (\cup_n (K_n \cap H)) = H$ and $H \cap N$ is μ-negligible.

For each n, there exists a μ-negligible set $N_n \subset K_n \cap H$ such that for $t \in (K_n \cap H) \setminus N_n$ we have $f_{K_n}(t) = f_H(t)$.

For $t \in (K_n \cap H) \setminus N_n$ we have then $f(t) = f_{K_n}(t) = f_H(t)$. Hence we have $f(t) = f_H(t)$ for the points $t \in H$ which do not belong to the μ-negligible set $(H \cap N) \cup \cap_n N_n$.

Thus the proposition is proved.

11.24 COROLLARY. *Let $(U_\alpha)_{\alpha \in A}$ be a covering of T consisting of open sets, and for each $\alpha \in A$, let f_α be a mapping of U_α into a set G such that for every pair of indices (α, β) we have $f_\alpha(t) = f_\beta(t)$, μ-almost everywhere on $U_\alpha \cap U_\beta$. There exists then a function $f: T \to G$ such that for every $\alpha \in A$ we have $f(t) = f_\alpha(t)$, μ-almost everywhere on U_α.*

For every compact set $K \subset T$ choose a finite family $U_{\alpha_1}, \ldots, U_{\alpha_n}$ which covers K. Let N_{ij} be the μ-negligible set of $U_{\alpha_i} \cap U_{\alpha_j}$ such that for $t \in U_{\alpha_i} \cap U_{\alpha_j} \setminus N_{ij}$ we have $f_{\alpha_i}(t) = f_{\alpha_j}(t)$. The set $N = \cup N_{ij}$ is μ-negligible and for every $1 \leqslant i, j \leqslant n$ and $t \in U_{\alpha_i} \cap U_{\alpha_j} \setminus N$ we have $f_{\alpha_i}(t) = f_{\alpha_j}(t)$. We define then the function $g_K : K \to G$ by

$$g_K(t) = f_{\alpha_i}(t), \text{ if } t \in U_{\alpha_i} \setminus N \,,$$
$$g_K(t) = a \in G, \text{ if } t \in N \,.$$

It is easily verified that for every $\alpha \in A$, we have $g_K(t) = f_\alpha(t)$ μ-almost everywhere on $K \cap U_\alpha$.

The functions g_K satisfy the hypothesis of proposition 11.23. Consequently, there exists a function $f: T \to G$ such that for every compact set $K \subset T$

we have $f(t) = g_K(t)$, μ-almost everywhere on K. Then for every $\alpha \in A$ and every compact set $K \subset T$ we have $f(t) = g_K(t) = f_\alpha(t)$, μ-almost everywhere on $K \cap U_\alpha$, whence it follows that $f(t) = f_\alpha(t)$, μ-almost everywhere on U_α.

11.25 PROPOSITION. *Let E, F and G be three Banach spaces and $u: E \times F \to G$ a bilinear continuous mapping such that $|x| = \sup_{|y| \leqslant 1} |u(x, y)|$ for every $x \in E$. For every μ-measurable function $\boldsymbol{f}: T \to E$ and $0 < a < 1$, there exists a μ-measurable function $\boldsymbol{g}: T \to F$ such that*

$$a|\boldsymbol{f}(t)| \leqslant |u(\boldsymbol{f}(t), \boldsymbol{g}(t))| \text{ and } |\boldsymbol{g}(t)| = 1$$

μ-almost everywhere.

Let $K \subset T$ be a compact set. Assume first that the restriction of $\boldsymbol{f}$ to K is continuous and $\boldsymbol{f}(t) \neq 0$ for $t \in K$. For every point $t_0 \in K$ there exists an element $y_0 \in F$ with $|y_0| = 1$ and

$$a|\boldsymbol{f}(t_0)| < |u(\boldsymbol{f}(t), y_0)| .$$

Since the restrictions to K of the functions from this inequality are continuous, there exists a neighbourhood V_0 of t_0 such that for $t \in V_0 \cap K$ we have

$$a|\boldsymbol{f}(t)| < |u(\boldsymbol{f}(t), y_0)| .$$

Since K is a compact set, there exists a finite family $V_1, \ldots, V_n$ of open sets which cover K and a finite family $y_1, \ldots, y_n$ of elements of F such that for each i we have

$$|y_i| = 1 \text{ and } a|\boldsymbol{f}(t)| < |u(\boldsymbol{f}(t), y_i)| \text{ for } t \in V_i \cap K .$$

Choose a family $A_1, \ldots, A_n$ of mutually disjoint μ-measurable sets with $A_i \subset V_i$ for each i and $\cup_{1 \leqslant i \leqslant n} A_i = K$. The step function $\boldsymbol{g} = \Sigma_{1 \leqslant i \leqslant n} \varphi_{A_i} y_i$ is μ-measurable and for $t \in K$ we have

$$|\boldsymbol{g}(t)| = 1 \quad \text{and} \quad a|\boldsymbol{f}(t)| \leqslant |u(\boldsymbol{f}(t), \boldsymbol{g}(t))| .$$

Assume now that $\boldsymbol{f}$ is measurable and let $A_0 = \{t \in K; \boldsymbol{f}(t) = 0\}$. Since $K \setminus A_0$ is μ-integrable, there exists a μ-negligible set $N \subset K \setminus A$ and a sequence (K_n) of mutually disjoint compact sets, with union $K \setminus A_0 \setminus N$ such that $\boldsymbol{f}$ restricted to each K_n is continuous and different from 0.

For each n, there exists a μ-measurable step function $\boldsymbol{g}_n: T \to F$ with $\boldsymbol{g}_n(t) = 0$ for each $t \notin K_n$ and

$$|\boldsymbol{g}_n(t)| = 1 \quad \text{and} \quad a|\boldsymbol{f}(t)| < |u(\boldsymbol{f}(t), \boldsymbol{g}_n(t))| \text{ for } t \in K_n .$$

The function $\boldsymbol{g}_K = \Sigma_{1 \leqslant n < \infty} \boldsymbol{g}_n + \varphi_A y$, where $y \in F$ and $|y| = 1$, is μ-measurable, takes on an infinite set of values and for $t \in K \setminus N$ we have

$$|\boldsymbol{g}_K(t)| = 1 \text{ and } a|\boldsymbol{f}(t)| \leqslant |u(\boldsymbol{f}(t), \boldsymbol{g}_K(t))| .$$

Let now $\mathscr{K}$ be a locally countable family of mutually disjoint compact sets with union differing from T by a μ-negligible set (proposition 11.22). The function

$$\boldsymbol{g} = \Sigma_{K \in \mathscr{K}} \boldsymbol{g}_K$$

is μ-measurable and fulfills the required conditions.

11.26 Corollary. *Let X and Y be two Banach spaces. For each μ-measurable function $U : T \to \mathscr{L}(X, Y)$ and for every $0 < a < 1$, there exists a μ-measurable function $\boldsymbol{g} : T \to X$ such that*

$$a|U(t)| < |U(t)\boldsymbol{g}(t)| \text{ and } |\boldsymbol{g}(t)| = 1$$

μ-almost everywhere.

11.27 Corollary. *Let E be a Banach space, $\boldsymbol{f} : T \to E$ a μ-measurable function and $0 < a < 1$. There exists a μ-measurable function $\boldsymbol{f}' : T \to E'$ with*

$$a|\boldsymbol{f}(t)| < \langle \boldsymbol{f}(t), \boldsymbol{f}'(t) \rangle \text{ and } |\boldsymbol{f}'(t)| = 1$$

μ-almost everywhere.

11.28 Corollary. *Let E be a Banach space, $\boldsymbol{f}' : T \to E'$ a μ-measurable function and $0 < a < 1$. There exists a μ-measurable function $\boldsymbol{f} : T \to E$ with*

$$a|\boldsymbol{f}'(t)| < \langle \boldsymbol{f}(t), \boldsymbol{f}'(t) \rangle \text{ and } |\boldsymbol{f}(t)| = 1$$

μ-almost everywhere.

6 *Borel functions*

We have seen that lower and upper semicontinuous functions are measurable with respect to every measure. A larger class of functions which have this property are the Borel functions. A Borel step function is a step-function such that the sets on which it is constant are Borel sets.

11.29 Definition. *Let E be a metric space. We denote by $\mathscr{B}_E(T)$ the smallest class of functions $f : T \to E$ which contains the Borel step functions and all the* *functions which have the property that for each compact set $K \subset T$, there exists a sequence (f_n) of functions from $\mathscr{B}_E(T)$ converging to f on K.*

The functions of $\mathscr{B}_E(T)$ are called *Borel functions.*

It is easy to see that $\mathscr{B}_E(T)$ is a vector space and a module over $\mathscr{B}(T) = \mathscr{B}_R(T)$.

The class of all measurable functions with respect to a measure μ contains the Borel step functions and in addition has the stated property. If follows that this class contains $\mathscr{B}_E(T)$; therefore, the Borel functions are measurable with respect to every measure.

By lemma 8.52, every continuous function is a Borel function.

11.30 PROPOSITION. *Let $\mu \geqslant 0$ be a measure on T. For every μ-measurable function $f: T \to E$, there exists a Borel function $g \in \mathscr{B}_E(T)$ equal to f, μ-almost everywhere.*

For each compact set $K \subset T$, there exists a sequence (g_n) of Borel step functions converging to f, μ-almost everywhere on K (theorem 10.3). If we set $g_K(t) = \lim g_n(t)$ when the limit exists and $g_K(t) = 0$ in the contrary case, then g_K is a Borel function equal to f, μ-almost everywhere on K.

By proposition 11.23, there exists a function $g : T \to E$, such that for each compact set $K \subset T$ we have $g(t) = g_K(t)$, μ-almost everywhere.

It follows that g is a Borel function equal to f, μ-almost everywhere.

11.31 PROPOSITION. *Let $\mu \geqslant 0$ be a measure on T. For every μ-measurable step function $\boldsymbol{f}: T \to E$ there exists a Borel step function $\boldsymbol{g}: T \to E$ equal to f, μ-almost everywhere.*

Let $A \subset T$ be a μ-measurable set. We show that there exists a Borel set $B \subset A$ such that $\mu^*(A \setminus B) = 0$.

Let $(K_j)_{j \in J}$ be a locally countable family of disjoint compact subsets such that the set $N = T \setminus \cup_{j \in J} K_j$ is μ-negligible. Since A is μ-measurable, for each $j \in J$ the set $A \cap K_j$ is μ-integrable; therefore, it is the union of a negligible set N_j and a sequence $(C_{jn})_{n \in N}$ of disjoint compact sets. The set $\mathscr{B}_j = \cup_{1 \leqslant n < \infty} C_{jn}$ is a Borel set and differs from $A \cap K_j$ by a μ-negligible set. Show that the set $B = \cup_{j \in J} B_j$ is a Borel set and differs from A by a negligible set.

To do this, let $K \subset T$ be an arbitrary compact set.

Each point $t \in K$ has a neighbourhood V which is intersected at most by a countable family of sets K_j. As K can be covered by a finite number of sets V, it follows that the family of the sets K_j which intersect K is at most countable; denote them by $K_1, K_2, \ldots$ These sets together with N, cover the set K. As $(A \setminus B) \cap K_i$ is μ-negligible for each i, it follows that $(A \setminus B) \cap K$ is μ-negligible, therefore $A \setminus B$ is μ-negligible. On the other hand, for each i, $K_i \cap B$ is a Borel set, hence the set $K_i \cap B \cap K$ is a relatively compact Borel set. Then $K \cap B = \cup_{1 \leqslant i < \infty} (K_i \cap B \cap K)$ is a relatively compact Borel set, therefore B is a Borel set.

Now let $f: T \to E$ be a μ-measurable step function with values $x_1, x_2, \ldots, x_n$, respectively on the μ-measurable sets $A_1, A_2, \ldots, A_n$. For each i consider the Borel set B_i which differs from A_i by a negligible set. Then the function $g: T \to E$ which takes on the values $x_1, x_2, \ldots, x_n$ respectively on the sets $B_1, B_2, \ldots, B_n$, is a Borel step function and it is equal to f almost everywhere.

11.32 Corollary. *Let E be a compact metric space and $f: T \to E$ a μ-measurable function. There exists then a μ-negligible set $N \subset T$ and a sequence (g_n) of Borel step functions which converges uniformly on $T \setminus N$ to f.*

We use proposition 10.7.

11.33 Corollary. *If E is finite dimensional and if $f: T \to E$ is a bounded μ-measurable function, there exists a μ-negligible set $N \subset T$ and a sequence (f_n) of Borel step functions which converges uniformly on $T \setminus N$ to f.*

In particular, corollary 11.33 is valid for real or complex bounded measurable functions.

§12 The space $\mathscr{L}^\infty$

1 *The N_∞ seminorm*

Let μ be a positive measure on T and E a Banach space. For every function $f: T \to E$ we shall denote by $N_\infty(f, \mu)$ or $N_\infty(f)$, the infimum of the set of real numbers α which have the property that $|f(t)| \leqslant \alpha$, μ-almost everywhere.

The number $N_\infty(f)$ is called the *maximum in measure of the function* $|f|$, *or μ-maximum of the function* $|f|$.

If f and g are equal μ-almost everywhere, then $N_\infty(f) = N_\infty(g)$.

12.1 PROPOSITION. *For every function $f:T\to E$ we have $|f(t)|\leqslant N_\infty(f)$, μ-almost everywhere.*

If $N_\infty(f)=+\infty$, the assertion is evidently true. Assume that $N_\infty(f)<+\infty$. Let α_n be a decreasing sequence such that $\alpha_n>N_\infty(f)$ and $\alpha_n\to N_\infty(f)$. For each n we have $|f(t)|\leqslant\alpha_n$, μ-almost everywhere, therefore the set $A_n=\{t|f(t)>\alpha_n\}$ is μ-negligible. Then the set

$$\cup_{1\leqslant n<\infty}A_n=\{t\,|\,|f(t)|>\inf\alpha_n\}$$

is also μ-negligible. Remarking that $\inf\alpha_n=\lim\alpha_n=N_\infty(f)$ we deduce that $|f(t)|\leqslant N_\infty(f)$, μ-almost everywhere.

We list some properties of the function $N_\infty(f)$:

(1) $0\leqslant N_\infty(f)\leqslant+\infty$; $N_\infty(f)=0$ if and only if f is μ-negligible;

(2) $N_\infty(\alpha f)=|\alpha|N_\infty(f)$, α scalar;

(3) $N_\infty(f+g)\leqslant N_\infty(f)+N_\infty(g)$;

(4) $N_\infty(f)\leqslant N_\infty(g)$ if $|f|\leqslant|g|$.

The first part of property (1) is evident; from the inequality $|f(t)|\leqslant N_\infty(f)$, μ-almost everywhere, we deduce that if $N_\infty(f)=0$, then $f(t)=0$, μ-almost everywhere; conversely, if $f(t)=0$, μ-almost everywhere, then $N_\infty(f)=N_\infty(0)=0$.

Property (2) is evidently true for $\alpha=0$, since $\alpha f=0$ and $0\cdot N_\infty(f)=0$, even if $N_\infty(f)=+\infty$; if $\alpha\neq0$, we remark that $|\alpha f(t)|\leqslant\beta$, μ-almost everywhere, if and only if $|f(t)\leqslant\beta/|\alpha|$, μ-almost everywhere. Thus, $\beta\geqslant N_\infty(\alpha f)$ if and only if $\beta/|\alpha|\geqslant N_\infty(f)$, that is $\beta\geqslant|\alpha|N_\infty(f)$. It follows then the equality $N_\infty(\alpha f)=|\alpha|N_\infty(f)$.

Property (3) is obtained by remarking that $|f(t)|\leqslant N_\infty(f)$ and $|g(t)|\leqslant N_\infty(g)$, μ-almost everywhere, therefore

$$|f(t)+g(t)|\leqslant|f(t)|+|g(t)|\leqslant N_\infty(f)+N_\infty(g),$$

μ-almost everywhere, whence

$$N_\infty(f+g)\leqslant N_\infty(f)+N_\infty(g).$$

Property (4) is immediate.

From these properties we deduce also the inequality

(5) $|N_\infty(f)-N_\infty(g)|\leqslant N_\infty(f-g)$.

If $f:T\to E$ and φ is a scalar function, then

(6) $N_\infty(f\varphi)\leqslant N_\infty(f)N_\infty(\varphi)$.

In fact, $|\boldsymbol{f}(t)| \leqslant N_{\infty}(\boldsymbol{f})$ and $|\varphi(t)| \leqslant N_{\infty}(\varphi)$, μ-almost everywhere, therefore

$$|\boldsymbol{f}(t)\varphi(t)| = |\boldsymbol{f}(t)|\,|\varphi(t)| \leqslant N_{\infty}(\boldsymbol{f})N_{\infty}(\varphi),$$

μ-almost everywhere, whence the inequality (6) follows.

If E is a normed algebra, then the inequality

(7) $N_{\infty}(\boldsymbol{fg}) \leqslant N_{\infty}(\boldsymbol{f})\,N_{\infty}(\boldsymbol{g})$.

is true for every functions $\boldsymbol{f}, \boldsymbol{g} : T \to E$.

(8) $N_{\infty}(\boldsymbol{f}) = \sup_K N_{\infty}(\boldsymbol{f}\varphi_K)$, K compact.

If $N_{\infty}(\boldsymbol{f}) = 0$, then $N_{\infty}(\boldsymbol{f}\varphi_K) = 0$ and the equality is verified.

Assume that $N_{\infty}(\boldsymbol{f}) = a > 0$ and let $0 < \alpha < a$. The set $A = \{t \,|\, |\boldsymbol{f}(t)| > \alpha\}$ is not negligible. There exists a compact set $K \subset T$ such that $\mu^*(A \cap K) > 0$. Then $N_{\infty}(\boldsymbol{f}\varphi_K) > \alpha$ therefore $\sup_K N_{\infty}(\boldsymbol{f}\varphi_K) = N_{\infty}(\boldsymbol{f})$.

For a (finite of infinite) numerical function $f: T \to \overline{R}$ we define $N_{\infty}(f)$ as before. Proposition 12.1 remains true also in this case. It follows that if $N_{\infty}(f) < +\infty$, then $|\boldsymbol{f}(t)| < +\infty$, μ-almost everywhere.

The above properties remain true also in this case, if the sum $f+g$ or the difference $f-g$ is defined on T. Still in this case, for two functions f and g equal μ-almost everywhere, we have $N_{\infty}(f) = N_{\infty}(g)$.

Thus, for functions f defined on T with values in E or in $\overline{R}$, $N_{\infty}(f)$ depends only on the class $\tilde{f}$ of the functions equal to f, μ-almost everywhere. That is why we put by definition,

$$N_{\infty}(\tilde{f}) = N_{\infty}(f).$$

If $\boldsymbol{f}$ is a function with values in E or in $\overline{R}$, defined μ-almost everywhere on E, we put, $N_{\infty}(\boldsymbol{f}) = N_{\infty}(\tilde{\boldsymbol{f}})$.

In this case, $N_{\infty}(\boldsymbol{f})$ has the same properties as for the functions defined everywhere.

The Hölder inequality

$$\int^* fg\,\mathrm{d}\mu \leqslant N_p(f)\,N_q(g),\; f \geqslant 0,\; g \geqslant 0$$

was proved for $1 < p$, $q < +\infty$, $p^{-1} + q^{-1} = 1$. We shall prove now the inequality also in case $p = 1$ and $q = +\infty$.

12.2 Proposition. *For every positive functions f and g defined on T (with finite or $+\infty$ values), we have*

$$\int^* fg\,\mathrm{d}\mu \leqslant N_1(f)N_{\infty}(g).$$

In fact, we have $g(t) \leqslant N_\infty(g)$, μ-almost everywhere, therefore $f(t)g(t) \leqslant f(t)\, N_\infty(g)$ μ-almost everywhere. Then

$$\int^* fg\mathrm{d}\mu \leqslant \int^* f\, N_\infty(g)\mathrm{d}\mu = \int^* f\mathrm{d}\mu\, N_\infty(g) = N_1(f) N_\infty(g)\,.$$

2 *The space* $\mathscr{F}^\infty$

Let μ be a positive measure on T and E a Banach space. We shall denote by $\mathscr{F}_E^\infty(T, \mu)$ or $\mathscr{F}_E^\infty(\mu)$ or $\mathscr{F}_E^\infty$ the set of functions $\boldsymbol{f}: T \to E$ with $N_\infty(\boldsymbol{f}) < +\infty$. The set $\mathscr{F}_E^\infty$ is a vector space and $N_\infty(\boldsymbol{f})$ is a seminorm on this space.

If E is a normed algebra, then $\mathscr{F}_E^\infty$ is an algebra.

A function $\boldsymbol{f}$ of $\mathscr{F}_E^\infty$ can be characterized by the fact that there exists a *bounded* function $\boldsymbol{g}$ equal to $\boldsymbol{f}$, μ-almost everywhere.

In fact, we may take $\boldsymbol{g}(t) = \boldsymbol{f}(t)$ if $|\boldsymbol{f}(t)| \leqslant N_\infty(\boldsymbol{f})$ and $\boldsymbol{g}(t) = 0$ in the contrary case.

In particular, $\mathscr{F}_E^\infty$ contains all the constant functions.

12.3 PROPOSITION. *A sequence* $(\boldsymbol{f}_n)$ *of function of* $\mathscr{F}_E^\infty$ *converges to a function* $\boldsymbol{f} \in \mathscr{F}_E^\infty$ *for the topology defined by the seminorm* N_∞ *if and only if there exists a* μ*-negligible set* $A \subset T$ *such that the sequence* $(\boldsymbol{f}_n)$ *converges uniformly to* $\boldsymbol{f}$ *on* $T \setminus A$.

We remark first that for each n there exists a μ-negligible set $A_n \subset T$, such that $|\boldsymbol{f}_n(t) - \boldsymbol{f}(t)| \leqslant N_\infty(\boldsymbol{f}_n - \boldsymbol{f})$ for $t \notin A_n$. The set $A = \cup_{1 \leqslant n < \infty} A_n$ is μ-negligible and for every $t \notin A$ and every $n \in N$ we have

$$|\boldsymbol{f}_n(t) - \boldsymbol{f}(t)| \leqslant N_\infty(\boldsymbol{f}_n - \boldsymbol{f})\,.$$

Moreover, for every $n \in N$ we have

$$\sup_{t \notin A} |\boldsymbol{f}_n(t) - \boldsymbol{f}(t)| = N_\infty(\boldsymbol{f}_n - \boldsymbol{f}).$$

Thus it follows that if $N_\infty(\boldsymbol{f}_n - \boldsymbol{f}) \to 0$, then the sequence $(\boldsymbol{f}_n)$ converges uniformly to $\boldsymbol{f}$ on the set $T \setminus A$,and conversely, if $(\boldsymbol{f}_n)$ converges uniformly on on $T \setminus A$, then $N_\infty(\boldsymbol{f}_n - \boldsymbol{f}) \to 0$.

12.4 PROPOSITION. *The space* $\mathscr{F}_E^\infty$ *is complete.*

Let $(\boldsymbol{f}_n)$ be a Cauchy sequence of functions of $\mathscr{F}_E^\infty$. For every pair (m, n) of natural numbers, there exists a μ-negligible set $A_{m,n}$ such that for $t \notin A_{m,n}$

we have $|f_n(t)-f_m(t)| \leqslant N_{\infty}(f_n-f_m)$. Also, for every $n \in N$, there exists a μ-negligible set B_n such that for $t \notin B_n$ we have $|f_n(t)| \leqslant N_{\infty}(f_n)$. The union A of the sets $A_{m,n}$ and B_n is μ-negligible and for every $t \notin A$ we have $|f_n(t)-f_m(t)| \leqslant N_{\infty}(f_n-f_m)$ and $|f_n(t)| \leqslant N_{\infty}(f_n)$, for every n and m. Since $\lim_{n,m} N_{\infty}(f_n-f_m)=0$, it follows that $\lim_{n,m}|f_n(t)-f_m(t)|=0$, uniformly for $t \in T \setminus A$.

Thus, for $t \notin A$ the sequence $(f_n(t))$ is a Cauchy sequence of elements of the Banach space E, consequently it has a limit. Put $f(t)=\lim f_n(t)$, for $t \notin A$, $f(t)=0$, for $t \in A$.

From the inequality

$$|N_{\infty}(f_n)-N_{\infty}(f_m)| \leqslant N_{\infty}(f_n-f_m)$$

it follows that the numerical sequence $(N_{\infty}(f_n))$ is a Cauchy sequence, therefore it is convergent. We have, for $t \notin A$,

$$|f(t)| = \lim |f_n(t)| \leqslant \lim N_{\infty}(f_n) < +\infty$$

therefore

$$N_{\infty}(f) \leqslant \lim N_{\infty}(f_n) < +\infty ,$$

that is, $f \in \mathscr{F}_E^{\infty}$.

Since (f_n) converges uniformly to f on $T \setminus A$, it follows that $N_{\infty}(f_n-f) \to 0$, and the proposition is proved.

3 *The space $\mathscr{L}^{\infty}$*

For $1 \leqslant p < +\infty$, the space $\mathscr{L}_E^p$ is characterized by being the *subspace of the μ-measurable functions of* $\mathscr{F}_E^p$.

Similarly, in case $p=\infty$, we shall denote by $\mathscr{L}_E^{\infty}(T, \mu)$ or $\mathscr{L}_E^{\infty}(\mu)$ or $\mathscr{L}_E^{\infty}$, the set of the *μ-measurable functions of* $\mathscr{F}_E^{\infty}$.

We deduce immediately that $\mathscr{L}_E^{\infty}$ is a vector space and that N_{∞} is a seminorm on this space. The space $\mathscr{L}_E^{\infty}$ contains the subspace $\mathscr{N}_{\infty}$ of the μ-negligible functions $f: T \to E$.

The quotient space $\mathscr{L}_E^{\infty}/\mathscr{N}_{\infty}$ will be denoted by L_E^{∞}. The function $N_{\infty}(\tilde{f})$ is a norm on L_E^{∞}. This norm is denoted also by $\|\tilde{f}\|_{\infty}$.

In case $E=R$, we shall write $\mathscr{L}^{\infty}$ and L^{∞} instead of $\mathscr{L}_R^{\infty}$ and L_R^{∞}.

If $\boldsymbol{m}: \mathscr{K}_E(T) \to F$ is a dominated measure, we shall write sometimes $\mathscr{L}_E^{\infty}(\boldsymbol{m})$ instead of $\mathscr{L}_E^{\infty}(|\boldsymbol{m}|)$, and $N_{\infty}(f, \boldsymbol{m})$ instead of $N_{\infty}(f, |\boldsymbol{m}|)$.

12.5 PROPOSITION. *The space* $\mathscr{L}_E^\infty$ *is complete. The space* L_E^∞ *is a Banach space.*

Let (f_n) be a sequence of functions of $\mathscr{L}_E^\infty$ which converges to a function $f \in \mathscr{F}_E^\infty$ for the seminorm N_∞. There exists a μ-negligible set $N \subset T$ outside of which the sequence (f_n) converges uniformly to f. The functions f_n are μ-measurable, therefore the functions $g_n = f_n \varphi_{T \setminus N}$ are μ-measurable and converge to the function $g = f\varphi_{T \setminus N}$. It follows that g is μ-measurable; the function f is also μ-measurable since it is equal μ-almost everywhere to g. It follows that $f \in \mathscr{L}_E^\infty$. Thus $\mathscr{L}_E^\infty$ is closed in the complete space $\mathscr{F}_E^\infty$. We deduce that $\mathscr{L}_E^\infty$ is also complete. The quotient space L_E^∞ is then also complete.

12.6 PROPOSITION. *Every bounded Borel function* $f: T \to E$ *belongs to the space* $\mathscr{L}_E^\infty(\mu)$, *for every measure* μ.

In fact, the Borel functions are measurable for every measure μ.

In particular, $\mathscr{L}_E^\infty(\mu)$ contains the bounded continuous functions. We remark that if f is continuous and bounded, we have

$$N_\infty(f) \leqslant \| f \| = \sup_{t \in T} |f(t)|$$

and the inequality may be strict.

12.7 PROPOSITION. *We have the equality* $N_\infty(f) = \| f \|$ *for every bounded and continuous function* f *if and only if the support of the measure* μ *is equal to* T.

If the support $S(\mu)$ of the measure μ differs from T, the set $T \setminus S(\mu)$ is open and μ-negligible. There exists a continuous non-zero function f with the support contained in $T \setminus S(\mu)$. We have then $\| f \| > 0$ and $N_\infty(f) = 0$. Thus if $N_\infty(f) = \| f \|$ for every continuous bounded function f, then $S(\mu) = T$.

Conversely assume that $S(\mu) = T$. Let $f: T \to E$ be a continuous, bounded, non identically zero function and let $0 < \alpha < \| f \|$. The set $\{t \mid |f(t)| < \alpha\}$ is open and nonempty, therefore it has strictly positive measure. It follows that $N_\infty(f) > \alpha$. We deduce then that $N_\infty(f) = \| f \|$. The proposition is proved.

If $S(\mu) = T$, we may embed the Banach space $\mathscr{C}_E^\infty(T)$ of the continuous bounded functions $f: T \to E$, in the space $\mathscr{L}_E^\infty(T)$. The space $\mathscr{L}_E^\infty(T)$ is not separated, in general, so that the Banach space $\mathscr{C}_E^\infty(T)$ is not a closed subspace of $\mathscr{L}_E^\infty$. If we identify a function $f \in \mathscr{C}_E^\infty(T)$ with the class $\tilde{f} \in L_E^\infty$, the space $\mathscr{C}_E^\infty(T)$ may be isometrically embedded in the space L_E^∞, since, if $S(\mu) = T$, two different functions f and g of $\mathscr{C}_E^\infty(T)$ belong to different classes

$\tilde{f}$ and $\tilde{g}$ of L_E^{∞}. In this case, $\mathscr{C}_E^{\infty}(T)$ is a closed subspace of the Banach space L_E^{∞}. The space $\mathscr{C}_E^{\infty}$ differs, in general, from L_E^{∞}, since for an arbitrary function $f \in \mathscr{L}_E^{\infty}$, in general there is no continuous function g, equal to f μ-almost everywhere. It follows that the space $\mathscr{K}_E(T)$ of the continuous functions $f: T \to E$ with compact support is, in general, not dense in $\mathscr{L}_E^{\infty}$, since the closure of $\mathscr{K}_E(T)$ is contained in $\mathscr{C}_E^{\infty}(T)$.

12.8 PROPOSITION. *Every μ-measurable step function $f: T \to E$ belongs to the space $\mathscr{L}_E^{\infty}$.*

In fact, a step function has a finite number of values, therefore it is bounded.

We have seen that the space of the Borel step functions with compact support is dense in each space $\mathscr{L}_E^{\infty}$, $1 \leqslant p < +\infty$. But, in general, the space of all μ-measurable step functions (with arbitrary support) is not dense in $\mathscr{L}_E^{\infty}$, since an arbitrary function $f \in \mathscr{L}_E^{\infty}$ is not, in general, a uniform limit, outside a μ-negligible set, of a sequence of step functions.

Nevertheless, for functions with values in a finite-dimensional space we have:

12.9 PROPOSITION. *If E is finite dimensional, the space of the Borel step functions $f: T \to E$ is dense in the space $\mathscr{L}_E^{\infty}$.*

In fact, every function $f \in \mathscr{L}_E^{\infty}$ is equal almost everywhere to a bounded function $g \in \mathscr{L}_E^{\infty}$, and the μ-measurable bounded function g can be uniformly approximated almost everywhere by Borel step functions (corollary 11.33).

12.10 COROLLARY. *The space of the real (respectively complex) Borel step functions is dense in the space $\mathscr{L}^{\infty}$ (respectively $\mathscr{L}_C^{\infty}$).*

§13 The lifting property

1 *The lifting property of $\mathscr{L}^{\infty}$*

Let μ be a *positive* measure on T. For two functions f and g defined on T we write $f \sim g$ if f and g are equal μ-almost everywhere.

13.1 DEFINITION. *Let $\mathscr{F}$ be an algebra of real or complex functions defined on T.*

A mapping $\rho : \mathscr{F} \to \mathscr{F}$ *is called a lifting of* $\mathscr{F}$ *if it fulfills the following conditions.*

I. $\rho(f) \sim f$;
II. $f \sim g$ *implies* $\rho(f) = \rho(g)$;
III. $f =$ constant *implies* $\rho(f) = f$;
IV. $\rho(\alpha f + \beta g) = \alpha\rho(f) + \beta\rho(g)$;
V. $f \geqslant 0$ *implies* $\rho(f) \geqslant 0$;
VI. $\rho(fg) = \rho(f)\rho(g)$.

If $\mathscr{F}$ *has a lifting* ρ, *we say that* $\mathscr{F}$ *has the lifting property.*

An algebra $\mathscr{F}$ does not necessarily have a lifting. For example, the algebra of all μ-measurable functions has no lifting.

13.2 Proposition. *If* $\mu \neq 0$, *then the algebra* $\mathscr{L}^\infty(\mu)$ *has the lifting property.*

The proof is given in my book *Vector measures*, Pergamon Press, 1967, and in the book by A. and C. Ionescu Tulcea, *Topics in the theory of lifting*, Springer, 1969.

A lifting ρ of $\mathscr{L}^\infty(\mu)$ has, in addition, the following properties:

VII. $f(t) \leqslant g(t)$ μ-almost everwhere, implies $\rho(f) \leqslant \rho(g)$ everywhere;
VIII. $|\rho(f)(t)| \leqslant N_\infty(f)$, for every $t \in T$;
IX. $|\rho(f)| = \rho(|f|)$;
X. $\rho(\sup(f, g) = \sup\ (\rho(f), \rho(g))$,
$\rho\ (\inf\ (f, g) = \inf\ (\rho(f), \rho(g))$.

Each lifting ρ of $\mathscr{L}^\infty(\mu)$ is extended uniquely to a lifting of the space $\mathscr{L}_C^\infty(\mu)$ denoted also by ρ, if for each complex function $f = f_1 + if_2$ with $f_1, f_2 \in \mathscr{L}^\infty(\mu)$ we set

$$\rho(f) = \rho(f_1) + i\rho(f_2).$$

It is immediately seen that on $\mathscr{L}_C^\infty(\mu)$. ρ has all the properties I–VI and VII–X.

In the sequel every lifting of $\mathscr{L}^\infty(\mu)$ will be considered automatically extended to $\mathscr{L}_C^\infty(\mu)$.

For two sets $A, B \subset T$, we write $A \sim B$ if $\varphi_A \sim \varphi_B$, that is if $\mu^*(A \triangle B) = 0$.

Let ρ be a lifting of $\mathscr{L}^\infty(\mu)$. For each μ-measurable set $A \subset T$, $\rho(\varphi_A)$ is a characteristic function of a set denoted by $\rho(A)$:

$$\varphi_{\rho(A)} = \rho(\varphi_A).$$

The mapping ρ defined for the μ-measurable sets has the following properties:

I′. $\rho(A) \sim A$;
II′. $A \sim B$ implies $\rho(A) = \rho(B)$;
III′. $\rho(\varnothing) = \varnothing$ and $\rho(T) = T$;
IV′. $\rho(A \cup B) = \rho(A) \cup \rho(B)$;
V′. $\rho(A \cap B) = \rho(A) \cap \rho(B)$;

and also the following properties which are deduced from the preceding ones;

VI′. $A \subset B$ implies $\rho(A) \subset \rho(B)$;
VII′. $A \cap B = \varnothing$ implies $\rho(A) \cap \rho(B) = \varnothing$;
VIII′. $\rho(A \setminus B) = \rho(A) \setminus \rho(B)$.

Besides $\mathscr{L}^\infty(\mu)$, there exist also other algebras which have the lifting property.

We remark first that if ρ is a lifting of an algebra $\mathscr{F}$ and if $\mathscr{G}$ is a subalgebra of $\mathscr{F}$ which contains together with a function f, all functions of $\mathscr{F}$ equivalent to f, then the restriction of ρ to $\mathscr{G}$ is a lifting of $\mathscr{G}$.

We shall denote by $C(\mu)$ the set of locally countable families $\mathscr{K} = (K_j)_{j \in J}$ of disjoint nonempty compact subsets, such that the set $T \setminus \cup_{j \in J} K_j$ is μ-negligible.

If $\mathscr{K}_i = (K_i)_{i \in I} \in C(\mu)$ and $\mathscr{K}_2 = (K_j)_{j \in J} \in C(\mu)$, then the family $\mathscr{K}$ consisting of the sets $K_i \cap K_j$ with $i \in I$ and $j \in J$ which are non empty, belongs to $C(\mu)$.

Thus, the set $C(\mu)$ is ordered by the relation $\mathscr{K}' < \mathscr{K}''$ which means that each set of $\mathscr{K}''$ is contained in some set of $\mathscr{K}'$.

For each family $\mathscr{K} = (K_j)_{j \in J} \in C(\mu)$ denote by $\mathscr{M}_C^\infty(\mathscr{K})$ the set of the complex functions f defined on T, which have the property that $f\varphi_{K_j} \in \mathscr{L}_C^\infty(\mu)$, for every $j \in J$. The set $\mathscr{M}_C^\infty(\mathscr{K})$ is an algebra which contains $\mathscr{L}_C^\infty(\mu)$.

13.3 Proposition. *For each lifting ρ of $\mathscr{L}^\infty(\mu)$ there exists a lifting $\rho_{\mathscr{K}}$ of $\mathscr{M}_C^\infty(\mathscr{K})$ such that for every function $f \in \mathscr{M}_C^\infty(\mathscr{K})$ and every $j \in J$ we have*

$$\rho(\varphi_{K_j} f) = \varphi_{\rho(K_j)} \rho_{\mathscr{K}}(f).$$

Let ρ be a lifting of $\mathscr{L}_C^\infty(\mu)$ and A a compact set with $\mu(A) > 0$. Consider the subalgebra $\mathscr{L}_C^\infty(A, \mu)$ of functions $f \in \mathscr{L}_C^\infty(\mu)$ which vanish outside A. Let $\chi : \mathscr{L}_C^\infty(A, \mu) \to C$ be a continuous linear and multiplicative mapping with

$\chi(\varphi_A)=1$. From the theory of Banach algebras it is known that there exist such mappings.

For each function $f\in\mathcal{M}_C^\infty(\mathcal{K})$ we define the function $\rho_{\mathcal{K}}(f)$ by

$$\rho_{\mathcal{K}}(f)(t)=\rho(\varphi_{K_j}f)(t), \text{ if } t\in\rho(K_j) \text{ and } j\in J\,,$$
$$\rho_{\mathcal{K}}(f)(t)=\chi(f\varphi_A) \quad \text{if } t\in T\setminus\cup_{j\in J}\rho(K_j)\,.$$

It is immediately seen that $\rho_{\mathcal{K}}(f)\varphi_{K_j}\in\mathcal{L}_C^\infty(\mu)$ for each $j\in J$, consequently $\rho_{\mathcal{K}}(f)\in\mathcal{M}_C^\infty(\mu)$, and that $\rho_{\mathcal{K}}$ has all the properties I–VI.

Denote by $\mathcal{M}_C^\infty$ the set functions $f\colon T\to C$ which have the property that $f\varphi_K\in\mathcal{L}_C^\infty(\mu)$ for every compact set K; for every locally countable family $\mathcal{K}=(K_j)_{j\in J}$ of disjoint compact subsets such that $T\setminus\cup K_j$ is μ-negligible, $\mathcal{M}_C^\infty$ is a subalgebra of the algebra $\mathcal{M}_C^\infty(\mathcal{K})$.

13.4 COROLLARY. *There exists a mapping $\rho_1:\mathcal{M}_C^\infty\to\mathcal{M}_C^\infty$, called a lifting of $\mathcal{M}_C^\infty$, having properties* I–VI.

Let $\mathcal{K}=(K_j)_{j\in J}$ be a locally countable family of disjoint compact sets such that $T\setminus\cup K_j$ is μ-negligible. The restriction ρ_1 of $\rho_{\mathcal{K}}$ to $\mathcal{M}_C^\infty$ is a lifting of $\mathcal{M}_C^\infty$, since for $f\in\mathcal{M}_C^\infty$ and K compact we have $\rho_{\mathcal{K}}(f)\varphi_K\sim f\varphi_K\in\mathcal{M}_C^\infty$, therefore $\rho_{\mathcal{K}}(f)\in\mathcal{M}_C^\infty$.

An immediate application of the existence of a lifting of the space $\mathcal{L}_C^\infty$ is the following proposition, which extends theorem 5.3 to the case of measurable functions.

13.5 PROPOSITION. *Let ρ be a lifting of $\mathcal{M}_R^\infty$ and let $\mathcal{F}$ be a directed set (for the relation $\leqslant$) of functions $f\in\mathcal{M}_R^\infty$ such that $f\geqslant 0$ and $\rho(f)=f$. Then the upper envelope $f_\infty=\sup_{f\in\mathcal{F}}f$ is μ-measurable and*

$$\int^* \sup_{f\in\mathcal{F}} f\,d\mu=\sup_{f\in\mathcal{F}}\int^* f\,d\mu\,.$$

Assume first that f_∞ is bounded. Let $K\subset T$ be a compact set. Then

$$f_\infty(t)\varphi_{\rho(K)}(t)=\sup_{f\in\mathcal{F}}f(t)\varphi_{\rho(K)}(t), \text{ for every } t\in T\,.$$

Since f_∞ is bounded, each function $f\in\mathcal{F}$ is bounded, therefore we have $f\varphi_K\in\mathcal{L}^1(\mu)$ for each $f\in\mathcal{F}$. It follows that the set $\mathcal{F}_K=\{\tilde f\tilde\varphi_{\rho(K)};\,f\in\mathcal{F}\}$ of elements of $L^1(\mu)$ is directed and $\sup_{f\in\mathcal{F}_K}\|\tilde f\tilde\varphi_{\rho(K)}\|_1\leqslant N_1(f_\infty\varphi_K)<\infty$. By theorem 7.47, $\mathcal{F}_K$ has a supremum $\tilde g\in L^1(\mu)$ and converges to $\tilde g$ in $L^1(\mu)$. We can find an increasing sequence of functions of $\mathcal{F}$ such that $(\tilde f_n\tilde\varphi_{\rho(K)})$ tends to $\tilde{g}$ in $L^1(\mu)$ and almost everywhere. Then $f_\infty(t)\varphi_{\rho(K)}(t) \geqslant \lim_n f_n(t)\varphi_{\rho(K)}(t) = g(t)$, μ-almost everywhere. It follows that $g \in \mathscr{L}^\infty(\mu)$. From the inequality

$$g(t) \geqslant f(t)\varphi_{\rho(K)}(t) \ \mu\text{-almost everywhere,}$$

for each $f \in \mathscr{F}$, we deduce

$$\rho(g)(t) \geqslant \rho(f\varphi_{\rho(K)})(t) = f(t)\varphi_{\rho(K)}(t)$$

for each $t \in T$, and $f \in \mathscr{F}$, therefore

$$\rho(g)(t) \geqslant f_\infty(t)\varphi_{\rho(K)}(t), \text{ for every } t \in T.$$

Thus $f_\infty \varphi_{\rho(K)}$ coincides with g almost everywhere, therefore it is μ-measurable and

$$\sup_{f \in \mathscr{F}} \int f\varphi_{\rho(K)} \mathrm{d}\mu = \int g \,\mathrm{d}\mu = \int f_\infty \varphi_{\rho(K)} \mathrm{d}\mu .$$

Since K is arbitrary it follows that f_∞ is μ-measurable; taking in the preceding equality the supremum for all compact sets $K \subset T$, we deduce

$$\sup_{f \in \mathscr{F}} \int^* f \mathrm{d}\mu = \int^* f_\infty \mathrm{d}\mu .$$

Assume now that f_∞ is arbitrary. For each natural number p, consider the family $\mathscr{F}_p = \{\inf(f, p); f \in \mathscr{F}\}$. The set $\mathscr{F}_p$ is directed, consists of bounded functions and we have

$$\rho(\inf(f, p)) = \inf(\rho(f), \rho(p)) = \inf(f, p)$$

and

$$\inf(f_\infty, p) = \sup_{f \in \mathscr{F}} \inf(f, p) = \sup_{g \in \mathscr{F}_p} g .$$

Since $\inf(f_\infty, p)$ is bounded, from the first part of the proof we deduce that $\inf(f_\infty, p)$ is μ-measurable and

$$\int^* \inf(f_\infty, p) \mathrm{d}\mu = \sup_{f \in \mathscr{F}} \int^* \inf(f, p) \mathrm{d}\mu .$$

Taking then $p \to \infty$, we deduce that $f_\infty = \sup_p \inf(f_\infty, p))$ is μ-measurable and

$$\int^* f_\infty \mathrm{d}\mu = \sup_{f \in \mathscr{F}} \int^* f \mathrm{d}\mu .$$

Thus the proposition is proved.

2 *The lifting property of spaces of vector functions*

Let μ be a positive measure on T, E and F two Banach spaces and Z a subspace of F', *norming* for F, that is:

$$|y| = \sup_{z \in Z} |\langle y, z\rangle|/|z| \text{ for every } y \in F .$$

We shall denote by $\mathscr{L}^*(E, F)$ the set of linear mappings $U : E \to F$. For each $U \in \mathscr{L}^*(E, F)$ we shall write

$$|U| = \sup_{|x| \leq 1, |z| \leq 1} |\langle Ux, z\rangle| = \sup_{|x| \leq 1} |Ux| \leqslant +\infty .$$

For $U : T \to \mathscr{L}^*(E, F)$, $\boldsymbol{x} : T \to E$ and $\boldsymbol{z} : T \to Z$, we shall denote by $\langle U\boldsymbol{x},\boldsymbol{z}\rangle$ the function $t \to \langle U(t)\boldsymbol{x}(t), \boldsymbol{z}(t)\rangle$ and by $|U|$ the function $t \to |U(t)|$.

For two functions $U_1, U_2 : T \to \mathscr{L}^*(E, F)$ we shall write $U_1 \equiv U_2$ if

$$\langle U_1 x, z\rangle = \langle U_2 x, z\rangle, \text{ for every } x \in E \text{ and } z \in Z .$$

Remarks.

1. The relation $U_1 \equiv U_2$ depends not only on the spaces E and F but also on the space Z.

2. If Z is of countable type, then we have $U_1 \equiv U_2$ if and only if $U_1(t)x = U_2(t)x$, μ-almost everywhere, for each $x \in E$.

In particular, if $F = C$, then $\mathscr{L}^*(E, C) = E^*$; in this case, for two functions $\boldsymbol{f}, \boldsymbol{g} : T \to E^*$ we have $\boldsymbol{f} \equiv \boldsymbol{g}$ if and only if

$$\langle x, \boldsymbol{f}(t)\rangle = \langle x, \boldsymbol{g}(t)\rangle, \ \mu\text{-almost everywhere,}$$

for each $x \in E$.

3. If $E = C$, then $\mathscr{L}^*(C, F) = F$; in this case for two functions $\boldsymbol{f}, \boldsymbol{g} : T \to F$, the relation $\boldsymbol{f} \equiv \boldsymbol{g}$ means $\langle \boldsymbol{f}(t), z\rangle = \langle \boldsymbol{g}(t), z\rangle$, μ-almost everywhere, for each $z \in Z$.

4. If E and Z are of countable type, then $U_1 \equiv U_2$ if and only if $U_1(t) = U_2(t)$, μ-almost everywhere.

5. If F is of countable type, then there exists a subspace of countable type $Z \subset F'$ norming for F.

In fact, let (y_n) be a sequence dense in F; for each y_n there exists $z_n \in F'$ with $|z_n| = 1$ and $|\langle y_n, z_n\rangle| = |y_n|$. The space Z generated by the sequence (z_n) has the above property.

13.6 Definition. *Let $\mathscr{G}$ be a vector spaces of functions $U : T \to \mathscr{L}^*(E, F)$. A mapping $\rho : \mathscr{G} \to \mathscr{G}$ is called a linear lifting of $\mathscr{G}$ if it fulfills the following conditions:*

I. $\rho(U) \equiv U$;

II. $U_1 \equiv U_2$ *implies* $\rho(U_1) = \rho(U_2)$;

III. $U \equiv$*constant implies* $\rho(U) = U$;

IV. $\rho(\alpha U+\beta V)=\alpha\rho(U)+\beta\rho(V)$.

If $\mathscr{G}$ possesses a linear lifting we say that it has the linear lifting property.

A vector space does not necessarily have the linear lifting property. It will be shown later that if $\mu\neq 0$ and is not atomic, then the space $\mathscr{L}^p(\mu)$, $1\leqslant p<\infty$, has no positive linear lifting.

Let $\mathscr{F}$ be an algebra of scalar functions, such that $f\in\mathscr{F}$ implies $|f|\in\mathscr{F}$.

We shall denote by $\mathscr{F}_{E,F}$ the space of functions $U: T\to\mathscr{L}^*(E,F)$ which have the following two properties:

(1) $\langle Ux, z\rangle\in\mathscr{F}$ for every $x\in E$ and $z\in Z$;

(2) for every $x\in E$ there exists a function $\varphi_x\in\mathscr{F}$ with $|Ux|\leqslant\varphi_x$.

It follows that if $U\in\mathscr{F}_{E,F}$ and $\varphi\in\mathscr{F}$, then $U\varphi\in\mathscr{F}_{E,F}$.

We can take, for example, $\mathscr{F}=\mathscr{L}^\infty(\mu)$ or $\mathscr{F}=\mathscr{M}^\infty(\mathscr{K})$. We remark that, for the space $\mathscr{L}^\infty_{E,F}$, the property (2) means

$$N_\infty(Ux)<\infty \text{ for every } x\in E\,.$$

For the space $\mathscr{M}^\infty_{E,F}(\mathscr{K})$ with $\mathscr{K}=(K_j)_{j\in J}\in C(\mu)$, property (2) means

$$N_\infty(Ux\varphi_{K_j})<\infty \text{ for every } x\in E \text{ and } j\in J\,.$$

We shall show that in these two cases, $\mathscr{F}_{E,Z'}$ has the linear lifting property.

13.7 Proposition. *For each lifting ρ of $\mathscr{L}^\infty(\mu)$ there exists a unique linear lifting of $\mathscr{L}^\infty_{E,Z'}(\mu)$, also denoted by ρ, such that*

$$\langle\rho(U)x, z\rangle=\rho(\langle Ux, z\rangle), \text{ for } U\in\mathscr{L}^\infty_{E,Z'},\ x\in E \text{ and } z\in Z$$

Then we have

V. $\rho(U\varphi)=\rho(U)\rho(\varphi)$, *for* $U\in\mathscr{L}^\infty_{E,Z'}$, *and* $\varphi\in\mathscr{L}^\infty$;

VI. $|\rho(U)|\leqslant\rho(\varphi)$ *if* $U\in\mathscr{L}^\infty_{E,Z'}$, $\varphi\in\mathscr{L}^\infty$ *and* $|U|\leqslant\varphi$.

Let ρ be a lifting of $\mathscr{L}^\infty$. For each $x\in E$ and $z\in Z$ put $U_{x,z}=\rho(\langle Ux, z\rangle)$. For each $t\in T$ the mapping $(x,y)\to U_{x,z}(t)$ is a bilinear functional on $E\times Z$. For fixed x, the mapping $U_x(t): Z\to U_{x,z}(t)$ is a linear functional on Z. It is also continuous since

$$|U_x(t)z|=|\rho(\langle Ux, z\rangle)(t)|\leqslant N_\infty(Ux)|z|\,,$$

therefore $U_x(t)\in Z'$ and $|U_x(t)|\leqslant N_\infty(U_x)$. The mapping $\rho(U)(t): x\to U_x(t)$ of E into Z' is linear, hence $\rho(U)(t)\in\mathscr{L}^*(E,Z')$ and we have

$$\langle\rho(U)(t)x, z\rangle=\rho(\langle Ux, z\rangle)(t)$$

for every $t\in T$, $x\in E$ and $z\in Z$, that is,

$$\langle \rho(U)x, z\rangle = \rho(\langle Ux, z\rangle).$$

From this equality, the uniqueness of $\rho(U)$ follows. From the properties of the lifting ρ of $\mathscr{L}^\infty$ we easily deduce the properties of the linear lifting of ρ on $\mathscr{L}^\infty_{E,Z'}$, as well as properties V and VI. For example, property VI is deduced in the following way: if $|U|\leqslant\varphi$,

$$|\langle \rho(U)(t)x, z\rangle| = |\rho(\langle Ux, z\rangle)(t)| =$$

$$= \rho(|\langle Ux, z\rangle|)(t) \leqslant \rho(\varphi|x||z|)(t) = \rho(\varphi)(t)|x|\,|z|,$$

whence

$$|\rho(U)(t)| \leqslant \rho(\varphi)(t).$$

Taking, in the preceding proposition, $E=C$ we obtain

13.8 COROLLARY. *Denote by $\mathscr{L}^\infty_{Z_s}$ the space of functions $\boldsymbol{f}: T\to Z'$ with the following two properties:*

(1) $\langle \boldsymbol{f}, z\rangle \in \mathscr{L}^\infty(\mu)$, *for every* $z\in Z$;

(2) $N_\infty(\boldsymbol{f}) < \infty$.

For every lifting ρ of $\mathscr{L}^\infty$ there exists a unique linear lifting of $\mathscr{L}^\infty_{Z_s}$, also denoted by ρ, such that

$$\langle \rho(\boldsymbol{f}), z\rangle = \rho(\langle \boldsymbol{f}, z\rangle) \text{ for every } z\in Z.$$

We remark that the relation $\boldsymbol{f}\equiv\boldsymbol{g}$ means in this case $\langle \boldsymbol{f}(t), z\rangle = \langle \boldsymbol{g}(t), z\rangle$, μ-almost everywhere, for every $z\in Z$.

Taking, in the preceding proposition, $F=C$ we obtain

13.9 COROLLARY. *Denote by $\mathscr{L}^\infty_{E^*}$ the space of functions $\boldsymbol{f}: T\to E^*$ with the following property:*

$$\langle x, \boldsymbol{f}\rangle \in \mathscr{L}^\infty(\mu), \text{ for every } x\in E.$$

For every lifting ρ of $\mathscr{L}^\infty$, there exists a unique linear lifting of $\mathscr{L}^\infty_{E^*}$, also denoted by ρ, such that

$$\langle x, \rho(\boldsymbol{f})\rangle = \rho(\langle x, \boldsymbol{f}\rangle), \text{ for every } x\in E.$$

We remark that, in this case, $\boldsymbol{f}\equiv\boldsymbol{g}$ means $\langle x, \boldsymbol{f}(t)\rangle = \langle x, \boldsymbol{g}(t)\rangle$, μ-almost everywhere for each $x\in E$.

Combining the preceding results we deduce

13.10 COROLLARY. *For every lifting ρ of $\mathscr{L}^\infty(\mu)$, its extensions to the spaces $\mathscr{L}^\infty_{E,Z'}$ and $\mathscr{L}^\infty_{Z'_s}$ again denoted by ρ, satisfy the equality*

$$\rho(U)x=\rho(Ux)$$

for every $U\in\mathscr{L}^\infty_{E,Z'}$ and $x\in E$.

13.11 PROPOSITION. *Let $\mathscr{K}=(K_j)_{j\in J}\in C(\mu)$. For each lifting ρ of $\mathscr{L}^\infty(\mu)$, there exists a linear lifting $\rho_{\mathscr{K}}$ of the space $\mathscr{M}^\infty_{E,Z'}(\mathscr{K})$ such that*

$$\langle\rho_{\mathscr{K}}(U)x, z\rangle\,\varphi_{\rho(K_j)}=\rho(\langle Ux, z\rangle\varphi_{K_j})$$

for each $U\in\mathscr{M}^\infty_{E,Z'}(\mathscr{K})$, $x\in E$, $z\in Z$ and $j\in J$.

Let $U\in\mathscr{M}^\infty_{E,Z'}(\mathscr{K})$ and $j\in J$. For each $x\in E$ and $z\in Z$, put $U_{x,z}=\rho(\langle U_{x,z}\rangle\,\varphi_{K_j})$. For each $t\in T$ the mapping $(x, z)\to U_{x,z}(t)$ is a bilinear functional on $E\times Z$; there exists then a linear mapping $U_j(t): E\to Z^*$ which satisfies the equality

$$\langle U_j(t)\,x, z\rangle=U_{x,z}(t),\quad \text{for } x\in E \text{ and } z\in Z\,.$$

The linear functional $U_j(t)x$ is also continuous on Z, since

$$|\langle U_j(t)x, z\rangle|=|\rho(\langle Ux, z\rangle)(t)\varphi_{K_j}(t)|\leqslant N_\infty(Ux\varphi_{K_j})|z|,$$

consequently $U_j(t)\in\mathscr{L}^*(E, Z')$

Now let A be a compact set with $\mu(A)>0$, such that $\sup_{t\in A}|U(t)x|<\infty$ for each $x\in E$. For example, A may be taken equal to one of the sets $K_j\in\mathscr{K}$ which are not μ-negligible. For each $x\in E$ and $z\in Z$ consider the bilinear functional:

$$(x, z)\to(\mu(A))^{-1}\int_A\langle U(t)x, z\rangle\,d\mu$$

on $E\times Z$. There exists then a mapping $U_0: E\to Z^*$ such that

$$\langle U_0x, z\rangle=(\mu(A))^{-1}\int_A\langle U(t)x, z\rangle\,d\mu\,.$$

The linear functional U_0x is continuous on Z since

$$|\langle U_0x, z\rangle|\leqslant\sup_{t\in A}|U(t)x|\;|z|$$

therefore $U_0\in\mathscr{L}^*(E, Z')$.

We now define the function $\rho_{\mathscr{K}}(U)$ on T by

$\rho_{\mathscr{K}}(U)(t) = U_j(t)$, for $t \in \rho(K_j)$ and $j \in J$.

$\rho_{\mathscr{K}}(U)(t) = U_0$, for $t \in T \setminus \cup_{j \in J} \rho(K_j)$.

It is easy to see that $\rho_{\mathscr{K}}(U) \in \mathscr{M}^{\infty}_{E,Z'}(\mathscr{K})$ and that $\rho_{\mathscr{K}}$ is a linear lifting which fulfils the stated conditions.

Remark. The linear lifting $\rho_{\mathscr{K}}$ is not uniquely determined by ρ It depends, for example, on the set A, which was used in the proof.

Taking $E = C$, in proposition 13.11, we obtain

13.12 COROLLARY. *Denote by $\mathscr{M}^{\infty}_{Z'_s}(\mathscr{K})$ the space of functions $f: T \to Z'$ having the following two properties:*

(1) $\langle f, z \rangle \varphi_{K_j} \in \mathscr{L}^{\infty}(\mu)$, for every $z \in Z$ and $j \in J$;

(2) $N_{\infty}(f \varphi_{K_j}) < \infty$ for every $j \in J$.

For every lifting ρ of $\mathscr{L}^{\infty}$ there exists a linear lifting $\rho_{\mathscr{K}}$ of $\mathscr{M}^{\infty}_{Z'_s}(\mathscr{K})$ such that we have

$$\langle \rho_{\mathscr{K}}(f), z \rangle \varphi_{\rho(K_j)} = \rho(\langle f, z \rangle \varphi_{K_j})$$

for $f \in \mathscr{M}^{\infty}_{Z'_s}(\mathscr{K})$, $z \in Z$ and $j \in J$.

Taking $F = C$, in the preceding proposition, we obtain

13.13 COROLLARY. *Denote by $\mathscr{M}^{\infty}_{E^*}(\mathscr{K})$ the space of functions $f: T \to E$ having the following property:*

$\langle x, f \rangle \varphi_{K_j} \in \mathscr{L}^{\infty}(\mu)$ for every $x \in E$ and $j \in J$.

For every lifting ρ of $\mathscr{L}^{\infty}(\mu)$ there exists a linear lifting $\rho_{\mathscr{K}}$ of $\mathscr{M}_{E^*}(\mathscr{K})$ such that

$$\langle x, \rho_{\mathscr{K}}(f) \rangle \varphi_{\rho(K_j)} = \rho(\langle x, f \rangle \varphi_{K_j})$$

for $f \in \mathscr{M}^{\infty}_{E^*}(\mathscr{K})$, $x \in E$ and $j \in J$.

Combining the preceding results we deduce

13.14 COROLLARY. *Let ρ be a lifting of $\mathscr{L}^{\infty}(\mu)$. Then the linear lifting $\rho_{\mathscr{K}}$ of the space $\mathscr{M}^{\infty}_{E,F'}(\mathscr{K})$ or $\mathscr{M}^{\infty}_{F'_s}(\mathscr{K})$ correspoding to ρ, satisfies the equality*

$\rho_{\mathscr{K}}(U)x = \rho_{\mathscr{K}}(Ux)$, μ-almost everywhere

for every $U \in \mathscr{M}^{\infty}_{E,F'}(\mathscr{K})$ and $x \in E$.

In fact, for each $j \in J$, we have

$$\rho_{\mathcal{K}}(U)x\varphi_{\rho(K_j)} = \rho_{\mathcal{K}}(Ux)\varphi_{\rho(K_j)}.$$

3 *Functions with the lifting property*

Let μ be a positive measure on T, E and F two Banach spaces and Z a subspace of F', norming for F. Let ρ be a lifting of $\mathcal{L}^\infty(\mu)$.

13.15 Definition. *Let $U : T \to \mathcal{L}^*(E, F)$.*

(a) *We shall write $\rho(U) = U$, if for each $x \in E$ and $z \in Z$ we have $\langle Ux, z\rangle \in \mathcal{L}^\infty(\mu)$ and*

$$\rho(\langle Ux, z\rangle) = \langle Ux, z\rangle.$$

(b) *We shall write $\rho[U] = U$, if there exists a family $\mathcal{K} = (K_j)_{j\in J} \in C(\mu)$ such that for each $j \in J$, $x \in E$ and $z \in Z$ we have $\langle Ux, z\rangle \varphi_{K_j} \in \mathcal{L}^\infty(\mu)$ and*

$$\rho(\langle Ux, z\rangle \varphi_{K_j}) = \langle Ux, z\rangle \varphi_{\rho(K_j)}.$$

We remark that the relations $\rho(U) = U$ and $\rho[U] = U$ are defined with respect to a certain subspace $Z \subset F'$. We remark also that the relation $\rho[U] = U$ is defined with respect to a certain family $\mathcal{K} = (K_j)_{j\in J} \in C(\mu)$. If a μ-measurable set A is contained in a set K_j, then we also have

$$\rho(\langle Ux, z\rangle \varphi_A) = \langle Ux, z\rangle \varphi_{\rho(A)}, \text{ for } x \in E \text{ and } z \in Z.$$

In fact,

$$\begin{aligned}\rho(\langle Ux, z\rangle \varphi_A) &= \rho(\langle Ux, z\rangle \varphi_{K_j}\varphi_A) = \rho(\langle Ux, z\rangle \varphi_{K_j})\rho(\varphi_A)\\ &= \langle Ux, z\rangle \varphi_{\rho(K_j)}\varphi_{\rho(A)} = \langle Ux, z\rangle \varphi_{\rho(A)}\end{aligned}$$

since $A \subset K_j$ implies $\rho(A) \subset \rho(K_j)$.

It follows that if $U_1, U_2 : T \to \mathcal{L}^*(E, F)$ are two functions with $\rho[U_1] = U_1$ and $\rho[U_2] = U_2$, we can find a common family $\mathcal{K} = (K_j)_{j\in J} \in C(\mu)$ such that, for every $j \in J$, $x \in E$ and $z \in Z$, we have simultaneously

$$\rho(\langle U_i x, z\rangle \varphi_{K_j}) = \langle U_i x, z\rangle \varphi_{\rho(K_j)} \text{ for } i = 1, 2.$$

The functions with $\rho(U) = U$ and $\rho[U] = U$ have the following properties:

(1) $\rho(U)=U$ *implies* $\rho[U]=U$ (*for every family* $\mathscr{K}\in C(\mu)$);

(2) *if* $\rho[U]=U$, *then the function* $\langle Ux, z\rangle$ *is* μ-*measurable, for every* $x\in E$ *and* $z\in Z$;

(3) *if* $U_1\equiv U_1$, $\rho(U_1)=U_1$ and $\rho(U_2)=U_2$, then $U_1=U_2$;

(4) *if* $U_1\equiv U_2$, $\rho[U_1]=U_1$ and $\rho[U_2]=U_2$, then $U_1(t)=U_2(t)$, μ-*almost everywhere;*

(5) *if* $U_1(t)=U_2(t)$, μ-*almost everwhere and* $\rho[U_1]=U_1$ *then* $\rho[U_2]=U_2$.
In fact, let $\mathscr{K}=(K_j)_{j\in J}\in C(\mu)$ be such that

$$\rho(\langle U_1 x, z\rangle \varphi_{K_j})=\langle U_1 x, z\rangle \varphi_{\rho(K_j)}$$

for every $j\in J$, $x\in E$ and $z\in Z$. Let also $N\subset T$ be the μ-negligible set such that

$$U_1(t)=U_2(t), \text{ for } t\in T\setminus N.$$

Each set $K_j\cap\rho(K)\setminus N$ is μ-integrable; therefore, it is the union of a sequence (K_{jn}) of disjoint compact sets and of a μ-negligible set N_j. The sets K_{jn} which are not empty form a family of $C(\mu)$, and U_1 and U_2 coincide on each K_{jn} and on each $\rho(K_{jn})$, consequently

$$\rho(\langle U_2 x, z\rangle\varphi_{K_{jn}})=\rho(\langle U_1 x, z\rangle \varphi_{K_{jn}})= \\ =\langle U_1 x, z\rangle\varphi_{\rho(K_{jn})}=\langle U_2 x, z\rangle \varphi_{\rho(K_{jn})},$$

that is, $\rho[U_2]=U_2$.

(6) *If* $\rho(U)=U$, *then* $|U(t)|\leqslant N_\infty(U)<\infty$ *for each* $t\in T$.

In fact, for each $x\in E$ and $z\in U$ we have $\langle Ux, z\rangle\in\mathscr{L}^\infty$ and $\rho(\langle Ux, z\rangle)=\langle Ux, z\rangle$; therefore, for every $t\in T$ we have

$$|\langle U(t)x, z\rangle|\leqslant N_\infty(\langle Ux, z\rangle)\leqslant|x|\,|z|\,N_\infty(U),$$

whence $|U(t)|<N_\infty(U)$ for every $t\in T$.

Remark. If $\rho(U)=U$ and $N_\infty(Ux)<\infty$ for each $x\in E$, then $U\in\mathscr{L}^\infty_{E,F}$; in this case $\rho(U)$ represents the value of the linear lifting of $\mathscr{L}^\infty_{E,F'}$ corresponding to the lifting ρ of $\mathscr{L}^\infty$ (proposition 13.7).

(7) *If* $\rho[U]=U$ *with respect to a family* $\mathscr{K}=(K_j)_{j\in J}\in C(\mu)$, *then* $|U(t)|\leqslant N_\infty(U\varphi_{K_j})<\infty$, *for each* $j\in J$ *and* $t\in K_j$.

Other properties are given in the following propositions.

13.16 Proposition. *Let* $U: T\to\mathscr{L}^*(E, F)$ *be a function such that* $\langle Ux, z\rangle$ *is* μ-*measurable, for every* $x\in E$ *and* $z\in Z$.

(a) *If* $N_\infty(Ux) < \infty$ *for every* $x \in E$, *then there exists a function* $U' : T \to \mathscr{L}^*(E, Z')$ *with* $U' \equiv U$ *and* $\rho(U') = U'$.

(b) *If there exists a family* $\mathscr{K} = (K_j)_{j \in J} \in C(\mu)$ *such that* $N_\infty(Ux\varphi_{K_j}) < \infty$ *for each* $x \in E$ *and* $j \in J$, *then there exists a function* $U' : T \to \mathscr{L}^*(E, Z')$ *with* $U' \equiv U$ *and* $\rho[U'] = U'$.

In fact, from the hypotheses it follows that $U \in \mathscr{L}^\infty_{E,Z'}(\mu)$ in the first case and $U \in \mathscr{M}^\infty_{E,Z'}(\mathscr{K})$ in the second case. If ρ is the linear lifting of $\mathscr{L}^\infty_{E,Z'}$ corresponding to the lifting ρ of $\mathscr{L}^\infty(\mu)$, by proposition 13.7, then the function $U' = \rho(U)$ has the required properties in the first case. Also, if $\rho_{\mathscr{K}}$ is a linear lifting of $\mathscr{M}^\infty_{E,Z'}$ corresponding to ρ by proposition 13.11, then $U' = \rho_{\mathscr{K}}(U)$ has the required properties in the second case.

Remark. In the second case, every function equal to $\rho_{\mathscr{K}}(U)$ μ-almost everywhere has the desired properties.

13.17 COROLLARY. *Let* $f : T \to F$ *be a function such that* $\langle f, z\rangle$ *is* μ*-measurable for every* $z \in Z$.

(a) *If* $N_\infty(f) < \infty$, *there exists a function* $f' : T \to Z'$ *with* $f' \equiv f$ *and* $\rho(f') = f'$.

(b) *If there exists* $\mathscr{K} = (K_j)_{j \in J} \in C(\mu)$ *with* $N_\infty(f\varphi_{K_j}) < \infty$ *for each* $j \in J$, *then there exists* $f' : T \to Z'$ *with* $f' \equiv f$ *and* $\rho[f'] = f'$.

13.18 PROPOSITION. *Let* $U : T \to \mathscr{L}^*(E, F)$. *If* $\rho[U] = U$, *then the function* $t \to |U(t)|$ *is* μ*-measurable.*

Let $(K_j)_{j \in J} \in C(\mu)$ be a family such that

$$\rho(\varphi_{K_j}\langle Ux, z\rangle) = \varphi_{\rho(K_j)}\langle Ux, z\rangle$$

for every $j \in J$, $x \in E$ and $z \in Z$. Denote $T_0 = \cup_{j \in J} \rho(K_j)$. The set $T \setminus T_0$ is μ-negligible.

Denote by $\mathscr{F}$ the set of positive functions of the form

$$f = \Sigma_{1 \leqslant i \leqslant n} \varphi_{A_i} |\langle Ux_i, z_i\rangle| ,$$

where the sets A_i are μ-measurable, mutually disjoint, $\rho(A_i) = A_i$ and each A_i is contained in some set $\rho(K_j)$, $x_i \in E$ with $|x_i| \leqslant 1$ and $z_i \in Z$ with $|z_i| \leqslant 1$. In particular, $\mathscr{F}$ contains all functions of the form $\varphi_{\rho(K_j)} |\langle Ux, z\rangle|$ with $|x| \leqslant 1$ and $|z| \leqslant 1$.

Since

$$f(t) \leqslant \Sigma_{1 \leqslant i \leqslant n} \varphi_{A_i}(t) |U(t)| \leqslant |U(t)|, \text{ for } t \in T_0,$$

and, on the other hand, for each $t \in T_0$ and for $j \in J$ such that $t \in \rho(K_j)$ we have

$$|U(t)| = \sup_{|x| \leqslant 1, |z| \leqslant 1} \varphi_{\rho(K_j)}(t) |\langle U(t)x, z \rangle| ;$$

we deduce that

$$|U(t)| = \sup_{f \in \mathscr{F}} f(t) \text{ for } t \in T_0 .$$

For each function $f = \Sigma \varphi_{A_i} |\langle Ux, z \rangle|$ of $\mathscr{F}$ we have $\rho(f) = f$. In fact, if for every i we have $A_i \subset \rho(K_j)$, then

$$f = \Sigma_i \varphi_{A_i} \varphi_{\rho(K_j)} |\langle Ux_i, z_i \rangle| ;$$

therefore,

$$\rho(f) = \Sigma_i \varphi_{A_i} \varphi_{\rho(K_j)} |\langle Ux_i, z \rangle| = f .$$

We show that $\mathscr{F}$ is directed. Let $f_1, f_2 \in \mathscr{F}$. Let us remark that if A is μ-measurable and $\rho(A) = A$, then $\varphi_A f_1 + \varphi_{T \setminus A} f_2 \in \mathscr{F}$. Since there exists a μ-measurable set $B \subset T$ such that

$$\sup(f_1, f_2) = \varphi_B f_1 + \varphi_{T \setminus B} f_2 ,$$

we deduce

$$\sup(f_1, f_2) = \sup[\rho(f_1), \rho(f_2)] = \rho(\sup(f_1, f_2)) =$$
$$= \varphi_{\rho(B)} f_1 = \varphi_{\rho(T \setminus A)} f_2 = \varphi_A f_1 + \varphi_{T \setminus A} f_2 \in \mathscr{F} ,$$

where we have put $A = \rho(B)$; consequently, $\rho(A) = A$.

By proposition 13.5, it follows that the function f_∞ defined by the equality

$$f_\infty(t) = \sup_{f \in \mathscr{F}} f(t), \text{ for } t \in T,$$

is μ-measurable. Since

$$|U(t)| = f_\infty(t) \text{ for } t \in T_0$$

and $T \setminus T_0$ is μ-negligible, it follows that the function $t \to |U(t)|$ is μ-measurable and the proposition is proved.

13.19 COROLLARY. *If $\rho[U] = U$, then the function $t \to |U(t)x|$ is μ-measurable for every $x \in E$.*

In fact, we have $\rho[Ux]=Ux$ for every $x\in E$. In particular, proposition 13.18 and corollary 13.19 are true if $\rho(U)=U$.

13.20 PROPOSITION. *Let $U:T\to\mathcal{L}^*(E,F)$ be a function such that $t\to|U(t)|$ is μ-measurable.*

If $U'\equiv U$ and $\rho[U']=U'$, then

$$|U'(t)|\leqslant|U(t)|,\ \mu\text{-almost everywhere}.$$

Let $\mathscr{K}=(K_j)_{j\in J}\in C(\mu)$ be a family such that

$$\langle U'x, z\rangle\in\mathcal{L}^\infty(\mu) \text{ and } \rho(\langle Ux', z\rangle\varphi_{K_j})=\langle Ux', z\rangle\varphi_{\rho(K_j)}$$

for every $j\in J$, $x\in E$ and $z\in Z$.

We remark that

$$\varphi_{\rho(K_j)}|\langle U'x, z\rangle|=\rho(\varphi_{K_j}|\langle Ux, z\rangle|)$$

and

$$\varphi_{K_j}|\langle U'x, z\rangle|\sim\varphi_{K_j}|\langle Ux, z\rangle|.$$

Then

$$\varphi_{\rho(K_j)}|\langle U'x, z\rangle|=\rho(\varphi_{K_j}|\langle Ux, z\rangle|)\leqslant|x|\,|z|\,\rho(\varphi_{K_j}|U|)\ ;$$

therefore,

$$\varphi_{\rho(K_j)}|U'|\leqslant\rho(\varphi_{K_j}|U|)\sim\varphi_{\rho(K_j)}|U|,$$

whence

$$|U'(t)|\leqslant|U(t)|,\ \text{almost everywhere on } \cup\,\rho(K_j),$$

that is, almost everywhere, and the proposition is proved.

13.21 COROLLARY. *Let $U:T\to\mathcal{L}^*(E,F)$ be such that the function $t\to|U(t)|$ is μ-measurable. If $U'\equiv U$ and $\rho(U')=U'$, then*

$$|U'(t)|\leqslant|U(t)|,\ \mu\text{-almost everywhere}.$$

13.22 PROPOSITION. *Let $U:T\to\mathcal{L}^*(E,F)$ be a function such that $\langle Ux, z\rangle$ is μ-measurable for every $x\in E$ and $z\in Z$.*

(a) *If the function* $t \to |U(t)|$ *is* μ*-measurable and finite* μ*-almost everywhere, then there exists a function* $U' : T' \to \mathscr{L}^*(E, Z')$ *with* $U' \equiv U$, $\rho[U'] = U'$ *and*

$$|U'(t)| \leqslant |U(t)|, \text{ for every } t \in T.$$

(b) *If* $|U| \in \mathscr{L}^\infty(\mu)$, *then there exists a function* $U' : T \to \mathscr{L}^\infty(E, Z')$ *with* $U' \stackrel{\bullet}{\equiv} U$, $\rho(U') = U'$ *and*

$$|U'(t)| \leqslant |U(t)|, \ \mu\text{-almost everywhere}.$$

Let us prove (a): since $|U|$ is μ-measurable and finite μ-almost everywhere, we can find a family $\mathscr{K} = (K_j)_{j \in J} \in C(\mu)$ such that the restriction of $|U|$ to each K_j is finite and continuous; consequently, bounded. Then $N_\infty(Ux\varphi_{K_j}) < \infty$ for each $x \in E$ and $j \in J$. By proposition 13.16, there exists a function $U_1 : T \to \mathscr{L}^*(E, Z')$ with $U_1 \equiv U$ and $\rho[U_1] = U_1$. By proposition 13.20 we have $|U_1(t)| \leqslant |U(t)|$ almost everywhere. Modifying the function U_1 on a convenient negligible set, we obtain a function U' such that $|U'(t)| \leqslant |U(t)|$ for every $t \in T$; we have also $U' \equiv U$ and $\rho[U'] = U'$.

The point (b) is proved in a similar way.

Remark. All the preceding results can be applied, in particular, to functions $f: T \to F$, taking $E = C$, and to functions $\boldsymbol{f}: T \to E^*$, taking $F = C$, as well as to measurable scalar functions.

§14 Relationships between the $\mathscr{L}^p$ spaces

1 *The Hölder inequality*

Let μ be a positive measure on T.

14.1 THEOREM. *Let* f *and* g *be two numerical functions defined and finite* μ*-almost everwhere on* T. *If* f *is equivalent to a function* $f_1 \in \mathscr{L}^p$, *and* g *is equivalent to a function* $g_1 \in \mathscr{L}^q$, *where* $1 \leqslant p, q \leqslant +\infty$, $p^{-1} + q^{-1} = 1$, *then* fg *is* μ*-integrable and*

$$\int |fg| \, \mathrm{d}\mu \leqslant N_p(f) \, N_q(g).$$

We have $N_p(f_1) < +\infty$ and $N_q(g_1) < +\infty$, and f_1 and g_1 are μ-measurable.

The function fg is equal μ-almost everywhere to $f_1 g_1$. The function $f_1 g_1$ is μ-measurable as the product of two μ-measurable functions, and we have

$$\int^* |f_1 g_1| d\mu = \int^* |f_1| |g_1| d\mu \leqslant N_p(f_1) N_q(g_1) < +\infty ;$$

therefore $f_1 g_1$ is μ-integrable. Then fg is also μ-integrable and

$$\int |fg| d\mu = \int |f_1 g_1| d\mu \leqslant N_p(f_1) N_q(g_1) = N_p(f) N_q(g)$$

and the theorem is proved.

14. Proposition. *Let E, F, G, H be four Banach spaces, $(x, y) \to u(x, y)$ a continuous bilinear mapping of $E \times F$ into G such that $|u(x,y)| \leqslant |x| |y|$ and $\boldsymbol{m} : \mathscr{K}_G(T) \to H$ a dominated measure. If $\boldsymbol{f} \in \mathscr{L}_E^p(\boldsymbol{m})$ and $\boldsymbol{g} \in \mathscr{L}_F^q(|\boldsymbol{m}|)$, $(1 \leqslant p, q \leqslant +\infty, p^{-1} + q^{-1} = 1)$, then the function $u(\boldsymbol{f}, \boldsymbol{g})$ is $\boldsymbol{m}$-integrable and*

$$\left|\int u(\boldsymbol{f}, \boldsymbol{g}) d\boldsymbol{m}\right| \leqslant \int |u(\boldsymbol{f}, \boldsymbol{g})| d|\boldsymbol{m}| \leqslant N_p(\boldsymbol{f}, |\boldsymbol{m}|) N_q(\boldsymbol{g}, |\boldsymbol{m}|) .$$

In fact, since the function $\boldsymbol{f} : T \to E$ and $\boldsymbol{g} : T \to F$ are $|\boldsymbol{m}|$-measurable, the function $(\boldsymbol{f}, \boldsymbol{g}) : T \to E \times F$ is $|\boldsymbol{m}|$-measurable; the function $u : E \times F \to G$ being continuous, we deduce that the composed function $u(\boldsymbol{f}, \boldsymbol{g}) : T \to G$ is $|\boldsymbol{m}|$-measurable. On the other hand, by the Hölder inequality we have

$$\int^* |u(\boldsymbol{f}, \boldsymbol{g})| d|\boldsymbol{m}| \leqslant \int^* |\boldsymbol{f}| |\boldsymbol{g}| d|\boldsymbol{m}| \leqslant N_p(\boldsymbol{f}, |\boldsymbol{m}|) N_q(\boldsymbol{g}, |\boldsymbol{m}|) < +\infty$$

therefore the function $u(\boldsymbol{f}, \boldsymbol{g})$ is $|\boldsymbol{m}|$-integrable. It follows then that $u(\boldsymbol{f}, \boldsymbol{g})$ is $\boldsymbol{m}$-integrable and

$$\left|\int^* |u(\boldsymbol{f}, \boldsymbol{g}) d\boldsymbol{m}\right| \leqslant \int |u(\boldsymbol{f}, \boldsymbol{g})| d|\boldsymbol{m}| \leqslant N_p(\boldsymbol{f}, |\boldsymbol{m}|) N_q(\boldsymbol{g}, |\boldsymbol{m}|) .$$

Remark. The preceding proposition is true, in particular, if $H = G$ and $\boldsymbol{m}$ is a scalar measure.

14.3 Corollary. *Let E be a Banach algebra and $\boldsymbol{m} : \mathscr{K}_E(T) \to E$ a dominated measure. If $\boldsymbol{f} \in \mathscr{L}_E^p(\boldsymbol{m})$ and $\boldsymbol{g} \in \mathscr{L}_E^q(\boldsymbol{m})$, $(1 \leqslant p, q \leqslant +\infty, p^{-1} + q^{-1} = 1)$, then $\boldsymbol{fg} \in \mathscr{L}_E^1(\boldsymbol{m})$ and*

$$\left|\int \boldsymbol{fg} \, d\boldsymbol{m}\right| \leqslant \int |\boldsymbol{fg}| d|\boldsymbol{m}| \leqslant N_p(\boldsymbol{f}, \boldsymbol{m}) \leqslant N_q(\boldsymbol{g}, \boldsymbol{m}) .$$

In fact, the function $u(x, y) = xy$ is bilinear and continuous on $E \times E$ and $|u(x, y)| \leqslant |x| |y|$.

The measure $\boldsymbol{m}$ may be, in particular, a scalar measure.

14.4 COROLLARY. *Let E and F be two Banach spaces and* $\boldsymbol{m}:\mathscr{K}_E(T)\to F$ *a dominated measure. If* $\boldsymbol{f}\in\mathscr{L}_E^p(\boldsymbol{m})$ *and* $\varphi\in\mathscr{L}^q(\boldsymbol{m})$, $(1\leqslant p, q\leqslant +\infty,\ p^{-1}+q^{-1}=1)$, *then* $\boldsymbol{f}\varphi\in\mathscr{L}_E^1(\boldsymbol{m})$ *and*

$$|\int \boldsymbol{f}\varphi\,\mathrm{d}\boldsymbol{m}| \leqslant \int|\boldsymbol{f}\varphi|\,\mathrm{d}|\boldsymbol{m}| \leqslant N_p(\boldsymbol{f},\boldsymbol{m})\,N_q(\varphi,\boldsymbol{m})\,.$$

In fact, the function $u(\alpha, x)=\alpha x$ is bilinear and continuous on $R\times E$ and $|u(\alpha, x)|=|\alpha|\,|x|$.

Remark. If E is a complex Banach space, then we may take $\varphi\in\mathscr{L}_C^q(\boldsymbol{m})$. The measure $\boldsymbol{m}$ may be, in particular, a scalar measure.

14.5 COROLLARY. *Let E be a Banach space, E′ its dual and* μ *a scalar measure. If* $\boldsymbol{f}\in\mathscr{L}_E^p(\mu)$ *and* $\boldsymbol{g}\in\mathscr{L}_{E'}^p(\mu)$, $(1\leqslant p, q\leqslant +\infty,\ p^{-1}+q^{-1}=1)$, *then the scalar function* $\langle \boldsymbol{f}, \boldsymbol{g}\rangle$ *is* μ*-integrable and*

$$|\int\langle \boldsymbol{f}, \boldsymbol{g}\rangle\,\mathrm{d}\mu| \leqslant \int |\langle \boldsymbol{f}, \boldsymbol{g}\rangle|\,\mathrm{d}|\mu| \leqslant N_p(\boldsymbol{f},\mu)N_q(\boldsymbol{g},\mu)\,.$$

In fact, the function $u(z, z')=\langle z, z'\rangle$ defined on $E\times E'$ with scalar values is bilinear, continuous and $|u(z, z')|\leqslant|z|\,|z'|$.

Remark. Corollary 14.5 remains true if $E=C$ or $E=R$, for every dominated vector measure $\boldsymbol{m}:\mathscr{K}_G(T)\to H$, considered as defined on $\mathscr{K}(T)$ with values in $\mathscr{L}(G, H)$.

2 *Computation of the seminorms* N_p

Let μ be a positive measure on T, p and q two conjugate exponents, i.e.

$$1\leqslant p\leqslant +\infty,\quad 1\leqslant q\leqslant +\infty,\quad p^{-1}+q^{-1}=1\,.$$

14.6 THEOREM. *Let E, F and G be three Banach spaces and* $u:E\times F\to G$ *a continuous bilinear mapping such that* $|x|=\sup_{|y|\leqslant 1}|u(x, y)|$ *for every* $x\in E$.

For every μ*-measurable function* $\boldsymbol{f}: T\to E$ *we have*

$$N_p(\boldsymbol{f})=\sup\int^* |u(\boldsymbol{f}, \boldsymbol{g})\,\mathrm{d}\mu=\sup\int^* |\boldsymbol{f}|\,|\boldsymbol{g}|\,\mathrm{d}\mu$$

for $\boldsymbol{g}\in\mathscr{E}_F$ *with* $N_q(\boldsymbol{g})=1$, *where* $\mathscr{E}_F$ *is the space of Borel step functions* $\boldsymbol{g}: T\to F$ *with compact support.*

From the equality $|x| = \sup_{|y| \leqslant 1} |u(x, y)|$ it follows $|u(x, y)| \leqslant |x|\,|y|$. From the Hölder inequality it follows then, for $f \in \mathscr{L}^p_E$,

$$\sup \int^* |u(f, g)|\,d\mu \leqslant \sup \int^* |f|\,|g|\,d\mu \leqslant N_p(f), \quad g \in \mathscr{E}_F,\ N_q(g) = 1\,.$$

It remains to prove the inequality

$$N_p(f) \leqslant \sup \int^* |u(f, g)|\,d\mu, \quad g \in \mathscr{E}_F,\ N_q(g) = 1\,.$$

If $N_p(f) = 0$, the inequality is true. We shall assume hence that $N_p(f) > 0$. We shall consider several successive cases.

(i) $p = 1$, f a Borel step function of $\mathscr{E}_E$. We can write the function f in the form $f = \Sigma_{1 \leqslant i \leqslant n} \varphi_{A_i} x_i$ with $x_i \in E$ and A_i mutually disjoint relatively compact Borel sets of $\mathscr{B}$. We have then $|f| = \Sigma_{1 \leqslant i \leqslant n} \varphi_{A_i} |x_i|$ and

$$N_1(f) = \int |f|\,d\mu = \Sigma_{1 \leqslant i \leqslant n} \mu(A_i) |x_i|\,.$$

Let $\varepsilon > 0$. For each i there exists an element $y_i \in F$ with $|y_i| = 1$ such that $|u(x_i, y_i)| > (1 - \varepsilon)|x_i|$.

The function $g = \Sigma_{1 \leqslant i \leqslant n} \varphi_{A_i} y_i$ is a Borel step function of $\mathscr{E}_F$ and

$$N_\infty(g) = \sup_{1 \leqslant i \leqslant n} |y_i| = 1\,.$$

Then

$$u(f, g) = u(\Sigma_i \varphi_{A_i} x_i, \Sigma_j \varphi_{A_j} y_i) = \Sigma_{i,j} \varphi_{A_i} u(x_i, y_j) =$$
$$= \Sigma_i \varphi_{A_i} u(x_i, y_i)$$

and

$$|u(f, g)| = |\Sigma\, \varphi_{A_i} u(x_i, y_i)| = \Sigma\, \varphi_{A_i} |u(x_i, y_i)| \geqslant$$
$$\geqslant \Sigma\, \varphi_{A_i} (1 - \varepsilon) |x_i|\ ;$$

therefore,

$$\int |u(f, g)|\,d\mu \geqslant (1 - \varepsilon)\, \Sigma\, \mu(A_i) |x_i| = (1 - \varepsilon) N_1(f)\,,$$

whence

$$\sup \int |u(f, g)|\,d\mu \geqslant N_1(f), \quad g \in \mathscr{E}_F,\ N_\infty(g) = 1\,.$$

(ii) $1 < p < +\infty$, $f = \Sigma_{1 \leqslant i \leqslant n} \varphi_{A_i} x_i$, a Borel step function of $\mathscr{E}_E$ with A_i mutually disjoint and $N_p(f) = 1$.

We have, therefore,

$$[N_p(f)]^p = \int |f|^p\,d\mu = \int \Sigma_{1 \leqslant i \leqslant n} \varphi_{A_i} |x_i|^p\,d\mu = \Sigma_{1 \leqslant i \leqslant n} \mu(A_i) |x_i|^p = 1\,.$$

Let $\varepsilon > 0$. For each i there exists $z_i \in F$ with $|z_i| = 1$ and $|u(x_i, z_i)| \geqslant (1-\varepsilon)|x_i|$. Denote $y_i = |x_i|^{p/q} z_i$. We have

$$|y_i|^q = |x_i|^p \text{ and } |u(x_i, y_i)| \geqslant (1-\varepsilon)|x_i|\,|y_i|\,.$$

The function $\boldsymbol{g} = \Sigma_{1 \leqslant i \leqslant n} \varphi_{A_i} y_i$ is a Borel step function of $\mathscr{E}_F$ and

$$N_q(\boldsymbol{g}) = (\int |\Sigma_{1 \leqslant i \leqslant n} \varphi_{A_i} y_i|^q \mathrm{d}\mu)^{1/q} = (\int \Sigma_{1 \leqslant i \leqslant n} \varphi_{A_i} |y_i|^q \mathrm{d}\mu)^{1/q} =$$
$$= (\Sigma_{1 \leqslant i \leqslant n} \mu(A_i) |y_i|^q)^{1/q} = (\Sigma_{1 \leqslant i \leqslant n} \mu(A_i) |x_i|^p)^{1/q} = 1\,.$$

Then, as for case (i),

$$|u(\boldsymbol{f}, \boldsymbol{g})| = \Sigma\, \varphi_{A_i} |u(x_i, y_i)| \geqslant (1-\varepsilon) \Sigma\, \varphi_{A_i} |x_i|\,|y_i|$$

therefore

$$\int |u(\boldsymbol{f}, \boldsymbol{g})| \mathrm{d}\mu \geqslant (1-\varepsilon) \Sigma\, \mu(A_i) |x_i|\,|y_i| = (1-\varepsilon) \Sigma\, \mu(A_i) |x_i|\,|x_i|^{p/q}$$
$$= (1-\varepsilon) \Sigma\, \mu(A_i) |x_i|^p = 1 - \varepsilon\,,$$

whence

$$\sup \int |u(\boldsymbol{f}, \boldsymbol{g})| \mathrm{d}\mu \geqslant 1, \ \boldsymbol{g} \in \mathscr{E}_F,\ N_q(\boldsymbol{g}) = 1\,.$$

(iii) $1 < p < +\infty$, $\boldsymbol{f} = \Sigma_{1 \leqslant i \leqslant n} \varphi_{A_i} x_i$, a Borel step function of $\mathscr{E}_E$ with $N_p(\boldsymbol{f}) = a > 0$. Then the function $\boldsymbol{f}_1 = a^{-1} \boldsymbol{f}$ is a Borel function of $\mathscr{E}_E$ with $N_p(\boldsymbol{f}_1) = a^{-1} N_p(\boldsymbol{f}) = 1$, hence, for $\boldsymbol{g} \in \mathscr{E}_F$ with $N_g(\boldsymbol{g}) = 1$, we have

$$\int |u(\boldsymbol{f}, \boldsymbol{g})|\, \mathrm{d}\mu \;=\; \int |u(a\boldsymbol{f}_1, \boldsymbol{g})| \mathrm{d}\mu \;=\; a \int |u(\boldsymbol{f}_1, \boldsymbol{g})| \mathrm{d}\mu\,,$$

whence

$$\sup \int |u(\boldsymbol{f}, \boldsymbol{g})| \mathrm{d}\mu = a \sup \int |u(\boldsymbol{f}_1, \boldsymbol{g})| \mathrm{d}\mu = a = N_p(\boldsymbol{f}),\ \ g \in \mathscr{E}_F,\ N_q(g) = 1\,.$$

(iv) $1 \leqslant p < +\infty$, $\boldsymbol{f} \in \mathscr{L}^p_E$ an arbitrary function with $N_p(\boldsymbol{f}) > 0$. Let $\varepsilon > 0$. There exists a Borel step function $\boldsymbol{f}_1 \in \mathscr{E}_E$ with $N_p(\boldsymbol{f} - \boldsymbol{f}_1) < \varepsilon$ and $N_p(\boldsymbol{f}_1) > 0$. By the above considerations, there exists a Borel step function $\boldsymbol{g} \in \mathscr{E}_F$ with $N_q(\boldsymbol{g}) = 1$ such that

$$\int |u(\boldsymbol{f}_1, \boldsymbol{g})| \mathrm{d}\mu \geqslant N_p(\boldsymbol{f}_1) - \varepsilon \geqslant N_p(\boldsymbol{f}) - 2\varepsilon\,.$$

But

$$\int |u(\boldsymbol{f}, \boldsymbol{g})| \mathrm{d}\mu \;=\; \int |u(\boldsymbol{f}_1, \boldsymbol{g}) + u(\boldsymbol{f} - \boldsymbol{f}_1, \boldsymbol{g})| \mathrm{d}\mu \geqslant$$
$$\geqslant \int |u(\boldsymbol{f}_1, \boldsymbol{g})| \mathrm{d}\mu - \int |u(\boldsymbol{f} - \boldsymbol{f}_1, \boldsymbol{g})|\, \mathrm{d}\mu$$

and

$$\int |u(\boldsymbol{f} - \boldsymbol{f}_1, \boldsymbol{g})| \mathrm{d}\mu \leqslant N_p(\boldsymbol{f} - \boldsymbol{f}_1) N_q(\boldsymbol{g}) < \varepsilon\,,$$

therefore,

$$\int |u(f, g)| \mathrm{d}\mu \geqslant N_p(f) - 3\varepsilon ,$$

whence

$$\sup \int |u(f, g)| \mathrm{d}\mu \geqslant N_p(f), \quad g \in \mathscr{E}_F, \ N_q(g) = 1 \ ;$$

consequently,

$$N_p(f) = \sup \int |u(f, g)| \mathrm{d}\mu = \sup \int |f| \, |g| \mathrm{d}\mu$$

for $g \in \mathscr{E}_F$ with $N_q(g) = 1$.

(v) $1 \leqslant p < +\infty$, $f : T \to E$ an arbitrary measurable function with $N_p(f) > 0$. Let $0 < \alpha < N_p(f)$. Since

$$N_p(f) = \sup N_p(f\varphi_K), \ K \text{ compact} ,$$

there exists a compact set $K \subset T$ on which f does not vanish with

$$\alpha < N_p(f\varphi_K) .$$

Since f is measurable, there exists an increasing sequence (K_n) of compact sets and a μ-negligible set $N \subset K$ such that $\cup_{1 \leqslant n < \infty} K_n = K \setminus N$ and the restriction of f to each K_n is continuous. For each n, the function $f\varphi_{K_n}$ belongs to the space $\mathscr{L}^p_E$. The sequence $(|f|\varphi_{K_n})$ is increasing and $|f|\varphi_{K \setminus N} = \sup |f| \varphi_{K_n}$, therefore

$$N_p(f\varphi_K) = N_p(f\varphi_{K \setminus N}) = \sup N_p(f\varphi_{K_n}) .$$

There exists n such that $\alpha < N_p(f\varphi_{K_n})$. Since $f\varphi_{K_n} \in \mathscr{L}^p_E$, there exists a real Borel step function φ with $N_q(\varphi) = 1$ and

$$\alpha < \int |f| \varphi_{K_n} |\varphi| \mathrm{d}\mu .$$

Let $0 < a < 1$. Since the restriction of f to K_n is continuous and does not vanish, there exists a measurable step function $g_0 : T \to F$ with

$$|g_0(t)| = 1 \ \text{and} \ a|f(t)| < |u(f(t), g_0(t)|, \ \text{for} \ t \in K_n$$

and $g_0(t) = 0$ for $t \notin K_n$ (see the proof of proposition 11.25). Then $g_0 \varphi$ is a step function of $\mathscr{E}_F$ with $N_q(g_0 \varphi) = 1$ and

$$\int |f| \varphi_{K_n} |\varphi| \mathrm{d}\mu \leqslant a^{-1} \int |u(f, g_0 \varphi)| \mathrm{d}\mu \leqslant a^{-1} \sup \int^* |u(f, g)| \mathrm{d}\mu, \ N_q(g) = 1.$$

It follows that

$$\alpha < a^{-1} \sup \int^* |u(f, g)| \mathrm{d}\mu, \ N_q(g) = 1, \ g \in \mathscr{E}_F .$$

Taking $a \to 1$ and $\alpha \to N_p(f)$, we deduce

$$N_p(\boldsymbol{f}) \leqslant \sup \int^* |u(\boldsymbol{f}, \boldsymbol{g})| \mathrm{d}\mu, \quad N_q(\boldsymbol{g}) = 1, \ \boldsymbol{g} \in \mathscr{E}_F$$

and the theorem is proved in this case.

(vi) $p = +\infty$, $\boldsymbol{f}: T \to E$ is an arbitrary measurable function with $N_\infty(\boldsymbol{f}) > 0$. Let α be a number such that $0 < \alpha < N_\infty(\boldsymbol{f})$. The set $B = \{t \mid |\boldsymbol{f}(t)| > \alpha\}$ is μ-measurable (since $|\boldsymbol{f}|$ is μ-measurable) and it is not μ-negligible. There exists a compact set $K \subset B$ which is not μ-negligible: $\mu(K) > 0$. Since $\boldsymbol{f}$ is μ-measurable, there exists a compact set $K_1 \subset K$ with $\mu(K_1) > 0$ such that the restriction of $\boldsymbol{f}$ to K_1 is continuous.

Let $\varepsilon > 0$. Since the restriction of $\boldsymbol{f}$ to K_1 is continuous, there exists a finite family of relatively compact open sets $(G_i)_{1 \leqslant i \leqslant n}$ which cover K_1 and such that the oscillation of $\boldsymbol{f}$ on each $G_i \cap K$ is $< \varepsilon$. There exists then a partition $(A_i)_{1 \leqslant i \leqslant n}$ of K_1 consisting of relatively compact Borel sets such that $A_i \subset G_i$ for each i; therefore, the oscillation of $\boldsymbol{f}$ on each A_i is $< \varepsilon$. At least one of the sets A_i has measure > 0. Denote by A one of the sets A_i with $\mu(A) > 0$. Let $x \in \boldsymbol{f}(A)$. Since $A \subset B$ we have $|x| > \alpha$ and $|\boldsymbol{f}(t) - x| < \varepsilon$ for every $t \in A$. Let $y \in F$ with $|y| = 1$ and $|u(x, y)| \geqslant (1 - \varepsilon)|x|$. The function $\boldsymbol{g} = (\mu(A))^{-1} \varphi_A y$ is a Borel step function of $\mathscr{E}_F$ and

$$N_1(\boldsymbol{g}) = \int |\boldsymbol{g}| \mathrm{d}\mu = 1 .$$

Then

$$\begin{aligned} \int |u(\boldsymbol{f}, \boldsymbol{g})| \mathrm{d}\mu &= (\mu(A))^{-1} \int \varphi_A |u(\boldsymbol{f}, \boldsymbol{g})| \mathrm{d}\mu = \\ &= (\mu(A))^{-1} \int |u(x, y)\varphi_A + u(\boldsymbol{f} - x, y)\varphi_A| \mathrm{d}\mu \geqslant \\ &\geqslant (\mu(A))^{-1} \int |u(x, y)| \varphi_A \mathrm{d}\mu - (\mu(A))^{-1} \int |u(\boldsymbol{f} - x, y)| \varphi_A \mathrm{d}\mu \end{aligned}$$

But

$$(\mu(A))^{-1} \int |u(x, y)| \varphi_A \mathrm{d}\mu \geqslant (\mu(A))^{-1} \int (1 - \varepsilon)|x| \varphi_A \mathrm{d}\mu = (1 - \varepsilon)|x|$$

and

$$\begin{aligned} (\mu(A))^{-1} \int |u(\boldsymbol{f} - x, y)| \varphi_A \mathrm{d}\mu &\leqslant (\mu(A))^{-1} \int |\boldsymbol{f} - x| \, |y| \, \varphi_A \mathrm{d}\mu \leqslant \\ &\leqslant (\mu(A))^{-1} \int \varepsilon \varphi_A \mathrm{d}\mu = \varepsilon , \end{aligned}$$

therefore

$$\int |u(\boldsymbol{f}, \boldsymbol{g})| \mathrm{d}\mu \geqslant (1 - \varepsilon)|x| - \varepsilon > (1 - \varepsilon)\alpha - \varepsilon = \alpha - \varepsilon(\alpha + 1),$$

whence

$$\sup \int^* |u(\boldsymbol{f}, \boldsymbol{g})| \mathrm{d}\mu \geqslant \alpha, \ \boldsymbol{g} \in \mathscr{E}_F, \ N_1(\boldsymbol{g}) = 1 .$$

Since $\alpha < N_\infty(\boldsymbol{f})$ is arbitrary, we deduce

$$\sup \int^* |u(f, g)| \, d\mu \leqslant N_{\infty}(f), \quad g \in \mathscr{E}_F, \, N_1(g) = 1 \,,$$

consequently

$$N_{\infty}(f) = \sup \int^* |u(f, g)| \, d\mu = \sup \int^* |f| \, |g| \, d\mu \,,$$

for $g \in \mathscr{E}_F$ with $N_1(g) = 1$ and the theorem is completely proved.

Remarks.

1. *For every μ-measurable function $f: T \to E$ and for $1 \leqslant p \leqslant +\infty$ we have also*

$$N_p(f) = \sup \int^* |u(f, g)| \, d\mu = \sup \int^* |f| \, |g| \, d\mu \,,$$

for $g \in \mathscr{E}_F$ with $N_q(g) < 1$ (instead of $N_q(g) = 1$), $p^{-1} + q^{-1} = 1$.

Let $\alpha < N_p(f)$ and $0 < \varepsilon < 1$. There exists a function $g_1 \in \mathscr{E}_F$ with $N_q(g_1) = 1$ such that

$$\alpha < \int |u(f, g_1)| \, d\mu \,.$$

The function $g = (1 - \varepsilon) g_1$ belongs to $\mathscr{E}_F$, $N_q(g) = 1 - \varepsilon < 1$ and

$$\int^* |u(f, g)| \, d\mu = (1 - \varepsilon) \int^* |u(f, g_1)| \, d\mu \geqslant (1 - \varepsilon)\alpha \,,$$

whence

$$\sup \int^* |u(f, g)| \, d\mu \geqslant N_p(f), \text{ for } g \in \mathscr{E}_F, \, N_q(g) < 1 \,.$$

2. *For every μ-measurable function $f: T \to E$ and for $1 \leqslant p \leqslant +\infty$ we have*

$$N_p(f) = \sup \int^* |u(f, g)| \, d\mu = \sup \int^* |f| \, |g| \, d\mu$$

for $g \in \mathscr{E}_F$ with $N_q(g) \leqslant 1$, $p^{-1} + q^{-1} = 1$.

(3) *For every μ-measurable function $f: T \to E$ and for $1 \leqslant p \leqslant +\infty$ we have*

$$N_p(f) = \sup \int^* |u(f, g)| \, d\mu = \sup \int^* |f| \, |g| \, d\mu$$

for $f \in \mathscr{L}^q_F$ (instead of $f \in \mathscr{E}_F$) with $N_q(g) = 1$ or $N_q(g) < 1$ or $N_q(g) \leqslant 1$, $p^{-1} + q^{-1} = 1$.

14.7 Proposition. *Let E, F and G be three Banach spaces and $u: E \times F \to G$ a continuous bilinear mapping with $|x| = \sup_{|y| \leqslant 1} |u(x, y)|$, for every $x \in E$.*

Let $f: T \to E$ be a μ-measurable function. If for every $y \in E$, the function $t \to u(f(t), y)$ is μ-negligible, then f is μ-negligible.

In fact, for every function $g = \Sigma_{1 \leqslant i \leqslant n} \varphi_{A_i} y_i$ of $\mathscr{E}_F$, the function $t \to u(f(t), g(t))$ is μ-negligible, therefore

$$N_1(f) = \sup \int^* |u(f, g)| \, d\mu = 0, \text{ for } g \in \mathscr{E}_F \text{ with } N_\infty(g) < 1 .$$

It follows that f is μ-negligible.

14.8 Corollary. *Let E and F be two Banach spaces and $U: T \to \mathscr{L}(E, F)$ a μ-measurable function. If for every $x \in E$, the function $t \to U(t)x$ is μ-negligible, then U is μ-negligible.*

14.9 Corollary. *Let E be a Banach space and Z a subspace of E such that $|x| = \sup_{z \in Z} |\langle x, z \rangle| / |z|$ for every $x \in E$. Let $f: T \to E$ be a μ-measurable function. If the function $t \to \langle f(t), z \rangle$ is μ-negligible for every $z \in Z$, then f is μ-negligible.*

We take $F = Z$ in proposition 14.7.

In particular, corollary 14.9 is true if we take $Z = E'$.

14.10 Corollary. *Let E be a Banach space, E' its dual and $f: T \to E'$ a μ-measurable function. If the function $t \to \langle x, f'(t) \rangle$ is μ-negligible for every $x \in E$, then f' is μ-negligible.*

For every Banach space X denote by $\mathscr{H}_X$ the set of the functions of the form $g = \Sigma_{1 \leqslant i \leqslant n} \varphi_i x_i$ with $\varphi_i \in \mathscr{K}(T)$ and $x_i \in E$. We have $\mathscr{H}_X \subset \mathscr{K}_X(T)$ and $\mathscr{H}_X$ is a rich subspace of $\mathscr{K}_X(T)$ (definition 1.7).

14.11 Theorem. *Let E, F and G be three Banach spaces and $u: E \times F \to G$ a continuous bilinear mapping such that $|x| = \sup_{|y| \leqslant 1} |u(x, y)|$ for every $x \in E$. For every μ-measurable function $f: T \to E$ we have*

$$N_p(f) = \sup \int^* |u(f, g)| \, d\mu = \sup \int^* |f| \, |g| \, d\mu$$

for $g \in \mathscr{H}_F(T)$ with $N_q(g) < 1$ (or $N_q(g) = 1$), $1 \leqslant p \leqslant +\infty$, $p^{-1} + q^{-1} = 1$.

If $g \in \mathscr{H}_F(T)$ with $N_q(g) < 1$, we have

$$\int^* |u(f, g)| \, d\mu \leqslant \int^* |f| \, |g| \, d\mu \leqslant N_p(f) .$$

If $N_p(\boldsymbol{f})=0$, the conclusion of the theorem is evident. Assume that $N_p(\boldsymbol{f})>0$. Let $0<\alpha<N_p(\boldsymbol{f})$. For each n define the function $\boldsymbol{f}_n$ by $\boldsymbol{f}_n(t)=\boldsymbol{f}(t)$ if $|\boldsymbol{f}(t)|\leqslant n$, and $\boldsymbol{f}_n(t)=0$, if $|\boldsymbol{f}(t)|>n$. The function $\boldsymbol{f}_n$ is μ-measurable, the sequence $(|\boldsymbol{f}_n|)$ is increasing and $\sup|\boldsymbol{f}_n|=\boldsymbol{f}$, therefore

$$N_p(\boldsymbol{f})=\sup N_p(\boldsymbol{f}_n)\,.$$

There exists hence a natural number n with $\alpha<N_p(\boldsymbol{f}_n)$.

There exists then a step function $\boldsymbol{h}=\Sigma_{1\leqslant i\leqslant k}\varphi_{A_i}y_i$ of $\mathscr{E}_F$ with disjoint sets A_i, $y_i\neq 0$ and $N_q(\boldsymbol{h})<1$, such that

$$\alpha<\int^* |u(\boldsymbol{f}_n,\boldsymbol{h})|\,\mathrm{d}\mu\,.$$

We remark that the function $u(\boldsymbol{f}_n,\boldsymbol{h})$ is μ-measurable, bounded, $(|u(\boldsymbol{f}_n(t),\boldsymbol{h}(t))|\leqslant n\,||\boldsymbol{h}||)$ and with compact support (contained in the support of $\boldsymbol{h}$), therefore is μ-integrable. Let $\eta>0$ be a number such that we have simultaneously

$$\Sigma_{1\leqslant i\leqslant k}\eta|y_i|<n^{-1}\left(\int|u(\boldsymbol{f}_n,\boldsymbol{h})|\,\mathrm{d}\mu-\alpha\right)$$

and

$$\Sigma_{1\leqslant\ \leqslant k}\eta|y_i|^q<1-[N_q(\boldsymbol{h})]^q\,.$$

For each i there exist a compact set $K_i\subset A_i$ with $\mu(A_i\setminus K_i)<\eta$, a relatively compact open set $U_i\supset K_i$ such that $\mu(U_i\setminus K_i)<\eta$ and a continuous function $\varphi_i: T\to[0,1]$ with compact support, equal to 1 on K_i and vanishing outside U_i. In addition, since the sets K_i are disjoint, we can choose the sets U_i disjoint. We have

$$\varphi_i\leqslant\varphi_{U_i}=\varphi_{K_i}+\varphi_{U_i\setminus K_i}\leqslant\varphi_{A_i}+\varphi_{U_i\setminus K_i}\,,$$

therefore

$$\int\varphi_i\,\mathrm{d}\mu\leqslant\mu(A_i)+\mu(U_i\setminus K_i)\leqslant\mu(A_i)+\eta\,.$$

The function $\boldsymbol{g}=\Sigma_{1\leqslant i\leqslant k}\varphi_i y_i$ is continuous with compact support. We have $\boldsymbol{g}(t)=\boldsymbol{h}(t)$ for $t\in K=\cup_{1\leqslant i\leqslant k}K_i$ and $|\boldsymbol{g}(t)|\leqslant|y_i|$ for $t\in U_i$. Then, if $q<+\infty$,

$$[N_q(\boldsymbol{g})]^q=\int|\boldsymbol{g}(t)|^q\,\mathrm{d}\mu=\Sigma|y_i|^q\int\varphi_i\,\mathrm{d}\mu\leqslant\Sigma\,\mu(A_i)|y_i|^q+\Sigma\,\eta|y_i|^q<1\,,$$

therefore $N_q(\boldsymbol{g})<1$; if $q=+\infty$, we have $N_\infty(\boldsymbol{g})\leqslant\sup|y_i|=N_\infty(\boldsymbol{h})<1$. Thus, for every q we have $N_q(\boldsymbol{g})<1$. On the other hand,

$$\int_{T\setminus K}|u(\boldsymbol{f}_n,\boldsymbol{h})|\,\mathrm{d}\mu\leqslant n\,\Sigma\,\mu(A_i\setminus K_i)|y_i|\ <\int|u(\boldsymbol{f}_n,\boldsymbol{h})|\,\mathrm{d}\mu+\alpha\,,$$

hence

$$\int^* |u(f, g)| d\mu \geqslant \int^* |u(f_n, g)| d\mu \geqslant \int_K^* |u(f_n, g)| d\mu = \int_K |u(f_n, h)| d\mu =$$
$$= \int |u(f_n, h)| d\mu - \int_{T \setminus K} |u(f_n, h)| d\mu > \alpha .$$

It follows that

$$N_p(f) = \sup \int^* |u(f, g)| d\mu, \text{ for } g \in \mathcal{H}_F, N_q(g) < 1 .$$

Let us show that the supremum can be taken for the functions $g \in \mathcal{H}_F$ with $N_q(g) = 1$.

Assume $N_p(f) > 0$ and let $0 < \alpha < N_p(f)$. Let $h \in \mathcal{H}_F$ with $N_q(h) < 1$ and

$$\alpha < \int^* |u(f, h)| d\mu .$$

We deduce first that $N_q(h) > 0$. Put $g = h/N_q(h)$. The function g belongs tl the space $\mathcal{H}_F$ and $N_q(g) = 1$. We have then

$$\alpha < \int^* |u(f, h)| d\mu = N_q(h) \int^* |u(f, g)| d\mu \leqslant \int^* |u(f, g)| d\mu ,$$

whence it follows that

$$N_p(f) = \sup \int^* |u(f, g)| d\mu, \text{ for } g \in \mathcal{H}_F, N_q(g) = 1 .$$

14.12 COROLLARY. *Let E and F be two Banach spaces and $U : T \to \mathcal{L}(E, F)$ a μ-measurable function. We have*

$$N_p(U) = \sup \int^* |U(t) g(t)| d\mu(t) = \sup \int^* |U(t)| \, |g(t)| d\mu(t)$$

for $g \in \mathcal{E}_E$ or $g \in \mathcal{H}_E$ with $N_q(g) < 1$ or $N_q(g) = 1$, $1 \leqslant p \leqslant +\infty$, $p^{-1} + q^{-1} = 1$.

14.13 COROLLARY. *Let E be a Banach space and $f : T \to E$ a μ-measurable function. We have*

$$N_p(f) = \sup \int^* |f(t) \varphi(t)| d\mu(t) = \sup \int^* |f(t)| \, |\varphi(t)| d\mu(t)$$

for $\varphi \in \mathcal{E}$ or $\varphi \in \mathcal{K}(T)$ with $N_q(\varphi) < 1$ or $N_q(\varphi) = 1, 1 \leqslant p \leqslant +\infty, p^{-1} + q^{-1} = 1$.

14.14 PROPOSITION. *Let E be a Banach space and $f : T \to E$ a μ-measurable function. If for every function $\varphi \in \mathcal{K}(T)$ the function $f\varphi$ is μ-negligible, then f is μ-negligible.*

14.15 THEOREM. *Let E be a Banach space, E' its dual, $1 \leqslant p \leqslant \infty$ and $p^{-1} + q^{-1} = 1$. For every function $f \in \mathcal{L}_E^p$ we have*

$$N_p(f) = \sup \left| \int \langle f, f' \rangle d\mu \right| = \sup \int |\langle f, f' \rangle| d\mu =$$
$$= \sup \int |f| \, |f'| d\mu$$

when f' *runs over the set of functions of* $\mathscr{E}_{E'}$ *or* $\mathscr{K}_{E'}$ *with* $N_p(f')\leqslant 1$ *(or* $N_p(f')=1$*).*

For every function $f'\in\mathscr{L}^q_{E'}$, *we have*

$$N_q(f')=\sup|\textstyle\int\langle f,f'\rangle\,d\mu|=\sup\int|\langle f,f'\rangle|\,d\mu= \\ =\sup\int|f|\,|f'|\,d\mu$$

when f *runs over the set of functions of* $\mathscr{E}_E$ *or* $\mathscr{K}_E$ *with* $N_p(f)\leqslant 1$ *(or* $N_p(f)=1$*).*

In order to prove the first set of equalities, we consider the bilinear functional $u(x, y)=\langle x, y\rangle$ defined on $E\times E'$ and we remark that for every $x\in E$ we have $|x|=\sup_{|y|\leqslant 1}|\langle x, y\rangle|$. For every $x\in E$ and every $\varepsilon>0$ there exists an element $y\in E'$ with $|y|\leqslant 1$ such that $\langle x, y\rangle$ is positive and $\langle x, y\rangle\geqslant(1-\varepsilon)|x|$. Then we follow the proof of theorems 14.6 and 14.11 to prove that for $f\in\mathscr{L}^p_E$, we have

$$\sup|\textstyle\int\langle f, g\rangle|\,d\mu\geqslant N_p(f),$$

for $g\in\mathscr{E}_E$, or $g\in\mathscr{K}_E$, with $N_p(g)\leqslant 1$.

In the same way we prove the second set of equalities.

Remark. In the first set of equalities we can replace E' by a subspace Z with the property that $|x|=\sup_{z\in Z}|\langle x, z\rangle|/|z|$ for $x\in E$.

14.16 COROLLARY. *If* $f\in\mathscr{L}^p(\mu)$, $1\leqslant p\leqslant\infty$, *then*

$$N_p(f)=\sup\int|fg|\,d\mu=\sup\int|f|\,g\,d\mu$$

when g *runs over the set of functions of* $\mathscr{E}(T)$ *or of* $\mathscr{K}(T)$ *with* $N_q(g)\leqslant 1$ *(or* $N_q(g)=1$*),* $p^{-1}+q^{-1}=1$.

14.17 PROPOSITION. *Let E be a Banach space and* $f: T\to E$ *a* μ*-measurable function. If for every compact set* $K\subset T$ *the function* $f\varphi_K$ *is* μ*-integrable and*

$$\int_K f\,d\mu=0$$

or if, for every function $\varphi\in\mathscr{K}(T)$, *the function* $f\varphi$ *is* μ*-integrable and*

$$\int f\varphi\,d\mu=0$$

then f *is* μ*-negligible.*

Assume first that for every compact set $K\subset T$ the function $f\varphi_K$ is μ-integrable and

$$\int_K \boldsymbol{f}\,\mathrm{d}\mu = 0 .$$

Let $A \subset T$ be a relatively compact Borel set (contained in a compact set B). Show that $\boldsymbol{f}\varphi_A$ is μ-integrable and

$$\int_A \boldsymbol{f}\,\mathrm{d}\mu = 0 .$$

Since A is μ-integrable, there exists an increasing sequence (K_n) of compact sets such that the set $A \setminus \cup K_n$ is μ-negligible. The sequence $(\boldsymbol{f}\varphi_{K_n})$ consists of integrable functions of $\mathscr{L}_E^1(\mu)$ and tends almost everywhere to $\boldsymbol{f}\varphi_A$. Since $|\boldsymbol{f}\varphi_{K_n}| \leqslant |\boldsymbol{f}\varphi_B|$ for every $n \in N$ and $\boldsymbol{f}\varphi_B$ is μ-integrable, by Lebesgue's theorem if follows that $\boldsymbol{f}\varphi_A$ is μ-integrable and

$$\int \boldsymbol{f}\varphi_A \mathrm{d}\mu = \lim \int \boldsymbol{f}\varphi_{K_n} \mathrm{d}\mu = 0 .$$

It follows then that for every $x' \in E'$ we have

$$\int \langle \boldsymbol{f}, x'\varphi_A \rangle \mathrm{d}\mu = \langle \int \boldsymbol{f}\varphi_A \mathrm{d}\mu, x' \rangle = 0 ,$$

therefore, for every step function $\boldsymbol{f}'$ of $\mathscr{E}_E$, and every compact set $K \subset T$ we have $\varphi_K \boldsymbol{f}' \in \mathscr{E}_{E'}$, and

$$\int \langle \boldsymbol{f}\varphi_K, \boldsymbol{f}' \rangle \mathrm{d}\mu = \int \langle \boldsymbol{f}, \varphi_K \boldsymbol{f}' \rangle \mathrm{d}\mu = 0 .$$

Since $\boldsymbol{f}\varphi_K \in \mathscr{L}_E^1(\mu)$, by theorem 14.15, we deduce that $N_p(\boldsymbol{f}\varphi_K) = 0$, hence $\boldsymbol{f}\varphi_K$ is μ-negligible, consequently $\boldsymbol{f}$ is μ-negligible.

Assume now that for every function $\varphi \in \mathscr{K}(T)$ we have

$$\int \boldsymbol{f}\varphi\,\mathrm{d}\mu = 0 .$$

For every $x' \in E'$ we have then

$$\int \langle \boldsymbol{f}, \varphi x' \rangle \mathrm{d}\mu = \langle \int \boldsymbol{f}\varphi\,\mathrm{d}\mu, x' \rangle = 0 ;$$

therefore, for every function $\boldsymbol{g} = \Sigma\,\varphi_i x_i'$ of $\mathscr{H}_{E'}$ and every function $\psi \in \mathscr{K}(T)$, we have $\boldsymbol{g}\,\psi \in \mathscr{H}_{E'}$, and

$$\int \langle \boldsymbol{f}\psi, \boldsymbol{g} \rangle \mathrm{d}\mu = \int \langle \boldsymbol{f}, \psi \boldsymbol{g} \rangle \mathrm{d}\mu = 0 .$$

Since $\boldsymbol{f}\psi \in \mathscr{L}_E^1(\mu)$, by theorem 14.15 we deduce that $\boldsymbol{f}\psi$ is μ-negligible for every $\psi \in \mathscr{K}(T)$. For every compact set $K \subset T$ there exists a function $\psi \in \mathscr{K}(T)$ with $\psi(t) = 1$ for $t \in K$. Then $|\boldsymbol{g}\varphi| \leqslant |\boldsymbol{g}\psi|$, therefore $\boldsymbol{g}\varphi_K$ is μ-negligible, consequently $\boldsymbol{g}$ is μ-negligible.

14.18 Corollary. *If* $\boldsymbol{f} \in \mathscr{L}_E^p$, $1 \leqslant p \leqslant \infty$, *and if*

$$\int_K \boldsymbol{f}\,\mathrm{d}\mu = 0, \textit{ for every compact set } K \subset T ,$$

or if

$$\int f\varphi \, d\mu = 0, \text{ for every function } \varphi \in \mathscr{K}(T),$$

then f *is* μ*-negligible.*

In fact, if $f \in \mathscr{L}_E^p$, then $f\varphi_K$ and $f\varphi$ are μ-integrable for every compact set $K \subset T$ and every function $\varphi \in \mathscr{K}(T)$.

14.19 PROPOSITION. *Let* E, F *and* G *be three Banach spaces and* $u: E \times F \to G$ *a continuous bilinear mapping with* $|x| = \sup_{|y| \leq 1} |u(x, y)|$ *for every* $x \in E$. *Let* $f: T \to E$ *be a* μ*-measurable function.*

If for every compact set $K \subset T$ *and every* $y \in F$ *the function* $t \to u(f(t), y)\varphi_K(t)$ *is* μ*-negligible and*

$$\int_K u(f(t), y) \, d\mu(t) = 0,$$

or if for every function $\varphi \in \mathscr{K}(T)$ *and every* $y \in F$, *the function* $t \to u(f(t), y)\,\varphi(t)$ *is* μ*-integrable and*

$$\int u(f(t), y)\,\varphi(t) \, d\mu(t) = 0$$

then f *is* μ*-negligible.*

In fact, by proposition 14.17, the function $t \to u(f(t), y)$ is μ-negligible for every $y \in F$. By proposition 14.7, the function f is μ-negligible.

14.20 COROLLARY. *Let* E *and* F *be two Banach spaces and* $U: T \to \mathscr{L}(E, F)$ *a* μ*-measurable function.*

If for every $x \in E$ *and every compact set* $K \subset T$ *the function* $t \to U(t)x\varphi_K(t)$ *is* μ*-integrable and*

$$\int_K U(t)x \, d\mu = 0$$

or if for every $x \in E$ *and every* $\varphi \in \mathscr{K}(T)$ *the function* $Ux\varphi$ *is* μ*-integrable and*

$$\int Ux\varphi \, d\mu = 0,$$

then U *is* μ*-negligible.*

14.21 COROLLARY. *Let* E *be a Banach space,* Z *a subspace of* E' *with* $|x| = \sup_{z \in Z} |\langle x, z\rangle| / |z|$ *for every* $x \in E$ *and* $f: T \to E$ *a* μ*-measurable function.*

If for every $z \in Z$ *and every compact set* $K \subset T$, *the function* $\langle f, z\rangle\,\varphi_K$ *is* μ*-integrable and*

$$\textstyle\int_K \langle f, z\rangle \mathrm{d}\mu = 0\,,$$

or if for every $z \in Z$ and every $\varphi \in \mathscr{K}(T)$, the function $\langle f, z\rangle\varphi$ is μ-integrable and

$$\textstyle\int \langle f, z\rangle \varphi \,\mathrm{d}\mu = 0\,,$$

then f is μ-negligible.

14.22 COROLLARY. *Let E be a Banach space, E' its dual and $f' : T \to E'$ a μ-measurable function.*

If for every $x \in E$ and every compact set $K \subset T$ the function $\langle x, f'\rangle \varphi_K$ is μ-integrable and

$$\textstyle\int_K \langle x, f'\rangle \mathrm{d}\mu = 0$$

or if for every $x \in E$ and $\varphi \in \mathscr{K}(T)$, the function $\langle x, f'\rangle \varphi$ is μ-integrable and

$$\textstyle\int \langle x, f'\rangle \varphi \,\mathrm{d}\mu = 0\,,$$

then f' is μ-negligible.

14.23 THEOREM. *Let E, F and G be three Banach spaces and $u : E \times F \to G$ a continuous bilinear mapping with $|x| = \sup_{|y| \leqslant 1} |u(x, y)|$ for each $x \in E$. A μ-measurable mapping $f : T \to E$ belongs to the space $\mathscr{L}_E^p(\mu)$, $1 \leqslant p \leqslant +\infty$, if and only if for every $g \in \mathscr{L}_F^p(\mu)$, $p^{-1} + q^{-1} = 1$, the function $t \to u(f(t), g(t))$ is μ-integrable.*

If $f \in \mathscr{L}_E^p$, then $u(f, g) \in \mathscr{L}_G^1$, for every function $g \in \mathscr{L}_F^q$ even if $p = 1$.

Conversely, assume that $u(f, g) \in \mathscr{L}_G^1$ for every $g \in \mathscr{L}_F^p$. Since f is μ-measurable, in order to show that $f \in \mathscr{L}_E^p$, it is sufficient to show that

$$N_p(f) = \sup \textstyle\int^* |u(f, g)| \,\mathrm{d}\mu < +\infty, \text{ for } N_q(g) \leqslant 1\,.$$

Assume first that f is a measurable *real* function and that $fg \in \mathscr{L}^1$ for every function $g \in \mathscr{L}^q$. Assume the contrary, that

$$N_p(f) = \sup \textstyle\int |fg| \,\mathrm{d}\mu = +\infty \text{ for } N_q(g) \leqslant 1,\ g \in \mathscr{L}^q\,.$$

For each n there exists a function $g_n \in \mathscr{L}^q$ with $N_q(g_n) \leqslant 1$ and

$$\textstyle\int |fg_n| \,\mathrm{d}\mu > n^3\,.$$

The function

$$g = \textstyle\sum_{1 \leqslant n < \infty} n^{-2} |g_n|$$

with finite or $+\infty$ values, is μ-measurable and

$$N_q(g) < \Sigma_{1\leqslant n<\infty} n^{-2} N_q(g_n) \leqslant \Sigma_{1\leqslant n<\infty} n^{-2} < +\infty ,$$

therefore g is finite μ-almost everywhere. Modifying g on the μ-negligible set on which it has the value $+\infty$, we obtain $g\in\mathscr{L}^q$.

Since $g \geqslant n^{-2}|g_n|$ for every n, we have

$$\int^* |f| g \,\mathrm{d}\mu \geqslant n^{-2}\int |f|\,|g_n|\,\mathrm{d}\mu = n^{-2}\int |fg_n|\,\mathrm{d}\mu > n,$$

therefore $\int^* |f| g\,\mathrm{d}\mu = +\infty$, consequently fg would not be integrable, which contradicts the hypothesis. Thus, $f\in\mathscr{L}^p$.

Consider now the general case, when $\boldsymbol{f}: T\to E$ is μ-measurable and $u(\boldsymbol{f}, \boldsymbol{g})\in\mathscr{L}^1_G$ for every $\boldsymbol{g}\in\mathscr{L}^q_F$.

Let $0<a<1$. There exists a μ-measurable function $\boldsymbol{h}: T\to F$ with

$$a|\boldsymbol{f}(t)| < |u(\boldsymbol{f}(t), \boldsymbol{h}(t))| \text{ and } |\boldsymbol{h}(t)| = 1$$

μ-almost everywhere. Let $\varphi\in\mathscr{L}^q$. Then $\varphi\boldsymbol{h}\in\mathscr{L}^q_F$, therefore the function $u(\boldsymbol{f}, \boldsymbol{h}\varphi)$ is μ-integrable. But

$$a|\boldsymbol{f}(t)|\,|\varphi(t)| \leqslant |u(\boldsymbol{f}(t), \boldsymbol{h}(t))|\,|\varphi(t)| ,$$

From the first part of the theorem we deduce that $a|\boldsymbol{f}|\in\mathscr{L}^p$, consequently $\boldsymbol{f}\in\mathscr{L}^p$, that is, $N_p(\boldsymbol{f}) < +\infty$. It follows that $\boldsymbol{f}\in\mathscr{L}^p_E$ and the theorem is proved.

14.24 COROLLARY. *Let E and F be two Banach spaces. A μ-measurable function $U: T\to X = \mathscr{L}(E, F)$ belongs to the space $\mathscr{L}^p_X$, if and only if for every function $\boldsymbol{f}\in\mathscr{L}^q_E$, the function $t\to U(t)\boldsymbol{f}(t)$ is μ-integrable,* $(1\leqslant p\leqslant +\infty,$ $p^{-1}+q^{-1}=1)$.

14.25 COROLLARY. *Let E be a Banach space. A μ-measurable function $\boldsymbol{f}: T\to E$ belongs to the space $\mathscr{L}^p_E$, $1\leqslant p\leqslant +\infty$, if and only if for every real function $\varphi\in\mathscr{L}^q$, $p^{-1}+q^{-1}=1$, the function $\boldsymbol{f}\varphi$ is μ-integrable.*

14.26 COROLLARY. *Let E be a Banach space. A μ-measurable function $\boldsymbol{f}: T\to E$ belongs to the space $\mathscr{L}^p_E$, $1\leqslant p\leqslant +\infty$, if and only if for every function $\boldsymbol{f}'\in\mathscr{L}^q_{E'}$, $p^{-1}+q^{-1}=1$, the scalar function $t\to\langle \boldsymbol{f}(t), \boldsymbol{f}'(t)\rangle$ is μ-integrable.*

14.27 COROLLARY. *Let E be a Banach space. A μ-measurable function $\boldsymbol{f}': T\to E'$ belongs to the space $\mathscr{L}^p_{E'}$, $1\leqslant p\leqslant +\infty$, if and only if for every func-*

tion $f \in \mathscr{L}^q_E$, $p^{-1}+q^{-1}=1$, *the scalar function* $t \to \langle f(t), f'(t)\rangle$ *is* μ*-integrable.*

14.28 COROLLARY. *A real function f belongs to* $\mathscr{L}^p$, $1 \leqslant p \leqslant +\infty$, *if and only if* $fg \in \mathscr{L}^1$ *for every* $g \in \mathscr{L}^q$, $p^{-1}+q^{-1}=1$.

In fact, f is μ-measurable if and only if fg is μ-measurable, for every μ-measurable step-function g.

14.29 PROPOSITION. *For every function* $g \in \mathscr{L}^q_{E'}$, *the mapping*

$$U(f) = \int \langle f, g\rangle \,\mathrm{d}\mu, \quad f \in \mathscr{L}^p_E$$

is a continuous linear functional on $\mathscr{L}^p_E$ *and*

$$\|U\| = N_q(g)\,.$$

The linearity of the mapping U is immediate. The continuity follows from the equality

$$\|U\| = \sup|U(f)| = \sup\left|\int \langle f, g\rangle \,\mathrm{d}\mu\right| = N_q(g) < \infty, \quad \text{for } N_p(f) < 1\,.$$

Remark. From the equality $\|U\| = N_q(g)$ it follows that if $N_q(g)=0$, then $U=0$, therefore for two equivalent functions g and g' of $\mathscr{L}^q_{E'}$ corresponds the same functional U on $\mathscr{L}^p_E$.

On the other hand, the dual of $\mathscr{L}^p_E$ can be identified with the dual of L^p_E. From the above proposition it follows that the space $L^q_{E'}$, can be embedded isometrically in the dual of the space L^p_E.

Later we shall show that under additional conditions on E, each continuous linear functional U on the space $\mathscr{L}^p_E$ with $1 \leqslant p < +\infty$ can be obtained as above, from a function $g \in \mathscr{L}^q_{E'}$. Therefore, the dual of the space L^p_E with $1 \leqslant p < +\infty$ can be identified with the space $L^q_{E'}$.

The dual of the space $L^\infty_{E'}$ is not isomorphic to L^1_E except when the measure μ has finite support.

14.30 COROLLARY. *If E is a Hilbert space, then* L^2_E *is a Hilbert space for the scalar product*

$$\langle \tilde{f}, \tilde{g}\rangle = \int \langle f(t), g(t)\rangle \,\mathrm{d}\mu(t)\,.$$

For every continuous linear functional U on $\mathscr{L}^2_E$ *there exists a function* $g \in \mathscr{L}^2_E$, *determined μ-almost everywhere, such that*

$$U(f)=\int \langle f(t), g(t)\rangle \,\mathrm{d}\mu(t), \text{ for } f\in\mathscr{L}_E^2$$

and

$$\| U \| = N_2(g)\,.$$

The fact that $\langle \tilde{f}, \tilde{g}\rangle$ is a scalar product on L_E^2 is immediately seen without difficulty. We have then $\| f \|_2 = N_2(f) = (\langle \tilde{f}, \tilde{f}\rangle)^{1/2}$, therefore the Banach space L_E^2 is a Hilbert space.

Now let U be a continuous linear functional on $\mathscr{L}_E^2$.

If $\tilde{f}=\tilde{g}$ we have $U(f)=U(g)$. We define the functional U' on L_E^2 by setting

$$U'(\tilde{f}) = U(f)\,.$$

This functional is continuous since

$$\| U' \| = \sup |U(\tilde{f})| = \sup |U(f)| = \| U \| < +\infty, \text{ for } N_2(f)\leqslant 1\,.$$

As L_E^2 is a Hilbert space, there exists an element $\tilde{g}\in L_E^2$ such that

$$U'(\tilde{f}) = \langle \tilde{f}, \tilde{g}\rangle \text{ for every } \tilde{f}\in L_E^2$$

and $\| U' \| = \| \tilde{g} \|_2$. It follows then that for an arbitrary function $g\in\tilde{g}$, we have

$$U(f) = \langle \tilde{f}, \tilde{g}\rangle = \int \langle f(t), g(t)\rangle \,\mathrm{d}\mu(t), \text{ for } f\in\mathscr{L}_E^2$$

and

$$\| U \| = N_2(g)\,.$$

3 *Relationships between $\mathscr{L}^r$ and $\mathscr{L}^s$*

Let E be a Banach space and μ a positive measure on T.

14.31 PROPOSITION. *If $1\leqslant r<p<s\leqslant +\infty$, then*

$$\mathscr{F}_E^r \cap \mathscr{F}_E^s \subset \mathscr{F}_E^p \text{ and } \mathscr{L}_E^r \cap \mathscr{L}_E^s \subset \mathscr{L}_E^p\,.$$

Assume first that $s<+\infty$. There exists a number t such that $0<t<1$ and $p=tr+(1-t)s$. Denote $\alpha=t^{-1}$ and $\beta=(1-t)^{-1}$. We have $0<\alpha, \beta<+\infty$, $\alpha^{-1}+\beta^{-1}=1$ and $p=r\alpha^{-1}+s\beta^{-1}$. Using the Hölder inequality, for every function $f: T\to E$ we have

$$\int^* |f|^p \, d\mu = \int^* |f|^{r/\alpha} |f|^{s/\beta} d\mu \leqslant N_\alpha(|f|^{r/\alpha}) N_\beta(|f|^{s/\beta}) = \\ = (\int^* |f|^r d\mu)^{1/\alpha} (\int^* |f|^s d\mu)^{1/\beta} .$$

If $N_r(f) < +\infty$ and $N_s(f) < \infty$, it follows that $N_p(f) < +\infty$, therefore

$$\mathscr{F}_E^r \cap \mathscr{F}_E^s \subset \mathscr{F}_E^p .$$

Assume now that $s = +\infty$. If $f \in \mathscr{F}_E^r \cap \mathscr{F}_E^{+\infty}$, we then have

$$\int^* |f|^p \, d\mu = \int^* |f|^r |f|^{p-r} d\mu \leqslant N_1(|f|^r) N_\infty(|f|^{p-r}) = \\ = \int^* |f|^r d\mu [N_\infty(f)]^{p-r} < +\infty ,$$

consequently $f \in \mathscr{F}_E^p$. Thus, the inclusion $\mathscr{F}_E^r \cap \mathscr{F}_E^s \subset \mathscr{F}_E^p$ is valid for $1 < s \leqslant +\infty$.

If now $f \in \mathscr{L}_E^r \cap \mathscr{L}_E^s$, then f is μ-measurable and $f \in \mathscr{F}_E^p$, hence $f \in \mathscr{L}_E^p$. Thus the proposition is proved.

14.32 Corollary. *For every function $f: T \to E$, the set $I_f = \{p \mid 1 \leqslant p \leqslant +\infty, N_p(f) < +\infty\}$ is either empty or an interval of $\overline{R}$. If I_f is not empty (and is not reduced to a point), then the function* lg $N_p(f)$ *is a convex function of p^{-1}, and the mapping $p \to N_p(f)$ is continuous on I_f.*

The first part of the corollary follows from the preceding proposition. Let us prove the convexity of the mapping $p^{-1} \to \lg N_p(f)$. This means that if $r, s, p \in I_f$ and if

$$p^{-1} = tr^{-1} + (1-t)/s \text{ with } 0 < t < 1 ,$$

then

$$\lg N_p(f) \leqslant t \lg N_r(f) + (1-t) \lg N_s(f) .$$

To do that, assume first that $+\infty \notin I_f$, therefore $1 \leqslant r < p < s < +\infty$.

Put $\alpha = r/tp$ and $\beta = s/(1-t)p$. We have $\alpha^{-1} + \beta^{-1} = 1$ and $1 \leqslant \alpha, \beta < +\infty$. Applying the Hölder inequality we obtain

$$\int^* |f|^p d\mu = \int^* |f|^{tp} |f|^{(1-t)p} d\mu = \int^* |f|^{r/\alpha} |f|^{s/\beta} d\mu \leqslant$$

$$\leqslant N_\alpha(|f|^{r/\alpha}) N_\beta(|f|^{s/\beta}) = (\int^* |f|^r d\mu)^{1/\alpha} (\int^* |f|^s d\mu)^{1/\beta} =$$

$$= (\int^* |f|^r d\mu)^{(1/r)tp} (\int^* |f|^s d\mu)^{(1/s)(1-t)p} ,$$

that is

$$N_p(f)^p \leqslant N_r(f)^{tp} N_s(f)^{(1-t)p} ,$$

whence

$$N_p(f) \leqslant N_r(f)^t N_s(f)^{(1-t)}$$

therefore

$$\lg N_p(f) \leqslant t \lg N_r(f) + (1-t) \lg N_s(f) .$$

If $+\infty \in I_f$ and $s = +\infty$, and if $p^{-1} = tr^{-1} + (1-t)/s = tr^{-1}$, $0 < t < 1$, then, as in the preceding proposition, we have

$$\int^* |f|^p \mathrm{d}\mu \leqslant \int^* |f|^r \mathrm{d}\mu [N_\infty(f)]^{p-r} < +\infty ,$$

that is,

$$N_p(f)^p \leqslant N_r(f)^r N_\infty(f)^{p-r} ,$$

whence

$$N_p(f) \leqslant N_r(f)^{r/p} N_\infty(f)^{(p-r)/p} = N_r(f)^t N_\infty(f)^{1-t} ,$$

consequently, also,

$$\lg N_p(f) \leqslant t \lg N_r(f) + (1-t) \lg N_\infty(f) .$$

Thus, the function $\lg N_p(f)$ is convex with respect to the variable p^{-1}, (when p runs over I_f). We deduce then the continuity of $\lg N_p(f)$ as a function of p^{-1}, hence as a function of p too. Then $N_p(f) = \exp \lg N_p(f)$ is also a continuous function of p. Thus the proposition is proved.

14.33 Proposition. *If the measure μ is bounded and if $1 \leqslant r < s \leqslant +\infty$, then $\mathscr{F}_E^s \subset \mathscr{F}_E^r$ and $\mathscr{L}_E^s \subset \mathscr{L}_E^r$, and the topology of the convergence in the mean of order s is finer than the topology of the convergence in the mean of order p (on $\mathscr{F}_E^s$).*

The proposition is evidently true if $\mu = 0$. Assume therefore that $\mu \neq 0$, that is, $0 < \mu(T) < +\infty$. Let p be a number such that

$$r^{-1} = s^{-1} + p^{-1} .$$

It follows that $1 \leqslant r \leqslant p < +\infty$. Denote $\alpha^{-1} = rs^{-1}$ and $\beta^{-1} = rp^{-1}$. We have $1 \leqslant \alpha, \beta \leqslant +\infty$ and $\alpha^{-1} + \beta^{-1} = 1$. For every function $f : T \to E$, we apply the Hölder inequality and obtain:

$$\int^* |f|^r \mathrm{d}\mu = \int^* |f|^r 1 \,\mathrm{d}\mu \leqslant N_\alpha(|f|^r) N_\beta(1) ,$$

whence

$$(\int^* |f|^r \mathrm{d}\mu)^{1/r} \leqslant N_\alpha(|f|^r)^{1/r} N_\beta(1)^{1/r},$$

that is,

$$N_r(f) \leqslant N_\alpha(|f|^r)^{1/r} \mu(T)^{1/\beta r}.$$

If $s < +\infty$, then $\alpha < +\infty$, therefore

$$N_r(f) \leqslant (\int^* |f|^{\alpha r} \mathrm{d}\mu)^{1/\alpha r} \mu(T)^{1/\beta r} = (\int^* |f|^s \mathrm{d}\mu)^{1/s} \mu(T)^{1/p} = N_s(f)\mu(T)^{1/r - 1/s},$$

whence

$$\mu(T)^{-1/r} N_r(f) \leqslant \mu(T)^{-1/s} N_s(f).$$

This relation remains valid also for $s = +\infty$, since in this case $\alpha = +\infty$, hence $\beta = 1$ and

$$N_r(f) \leqslant N_\infty(|f|^r)^{1/r} \mu(T)^{1/r} = N_\infty(f)\mu(T)^{1/r},$$

whence

$$\mu(T)^{-1/r} N_r(f) \leqslant N_\infty(f) = \mu(T)^{-1/\infty} N_\infty(f).$$

From the inequality

$$\mu(T)^{-1/r} N_r(f) \leqslant \mu(T)^{-1/s} N_r(f)$$

it follows that $\mathscr{F}^s_E \subset \mathscr{F}^r_E$ and also that $\mathscr{L}^s_E \subset \mathscr{L}^r_E$. If (f_n) is a sequence of functions of $\mathscr{L}^s_E$, converging in the mean of order s to a function $f \in \mathscr{F}^s_E$, then from the preceding inequality it follows that (f_n) converges to f also in the mean of order r, hence on $\mathscr{F}^s_E$, the topology of the convergence in the mean of order s is finer than the topology of the convergence in the mean of order r.

14.34 Corollary. *If μ is bounded, and $f: T \to E$, then the set $I_f = \{p \mid 1 \leqslant p \leqslant +\infty, N_p(f) < +\infty\}$ is either empty or an interval of $\bar{R}$, which contains the point 1. If I_f is not empty (and is not reduced to a point), the function $p \to \mu(T)^{-1/p} N_p(f)$ is increasing on I_f.*

14.35 Proposition. *Let T be a discrete space and μ the measure defined by the mass $+1$ in each point of T. If $1 \leqslant r < s \leqslant +\infty$, then $\mathscr{F}^r_E \subset \mathscr{F}^s_E$ and $\mathscr{L}^r_E \subset \mathscr{L}^s_E$ and the topology of the convergence in the mean of order r is finer than the topology of the convergence in the mean of order s (on $\mathscr{F}^r_E$).*

Assume first that $s = +\infty$, and show that $N_\infty(f) \leqslant N_r(f)$ for every function $f: T \to E$. In fact, if

$$\alpha < N_\infty(f) = \|f\| = \sup_{t\in T}|f(t)|,$$

there exists a point $t_0 \in T$ with $|f(t_0)| > \alpha$. Then

$$N_r(f) = (\int^* |f(t)|^r d\mu)^{1/r} = (\Sigma_{t\in T}|f(t)|^r)^{1/r} \geqslant$$

$$\geqslant (|f(t_0)|^r)^{1/r} = |f(t_0)| > \alpha,$$

whence $N_r(f) \geqslant N_\infty(f)$.

Assume now that $1 \leqslant r < s < +\infty$. Then

$$\int^* |f|^s d\mu = \int |f|^r |f|^{s-r} d\mu \leqslant N_1(|f|^r(N_\infty(|f|^{s-r})\ ;$$

but

$$N_\infty(|f|^{s-r}) = N_\infty(|f|^r)^{(s-r)/r} \leqslant N_1(|f|^r)^{(s-r)/r},$$

therefore

$$\int^* |f|^s d\mu \leqslant N_1(|f|^r) N_1(|f|^r)^{(s-r)/r} = N_1(|f|^r)^{s/r} = (\int^* |f|^r d\mu)^{s/r},$$

whence

$$(\int^* |f|^s d\mu)^{1/s} \leqslant (\int^* |f|^r d\mu)^{1/r},$$

that is,

$$N_s(f) \leqslant N_r(f).$$

Thus, the inequality

$$N_s(f) \leqslant N_r(f)$$

is true for every s. From this inequality the proposition follows.

14.36 Corollary. *If T is discrete and if μ is a measure defined by the mass $+1$ in each point of T, then for every function $f: T \to E$, the set $I = \{p \mid 1 \leqslant p \leqslant +\infty,\ N_p(f) < +\infty\}$ is either empty or an interval of R, which contains the point $+\infty$. If I_f is not empty (and is not reduced to a point), the function $p \to N_p(f)$ is decreasing on I_f.*

PART II

Chapter IV

MEASURES DEFINED BY DENSITIES

§15 Locally integrable functions. Measure defined by densities

1 *Locally integrable functions*

Let μ be a positive measure and E a Banach space.

15.1 DEFINITION. *We say that a function $\boldsymbol{f}$ defined on T with values in E or in $\overline{R}$ is locally integrable with respect to μ, or locally μ-integrable, if the function $\boldsymbol{f}\varphi_K$ is μ-integrable for every compact set $K \subset T$.*

To say that $\boldsymbol{f}$ is *locally μ-integrable* means that $\boldsymbol{f}$ is *μ-measurable* and

$$\int_K^* |\boldsymbol{f}|\,\mathrm{d}\mu < \infty, \text{ for every compact set } K \subset T.$$

Examples

1. Every function $\boldsymbol{f}$ of $\mathscr{L}_E^p(\mu)$ with $1 \leqslant p \leqslant \infty$ is locally μ-integrable.

In fact, for every compact set $K \subset T$ the function φ_K belongs to the space $\mathscr{L}^q(\mu)$ with $p^{-1} + q^{-1} = 1$, therefore the function $\boldsymbol{f}\varphi_K$ is μ-integrable.

2. Every μ-measurable function which is essentially bounded on every compact set is locally integrable.

If $\boldsymbol{m}$ is a dominated measure, we say that a function is locally $\boldsymbol{m}$-integrable if it is locally $|\boldsymbol{m}|$-integrable.

Every function equal almost everywhere to a locally μ-integrable function is also locally μ-integrable.

A function $\boldsymbol{f}$ defined μ-almost everywhere on T with values in E or in $\overline{R}$ will be also called locally μ-integrable if it is equal μ-almost everywhere to a locally μ-integrable function $\boldsymbol{f}'$ defined on the whole space T.

If T is compact, or if the measure μ has compact support, the notion of local integrability is equivalent to that of integrability.

If $\boldsymbol{f}$ is locally μ-integrable, then the function $|\boldsymbol{f}|$ is locally integrable.

Conversely, if $\boldsymbol{f}$ is *μ-measurable* and if $|\boldsymbol{f}|$ is locally μ-integrable, then $\boldsymbol{f}$ is locally μ-integrable.

Thus, to say that $\boldsymbol{f}$ is locally μ-integrable means that $\boldsymbol{f}$ is μ-measurable and that the positive function $|\boldsymbol{f}|$ is locally μ-integrable.

Remark. If the functions f and g have values in $\bar{R}$ and are locally μ-integrable, then they are finite almost everywhere; therefore, the sum $f+g$ is defined μ-almost everywhere.

15.2 PROPOSITION. *A function $\boldsymbol{f}$ defined on T with values in E or in $\bar{R}$ is locally μ-integrable if and only if for every function $h \in \mathscr{K}(T)$, the function $\boldsymbol{f}h$ is μ-integrable.*

Assume that $\boldsymbol{f}$ is locally μ-integrable. Let $h \in \mathscr{K}(T)$ and let K be the support of h. The function $\boldsymbol{f}\varphi_K$ is μ-integrable, and $h \in \mathscr{L}^\infty(\mu)$, therefore the function $\boldsymbol{f}h = \boldsymbol{f}\varphi_K h$ is μ-integrable.

Conversely, assume that $\boldsymbol{f}h$ is μ-integrable for every $h \in \mathscr{K}(T)$ and let $K \subset T$ be a compact set. Let h_0 be a function of $\mathscr{K}(T)$ equal to 1 on K. The function $\boldsymbol{f}h_0$ is μ-integrable and $\varphi_K \in \mathscr{L}^\infty(\mu)$, therefore the function $\boldsymbol{f}\varphi_K = \boldsymbol{f}h_0\varphi_K$ is μ-integrable. It follows that $\boldsymbol{f}$ is locally μ-integrable.

15.3 PROPOSITION. *Let E and F be two Banach spaces and $\boldsymbol{f} : T \to \mathscr{L}(E, F)$ a locally μ-integrable function. Then for every function $\boldsymbol{h} \in \mathscr{K}_E(T)$ the function $\boldsymbol{f}\boldsymbol{h} : T \to F$ is μ-integrable.*

In fact, the function $\boldsymbol{f}$ is μ-measurable, hence the function $\boldsymbol{f}\boldsymbol{h}$ is μ-measurable. On the other hand, the function $|\boldsymbol{f}|$ is locally μ-integrable and $|\boldsymbol{h}| \in \mathscr{K}(T)$; consequently, the function $|\boldsymbol{f}|\,|\boldsymbol{h}|$ is μ-integrable. Then

$$\int^* |\boldsymbol{f}\boldsymbol{h}|\,\mathrm{d}\mu \leqslant \int^* |\boldsymbol{f}|\,|\boldsymbol{h}|\,\mathrm{d}\mu < \infty\,.$$

By the integrability criterion, the function $\boldsymbol{f}\boldsymbol{h}$ is μ-integrable.

15.4 PROPOSITION. *If $\boldsymbol{f}, \boldsymbol{g} : T \to E$ are locally μ-integrable and α is a scalar, then $\boldsymbol{f}+\boldsymbol{g}$ and $\alpha\boldsymbol{f}$ are locally μ-integrable.*

This property follows immediately from definition 15.1.

In the case of scalar functions, this property can be stated more precisely:

15.5 PROPOSITION. *A real function f is locally μ-integrable if and only if the functions f^+ and f^- are locally μ-integrable.*

If f^+ and f^- are locally μ-integrable, then the function $f=f^+-f^-$ is locally μ-integrable. Conversely, if f is locally μ-integrable then the function $|f|$ is locally μ-integrable, hence the function $f^+=\frac{1}{2}(|f|+f)$ and $f^-=\frac{1}{2}(|f|-f)$ are locally μ-integrable.

15.6 PROPOSITION. *A complex function $f=f_1+if_2$ is locally μ-integrable if and only if f_1 and f_2 are locally μ-integrable.*

In fact, for every compact set $K \subset T$ we have $f\varphi_K=f_1\varphi_K+if_2\varphi_K$ and the function $f\varphi_K$ is μ-integrable if and only if the functions $f_1\varphi_K$ and $f_2\varphi_K$ are μ-integrable.

15.7 PROPOSITION. *Let μ and v be two positive measures. If $\mu \leqslant v$ and $\boldsymbol{f}: T \to E$ is a locally v-integrable function, then $\boldsymbol{f}$ is locally μ-integrable.*

This property follows immediately from definition 15.1.

15.8 PROPOSITION. *Let $\boldsymbol{m}$ and $\boldsymbol{n}$ be two dominated measures and α a scalar. If the function $\boldsymbol{f}: T \to E$ is locally integrable with respect to $\boldsymbol{m}$ and with respect to $\boldsymbol{n}$, then $\boldsymbol{f}$ is locally integrable with respect to $\boldsymbol{m}+\boldsymbol{n}$ and with respect to $\alpha\,\boldsymbol{m}$.*

This property follows from the relations $|\boldsymbol{m}+\boldsymbol{n}| \leqslant |\boldsymbol{m}|+|\boldsymbol{n}|$ and $|\alpha \boldsymbol{m}| = |\alpha|\,|\boldsymbol{m}|$, using the preceding proposition.

In the case of scalar measures, this property can be stated more precisely:

15.9 PROPOSITION. *Let μ and v be two positive measures. A function $\boldsymbol{f}: T \to E$ is locally integrable with respect to $\mu+v$ if and only if $\boldsymbol{f}$ is locally integrable with respect to μ and with respect to v.*

In fact, $\mu \leqslant \mu+v$ and $v \leqslant \mu+v$.

15.10 PROPOSITION. *Let μ be a real measure. A function $\boldsymbol{f}: T \to E$ is locally μ-integrable, if and only if $\boldsymbol{f}$ is locally integrable with respect to μ^+ and with respect to μ^-.*

In fact, $|\mu| = \mu^+ + \mu^-$, and the local integrability with respect to μ is equivalent with the local integrability with respect to $|\mu|$.

15.11 PROPOSITION. *Let μ and ν be two real measures. A function $\boldsymbol{f}: T \to E$ is locally integrable with respect to the complex measure $\mu + i\nu$ if and only if $\boldsymbol{f}$ is locally integrable with respect to μ and with respect to ν.*

We use the relations $|\mu + i\nu| \leqslant |\mu| + |\nu|$ and $|\mu| \leqslant |\mu + i\nu|$, $|\nu| \leqslant |\mu + i\nu|$ and we apply proposition 15.7.

2 *Measures defined by locally integrable densities*

Let X, E and F be three Banach spaces. Assume that there exists a bilinear mapping $(u, v) \to uv$ of $X \times E$ into F such that $|uv| \leqslant |u|\,|v|$ for every $u \in X$ and $v \in E$.

We may have, for example, $X \subset \mathscr{L}(E, F)$ or $E \subset \mathscr{L}(X, F)$.

Let $\boldsymbol{m} : \mathscr{K}(T) \to X$ be a *dominated measure* and $\boldsymbol{g} : T \to E$ *a locally $\boldsymbol{m}$-integrable function.*

For every *real* function $\varphi \in \mathscr{K}(T)$, the function $\boldsymbol{g}\varphi : T \to E$ is $\boldsymbol{m}$-integrable. Denote

$$\boldsymbol{n}(\varphi) = \int \varphi \boldsymbol{g}\, \mathrm{d}\boldsymbol{m}, \text{ for } \varphi \in \mathscr{K}(T).$$

The function $|\boldsymbol{g}|$ is also locally $|\boldsymbol{m}|$-integrable, therefore for every real function $\varphi \in \mathscr{K}(T)$, the function $\varphi|\boldsymbol{g}|$ is locally $|\boldsymbol{m}|$-integrable. Denote

$$\nu(\varphi) = \int \varphi |\boldsymbol{g}|\, \mathrm{d}|\boldsymbol{m}|, \text{ for } \varphi \in \mathscr{K}(T).$$

It is immediately seen that ν is a positive *linear functional* on $\mathscr{K}(T)$, therefore it is a *positive measure* on T.

Also, the mapping $\boldsymbol{n} : \mathscr{K}(T) \to F$ is linear and dominated by ν:

$$|\boldsymbol{n}(\varphi)| = |\int \varphi \boldsymbol{g}\, \mathrm{d}\boldsymbol{m}| \leqslant \int |\varphi|\,|\boldsymbol{g}|\, \mathrm{d}|\boldsymbol{m}| = \nu(|\varphi|).$$

Therefore $\boldsymbol{n}$ is a dominated vector measure and we have $|\boldsymbol{n}| \leqslant \nu$.

15.12 DEFINITION. *Let $\boldsymbol{m} : \mathscr{K}(T) \to X$ be a dominated measure, $\boldsymbol{g} : T \to E$ a locally $\boldsymbol{m}$-integrable function and $\boldsymbol{n} : \mathscr{K}(T) \to F$ the measure defined by the equality*

$$\boldsymbol{n}(\varphi) = \int \varphi \boldsymbol{g}\, \mathrm{d}\boldsymbol{m}, \textit{ for } \varphi \in \mathscr{K}(T).$$

We say that $\boldsymbol{n}$ *is the measure with density* $\boldsymbol{g}$ *and with basis* $\boldsymbol{m}$, *or that* $\boldsymbol{n}$ *is the product of the measure* $\boldsymbol{m}$ *by the locally* $\boldsymbol{m}$*-integrable function* $\boldsymbol{g}$ *and we write*

$$\boldsymbol{n} = \boldsymbol{g}\boldsymbol{m} .$$

We have, therefore,

$$\int \varphi \, \mathrm{d}(\boldsymbol{g}\, \boldsymbol{m}) = \int \varphi \boldsymbol{g} \, \mathrm{d}\boldsymbol{m}, \text{ for } \varphi \in \mathscr{K}(T) .$$

By this definition, the measure ν defined by the equality

$$\nu(\varphi) = \int \varphi |\boldsymbol{g}| \, \mathrm{d}|\boldsymbol{m}|, \text{ for } \varphi \in \mathscr{K}(T)$$

is with basis $|\boldsymbol{m}|$ and density $|\boldsymbol{g}|$, that is,

$$\nu = |\boldsymbol{g}|\, |\boldsymbol{m}| .$$

The inequality $|\boldsymbol{n}| \leqslant \nu$ is written now

$$|\boldsymbol{g}\boldsymbol{m}| \leqslant |\boldsymbol{g}|\, |\boldsymbol{m}| .$$

We shall show that in certain cases, we have even the equality $|\boldsymbol{g}\boldsymbol{m}| = |\boldsymbol{g}|\, |\boldsymbol{m}|$. Namely:

(1) If μ is a *scalar* measure and $\boldsymbol{g} : T \to E$ is a locally μ-integrable function, we have (theorem 18.17)

$$|\boldsymbol{g}\mu| = |\boldsymbol{g}|\, |\mu| .$$

(2) If $\boldsymbol{m} : \mathscr{K}(T) \to E$ is a *dominated measure*, and g is a *locally* $\boldsymbol{m}$-integrable *scalar* function, we have (theorem 18.20).

$$|g\boldsymbol{m}| = |g|\, |\boldsymbol{m}| .$$

Remark. The equality $\boldsymbol{n} = \boldsymbol{g}\boldsymbol{m}$ is often written

$$\mathrm{d}\boldsymbol{n}(t) = \boldsymbol{g}(t)\, \mathrm{d}\boldsymbol{m}(t) ;$$

thus the equality $\nu = |\boldsymbol{g}|\, |\boldsymbol{m}|$ is written

$$\mathrm{d}\nu(t) = |\boldsymbol{g}(t)|\, \mathrm{d}|\boldsymbol{m}|(t) .$$

15.13 Proposition. *Let* μ *be a scalar measure,* $\boldsymbol{g} : T \to \mathscr{L}(E, F)$ *a locally* μ*-integrable function and* $\boldsymbol{g}\mu : \mathscr{K}(T) \to \mathscr{L}(E, F)$ *the product measure. We have*

$$\int \boldsymbol{f} \mathrm{d}(\boldsymbol{g}\mu) = \int \boldsymbol{f}\boldsymbol{g}\, \mathrm{d}\mu, \textit{ for every } \boldsymbol{f} \in \mathscr{K}_E(T) .$$

We remark first that if $\boldsymbol{f}\in\mathscr{K}_E(T)$, then the function $\boldsymbol{fg}$ is μ-integrable (proposition 15.3), hence in the above equality both integrals make sense.

Put

$$\boldsymbol{m}(\boldsymbol{f}) = \int \boldsymbol{fg}\,\mathrm{d}\mu, \text{ for } \boldsymbol{f}\in\mathscr{K}_E(T).$$

It is easy to see that $\boldsymbol{m}:\mathscr{K}_E(T)\to F$ is a linear mapping dominated by $|\boldsymbol{g}|\,|\mu|$, consequently it is a dominated measure.

For every $\varphi\in\mathscr{K}(T)$ and $x\in E$ we have (corollary 8.4)

$$\boldsymbol{m}(\varphi)x = \boldsymbol{m}(\varphi x) = \int \varphi x\boldsymbol{g}\,\mathrm{d}\mu = x\int \boldsymbol{g}\,\mathrm{d}\mu = x\int \varphi\,\mathrm{d}(\boldsymbol{g}\mu),$$

therefore

$$\boldsymbol{m}(\varphi) = \int \varphi\,\mathrm{d}(\boldsymbol{g}\mu), \text{ for } \varphi\in\mathscr{K}(T).$$

It follows that $\boldsymbol{m}=\boldsymbol{g}\mu$, consequently

$$\int \boldsymbol{f}\mathrm{d}(\boldsymbol{g}\mu) = \int \boldsymbol{fg}\,\mathrm{d}\mu, \text{ for } \boldsymbol{f}\in\mathscr{K}_E(T)$$

and the proposition is proved.

In particular, the proposition is true if $\boldsymbol{g}$ is a locally μ-integrable *scalar* function.

3 *Simply measurable operator-valued functions*

Let E and F be two Banach spaces and μ a positive measure on T. Denote by $\mathscr{L}^*(E, F)$ the space of all linear mappings of E into F.

15.14 Definition. *We say that a function $\boldsymbol{f}:T\to\mathscr{L}^*(E,F)$ is simply μ-measurable if for every $x\in E$ the function $t\to\boldsymbol{f}(t)x$ (with values in F) is μ-measurable.*

If $\boldsymbol{m}$ is a dominated measure, we say that a function is simply $\boldsymbol{m}$-measurable if it is simply $|\boldsymbol{m}|$-measurable.

If E is the scalar field—that is, if $\mathscr{L}^*(E, F)=F$—then the notion of simple measurability is identical to that of measurability.

Every μ-measurable function $\boldsymbol{f}: T\to\mathscr{L}(E, F)$ is also simply μ-measurable.

We shall say that a function $\boldsymbol{g}$ defined μ-almost everywhere on T, with values in $\mathscr{L}^*(E, F)$, is simply μ-measurable if it is equal almost everywhere to a simply μ-measurable function defined on the whole space T.

If $\boldsymbol{f}$, $\boldsymbol{g}$ are simply μ-measurable and α is a scalar, then the functions $\boldsymbol{f}+\boldsymbol{g}$ and $\alpha\boldsymbol{f}$ are simply μ-measurable.

15.15 PROPOSITION. *If the function $\boldsymbol{f}: T \to \mathscr{L}^*(E, F)$ is simply μ-measurable, then for every μ-measurable function $\boldsymbol{h}: T \to E$, the function $t \to \boldsymbol{f}(t)\, \boldsymbol{h}(t)$ is μ-measurable.*

Let $K \subset T$ be a compact set. Since $\boldsymbol{h}$ is μ-measurable, it can be approximated on K by μ-measurable step functions of the form $\boldsymbol{h}_n = \Sigma\, \varphi_{A_i} x_i$, where $x_i \in E$ and (A_i) is a finite partition of K into μ-measurable sets. Then the function $\boldsymbol{f}\boldsymbol{h}$ can be approximated on K by the μ-measurable functions $\boldsymbol{f}\boldsymbol{h}_n = \Sigma\, \varphi_{A_i} \boldsymbol{f} x_i$. It follows that $\boldsymbol{f}\boldsymbol{h}\varphi_K$ is μ-measurable.

By the localization principle, the function $\boldsymbol{f}\boldsymbol{h}$ is μ-measurable.

If the function $\boldsymbol{f}: T \to \mathscr{L}(E, F)$ is μ-measurable, then the positive function $t \to |\boldsymbol{f}(t)|$ is μ-measurable. Under certain conditions this function is μ-measurable even if $\boldsymbol{f}$ is simply μ-measurable.

15.16 PROPOSITION. *If the function $\boldsymbol{f}: T \to \mathscr{L}^*(E, F)$ is simply μ-measurable, and if for each compact set $K \subset T$ thêre exists a countable set $H \subset E$ such that*

$$|\boldsymbol{f}(t)| = \sup_{x \in H} |\boldsymbol{f}(t)x|/|x|, \quad \mu\text{-almost everywhere on } K,$$

then the function $t \to |\boldsymbol{f}(t)|$ is μ-measurable.

In fact, the functions $t \to \boldsymbol{f}(t)x$ are μ-measurable, hence the functions $t \to |\boldsymbol{f}(t)x|$ and $t \to |\boldsymbol{f}(t)x|/|x|$ are also μ-measurable. Then the function $|\boldsymbol{f}|\varphi_K$ is μ-measurable as supremum of a countable family of μ-measurable functions. By the localization principle, the function $|\boldsymbol{f}|$ is μ-measurable.

15.17 COROLLARY. *If $\boldsymbol{f}: T \to \mathscr{L}^*(E, F)$ is simply μ-measurable and if E is of countable type, then the function $t \to |\boldsymbol{f}(t)|$ is μ-measurable.*

15.18 COROLLARY. *If $\boldsymbol{f}: T \to \mathscr{L}(E, F)$ is simply μ-measurable and if for every compact set $K \subset T$ there exists a μ-negligible set $N \subset K$ and a countable set $H \subset \mathscr{L}(E, F)$ such that $\boldsymbol{f}(K \setminus N) \subset \bar{H}$, then $\boldsymbol{f}$ is μ-measurable.*

For the proof see theorem 10.10.

15.19 Corollary. *If $\boldsymbol{f}: T \to \mathscr{L}(E, F)$ is simply μ-measurable and if there exists a countable set $H \subset \mathscr{L}(E, F)$ such that $\boldsymbol{f}(T) \subset \bar{H}$, then $\boldsymbol{f}$ is μ-measurable.*

4 *Simply locally integrable operator-valued functions*

Let E and F be two Banach spaces and μ a positive measure.

15.20 Definition. *We say that a function $\boldsymbol{f}: T \to \mathscr{L}^*(E, F)$ is simply locally μ-integrable if $\boldsymbol{f}$ is simply μ-measurable and the positive function $|\boldsymbol{f}|$ is locally μ-integrable.*

This means that for every $x \in E$, the function $t \to \boldsymbol{f}(t)x$ (with values in F) is μ-measurable and the function $t \to |\boldsymbol{f}(t)|$ is μ-measurable and that for every compact set $K \subset T$ we have

$$\int_K^* |\boldsymbol{f}| \, \mathrm{d}\mu < \infty .$$

It follows that $|\boldsymbol{f}(t)| < \infty$ almost everywhere, therefore $\boldsymbol{f}(t) \in \mathscr{L}(E, F)$ almost everywhere.

If $\boldsymbol{f}$ is simply locally μ-integrable, then the function $\boldsymbol{f}(t)x$ is locally μ-integrable, for every $x \in E$.

If $\boldsymbol{m}$ is a dominated measure, we shall say that a function is simply locally $\boldsymbol{m}$-integrable if it is simply locally $|\boldsymbol{m}|$-integrable.

Every locally μ-integrable function $\boldsymbol{f}: T \to \mathscr{L}(E, F)$ is simply locally μ-integrable.

In case E is the field of scalars, to say that the function $\boldsymbol{f}: T \to \mathscr{L}^*(E, F)$ is simply locally μ-integrable means that the function $\boldsymbol{f}: T \to F$ is locally μ-integrable.

If $\boldsymbol{f}$ is a function defined μ-almost everywhere on T with values in $\mathscr{L}^*(E, F)$, we say that $\boldsymbol{f}$ is simply locally μ-integrable if it is equal μ-almost everywhere to a simply locally μ-integrable function defined on the whole space T.

If $\boldsymbol{f}$ is simply locally μ-integrable and α is a scalar, then $\alpha \boldsymbol{f}$ is simply locally μ-integrable.

The sum $\boldsymbol{f}+\boldsymbol{g}$ of two simply locally μ-integrable functions is not simply locally μ-integrable in general, since, from the fact that $|\boldsymbol{f}|$ and $|\boldsymbol{g}|$ are μ-measurable, it does not follow that $|\boldsymbol{f}+\boldsymbol{g}|$ is μ-measurable.

If $0 \leqslant \mu \leqslant \nu$ and if $\boldsymbol{f}$ is simply locally ν-integrable, then $\boldsymbol{f}$ is simply locally μ-integrable.

15.21 Proposition. *If the function* $\boldsymbol{f}: T \to \mathscr{L}^*(E, F)$ *is simply locally* μ-*integrable, then the function* $\boldsymbol{fh}$ *is* μ-*integrable for every function* $\boldsymbol{h} \in \mathscr{K}_E(T)$.

Let $\boldsymbol{h} \in \mathscr{K}_E(T)$. Since the function $\boldsymbol{f}$ is simply μ-measurable, the function $t \to \boldsymbol{f}(t)\boldsymbol{h}(t)$ is μ-measurable. Since the function $|\boldsymbol{f}|$ is locally μ-integrable we have

$$\int^* |\boldsymbol{fh}| \,\mathrm{d}\mu \leqslant \int^* |\boldsymbol{f}|\, |\boldsymbol{h}| \,\mathrm{d}\mu < \infty .$$

Thus the function $\boldsymbol{fh}$ is μ-integrable.

Remark. Assume that the function $\boldsymbol{f}: T \to \mathscr{L}^*(E, F)$ is simply μ-measurable and that $\int_K^* |\boldsymbol{f}| \,\mathrm{d}\mu < \infty$ for every compact set $K \subset T$.

In general, the function $\boldsymbol{f}$ is not simply locally μ-integrable, since the function $|\boldsymbol{f}|$ is not necessarily μ-measurable.

Under the conditions of proposition 15.16, it follows that the function $\boldsymbol{f}$ is simply locally μ-integrable.

Under the conditions of proposition 15.18 it follows that $\boldsymbol{f}$ is even locally μ-integrable.

5 *Measures defined by simply locally integrable densities*

Let E, F and G be three Banach spaces, $\boldsymbol{m}: \mathscr{K}(T) \to \mathscr{L}(F, G)$ a *dominated measure* and $\boldsymbol{g}: T \to \mathscr{L}^*(E, F)$ a *simply locally* $\boldsymbol{m}$-*integrable* function.

Then the function $|\boldsymbol{g}|$ is locally $\boldsymbol{m}$-integrable, hence we may consider the measure $|\boldsymbol{g}|\,|\boldsymbol{m}|$.

For every function $\boldsymbol{f} \in \mathscr{K}_E(T)$, the function $\boldsymbol{gf}: T \to F$ is $\boldsymbol{m}$-integrable. In particular, for every function $\varphi \in \mathscr{K}(T)$ and $x \in E$ we have $\varphi x \in \mathscr{K}_E(T)$, hence the function $\boldsymbol{g} x \varphi$ is $\boldsymbol{m}$-integrable. Put

$$\boldsymbol{n}(\varphi)x = \int \boldsymbol{g} x \varphi \,\mathrm{d}\boldsymbol{m}, \text{ for } \varphi \in \mathscr{K}(T) \text{ and } x \in E .$$

It is immediately seen that for each $\varphi \in \mathscr{K}(T)$ the mapping $\boldsymbol{n}(\varphi)$: $x \to \boldsymbol{n}(\varphi)x$ of E into G is linear and continuous:

$$|\boldsymbol{n}(\varphi)x| \leqslant \int |\boldsymbol{g}|\, |x| \,\mathrm{d}|\boldsymbol{m}| = |x| \int |\varphi| \,\mathrm{d}(|\boldsymbol{g}|\, |\boldsymbol{m}|),$$

consequently

$$|\boldsymbol{n}(\varphi)| \leqslant \int |\varphi| \,\mathrm{d}(|\boldsymbol{g}|\, |\boldsymbol{m}|) .$$

It is also easy to see that the mapping $\boldsymbol{n}:\mathscr{K}(T)\to\mathscr{L}(E, G)$ is linear and dominated by the measure $|\boldsymbol{g}|\,|\boldsymbol{m}|$, hence $\boldsymbol{n}$ is a *dominated measure* and $|\boldsymbol{n}|\leqslant |\boldsymbol{g}|\,|\boldsymbol{m}|$.

We see that if, for example, $\boldsymbol{m}$ is a *scalar* measure and $\boldsymbol{g}$ is *locally-$\boldsymbol{m}$-integrable*, or if $\boldsymbol{m}$ is a dominated measure and $\boldsymbol{g}$ is a *scalar* locally $\boldsymbol{m}$-integrable function, the measure $\boldsymbol{n}$ defined before is equal to the product measure $\boldsymbol{g}\,\boldsymbol{m}$.

In fact, in each of these cases, the function $\boldsymbol{g}\varphi$ is $\boldsymbol{m}$-integrable, hence

$$\boldsymbol{n}(\varphi)x = \int \boldsymbol{g}x\,\varphi\,\mathrm{d}\boldsymbol{m} = x\int \boldsymbol{g}\varphi\,\mathrm{d}\boldsymbol{m} = x\int \varphi\,\mathrm{d}(\boldsymbol{g}\,\boldsymbol{m}),$$

whence $\boldsymbol{n}=\boldsymbol{g}\,\boldsymbol{m}$.

By extension, we are led to the following definition:

15.22 DEFINITION. *Let $\boldsymbol{m}:\mathscr{K}(T)\to\mathscr{L}(F, G)$ be a dominated measure, $\boldsymbol{g}:T\to\mathscr{L}^*(E,F)$ a simply locally $\boldsymbol{m}$-integrable function and $\boldsymbol{n}:\mathscr{K}(T)\to\mathscr{L}(E, G)$ the measure defined by the equality*

$$\boldsymbol{n}(\varphi)x = \int \boldsymbol{g}x\varphi\,\mathrm{d}\boldsymbol{m},\ \textit{for}\ \varphi\in\mathscr{K}(T)\ \textit{and}\ x\in E.$$

We say that $\boldsymbol{n}$ is the measure with density $\boldsymbol{g}$ and with basis $\boldsymbol{m}$, or that $\boldsymbol{n}$ is the product of the measure $\boldsymbol{m}$ by the simply locally $\boldsymbol{m}$-integrable function $\boldsymbol{g}$ and we write

$$\boldsymbol{n} = \boldsymbol{g}\,\boldsymbol{m}.$$

We have, therefore,

$$\int \varphi\,\mathrm{d}(\boldsymbol{g}\,\boldsymbol{m})x = \int \boldsymbol{g}x\varphi\,\mathrm{d}\boldsymbol{m},\ \text{for}\ \varphi\in\mathscr{K}(T)\ \text{and}\ x\in E.$$

The inequality $|\boldsymbol{n}|\leqslant|\boldsymbol{g}|\,|\boldsymbol{m}|$ is now written

$$|\boldsymbol{g}\,\boldsymbol{m}|\leqslant|\boldsymbol{g}|\,|\boldsymbol{m}|.$$

In certain cases we even have the equality $|\boldsymbol{g}\,\boldsymbol{m}|=|\boldsymbol{g}|\,|\boldsymbol{m}|$.

In case $\boldsymbol{g}$ is *locally $\boldsymbol{m}$-integrable* and $E=R$, we have $\mathscr{L}(E, F)=F$ and the product $\boldsymbol{g}\,\boldsymbol{m}$ in definition 15.22 coincides with the product $\boldsymbol{g}\,\boldsymbol{m}$ is definition 15.12.

Remark. For every $x\in E$, the function $\boldsymbol{g}x:T\to F$ is locally $\boldsymbol{m}$-integrable; therefore, we can consider the product $(\boldsymbol{g}x)\boldsymbol{m}$. We have then

$$\int \varphi \, \mathrm{d}[(\boldsymbol{g}x)\boldsymbol{m}] = \int \varphi \boldsymbol{g} x \, \mathrm{d}\boldsymbol{m} = \int \varphi \, \mathrm{d}(\boldsymbol{g}\boldsymbol{m}) \cdot x$$

for $\varphi \in \mathscr{K}(T)$ and $x \in E$, thus $(\boldsymbol{g}x)\boldsymbol{m} = x \cdot (\boldsymbol{g}\boldsymbol{m})$.

15.23 PROPOSITION. *Let* $\boldsymbol{m} : \mathscr{K}(T) \to \mathscr{L}(F, G)$ *be a dominated measure,* $\boldsymbol{g} : T \to \mathscr{L}^*(E, F)$ *a simply locally* $\boldsymbol{m}$*-integrable function and the measure* $\boldsymbol{g}\boldsymbol{m} : \mathscr{K}(T) \to \mathscr{L}(E, G)$. *We have then*

$$\int \boldsymbol{f} \mathrm{d}(\boldsymbol{g}\boldsymbol{m}) = \int \boldsymbol{f}\boldsymbol{g} \, \mathrm{d}\boldsymbol{m}, \text{ for } \boldsymbol{f} \in \mathscr{K}_E(T).$$

We remark first that for $\boldsymbol{f} \in \mathscr{K}_E(T)$, the function $\boldsymbol{f}\boldsymbol{g} : T \to F$ is $\boldsymbol{m}$-integrable, hence in the preceding equality both integrals make sense.

For every $\varphi \in \mathscr{K}(T)$ and every $x \in E$ we have $\varphi x \in \mathscr{K}_E(T)$. Denoting $\boldsymbol{n} = \boldsymbol{g}\boldsymbol{m}$, we have

$$\boldsymbol{n}(\varphi)x = \boldsymbol{n}(\varphi x) = \int \varphi x \boldsymbol{g} \, \mathrm{d}\boldsymbol{m}.$$

Then, for every function of the form

$$\boldsymbol{f} = \Sigma_{1 \leqslant i \leqslant n} \varphi_i x_i \text{ with } \varphi_i \in \mathscr{K}(T) \text{ and } x_i \in E,$$

we have

$$\boldsymbol{n}(\boldsymbol{f}) = \int \boldsymbol{f}\boldsymbol{g} \, \mathrm{d}\boldsymbol{m}.$$

Now let $\boldsymbol{f} \in \mathscr{K}_E(T)$ be an arbitrary function. There exists then a sequence $(\boldsymbol{f}_k)$ of linear combinations of the preceding form, with supports contained in the same compact set, and uniformly converging to $\boldsymbol{f}$. Then

$$\lim_k \boldsymbol{n}(\boldsymbol{f}_k) = \boldsymbol{n}(\boldsymbol{f}).$$

On the other hand, the functions $\boldsymbol{g}\boldsymbol{f}_k$ and $\boldsymbol{g}\boldsymbol{f}$ are $\boldsymbol{m}$-integrable, have supports contained in the same compact set and the sequence $(\boldsymbol{g}\boldsymbol{f}_k)$ tends uniformly to $\boldsymbol{g}\boldsymbol{f}$, therefore,

$$\lim_k \int \boldsymbol{f}_k \boldsymbol{g} \, \mathrm{d}\boldsymbol{m} = \int \boldsymbol{f}\boldsymbol{g} \, \mathrm{d}\boldsymbol{m}.$$

As for each k we have

$$\boldsymbol{n}(\boldsymbol{f}_k) = \int \boldsymbol{f}_k \boldsymbol{g} \, \mathrm{d}\boldsymbol{m},$$

passing to the limit we obtain

$$\boldsymbol{n}(\boldsymbol{f}) = \int \boldsymbol{f}\boldsymbol{g} \, \mathrm{d}\boldsymbol{m},$$

and the proposition is proved.

6 Weakly measurable operator-valued functions

Let E and F be two Banach spaces, Z a vector subspace of F' and μ a *positive* measure on T. We say that Z is a *norming* set (for F) if

$$|y| = \sup_{z\in Z} |\langle y, z\rangle|/|z| \text{ for every } y\in F .$$

15.24 DEFINITION. *We say that $\boldsymbol{\mu} : T\to \mathscr{L}^*(E, F)$ is a Z-weakly μ-measurable function if for every $x\in E$ and every $z\in Z$ the scalar function $t\to\langle \boldsymbol{u}(t)\, x, z\rangle$ is μ-measurable.*

The F'-weakly measurable functions will be called, simply, weakly measurable functions.

To say that the function $\boldsymbol{u}$ is Z-weakly μ-measurable means that for every $z\in Z$, the composed function $z\circ \boldsymbol{u} : T\to\mathscr{L}^*(E, C)$ is simply μ-measurable.

Evidently, a weakly measurable function is Z-weakly measurable for every subspace $Z\subset F'$.

If $\boldsymbol{m}$ is a dominated measure, we say that a function is weakly $\boldsymbol{m}$-measurable if it is weakly $|\boldsymbol{m}|$-measurable.

If E is the scalar field, to say that the operator function $\boldsymbol{u} : T\to\mathscr{L}(E, F)$ is Z-weakly μ-measurable means that the vector function $\boldsymbol{u} : T\to F$ is Z-weakly μ-measurable, that is, $t\to\langle \boldsymbol{u}(t), z\rangle$ is μ-measurable for every $z\in Z$.

If F is the scalar field, to say that the function $\boldsymbol{u} : T\to\mathscr{L}^*(E, F) = E^*$ is E-weakly μ-measurable means that the function $\boldsymbol{u}$ is simply μ-measurable, that is, $t\to\langle x, \boldsymbol{u}(t)\rangle$ is μ-measurable for every $x\in E$.

If $\boldsymbol{u}$ and $\boldsymbol{v}$ are Z-weakly μ-measurable and α is a scalar, then $\boldsymbol{u}+\boldsymbol{v}$ and $\alpha\boldsymbol{u}$ are Z-weakly μ-measurable.

To say that $\boldsymbol{u} : T\to\mathscr{L}^*(E, F)$ is Z-weakly μ-measurable means that for every $x\in E$, the function $t\to \boldsymbol{u}(t)x$ with values in F is Z-weakly μ-measurable.

15.25 PROPOSITION. *If the function $\boldsymbol{u} : T\to\mathscr{L}^*(E, F)$ is Z-weakly μ-measurable, then for every μ-measurable functions $\boldsymbol{f} : T\to E$ and $\boldsymbol{h} : T\to Z$, the scalar function $t\to\langle \boldsymbol{u}(t)\boldsymbol{f}(t), \boldsymbol{h}(t)\rangle$ is μ-measurable.*

Let $z\in Z$. For every Borel step function $\boldsymbol{f} = \Sigma\, \varphi_{A_i} x_i$ with compact support we have

$$\langle \boldsymbol{u}\boldsymbol{f}, z\rangle = \Sigma\, \varphi_{A_i}\langle \boldsymbol{u}x_i, z\rangle \ ;$$

consequently, the function $\langle \boldsymbol{u}\boldsymbol{f}, z\rangle$ is μ-measurable.

Assume that $\boldsymbol{f}: T \to E$ is an arbitrary μ-measurable function. Let $K \subset T$ be a compact set. There exists a sequence of Borel step functions $\boldsymbol{f}_n : T \to E$, with supports contained in K, such that $\boldsymbol{f}_n(t) \to \boldsymbol{f}(t)$, μ-almost everywhere on K. Then

$$\langle \boldsymbol{u}(t)\boldsymbol{f}_n(t), z\rangle \to \langle \boldsymbol{u}(t)\boldsymbol{f}(t), z\rangle$$

μ-almost everywhere on K. Since the functions $\langle \boldsymbol{u}\boldsymbol{f}_n, z\rangle$ are μ-measurable, it follows that the function $\langle \boldsymbol{u}\boldsymbol{f}, z\rangle \varphi_K$ is μ-measurable.

By the localization principle, the function $\langle \boldsymbol{u}\boldsymbol{f}, z\rangle$ is μ-measurable.

We show then that the function $\langle \boldsymbol{u}\boldsymbol{f}, \boldsymbol{h}\rangle$ is μ-measurable, first for step functions of the form $\boldsymbol{h} = \Sigma \varphi_{A_i} z_i$ with A_i Borel sets and $z_i \in Z$ and then, for arbitrary μ-measurable functions $\boldsymbol{h} : T \to Z$.

Remark. If the function $\boldsymbol{u} : T \to \mathscr{L}^*(E, F)$ is weakly μ-measurable and the function $\boldsymbol{f}: T \to E$ is μ-measurable, it does not follow that the function $t \to \boldsymbol{u}(t)\boldsymbol{f}(t)$ and $t \to |\boldsymbol{u}(t)\boldsymbol{f}(t)|$ are μ-measurable.

Under certain conditions, the last function is μ-measurable.

15.26 PROPOSITION. *Let $\boldsymbol{f}: T \to F$ be a Z-weakly μ-measurable function. If for each compact set $K \subset T$ there exists a countable set $S \subset Z$ such that*

$$|\boldsymbol{f}(t)| = \sup_{s \in S} |\langle \boldsymbol{f}(t), s\rangle|/|s|, \text{ almost for every } t \in K,$$

then the function $|\boldsymbol{f}|$ is μ-measurable.

In fact, for each $s \in S$, the function $t \to \langle \boldsymbol{f}(t), s\rangle$ is μ-measurable, hence the function $t \to |\langle \boldsymbol{f}(t), s\rangle|/|s|$ is μ-measurable. From the hypothesis it follows that for every compact set $K \subset T$ the function $|\boldsymbol{f}|\varphi_K$ is equal almost everywhere to the supremum of a countable family of μ-measurable functions, hence $|\boldsymbol{f}|\varphi_K$ is μ-measurable, consequently $|\boldsymbol{f}|$ is μ-measurable.

15.27 COROLLARY. *If there exists a countable set $S \subset Z$, norming for F, and if $\boldsymbol{f}: T \to F$ is Z-weakly μ-measurable, then the function $|\boldsymbol{f}|$ is μ-measurable.*

15.28 PROPOSITION. *Assume that there exists a countable set $S \subset Z$ norming for F. If the function $\boldsymbol{u} : T \to \mathscr{L}^*(E, F)$ is Z-weakly μ-measurable, then for every μ-measurable function $\boldsymbol{f}: T \to E$ the function $t \to |\boldsymbol{u}(t)\boldsymbol{f}(t)|$ is μ-measurable.*

In fact, for each μ-measurable function $\boldsymbol{f}: T \to E$ and every $z \in Z$, the function

$t \to \langle \boldsymbol{u}(t)\boldsymbol{f}(t), z\rangle$ is μ-measurable; hence, the function $t \to \boldsymbol{u}(t)\boldsymbol{f}(t)$ is Z-weakly μ-measurable and we apply proposition 15.26.

In particular, the conditions of proposition 15.28 are fulfilled in the following cases:

(1) Z is a Banach space of countable type and $F = Z'$;

(2) F' is of countable type and $Z = F'$;

(3) F is of countable type and $Z = F'$.

In the last case we have even more (corollary 10.13).

15.29 PROPOSITION. *If F is of countable type, then every weakly μ-measurable function $\boldsymbol{u}: T \to \mathscr{L}^*(E, F)$ is simply μ-measurable.*

Under certain conditions, even the function $|\boldsymbol{u}|$ is μ-measurable.

15.30 PROPOSITION. *Let $\boldsymbol{u}: T \to \mathscr{L}^*(E, F)$ be a Z-weakly μ-measurable function. Assume that Z is norming for F. If there exists a lifting ρ of $\mathscr{L}^\infty(\mu)$ such that $\rho[\boldsymbol{u}] = \boldsymbol{u}$, then:*

(1) the function $t \to |\boldsymbol{u}(t)|$ is μ-measurable;

(2) the function $t \to |\boldsymbol{u}(t)x|$ is μ-measurable, for every $x \in E$;

(3) if, in addition, $\boldsymbol{u}$ has values in $\mathscr{L}(E, F)$, the function $t \to |\boldsymbol{u}(t)\boldsymbol{f}(t)|$ is μ-measurable, for every μ-measurable function $\boldsymbol{f}: T \to E$.

The assertions (1) and (2) follow from proposition 13.18 and corollary 13.19. The assertion (3) follows from the following lemma.

15.31 LEMMA. *Let $\boldsymbol{u} \to \mathscr{L}(E, F)$ be a function such that the function $t \to |\boldsymbol{u}(t)x|$ is μ-measurable for every $x \in E$. Then for every μ-measurable function $\boldsymbol{f}: T \to E$, the function $t \to |\boldsymbol{u}(t)\boldsymbol{f}(t)|$ is μ-measurable.*

Let first $\boldsymbol{f} = \Sigma\, \varphi_{A_i} x_i$ be a Borel step function with $x_i \in E$. We may assume that the sets A_i are disjoint. In this case we have

$$|\boldsymbol{u}(t)\boldsymbol{f}(t)| = \Sigma\, \varphi_{A_i}(t) |\boldsymbol{u}(t)x_i|,$$

therefore $|\boldsymbol{u}\boldsymbol{f}|$ is μ-measurable.

Let now: $\boldsymbol{f}: T \to E$ be an arbitrary μ-measurable function and $K \subset T$ a compact set. There exists a μ-negligible set $N \subset K$ and a sequence $(\boldsymbol{f}_n)$ of Borel step functions converging to $\boldsymbol{f}$ on $K \setminus N$. Then

$$\boldsymbol{u}(t)\boldsymbol{f}_n(t) \to \boldsymbol{u}(t)\boldsymbol{f}(t), \text{ for } t \in K \setminus N$$

consequently

$$|\boldsymbol{u}(t)\boldsymbol{f}_n(t)| \to |\boldsymbol{u}(t)\boldsymbol{f}(t)|, \text{ for } t \in K \setminus N .$$

It follows that the function $|\boldsymbol{uf}|$ is μ-measurable.

15.32 PROPOSITION. *Assume that E is of countable type and that there exists a countable set $S \subset Z$, norming for F.*

Then for every Z-weakly μ-measurable function $\boldsymbol{u} : T \to \mathscr{L}^*(E, F)$, the function $|\boldsymbol{u}|$ is μ-measurable.

By proposition 15.16, it follows that for every $x \in E$ the function $t \to |\boldsymbol{u}(t)x|$ is μ-measurable. If (x_n) is a sequence dense in E, we have

$$|\boldsymbol{u}(t)| = \sup_n |\boldsymbol{u}(t)x_n| / |x_n| ,$$

therefore $|\boldsymbol{u}|$ is μ-measurable.

In particular, the conditions of proposition 15.32 are fulfilled in the following cases:

(1) E is of countable type, Z is a Banach space of countable type and $F = Z'$;
(2) E is of countable type, F' is of countable type and $Z = F'$;
(3) E is of countable type, F is of countable type and $Z = F'$.

Remark. In the conditions of proposition 15.32, the function $t \to |\boldsymbol{u}(t)\boldsymbol{f}(t)|$ is also μ-measurable for every μ-measurable function $\boldsymbol{f}: T \to E$.

In case (3), the function $\boldsymbol{u}$ is even simply μ-measurable.

7 *Weakly locally integrable operator-valued functions*

Let E and F be two Banach spaces, $Z \subset F'$ a vector space and μ a positive measure.

15.33 DEFINITION. *We say that a function $\boldsymbol{u} : T \to \mathscr{L}^*(E, F)$ is Z-weakly locally μ-integrable if $\boldsymbol{u}$ is Z-weakly μ-measurable and if the function $t \to |\boldsymbol{u}(t)|$ is locally μ-integrable.*

If $\boldsymbol{m}$ is a dominated measure, we say that the function $\boldsymbol{u}$ is Z-weakly locally $\boldsymbol{m}$-integrable if $\boldsymbol{u}$ is Z-weakly locally $|\boldsymbol{m}|$-integrable.

From definition 15.33, it follows that if $\boldsymbol{u}$ is weakly locally μ-integrable we have $|\boldsymbol{u}(t)| < \infty$ therefore $\boldsymbol{u}(t) \in \mathscr{L}(E, F)$, μ-almost everywhere.

Every simply locally μ-integrable function is weakly locally μ-integrable.

If $0 \leqslant \mu \leqslant \nu$ and $\boldsymbol{u}$ is Z-weakly locally ν-integrable, then $\boldsymbol{u}$ is Z-weakly locally μ-integrable.

The sum of two Z-weakly locally μ-integrable functions may not be Z-weakly locally μ-integrable.

15.34 PROPOSITION. *If the function $\boldsymbol{u} : T \to \mathscr{L}^*(E, F)$ is Z-weakly locally μ-integrable, then for every function $\boldsymbol{f} \in \mathscr{K}_E(T)$ and every $z \in Z$, the numerical function $t \to \langle \boldsymbol{u}(t)\boldsymbol{f}(t), z \rangle$ is μ-integrable.*

In fact, the function $t \to \langle \boldsymbol{u}(t)\boldsymbol{f}(t), z \rangle$ is μ-measurable; since $|\boldsymbol{u}|$ is locally μ-integrable and $|\boldsymbol{f}| \in \mathscr{K}(T)$ we have

$$\int^* |\langle \boldsymbol{u}(t)\boldsymbol{f}(t), z \rangle| \, d\mu \leqslant |z'| \int^* |\boldsymbol{u}| \, |\boldsymbol{f}| \, d\mu < \infty .$$

15.35 COROLLARY. *If $\boldsymbol{u} : T \to \mathscr{L}^*(E, F)$ is Z-weakly locally μ-integrable, then for every $x \in E$ and $z \in Z$ the function $t \to \langle \boldsymbol{u}(t)x, z \rangle$ is locally μ-integrable.*

In fact, for every function $\varphi \in \mathscr{K}(T)$ and $x \in E$ we have $\varphi x \in \mathscr{K}_E(T)$.

Remark. Assume that the function $\boldsymbol{u} : T \to \mathscr{L}^*(E, F)$ is weakly μ-measurable and that $\int_K^* |\boldsymbol{u}| \, d\mu < \infty$ for every compact set $K \subset T$.

In general, the function $\boldsymbol{u}$ is not weakly locally μ-integrable since the function $|\boldsymbol{u}|$ is not necessarily μ-measurable.

Under the conditions of propositions 15.30 and 15.32, it follows that $\boldsymbol{u}$ is weakly locally μ-integrable. In case E and F are of countable type and $Z = F'$, if follows that $\boldsymbol{u}$ is even simply locally μ-integrable.

8 *Measures defined by weakly locally integrable densities*

Let E and F two Banach spaces, Z a vector subspace of F' and μ a scalar measure. We shall assume that Z is norming for F.

15.36 PROPOSITION. *For every Z-weakly locally μ-integrable function $\boldsymbol{u} : T \to \mathscr{L}^*(E, F)$ there exists a measure $\boldsymbol{m} : \mathscr{K}(T) \to \mathscr{L}(E, Z')$ such that*

$$\langle \boldsymbol{m}(\varphi)x, z \rangle = \int \langle \boldsymbol{u}(t)\varphi(t)x, z \rangle \, d\mu(t)$$

for every $\varphi \in \mathscr{K}(T)$, $x \in E$ *and* $z \in Z$. *The measure* $\boldsymbol{m}$ *is dominated and*

$$|\boldsymbol{m}| \leqslant |\boldsymbol{u}|\,|\mu|\,.$$

The measure $\boldsymbol{m}$ *has values in* $\mathscr{L}(E, F)$ *in each of the following cases*:

(1) $F = Z'$;

(2) $\boldsymbol{u}$ *is simply* μ*-measurable, in particular if* F *is of countable type*;

(3) *for every* $x \in E$ *there exists a locally countable family* $(K_j)_{j \in J}$ *of disjoint compact subsets with* $T \setminus \cup K_j$ μ*-negligible such that the convex closed (in the topology* $\sigma(F, Z)$*) balanced hull of the set* $\{\boldsymbol{u}(t)x;\ t \in K_j\}$ *is compact for the topology* $\sigma(F, Z)$.

For $\varphi \in \mathscr{K}(T)$, $x \in E$ and $z \in Z$ we have $\varphi x \in \mathscr{K}_E(T)$; therefore, (proposition 15.34), the function $t \to \langle \boldsymbol{u}(t)\varphi(t)x, z\rangle$ is μ-integrable. Put

$$M(\varphi, x, z) = \int \langle \boldsymbol{u}\,\varphi x, z\rangle \mathrm{d}\mu\,.$$

The mapping $M(z, x) : z \to M(\varphi, x, z)$ is a linear functional on Z; it is also continuous:

$$|M(\varphi, x, z)| \leqslant \int |\boldsymbol{u}|\,|\varphi|\,|x|\,|z|\,\mathrm{d}|\mu|\ ;$$

consequently, $M(\varphi, x) \in Z'$ and

$$|M(\varphi, x)| \leqslant \int |\boldsymbol{u}|\,|\varphi|\,|x|\,\mathrm{d}|\mu|\,.$$

The mapping $\boldsymbol{m}(\varphi) : x \to M(\varphi, x)$ of E into Z' is linear and continuous, hence $\boldsymbol{m}(\varphi) \in \mathscr{L}(E, Z')$ and

$$|\boldsymbol{m}(\varphi)| \leqslant \int |\boldsymbol{u}|\,|\varphi|\,\mathrm{d}|\mu|\,.$$

The mapping $\boldsymbol{m} : \varphi \to \boldsymbol{m}(\varphi)$ of $\mathscr{K}(T)$ into $\mathscr{L}(E, Z')$ is linear and dominated by the positive measure $|\boldsymbol{u}|\,|\mu|$, hence $\boldsymbol{m}$ is a dominated measure and $|\boldsymbol{m}| \leqslant |\boldsymbol{u}|\,|\mu|$. It follows then that

$$\langle \boldsymbol{m}(\varphi)x, z\rangle = \int \langle \boldsymbol{u}\,\varphi x, z\rangle \mathrm{d}\mu, \text{ for } \varphi \in \mathscr{K}(T),\ x \in E \text{ and } z \in Z\,.$$

It remains to consider the cases when $\boldsymbol{m}$ takes on values in $\mathscr{L}(E, F)$. The case $F = Z'$ is trivial. Assume now that $\boldsymbol{u}$ is simply μ-measurable and consider the measure $\boldsymbol{m}_1 : \mathscr{K}(T) \to \mathscr{L}(E, F)$ defined by the equality (definition 15.22):

$$\boldsymbol{m}_1(\varphi)x = \int \boldsymbol{u}\,x\,\varphi\,\mathrm{d}\mu, \text{ for } \varphi \in \mathscr{K}(T) \text{ and } x \in E.$$

Then for every $\varphi \in \mathscr{K}(T)$, $x \in E$ and $z \in Z$ we have

$$\langle \boldsymbol{m}_1(\varphi)x, z\rangle = \langle \int \boldsymbol{u}x\varphi\, \mathrm{d}\mu, z\rangle = \int \langle \boldsymbol{u}x\varphi, z\rangle \mathrm{d}\mu = \langle \boldsymbol{m}(\varphi)x, z\rangle ,$$

whence $\boldsymbol{m} = \boldsymbol{m}_1$, consequently $\boldsymbol{m}$ has values in $\mathscr{L}(E, F)$.

Finally, assume that condition (3) is fulfilled. In this case, we shall be able to prove that $\boldsymbol{m}$ has values in $\mathscr{L}(E, F)$ after proving proposition 16.7, but the proof of the present proposition is given here. Let $\varphi \in \mathscr{K}(T)$, $x \in E$, $z \in Z$ and a compact set $K \subset T$. Let us show that we have

$$\langle \int_K \varphi\, \mathrm{d}\boldsymbol{m}x, z\rangle = \int_K \langle \boldsymbol{u}\varphi x, z\rangle \mathrm{d}\mu .$$

There exists a sequence (φ_n) of functions of $\mathscr{K}(T)$ such that

$$\lim_n \int |\varphi_n - \varphi\varphi_K| \mathrm{d}(|\boldsymbol{u}|\, |\mu|) = 0 .$$

The functions φ_n and $\varphi\varphi_K$ are integrable with respect to every measure. Since $|\boldsymbol{m}| \leqslant |\boldsymbol{u}|\, |\mu|$, we deduce that

$$\lim_n \int |\varphi_n - \varphi\varphi_K| \mathrm{d}|\boldsymbol{m}| = 0,$$

hence

$$\lim_n \int \varphi_n \mathrm{d}\boldsymbol{m} = \int \varphi\varphi_K \mathrm{d}\boldsymbol{m} ,$$

consequently

$$\lim_n \langle \int \varphi_n \mathrm{d}\boldsymbol{m}x, z\rangle = \langle \int \varphi\varphi_K \mathrm{d}\boldsymbol{m}x, z\rangle .$$

From proposition 16.7, we deduce that

$$\lim_n \int^* |\boldsymbol{u}|\, |\varphi_n - \varphi\varphi_K|\, \mathrm{d}|\mu| = 0 .$$

The functions $\langle \boldsymbol{u}\varphi_n x, z\rangle$ and $\langle \boldsymbol{u}\varphi\varphi_K x, z\rangle$ are μ-integrable and we have

$$|\int \langle \boldsymbol{u}\varphi_n x, z\rangle \mathrm{d}\mu - \int \langle \boldsymbol{u}\varphi\varphi_K x, z\rangle \mathrm{d}\mu| \leqslant |x|\, |z| \int^* |\boldsymbol{u}|\, |\varphi_n - \varphi\varphi_K| \mathrm{d}|\mu| ;$$

therefore,

$$\lim_n \int \langle \boldsymbol{u}\varphi_n x, z\rangle \mathrm{d}\mu = \int \langle \boldsymbol{u}\varphi\varphi_K x, z\rangle \mathrm{d}\mu .$$

As for each n we have

$$\langle \int \varphi_n \mathrm{d}\boldsymbol{m}x, z\rangle = \int \langle \boldsymbol{u}\varphi_n x, z\rangle \mathrm{d}\mu ,$$

passing to the limit we obtain

$$\langle \int \varphi\varphi_K \mathrm{d}\boldsymbol{m}x, z\rangle = \int \langle \boldsymbol{u}\varphi\varphi_K x, z\rangle \mathrm{d}\mu .$$

Let now $x \in E$ and let $\mathscr{K} = (K_j)_{j \in J}$ be a family of disjoint compact subsets such that $T \setminus \cup K_j$ is μ-negligible and such that for each $j \in J$, the convex closed balanced hull A_j of the set $\{\boldsymbol{u}(t)x; t \in K_j\}$ is compact for the topology $\sigma(F, Z)$. Let $\varphi \in \mathscr{K}(T)$ with $\|\varphi\| \leqslant 1$. Then $\boldsymbol{u}(t)\varphi(t)x \in A_j$ for every $t \in K_j$ and every $j \in J$.

Let (K_n) be the sequence of sets of the family $\mathscr{K}$ which intersects the support of φ.

For each K_n the set A_n is compact and convex in the algebraic dual Z^* of Z for the topology $\sigma(Z^*, Z)$. There exists therefore a family $(z_i)_{i \in I}$ of elements of Z such that

$$A_n = \cap_{i \in I} \{y; y \in Z^*, |\langle y, z_i \rangle| \leqslant 1\} .$$

Then

$$|\langle \boldsymbol{u}(t)\varphi(t)x, z_i \rangle| \leqslant 1, \text{ for } i \in I \text{ and } t \in K_n .$$

It follows that for every n we have

$$|\langle \textstyle\int_{K_n} \varphi \, \mathrm{d}\boldsymbol{m}\, x, z_i \rangle| \leqslant \textstyle\int_{K_n} |\langle \boldsymbol{u}(t)\varphi(t)x, z_i \rangle| \, \mathrm{d}|\mu| \leqslant |\mu|(K_n) ,$$

consequently

$$\textstyle\int_{K_n} \varphi \, \mathrm{d}\boldsymbol{m}\, x \in |\mu|(K_n) A_n \subset F .$$

It follows then that

$$\textstyle\int \varphi \, \mathrm{d}\boldsymbol{m}\, x = \Sigma \int_{K_n} \varphi \, \mathrm{d}\boldsymbol{m}\, x \in F$$

and as $x \in E$ is arbitrary, we deduce that $\boldsymbol{m}(\varphi) \in \mathscr{L}(E, F)$. This relation is deduced then for every $\varphi \in \mathscr{K}(T)$. Thus the proposition is proved.

Remarks

1. The measure $\boldsymbol{m}$ depends on the subspace Z. If Z_1 and Z_2 are two subspaces of F' with $Z_1 \subset Z_2$ and if $\boldsymbol{m}_1 : \mathscr{K}(T) \to \mathscr{L}(E, Z_1')$ and $\boldsymbol{m}_2 : \mathscr{K}(T) \to \mathscr{L}(E, Z_2')$ are the corresponding measures, then for each $\varphi \in \mathscr{K}(T)$ and $x \in E$, the functional $\boldsymbol{m}_1(\varphi)x \in Z_1'$ is the restriction to the subspace Z_1 of the functional $\boldsymbol{m}_2(\varphi)x \in Z_2'$.

2. The existence of the measure $\boldsymbol{m} : \mathscr{K}(T) \to \mathscr{L}(E, Z')$ was deduced by making no hypothesis on Z. The fact that Z is norming has been used in order to embed F isometrically in Z' and to deduce that $\boldsymbol{m}$ has values in $\mathscr{L}(E, F)$, in cases (2) and (3). If we assume only that Z separates the points of F, then we may embed F algebraically in Z' and we can also deduce that $\boldsymbol{m}$ has values in $\mathscr{L}(E, F)$. But in this case, the modulus of $\boldsymbol{m}$ differs according to whether we consider $\boldsymbol{m}$ with values in $\mathscr{L}(E, F)$ or in $\mathscr{L}(E, Z')$ since the norms on these space are different.

In the sequel we shall assume always that Z is norming for F.

If $\boldsymbol{u}$ is locally μ-integrable, then the measure $\boldsymbol{m}$ is the product of the measure μ by the function $\boldsymbol{u}$ in the sense of definition 15.12, that is $\boldsymbol{m} = \boldsymbol{u}\mu$.

In fact, in this case, for every $\varphi \in \mathscr{K}(T)$ the function $\boldsymbol{u}\varphi$ is μ-integrable and for every $x \in E$ and $z \in F$ we have

$$\langle \boldsymbol{m}(\varphi)x, z\rangle = \int \langle \boldsymbol{u}\varphi x, z\rangle \mathrm{d}\mu = \langle \int \boldsymbol{u}\varphi x \mathrm{d}\mu, z\rangle = \langle x \int \boldsymbol{u}\varphi \mathrm{d}\mu, z\rangle;$$

therefore,

$$\boldsymbol{m}(\varphi) = \int \boldsymbol{u}\varphi \,\mathrm{d}\mu$$

that is, $\boldsymbol{m} = \boldsymbol{u}\mu$.

Also, if $\boldsymbol{u}$ is simply locally μ-integrable, then the measure $\boldsymbol{m}$ is the product of the measure μ by the function $\boldsymbol{u}$ in the sense of definition 15.22, that is

$$\boldsymbol{m}(\varphi)x = \int \boldsymbol{u}x\varphi \,\mathrm{d}\mu, \text{ for } \varphi \in \mathscr{K}(T) \text{ and } x \in E,$$

consequently, in this case too we have $\boldsymbol{m} = \boldsymbol{u}\mu$.

By extension we are led to the following definition:

15.37 Definition. *Let μ be a scalar measure, $\boldsymbol{u} : T \to \mathscr{L}^*(E, F)$ a Z-weakly locally μ-integrable function and $\boldsymbol{m} : \mathscr{K}(T) \to \mathscr{L}(E, Z')$ the dominated measure defined by the equality*

$$\langle \boldsymbol{m}(\varphi)x, z\rangle = \int \langle \boldsymbol{u}\varphi x, z\rangle \mathrm{d}\mu, \textit{ for } \varphi \in \mathscr{K}(T), x \in E, \textit{ and } z \in Z.$$

We say that $\boldsymbol{m}$ is the measure with density $\boldsymbol{u}$ and basis μ, or that $\boldsymbol{m}$ is the product of the measure μ by the Z-weakly locally μ-integrable function $\boldsymbol{u}$, and we write

$$\boldsymbol{m} = \boldsymbol{u}\mu .$$

We have therefore

$$\langle \int \varphi \mathrm{d}(\boldsymbol{u}\mu) x, z\rangle = \int \langle \boldsymbol{u}\varphi x, z\rangle \mathrm{d}\mu$$

for $\varphi \in \mathscr{K}(T)$, $x \in E$ and $z \in Z$.

The inequality $|\boldsymbol{m}| \leqslant |\boldsymbol{u}|\,|\mu|$ is written

$$|\boldsymbol{u}\mu| \leqslant |\boldsymbol{u}|\,|\mu| .$$

In certain cases we have even the equality $|\boldsymbol{u}\mu| = |\boldsymbol{u}|\,|\mu|$.

We shall show later that every dominated measure is the product of a positive measure μ by a weakly locally μ-integrable function.

If the function $\boldsymbol{u} : T \to \mathscr{L}^*(E, F)$ is *simply locally μ-integrable* then we have

$$x \int \varphi \mathrm{d}(\boldsymbol{u}\mu) = \int \boldsymbol{u}\varphi x \mathrm{d}\mu ,$$

for $\varphi \in \mathscr{K}(T)$ and $x \in E$; therefore, $\boldsymbol{u}\mu$ coincides with the product of $\boldsymbol{u}$ by μ in the sense of definition 15.22.

If the function $\boldsymbol{u} : T \to \mathscr{L}(E, F)$ is *locally μ-integrable*, then we have

$$\int \varphi \mathrm{d}(\boldsymbol{u}\mu) = \int \boldsymbol{u}\varphi \mathrm{d}\mu, \text{ for } \varphi \in \mathscr{K}(T) ,$$

therefore $\boldsymbol{u}\mu$ is the product of $\boldsymbol{u}$ by μ also in the sense of definition 15.12.

Propositions 15.13 and 15.23 are extended also for weakly locally μ-integrable functions.

15.38 PROPOSITION. *If $\boldsymbol{u} : T \to \mathscr{L}^*(E, F)$ is a Z-weakly locally μ-integrable function, then we have*

$$\langle \int \boldsymbol{f} \mathrm{d}(\boldsymbol{u}\mu), z \rangle = \int \langle \boldsymbol{u}\boldsymbol{f}, z \rangle \mathrm{d}\mu, \textit{for } \boldsymbol{f} \in \mathscr{K}_E(T) \textit{ and } z \in Z .$$

Denote $\boldsymbol{m} = \boldsymbol{u}\mu$. For $\varphi \in \mathscr{K}(T)$ and $x \in E$ we have $\varphi x \in \mathscr{K}_E(T)$ and $\boldsymbol{m}(\varphi x) = \boldsymbol{m}(\varphi)x$; therefore, for every $z \in Z$ we have

$$\langle \boldsymbol{m}(\varphi x), z \rangle = \int \langle \boldsymbol{u}\varphi x, z \rangle \mathrm{d}\mu .$$

It follows that for every function $\boldsymbol{f} = \Sigma \varphi_i x_i$ with $\varphi_i \in \mathscr{K}(T)$ and $x_i \in E$ we have

$$\langle \boldsymbol{m}(\boldsymbol{f}), z \rangle = \int \langle \boldsymbol{u}\boldsymbol{f}, z \rangle \mathrm{d}\mu .$$

Now let $\boldsymbol{f} \in \mathscr{K}(T)$ be arbitrary. There exists then a sequence $(\boldsymbol{f}_n)$ of linear combinations of the preceding form with supports contained in the same compact set K and uniformly convergent to $\boldsymbol{f}$. Then

$$\lim_n \boldsymbol{m}(\boldsymbol{f}_n) = \boldsymbol{m}(\boldsymbol{f}) .$$

On the other hand, the functions $\langle \boldsymbol{u}\, \boldsymbol{f}_n, z \rangle$ have the supports contained in K and tend uniformly to $\langle \boldsymbol{u}\boldsymbol{f}, z \rangle$, for each $z \in Z$, therefore

$$\lim_n \int \langle \boldsymbol{u}\boldsymbol{f}_n, z \rangle \mathrm{d}\mu = \int \langle \boldsymbol{u}\, \boldsymbol{f}, z \rangle \mathrm{d}\mu .$$

As for each n we have

$$\langle \boldsymbol{m}(\boldsymbol{f}_n), z \rangle = \int \langle \boldsymbol{u}\boldsymbol{f}_n, z \rangle \mathrm{d}\mu ,$$

passing to the limit we obtain the desired equality.

Thus the proposition is proved.

Remark. Let $\boldsymbol{u}: T \to \mathscr{L}^*(E, F)$ be a *Z-weakly locally μ-integrable* function. We can integrate vector functions $\boldsymbol{f}: T \to E$ with respect to the measure $\boldsymbol{u}\mu$: $\mathscr{K}(T) \to \mathscr{L}(E, Z')$. Nevertheless, in general, we do not have the equality

$$\int \boldsymbol{f} \mathrm{d}(\boldsymbol{u}\mu) = \int \boldsymbol{f}\boldsymbol{u} \,\mathrm{d}\mu, \text{ for } \boldsymbol{f} \in \mathscr{K}_E(T),$$

since $\boldsymbol{uf}$ is not μ-measurable. As we have seen, this equality is true if $\boldsymbol{u}$ is simply locally μ-integrable.

Also, we can integrate scalar functions φ with respect to $\boldsymbol{u}\mu$, but, in general, we do not have the equality

$$\int \varphi \,\mathrm{d}(\boldsymbol{u}\mu) = \int \varphi \boldsymbol{u} \,\mathrm{d}\mu, \text{ for } \varphi \in \mathscr{K}(T).$$

This equality is true if $\boldsymbol{u}$ is locally μ-integrable.

§16 Integration with respect to a measure defined by densities

1 *Directed families of measurable functions*

We prove first the following proposition which will be used in the sequel. Let μ be a positive measure on T.

16.1 PROPOSITION. *Let $(f_i)_{i \in I}$ be a family of positive (finite or infinite) functions, directed for the relation $\leqslant$. If the mapping $t \to (f_i(t))_{i \in I}$ of T into $\bar{R}^I$ is μ-measurable, then the function $\sup_{i \in I} f_i$ is μ-measurable and*

$$\int^* \sup_{i \in I} f_i \mathrm{d}\mu = \sup_{i \in I} \int^* f_i \mathrm{d}\mu .$$

(a) Let $K \subset T$ be a compact set and $\varepsilon > 0$. There exists a compact set $K_1 \subset K$ such that the restriction to K_1 of the mapping $t \to (f_i(t))_{i \in I}$ is continuous and $\mu(K \setminus K_1) < \varepsilon/2$. Then for every $i \in I$, the restriction of the function f_i to K_1 is continuous, therefore the restriction to K_1 of the function $f = \sup_{i \in I} f_i$ is lower semicontinuous. Denoting by g a function equal to f on K_1, and constant on $T \setminus K_1$, g is μ-measurable. There exists therefore a compact set $K_2 \subset K_1$ such that $\mu(K_1 \setminus K_2) < \varepsilon/2$ and the restriction of g to K_2 is continuous. This means that the restriction of f to K_2 is continuous and $\mu(K \setminus K_2) < \varepsilon$. Thus the function $f = \sup_{i \in I} f_i$ is μ-measurable.

(b) If the set $A = \{t; f(t) = +\infty\}$ is not μ-negligible, then

$$\int^* f \mathrm{d}\mu = \sup_{i \in I} \int^* f_i \mathrm{d}\mu = +\infty .$$

In fact, A is μ-measurable since f is μ-measurable; hence $A \cap K$ is μ-integrable for every compact set $K \subset T$. There exists then a compact set $K \subset A$ with $\mu(K) > 0$, such that the restriction of the function $t \to (f_i(t))_{i \in I}$ to K is continuous. We have then $\lim_{i \in I} f_i(t) = \sup_{i \in I} f_i(t) = f(t) = +\infty$ for $t \in K$. The restrictions of the functions $g_i(t) = [1 + f_i(t)]^{-1}$ to K are also continuous ($g_i(t) = 0$ if $f_i(t) = +\infty$), the functions g_i form a directed family for the relation $\geqslant$ and $\inf_{i \in I} g_i(t) = \lim_{i \in I} g_i(t) = 0$. By the Dini theorem, the functions g_i tend uniformly to 0 on K. For every number $a > 0$ there exists an index $i \in I$ such that for every function $g_j \leqslant g_i$ we have $g_j(t) \leqslant (1+a)^{-1}$ for $t \in K$. This means that $f_j(t) \geqslant a$ for $t \in K$, that is $f_j \varphi_K \geqslant a \varphi_K$; therefore,

$$\int^* f_j \varphi_K \mathrm{d}\mu \geqslant a\mu(K).$$

It follows that

$$\sup_{i \in I} \int^* f_i \varphi_K \mathrm{d}\mu = +\infty = \int^* f \varphi_K \mathrm{d}\mu,$$

whence

$$\sup_{i \in I} \int^* f_i \mathrm{d}\mu = +\infty = \int^* f \mathrm{d}\mu.$$

(c) Assume now that A is μ-negligible, therefore $f(t) < +\infty$, μ-almost everywhere.

Let $K \subset T$ be a compact set. There exists a μ-negligible set $N \subset K$ and an increasing sequence (K_n) of compact sets whose union is $K \setminus N$ and such that the restrictions of the functions $t \to (f_i(t))_{i \in I}$ and $f(t) = \sup_{i \in I} f_i(t)$ to each K_n are continuous, and $f(t) < +\infty$ for $t \in K \setminus N$.

It follows that the restriction of each function f_i to each set K_n is continuous. By the Dini theorem, f_i converges uniformly to f on each K_n.

For each K_n and each $\varepsilon > 0$, there exists an index i such that for every function $f_j \geqslant f_i$ of the family we have $|f(t) - f_j(t)| < \varepsilon$ for $t \in K_n$, that is, $|f\varphi_{K_n} - f_j \varphi_{K_n}| \leqslant \varepsilon \varphi_{K_n}$. The functions $f\varphi_{K_n}$ and $f_j \varphi_{K_n}$ are μ-integrable, therefore

$$\left|\int f\varphi_{K_n} \mathrm{d}\mu - \int f_j \varphi_{K_n} \mathrm{d}\mu\right| \leqslant \int |f\varphi_{K_n} - f_j \varphi_{K_n}| \mathrm{d}\mu \leqslant \varepsilon \mu(K_n).$$

This means that

$$\int f \varphi_{K_n} \mathrm{d}\mu = \lim_{i \in I} \int f_i \varphi_{K_n} \mathrm{d}\mu = \sup_{i \in I} \int f_i \varphi_{K_n} \mathrm{d}\mu.$$

The sequence $(f\varphi_{K_n})$ is increasing and $\sup_n f\varphi_{K_n} = f\varphi_{K \setminus N}$; also, for each $i \in I$, the sequence $(f_i \varphi_{K_n})$ is increasing and

$$\sup_n f_i \varphi_{K_n} = f_i \varphi_{K \setminus N}.$$

From the monotone convergence theorem for positive functions, we deduce that

$$\int^* f\varphi_K \mathrm{d}\mu = \int^* f\varphi_{K\setminus N} \mathrm{d}\mu = \sup_n \int f\varphi_{K_n} \mathrm{d}\mu ,$$

$$\int^* f_i\varphi_K \mathrm{d}\mu = \int^* f_i\varphi_{K\setminus N} \mathrm{d}\mu = \sup_n \int^* f_i\varphi_{K_n} \mathrm{d}\mu .$$

Then

$$\int^* f\varphi_K \mathrm{d}\mu = \sup_n \int^* f\varphi_{K_n} \mathrm{d}\mu = \sup_n \sup_i \int^* f_i\varphi_{K_n} \mathrm{d}\mu =$$

$$= \sup_i \sup_n \int^* f_i\varphi_{K_n} \mathrm{d}\mu = \sup_i \int^* f_i\varphi_K \mathrm{d}\mu .$$

We deduce then that

$$\int^* f\mathrm{d}\mu = \sup_K \int^* f\varphi_K \mathrm{d}\mu = \sup_K \sup_i \int^* f_i\varphi_K \mathrm{d}\mu =$$

$$= \sup_i \sup_K \int^* f_i\varphi_K \mathrm{d}\mu = \sup_i \int^* f_i \mathrm{d}\mu ,$$

where K runs over the set of compact subsets of T.

Thus the proposition is proved.

2 *The upper integral with respect to a positive measure defined by densities*

Let μ be a positive measure on T, $g \geqslant 0$ a locally μ-integrable function and $\nu = g\mu$.

For every $f \in \mathcal{K}(T)$, $\nu(f)$ is defined by the equality

$$\int f\mathrm{d}\nu = \int fg\,\mathrm{d}\mu .$$

We shall show that this equality holds also for the upper integrals of an arbitrary function $f \geqslant 0$.

16.2 Lemma. *If $f \geqslant 0$ is lower semicontinuous, then*

$$\int^* f\mathrm{d}\nu = \int^* fg\,\mathrm{d}\mu .$$

Let $(h_\alpha)_{\alpha\in A}$ be the directed family of functions of $\mathcal{K}_+(T)$ such that $h_\alpha \geqslant f$. For each function h_α we have

$$\int h_\alpha \mathrm{d}\nu = \int h_\alpha g\,\mathrm{d}\mu .$$

We have, also, $\sup_\alpha h_\alpha = f$ and $\sup_\alpha h_\alpha g = fg$.

The mapping $t \to (h_\alpha(t) g(t))_{\alpha\in A}$ of T into $\bar{R}^A$ is μ-measurable. In fact, g

being μ-measurable, for each compact set $K \subset T$ and every $\varepsilon > 0$, there exists a compact set $K' \subset K$ such that the restriction of g to K' is continuous and $\mu(K \setminus K') < \varepsilon$; then the restrictions of the functions $h_\alpha g$ to K' are continuous, therefore also the restriction of the mapping $t \to (h_\alpha(t) g(t))_{\alpha \in A}$ to K' is continuous. This means that the function $t \to (h_\alpha(t) g(t))_{\alpha \in A}$ is μ-measurable.

From proposition 16.1 it follows then that

$$\int^* fg\,\mathrm{d}\mu = \sup_\alpha \int h_\alpha g\,\mathrm{d}\mu = \sup_\alpha \int h_\alpha \mathrm{d}\nu = \int^* f\mathrm{d}\nu\,.$$

16.3 Lemma. *For every function $f \geqslant 0$ we have*

$$\int^* f\mathrm{d}\nu \geqslant \int^* fg\mathrm{d}\mu\,.$$

Let $K \subset T$ be a compact set and let $h \in \mathscr{I}_+$ with $h \geqslant f\varphi_K$. From lemma 16.2 it follows that

$$\int^* h\mathrm{d}\nu = \int^* hg\,\mathrm{d}\mu \geqslant \int^* fg\,\varphi_K \mathrm{d}\mu\,.$$

Then

$$\int^* f\varphi_K \mathrm{d}\nu = \inf_h \int^* h\mathrm{d}\nu \geqslant \int^* fg\,\varphi_K \mathrm{d}\mu, \text{ for } h \in \mathscr{I}_+,\, h \geqslant f\varphi_K$$

whence

$$\int^* f\mathrm{d}\mu \geqslant \int^* fg\,\mathrm{d}\mu\,.$$

16.4 Lemma. *If $f \geqslant 0$ is ν-integrable, then fg is μ-integrable and we have*

$$\int f\mathrm{d}\nu = \int fg\,\mathrm{d}\mu\,.$$

Since $f \geqslant 0$ is ν-integrable, there exists a sequence (f_n) of positive functions of $\mathscr{K}(T)$ such that

$$\lim \int |f_n - f|\,\mathrm{d}\nu = 0\,.$$

Since, by lemma 16.3, we have

$$\int |f_n - f|\,\mathrm{d}\nu = \int^* |f_n - f|\,\mathrm{d}\nu \geqslant \int^* |f_n - f| g\,\mathrm{d}\mu = \int^* |f_n g - fg|\,\mathrm{d}\mu\,,$$

we deduce that

$$\lim \int^* |f_n g - fg|\,\mathrm{d}\mu = 0\,.$$

But $f_n g$ are μ-integrable, therefore fg is μ-integrable. From the equalities

$$\lim \int f_n \mathrm{d}\nu = \int f \,\mathrm{d}\nu, \quad \lim \int f_n g \,\mathrm{d}\mu = \int fg \,\mathrm{d}\mu$$

and

$$\int f_n \mathrm{d}\nu = \int f_n g \,\mathrm{d}\mu ,$$

we deduce

$$\int f \mathrm{d}\nu = \int fg \,\mathrm{d}\mu .$$

16.5 LEMMA. *If $f \geqslant 0$ is ν-measurable and has compact support, then fg is μ-measurable and*

$$\int^* f \,\mathrm{d}\nu = \int^* fg \,\mathrm{d}\mu .$$

If for each n we put $f_n = \inf(n, f)$, the function f_n is ν-integrable, the sequence (f_n) is increasing and $f = \sup f_n$. By lemma 16.4 it follows that $f_n g$ are μ-integrable and

$$\int f_n \mathrm{d}\nu = \int f_n g \,\mathrm{d}\mu .$$

But

$$\int^* f \mathrm{d}\nu = \sup \int^* f_n \mathrm{d}\nu \ \text{ and } \ \int^* fg \,\mathrm{d}\mu = \sup \int^* f_n g \,\mathrm{d}\mu ,$$

whence

$$\int^* f \mathrm{d}\nu = \int^* fg \,\mathrm{d}\mu .$$

The functions $f_n g$ are μ-measurable since they are μ-integrable and $fg = \sup f_n g$. It follows that fg is μ-measurable.

16.6 PROPOSITION. *For every function $f \geqslant 0$ we have*

$$\int^* f \mathrm{d}(g\mu) = \int^* fg \,\mathrm{d}\mu .$$

Denote $\nu = g\mu$. We prove first the proposition in case f has *compact support* K.

Since the inequality $\int^* f \mathrm{d}\nu \geqslant \int^* fg \,\mathrm{d}\mu$ has been proved in lemma 16.3, it remains to prove the converse inequality

$$\int^* f \mathrm{d}\nu \leqslant \int^* fg \,\mathrm{d}\mu .$$

In order to do this it is sufficient to show that if $h \in \mathscr{I}_+$ and $fg \leqslant h$, then $\int^* f \mathrm{d}\nu \leqslant \int^* h \,\mathrm{d}\mu$.

Let $h \in \mathscr{I}_+$ with $fg \leqslant h$. We shall show that we can find a ν-measurable function f' with compact support such that $f \leqslant f'$ and $f'g \leqslant h$. It will follow then from lemma 16.5 that

$$\int^* f \mathrm{d}\nu \leqslant \int^* f' \mathrm{d}\nu = \int^* f' g \mathrm{d}\mu \leqslant \int^* h \mathrm{d}\mu$$

and the proposition will be proved.

Let $S = \{t; g(t) \neq 0\}$. Since g is μ-measurable, the set S is μ-measurable, therefore $K \cap S$ is μ-integrable.

There exists then a μ-negligible set $N \subset K \cap S$ and a sequence (K_n) of mutually disjoint compact sets such that $\cup K_n = (K \cap S) \setminus N$ and such that the restriction of g to each K_n is continuous. Let $\varphi \in \mathscr{I}_+$ with $\varphi \geqslant f$. Define the function f' on T by the equality

$$f'(t) = h(t)/g(t) \quad \text{if } t \in (K \cap S) \setminus N,$$

$$f'(t) = \varphi(t) \quad \text{if } t \in (K \setminus S) \cup N,$$

$$f'(t) = 0 \quad \text{if } t \in T \setminus K.$$

Then $f' \geqslant f$ since:

for $t \in (K \cap S) \setminus N = \cup K_n$ we have $f'(t) g(t) = h(t) \geqslant f(t) g(t)$, and $g(t) > 0$;

for $t \in (K \setminus S) \cup N$ we have $f'(t) = \varphi(t) \geqslant f(t)$;

for $t \in T \setminus K$ we have $f'(t) = f(t) = 0$.

The function f' is ν-measurable since:

the restriction of $f' = h/g$ to each K_n is lower semicontinuous;

the restriction of $f' = \varphi$ to $(K \setminus S) \cup N$ is lower semicontinuous;

the restriction of f' to $T \setminus K$ is identically zero.

From the definition of f' it follows that $h \geqslant f'g$, μ-almost everywhere, since:

for $t \in (K \cap S) \setminus N$ we have $h(t) = f'(t) g(t)$;

for $t \in K \setminus S$ we have $g(t) = 0$;

for $t \in T \setminus K$ we have $f'(t) = 0$.

Then, as we have remarked,

$$\int^* f \mathrm{d}\nu \leqslant \int^* f' \mathrm{d}\nu = \int^* f' g \mathrm{d}\mu \leqslant \int^* h \mathrm{d}\mu$$

and since the function $h \in \mathscr{I}_+$ with $h \geqslant fg$ is arbitrary, we deduce

$$\int^* f \mathrm{d}\nu \leqslant \int^* fg \mathrm{d}\mu ;$$

consequently,

$$\int^* f \mathrm{d}\nu = \int^* fg \mathrm{d}\mu .$$

Now let $f \geqslant 0$ be an arbitrary function. For every compact set $K \subset T$, the function $f\varphi_K$ has compact support; from the first part of the proof it follows that

$$\int^* f\varphi_K \mathrm{d}\nu = \int^* fg\,\varphi_K \mathrm{d}\mu\ .$$

Taking the supremum for all the compact sets $K \subset T$, we obtain

$$\int^* f\mathrm{d}\nu = \int^* fg\,\mathrm{d}\mu$$

and the proposition is proved.

3 *Integration with respect to a positive measure defined by densities*

Let μ be a positive measure on T, $g \geqslant 0$ a locally μ-integrable function and $\nu = g\mu$.

16.7 PROPOSITION. *A function f defined on T with values in a Banach space E or in $\bar{R}$ is $g\mu$-negligible if and only if fg is μ-negligible.*

In fact,

$$\int^* |f|\,\mathrm{d}\nu = \int^* |f|\,\mathrm{d}(g\mu) = \int^* |f|\,g\,\mathrm{d}\mu = \int^* |fg|\,\mathrm{d}\mu\ ;$$

therefore, $\nu^*(|f|) = 0$ if and only if $\mu^*(|f|g) = 0$.

Remark. Denote $S_g = \{t;\, g(t) \neq 0\}$. To say that fg is μ-negligible means that f is μ-negligible on the set S_g (that is, $f(t) = 0$ for almost every $t \in S_g$).

16.8 COROLLARY. *A set $A \subset T$ is $g\mu$-negligible if and only if the set $A \cap S_g$ is μ-negligible.*

16.9 PROPOSITION. *A function f defined on T with values in a Banach space E or in $\bar{R}$ is $g\mu$-measurable if and only if fg is μ-measurable.*

Assume that f is $g\mu$-measurable. Let $K \subset T$ be a compact set. There exists a $g\mu$-negligible set $M_0 \subset K$ and a sequence (K_n) of mutually disjoint compact sets whose union is $K \setminus M_0$ such that the restriction of f to each K_n is continuous. Since g is μ-measurable, each K_n is the union of a μ-negligible set N_n and of a sequence of mutually disjoint compact sets $(K_{nm})_{m \in N}$ such that the restriction of g to each K_{nm} is continuous. Denote $S = S_g = \{t;\, g(t) \neq 0\}$.

Since M_0 is $g\mu$-negligible, the set $S \cap M_0$ is μ-negligible.

Since S is μ-measurable and $S \cap M_0$ is μ-negligible, the set $M_0 \setminus S$ is μ-measurable and relatively compact, therefore it is μ-integrable. There exists then a μ-negligible set $N_0 \subset M_0 \setminus S$ and a sequence $(K_{0m})_{m \in N}$ of mutually disjoint compact sets whose union is $(M_0 \setminus S) \setminus N_0$.

The restriction of g to each K_{0m} vanishes, therefore it is continuous. Then the set $N = (\cup_{n \geq 0} N_n)$ is μ-negligible and $K \setminus N$ is the union of the compact sets K_{nm}, $0 \leqslant n < \infty$, $1 \leqslant m < \infty$, such that the restriction of the function fg to each K_{nm} is continuous (on the sets K_0 we have $fg = 0$). It follows that fg is μ-measurable.

Conversely assume that fg is μ-measurable. Let $K \subset T$ be a compact set. The set $A = K \setminus S$ is disjoint from S; therefore, by corollary 16.8, A is $g\mu$-negligible. The functions g and fg are μ-measurable and the set $K \cap S$ is μ-integrable, hence it can be written as a union of a μ-negligible set M and a sequence (K_n) of compact sets such that the restrictions of the functions g and fg to each K_n are continuous. Since M is μ-negligible and $M \subset S$, it follows that M is $g\mu$-negligible.

On each K_n we have $g_n(t) > 0$, therefore the restriction of the function $f = fg/g$ to each K_n is continuous. The set $N = A \cup M$ is $g\mu$-negligible and the union of the sets (K_n) is equal to $K \setminus N$. It follows that f is $g\mu$-measurable.

16.10 Corollary. *A set $A \subset T$ is $g\mu$-measurable if and only if $A \cap S_g$ is μ-measurable.*

Assume first that A is $g\mu$-measurable. Then φ_A is $g\mu$-measurable, hence $\varphi_A g$ is μ-measurable.

The function h defined by: $h(t) = 1/|g(t)|$ if $g(t) \neq 0$ and $h(t) = 0$ if $g(t) = 0$, is μ-measurable, hence $\varphi_A gh$ is μ-measurable. But $gh = \varphi_{S_g}$, consequently $\varphi_A \varphi_{S_g}$ is μ-measurable, that is, $A \cap S_g$ is μ-measurable.

Conversely, assume that $A \cap S_g$ is μ-measurable. Then $\varphi_A \varphi_{S_g}$ is μ-measurable, hence $\varphi_A g = \varphi_A \varphi_{S_g} g$ is μ-measurable. It follows that φ_A is $g\mu$-measurable, that is, A is $g\mu$-measurable.

16.11 Theorem. *A function f defined on T with values in a Banach space E or in $\bar{R}$ is $g\mu$-integrable if and only if the function fg is μ-integrable. In this case we have*

$$\int f \mathrm{d}(g\mu) = \int fg \, \mathrm{d}\mu .$$

Put $\nu = g\mu$. From the equality

$$\int^* |f| \mathrm{d}\nu = \int^* |f| g \, \mathrm{d}\mu ,$$

it follows that $\int^* |f| \mathrm{d}\nu < \infty$ if and only if $\int^* |f| g \, \mathrm{d}\mu < \infty$.

Also, f is ν-measurable if and only if fg is μ-measurable. By the integrability criterion, it follows that f is ν-integrable if and only if fg is μ-integrable.

Now let f be a ν-integrable function with values in E (or in $\bar{R}$) and (f_n) a sequence of functions of $\mathscr{K}_E(T)$ (or of $\mathscr{K}(T)$) such that

$$\lim \int |f - f_n| \, \mathrm{d}\nu = 0 .$$

Then

$$\lim \int f_n \, \mathrm{d}\nu = \int f \mathrm{d}\nu .$$

But, for the positive functions $|f - f_n|$, we have (proposition 16.6)

$$\int |f - f_n| \mathrm{d}\nu = \int |f - f_n| g \, \mathrm{d}\mu ,$$

therefore

$$\lim \int |fg - f_n g| \, \mathrm{d}\mu = 0 ,$$

whence

$$\lim \int f_n g \, \mathrm{d}\mu = \int fg \, \mathrm{d}\mu .$$

Since for the functions f_n we have (proposition 15.13)

$$\int f_n \mathrm{d}\nu = \int f_n g \, \mathrm{d}\mu ,$$

passing to the limit we deduce

$$\int f \mathrm{d}\nu = \int f g \, \mathrm{d}\mu .$$

16.12 COROLLARY. *A set $A \subset T$ is $g\mu$-integrable if and only if $\varphi_A g$ is μ-integrable. In this case we have*

$$(g\mu)(A) = \int_A g \, \mathrm{d}\mu .$$

16.13 COROLLARY. *The measure $g\mu$ is bounded if and only if g is μ-integrable. In this case we have*

$$\|g\mu\| = \int g \, \mathrm{d}\mu .$$

In fact, T is $g\mu$-integrable if and only if $g=\varphi_T g$ is μ-integrable, and we have

$$\|g\mu\| = (g\mu)(T) = \int g \,\mathrm{d}\mu .$$

16.14 PROPOSITION. *Let μ be a positive measure, $g_1 \geqslant 0$ a locally μ-integrable function and E a Banach space. A function $\boldsymbol{g}_2 : T \to E$ is locally $g_1\mu$-integrable if and only if the function $\boldsymbol{g}_2 g_1$ is locally μ-integrable. In this case we have*

$$\boldsymbol{g}_2(g_1\mu) = (\boldsymbol{g}_2 g_1)\mu .$$

Assume first that $\boldsymbol{g}_2$ is locally $g_1\mu$-integrable. For every function $\varphi \in \mathscr{K}(T)$, the function $\boldsymbol{g}_2\varphi$ is $g_1\mu$-integrable; therefore, $\boldsymbol{g}_2\varphi g_1$ is μ-integrable and we have

$$\int \boldsymbol{g}_2 \varphi \,\mathrm{d}(g_1\mu) = \int \boldsymbol{g}_2 \varphi g_1 \,\mathrm{d}\mu .$$

It follows that $\boldsymbol{g}_2 g_1$ is locally μ-integrable and

$$\boldsymbol{g}_2(g_1\mu) = (\boldsymbol{g}_2 g_1)\mu .$$

Conversely, if $\boldsymbol{g}_2 g_1$ is locally μ-integrable, then for every $\varphi \in \mathscr{K}(T)$ the function $\boldsymbol{g}_2 g_1 \varphi$ is μ-integrable, hence $\boldsymbol{g}_2\varphi$ is $g_1\mu$-integrable, consequently $\boldsymbol{g}_2$ is locally $g_1\mu$-integrable.

Remark. Let A be a set of T. The function φ_A is locally μ-integrable if and only if A is μ-measurable.

Taking $g=\varphi_A$ in the preceding propositions we obtain:

16.15 PROPOSITION. *Let A be a μ-measurable set.*

(1) *For every function $f \geqslant 0$ defined on T we have*

$$\int^* f \,\mathrm{d}(\varphi_A\mu) = \int^* f\varphi_A \,\mathrm{d}\mu = \int_A^* f \,\mathrm{d}\mu .$$

(2) *A function $\boldsymbol{f}$ defined on T with values in E or in $\overline{R}$ is $\varphi_A\mu$-negligible if and only if $\boldsymbol{f}$ is μ-negligible on A;*

(3) *A function $\boldsymbol{f}$ defined on T with values in E or in $\overline{R}$ is $\varphi_A\mu$-measurable if and only if the restriction of $\boldsymbol{f}$ to A is μ-measurable.*

(4) *A function $\boldsymbol{f}$ defined on T with values in E or in $\overline{R}$ is $\varphi_A\mu$-integrable if and only if $\boldsymbol{f}\varphi_A$ is μ-integrable, and we have*

$$\int \boldsymbol{f} \,\mathrm{d}(\varphi_A\mu) = \int_A \boldsymbol{f} \,\mathrm{d}\mu .$$

If two functions $\boldsymbol{f}$ and $\boldsymbol{g}$ defined on T with values in E or in $\overline{R}$ are equal on A, then one of them is $\varphi_A \mu$-negligible, $\varphi_A \mu$-measurable or $\varphi_A \mu$-integrable if and only if the another function has the same property. If they are $\varphi_A \mu$-integrable, then

$$\int f \mathrm{d}(\varphi_A \mu) = \int \boldsymbol{g} \mathrm{d}(\varphi_A \mu),$$

that is

$$\int_A \boldsymbol{f} \mathrm{d}\mu = \int_A \boldsymbol{g} \, \mathrm{d}\mu .$$

Now let $B \supset A$ and let $\boldsymbol{f}$ be a function defined on B with values in E or in $\overline{R}$. We say that $\boldsymbol{f}$ is μ-negligible on A (respectively μ-measurable on A, μ-integrable on A), if an arbitrary extension $\boldsymbol{f}'$ to T of the restriction of $\boldsymbol{f}$ to A is μ-negligible (respectively μ-measurable, μ-integrable). If $\boldsymbol{f}$ is integrable on A, we denote

$$\int_A \boldsymbol{f} \, \mathrm{d}\mu = \int_A \boldsymbol{f}' \mathrm{d}\mu = \int \boldsymbol{f}' \varphi_A \mathrm{d}\mu$$

and we say that $\int_A \boldsymbol{f} \mathrm{d}\mu$ is the integral of $\boldsymbol{f}$ on the set A. We define also $\int_A^* f \mathrm{d}\mu = \int_A^* f' \mathrm{d}\mu$ if $f \geqslant 0$.

We say that the function $\boldsymbol{f}$ is locally μ-integrable on A if $\boldsymbol{f}'$ is locally $\varphi_A \mu$-integrable, that is, if for every compact set $K \subset T$ the function $\boldsymbol{f} \varphi_{K \cap A}$ is μ-integrable.

4 *Integration with respect to measures defined by weakly locally integrable densities*

Let E and F be two Banach spaces and $Z \subset F'$ a norming subspace for F. Let μ be a *scalar* measure on T.

16.16 THEOREM. *Let $\boldsymbol{u} : T \to \mathscr{L}^*(E, F)$ be a Z-weakly locally μ-integrable function and $\boldsymbol{f}: T \to E$ a function.*

(1) *If $\boldsymbol{f}$ is $|\boldsymbol{u}| \, |\mu|$-negligible, then $\boldsymbol{f}$ is $\boldsymbol{u}\mu$-integrable and $\boldsymbol{uf}$ is μ-negligible.*

(2) *If $\boldsymbol{f}$ is $|\boldsymbol{u}| \, |\mu|$-measurable, then $\boldsymbol{f}$ is $\boldsymbol{u}\mu$-measurable and $\langle \boldsymbol{uf}, z \rangle$ is μ-measurable for every $z \in Z$.*

(3) *If $\boldsymbol{f}$ is $|\boldsymbol{u}| \, |\mu|$-integrable, then $\boldsymbol{f}$ is $\boldsymbol{u}\mu$-integrable and $\langle \boldsymbol{uf}, z \rangle$ is μ-integrable for every $z \in Z$ and*

$$\langle \int \boldsymbol{f} \mathrm{d}(\boldsymbol{u}\mu), z \rangle = \int \langle \boldsymbol{uf}, z \rangle \mathrm{d}\mu .$$

The first part of each of the three propositions concerning the measure

$u\mu$ follows from the inequality $|u\mu| \leqslant |u|\,|\mu|$.

From the inequality

$$\int^* |uf|\,d|\mu| \leqslant \int^* |u|\,|f|\,d|\mu| = \int^* |f|\,d(|u|\,|\mu|),$$

it follows that if f is $|u|\,|\mu|$-negligible, then uf is μ-negligible. If $\int^* |f|\,d(|u|\,|\mu|) < \infty$, then $\int^* |uf|\,d|\mu| < \infty$.

Assume that f is $|u|\,|\mu|$-measurable. Then the function $f|u|$ is μ-measurable. Since the function $|u|$ is μ-measurable, the function $h: T \to E$ defined by

$$h(t) = f(t)u(t)/|u(t)| = f(t) \text{ if } u(t) \neq 0 \text{ and } h(t) = 0 \text{ if } u(t) = 0,$$

is μ-measurable. It follows that for every $z \in Z$ the function $\langle u\,h, z\rangle$ is μ-measurable. Since $u(t)h(t) = u(t)f(t)$, it follows that for every $z \in Z$ the function $\langle uf, z\rangle$ is μ-measurable.

From these facts it follows that if f is $|u|\,|\mu|$-integrable, then $\langle uf, z\rangle$ is μ-integrable, for every $z \in Z$.

Assume that f is $|u|\,|\mu|$-integrable.

Let (f_n) be a sequence of functions of $\mathscr{K}_E(T)$ such that

$$\lim \int |f - f_n|\,d(|u|\,|\mu| = 0.$$

The functions f and f_n are $u\mu$-integrable. From the inequalities

$$|\int f\,d(u\mu) - \int f_n\,d(u\mu)| \leqslant \int |f - f_n|\,d|u\mu| \leqslant \int |f - f_n|\,d(|u|\,|\mu|),$$

it follows that

$$\lim \int f_n\,d(u\mu) = \int f\,d(u\mu)$$

and for every $z \in Z$

$$\lim \langle \int f_n\,d(u\mu), z\rangle = \langle \int f\,d(u\mu), z\rangle.$$

On the other hand, for every $z \in Z$, the functions $\langle uf_n, z\rangle$ and $\langle uf, z\rangle$ are μ-integrable. From the inequalities

$$|\int \langle uf_n, z\rangle\,d\mu - \int \langle uf, z\rangle\,d\mu| \leqslant |z| \int |f_n - f|\,|u|\,d|\mu| =$$
$$= |z| \int |f_n - f|\,d(|u|\,|\mu|,$$

we deduce

$$\lim \int \langle uf_n, z\rangle\,d\mu = \int \langle uf, z\rangle\,d\mu.$$

But for the functions $f_n \in \mathscr{K}_E(T)$ we have

$$\langle \int \boldsymbol{f}_n \mathrm{d}(\boldsymbol{u}\mu), z \rangle = \int \langle \boldsymbol{u}\boldsymbol{f}_n, z \rangle \mathrm{d}\mu .$$

Passing to the limit we obtain:

$$\langle \int \boldsymbol{f} \mathrm{d}(\boldsymbol{u}\mu), z \rangle = \int \langle \boldsymbol{u}\boldsymbol{f}, z \rangle \mathrm{d}\mu .$$

16.17 Corollary. *Let $\boldsymbol{u} : T \to \mathscr{L}^*(E, F)$ be a Z-weakly locally μ-integrable function. Assume that we have*

$$|\boldsymbol{u}\mu| = |\boldsymbol{u}|\,|\mu| .$$

Let $\boldsymbol{f} : T \to E$.

(1) *If $\boldsymbol{f}$ is $\boldsymbol{u}\mu$-negligible, the $\boldsymbol{u}\boldsymbol{f}$ is μ-negligible.*
(2) *If $\boldsymbol{f}$ is $\boldsymbol{u}\mu$-measurable, then the function $\langle \boldsymbol{u}\boldsymbol{f}, z \rangle$ is μ-measurable for every $z \in Z$.*
(3) *If $\boldsymbol{f}$ is $\boldsymbol{u}$ μ-integrable, then the function $\langle \boldsymbol{u}\boldsymbol{f}, z \rangle$ is μ-integrable for every $z \in Z$ and*

$$\langle \int \boldsymbol{f} \mathrm{d}(\boldsymbol{u}\mu), z \rangle = \int \langle \boldsymbol{u}\boldsymbol{f}, z \rangle \mathrm{d}\mu .$$

The scalar functions have additional properties, under certain conditions of countability for E and F.

16.18 Proposition. *Let $\boldsymbol{u} : T \to \mathscr{L}^*(E, F)$ be a Z-weakly locally μ-integrable function and f a scalar function defined on T.*

(1) *If f is $|\boldsymbol{u}|\,|\mu|$-negligible, then f is $\boldsymbol{u}\mu$-negligible and $\boldsymbol{u}f$ is μ-negligible.*
(2) *If f is $|\boldsymbol{u}|\,|\mu|$-measurable, then f is $\boldsymbol{u}\mu$-measurable and $\langle \boldsymbol{u}xf, z \rangle$ is μ-measurable for every $x \in E$ and $z \in Z$.*
(3) *If f is $|\boldsymbol{u}|\,|\mu|$-integrable, then f is $\boldsymbol{u}\mu$-integrable and $\langle \boldsymbol{u}fx, z \rangle$ is μ-integrable for every $x \in E$ and $z \in Z$ and we have*

$$\langle \int f \mathrm{d}(\boldsymbol{u}\mu)x, z \rangle = \int \langle \boldsymbol{u}fx, z \rangle \mathrm{d}\mu ,$$

for $x \in E$ and $z \in Z$.

Conversely, assume that E is of countable type and that there exists a countable set $S \subset Z$ norming for E.

In this case:

(1′) *If $\langle \boldsymbol{u}fx, z \rangle$ is μ-negligible for every $x \in E$ and every $z \in Z$, then f is negligible with respect to $|\boldsymbol{u}|\,|\mu|$ and with respect to $\boldsymbol{u}\mu$.*
(2′) *If f is real and if $\langle \boldsymbol{u}fx, z \rangle$ is μ-measurable for every $x \in E$ and $z \in Z$, then f is measurable with respect to $|\boldsymbol{u}|\,|\mu|$ and with respect to $\boldsymbol{u}\mu$.*

If f is $|\boldsymbol{u}|\,|\mu|$-negligible (measurable, integrable), then from the inequality $|\boldsymbol{u}\mu| \leqslant |\boldsymbol{u}|\,|\mu|$ it follows that f is $\boldsymbol{u}\mu$-negligible (measurable, integrable).

If f is $|\boldsymbol{u}|\,|\mu|$-negligible, it is shown as in theorem 16.11 that $\boldsymbol{u}f$ is μ-negligible and property (1) is proved.

If f is $|\boldsymbol{u}|\,|\mu|$-measurable (integrable), then for every $x \in E$ the function $fx : T \to E$ is $|\boldsymbol{u}|\,|\mu|$-measurable (integrable) and from theorem 16.16 we deduce that the function $\langle \boldsymbol{u}fx, z \rangle$ is μ-measurable (integrable) for every $z \in Z$ and thus properties (2) and (3) are also proved.

Conversely, assume that E and Z have the stated countability property.

Let (x_n) be a sequence dense in E. For every $t \in T$ we have

$$|\boldsymbol{u}(t)|\,|f(t)| = |\boldsymbol{u}(t)f(t)| = \sup_{s\in S,\, n\in N} |\langle \boldsymbol{u}(t)f(t)x_n, s\rangle| / |x_n|\,|s|\,.$$

From this equality it follows that if $\langle \boldsymbol{u}fx, z\rangle$ is μ-negligible or μ-measurable for every $x \in E$ and every $z \in Z$, then $|f|\,|\boldsymbol{u}|$ is μ-negligible respectively μ-measurable; therefore, $|f|$ is $|\boldsymbol{u}|\,|\mu|$-negligible respectively $|\boldsymbol{u}|\,|\mu|$-measurable.

In this way property (1′) is proved.

Assume now that f is real and that $\langle \boldsymbol{u}fx, z\rangle$ is μ-measurable for every $x \in E$ and every $z \in Z$.

From these facts it follows that $|f|$ is $|\boldsymbol{u}|\,|\mu|$-measurable, and from property (2) we deduce that $|f|\langle \boldsymbol{u}x, z\rangle$ is μ-measurable for every $x \in E$ and $z \in Z$.

Since the functions $f\langle \boldsymbol{u}x, z\rangle$ and $|f|\langle \boldsymbol{u}x, z\rangle$ are μ-measurable, the functions

$$f^+\langle \boldsymbol{u}x, z\rangle = \tfrac{1}{2}(|f|+f)\langle \boldsymbol{u}x, z\rangle \quad \text{and}$$
$$f^-\langle \boldsymbol{u}x, z\rangle = \tfrac{1}{2}(|f|-f)\langle \boldsymbol{u}x, z\rangle$$

are μ-measurable for every $x \in E$ and $z \in Z$.

Repeating the same argument with f^+ and f^- instead of f, we deduce that the functions $|f^+| = f^+$ and $|f^-| = f^-$ are $|\boldsymbol{u}|\,|\mu|$-measurable.

Then the function $f = f^+ - f^-$ is $|\boldsymbol{u}|\,|\mu|$-measurable and thus property (2′) is also proved.

16.19 Corollary. *Assume that E is of countable type and that there exists a countable set $S \subset Z$ norming for E.*

Let $\boldsymbol{u} : T \to \mathscr{L}^*(E, F)$ be a Z-weakly locally μ-integrable function such that we have

$$|\boldsymbol{u}\mu| = |\boldsymbol{u}|\,|\mu|\,.$$

(1) *A scalar function f is* $\boldsymbol{u}\mu$*-negligible if and only if* $\langle \boldsymbol{u}fx, z\rangle$ *is* μ*-negligible for every* $x \in E$ *and every* $z \in Z$;

(2) *A real function f is* $\boldsymbol{u}\mu$*-measurable, if and only if* $\langle \boldsymbol{u}fx, z\rangle$ *is* μ*-measurable for every* $x \in E$ *and every* $z \in Z$.

16.20 PROPOSITION. *Assume that there exists a countable set* $S \subset Z$ *norming for* F.

Let $\boldsymbol{u}: T \to \mathscr{L}^*(E, F)$ *be a Z-weakly locally* μ*-integrable function and* $\boldsymbol{f}: T \to E$ *be a function.*

If $\boldsymbol{f}$ *is locally* $|\boldsymbol{u}|\,|\mu|$*-integrable, then* $\boldsymbol{f}$ *is locally* $\boldsymbol{u}\mu$*-integrable and the function* $\boldsymbol{uf}$ *is Z-weakly locally* μ*-integrable and we have*

$$\boldsymbol{f}(\boldsymbol{u}\mu) = (\boldsymbol{u}\boldsymbol{f})\mu\,.$$

From the inequality $|\boldsymbol{u}\mu| \leqslant |\boldsymbol{u}|\,|\mu|$ it follows that if $\boldsymbol{f}$ is locally integrable with respect to $|\boldsymbol{u}|\,|\mu|$, then $\boldsymbol{f}$ is locally integrable also with respect to $\boldsymbol{u}\mu$.

Since $\boldsymbol{f}$ is locally $|\boldsymbol{u}|\,|\mu|$-integrable, the function $\boldsymbol{f}$ is $|\boldsymbol{u}|\,|\mu|$-measurable; therefore, for every $z \in Z$ the function $\langle \boldsymbol{uf}, z\rangle$ is μ-measurable, that is, $\boldsymbol{uf}$ is Z-weakly μ-measurable.

The function $|\langle \boldsymbol{uf}, z\rangle|$ is μ-measurable for every $z \in Z$. From the equality

$$|\boldsymbol{u}(t)\boldsymbol{f}(t)| = \sup_{s\in S} |\langle \boldsymbol{u}(t)\boldsymbol{f}(t), s\rangle|/|s|, \text{ for } t \in T\,,$$

we deduce that $|\boldsymbol{uf}|$ is μ-measurable. For every compact set $K \subset T$ we have

$$\textstyle\int_K^* |\boldsymbol{uf}|\,\mathrm{d}|\mu| \leqslant \int_K^* |\boldsymbol{u}|\,|\boldsymbol{f}|\,\mathrm{d}|\mu| = \int_K^* |\boldsymbol{f}|\,\mathrm{d}(|\boldsymbol{u}|\,|\mu|) < \infty\,,$$

therefore $\boldsymbol{uf}$ is Z-weakly locally μ-integrable.

Let $\varphi \in \mathscr{K}(T)$. Then $\boldsymbol{f}\varphi$ is $|\boldsymbol{u}|\,|\mu|$-integrable, therefore $\langle \boldsymbol{uf}\varphi, z\rangle$ is μ-integrable for every $z \in Z$ and we have

$$\langle \textstyle\int \boldsymbol{f}\varphi\,\mathrm{d}(\boldsymbol{u}\mu), z\rangle = \int \langle \boldsymbol{uf}\varphi, z\rangle\,\mathrm{d}\mu\,,$$

that is,

$$\langle \textstyle\int \varphi\,\mathrm{d}[\boldsymbol{f}(\boldsymbol{u}\mu)], z\rangle = \langle \int \varphi\,\mathrm{d}[(\boldsymbol{uf})\mu], z\rangle\,,$$

consequently

$$\boldsymbol{f}(\boldsymbol{u}\mu) = (\boldsymbol{f}\boldsymbol{u})\mu\,.$$

16.21 PROPOSITION. *Let* $\boldsymbol{u}: T \to \mathscr{L}^*(E, F)$ *be a Z-weakly locally* μ*-integrable function and f a scalar function defined on* T.

If f is locally $|\boldsymbol{u}|\,|\mu|$-integrable, then f is locally $\boldsymbol{u}\mu$-integrable and $\boldsymbol{u}f$ is Z-weakly locally μ-integrable and we have

$$f(\boldsymbol{u}\mu)=(f\boldsymbol{u})\mu\,.$$

Assume that f is locally $|\boldsymbol{u}|\,|\mu|$-integrable. From the inequality $|\boldsymbol{u}\mu|\leqslant|\boldsymbol{u}|\,|\mu|$ we deduce that f is locally $\boldsymbol{u}\mu$-integrable.

Since f is $|\boldsymbol{u}|\,|\mu|$-measurable, it follows that $\langle \boldsymbol{u}fx, z\rangle$ is μ-measurable for every $x\in E$ and every $z\in Z$, that is, $\boldsymbol{u}f$ is *Z-weakly μ-measurable.*

Since f is locally $|\boldsymbol{u}|\,|\mu|$-integrable, the function $|f|$ is locally $|\boldsymbol{u}|\,|\mu|$-integrable, therefore the function $|\boldsymbol{u}f|=|\boldsymbol{u}|\,|f|$ is locally $|\mu|$-integrable. It follows that $\boldsymbol{u}f$ is Z-weakly locally μ-integrable.

Let $h\in\mathscr{K}(T)$. Then fh is $|\boldsymbol{u}|\,|\mu|$-integrable; therefore, fh is $\boldsymbol{u}\mu$-integrable and $\langle \boldsymbol{u}fhx, z\rangle$ is μ-integrable for every $x\in E$ and $z\in Z$ and we have

$$\langle \textstyle\int hd[f(\boldsymbol{u}\mu)]\,x, z\rangle=\langle\int f\mathrm{d}h(\boldsymbol{u}\mu)x, z\rangle=\int\langle \boldsymbol{u}fx, z\rangle\mathrm{d}\mu= \\ =\langle\int h\mathrm{d}[(\boldsymbol{u}f)\mu]x, z\rangle$$

whence

$$\textstyle\int h\mathrm{d}[f(\boldsymbol{u}\mu)]=\int h\mathrm{d}[(\boldsymbol{u}f)\mu]$$

that is

$$f(\boldsymbol{u}\mu)=(f\boldsymbol{u})\mu\,.$$

5 *Integration with respect to measures defined by simply locally integrable densities*

Let E, F and G be three Banach spaces.

16.22 Theorem. *Let $\boldsymbol{m}:\mathscr{K}(T)\to\mathscr{L}(F, G)$ be a dominated measure, $\boldsymbol{g}:T\to\mathscr{L}^*(E, F)$ a simply locally $\boldsymbol{m}$-integrable function, the measure $\boldsymbol{g}\boldsymbol{m}:\mathscr{K}(T)\to\mathscr{L}(E, G)$ and a function $\boldsymbol{f}:T\to E$.*

(1) *If $\boldsymbol{f}$ is $|\boldsymbol{g}|\,|\boldsymbol{m}|$-negligible, then $\boldsymbol{f}$ is $\boldsymbol{g}\boldsymbol{m}$-negligible and $\boldsymbol{f}\boldsymbol{g}$ is $\boldsymbol{m}$-negligible.*

(2) *If $\boldsymbol{f}$ is $|\boldsymbol{g}|\,|\boldsymbol{m}|$-measurable, then $\boldsymbol{f}$ is $\boldsymbol{g}\boldsymbol{m}$-measurable and $\boldsymbol{f}\boldsymbol{g}$ is $\boldsymbol{m}$-measurable.*

(3) *If $\boldsymbol{f}$ is $|\boldsymbol{g}|\,|\boldsymbol{m}|$-integrable, then $\boldsymbol{f}$ is $\boldsymbol{g}\boldsymbol{m}$-integrable and $\boldsymbol{f}\boldsymbol{g}$ is $\boldsymbol{m}$-integrable and we have*

$$\textstyle\int \boldsymbol{f}\mathrm{d}(\boldsymbol{g}\,\boldsymbol{m})=\int \boldsymbol{f}\boldsymbol{g}\,\mathrm{d}\boldsymbol{m}\,.$$

The proof is the same as for theorem 16.16 without using Z.

16.23 COROLLARY. *Let $\boldsymbol{m}:\mathscr{K}(T)\to\mathscr{L}(F,G)$ be a dominated measure and $\boldsymbol{g}:T\to\mathscr{L}^*(E,F)$ a simply locally $\boldsymbol{m}$-integrable function such that*

$$|\boldsymbol{g}\,\boldsymbol{m}|=|\boldsymbol{g}|\,|\boldsymbol{m}|\,.$$

If a function $\boldsymbol{f}:T\to E$ is $\boldsymbol{g}\,\boldsymbol{m}$-negligible (measurable, integrable) then $\boldsymbol{f}\boldsymbol{g}$ is $\boldsymbol{m}$-negligible (measurable, integrable).

For scalar functions we have the following proposition:

16.24 PROPOSITION. *Let $\boldsymbol{m}:\mathscr{K}(T)\to\mathscr{L}(F,G)$ be a dominated measure, $\boldsymbol{g}:T\to\mathscr{L}^*(E,F)$ a simply locally $\boldsymbol{m}$-integrable function, the measure $\boldsymbol{g}\,\boldsymbol{m}$: $\mathscr{K}(T)\to\mathscr{L}(E,G)$ and f a scalar function.*

(1) *If f is $|\boldsymbol{g}|\,|\boldsymbol{m}|$-negligible, then f is $\boldsymbol{g}\,\boldsymbol{m}$-negligible and $f\boldsymbol{g}$ is $\boldsymbol{m}$-negligible.*
(2) *If f is $|\boldsymbol{g}|\,|\boldsymbol{m}|$-measurable, then f is $\boldsymbol{g}\,\boldsymbol{m}$-measurable and $f\boldsymbol{g}x$ is m-measurable, for every $x\in E$.*
(3) *If f is $|\boldsymbol{g}|\,|\boldsymbol{m}|$-integrable, then f is $\boldsymbol{g}\,\boldsymbol{m}$-integrable and $f\boldsymbol{g}x$ is $\boldsymbol{m}$-integrable for every $x\in E$ and we have*

$$\int f\,\mathrm{d}(\boldsymbol{g}\,\boldsymbol{m})x=\int f\boldsymbol{g}x\,\mathrm{d}\boldsymbol{m},\ \textit{for}\ x\in E\,.$$

Conversely, assume that E is of countable type. In this case:

(1′) *If $f\boldsymbol{g}x$ is $\boldsymbol{m}$-negligible for every $x\in E$, then f is $\boldsymbol{g}\,\boldsymbol{m}$-negligible and $|\boldsymbol{g}|\,|\boldsymbol{m}|$-negligible.*
(2′) *If f is real and if $f\boldsymbol{g}x$ is $\boldsymbol{m}$-measurable for every $x\in E$, then f is $|\boldsymbol{g}|\,|\boldsymbol{m}|$-measurable and $\boldsymbol{g}\,\boldsymbol{m}$-measurable.*

The proof is the same as that of proposition 16.18 without using Z.

16.25 COROLLARY. *Let $\boldsymbol{m}:\mathscr{K}(T)\to\mathscr{L}(F,G)$ be a dominated measure and $\boldsymbol{g}:T\to\mathscr{L}^*(E,F)$ a simply locally $\boldsymbol{m}$-integrable function. Assume that E is of countable type and that*

$$|\boldsymbol{g}\,\boldsymbol{m}|=|\boldsymbol{g}|\,|\boldsymbol{m}|\,.$$

Then:

(1) *A scalar function f is $\boldsymbol{g}\,\boldsymbol{m}$-negligible if and only if $f\boldsymbol{g}x$ is $\boldsymbol{m}$-negligible for every $x\in E$.*
(2) *A real function f is $\boldsymbol{g}\,\boldsymbol{m}$-measurable if and only if $f\boldsymbol{g}x$ is $\boldsymbol{m}$-measurable for every $x\in E$.*

16.26 PROPOSITION. *Let* $\boldsymbol{m}:\mathscr{K}(T)\to\mathscr{L}(F,G)$ *be a dominated measure,* $\boldsymbol{g}:T\to\mathscr{L}^*(E,F)$ *a simply locally* $\boldsymbol{m}$*-integrable function and* $\boldsymbol{f}:T\to E$ *a function.*

If $\boldsymbol{f}$ *is locally* $|\boldsymbol{g}|\,|\boldsymbol{m}|$*-integrable, then* $\boldsymbol{f}$ *is locally* $\boldsymbol{g}\,\boldsymbol{m}$*-integrable and* $\boldsymbol{f}\boldsymbol{g}$ *is locally* $\boldsymbol{m}$*-integrable and we have*

$$\boldsymbol{f}(\boldsymbol{g}\boldsymbol{m})=(\boldsymbol{f}\boldsymbol{g})\boldsymbol{m}\,.$$

If $\boldsymbol{f}$ is locally $|\boldsymbol{g}|\,|\boldsymbol{m}|$-integrable, from the inequality $|\boldsymbol{g}\,\boldsymbol{m}|\leqslant|\boldsymbol{g}|\,|\boldsymbol{m}|$ it follows that $\boldsymbol{f}$ is locally $\boldsymbol{g}\,\boldsymbol{m}$-integrable. On the other hand, for every $\varphi\in\mathscr{K}(T)$ the function $\boldsymbol{f}\varphi$ is $|\boldsymbol{g}|\,|\boldsymbol{m}|$-integrable therefore $\boldsymbol{g}\boldsymbol{f}\varphi$ is $\boldsymbol{m}$-integrable and we have

$$\int \boldsymbol{f}\varphi\,\mathrm{d}(\boldsymbol{g}\boldsymbol{m})=\int \boldsymbol{f}\boldsymbol{g}\varphi\,\mathrm{d}\boldsymbol{m}\,.$$

It follows that $\boldsymbol{f}\boldsymbol{g}$ is locally $\boldsymbol{m}$-integrable and that

$$\int \varphi\,\mathrm{d}[\boldsymbol{f}(\boldsymbol{g}\boldsymbol{m})]=\int\varphi\,\mathrm{d}[(\boldsymbol{f}\boldsymbol{g})\boldsymbol{m}],\ \text{for}\ \varphi\in\mathscr{K}(T)\,,$$

consequently

$$\boldsymbol{f}(\boldsymbol{g}\,\boldsymbol{m})=(\boldsymbol{f}\boldsymbol{g})\boldsymbol{m}.$$

For scalar functions, the preceding proposition can be stated in the following way:

16.27 PROPOSITION. *Let* $\boldsymbol{m}:\mathscr{K}(T)\to\mathscr{L}(F,G)$ *be a dominated measure,* $\boldsymbol{g}:T\to\mathscr{L}^*(E,F)$ *a simply locally* $\boldsymbol{m}$*-integrable function and* f *a scalar function defined on* T.

If f *is locally* $|\boldsymbol{g}|\,|\boldsymbol{m}|$*-integrable, then* f *is locally* $\boldsymbol{g}\boldsymbol{m}$*-integrable and* $f\boldsymbol{g}$ *is simply locally* $\boldsymbol{m}$*-integrable and we have*

$$f(\boldsymbol{g}\,\boldsymbol{m})=(f\boldsymbol{g})\boldsymbol{m}\,.$$

The proof is the same as for proposition 16.21 without using Z.

6 *Integration with respect to measures defined by locally integrable densities*

Let E, F and Y be three Banach spaces.

We shall consider three cases according to whether m, g or f are scalar valued.

Consider first the case where m is *scalar.*

It will be shown in the sequel (theorem 18.17) that if μ is a scalar measure and $\boldsymbol{g}:T\to X$ is locally μ-integrable we have $|\boldsymbol{g}\,\mu|=|\boldsymbol{g}|\,|\mu|$.

Theorems 16.16 and 16.22 and their corollaries are stated now for this particular case:

16.28 THEOREM. *Let μ be a scalar measure, $\boldsymbol{g}: T \to \mathscr{L}(E, F)$ a locally μ-integrable function, the measure $\boldsymbol{g}\mu : \mathscr{K}(T) \to \mathscr{L}(E, F)$ and a function $\boldsymbol{f}: T \to E$.*

If $\boldsymbol{f}$ is $|\boldsymbol{g}|\,|\mu|$-negligible (measurable, integrable), then $\boldsymbol{f}\boldsymbol{g}$ is μ-negligible (measurable, integrable).

If $\boldsymbol{f}$ is $|\boldsymbol{g}|\,|\mu|$-integrable, then $\boldsymbol{f}$ is $\boldsymbol{g}\mu$-integrable and we have

$$\int \boldsymbol{f} \mathrm{d}(\boldsymbol{g}\mu) = \int \boldsymbol{f}\boldsymbol{g} \,\mathrm{d}\mu .$$

From proposition 16.20 and 16.26 we deduce:

16.29 PROPOSITION. *Let μ be a scalar measure, $\boldsymbol{g}: T \to \mathscr{L}(E, F)$ a locally μ-integrable function and $\boldsymbol{f}: T \to E$ a function.*

If $\boldsymbol{f}$ is a locally $|\boldsymbol{g}|\,|\mu|$-integrable, then $\boldsymbol{f}$ is locally $\boldsymbol{g}\mu$-integrable and $\boldsymbol{f}\boldsymbol{g}$ is locally μ-integrable and we have

$$\boldsymbol{f}(\boldsymbol{g}\mu) = (\boldsymbol{f}\boldsymbol{g})\mu .$$

Consider now the case where g is *scalar*.

It will be shown later (theorem 18.20) that if $\boldsymbol{m} : \mathscr{K}(T) \to X$ is a dominated measure and g is a locally $\boldsymbol{m}$-integrable *scalar* function, then $|g\boldsymbol{m}| = |g|\,|\boldsymbol{m}|$.

Theorems 16.22 and 16.28 can now be completed:

16.30 THEOREM. *Let $\boldsymbol{m} : \mathscr{K}(T) \to \mathscr{L}(E, F)$ be a dominated measure, g a locally $\boldsymbol{m}$-integrable scalar function, the measure $g\boldsymbol{m} : \mathscr{K}(T) \to \mathscr{L}(E, F)$ and the function $\boldsymbol{f}: T \to E$.*

The function $\boldsymbol{f}$ is $|g|\,|\boldsymbol{m}|$-negligible (measurable, integrable) if and only if $\boldsymbol{f}g$ is $\boldsymbol{m}$-negligible (measurable, integrable).

If $\boldsymbol{f}$ is $|g|\,|\boldsymbol{m}|$-integrable, then $\boldsymbol{f}$ is $g\boldsymbol{m}$-integrable and we have

$$\int \boldsymbol{f} \mathrm{d}(g\boldsymbol{m}) = \int \boldsymbol{f} g \,\mathrm{d}\boldsymbol{m} .$$

Assume that $\boldsymbol{f}g$ is $\boldsymbol{m}$-negligible. Then the function $|\boldsymbol{f}g| = |\boldsymbol{f}|\,|g|$ is $|\boldsymbol{m}|$-negligible, therefore $|\boldsymbol{f}|$ is $|g|\,|\boldsymbol{m}|$-negligible, that is, $\boldsymbol{f}$ is $|g|\,|\boldsymbol{m}|$-negligible.

Assume that $\boldsymbol{f}g$ is $\boldsymbol{m}$-measurable. The scalar function h defined on T by $h(t) = |g(t)|/g(t)$ if $g(t) \neq 0$ and $h(t) = 0$ if $g(t) = 0$ is $\boldsymbol{m}$-measurable, consequently the function $f\,|g| = fgh$ is $|\boldsymbol{m}|$-measurable. It follows that f is $|g|\,|\boldsymbol{m}|$-measurable.

If fg is $\boldsymbol{m}$-integrable, then fg is $\boldsymbol{m}$-measurable and the function $|fg| = |f|\,|g|$ is $\boldsymbol{m}$-integrable. It follows that f is $|g|\,|\boldsymbol{m}|$-measurable and $|f|$ is $|g|\,|\boldsymbol{m}|$-integrable, hence f is $|g|\,|\boldsymbol{m}|$-integrable. The other assertions follow from theorem 16.28, considering $g : T \to \mathscr{L}(E, F)$.

16.31 PROPOSITION. *Let $\boldsymbol{m} : \mathscr{K}(T) \to \mathscr{L}(E, F)$ be a dominated measure, g a locally $\boldsymbol{m}$-integrable scalar function and the measure $g\boldsymbol{m} : \mathscr{K}(T) \to \mathscr{L}(E, F)$.*

(1) *A function $f : T \to \mathscr{L}^*(H, E)$ is simply locally $|g|\,|\boldsymbol{m}|$-integrable if and only if fg is simply locally $\boldsymbol{m}$-integrable. If f is simply locally $|g|\,|\boldsymbol{m}|$-integrable, then f is simply locally $g\boldsymbol{m}$-integrable and we have*

$$f\,(g\boldsymbol{m}) = (fg)\boldsymbol{m}\,.$$

(2) *A function $f : T \to E$ is locally $|g|\,|\boldsymbol{m}|$-integrable if and only if fg is locally $\boldsymbol{m}$-integrable. If f is locally $|g|\,|\boldsymbol{m}|$-integrable, we have*

$$f\,(g\boldsymbol{m}) = (fg)\boldsymbol{m}\,.$$

If $f : T \to \mathscr{L}^*(H, E)$ is simply locally $|g|\,|\boldsymbol{m}|$-integrable, then from the inequality $|g\boldsymbol{m}| \leqslant |g|\,|\boldsymbol{m}|$ it follows that f is simply locally $g\boldsymbol{m}$-integrable. It follows also that $|f|$ is $|g|\,|\boldsymbol{m}|$-integrable and fx is $|g|\,|\boldsymbol{m}|$-measurable for every $x \in H$. Then the function $|fg| = |f|\,|g|$ is locally $\boldsymbol{m}$-integrable and fxg is $\boldsymbol{m}$-measurable for every $x \in H$, hence fg is simply locally $\boldsymbol{m}$-integrable. For every $\varphi \in \mathscr{K}(T)$ and $x \in H$ the function $\varphi x f$ is $g\boldsymbol{m}$-integrable and $\varphi x f g$ is $\boldsymbol{m}$-integrable and we have

$$\int \varphi x f \,\mathrm{d}(g\boldsymbol{m}) = \int \varphi x f g \,\mathrm{d}\boldsymbol{m}\,,$$

whence

$$\int \varphi \,\mathrm{d}[f\,(g\boldsymbol{m})] \cdot x = \int \varphi \,\mathrm{d}[(fg)\boldsymbol{m}] \cdot x\,,$$

hence

$$f\,(g\boldsymbol{m}) = (fg)\boldsymbol{m}\,.$$

If fg is simply locally $\boldsymbol{m}$-integrable, then the function $|fg| = |f|\,|g|$ is locally $|\boldsymbol{m}|$-integrable and fgx is $|\boldsymbol{m}|$-measurable for every $x \in H$. Then $|f|$ is locally $|g|\,|\boldsymbol{m}|$-integrable and f is $|g|\,|\boldsymbol{m}|$-measurable, consequently f is simply locally $|g|\,|\boldsymbol{m}|$-integrable.

The second part follows from the first part, taking $H = R$.

16.32 PROPOSITION. *Let μ be a scalar measure and g a locally μ-integrable scalar function.*

A function $\boldsymbol{f}: T \to \mathscr{L}^*(E, F)$ is Z-weakly locally $|g|\,|\mu|$-integrable if and only if $\boldsymbol{f}g$ is Z-weakly locally μ-integrable. If $\boldsymbol{f}$ is Z-weakly locally $|g|\,|\mu|$-integrable, then $\boldsymbol{f}$ is Z-weakly locally $g\mu$-integrable and we have

$$\boldsymbol{f}(g\mu) = (\boldsymbol{f}g)\mu\,.$$

The proof is similar to that of proposition 16.31.

Consider, finally, the case where f is *scalar*.

Proposition 16.24 can be stated more precisely:

16.33 THEOREM. *Let $\boldsymbol{m}: \mathscr{K}(T) \to \mathscr{L}(E, F)$ be a dominated measure, $\boldsymbol{g}: T \to E$ a locally $\boldsymbol{m}$-integrable function, the product measure $\boldsymbol{g}\boldsymbol{m}: \mathscr{K}(T) \to F$ and f a scalar function defined on T.*

(1) *If the function f is $|\boldsymbol{g}|\,|\boldsymbol{m}|$-negligible, then f is $\boldsymbol{g}\boldsymbol{m}$-negligible and $f\boldsymbol{g}$ is $\boldsymbol{m}$-negligible.*

(1′) *If $f\boldsymbol{g}$ is $\boldsymbol{m}$-negligible, then f is $|\boldsymbol{g}|\,|\boldsymbol{m}|$-negligible and $\boldsymbol{g}\,\boldsymbol{m}$-negligible.*

(2) *If f is $|\boldsymbol{g}|\,|\boldsymbol{m}|$-measurable (integrable), then f is $\boldsymbol{g}\,\boldsymbol{m}$-measurable (integrable) and $f\boldsymbol{g}$ is $\boldsymbol{m}$-measurable (integrable). If f is $|\boldsymbol{g}|\,|\boldsymbol{m}|$-integrable, we have*

$$\int f\,\mathrm{d}(\boldsymbol{g}\boldsymbol{m}) = \int f\boldsymbol{g}\,\mathrm{d}\boldsymbol{m}\,.$$

(2′) *If f is real and if $f\boldsymbol{g}$ is $\boldsymbol{m}$-measurable (integrable), then f is measurable (integrable) with respect to $|\boldsymbol{g}|\,|\boldsymbol{m}|$ and with respect to $\boldsymbol{g}\,\boldsymbol{m}$.*

The proof is the same as that of proposition 16.18.

16.34 COROLLARY. *Let $\boldsymbol{m}: \mathscr{K}(T) \to \mathscr{L}(E, F)$ be a dominated measure and $\boldsymbol{g}: T \to E$ a locally $\boldsymbol{m}$-integrable function such that*

$$|\boldsymbol{g}|\,|\boldsymbol{m}| = |\boldsymbol{g}|\,|\boldsymbol{m}|\,.$$

A real function f is $\boldsymbol{g}\,\boldsymbol{m}$-negligible (measurable, integrable), if and only if $f\boldsymbol{g}$ is $\boldsymbol{m}$-negligible (measurable, integrable).

From proposition 16.27 we deduce:

16.35 PROPOSITION. *Let* $\boldsymbol{m}: \mathscr{K}(T) \to \mathscr{L}(E, F)$ *be a dominated measure,* $\boldsymbol{g}: T \to E$ *a locally integrable function and* f *a scalar function.*

If f *is locally* $|\boldsymbol{g}|\,|\boldsymbol{m}|$*-integrable, then* f *is locally* $\boldsymbol{g}\boldsymbol{m}$*-integrable and* $f\boldsymbol{g}$ *is locally* $\boldsymbol{m}$*-integrable and we have*

$$f(\boldsymbol{g}\boldsymbol{m}) = (f\boldsymbol{g})\boldsymbol{m}\,.$$

Conversely, if f *is real and the function* $f\boldsymbol{g}$ *is locally* $\boldsymbol{m}$*-integrable, then* f *is locally integrable with respect to* $|\boldsymbol{g}|\,|\boldsymbol{m}|$ *and with respect to* $\boldsymbol{g}\boldsymbol{m}$.

§17 Properties of measures defined by densities

1 *Algebraic properties*

We give first the algebraic properties of the measures defined by densities.

17.1 PROPOSITION. *Let* $\boldsymbol{m}$ *be a dominated measure,* $\boldsymbol{f}$ *and* $\boldsymbol{g}$ *two functions defined on* T *and* α *a scalar. If the measures* $\boldsymbol{f}\boldsymbol{m}$, $\boldsymbol{g}\boldsymbol{m}$ *and* $(\boldsymbol{f}+\boldsymbol{g})\boldsymbol{m}$ *are defined, then we have*

$$(\boldsymbol{f}+\boldsymbol{g})\boldsymbol{m} = \boldsymbol{f}\boldsymbol{m} + \boldsymbol{g}\boldsymbol{m} \quad \text{and} \quad (\alpha\boldsymbol{f})\boldsymbol{m} = \alpha(\boldsymbol{f}\boldsymbol{m})\,.$$

The proof is immediate.

Remark. If $\boldsymbol{f}$ and $\boldsymbol{g}$ are, for example, simply locally $\boldsymbol{m}$-integrable it does not follow that $\boldsymbol{f}+\boldsymbol{g}$ is simply locally $\boldsymbol{m}$-integrable. That is why in the statement of the proposition we have imposed the condition that the measure $(\boldsymbol{f}+\boldsymbol{g})\boldsymbol{m}$ should also be defined.

17.2 PROPOSITION. *Let* $\boldsymbol{m}$ *and* $\boldsymbol{n}$ *be two dominated measures,* α *a scalar and* $\boldsymbol{g}$ *a function defined on* T. *If the measures* $\boldsymbol{g}\boldsymbol{m}$ *and* $\boldsymbol{g}\boldsymbol{n}$ *are defined, then the measures* $\boldsymbol{g}(\boldsymbol{m}+\boldsymbol{n})$ *and* $\boldsymbol{g}(\alpha\boldsymbol{m})$ *are defined and we have*

$$\boldsymbol{g}(\boldsymbol{m}+\boldsymbol{n}) = \boldsymbol{g}\boldsymbol{m} + \boldsymbol{g}\boldsymbol{n} \quad \text{and} \quad \boldsymbol{g}(\alpha\boldsymbol{m}) = \alpha(\boldsymbol{g}\boldsymbol{m}).$$

The proof is immediate.

2 *Measures with locally integrable densities and positive bases*

Let E be a Banach space and μ a *positive* measure.

17.3 THEOREM. *For every locally μ-integrable function $\boldsymbol{g}: T \to E$ we have*

$$|\boldsymbol{g}\mu| = |\boldsymbol{g}|\mu .$$

Assume first that $|\boldsymbol{g}(t)| \equiv 1$. Let φ be a positive function of $\mathscr{K}(T)$ and $0 < a < 1$; let K be the support of φ. There exists a μ-measurable function $\boldsymbol{g}'$, defined on T, with values in the dual E' and E, such that

$$a|\boldsymbol{g}(t)| \leqslant \langle \boldsymbol{g}(t), \boldsymbol{g}'(t)\rangle \text{ and } |\boldsymbol{g}'(t)| = 1$$

μ-almost everywhere on K (corollary 11.27).

Let $\varepsilon > 0$. There exists a compact set $K' \subset K$ such that $\mu(K \setminus K') < \varepsilon$ and the restriction of $\boldsymbol{g}'$ to K' is continuous. The restriction of $\boldsymbol{g}'$ to K' can be extended to a *continuous* function $\boldsymbol{f}': T \to E'$ with compact support such that $\|\boldsymbol{f}'\| \leqslant (2-a)\|\boldsymbol{g}'\|$. We have thus,

$$\boldsymbol{f}'(t) = \boldsymbol{g}'(t), \text{ for } t \in K \setminus K'$$

and

$$|\boldsymbol{f}'(t)| \leqslant (2-a), \text{ for } t \in T .$$

Then

$$a\mu(\varphi) = a \int |\boldsymbol{g}(t)| \varphi(t) \mathrm{d}\mu(t) \leqslant \int \langle \boldsymbol{g}(t), \boldsymbol{g}'(t)\rangle \varphi(t) \mathrm{d}\mu(t) = \\ = |\int \langle \boldsymbol{g}, \boldsymbol{f}'\rangle \varphi \mathrm{d}\mu + \int \langle \boldsymbol{g}, \boldsymbol{g}' - \boldsymbol{f}'\rangle \varphi \mathrm{d}\mu| \leqslant |\int \langle \boldsymbol{g}, \boldsymbol{f}'\varphi\rangle \mathrm{d}\mu| + \\ + |\int_{K \setminus K'} \langle \boldsymbol{g}, \boldsymbol{g}' - \boldsymbol{f}'\rangle \varphi \mathrm{d}\mu| .$$

Remarking that

$$|\int \langle \boldsymbol{g}, \boldsymbol{f}'\varphi\rangle \mathrm{d}\mu| = |\int \boldsymbol{f}'\varphi \mathrm{d}(\boldsymbol{g}\mu)| \leqslant \int |\boldsymbol{f}'| \varphi \mathrm{d}|\boldsymbol{g}\mu| \leqslant (2-a) \int \varphi \mathrm{d}|\boldsymbol{g}\mu| = \\ = (2-a)|\boldsymbol{g}\mu|(\varphi)$$

and

$$|\int_{K \setminus K'} \langle \boldsymbol{g}, \boldsymbol{g}' - \boldsymbol{f}'\rangle \varphi \mathrm{d}\mu| \leqslant \int_{K \setminus K'} |\boldsymbol{g}|(|\boldsymbol{g}'| + |\boldsymbol{f}'|) \varphi \mathrm{d}\mu = \\ = \int_{K \setminus K'} (1 + |\boldsymbol{f}'|) \varphi \mathrm{d}\mu \leqslant (3-a)\|\varphi\| \mu(K \setminus K') \leqslant (3-a)\|\varphi\| \varepsilon ,$$

we deduce

$$a\mu(\varphi) \leqslant (2-a)|\boldsymbol{g}\mu|(\varphi) + (3-a)\|\varphi\| \varepsilon .$$

Letting $a \to 1$ and $\varepsilon \to 0$, we obtain

$$\mu(\varphi) \leqslant |\boldsymbol{g}\mu|(\varphi) .$$

Since the function $\varphi \in \mathscr{K}_+(T)$ is arbitrary it follows $\mu \leqslant |\boldsymbol{g}\mu|$. As we have also $|\boldsymbol{g}\mu| \leqslant |\boldsymbol{g}|\mu = \mu$, we deduce

$|g\mu| = |g|\mu$.

Let now $g : T \to E$ be an arbitrary μ-integrable function. Put

$h(t) = g(t)/|g(t)|$ if $g(t) \neq 0$,
$h(t) = x_0 \in E$ with $|x_0| = 1$, if $g(t) = 0$.

We have $|h(t)| = 1$ and $h|g| = g$. Since g is μ-measurable, the function h is $\lambda = |g|\mu$-measurable.

From the first part it follows that $|h\lambda| = \lambda$.

Then

$$|g\mu| = |h|g|\mu| = |h\lambda| = \lambda = |g|\mu$$

and the theorem is proved.

Remark. It will be shown in the sequel (theorem 18.17) that the equality $|g\mu| = |g|\,|\mu|$ remains valid for every *scalar* measure μ and every locally μ-integrable function g. It will follow that propositions 17.4 and 17.5 below remain valid also if μ is a scalar measure.

17.4 Proposition. *Let $g : T \to E$ be a locally μ-integrable function. We have $g\mu = 0$ if and only if g is μ-negligible.*

If g is μ-negligible, then for every $\varphi \in \mathscr{K}(T)$ the function φg is μ-negligible, therefore

$$\int \varphi \, \mathrm{d}(g\mu) = \int \varphi g \, \mathrm{d}\mu = 0$$

that is, $g\mu = 0$.

Conversely, if $g\mu = 0$, then $|g|\,|\mu| = |g\mu| = 0$. It follows that the function φ_T is $|g\mu|$-negligible, hence $|g|\,|\mu|$-negligible; consequently, the function $\varphi_T|g| = |g|$ is μ-negligible, that is, g is μ-negligible.

17.5 Proposition. *Let $g_1, g_2 : T \to E$ be two locally μ-integrable functions. We have $g_1\mu = g_2\mu$ if and only if $g_1(t) = g_2(t)$, μ-almost everywhere.*

If $g_1 = g_2$, μ-almost everywhere, it follows immediately that $g_1\mu = g_2\mu$.

Conversely, if $g_1\mu = g_2\mu$, then $(g_1 - g_2)\mu = 0$, hence $g_1 - g_2$ is μ-negligible, consequently $g_1 = g_2$, μ-almost everywhere.

The measures defined by *real densities* with respect to *positive bases* have additional properties.

17.6 PROPOSITION. *If μ is a positive measure and g is a locally μ-integrable real function, then*

$$(g\mu)^+ = g^+\mu \text{ and } (g\mu)^- = g^-\mu .$$

In fact,

$$(g\mu)^+ = \tfrac{1}{2}(|g\mu| + g\mu) = \tfrac{1}{2}(|g|\mu + g\mu) = \tfrac{1}{2}(|g| + g)\mu = g^+\mu$$

and

$$(g\mu)^- = \tfrac{1}{2}(|g\mu| - g\mu) = \tfrac{1}{2}(|g|\mu - g\mu) = \tfrac{1}{2}(|g| - g)\mu = g^-\mu .$$

17.7 PROPOSITION. *Let μ be a positive measure and g a locally μ-integrable real function. We have $g\mu \geqslant 0$ if and only if $g(t) \geqslant 0$, μ-almost everywhere.*

Assume that $g\mu \geqslant 0$. Then $|g|\mu = |g\mu| = g\mu$, hence, by proposition 17.5, we have $|g(t)| = g(t)$ μ-almost everywhere, therefore $g(t) \geqslant 0$ μ-almost everwhere.

Conversely, if $g(t) \geqslant 0$ μ-almost everywhere, it follows immediately that $g\mu \geqslant 0$.

17.8 PROPOSITION. *Let μ be a positive measure, g_1 and g_2 two locally μ-integrable real functions. We have $g_1\mu \leqslant g_2\mu$ if and only if $g_1(t) \leqslant g_2(t)$, μ-almost everywhere.*

In fact, $g_1\mu \leqslant g_2\mu$ if and only if $(g_2 - g_1)\mu = g_2\mu - g_1\mu \geqslant 0$, therefore if and only if $g_2(t) - g_1(t) \geqslant 0$, μ-almost everywhere, that is, if and only if $g_1(t) \leqslant g_2(t)$, μ-almost everywhere.

17.9 PROPOSITION. *Let μ be a positive measure. If g_1 and g_2 are two locally μ-integrable real functions, then the functions* $\sup(g_1, g_2)$ *and* $\inf(g_1, g_2)$ *are locally μ-integrable and*

$$\sup(g_1, g_2)\mu = \sup(g_1\mu, g_2\mu),$$

$$\inf(g_1, g_2)\mu = \inf(g_1\mu, g_2\mu).$$

Put $f = \sup(g_1, g_2)$ and $g = \inf(g_1, g_2)$. Since g_1 and g_2 are μ-measurable, the functions f and g are μ-measurable.

Let $K \subset T$ be a compact set. From the inequality $|f| \leqslant |g_1| + |g_2|$ it follows

$$\int_K^* |f| \mathrm{d}\mu \leqslant \int_K^* |g_1| \mathrm{d}\mu + \int_K^* |g_2| \mathrm{d}\mu < \infty$$

hence f is locally μ-integrable.

From the inequality $|g| \leqslant |g_1| + |g_2|$ it follows that g is locally μ-integrable.

From the equality

$$\sup(x, y) = \tfrac{1}{2}(x+y+|x+y|)$$

true in every ordered vector space, we deduce

$$\sup(g_1, g_2)\mu = \tfrac{1}{2}(g_1+g_2+|g_1-g_2|)\mu =$$
$$= \tfrac{1}{2}(g_1\mu+g_2\mu+|g_1-g_2|\mu) =$$
$$= \tfrac{1}{2}(g_1\mu+g_2\mu+|g_1\mu-g_2\mu|) = \sup(g_1\mu, g_2\mu)\,.$$

From the inequality

$$\inf(x, y) = -\sup(-x, -y)$$

we deduce

$$\inf(g_1 g_2)\mu = -\sup(-g_1, -g_2)\mu = -\sup(-g_1\mu, -g_2\mu) = \inf(g_1\mu, g_2\mu).$$

17.10 PROPOSITION. *Let μ be a positive measure and (g_n) an increasing (μ-almost everywhere) sequence of locally μ-integrable real functions. The sequence of measures $(g_n\mu)$ is dominated if and only if the function* $\sup g_n$ *is locally μ-integrable. In this case we have*

$$\sup(g_n\mu) = (\sup g_n)\mu\,.$$

Modifying the functions g_n on a μ-negligible set, we may assume that the sequence (g_n) is increasing everywhere on T. Assume that the sequence of measures $(g_n\mu)$ is dominated by a measure λ. For every function $f \geqslant 0$ of $\mathscr{K}_+(T)$ we have then

$$\sup \int f g_n \mathrm{d}\mu \leqslant \lambda(f) < +\infty\,.$$

It follows that the function $\sup f g_n = f \sup g_n$ is μ-integrable and

$$\int f \sup g_n \mathrm{d}\mu = \int \sup f g_n \mathrm{d}\mu = \sup \int f g_n \mathrm{d}\mu\,.$$

This means that the function $\sup g_n$ is locally μ-integrable and that $(\sup g_n)\mu = \sup(g_n\mu)$.

Conversely if the function $\sup g_n$ is locally μ-integrable, then the measures $g_n\mu$ are dominated by the measure $(\sup g_n)\mu$.

The measures defined by *positive densities* with respect to *real* measures have the following properties:

17.11 PROPOSITION. *Let μ and ν be two real measures. If $\mu \leqslant \nu$ and if $g \geqslant 0$ is a function locally integrable with respect to μ and ν, then*

$$g\mu \leqslant g\nu .$$

The proof is immediate.

Remark. From the equality $|g\mu| = |g|\,|\mu|$ which will be proved in the sequel (theorem 18.17) it will follow that if μ is a *real* measure and $g \geqslant 0$ is a locally μ-integrable function, then we have

$$(g\mu)^+ = g\mu^+ \text{ and } (g\mu)^- = g\mu^- .$$

17.12 PROPOSITION. *Let μ and ν be two real measures on T. If the function $\boldsymbol{g} : T \to E$ is locally integrable for the measures μ and ν, then $\boldsymbol{g}$ is locally integrable for the measures* $\sup(\mu, \nu)$ *and* $\inf(\mu, \nu)$. *If g is positive we have*

$$g(\sup(\mu, \nu)) = \sup(g\mu, g\nu) ,$$

$$g(\inf(\mu, \nu)) = \inf(g\mu, g\nu) .$$

From the equality

$$\sup(\mu, \nu) = \tfrac{1}{2}(\mu + \nu + |\mu - \nu|), \ \inf(\mu, \nu) = -\sup(-\mu, -\nu),$$

true in every ordered vector space, we deduce

$$|\sup(\mu, \nu)| \leqslant |\mu| + |\nu|, \ |\inf(\mu, \nu)| \leqslant |\mu| + |\nu| ;$$

hence, if $\boldsymbol{g}$ is locally integrable for μ and ν, then $\boldsymbol{g}$ is locally integrable also for $\sup(\mu, \nu)$ and $\inf(\mu, \nu)$.

If g is positive, then we have

$$g(\sup(\mu, \nu)) = g(\tfrac{1}{2}(\mu + \nu + |\mu - \nu|)) = \tfrac{1}{2}(g\mu + g\nu + g|\mu - \nu|) = $$
$$= \tfrac{1}{2}(g\mu + g\nu + |g\mu - g\nu|) = \sup(g\mu, g\nu))$$

and

$$g(\inf(\mu, \nu)) = g(-\sup(-\mu, -\nu)) = -\sup(-g\mu, -g\nu) = \inf(g\mu, g\nu) .$$

3 *Measures with operator valued densities*

Let E and F be two Banach spaces and $Z \subset F'$ a norming subspace for F.

Let μ be a *positive* measure on T.

Let $\boldsymbol{u}, \boldsymbol{v} : T \to \mathscr{L}^*(E, F)$ be two functions. We recall that we have used the notation $\boldsymbol{u} \equiv \boldsymbol{v}$ if for every $x \in E$ and every $z \in Z$ we have

$$\langle \boldsymbol{u}(t)x, z \rangle = \langle \boldsymbol{v}(t)x, z \rangle, \ \mu\text{-almost everywhere.}$$

From proposition 17.5 it follows

17.13 PROPOSITION. *Let* $\boldsymbol{u}, \boldsymbol{v} : T \to \mathscr{L}^*(E, F)$ *be two* Z*-weakly locally* μ*-integrable functions. We have* $\boldsymbol{u}\mu = \boldsymbol{v}\mu$ *if and only if* $\boldsymbol{u} \equiv \boldsymbol{v}$.

In fact, if $\boldsymbol{u}\mu = \boldsymbol{v}\mu$, then for every function $\varphi \in \mathscr{K}(T)$ we have

$$\int \varphi \, \mathrm{d}(\boldsymbol{u}\mu) = \int \varphi \, \mathrm{d}(\boldsymbol{v}\mu) \ ;$$

hence, for every $x \in E$ and every $z \in Z$ we have

$$\langle \int \varphi \, \mathrm{d}(\boldsymbol{u}\mu)x, z \rangle = \langle \int \varphi \, \mathrm{d}(\boldsymbol{v}\mu)x, z \rangle$$

that is,

$$\int \langle \boldsymbol{u}x, z \rangle \varphi \, \mathrm{d}\mu = \int \langle \boldsymbol{v}x, z \rangle \varphi \, \mathrm{d}\mu, \ \text{for } \varphi \in \mathscr{K}(T) \ ;$$

whence, by proposition 17.5

$$\langle \boldsymbol{u}(t)x, z \rangle = \langle \boldsymbol{v}(t)x, z \rangle, \ \mu\text{-almost everywhere, that is } \boldsymbol{u} \equiv \boldsymbol{v}.$$

Conversely, if $\boldsymbol{u} \equiv \boldsymbol{v}$, then reasoning in a converse way, we deduce that $\boldsymbol{u}\mu = \boldsymbol{v}\mu$.

Remark. We may have $\boldsymbol{u}\mu = \boldsymbol{v}\mu$ even if $\boldsymbol{u}(t) \neq \boldsymbol{v}(t)$ for every $t \in T$.

Examples.

(1) Take $T = [0, 1]$, μ the Lebesgue measure on T, $E = \mathscr{C}(T)$ the space of continuous real functions on T with the topology defined by the norm $\|f\| = \sup_{t \in T} |f(t)|$, Z the space of the real functions Z defined on T such that

$$\|z\| = \Sigma_{t \in T} |z(t)| < \infty$$

and $F = Z'$. Then F is the space of the measurable real functions y defined on T such that

$$\|y\| = \sup_{t\in T}|y(t)| < \infty .$$

The space E is of countable type but Z is not of countable type. Every function $z\in Z$ vanishes outside a countable set hence it is μ-negligible.

For every $t\in T$ consider the mapping $\boldsymbol{u}(t) : E\to F$ defined by $\boldsymbol{u}(t)f=\varphi_{\{t\}}f$ for $f\in E$. We have $\|\boldsymbol{u}(t)\| =1$ for every $t\in T$ and

$$\langle z, \boldsymbol{u}(t)f\rangle = \langle z, \varphi_{\{t\}}f\rangle = \Sigma_{r\in T}\, z(r)\varphi_{\{t\}}(r)f(r) = z(t)f(t)$$

for every $z\in Z$ and $f\in E$.

For every function $x : T\to E$ and every $t\in T$ we put $x(t)=f_t$. Then for $z\in Z$ we have

$$\int \langle z, \boldsymbol{u}(t)x(t)\rangle \mathrm{d}\mu(t) = \int \langle z, \boldsymbol{u}(t)f_t\rangle \mathrm{d}\mu(t) = \int z(t)f_t(t)\mathrm{d}\mu(t) = 0 .$$

On the other hand, for $\boldsymbol{v}(t)\equiv 0$ we have also

$$\int \langle z, \boldsymbol{v}(t)x(t)\rangle \mathrm{d}\mu(t) = 0,$$

hence

$$\int\langle z, \boldsymbol{u}(t)x(t)\rangle \mathrm{d}\mu(t) = \int \langle z, \boldsymbol{v}(t)x(t)\rangle \mathrm{d}\mu(t)$$

for every $z\in Z$ and every $x : T\to Z$, but $\boldsymbol{u}(t)\neq \boldsymbol{v}(t)$ for every $t\in T$.

(2) Take $T=[0, 1]$, μ the Lebesgue measure on T, E the space of function $f: T\to R$ with $\|f\| =\Sigma_{t\in T}|f(t)| < \infty$, $Z=\mathscr{C}(T)$, hence $F=Z'$ is the space of the real measures on T. The space Z is of countable type, but E is not of countable type.

For every $t\in T$ consider the mapping $\boldsymbol{u}(t) : E\to F$ defined by $\boldsymbol{u}(t)=f(t)\varepsilon_t$ for $f\in E$, where ε_t is the measure defined by the unit mass placed at the point t. We have $\|\boldsymbol{u}(t)\|\equiv 1$ and $\langle z, \boldsymbol{u}(t)f\rangle = \langle z,f(t)\varepsilon_t\rangle = z(t)f(t)$ for every $z\in Z$ and $f\in E$.

It can be seen, as in example 1, that taking $\boldsymbol{v}(t)\equiv 0$ we have

$$\int\langle z, \boldsymbol{u}(t)x(t)\rangle \mathrm{d}\mu(t) = \int \langle z, \boldsymbol{v}(t)x(t)\rangle \mathrm{d}\mu(t)$$

for every $z\in Z$ and $x : T\to E$, but $\boldsymbol{u}(t)\neq \boldsymbol{v}(t)$ for every $t\in T$.

We give now sufficient conditions such that $\boldsymbol{u}\mu= \boldsymbol{v}\mu$ implies $\boldsymbol{u}(t)= \boldsymbol{v}(t)$, μ-almost everywhere.

17.14 PROPOSITION. *Let $\boldsymbol{u},\boldsymbol{v} : T\to \mathscr{L}^*(E, F)$ be two Z-weakly locally μ-*

integrable functions. If there exists a lifting ρ *of* $\mathscr{L}^\infty(\mu)$ *such that we have* $\rho[\boldsymbol{u}]=\boldsymbol{u}$ *and* $\rho[\boldsymbol{v}]=\boldsymbol{v}$ *and if* $\boldsymbol{u}\mu=\boldsymbol{v}\mu$, *then* $\boldsymbol{u}(t)=\boldsymbol{v}(t)$, μ*-almost everywhere.*

Assume that $\boldsymbol{u}\mu=\boldsymbol{v}\mu$. By proposition 17.13 it follows that $\boldsymbol{u}\equiv\boldsymbol{v}$. For every compact set $K\subset T$, every $x\in E$ and every $z\in Z$ we have then

$$\varphi_K(t)\langle \boldsymbol{u}(t)x, z\rangle = \varphi_K(t)\langle \boldsymbol{v}(t)x, z\rangle,\ \mu\text{-almost everywhere}$$

hence

$$\varphi_{\rho(K)}\langle \boldsymbol{u}x, z\rangle = \varphi_{\rho(K)}\langle \boldsymbol{v}x, z\rangle\ .$$

Since $\rho[\boldsymbol{u}]=\boldsymbol{u}$ and $\rho[\boldsymbol{v}]=\boldsymbol{v}$, there exists a locally countable family $(K_j)_{j\in J}$ of disjoint nonempty compact sets such that the set $T\setminus\cup K_j$ is μ-negligible, and such that

$$\rho(\varphi_{K_j}\langle \boldsymbol{u}x, z\rangle) = \varphi_{\rho(K_j)}\langle \boldsymbol{u}x, z\rangle$$

and

$$\rho(\varphi_{K_i}\langle \boldsymbol{v}x, z\rangle) = \varphi_{\rho(K_j)}\langle \boldsymbol{v}x, z\rangle$$

for every $j\in J$, $x\in E$ and $z\in Z$; therefore,

$$\varphi_{\rho(K_j)}\langle \boldsymbol{u}x, z\rangle = \varphi_{\rho(K_j)}\langle \boldsymbol{v}x, z\rangle,\ \text{for } j\in J, x\in E \text{ and } z\in Z.$$

The set $T'=\cup\rho(K_j)$ differs from T by a μ-negligible set and for $t\in T'$ we have

$$\langle \boldsymbol{u}(t)x, z\rangle = \langle \boldsymbol{v}(t)x, z\rangle \text{ for every } x\in E \text{ and } z\in Z.$$

It follows then that for $t\in T'$ we have

$$\boldsymbol{u}(t)x = \boldsymbol{v}(t)x,\ \text{for every } x\in E\ ;$$

consequently, $\boldsymbol{u}(t)=\boldsymbol{v}(t)$ for $t\in T'$, that is, μ-almost everywhere.

Remark. Proposition 17.14 is true in particular if $\boldsymbol{u}$ and $\boldsymbol{v}$ are simply locally μ-integrable.

17.15 PROPOSITION. *Let* $\boldsymbol{u}, \boldsymbol{v}: T\to\mathscr{L}^*(E, F)$ *be two Z-weakly locally* μ*-integrable functions. If E is of countable type, if there exists a countable set* $S\subset Z$ *norming for F and if* $\boldsymbol{u}\mu=\boldsymbol{v}\mu$, then $\boldsymbol{u}(t)=\boldsymbol{v}(t)$, μ*-almost everywhere.*

Since $\boldsymbol{u}\mu=\boldsymbol{v}\mu$, we have $\boldsymbol{u}\equiv\boldsymbol{v}$.

Let $x\in E$ and $s\in S$ and let $N(x, s)$ be a μ-negligible set such that for

$t \notin N(x, s)$ we have

$$\langle \boldsymbol{u}(t)x, s\rangle = \langle \boldsymbol{v}(t)x, s\rangle .$$

The set $N(x) = \cup_{s \in S} N(x, s)$ is μ-negligible and for $t \notin N(x)$ we have

$$\boldsymbol{u}(t)x = \boldsymbol{v}(t)x .$$

Let (x_n) be a sequence dense in E. The set $N = \cup N(x_n)$ is μ-negligible and for $t \notin N$ we have $\boldsymbol{u}(t) = \boldsymbol{v}(t)$.

In particular, the conditions of the proposition 17.15 are fulfilled in the following cases:

(1) Z is a Banach space of countable type and $F = Z'$;

(2) F' is of countable type and $Z = F'$;

(3) F is of countable type and $Z = F'$.

In the last case $\boldsymbol{u}$ and $\boldsymbol{v}$ are simply locally μ-integrable.

For simply locally integrable functions the following properties are deduced:

17.16 PROPOSITION. *Let $\boldsymbol{f}, \boldsymbol{g} : T \to \mathscr{L}^*(E, F)$ be two simply locally μ-integrable functions. We have $\boldsymbol{f}\mu = \boldsymbol{g}\mu$ if and only if for every $x \in E$ we have $\boldsymbol{f}(t)x = \boldsymbol{g}(t)x$, μ-almost everywhere.*

The proof is the same as for proposition 17.13, without using Z.

17.17 PROPOSITION. *Let $\boldsymbol{f}, \boldsymbol{g} : T \to \mathscr{L}^*(E, F)$ be two simply locally μ-integrable functions. If E is of countable type and if $\boldsymbol{f}\mu = \boldsymbol{g}\mu$, then, $\boldsymbol{f}(t) = \boldsymbol{g}(t)$ μ-almost everywhere.*

Let (x_n) be a sequence dense in E. For each n there exists a μ-negligible set $N(x_n)$ such that

$$\boldsymbol{f}(t)x_n = \boldsymbol{g}(t)x_n, \text{ for } t \notin N(x_n) .$$

The set $N = \cup N(x_n)$ is μ-negligible and we have

$$\boldsymbol{g}(t)x_n = \boldsymbol{g}(t)x_n , \text{ for } t \notin N \text{ and every } x_n .$$

Since (x_n) is dense in E, we deduce

$$\boldsymbol{f}(t) = \boldsymbol{g}(t), \text{ for } t \notin N .$$

§18 Absolutely continuous measures

1 *Absolutely continuous positive measures*

Let μ and ν be two positive measures on T.

18.1 DEFINITION. *We say that the measure ν is absolutely continous with respect to the measure μ if every μ-negligible set is ν-negligible.*

To say that ν is absolutely continuous with respect to μ means that $\mu^*(A)=0$ implies $\nu^*(A)=0$.

18.2 PROPOSITION. *The measure ν is absolutely continuous with respect to μ if and only if every μ-negligible function $\boldsymbol{f}: T \to E$ is ν-negligible.*

In fact, to say that $\boldsymbol{f}$ is μ-negligible (respectively ν-negligible), means that the set $A=\{t; \boldsymbol{f}(t)=0\}$ is μ-negligible (respectively ν-negligible).

18.3 DEFINITION. *The measure ν is absolutely continuous with respect to μ if and only if, for every compact set $K \subset T$ and every $\varepsilon>0$, there exists a number $\delta>0$ such that for every set $A \subset K$ with $\mu^*(A)<\delta$ we have $\nu^*(A)<\varepsilon$.*

Assume first that ν is absolutely continuous with respect to μ and show that the stated condition is satisfied. If this condition were not satisfied, there would exist a compact set $K_0 \subset T$ and a number $\varepsilon_0>0$ such that, for every natural number n, there exists a set $A_n \subset K_0$ with $\mu(A_n)<1/2^n$ and $\nu^*(A_n) \geqslant \varepsilon_0$. Since the outer measure of a relatively compact set is equal to the infimum of the measures of the open sets containing it, for each n there exists an open set $G_n \supset A_n$ such that

$$\mu^*(G_n) \leqslant \mu^*(A_n)+1/2^n \leqslant 2/2^n.$$

Denote

$$V_n = \cup_{p \geqslant 0} G_{n+p} \text{ and } V=\cap_{1 \leqslant n<\infty} V_n .$$

We have

$$\mu^*(V_n) \leqslant \Sigma_{p \geqslant 0} \mu^*(G_{n+p}) \leqslant \Sigma_{p \geqslant 0} 2/2^{n+p} \leqslant 4/2^{n+p}$$

and

$$\mu^*(V) \leqslant \mu^*(V_n) \leqslant 4/2^{n+p}, \text{ therefore } \mu^*(V) = 0.$$

On the other hand, for each n we have $V_n \supset G_n \supset A_n$ hence $\nu^*(V_n) \geqslant \nu^*(A_n) \geqslant \varepsilon_0$.

Since the sequence (V_n) is decreasing, we have

$$\nu^*(V) = \inf \nu^*(V_n) \geqslant \varepsilon_0 .$$

Thus, V is μ-negligible but it is not ν-negligible, which contradicts the assumption that ν is absolutely continuous with respect to μ.

It follows that if ν is absolutely continuous with respect to μ the stated condition is fulfilled.

Conversely, assume that this condition is fulfilled and show that ν is absolutely continuous with respect to μ.

Let A be a μ-negligible set and $K \subset T$ a compact set. For every $\varepsilon > 0$ we have $\mu^*(A \cap K) = 0 < \delta$, hence $\nu^*(A \cap K) < \varepsilon$. It follows that $\nu^*(A \cap K) = 0$, consequently $\nu^*(A) = 0$, that is, A is ν-negligible.

Thus, ν is absolutely continuous with respect to μ.

Thus the proposition is proved.

18.4 PROPOSITION. *The measure ν is absolutely continuous with respect to μ if and only if for every function $f \geqslant 0$, which is μ-integrable and ν-integrable and for every number $\varepsilon > 0$, there exists a number $\delta > 0$ such that, for every function h with $0 \leqslant h \leqslant f$ and $\int^* h \, \mathrm{d}\mu < \delta$ we have $\int^* h \, \mathrm{d}\nu < \varepsilon$.*

Assume that ν is absolutely continuous with respect to μ. If the stated condition were not satisfied, there would exist a function $f_0 \geqslant 0$ with compact support, integrable with respect to μ and ν, and a number $\varepsilon_0 > 0$ such that for every natural number n, there exists a function g_n with $0 \leqslant g_n \leqslant f_0$,

$$\int^* g_n \mathrm{d}\mu < 1/2^n \text{ and } \int^* g_n \mathrm{d}\nu \geqslant \varepsilon_0 .$$

Since the functions g_n have compact support, for each n there exists a lower semicontinuous function $g'_n \geqslant g_n$ with $\mu^*(g'_n) \leqslant \mu^*(g_n) + 1/2^n \leqslant 2/2^n$. Then the functions $g''_n = \inf(f_0, g'_n)$ are integrable with respect to μ and ν since they are measurable with respect to μ and ν and

$$\mu^*(g''_n) \leqslant \mu^*(g'_n) \leqslant 2/2^n \text{ and } \nu^*(g''_n) \leqslant \nu^*(f_0) < +\infty .$$

For each n put $h_n = \sup_{p \geqslant 0} g''_{n+p}$ and $h = \inf h_n$. We have

$$\mu^*(h_n) \leqslant \Sigma_{p \geqslant 0}(g''_{n+p}) \leqslant \Sigma_{p \geqslant 0} 2/2^{n+p} = 4/2^n .$$

Then

$$\mu^*(g) \leqslant \mu^*(h_n) \leqslant \Sigma_{p \geqslant 0}\, \mu^*(g''_{n+p}) \leqslant 4/2^n \text{ for every } n, \text{ hence } \mu^*(h) = 0\,.$$

On the other hand, $h_n \geqslant g''_n \geqslant g'_n \geqslant g_n$, hence $\nu^*(h_n) \geqslant \varepsilon_0$.

The sequence (h_n) is decreasing and it consists of ν-integrable functions, hence

$$\nu(h) = \inf \nu(h_n) \geqslant \varepsilon_0\,.$$

Thus, h is μ-negligible but it is not ν-negligible, which contradicts the assumption that ν is absolutely continuous with respect to μ.

Thus, if ν is absolutely continuous with respect to μ, the condition stated in the proposition is fulfilled.

Conversely, assume that this condition is fulfilled and show that ν is absolutely continuous with respect to μ.

Let A be a μ-negligible set and K a compact set. Taking $f = \varphi_K$ and $h = \varphi_{A \cap K}$, for every $\varepsilon > 0$ we have $\mu^*(h) \leqslant \mu^*(A \cap K) = 0 < \delta$, hence $\nu^*(A \cap K) = \nu^*(h) < \varepsilon$. Thus, $\nu^*(A \cap K) = 0$; consequently $\nu^*(A) = 0$, that is, A is ν-negligible. It follows that ν is absolutely continuous with respect to μ and thus the proposition is proved.

18.5 Proposition. *The measure ν is absolutely continuous with respect to μ if and only if every μ-negligible compact set is ν-negligible.*

If ν is absolutely continuous with respect to μ, every μ-negligible set is ν-negligible, in particular every μ-negligible compact set is ν-negligible.

Conversely, assume that every μ-negligible compact set is ν-negligible. Let A be a μ-negligible set and K a compact set.

Since

$$\mu^*(A \cap K) = \inf_{G \supset A} \mu^*(G) = 0, \; G \text{ open}\,,$$

there exists a sequence (G_n) of open sets, such that denoting $B = (\cap G_n) \cap K$ we have $B \supset A \cap K$ and $\mu^*(B) = 0$.

The set B is a relatively compact Borel set, therefore, it is integrable with respect to every measure, in particular, for the measure ν. There exists, hence, a ν-negligible set $N \subset B$ and a partition of $B \setminus N$ consisting of a sequence (K_n) of compact sets. Since $K_n \subset B$ and $\mu^*(B) = 0$ we have $\mu^*(K_n) = 0$ hence $\nu^*(K_n) = 0$. It follows then that $\nu^*(B) = 0$, hence $\nu^*(A \cap B) = 0$, consequently $\nu^*(A) = 0$.

2 The Lebesgue–Nikodym theorem

18.6 THEOREM (*Lebesgue-Nikodym*). *Let μ be a positive measure and ν a scalar measure. The measure ν is absolutely continuous with respect to μ if and only if ν is with basis μ.*

If ν is with basis μ, $\nu=g\mu$, and if A is a μ-negligible set, then $|g|\varphi_A$ is μ-negligible, hence φ_A is $|g|\mu$-negligible. From the equality $|\nu|=|g\mu|=|g|\mu$ it follows that φ_A is ν-negligible consequently ν is absolutely continuous with respect to μ.

Conversely, assume that ν is absolutely continuous with respect to μ and show that ν is with basis μ. We shall consider successively several cases.

(a) Assume first that μ and ν are *positive and bounded*. Then the measure $\lambda=\mu+\nu$ is positive and bounded, hence

$$\varphi_T\in\mathscr{L}^2(\lambda)\subset\mathscr{L}^1(\lambda)=\mathscr{L}^1(\mu)\cap\mathscr{L}^1(\nu).$$

Since $\nu\leqslant\lambda$, for every function $f\in\mathscr{L}^2(\lambda)$ we have $f\in\mathscr{L}^2(\nu)$ and

$$|\nu(f)|\leqslant\nu(|f|)\leqslant\lambda(|f|)=\int|f|\varphi_T\,\mathrm{d}\lambda=N_2(f,\lambda)N_2(\varphi_T,\lambda).$$

It follows that the linear mapping $f\to\nu(f)$ is continuous on the space $\mathscr{L}^2(\lambda)$. The mapping $\tilde{f}\to\nu(\tilde{f})$ is then a continuous linear functional in the Hilbert space $L^2(\lambda)$. There exists, hence, an element $\tilde{g}\in L^2(\lambda)$ such that for every $\tilde{f}\in L^2(\lambda)$ we have $\nu(\tilde{f})=(\tilde{f}|\tilde{g})$.

Choosing an arbitrary representative $g\in\tilde{g}$, we have

$$\nu(f)=(\tilde{f}|\tilde{g})=\int fg\,\mathrm{d}\lambda,\ \text{for } f\in\mathscr{L}^2(\lambda), \tag{1}$$

that is,

$$\nu(f)=\mu(fg)+\nu(fg),$$

whence

$$\nu(f(1-g))=\mu(fg)\ \text{for } f\in\mathscr{L}^2(\lambda). \tag{2}$$

We shall show that we have $\nu(h)=\mu(hg/(1-g))$ for $h\in\mathscr{K}(T)$, whence it will follow that ν is of basis μ, more precisely, $\nu=g(1-g)^{-1}\mu$.

For this purpose, we shall show first that we may choose the function $g\in\tilde{g}$ such that we have $g(t)\geqslant 0$ and $g(t)<1$ for every $t\in T$. First, $g(t)\geqslant 0$, λ-almost everywhere since the measure $\nu=g\lambda$ is positive; modifying the function g on a λ-negligible set (which is also ν-negligible and μ-negligible) we can obtain $g \geqslant 0$. Denote then $A = \{t | g(t) \geqslant 1\}$. The set A is λ-measurable. Since λ is bounded, A is integrable, hence $\varphi_A \in \mathscr{L}^2(\lambda)$. Writing equality (2) for the function $f = \varphi_A$, we obtain

$$\nu(\varphi_A(1-g)) = \mu(\varphi_A g) .$$

In this equality, the left-hand side is negative and the right-hand side is positive, therefore

$$\nu(\varphi_A(1-g)) = \mu(\varphi_A g) = 0 .$$

Thus, $\varphi_A g$ is μ-negligible, hence ν-negligible (since ν is absolutely continuous with respect to μ); consequently, it is also λ-negligible. Setting $g(t) = 0$ for $t \in A$ we obtain $0 \leqslant g(t) < 1$ for every $t \in T$, and equality (2) remains true also for the modified function g.

For every n, the function g^n is λ-measurable and bounded. For every function $h \geqslant 0$ from $\mathscr{K}(T)$, the function $f = h(1 + g + \ldots + g^{n-1})$ is λ-measurable and bounded, hence it belongs to the space $\mathscr{L}^2(\lambda)$. Writing equality (2) for this function, we obtain

$$\int h(1+g+\ldots+g^{n-1})(1-g)\mathrm{d}\nu = \int h(1+g+\ldots+g^{n-1})g\,\mathrm{d}\mu ,$$

that is,

$$\int h(1-g^n)\mathrm{d}\nu = \int h(g+g^2+\ldots+g^n)\mathrm{d}\mu =$$
$$= \int hg\,\mathrm{d}\mu + \int hg^2\,\mathrm{d}\mu + \ldots + \int hg^n\,\mathrm{d}\mu .$$

But

$$\int h(1-g^n)\,\mathrm{d}\nu \leqslant \int h\mathrm{d}\nu < +\infty ,$$

hence

$$\Sigma_{1 \leqslant k \leqslant n} \int hg^k\,\mathrm{d}\mu \leqslant \int h\mathrm{d}\nu < +\infty .$$

It follows that the function $\Sigma_{1 \leqslant n < \infty} hg^n = hg/(1-g)$ is μ-integrable and

$$\int hg(1-g)^{-1}\,\mathrm{d}\mu = \lim_n \int \Sigma_{1 \leqslant k \leqslant n} hg^k\,\mathrm{d}\mu = \lim_n \int h(1-g^n)\mathrm{d}\nu .$$

On the other hand, the sequence $(1-g^n)h$ is increasing and

$$\sup(1-g^n)h = \lim(1-g^n)h = h ,$$

consequently

$$\int h\mathrm{d}\nu = \sup \int (1-g^n)h\mathrm{d}\nu = \lim \int (1-g^n)h\mathrm{d}\nu .$$

It follows that

$$\int h \mathrm{d}\nu = \int hg(1-g)^{-1} \mathrm{d}\mu .$$

This equality remains true also for every function $h \in \mathscr{K}(T)$ (not necessarily positive).

Setting $g_0 = g/(1-g)$ we have $g_0 \geqslant 0$, the function g_0 is μ-integrable and $\nu = g_0 \mu$.

(b) Now let μ and ν be two arbitrary *positive* measures, ν absolutely continuous with respect to μ.

Let $(K_\alpha)_{\alpha \in A}$ be a locally countable family of mutually disjoint compact subsets such that the set $N = T \setminus \cup K_\alpha$ is μ-negligible. Then N is also ν-negligible, hence writing $M = \cup K_\alpha$ we have $\mu = \varphi_M \mu$ and $\nu = \varphi_M \nu$. For each $\alpha \in A$, the measures $\mu_\alpha = \varphi_{K_\alpha} \mu$ and $\nu_\alpha = \varphi_{K_\alpha} \nu$ are bounded and ν_α is absolutely continuous with respect to μ_α. From the first part it follows that there exists a μ_α-integrable function $g'_\alpha \geqslant 0$ such that we have $\nu_\alpha = g'_\alpha \mu_\alpha$. From the associativity property we deduce that the function $g_\alpha = g'_\alpha \varphi_{K_\alpha}$ is μ-integrable and $g'_\alpha \mu_\alpha = g'_\alpha(\varphi_{K_\alpha}\mu) = g_\alpha \mu$, consequently $\nu_\alpha = g_\alpha \mu$.

Define the function g on T by $g(t) = g_\alpha(t)$ if $t \in K_\alpha$ and $g(t) = 0$ if $t \notin M$.

The function g is positive, locally μ-integrable and for each $\alpha \in A$ we have $g\varphi_{K_\alpha} = g_\alpha \varphi_{K_\alpha}$. If we put $\nu' = g\mu$, for every $\alpha \in A$ we have

$$\varphi_{K_\alpha} \nu' = (\varphi_{K_\alpha} g)\mu = g_\alpha \mu = \nu_\alpha = \varphi_{K_\alpha} \nu .$$

It follows that $\varphi_M \nu' = \varphi_M \nu$. Since $N = T \setminus M$ is μ-negligible it follows that N is ν-negligible, hence $\nu' = \nu$, consequently $\nu = g\mu$.

(c) Assume now that ν is a real measure absolutely continuous with respect to μ. Then ν^+ and ν^- are positive measures absolutely continuous with respect to μ. There exists two locally μ-integrable functions $g_1 \geqslant 0$ and $g_2 \geqslant 0$ such that

$$\nu^+ = g_1 \mu \text{ and } \nu^- = g_2 \mu .$$

Then the function $g = g_1 - g_2$ is locally μ-integrable, and

$$\nu = \nu^+ - \nu^- = g_1 \mu - g_2 \mu = (g_1 - g_2)\mu = g\mu .$$

(d) If ν is a complex measure absolutely continuous with respect to μ and if $\nu = \nu_1 + i\nu_2$, then ν_1 and ν_2 are real measures, absolutely continuous with respect to μ. There exists then two locally μ-integrable real functions g_1 and g_2 such that

$$\nu_1 = g_1 \mu \text{ and } \nu_2 = g_2 \mu .$$

Then the function $g=g_1+ig_2$ is locally μ-integrable and

$$\nu=\nu_1+i\nu_2=g_1\mu+ig_2\mu=(g_1+ig_2)\mu=g\mu\,.$$

Thus the theorem is completely proved.

18.7 Proposition. *If μ is a scalar measure, then there exists a locally μ-integrable scalar function φ with $|\varphi(t)|\equiv 1$ and $\mu=\varphi|\mu|$. The function φ^{-1} is also locally μ-integrable and we have $|\mu|=\varphi^{-1}\mu$.*

In fact, μ is absolutely continuous with respect to $|\mu|$, hence, there exists a locally μ-integrable scalar function φ with $\mu=\varphi|\mu|$. By theorem 17.3, we have

$$|\mu|=|\varphi|\mu||=|\varphi|\,|\mu|,$$

hence by proposition 17.5 we have $|\varphi(t)|=1$ μ-almost everywhere. Modifying the function φ on a μ-negligible set we can obtain $|\varphi(t)|\equiv 1$. The function φ^{-1} is μ-measurable and $|[\varphi(t)]^{-1}|\equiv 1$, therefore φ^{-1} is locally μ-integrable. From proposition 16.26, we deduce then

$$\varphi^{-1}\mu=\varphi^{-1}(\varphi|\mu|)=(\varphi^{-1}\varphi)|\mu|=|\mu|\,.$$

18.8 Corollary. *If μ is a real measure there exist two disjoint μ-measurable sets A and B with $A\cup B=T$ such that*

$$\mu^+=\varphi_A|\mu|=\varphi_A\mu \text{ and } \mu^-=\varphi_B|\mu|=-\varphi_B\mu\,.$$

By proposition 18.7, there exists a locally μ-integrable real function φ with $\mu=\varphi|\mu|$, $|\mu|=\varphi^{-1}\mu$ and $|\varphi(t)|\equiv 1$. The function φ takes on only the values 1 and -1. Put $A=\{t;\,\varphi(t)=1\}$ and $B=\{t;\,\varphi(t)=-1\}$. The sets A and B are μ-measurable, disjoint, $A\cup B=T$ and $\varphi=\varphi_A-\varphi_B$, hence $\mu=\varphi_A|\mu|-\varphi_B|\mu|$ and $|\mu|=\varphi_A|\mu|+\varphi_B|\mu|$. It follows that $\mu^+=\varphi_A|\mu|$ and $\mu^-=\varphi_B|\mu|$.

In this case we have $\varphi^{-1}=\varphi=\varphi_A-\varphi_B$; consequently $|\mu|=\varphi\mu$, whence $\mu^+=\varphi_A|\mu|=\varphi_A\varphi\mu=\varphi_A\mu$ and $\mu^-=\varphi_B|\mu|=\varphi_B\varphi\mu=-\varphi_B\mu$.

3 *Equivalent measures*

18.9 Definition. *We say that two dominated measures $\boldsymbol{m},\boldsymbol{n}:\mathscr{K}(T)\to E$ are equivalent if the negligible sets are the same for $\boldsymbol{m}$ and $\boldsymbol{n}$.*

To say that $\boldsymbol{m}$ and $\boldsymbol{n}$ are equivalent means that the positive measures $|\boldsymbol{m}|$ and $|\boldsymbol{n}|$ are equivalent. It follows that if $\boldsymbol{m}$ and $\boldsymbol{n}$ are equivalent, the measurable functions are the same for $\boldsymbol{m}$ and $\boldsymbol{n}$.

To say that two positive measures μ and ν are equivalent means that each of them is absolutely continuous with respect to the other one.

18.10 PROPOSITION. *Two positive measures μ and ν are equivalent if and only if there exists a locally μ-integrable function g such that we gave $\nu = g\mu$ and $g(t) > 0$, μ-almost everywhere. We have then $g^{-1}\nu = \mu$.*

Assume that μ and ν are equivalent. Then ν is absolutely continuous with respect to μ and μ is absolutely continuous with respect to ν. There exists then a locally μ-integrable function $g \geqslant 0$ such that $\nu = g\mu$ and a locally μ-integrable function $h \geqslant 0$ such that $\mu = h\nu$. The function hg is locally μ-integrable and $\mu = h(g\mu) = (hg)\mu$. It follows that $h(t)g(t) = 1$, μ-almost everywhere, consequently $g(t) > 0$ and $h(t) = [g(t)]^{-1}$, μ-almost everywhere.

Conversely, assume that $\nu = g\mu$, where g is locally μ-integrable and $g(t) > 0$, μ-almost everwhere. The function $g^{-1}g$ is defined μ-almost everywhere and it is locally μ-integrable; then g^{-1} is locally $g\mu = \nu$-integrable and $g^{-1}\nu = g^{-1}(g\mu) = \mu$.

Thus ν is absolutely continuous with respect to μ and μ is absolutely continuous with respect to ν, consequently μ and ν are equivalent.

18.11 PROPOSITION. *If T is countable at infinity, then for every positive measure μ on T there exists a continuous function g on T such that $g(t) > 0$ for every $t \in T$ and the measure $\nu = g\mu$ is bounded.*

Let (K_n) be a sequence of compact sets whose union is T. For each n, let $f_n : T \to [0, 1]$ be a function of $\mathscr{K}(T)$ such that $f_n(t) = 1$ for $t \in K_n$. For each n, set

$$a_n = 1/(2^n \mu(f_n)) \text{ if } \mu(f_n) > 1 \,,$$

$$a_n = 1/2^n, \qquad \text{if } \mu(f_n) \leqslant 1 \,.$$

It follows that

$$\Sigma\, a_n \leqslant \Sigma\, 2^{-n} < \infty$$

and

$$\Sigma\, a_n \mu(f) \leqslant \Sigma\, 2^{-n} < \infty \,.$$

The series $\Sigma a_n f_n$ is uniformly convergent on T, hence its sum $g = \Sigma a_n f_n$ is a continuous function and $g(t) > 0$ for each $t \in T$.

For the measure $\nu = g\mu$ we have

$$\nu^*(\varphi_T) = \mu^*(g\varphi_T) = \mu^*(g) = \mu^*(\Sigma a_n f_n) \leqslant \Sigma a_n \mu(f_n) < \infty ,$$

consequently the measure $g\mu$ is bounded.

18.12 Proposition. *For every sequence (μ_n) of positive bounded measures on T, there exists a positive bounded measure μ on T having the following properties:*

(1) *every measure μ_n is absolutely continuous with respect to μ;*

(2) *$\mu(A) = 0$ if and only if $\mu_n(A) = 0$ for every n;*

(3) *every positive measure μ' on T with properties* (1) *and* (2) *is equivalent to μ.*

We may assume that $\mu_n \neq 0$ for every n, omitting from the sequence, if necessary, the measures equal to zero. For every function φ from $\mathscr{K}(T)$ we have $|\mu_n(\varphi)| \leqslant \mu_n(|\varphi|) \leqslant ||\mu_n|| \, ||\varphi||$, consequently

$$\Sigma (2^n ||\mu_n||)^{-1} |\mu_n(\varphi)| \leqslant \Sigma 2^{-n} ||\varphi|| < \infty .$$

Put

$$\mu(\varphi) = \Sigma (2^n \|\mu_n\|)^{-1} \mu_n(\varphi) .$$

It is easy to see that μ is a positive linear functional on $\mathscr{K}(T)$, hence it is a positive measure on T.

For every n we have $(2^n \|\mu_n\|)^{-1} \mu_n \leqslant \mu$, hence μ_n is absolutely continuous with respect to μ.

For every function $\varphi \in \mathscr{K}(T)$ with $||\varphi|| \leqslant 1$ we have

$$|\mu(\varphi)| = |\Sigma (2^n ||\mu_n||)^{-1} \mu_n(\varphi)| \leqslant \Sigma (2^n ||\mu_n||)^{-1} |\mu_n(\varphi)| \leqslant$$

$$\leqslant \Sigma (2^n ||\mu_n||)^{-1} ||\mu_n|| \, ||\varphi|| \leqslant \Sigma 2^{-n} ,$$

whence

$$\|\mu\| \leqslant \Sigma 2^{-n} < \infty$$

consequently μ is bounded.

We have then $\mu^*(A) = 0$ if and only if $\mu_n^*(A) = 0$ for every n (see proposition 19.8; paragraph 19 may be read independently of paragraph 18). Finally, if μ' fulfills condition (2), then the negligible sets are the same for μ and μ', consequently μ and μ' are equivalent.

4 *Absolutely continuous vector measures*

From proposition 18.5 we deduce that a positive measure ν is absolutely continuous with respect to a positive measure μ if and only if every relatively compact μ-negligible Borel set is ν-negligible.

For vector measures, this property is taken as a definition.

18.13 DEFINITION. *We say that a vector measure* $\mathbf{m}$ *is absolutely continuous with respect to a positive measure* μ *if for every relatively compact Borel set* A *with* $\mu(A)=0$ *we have* $\mathbf{m}(A)=0$.

For the dominated measures, the study of the absolute continuity is reduced to that for positive measures.

18.14 PROPOSITION. *A dominated measure* $\mathbf{m}$ *is absolutely continuous with respect to a positive measure* μ *if and only if the positive measure* $|\mathbf{m}|$ *is absolutely continuous with respect to* μ.

If $|\mathbf{m}|$ is absolutely continuous with respect to μ, then for every relatively compact Borel set A with $\mu(A)=0$ we have $|\mathbf{m}|(A)=0$, hence $\mathbf{m}(A)=0$ since $|\mathbf{m}(A)| \leqslant |\mathbf{m}|(A)$; it follows that $\mathbf{m}$ is absolutely continuous with respect to μ.

Conversely, assume that $\mathbf{m}$ is absolutely continuous with respect to μ and let A be a relatively compact Borel set with $\mu(A)=0$. For every Borel set $B \subset A$ we have $\mu(B)=0$, hence $\mathbf{m}(B)=0$. Then (proposition 8.60)

$$|\mathbf{m}|(A) = \sup \Sigma |\mathbf{m}(A_i)| = 0\,,$$

where the supremum is taken for all the finite partitions (A_i) of A with Borel sets. Thus $|\mathbf{m}|$ is absolutely continuous with respect to μ and the proposition is proved.

If a measure $\mathbf{m}$ is of the form $\mathbf{m}=\mathbf{u}\mu$, where μ is a positive measure, then, evidently, $\mathbf{m}$ is absolutely continuous with respect to μ.

Conversely, we shall prove that if $\mathbf{m}$ is absolutely continuous with respect to μ, then $\mathbf{m}$ is of basis μ.

We shall prove first that every dominated measure has as a basis its modulus.

18.15 THEOREM. *Let* X *be a Banach space,* $\mathbf{m}:\mathscr{K}(T)\to X$ *a dominated measure and* μ *its modulus.*

Let E and F be two Banach spaces with $X \subset \mathscr{L}(E, F)$ and let Z be a subspace of F′ norming for F. There exists then a function $U_{\boldsymbol{m}} : T \to \mathscr{L}(E, Z')$ having the following properties:

(1) $|U_{\boldsymbol{m}}(t)| = 1$, *μ-almost everywhere;*

(2) $U_{\boldsymbol{m}}$ *is Z-weakly locally μ-integrable and we have* $\boldsymbol{m} = U_{\boldsymbol{m}}\mu$, *hence*

$$\langle \int f \mathrm{d}\boldsymbol{m}, z\rangle = \int \langle U_{\boldsymbol{m}}(t) f(t), z\rangle \mathrm{d}\mu, \text{ for } f \in \mathscr{L}^1_E(\mu) \text{ and } z \in Z;$$

(3) *if ρ is a lifting of* $\mathscr{L}^\infty(\mu)$, *we can choose* $U_{\boldsymbol{m}}$ *uniquely such that* $\rho(U_{\boldsymbol{m}}) = U_{\boldsymbol{m}}$, *that is,*

$$\rho(\langle U_{\boldsymbol{m}} x, z\rangle) = \langle U_{\boldsymbol{m}} x, z\rangle \text{ for every } x \in E \text{ and } z \in Z;$$

(4) *we can choose* $U_{\boldsymbol{m}}(t) \in \mathscr{L}(E, F)$ *for every* $t \in T$ *in each of the following cases:*
 (a) $F = Z'$;
 (b) *for each* $x \in E$, *the convex balanced hull of the set* $A_x = \{\boldsymbol{m}(\varphi x); \varphi \in \mathscr{K}(T), \mu(|\varphi|) \leqslant 1\}$ *is relatively compact in F for the topology* $\sigma(F, Z)$.

Proof. Let ρ be lifting of $\mathscr{L}^\infty(\mu)$. For every $x \in E$ and every $z \in Z$ we define the scalar measure $\boldsymbol{m}_{x,z}$ by the equality

$$\boldsymbol{m}_{x,z}(\varphi) = \langle \boldsymbol{m}(\varphi x), z\rangle, \text{ for } \varphi \in \mathscr{K}(T).$$

We have

$$|\boldsymbol{m}_{x,z}| \leqslant |x|\,|z|\,\mu.$$

There exists then a bounded μ-measurable scalar function $g_{x,z}$ defined on T such that

$$\boldsymbol{m}_{x,z} = g_{x,z}\mu.$$

Modifying the function $g_{x,z}$ on a μ-negligible set, we may assume that

$$\rho(g_{x,z}) = g_{x,z}.$$

From this equality and from the relations $|\boldsymbol{m}_{x,z}| = |g_{x,z}|\mu \leqslant |x|\,|z|\mu$ we deduce that

$$|g_{x,z}(t)| \leqslant |x|\,|z|, \text{ for every } t \in T.$$

The mapping $(x, z) \to \boldsymbol{m}_{x,z}$ of $E \times Z$ into the space of scalar measures is bilinear. Since $\rho(g_{x,z})=g_{x,z}$, we deduce that the mapping $(x, z)\to g_{x,z}$ of $E\times Z$ into the space of the μ-measurable bounded scalar functions is bilinear, hence for each $t\in T$ the mapping $(x, z)\to g_{x,z}(t)$ is a *bilinear functional* on $E\times Z$.

For fixed $t\in T$ and $x\in E$, the mapping $g_x(t):z\to g_{x,z}(t)$ is a continuous linear functional on Z, hence $g_x(t)\in Z'$ and we have

$$|g_x(t)|\leqslant |x|\,.$$

For fixed $t\in T$, the mapping $U_{\boldsymbol{m}}(t):x\to g_x(t)$ of E into Z' is linear and continuous, hence $U_{\boldsymbol{m}}(t)\in\mathscr{L}(E, Z')$ and we have

$$|U_{\boldsymbol{m}}(t)|\leqslant 1\,.$$

Then

$$\langle U_{\boldsymbol{m}}(t)x, z\rangle = g_{x,z}(t), \text{ for } t\in T,\ x\in E \text{ and } z\in Z\,.$$

Using the equality $\rho(g_{x,z})=g_{x,z}$, we deduce that

$$\rho(U_{\boldsymbol{m}})= U_{\boldsymbol{m}}\,.$$

The uniqueness of $U_{\boldsymbol{m}}$ follows now from property (3) of section 3, §12.

From proposition 13.18, we deduce that the function $t\to|U_{\boldsymbol{m}}(t)|$ is μ-measurable; since it is also bounded, this function is locally μ-integrable, hence $U_{\boldsymbol{m}}$ is Z-weakly locally μ-integrable. For $\varphi\in\mathscr{K}(T)$, $x\in E$ and $z\in Z$ we have

$$\langle \int\varphi x\,\mathrm{d}\boldsymbol{m}, z\rangle = \int\varphi\,\mathrm{d}\boldsymbol{m}_{x,z} = \int\varphi g_{x,z}\,\mathrm{d}\mu = \int\langle U_{\boldsymbol{m}}(t)\varphi(t)x, z\rangle\,\mathrm{d}\mu\,,$$

consequently $\boldsymbol{m}=U_{\boldsymbol{m}}\mu$.

If $\boldsymbol{f}\in\mathscr{L}^1_E(\mu)$, then $\boldsymbol{f}\in\mathscr{L}^1_E(|U_{\boldsymbol{m}}|\mu)$ since $|U_{\boldsymbol{m}}|\mu\leqslant\mu$. By theorem 16.16 it follows then that

$$\langle\int \boldsymbol{f}\,\mathrm{d}\boldsymbol{m}, z\rangle = \int\langle U_{\boldsymbol{m}}(t)\boldsymbol{f}(t), z\rangle\,\mathrm{d}\mu \text{ for } \boldsymbol{f}\in\mathscr{L}^1_E(\mu) \text{ and } z\in Z.$$

Taking $\boldsymbol{f}=\varphi x$ with $\varphi\in\mathscr{K}(T)$ and $x\in E$, we deduce

$$|\boldsymbol{m}(\varphi)|\leqslant\int|U_{\boldsymbol{m}}|\,|\varphi|\,\mathrm{d}\mu=\int|\varphi|\,\mathrm{d}(|U_{\boldsymbol{m}}|\mu)\,,$$

that is, $|\boldsymbol{m}|\leqslant|U_{\boldsymbol{m}}|\mu$. Since $|\boldsymbol{m}|=\mu$, we deduce that $|U_{\boldsymbol{m}}(t)|\geqslant 1$, μ-almost everywhere, hence

$$|U_{\boldsymbol{m}}(t)|=1 \ \ \mu\text{-almost everywhere}\,.$$

Thus, the first three points of the theorem are proved. Let us prove point (4). The case $F = Z'$ is trivial. Assume now that condition (b) of point (4) is fulfilled and let $x \in E$. Denote by A the closure in the topology $\sigma(F, Z)$ of the convex balanced hull of the set $\{\boldsymbol{m}(\varphi x); \varphi \in \mathscr{K}(T), \mu(|\varphi|) \leqslant 1\}$; by hypothesis (b), the set A is compact for the topology $\sigma(F, Z)$. The set A is also compact in the algebraic dual Z^* for the topology $\sigma(Z^*, Z)$; consequently, there exists a family $(z_i)_{i \in I}$ of elements of Z such that

$$A = \cap_{i \in I} \{y; y \in Z^*, |\langle y, z_i \rangle| \leqslant 1\} .$$

Then, for every $i \in I$ and every $\varphi \in \mathscr{K}(T)$ with $\mu(|\varphi|) \leqslant 1$, we have

$$|\int \langle \varphi x \, \mathrm{d}\boldsymbol{m}, z_i \rangle| \leqslant 1 .$$

It follows that

$$|\int \langle U_{\boldsymbol{m}}(t) x, z_i \rangle \varphi(t) \mathrm{d}\mu| \leqslant 1$$

for every $i \in I$ and $\varphi \in \mathscr{K}(T)$ with $\int |\varphi| \mathrm{d}\mu \leqslant 1$, hence

$$|\langle U_{\boldsymbol{m}}(t) x, z_i \rangle| \leqslant 1, \quad \nu\text{-almost everywhere}$$

for each $i \in I$. Since

$$\rho(|\langle U_{\boldsymbol{m}} x, z_i \rangle|) = |\rho(\langle U_{\boldsymbol{m}} x, z_i \rangle)| = |\langle U_{\boldsymbol{m}} x, z_i \rangle| ,$$

we deduce that

$$|\langle U_{\boldsymbol{m}}(t) x, z_i \rangle| \leqslant 1, \text{ for every } i \in I \text{ and every } t \in T ,$$

consequently

$$U_{\boldsymbol{m}}(t) x \in A \subset F, \text{ for every } t \in T.$$

As $x \in E$ is arbitrary, it follows that $U_{\boldsymbol{m}}(t) \in \mathscr{L}(E, F)$, for every $t \in T$.

Remark. The function $U_{\boldsymbol{m}}$ depends on the spaces E and F as well as on the space Z. If we take $Z = F'$, then $U_{\boldsymbol{m}}(t) \in \mathscr{L}(E, F'')$ for each $t \in T$. In this case if $Z_1 \subset F'$ is an arbitrary subspace, norming for F, and if $U_1 : T \to \mathscr{L}(E, Z_1')$ is the corresponding function, then for every $t \in T$ and every $x \in E$, the functional $U_1(t) x \in Z_1'$ is the restriction to Z_1 of the functional $U_{\boldsymbol{m}}(t) x \in F''$. We denote the function U_1 still by $U_{\boldsymbol{m}}$.

We are able now to give a generalization of the Lebesgue–Nikodym theorem.

18.16 THEOREM. *Let ν be a scalar measure, X a Banach space, $\boldsymbol{m} : \mathscr{K}(T) \to X$ a dominated measure, absolutely continuous with respect to ν. Let E and F be two Banach spaces with $X \subset \mathscr{L}(E, F)$ and let Z be a subspace of F' norming for F. There exists then a function $V_{\boldsymbol{m}} : T \to \mathscr{L}(E, Z')$ having the following properties:*

(1) *The function $|V_{\boldsymbol{m}}|$ is locally μ-integrable and we have $|\boldsymbol{m}| = |V_{\boldsymbol{m}}|\,|\nu|$ that is,*

$$\int \varphi \, \mathrm{d}|\boldsymbol{m}| = \int |V_{\boldsymbol{m}}| \varphi \, \mathrm{d}|\nu|, \textit{ for } \varphi \in \mathscr{L}^1(\boldsymbol{m}).$$

(2) *The function $V_{\boldsymbol{m}}$ is Z-weakly locally μ-integrable and we have $\boldsymbol{m} = V_{\boldsymbol{m}} \nu$, hence*

$$\langle \textstyle\int \boldsymbol{f} \mathrm{d}\boldsymbol{m}, z \rangle = \int \langle V_{\boldsymbol{m}}(t) \boldsymbol{f}(t), z \rangle \mathrm{d}\nu \textit{ for } \boldsymbol{f} \in \mathscr{L}^1_E(\boldsymbol{m}) \textit{ and } z \in Z.$$

(3) *If ρ is a lifting of $\mathscr{L}^\infty(\nu)$, we can choose $V_{\boldsymbol{m}}$ uniquely ν-almost everywhere such that $\rho[V_{\boldsymbol{m}}] = V_{\boldsymbol{m}}$.*
If, in addition, there exists $\alpha > 0$ with $|\boldsymbol{m}| \leqslant \alpha |\nu|$, then we can choose $V_{\boldsymbol{m}}$ uniquely such that $\rho(V_{\boldsymbol{m}}) = V_{\boldsymbol{m}}$, that is

$$\rho(\langle V_{\boldsymbol{m}}, x, z \rangle) = \langle V_{\boldsymbol{m}} x, z \rangle, \textit{ for } x \in E \textit{ and } z \in Z.$$

(4) *We can choose $V_{\boldsymbol{m}}(t) \in \mathscr{L}(E, F)$ for every $t \in T$ in each of the following cases:*
   (a) *$F = Z'$;*
   (b) *for every $x \in E$, the convex balanced hull of the set $A = \{m(\varphi x); \varphi \in \mathscr{K}(T), \int |\varphi| \mathrm{d}|\boldsymbol{m}| \leqslant 1\}$ is relatively compact for the topology $\sigma(F, Z)$;*
   (c) *for each $x \in E$, the convex balanced hull of the set $B_x = \{\boldsymbol{m}(\varphi x); \varphi \in \mathscr{K}(T), \int |\varphi| \mathrm{d}|\nu| \leqslant 1\}$ is relatively compact in F for the topology $\sigma(F, Z)$.*

Denote $\mu = |\boldsymbol{m}|$. Since μ is absolutely continuous with respect to $|\nu|$, there exists a locally ν-integrable function $g \geqslant 0$ with $\mu = g|\nu|$ (theorem 18.6).

By proposition 18.7, we deduce that there exists a locally ν-integrable scalar function φ with $|\varphi(t)| \equiv 1$, $\nu = \varphi|\nu|$ and $|\nu| = \varphi^{-1}\nu$. The function $g\varphi^{-1}$ is then locally ν-integrable and $\mu = g\varphi^{-1}\nu$.

Let $U_{\boldsymbol{m}} : T \to \mathscr{L}(E, Z')$ be the function which corresponds to the measure $\boldsymbol{m}$ by theorem 18.15 and put $V = U_{\boldsymbol{m}} g \varphi^{-1}$. Since $|U_{\boldsymbol{m}}(t)| = 1$, μ-almost everywhere, the function $|U_{\boldsymbol{m}}| - 1$ is μ-negligible, hence the function $(|U_{\boldsymbol{m}}| - 1)g$ is ν-negligible, consequently $|U_{\boldsymbol{m}}(t)|\, g(t) = g(t)$, ν-almost everywhere. It follows then that $|V(t)| = g(t)$, ν-almost everywhere, consequently $|V|$ is locally ν-integrable and we have $\mu = |V|\,|\nu|$.

Taking $V_{\boldsymbol{m}} = V$, point (1) is proved.

Since $U_{\boldsymbol{m}}$ is Z-weakly locally μ-measurable, it follows that for every $x \in E$ and every $z \in Z$ the function $\langle U_{\boldsymbol{m}} x, z\rangle$ is μ-measurable; then the function $\langle U_{\boldsymbol{m}} x, z\rangle g$ is ν-measurable, hence the function $\langle Vx, z\rangle = \langle U_{\boldsymbol{m}} x, z\rangle g \varphi^{-1}$ is ν-measurable. This means that V is Z-weakly ν-measurable. Since $|V|$ is locally ν-integrable, it follows that V is Z-weakly locally ν-integrable.

For every $\boldsymbol{f} \in \mathscr{L}^1_E(\mu)$ and every $z \in Z$ we have

$$\langle \textstyle\int \boldsymbol{f} \, \mathrm{d}\boldsymbol{m}, z\rangle = \int \langle U_{\boldsymbol{m}} \boldsymbol{f}, z\rangle \mathrm{d}\mu = \int \langle U_{\boldsymbol{m}} g \varphi^{-1} \boldsymbol{f}, z\rangle \mathrm{d}\nu = \int \langle V\boldsymbol{f}, z\rangle \mathrm{d}\nu \,.$$

Taking $V_{\boldsymbol{m}} = V$, point (2) is as proved. Now let ρ be a lifting of $\mathscr{L}^\infty(\nu)$. From proposition 13.22, we deduce that there exists a function $V_{\boldsymbol{m}} : T \to \mathscr{L}^*(E, Z')$ with $\rho[V_{\boldsymbol{m}}] = V_{\boldsymbol{m}}$, $\langle V_{\boldsymbol{m}} x, z\rangle = \langle Vx, z\rangle$ for every $x \in E$ and $z \in Z$ (that is $V_{\boldsymbol{m}} \equiv V$) and $|V_{\boldsymbol{m}}(t)| \leqslant |V(t)|$ for every $t \in T$. The function $|V_{\boldsymbol{m}}|$ is ν-measurable, consequently, from the inequality $|V_{\boldsymbol{m}}| \leqslant |V|$, it follows that $|V_{\boldsymbol{m}}|$ is Z-weakly locally ν-integrable.

From the relation $V_{\boldsymbol{m}} \equiv V$ we deduce that $V_{\boldsymbol{m}}$ is Z-weakly locally ν-integrable and

$$\langle \textstyle\int \boldsymbol{f} \, \mathrm{d}\boldsymbol{m}, z\rangle = \int \langle V_{\boldsymbol{m}} \boldsymbol{f}, z\rangle \mathrm{d}\nu, \text{ for } \boldsymbol{f} \in \mathscr{L}^1_E(\boldsymbol{m}) \text{ and } z \in Z \,.$$

Taking $\varphi \in \mathscr{K}(T)$, $x \in E$ and $z \in Z$ we deduce

$$|\langle \boldsymbol{m}(\varphi) x, z\rangle| \leqslant \textstyle\int |V_{\boldsymbol{m}}|\,|\varphi|\,|x|\,|z|\,\mathrm{d}|\nu| \;;$$

hence, taking the supremum for $x \in E$ with $|x| \leqslant 1$ and $z \in Z$ with $|z| \leqslant 1$,

$$|\boldsymbol{m}(\varphi)| \leqslant \textstyle\int |V_{\boldsymbol{m}}|\,|\varphi|\,\mathrm{d}|\nu| = \int |\varphi|\,\mathrm{d}(|V_{\boldsymbol{m}}|\,|\nu|) \,.$$

It follows that $|\boldsymbol{m}| \leqslant |V_{\boldsymbol{m}}|\,|\nu| \leqslant |V|\,|\nu| = \mu$, consequently $\mu = |V_{\boldsymbol{m}}|\,|\nu|$. Assume finally that there exists $\alpha > 0$ with $\mu \leqslant \alpha|\nu|$.

Then, the function g belongs to the space $\mathscr{L}^\infty(\mu)$, consequently, $|V| \in \mathscr{L}^\infty(\nu)$. From corollary 13.21, it follows that there exists a function $V_{\boldsymbol{m}} : T \to \mathscr{L}^\infty(E, Z')$ with $\rho(V_{\boldsymbol{m}}) = V_{\boldsymbol{m}}$, $V_{\boldsymbol{m}} \equiv V$ and $|V_{\boldsymbol{m}}(t)| \leqslant |V(t)|$, ν-almost everywhere. It can be proved then that $V_{\boldsymbol{m}}$ has all the properties required for points (1) and (2) as in the case $\rho[V_{\boldsymbol{m}}] = V_{\boldsymbol{m}}$.

In this way point (3) has also been proved.

If conditions (a) and (b) of point (4) are fulfilled, then $U_{\boldsymbol{m}}$ has values in $\mathscr{L}(E, F)$ hence also $V_{\boldsymbol{m}}$ has values in $\mathscr{L}(E, F)$. If the condition (c) is fulfilled, it is proved that $V_{\boldsymbol{m}}$ has values in $\mathscr{L}(E, F)$ in the same way as in theorem 18.15.

Thus the theorem 18.16 is completely proved.

Remark. The function $V_{\boldsymbol{m}}$ depends on the spaces E and F as well as on the chosen space Z.

5 *The case* $|\boldsymbol{g}\,\boldsymbol{m}| = |\boldsymbol{g}|\,|\boldsymbol{m}|$

For a given measure $\boldsymbol{g}\boldsymbol{m}$, we have, in general, only the inequality $|\boldsymbol{g}\,\boldsymbol{m}| \leqslant |\boldsymbol{g}|\,|\boldsymbol{m}|$.

But if the measure $\boldsymbol{m}$ is scalar or if the function $\boldsymbol{g}$ is scalar we have the equality $|\boldsymbol{g}\,\boldsymbol{m}| = |\boldsymbol{g}|\,|\boldsymbol{m}|$.

We shall consider first the case of measures with scalar bases.

Let E and F be two Banach spaces and $Z \subset F'$ a subspace, norming for F.

18.17 THEOREM. *Let μ be a scalar measure and $\boldsymbol{g}: T \to \mathscr{L}^*(E, F)$ a Z-weakly locally μ-integrable function. We have*

$$|\boldsymbol{g}\mu| = |\boldsymbol{g}|\,|\mu|$$

in each of the following cases:

(1) *there exists a lifting ρ of $\mathscr{L}^\infty(\mu)$ such that $\rho[\boldsymbol{g}] = \boldsymbol{g}$;*

(2) *E is of countable type and there exists a countable set $S \subset Z$, norming for F;*

(3) *E is of countable type and $\boldsymbol{g}$ is simply locally μ-integrable;*

(4) *$\boldsymbol{g}$ is locally μ-integrable.*

Denote $\boldsymbol{m} = \boldsymbol{g}\mu$. The measure $\boldsymbol{m}$ has values in $\mathscr{L}(E, Z')$ and is absolutely continuous with respect to μ. By theorem 18.16 there exists a Z-weakly locally μ-integrable function $V: T \to \mathscr{L}(E, Z')$ such that $\boldsymbol{m} = V\mu$ and $|\boldsymbol{m}| = |V|\,|\mu|$.

In addition, for every lifting ρ of $\mathscr{L}^\infty(\mu)$ we may choose V such that $\rho[V] = V$. Then we have also $\rho[Vx] = Vx$ for every $x \in E$.

Assume first that there exists a lifting ρ of $\mathscr{L}^\infty(\mu)$ such that $\rho[\boldsymbol{g}] = \boldsymbol{g}$ (condition (1)). Then choose V such that $\rho[V] = V$. From the equality $\boldsymbol{m} = V\mu = \boldsymbol{g}\mu$ we deduce then (proposition 17.14) that $V(t) = \boldsymbol{g}(t)$, μ-almost everywhere, hence $|V(t)| = |\boldsymbol{g}(t)|$, μ-almost everywhere, consequently

$$|\boldsymbol{g}\mu| = |\boldsymbol{m}| = |V|\,|\mu| = |\boldsymbol{g}|\,|\mu|\,.$$

Assume then that condition (2) is fulfilled. From the equality $\boldsymbol{m} = V\mu = \boldsymbol{g}\mu$ we deduce then that $V(t) = \boldsymbol{g}(t)$, μ-almost everywhere (proposition 17.15), hence

$$|\boldsymbol{g}\mu| = |\boldsymbol{m}| = |V|\,|\mu| = |\boldsymbol{g}|\,|\mu|\,.$$

Assume that $\boldsymbol{g}$ is locally μ-integrable (condition (4)).

Let φ be a locally $|\mu|$-integrable scalar function such that $|\varphi(t)|\equiv 1$ and $\mu = \varphi|\mu|$ (proposition 18.7). The function $\boldsymbol{g}\varphi$ is locally $|\mu|$-integrable and we have (proposition 16.29)

$$\boldsymbol{g}\mu = \boldsymbol{g}(\varphi|\mu|) = (\boldsymbol{g}\varphi)|\mu|\,.$$

For the function $\boldsymbol{g}\varphi$ we have $|\boldsymbol{g}\varphi| = |\boldsymbol{g}|\,|\varphi| = |\boldsymbol{g}|$ and for the measure $(\boldsymbol{g}\varphi)|\mu|$ we have (theorem 17.3)

$$|(\boldsymbol{g}\varphi)|\mu|| = |\boldsymbol{g}\varphi|\,|\mu| = |\boldsymbol{g}|\,|\mu|\,,$$

whence

$$|\boldsymbol{g}\mu| = |(\boldsymbol{g}\varphi)|\mu|\,| = |\boldsymbol{g}|\,|\mu|\,.$$

Assume, finally, condition (3). Let $x \in F$. The function $\boldsymbol{g}x$ is locally μ-integrable, hence by the already proved case (4), we have

$$|\boldsymbol{g}(x)\mu| = |\boldsymbol{g}x|\,|\mu|\,.$$

Let ρ be a lifting of $\mathscr{L}^\infty(\mu)$ and choose V such that $\rho[V] = V$. Then we have also $\rho[Vx] = Vx$.

The function Vx is Z-weakly locally μ-integrable, hence by the already proved case (1) we have

$$|(Vx)\mu| = |Vx|\,|\mu|\,.$$

But we have $(\boldsymbol{g}x)\mu = (Vx)\mu$, since for every $\varphi \in \mathscr{K}(T)$ and every $z \in Z$ we have

$$\begin{aligned}\langle \textstyle\int \varphi \,\mathrm{d}(\boldsymbol{g}x)\mu, z\rangle &= \textstyle\int \langle \varphi \boldsymbol{g}x, z\rangle \mathrm{d}\mu = \langle \textstyle\int \varphi \,\mathrm{d}(\boldsymbol{g}\mu)x, z\rangle = \\ &= \langle \textstyle\int \varphi \,\mathrm{d}\boldsymbol{m}x, z\rangle = \textstyle\int \langle Vx\varphi, z\rangle \mathrm{d}\mu = \langle \textstyle\int \varphi \,\mathrm{d}(Vx)\mu, z\rangle\,,\end{aligned}$$

whence $\int \varphi \,\mathrm{d}(\boldsymbol{g}x)\mu = \int \varphi \,\mathrm{d}(Vx)\mu$, for every $\varphi \in \mathscr{K}(T)$, consequently $(\boldsymbol{g}x)\mu = (Vx)\mu$. Then

$$|\boldsymbol{g}x|\,|\mu| = |(\boldsymbol{g}x)\mu| = |(Vx)\mu| = |Vx|\,|\mu|\,,$$

hence $|\boldsymbol{g}(t)x| = |V(t)x|$, μ-almost everywhere. Letting x vary over a countable set dense in E, we deduce that $|\boldsymbol{g}(t)| = |V(t)|$, μ-almost everywhere; consequently,

$$|\boldsymbol{g}\mu| = |V\mu| = |V|\,|\mu| = |\boldsymbol{g}|\,|\mu|$$

and the theorem is completely proved.

Remark. In case (4) it is not necessary that the function $\boldsymbol{g}$ be operator-valued. Case (4) may be stated in the following way: Let X be an arbitrary Banach space.

If μ is a scalar measure and if $\boldsymbol{g}: T \to X$ is a locally μ-integrable function, then $|\boldsymbol{g}\mu| = |\boldsymbol{g}|\,|\mu|$.

18.18 COROLLARY. *Let μ be a real measure and $g \geqslant 0$ a locally μ-integrable function. Then*

$$(g\mu)^+ = g\mu^+ \text{ and } (g\mu)^- = g\mu^- .$$

In fact,

$$(g\mu)^+ = \tfrac{1}{2}(|g\mu| + g\mu) = \tfrac{1}{2}(g|\mu| + g\mu) = g\tfrac{1}{2}(|\mu| + \mu) = g\mu^+$$

and

$$(g\mu)^- = \tfrac{1}{2}(|g\mu| - g\mu) = \tfrac{1}{2}(g|\mu| - g\mu) = g\tfrac{1}{2}(|\mu| - \mu) = g\mu^- .$$

18.19 PROPOSITION. *Let μ be a scalar measure and $\boldsymbol{g}_1, \boldsymbol{g}_2 : T \to \mathscr{L}^*(E, F)$ two Z-weakly locally μ-integrable functions. We have $\boldsymbol{g}_1\mu = \boldsymbol{g}_2\mu$ if and only if $\boldsymbol{g}_1(t) = \boldsymbol{g}_2(t)$ μ-almost everywhere, in each of the following cases:*

(1) *there exists a lifting ρ of $\mathscr{L}^\infty(\mu)$ such that $\rho[\boldsymbol{g}_1] = \boldsymbol{g}_1$ and $\rho[\boldsymbol{g}_2] = \boldsymbol{g}_2$;*
(2) *E is of countable type and there exists a countable set $S \subset Z$ norming for F;*
(3) *E is of countable type and $\boldsymbol{g}_1$ and $\boldsymbol{g}_2$ are simply locally μ-integrable;*
(4) *$\boldsymbol{g}_1$ and $\boldsymbol{g}_2$ are locally μ-integrable.*

Assume that $\boldsymbol{g}_1\mu = \boldsymbol{g}_2\mu$.

Denote $\boldsymbol{g} = \boldsymbol{g}_1 - \boldsymbol{g}_2$. Then $\boldsymbol{g}$ is Z-weakly locally μ-integrable and we have

$$\boldsymbol{g}\mu = \boldsymbol{g}_1\mu - \boldsymbol{g}_2\mu = 0 .$$

In each of these cases $\boldsymbol{g}$ has the same properties as $\boldsymbol{g}_1$ and $\boldsymbol{g}_2$.

In fact, in case (1) we have $\rho[\boldsymbol{g}] = \boldsymbol{g}$, in case (2) the function $|\boldsymbol{g}|$ is μ-measurable (proposition 15.32), hence $\boldsymbol{g}$ is Z-weakly locally μ-integrable; in case (3) the function $\boldsymbol{g}$ is μ-measurable (corollary 15.17), consequently $\boldsymbol{g}$ is simply locally μ-integrable; in case (4) $\boldsymbol{g}$ is locally μ-integrable.

In each of these cases we have, therefore,

$$|\boldsymbol{g}\mu| = |\boldsymbol{g}|\,||\mu| ,$$

whence $|\boldsymbol{g}|\,|\mu|=0$. By proposition 17.4, we have $|\boldsymbol{g}(t)|=0$ μ-almost everywhere, hence $\boldsymbol{g}(t)=0$ μ-almost everywhere, consequently $\boldsymbol{g}_1(t)=\boldsymbol{g}_2(t)$ μ-almost everywhere.

The converse implication is evident.

Consider now the case of measures with vector bases.

18.20 THEOREM. *If $\boldsymbol{m}:\mathscr{K}(T)\to E$ is a dominated measure and g is a locally $\boldsymbol{m}$-integrable scalar function then*

$$|g\boldsymbol{m}|=|g|\,|\boldsymbol{m}|\,.$$

Let ρ be a lifting of $\mathscr{L}^{\infty}(|\boldsymbol{m}|)$. There exists an E'-weakly locally $|\boldsymbol{m}|$-integrable function $U:T\to E''=\mathscr{L}(R,E'')$ such that $\boldsymbol{m}=U|\boldsymbol{m}|$, $|U(t)|=1$, $|\boldsymbol{m}|$-almost everywhere and $\rho(U)=U$ (theorem 18.15).

There exists also (proposition 13.22) a locally $\boldsymbol{m}$-integrable scalar function g' equal to g, $|\boldsymbol{m}|$-almost everywhere, with $\rho[g']=g'$. It follows that the function g' is also locally $|\boldsymbol{m}|$-integrable, hence the function Ug' is E'-weakly locally $|\boldsymbol{m}|$-integrable and we have $\rho[Ug']=Ug'$ and (proposition 16.21)

$$g\boldsymbol{m}=g'\boldsymbol{m} \text{ and } g'(U|\boldsymbol{m}|)=(g'U)|\boldsymbol{m}|\,.$$

On the other hand, since $g\boldsymbol{m}$ is absolutely continuous with respect to $|\boldsymbol{m}|$, there exists an E'-weakly locally $|\boldsymbol{m}|$-integrable function $V:T\to E''$ with $g\boldsymbol{m}=V|\boldsymbol{m}|$, $|g\boldsymbol{m}|=|V|\,|\boldsymbol{m}|$ and $\rho[V]=V$ (theorem 18.16).

We deduce that $(g'U)|\boldsymbol{m}|=V|\boldsymbol{m}|$, hence $gU=g'U=V$, $|\boldsymbol{m}|$-almost everywhere. It follows that $|g|=|gU|=|V|$, $|\boldsymbol{m}|$-almost everywhere, consequently

$$|g\boldsymbol{m}|=|V|\,|\boldsymbol{m}|=|g|\,|\boldsymbol{m}|\,.$$

18.21 PROPOSITION. *Let $\boldsymbol{m}:\mathscr{K}(T)\to E$ be a dominated measure, g_1 and g_2 locally $\boldsymbol{m}$-integrable scalar functions. Then $g_1\boldsymbol{m}=g_2\boldsymbol{m}$ if and only if $g_1(t)=g_2(t)$, $\boldsymbol{m}$-almost everywhere.*

Assume that $g_1\boldsymbol{m}=g_2\boldsymbol{m}$, and denote $g=g_1-g_2$. Then g is locally $\boldsymbol{m}$-integrable and we have

$$g\boldsymbol{m}=g_1\boldsymbol{m}-g_2\boldsymbol{m}=0$$

and

$$|g\boldsymbol{m}| = |g|\,|\boldsymbol{m}|\,,$$

hence $|g|\,|\boldsymbol{m}|=0$. By proposition 17.4, we have $|g(t)|=0$, $\boldsymbol{m}$-almost everywhere.

6 Singular measures

18.22 DEFINITION. *We say that two dominated measures* $\boldsymbol{m}, \boldsymbol{n} : \mathscr{K}(T) \to F$ *are singular (with respect to each other) if* $\inf(|\boldsymbol{m}|, |\boldsymbol{n}|)=0$.

To say that $\boldsymbol{m}$ and $\boldsymbol{n}$ are singular means that for every measure $\lambda \geqslant 0$ with $\lambda \leqslant |\boldsymbol{m}|$ and $\lambda \leqslant |\boldsymbol{n}|$, we have $\lambda=0$.

18.23 DEFINITION. *We say that a dominated measure* $\boldsymbol{m}$ *is concentrated on a set* M *or that the set* M *carries the measure* $\boldsymbol{m}$, *if* $T \setminus M$ *is* $\boldsymbol{m}$*-negligible.*

To say that $\boldsymbol{m}$ is concentrated on M means that M is $\boldsymbol{m}$-measurable and that $\boldsymbol{m}=\varphi_M \boldsymbol{m}$.

If μ is a positive measure concentrated on M, every absolutely continuous measure with respect to μ is also concentrated on M.

Remark. Among the sets which carry a dominated measure $\boldsymbol{m}$, there is its support S. The support of $\boldsymbol{m}$ is the least *closed* set on which $\boldsymbol{m}$ is concentrated. But we might find subsets of the support S, and different from S, on which $\boldsymbol{m}$ is concentrated. In general, there is no least set which carries $\boldsymbol{m}$.

For example, the support of the Lebesgue measure is the whole line. The Lebesgue measure μ is concentrated on every set which is the complement of a finite subset of the line, since $\mu(\{t\})=0$ for every point $t \in R$.

18.24 PROPOSITION. *Two measures* $\boldsymbol{m}$ *and* $\boldsymbol{n}$ *are singular if and only if there exist two disjoint sets* M *and* N *which carry respectively* $\boldsymbol{m}$ *and* $\boldsymbol{n}$.

Denote $\lambda=|\boldsymbol{m}|+|\boldsymbol{n}|$. We have $|\boldsymbol{m}|=g\lambda$ and $|\boldsymbol{n}|=h\lambda$, where g and h are locally λ-integrable positive functions and

$$\inf(|\boldsymbol{m}|, |\boldsymbol{n}|) = \inf(g, h)\lambda\,.$$

The measures $|\boldsymbol{m}|$ and $|\boldsymbol{n}|$ are singular if and only if $\inf(g, h)=0$, λ-almost everywhere, that is, if and only if there exist two disjoint λ-measurable sets M and N such that $g=\varphi_M g$ and $h=\varphi_N h$, λ-almost everywhere.

But $g = \varphi_M g$, λ-almost everywhere, if and only if $g\lambda = (\varphi_M g)\lambda = \varphi_M(g\lambda)$, that is, $|\boldsymbol{m}| = \varphi_M |\boldsymbol{m}|$, which means that the set M carries the measure $|\boldsymbol{m}|$, consequently also the measure $\boldsymbol{m}$.

Also, $h = \varphi_N h$, λ-almost everywhere, if and only if the set N carries the measure $\boldsymbol{n}$.

Thus the proposition is proved.

18.25 Corollary. *Let* $\boldsymbol{m}$ *and* $\boldsymbol{n}$ *be two singular measures (with respect to each other). If* $\boldsymbol{m}'$ *is absolutely continuous with respect to* $\boldsymbol{m}$, *and* $\boldsymbol{n}'$ *is absolutely continuous with respect to* $\boldsymbol{n}$ *then* $\boldsymbol{m}'$ *and* $\boldsymbol{n}'$ *are singular (with respect to each other).*

In fact, there exist two disjoint sets M and N which carry $\boldsymbol{m}$ and $\boldsymbol{n}$ respectively. Thus $T \setminus M$ is $\boldsymbol{m}$-negligible and $T \setminus N$ is $\boldsymbol{n}$-negligible. It follows that $T \setminus M$ is $\boldsymbol{m}'$-negligible and $T \setminus N$ is $\boldsymbol{n}'$-negligible, hence M carries $\boldsymbol{m}'$ and N carries $\boldsymbol{n}'$. By the preceding proposition $\boldsymbol{m}'$ and $\boldsymbol{n}'$ are singular.

18.26 Corollary. *If* $\boldsymbol{m}$ *is at the same time absolutely continuous with respect to* $\boldsymbol{n}$ *and singular with respect to* $\boldsymbol{n}$, *then* $\boldsymbol{m} = 0$.

Since $\boldsymbol{m}$ and $\boldsymbol{n}$ are singular, there exist two disjoint sets M and N such that $T \setminus M$ is $\boldsymbol{m}$-negligible and $T \setminus N$ is $\boldsymbol{n}$-negligible. Since $\boldsymbol{m}$ is absolutely continuous with respect to $\boldsymbol{n}$, the set $T \setminus N$ is also $\boldsymbol{m}$-negligible. It follows that $T = (T \setminus M) \cup (T \setminus N)$ is $\boldsymbol{m}$-negligible, hence $\boldsymbol{m} = 0$.

18.27 Corollary. *If* μ *is a real measure, then* μ^+ *and* μ^- *are singular.*

In fact, there exist two μ-measurable disjoint sets A and B such that $\mu^+ = \varphi_A |\mu|$ and $\mu = \varphi_B |\mu|$. It follows that μ^+ is concentrated on A and μ^- is concentrated on B, consequently, μ^+ and μ^- are singular.

18.28 Theorem (*Lebesgue*). *Let* μ *be a positive measure. Every positive measure* ν *can be written uniquely in the form*

$$\nu = \mu' + \nu',$$

where μ' *is a positive measure absolutely continuous with respect to* μ *and* ν' *is a positive measure singular with respect to* μ.

Consider the positive measure $\lambda=\mu+\nu$.

Since the measure ν is absolutely continuous with respect to λ, there exists a locally λ-integrable function h such that $\nu=h\lambda$.

Since $0\leqslant\nu\leqslant\lambda$, we have $0\leqslant h(t)\leqslant 1$, λ-almost everywhere, consequently $0\leqslant h(t)\leqslant 1$, ν-almost everywhere (and also μ-almost everywhere, since $\mu\leqslant\lambda$).

The function h is also locally ν-integrable and locally μ-integrable. From the equality $\lambda=\mu+\nu$, it follows $\nu=h\lambda=h\mu+h\nu$.

Denote

$$A=\{t;0\leqslant h(t)<1\} \text{ and } B=\{t;h(t)=1\}.$$

The sets A and B are measurable with respect to each of the measures μ, ν and λ. From the equality $\nu=h\mu+h\nu$, it follows that $(1-h)\nu=h\mu$, consequently,

$$\int^* (1-h)\varphi_B \mathrm{d}\nu=\int^* h\varphi_B \mathrm{d}\mu\,,$$

whence $\mu^*(B)=0$. We deduce that the measure $\nu'=\varphi_B\nu$ is singular with respect to μ.

In fact, let ρ be a positive measure such that $\rho\leqslant\nu'$ and $\rho\leqslant\mu$. For every set $E\subset T$ we have

$$\rho^*(E)\leqslant\nu'^*(E)=\nu^*(E\cap B), \text{ and } \rho^*(E)\leqslant\mu^*(E).$$

It follows that if $E\cap B=\varnothing$ we have $\rho^*(E)=0$. Thus $\rho^*(E)=\rho^*(E\cap B)\leqslant\mu^*(E\cap B)=0$, hence $\rho=0$. This means that ν' and μ are singular.

The measure $\mu'=\varphi_A\nu$ is absolutely continuous with respect to μ. In fact, let $E\subset T$ be a set with $\mu^*(E)=0$.

From the equality $\nu=h\mu+h\nu$ we deduce $(1-h)\nu=h\mu$, hence

$$\int^* (1-h)\varphi_{A\cap E}\mathrm{d}\nu=\int^* h\varphi_{A\cap E}\mathrm{d}\mu=0\,.$$

But $1-h(t)>0$ for every $t\in A\cap E$. It follows then that $\nu^*(A\cap E)=0$, that is, $\mu'^*(E)=0$. Thus μ' is absolutely continuous with respect to μ. The measure ν is now written as

$$\nu=(\varphi_A+\varphi_B)\nu=\varphi_A\nu+\varphi_A\nu=\mu'+\nu'\,.$$

If μ'' is a positive measure absolutely continuous with respect to μ and ν'' is a positive measure singular with respect to μ, such that $\nu=\mu''+\nu''$, then $\mu''=\mu'$ and $\nu''=\nu'$.

In fact, from the equality $\mu'+\nu'=\mu''+\nu''$ we deduce $\mu'-\mu''=\nu''-\nu'$. Since ν' and ν'' are singular with respect to μ, there exists a set M which carries

μ and two sets N' and N'' which carry respectively ν' and ν'' such that $M \cap N' = \varnothing$ and $M \cap N'' = \varnothing$. The measure $\mu' - \mu''$ is absolutely continuous with respect to μ, hence it is concentrated on M. The measure $\nu'' - \nu'$ is concentrated on $T \setminus M$. It follows then that $\mu' - \mu'' = \nu'' - \nu' = 0$, consequently $\mu' = \mu''$ and $\nu' = \nu''$, that is, the decomposition $\nu = \mu' + \mu'$ is unique.

Thus the theorem is proved.

18.29 THEOREM. *Let μ be a positive measure and $\boldsymbol{m} : \mathscr{K}(T) \to E'$ a dominated measure. Then $\boldsymbol{m}$ can be written uniquely in the form*

$$\boldsymbol{m} = \boldsymbol{m}' + \boldsymbol{n}'$$

such that $\boldsymbol{m}' : \mathscr{K}(T) \to E'$ is a dominated measure, absolutely continuous with respect to μ, and $\boldsymbol{n}' : \mathscr{K}(T) \to E'$ is a singular measure with respect to μ.

By theorem 18.28, the positive measure $|\boldsymbol{m}|$ can be written in the form

$$|\boldsymbol{m}| = \mu' + \nu' ,$$

where μ' is a positive measure absolutely continuous with respect to μ and ν' is a positive measure singular with respect to μ. We have $\mu' \leqslant |\boldsymbol{m}|$ and $\nu' \leqslant |\boldsymbol{m}|$.

By theorem 18.15, there exists an E-weakly locally $\boldsymbol{m}$-integrable function $U : T \to E'$ with $|U(t)| = 1$, $|\boldsymbol{m}|$-almost everywhere such that $\boldsymbol{m} = U|\boldsymbol{m}|$.

Since $\mu' \leqslant |\boldsymbol{m}|$ and $\nu' \leqslant |\boldsymbol{m}|$, the function U is E-weakly locally integrable with respect to μ' and ν', therefore the measures $\boldsymbol{m}' = U\mu'$ and $\boldsymbol{n}' = U\nu'$ are well defined and we have $\boldsymbol{m} = \boldsymbol{m}' + \boldsymbol{n}'$.

From the inequalities $|\boldsymbol{m}'| \leqslant |U|\mu' = \mu'$ and $|\boldsymbol{n}'| = |U|\nu' = \nu'$ we deduce that $\boldsymbol{m}'$ is absolutely continuous with respect to μ' consequently also with respect to μ, and $\boldsymbol{n}'$ is singular with respect to μ.

We have then

$$\mu' + \nu' = |\boldsymbol{m}| = |\boldsymbol{m}' + \boldsymbol{n}'| \leqslant |\boldsymbol{m}'| + |\boldsymbol{n}'| \leqslant \mu' + \nu' ,$$

hence $|\boldsymbol{m}| = |\boldsymbol{m}'| + |\boldsymbol{n}'|$. It follows that $|\boldsymbol{m}'| = \mu'$ and $|\boldsymbol{n}'| = \nu'$.

It remains to show that the decomposition is unique.

Assume that we have also $\boldsymbol{m} = \boldsymbol{m}'' + \boldsymbol{n}''$, where $\boldsymbol{m}'' : \mathscr{K}(T) \to E'$ is a dominated measure, absolutely continuous with respect to μ and $\boldsymbol{n}'' : \mathscr{K}(T) \to E'$ is a dominated measure singular with respect to μ.

Then $\boldsymbol{m}' - \boldsymbol{m}'' = \boldsymbol{n}'' - \boldsymbol{n}'$, hence $|\boldsymbol{m}' - \boldsymbol{m}''| = |\boldsymbol{n}' - \boldsymbol{n}''|$. We have also $\lambda = |\boldsymbol{m} - \boldsymbol{m}''| \leqslant |\boldsymbol{m}'| + |\boldsymbol{m}''|$ and $\lambda = |\boldsymbol{n}'' - \boldsymbol{n}'| \leqslant |\boldsymbol{n}''| + |\boldsymbol{n}'|$. We deduce that

λ is at the same time absolutely continuous with respect to μ and singular with respect to μ. It follows that $\lambda=0$, that is $|\boldsymbol{m}'-\boldsymbol{m}''|=|\boldsymbol{n}''-\boldsymbol{n}'|=0$, hence $\boldsymbol{m}'=\boldsymbol{m}''$ and $\boldsymbol{n}'=\boldsymbol{n}''$.

Thus the theorem is proved.

7 *Diffuse measures. Atomic measures*

18.30 Definition. *We say that a dominated measure* $\boldsymbol{m}$ *on* T *is diffuse if for every* $t\in T$ *we have* $|\boldsymbol{m}|(\{t\})=0$.

To say that a measure $\boldsymbol{m}$ is diffuse means that every set whose complement is countable, carries $\boldsymbol{m}$.

If $\boldsymbol{n}$ is absolutely continuous with respect to a diffuse measure $\boldsymbol{m}$, then $\boldsymbol{n}$ also is diffuse.

18.31 Definition. *We say that a dominated measure* $\boldsymbol{m}$ *on* T *is atomic if there exists a family* $(\mu_\alpha)_{\alpha\in A}$ *of positive measures of the form* ε_t *such that*

$$|\boldsymbol{m}|(\varphi)=\Sigma_{\alpha\in A}\,\mu_\alpha(\varphi)\ \textit{for}\ \ \varphi\in\mathscr{K}(T).$$

To say that $\boldsymbol{m}$ is atomic means that there exists a positive function α defined on T such that

$$|\boldsymbol{m}|(\varphi)=\Sigma_{t\in T}\,\alpha(t)\,\varphi(t),\ \text{for}\ \ \varphi\in\mathscr{K}(T).$$

It follows that for every compact set $K\subset T$ we have

$$\Sigma_{t\in K}\,\alpha(t)<\infty,$$

hence if we denote $N=\{t;\,\alpha(t)\neq 0\}$, the set $N\cap K$ is at most countable for every compact set $K\subset T$. We deduce also that the measure $\boldsymbol{m}$ is concentrated on N.

18.32 Proposition. *Every atomic measure* $\boldsymbol{m}$ *is singular with respect to every diffuse measure* $\boldsymbol{n}$

We write

$$|\boldsymbol{m}|(\varphi)=\Sigma_{t\in T}\,\alpha(t)\,\varphi(t),\ \text{for}\ \ t\in T.$$

The measure $\boldsymbol{m}$ is concentrated on the set $N=\{t;\,\alpha(t)\neq 0\}$ which has the property that $N \cap K$ is at most countable for every compact set K (see the example following definition 5.8).

It follows that the set $N \cap K$ is $\boldsymbol{n}$-negligible, for every compact set $K \subset T$, hence N is $\boldsymbol{n}$-negligible, that is $\boldsymbol{n}$ is concentrated on $T \setminus N$. Thus $\boldsymbol{m}$ and $\boldsymbol{n}$ are singular.

18.33 COROLLARY. *If* $\boldsymbol{m}$ *is at the same time atomic and diffuse, then* $\boldsymbol{m}=0$.

18.34 PROPOSITION. *Every positive measure* μ *can be written uniquely in the form* $\mu=\mu_1+\mu_2$, *where* μ_1 *is a positive atomic measure and* μ_2 *is a positive diffuse measure.*

For every $t \in T$ put $\alpha(t)=\mu(\{t\})$.

Let $K \subset T$ be a compact set. For every finite subset $I \subset K$ we have

$$\Sigma_{t\in I}\ \varphi(t) = \mu(I) \leqslant \mu(K),$$

hence

$$\Sigma_{t\in K}\, \varphi(t) < \infty .$$

For every function $\varphi \in \mathscr{K}(T)$ we have

$$\Sigma_{t\in T}\, \alpha(t)\, |\varphi(t)| \leqslant \Sigma_{t\in K}\, \alpha(t)\, \|\varphi\| < \infty ,$$

where K is the support of φ.

Put

$$\mu_1(\varphi) = \Sigma_{t\in T}\, \alpha(t)\, \varphi(t) .$$

It is easy to see that μ_1 is linear and positive on $\mathscr{K}(T)$, hence it is a positive measure; μ_1 is an atomic measure, since

$$\mu_1(\varphi) = \Sigma_{t\in T}\, \alpha(t)\, \varepsilon_t(\varphi) .$$

We have, evidently, $\mu_1 \leqslant \mu$. Then the measure $\mu_2=\mu-\mu_1$ is positive and diffuse since for every $t \in T$ we have $\mu_2(\{t\})=\mu(\{t\})-\mu_1(\{t\})=0$.

The uniqueness of the decomposition follows from the fact that the atomic measures are singular with respect to the diffuse measures.

18.35 PROPOSITION. *Let E be a Banach space. Every dominated measure* $\boldsymbol{m}: \mathscr{K}(T) \to E'$ *can be written uniquely in the form*

$$\boldsymbol{m} = \boldsymbol{m}_1 + \boldsymbol{m}_2 ,$$

where $\boldsymbol{m}_1 : \mathscr{K}(T) \to E'$ *is an atomic dominated measure and* $\boldsymbol{m}_2 : \mathscr{K}(T) \to E'$ *is a diffuse dominated measure.*

By proposition 18.34, the positive measure $|\boldsymbol{m}|$ can be written

$$|\boldsymbol{m}| = \mu_1 + \mu_2 ,$$

whereμ_1 is an atomic positive measure and μ_2 is a diffuse positive measure.

The measure $\boldsymbol{m}$ can be written $\boldsymbol{m} = U|\boldsymbol{m}|$, where $U : T \to E'$ is an E-weakly locally $\boldsymbol{m}$-integrable function with $|U(t)| = 1$.

From the inequalities $\mu_1 \leqslant |\boldsymbol{m}|$ and $\mu_2 \leqslant |\boldsymbol{m}|$ we deduce that U is E-weakly locally integrable with respect to μ_1 and μ_2, hence we can define the measures $\boldsymbol{m}_1 = U\mu_1$ and $\boldsymbol{m}_2 = U\mu_2$. We have $\boldsymbol{m} = \boldsymbol{m}_1 + \boldsymbol{m}_2$.

On the other hand $|\boldsymbol{m}_1| \leqslant |U|\mu_1 = \mu_1$ and $|\boldsymbol{m}_2| \leqslant |U|\mu_2 = \mu_2$, therefore

$$\mu_1 + \mu_2 = |\boldsymbol{m}| = |\boldsymbol{m}_1 + \boldsymbol{m}_2| \leqslant |\boldsymbol{m}_1| + |\boldsymbol{m}_2| \leqslant \mu_1 + \mu_2 .$$

It follows that $|\boldsymbol{m}_1| + |\boldsymbol{m}_2| = \mu_1 + \mu_2$ whence $|\boldsymbol{m}_1| = \mu_1$ and $|\boldsymbol{m}_2| = \mu_2$. We deduce that $\boldsymbol{m}_1$ is atomic and $\boldsymbol{m}_2$ is diffuse.

Assume that $\boldsymbol{m}$ is written also in the form

$$\boldsymbol{m} = \boldsymbol{m}_1' + \boldsymbol{m}_2' ,$$

where $\boldsymbol{m}_1'$ is atomic and $\boldsymbol{m}_2'$ is diffuse. Then $\boldsymbol{m}_1 - \boldsymbol{m}_1' = \boldsymbol{m}_2' - \boldsymbol{m}_2$, hence $|\boldsymbol{m}_1 - \boldsymbol{m}_1'| = |\boldsymbol{m}_2' - \boldsymbol{m}_2|$. From the inequalities $\mu = |\boldsymbol{m}_1 - \boldsymbol{m}_1'| \leqslant |\boldsymbol{m}_1| + |\boldsymbol{m}_1'|$ and $\mu = |\boldsymbol{m}_2' - \boldsymbol{m}_2| \leqslant |\boldsymbol{m}_2| + |\boldsymbol{m}_2|$ we deduce that μ is atomic and diffuse, hence $\mu = 0$, that is $|\boldsymbol{m}_1' - \boldsymbol{m}_1| = |\boldsymbol{m}_2' - \boldsymbol{m}_2| = 0$, consequently $\boldsymbol{m}_1' = \boldsymbol{m}_1$ and $\boldsymbol{m}_2' = \boldsymbol{m}_2$.

8 *Linear operations on the space* $\mathscr{L}_E^p$

Let E and F be two Banach spaces and μ a positive measure on T. Consider the space $\mathscr{L}_E^p$ with $1 \leqslant p < \infty$. Let U be a continuous linear mapping of the space $\mathscr{L}_E^p(\mu)$ into F. Put

$$|||U||| = \sup \Sigma\, |U(\varphi_{A_i} x_i)| ,$$

the supremum being considered for all the Borel step functions $\boldsymbol{f} = \Sigma\, \varphi_{A_i} x_i$ with A_i mutually disjoint and $x_i \in E$, such that $N_p(\boldsymbol{f}, \mu) \leqslant 1$.

We have

$$\|U\| \leqslant |||U||| \leqslant \infty .$$

In fact

$$||U|| = \sup|U(\Sigma\,\varphi_{A_i}x_i)| = \sup|\Sigma\,U(\varphi_{A_i}x_i)| \leqslant$$
$$\leqslant \sup\Sigma\,|U(\varphi_{A_i}x_i)| = |||U|||\,.$$

The inequality $\|U\| \leqslant |||U|||$ may be strict.

Example. Consider $T=R$, μ the Lebesgue measure, $E=R$, $1<p<\infty$, $F=L^p(\mu)$ and the canonical mapping $U:\mathscr{L}^p(\mu)\to L^p(\mu)$ which to each function $f\in\mathscr{L}^p(\mu)$ associates its equivalence class $\tilde{f}$ in $L^p(\mu)$. Evidently, we have $\|U\|=1$. Let us show that $|||U|||=\infty$. For each natural number n choose n disjoint intervals $(A_i)_{1\leqslant i\leqslant n}$ and n numbers $(x_i)_{1\leqslant i\leqslant n}$ such that $\mu(A_i)=n^{-1}$ and $|x_i|=1$ for $1\leqslant i\leqslant n$.

For the step function $f=\Sigma_{1\leqslant i\leqslant n}\varphi_{A_i}x_i$ we have $N_p(f,\mu)=1$ and $\Sigma_{1\leqslant i\leqslant n}|U(\varphi_{A_i}x_i)| = \Sigma_{1\leqslant i\leqslant n}N_p(\varphi_{A_i}x_i,\mu) = \Sigma_{1\leqslant i\leqslant n}[\mu(A_i)]^{1/p}|x_i| = n(n^{-1})^{1/p}$ $=n^{1-1/p}$ hence $|||U|||\geqslant n^{1-(1/p)}$. Since $1-p^{-1}>0$, we have $\lim_n n^{1-(1/p)}=\infty$, consequently $|||U|||=\infty$.

But in certain cases we have $\|U\|=|||U|||$.

18.36 PROPOSITION. *In each of the following cases:*

(1) $p=1$,

(2) $1\leqslant p<\infty$ *and* $F=C$,

we have the equality $||U||=|||U|||$ *for every continuous linear mapping* $U:\mathscr{L}^p_E(\mu)\to F$.

(1) If $p=1$, then for every Borel step function $f=\Sigma\,\varphi_{A_i}x_i$ with $N_1(f,\mu)\leqslant 1$ such that the sets A_i are disjoint, we have

$$\Sigma|U(\varphi_{A_i}x_i)| \leqslant \Sigma\,\|U\|\,N_1(\varphi_{A_i}x_i,\mu) = \|U\|\,N_1(f,\mu) \leqslant \|U\|\,,$$

whence $|||U|||\leqslant||U||$, consequently $\|U\|=|||U|||$.

(2) Assume that $1\leqslant p<\infty$ and $F=C$. Let $f=\Sigma\,\varphi_{A_i}x_i$ be a Borel step function such that the sets A_i are mutually disjoint and $N_p(f,\mu)\leqslant 1$.

For each i there exists a complex number θ_i with $|\theta_i|=1$ such that

$$|U(\varphi_{A_i}x_i)| = \theta_i U(\varphi_{A_i}x_i) = U(\varphi_{A_i}\theta_i x_i)\ ;$$

then

$$\Sigma|U(\varphi_{A_i}x_i)| = U(\Sigma\,\varphi_{A_i}\theta_i x_i) \leqslant \|U\|\,N_p(\Sigma\,\varphi_{A_i}\theta_i x_i,\mu) =$$
$$= \|U\|\,N_p(\Sigma\,\varphi_{A_i}x_i,\mu) \leqslant \|U\|\,,$$

whence $|||U|||\leqslant\|U\|$, consequently $\|U\|=|||U|||$.

Remark. The set of the linear mappings $U : \mathscr{L}_E^p(\mu) \to F$ with $|||U||| < \infty$ is a vector space and $|||U|||$ is a norm on this space.

18.37 THEOREM. *There exists an isomorphism $U \leftrightarrow U'$ between the set of linear mappings $U : \mathscr{L}_E^p(\mu) \to F$ with $|||U||| < \infty$ and the set of linear mappings $U' : \mathscr{L}^p(\mu) \to \mathscr{L}(E, F)$ with $|||U'||| < \infty$ given by the equality*

$$U(\varphi x) = U'(\varphi)x, \text{ for } \varphi \in \mathscr{L}^p(\mu) \text{ and } x \in E,$$

and we have $|||U||| = |||U'|||$.

Let $U : \mathscr{L}_E^p(\mu) \to F$ be a linear mapping with $|||U||| < \infty$. For every $\varphi \in \mathscr{L}^p(\mu)$ and every $x \in E$ we have $\varphi x \in \mathscr{L}_E^p(\mu)$. For each $\varphi \in \mathscr{L}^p(\mu)$, the mapping $U'(\varphi) : x \to U(\varphi x)$ of E into F is linear and continuous:

$$|U'(\varphi)x| = |U(\varphi x)| \leqslant \|U\| \, N_p(\varphi x) \leqslant |x| \, \|U\| \, N_p(\varphi) < \infty,$$

hence $U'(\varphi) \in \mathscr{L}(E, F)$ and

$$|U'(\varphi)| \leqslant \|U\| \, N_p(\varphi).$$

The mapping $U' : \varphi \to U'(\varphi)$ of $\mathscr{L}^p(\mu)$ into $\mathscr{L}(E, F)$ is linear and continuous and we have $\|U'\| \leqslant \|U\|$. We have also $|||U'||| \leqslant |||U|||$. In fact, let $\varphi = \Sigma_{1 \leqslant i \leqslant n} \varphi_{A_i} \alpha_i$ be a step function of $\mathscr{L}^p(\mu)$ with A_i disjoint and $N_p(\varphi, \mu) \leqslant 1$ and let $\varepsilon > 0$. For each i there exists $x_i \in E$ with $|x_i| = 1$ and

$$|U'(\varphi_{A_i})| < |U'(\varphi_{A_i})x_i| + \varepsilon/n.$$

Then

$$N_p(\Sigma \varphi_{A_i} \alpha_i x_i) = N_p(\Sigma \varphi_{A_i} \alpha_i) \leqslant 1$$

and

$$\Sigma |U'(\varphi_{A_i} \alpha_i)| - \varepsilon < \Sigma |U'(\varphi_{A_i}) \alpha_i x_i| = \Sigma |U(\varphi_{A_i} x_i) \leqslant$$
$$\leqslant |||U||| \, N_p(\Sigma \varphi_{A_i} \alpha_i x_i) \leqslant |||U|||,$$

hence $|||U'||| \leqslant |||U|||$.

The mapping $U \to U'$ so defined is, evidently, linear. It is also injective. In fact, if for two mappings $U_1, U_2 : \mathscr{L}_E^p \to F$ we have

$$U_1(\varphi x) = U_2(\varphi x), \text{ for } \varphi \in \mathscr{L}^p \text{ and } x \in E,$$

then

$$U_1(\Sigma \varphi_i x_i) = U_2(\Sigma \varphi_i x_i)$$

for every function of the form $f = \Sigma \varphi_i x_i$ with $\varphi_i \in \mathscr{L}^p(\mu)$, and $x_i \in E$. As the set of functions of the preceding form is dense in $\mathscr{L}_E^p$ (since it contains the step functions) we deduce that $U_1 = U_2$. It follows that the mapping $U \to U'$ is injective.

Conversely, let $U' : \mathscr{L}^p(\mu) \to \mathscr{L}(E, F)$ be a linear mapping with $|||U'||| < \infty$.

For each function of the form $f = \Sigma \varphi_i x_i$ (finite sum, $1 \leqslant i \leqslant m$) with $\varphi_i \in \mathscr{L}^p(\mu)$ and $x_i \in E$ put

$$U(f) = \Sigma U'(\varphi_i) x_i .$$

The definition of $U(f)$ is independent of the representation of f in the above form. It is sufficient to prove this fact only for $f = 0$. Assume that $\Sigma \varphi_i(t) x_i \equiv 0$.

Let E_n be an n-dimensional subspace of E which contains all the vectors, $x_1, x_2, \ldots, x_m$ and let $y_1, y_2, \ldots, y_n$ be a basis of E_n. Each x_i can be written $x_i = \Sigma_{1 \leqslant j \leqslant n} \alpha_{ij} y_j$, hence

$$0 = \Sigma_{1 \leqslant i \leqslant n} \varphi_i x_i = \Sigma_{1 \leqslant i \leqslant n} (\Sigma_{1 \leqslant j \leqslant n} \varphi_i \alpha_{ij}) y_j .$$

Since $y_1, \ldots, y_n$ are linearly independent, we deduce that $\Sigma_{1 \leqslant i \leqslant m} \varphi_i(t) \alpha_{ij} = 0$ for each j, hence $U'(\Sigma_{1 \leqslant i \leqslant m} \varphi_i \alpha_{ij}) = 0$. Then

$$\Sigma_{1 \leqslant i \leqslant m} U'(\varphi_i) x_i = \Sigma_{i,j} U'(\varphi_i) \alpha_{ij} y_j = \Sigma_j U'(\Sigma_i \varphi_i \alpha_{ij}) y_j = 0 .$$

Now let $f = \Sigma \varphi_{A_i} x_i$ be a step function of $\mathscr{L}_E^p$ with A_i disjoint.

We have

$$|U(f)| = |\Sigma U'(\varphi_{A_i}) x_i| \leqslant \Sigma |U'(\varphi_{A_i} |x_i|)| \leqslant |||U'||| N_p(f) .$$

It follows that U is continuous for the seminorm N_p on the set of the functions of $\mathscr{L}_E^p(\mu)$ of the form $\Sigma \varphi_{A_i} x_i$. As this set is dense in $\mathscr{L}_E^p(\mu)$, we deduce that U can be extended to a linear continuous mapping of $\mathscr{L}_E^p(\mu)$ into F, denoted also by U.

For each step function $f = \Sigma \varphi_{A_i} x_i$ of $\mathscr{L}_E^p(\mu)$ with A_i disjoint and $N_p(f, \mu) \leqslant 1$, we have $|f| = \Sigma \varphi_{A_i} |x_i|$, hence $N_p(\Sigma \varphi_{A_i} |x_i|, \mu) \leqslant 1$ and

$$\Sigma |U(\varphi_{A_i} x_i)| = \Sigma |U'(\varphi_{A_i}) x_i| \leqslant \Sigma |U'(\varphi_{A_i} |x_i|)| \leqslant |||U'||| ,$$

whence $|||U||| \leqslant |||U'|||$. From the equality

$$U(\varphi x) = U'(\varphi) x, \quad \text{for } \varphi \in \mathscr{L}^p(\mu) \text{ and } x \in E ,$$

and from the first part of the proof, we deduce $|||U'||| \leqslant |||U|||$, consequently

$$|||U||| = |||U'||| .$$

18.38 COROLLARY. *There exists an isomorphism $U \leftrightarrow U'$ between the set of the linear continuous mappings $U : \mathscr{L}^1_E(\mu) \to F$ and the set of the linear continuous mappings $U' : \mathscr{L}^1(\mu) \to \mathscr{L}(E, F)$ given by the equality*

$$U(\varphi x) = U'(\varphi)x, \text{ for } \varphi \in \mathscr{L}^1(\mu) \text{ and } x \in E,$$

and we have $\| U \| = \| U' \|$.

We use the proposition 18.36.

18.39 COROLLARY. *There exists an isomorphism $U \leftrightarrow U'$ between the set of the linear continuous functionals $U : \mathscr{L}^p_E(\mu) \to C$ and the set of the linear mappings $U' : \mathscr{L}^p(\mu) \to E' = \mathscr{L}(E, C)$ with $|||U'||| < \infty$, given by the equality*

$$U(\varphi x) = U'(\varphi)x, \text{ for } \varphi \in \mathscr{L}^p(\mu) \text{ and } x \in E,$$

and we have $\| U \| = |||U'|||$.

We use the proposition 18.36.

Remark. In the sequel we shall identify the mappings U and U' which are in correspondence by the equality $U(\varphi x) = U'(\varphi)x$ and we shall write U instead of U'. With this notation we have

$$U(\varphi x) = U(\varphi)x, \text{ for } \varphi \in \mathscr{L}^p(\mu) \text{ and } x \in E.$$

18.40 THEOREM. *Let X be a Banach space. For every linear mapping $U : \mathscr{L}^p(\mu) \to X$, $1 \leqslant p < \infty$ with $|||U||| < \infty$, its restriction $\boldsymbol{m}$ to the space $\mathscr{K}(T)$ is a dominated measure, absolutely continuous with respect to μ and satisfying*

$$\int |\varphi| \, \mathrm{d}|\boldsymbol{m}| \leqslant |||U||| \, N_p(\varphi, \mu), \text{ for } \varphi \in \mathscr{K}(T).$$

Conversely, every dominated measure $\boldsymbol{m} : \mathscr{K}(T) \to X$ absolutely continuous with respect to μ for which there exists $\alpha > 0$ such that

$$\int |\varphi| \, \mathrm{d}|\boldsymbol{m}| \leqslant \alpha N_p(\varphi, \mu), \text{ for every } \varphi \in \mathscr{K}(T)$$

can be extended uniquely to a linear operation $U : \mathscr{L}^p(\mu) \to X$ with $|||U||| < \infty$. We have then $|\boldsymbol{m}| = g\mu$ with $N_q(g, \mu) = |||U|||$, $p^{-1} + q^{-1} = 1$, and for every pair of Banach spaces E and F with $X \subset \mathscr{L}(E, F)$ we have $\mathscr{L}^p_E(\mu) \subset \mathscr{L}^1_E(\boldsymbol{m})$,

$$U(\boldsymbol{f}) = \int \boldsymbol{f} \, \mathrm{d}\boldsymbol{m}, \text{ for } \boldsymbol{f} \in \mathscr{L}^p_E(\mu),$$

and

$$\int |\boldsymbol{f}|\,\mathrm{d}|\boldsymbol{m}| \leqslant |||U|||\,N_p(\boldsymbol{f}, \mu), \text{ for } \boldsymbol{f} \in \mathscr{L}^p_E(\mu)\,.$$

(a) Let $U : \mathscr{L}^p(\mu) \to X$ be a linear operation with $|||U||| < \infty$ and let $\boldsymbol{m}$ be the restriction of U to the space $\mathscr{K}(T)$. Let us show that $\boldsymbol{m}$ is dominated.

Let A be a μ-integrable set and (A_i) a finite partition of A in μ-integrable sets. We have $\varphi_A = \Sigma\,\varphi_{A_i}$, hence

$$\Sigma\,|U(\varphi_A)| \leqslant |||U|||\,N_p(\varphi_A, \mu)\,.$$

Taking the supremum for all the partitions of A in μ-integrable sets we obtain

$$v(A) = \sup \Sigma\,|U(\varphi_{A_i})| \leqslant |||U|||\,N_p(\varphi_A, \mu)\,.$$

It is easy to see that the set-function v is additive for the μ-integrable sets and that if $\Sigma_i \varphi_{A_i} c_i = \Sigma_j \varphi_{B_j} d_j$, where A_i are mutually disjoint μ-integrable sets and B_j are also mutually disjoint μ-integrable sets, then

$$\Sigma_i v(A_i) c_i = \Sigma_j v(B_j) d_j\,\cdot$$

For every μ-integrable step function of the form $\varphi = \Sigma\,\varphi_{A_i} c_i$, we define without ambiguity

$$v(\varphi) = \Sigma\,v(A_i) c_i\,.$$

It is easy to see that v is a linear functional on the set of the step functions of $\mathscr{L}^p(\mu)$. We shall show that v is continuous for the seminorm $N_p(\varphi, \mu)$. Let $\varphi = \Sigma_{1 \leqslant i \leqslant n} \varphi_{A_i} c_i$ be a step function of $\mathscr{L}^p(\mu)$ with A_i disjoint and let $\varepsilon > 0$.

For each i there exists a partition (B_{ij}) of A_i consisting of μ-integrable sets, such that

$$v(A_i) \leqslant \Sigma_j |U(\varphi_{B_{ij}})| + \varepsilon/(|c_i| n) \text{ if } c_i \neq 0\,,$$
$$v(A_i) \leqslant \Sigma_j |U(\varphi_{B_{ij}})| + \varepsilon/n, \text{ if } c_i = 0\,.$$

Then, for every i we have

$$v(A_i)\,|c_i| \leqslant \Sigma_j |U(\varphi_{B_{ij}})|\,|c_i| + \varepsilon/n = \Sigma_j |U(\varphi_{B_{ij}} c_i)| + \varepsilon/n\,,$$

hence

$$|v(\varphi)| = |\Sigma_i v(A_i) c_i| \leqslant \Sigma_i v(A_i) |c_i| \leqslant \Sigma_{i,j} |U(\varphi_{B_{ij}} c_i)| + \varepsilon \leqslant$$
$$\leqslant |||U|||\,N_p(\varphi_{B_{ij}} c_i, \mu) + \varepsilon = |||U|||\,N_p(\Sigma\,\varphi_{A_i} c_i, \mu) + \varepsilon\,.$$

ε being arbitrary, it follows

$$|v(\varphi)| \leqslant |||U|||\,N_p(\varphi, \mu)\,.$$

It follows that v can be extended to a linear continuous functional on $\mathscr{L}^p(\mu)$, also denoted by v, which satisfies the inequality

$$|v(\varphi)| \leqslant |||U||| N_p(\varphi, \mu) \text{ for } \varphi \in \mathscr{L}^p(\mu).$$

If $p=1$, for $\varphi \geqslant 0$ from $\mathscr{L}^1(\mu)$ we have $v(\varphi) \subset |||U||| \mu(\varphi)$, that is, $v \leqslant |||U||| \mu$. For every step function $\varphi = \Sigma \varphi_{A_i} \alpha_i$ of $\mathscr{L}^p(\mu)$ with A_i disjoint we have

$$|U(\varphi)| \leqslant \Sigma |U(\varphi_{A_i} \alpha_i)| \leqslant \Sigma |U(\varphi_{A_i})| \, |\alpha_i| \leqslant \Sigma v(A_i) |\alpha_i| = v(\Sigma \varphi_{A_i} |\alpha_i|) = v(|\varphi|).$$

Since the mappings U and v are continuous on $\mathscr{L}^p(\mu)$, the inequality

$$|U(\varphi)| \leqslant v(|\varphi|)$$

remains true for every $\varphi \in \mathscr{L}^p(\mu)$.

As the restriction of v to $\mathscr{K}(T)$ is a positive measure, it follows that the restriction $\boldsymbol{m}$ of U to $\mathscr{K}(T)$ is a dominated measure and we have $|\boldsymbol{m}| \leqslant v$.

From the inequality

$$\int |\varphi| \, \mathrm{d}|\boldsymbol{m}| \leqslant v(|\varphi|) \leqslant |||U||| N_p(\varphi, \mu),$$

it follows that $\boldsymbol{m}$ is absolutely continuous with respect to μ.

(b) Conversely, let $\boldsymbol{m} : \mathscr{K}(T) \to X$ be a dominated measure such that there exists $\alpha > 0$ with

$$\int |\varphi| \, \mathrm{d}|\boldsymbol{m}| \leqslant \alpha N_p(\varphi, \mu), \text{ for } \varphi \in \mathscr{K}(T).$$

Since $|\boldsymbol{m}|$ is absolutely continuous with respect to μ, there exists a locally μ-integrable function $g \geqslant 0$ with $|\boldsymbol{m}| = g\mu$.

For every function $\varphi \in \mathscr{K}(T)$ we have

$$\int |\varphi| g \, \mathrm{d}\mu = \int |\varphi| \, \mathrm{d}|\boldsymbol{m}| \leqslant \alpha N_p(\varphi, \mu),$$

hence

$$N_q(g, \mu) \leqslant \alpha < \infty, \quad p^{-1} + q^{-1} = 1.$$

It follows that if $\varphi \in \mathscr{L}^p(\mu)$, then $\varphi g \in \mathscr{L}^1(\mu)$, hence $\varphi \in \mathscr{L}^1(|\boldsymbol{m}|)$, consequently $\mathscr{L}^p(\mu) \subset \mathscr{L}^1(|\boldsymbol{m}|)$. We have then

$$\int |\varphi| \, \mathrm{d}|\boldsymbol{m}| \leqslant \alpha N_p(\varphi, \mu) \text{ for every } \varphi \in \mathscr{L}^p(\mu).$$

Put $U(\varphi) = \int \varphi \, \mathrm{d}\boldsymbol{m}$, for $\varphi \in \mathscr{L}^p(\mu)$.

The mapping $U : \mathscr{L}^p(\mu) \to X$ is linear and we shall show that $|||U||| < \infty$. Let $\varphi = \Sigma \varphi_{A_i} \alpha_i$ be a step function with A_i disjoint sets of $\mathscr{B}$. Then

$$\Sigma|U(\varphi_{A_i}\alpha_i)| = \Sigma|\int \varphi_{A_i}\alpha_i \mathrm{d}\boldsymbol{m}| \leqslant \Sigma \int \varphi_{A_i}|\alpha_i|\mathrm{d}|\boldsymbol{m}| = \\ = \int \Sigma\varphi_{A_i}|\alpha_i|\mathrm{d}|\boldsymbol{m}| = \int|\varphi|\mathrm{d}|\boldsymbol{m}| \leqslant \alpha N_p(\varphi, \mu),$$

hence $|||U||| \leqslant \alpha < \infty$.

(c) Let now $U : \mathscr{L}^p(\mu) \to X$ be a linear mapping with $|||U||| < \infty$ and let $\boldsymbol{m} : \mathscr{K}(T) \to X$ be the corresponding dominated measure. Let E and F be two Banach spaces with $X \subset \mathscr{L}(E, F)$.

From the first part of the proof we deduce

$$\int|\varphi|\mathrm{d}|\boldsymbol{m}| \leqslant |||U||| N_p(\varphi, \mu) \text{ for } \varphi \in \mathscr{L}^p(\mu).$$

From the second part of the proof, with $\alpha = |||U|||$, we deduce that there exists a locally μ-integrable function $g \geqslant 0$ with $|\boldsymbol{m}| = g\mu$ and $N_q(g, \mu) \leqslant |||U||| < \infty$. Let $\varphi = \Sigma\varphi_{A_i}\alpha_i$ be a step function with A_i disjoint sets of $\mathscr{B}$ and $N_p(\varphi, \mu) \leqslant 1$. Then, as before,

$$\Sigma|U(\varphi_{A_i}\alpha_i)| \leqslant \int|\varphi|\mathrm{d}|\boldsymbol{m}| = \int|\varphi|g\mathrm{d}\mu \leqslant N_q(g, \mu),$$

whence $|||U||| \leqslant N_q(g, \mu)$, consequently

$$|||U||| = N_q(g, \mu).$$

If $\boldsymbol{f} \in \mathscr{L}_E^p(\mu)$, then $\boldsymbol{f}g \in \mathscr{L}_E^1(\mu)$; therefore $\boldsymbol{f} \in \mathscr{L}_E^1(\boldsymbol{m})$, whence

$$\mathscr{L}_E^p(\mu) \subset \mathscr{L}_E^1(\boldsymbol{m})$$

and we have

$$\int|\boldsymbol{f}|\mathrm{d}|\boldsymbol{m}| = \int|\boldsymbol{f}|g\mathrm{d}\mu \leqslant N_p(\boldsymbol{f}, \mu)N_q(g, \mu) \leqslant |||U|||N_p(\boldsymbol{f}, \mu).$$

It follows that

$$|\int\boldsymbol{f}\mathrm{d}\boldsymbol{m}| \leqslant \int|\boldsymbol{f}|\mathrm{d}|\boldsymbol{m}| \leqslant |||U|||N_p(\boldsymbol{f}, \mu) \text{ for } \boldsymbol{f} \in \mathscr{L}_E^p(\mu),$$

hence the mapping $\boldsymbol{f} \to \int\boldsymbol{f}\mathrm{d}\boldsymbol{m}$ is continuous on $\mathscr{L}_E^p(\mu)$. Since we have

$$U(\boldsymbol{f}) = \int\boldsymbol{f}\mathrm{d}\boldsymbol{m}, \text{ for } \boldsymbol{f} \in \mathscr{K}_E(T),$$

and since the two operations U and $\int\boldsymbol{f}\mathrm{d}\boldsymbol{m}$ are continuous on $\mathscr{L}_E^p(\mu)$ it follows that

$$U(\boldsymbol{f}) = \int\boldsymbol{f}\mathrm{d}\boldsymbol{m}, \text{ for every } \boldsymbol{f} \in \mathscr{L}_E^p(\mu),$$

and the theorem is proved.

18.41 COROLLARY. *Let X be a Banach space. If $U : \mathscr{L}^1(\mu) \to X$ is a linear continuous mapping, then its restriction $\boldsymbol{m}$ to $\mathscr{K}(T)$ is a dominated measure, absolutely continuous with respect to μ satisfying $|\boldsymbol{m}| \leqslant \|U\|\mu$.*

Conversely, if $\boldsymbol{m}: \mathcal{K}(T) \to X$ *is a dominated measure such that* $|\boldsymbol{m}| \leqslant \alpha\mu$ *for some* $\alpha > 0$, *then* $\boldsymbol{m}$ *can be extended uniquely to a linear continuous mapping* $U: \mathcal{L}^1(\mu) \to X$. *We have then* $|\boldsymbol{m}| = g\mu$ *with* $N_\infty(g, \mu) = \|U\|$ *and for every pair of Banach spaces E and F with* $X \subset \mathcal{L}(E, F)$ *we have* $\mathcal{L}_E^1(\mu) \subset \mathcal{L}_E^1(\boldsymbol{m})$ *and*

$$U(\boldsymbol{f}) = \int \boldsymbol{f} \, \mathrm{d}\boldsymbol{m}, \text{ for } \boldsymbol{f} \in \mathcal{L}_E^1(\mu).$$

18.42 THEOREM. *Let* μ *be a scalar measure, E and F two Banach spaces, Z a subspace of F′ norming for F and* $U: \mathcal{L}_E^p(\mu) \to F$, $1 \leqslant p < \infty$, *a linear mapping with* $|||U||| < \infty$. *There exists then a function* $\boldsymbol{u}: T \to \mathcal{L}(E, Z')$ *with the following properties:*

(1) *the function* $|\boldsymbol{u}|$ *belongs to* $\mathcal{L}^p(\mu)$ *and we have*

$$N_q(\boldsymbol{u}, \mu) = |||U|||, \text{ where } p^{-1} + q^{-1} = 1;$$

(2) *the function* $\boldsymbol{u}$ *is Z-weakly locally* μ*-integrable and we have*

$$\langle U(\boldsymbol{f}), z \rangle = \int \langle \boldsymbol{u}(t) \boldsymbol{f}(t), z \rangle \, \mathrm{d}\mu, \text{ for } \boldsymbol{f} \in \mathcal{L}_E^p(\mu) \text{ and } z \in Z;$$

(3) *if* ρ *is a lifting of* $\mathcal{L}^\infty(\mu)$, *we can choose* $\boldsymbol{u}$ *uniquely* μ*-almost everywhere such that* $\rho[\boldsymbol{u}] = \boldsymbol{u}$.
If $p = 1$, *we can choose* $\boldsymbol{u}$ *uniquely such that* $\rho(\boldsymbol{u}) = \boldsymbol{u}$;

(4) *the function* $\boldsymbol{u}$ *has values in* $\mathcal{L}(E, F)$ *in each of the following cases:*
   (*a*) $F = Z'$;
   (*b*) *for each* $x \in E$, *the convex balanced hull of the set* $\{U(\varphi x); \varphi \in \mathcal{K}(T), \int |\varphi| \, \mathrm{d}|\mu| \leqslant 1\}$ *is relatively compact in F for the topology* $\sigma(E, Z)$.

Let $\boldsymbol{m}$ be the restriction of U to $\mathcal{K}(T)$. By theorem 18.40, $\boldsymbol{m}$ is a dominated measure and we have $|\boldsymbol{m}| = g|\mu|$ with $N_q(g, \mu) = |||U|||$.

If $p = 1$, then $|\boldsymbol{m}| \leqslant |||U||| \, |\mu|$. We have then $\mathcal{L}_E^p(\mu) \subset \mathcal{L}_E^1(\boldsymbol{m})$ and

$$U(\boldsymbol{f}) = \int \boldsymbol{f} \, \mathrm{d}\boldsymbol{m}, \text{ for } \boldsymbol{f} \in \mathcal{L}_E^p(\mu).$$

Let $V_{\boldsymbol{m}}: T \to \mathcal{L}(E, Z')$ be the function corresponding to $\boldsymbol{m}$ and μ in theorem 18.16. We have $|\boldsymbol{m}| = |V_{\boldsymbol{m}}| \, |\mu|$ hence $|V_{\boldsymbol{m}}(t)| = g(t)$, μ-almost everywhere, consequently

$$N_q(V_{\boldsymbol{m}}, \mu) = N_q(g, \mu) = |||U|||.$$

If we take $\boldsymbol{u} = V_{\boldsymbol{m}}$, then from theorem 18.16 we deduce that $\boldsymbol{u}$ has all the required properties.

18.43 COROLLARY. *For every linear continuous mapping* $U : \mathscr{L}^1_E(\mu) \to F$ *there exists a function* $\boldsymbol{u} : T \to \mathscr{L}(E, Z')$ *with the following properties*:

(1) $|\boldsymbol{u}| \in \mathscr{L}^\infty(\mu)$ and

$$N_\infty(\boldsymbol{u}, \mu) = \|U\| ;$$

(2) *the function* $\boldsymbol{u}$ *is Z-weakly locally* μ*-integrable and*

$$\langle U(\boldsymbol{f}), z\rangle = \int \langle \boldsymbol{u}(t)\boldsymbol{f}(t), z\rangle \,d\mu, \text{ for } \boldsymbol{f} \in \mathscr{L}^1_E(\mu) \text{ and } z \in Z;$$

(3) *if* ρ *is a lifting of* $\mathscr{L}^\infty(\mu)$, *we can choose* $\boldsymbol{u}$ *uniquely such that* $\rho(\boldsymbol{u}) = \boldsymbol{u}$;

(4) *the function* $\boldsymbol{u}$ *has values in* $\mathscr{L}(E, F)$ *in each of the following cases*:

(*a*) $F = Z'$;

(*b*) *for every* $x \in E$, *the convex balanced hull of the set* $\{U(\varphi x); \varphi \in \mathscr{K}(T), \int |\varphi| \,d|\mu| \leq 1\}$ *is relatively compact in F for the topology* $\sigma(F, Z)$.

The following theorem is a converse of theorem 18.42.

18.44 THEOREM. *Let E and F be two Banach spaces, Z a subspace of F' norming for F,* $\boldsymbol{u} : T \to \mathscr{L}^*(E, F)$ *a Z-weakly locally* μ*-integrable function and the space* $\mathscr{L}^p_E(\mu)$ *with* $1 \leq p < \infty$. *If*

(i) $N_q(\boldsymbol{u}, \mu) < \infty$,

then there exists a linear mapping $U : \mathscr{L}^p_E(\mu) \to Z'$ *such that*

$$\langle U(\boldsymbol{f}, z\rangle = \int \langle \boldsymbol{u}(t)\boldsymbol{f}(t), z\rangle \,d\mu \text{ for } \boldsymbol{f} \in \mathscr{L}^p_E(\mu) \text{ and } z \in Z .$$

In this case we have

$$|||U||| \leq N_q(\boldsymbol{u}, \mu) .$$

(ii) *We have* $U(\boldsymbol{f}) \in F$ *for every* $\boldsymbol{f} \in \mathscr{L}^p_E(\mu)$ *in each of the following cases*

(1) $F = Z'$;

(2) $\boldsymbol{u}$ *is simply* μ*-measurable*; *in particular, F is of countable type*;

(3) *for each* $x \in E$ *there exists a locally countable family* $\mathscr{K} = (K_j)_{j \in J}$, *of disjoint compact sets with* $T \setminus \cup K_j$ μ*-negligible, such that, for each* $j \in J$, *the convex balanced hull of the set* $\{\boldsymbol{u}(t)x; t \in K_j\}$ *is relatively compact in F for the topology* $\sigma(F, Z)$.

(iii) *In each of the following cases* $|\boldsymbol{u}|$ *is* μ*-measurable and we have*

$$|||U||| = N_q(\boldsymbol{u}, \mu) :$$

(1) *there exists a lifting* ρ *of* $\mathscr{L}^\infty(\mu)$ *such that* $\rho[\boldsymbol{u}]=\boldsymbol{u}$;
(2) *E is of countable type and there exists a sequence* (z_n) *in* Z, *norming for* F;
(3) *E is of countable type and* $\boldsymbol{u}$ *is simply locally* μ*-measurable*; *in this case we have*

$$U(\boldsymbol{f}) = \int \boldsymbol{u}(t)\boldsymbol{f}(t)\,\mathrm{d}\mu, \text{ for } \boldsymbol{f}\in\mathscr{L}_E^p(\mu)\ ;$$

(4) $\boldsymbol{u}$ *is locally* μ*-integrable*; *in this case we have*

$$U(\varphi) = \int \boldsymbol{u}(t)\varphi(t)\,\mathrm{d}\mu, \text{ for } \varphi\in\mathscr{L}_E^p(\mu)\,.$$

We remark first that for every $\boldsymbol{f}\in\mathscr{L}_E^p(\mu)$ and every $z\in Z$ the function $t\to\langle\boldsymbol{u}(t)\boldsymbol{f}(t), z\rangle$ is μ-integrable. In fact, this function is μ-measurable and

$$\int^* |\langle\boldsymbol{u}\boldsymbol{f}, z\rangle|\,\mathrm{d}|\mu| \leqslant \int^* |\boldsymbol{u}|\,|\boldsymbol{f}|\,|z|\,\mathrm{d}|\mu|| \leqslant |z|\,N_q(\boldsymbol{u},\mu)\,N_p(\boldsymbol{f},\mu) < \infty\,.$$

Put

$$M(\boldsymbol{f}, z) = \int \langle\boldsymbol{u}\boldsymbol{f}, z\rangle\,\mathrm{d}\mu\,.$$

The mapping $U(\boldsymbol{f}): z\to M(\boldsymbol{f}, z)$ is a linear functional on Z. It is continuous since

$$|M(\boldsymbol{f}, z)| \leqslant \int |\langle\boldsymbol{u}\boldsymbol{f}, z\rangle|\,\mathrm{d}|\mu| \leqslant |z|\,N_q(\boldsymbol{u},\mu)\,N_p(\boldsymbol{f},\mu)\,,$$

hence $U(\boldsymbol{f})\in Z'$ and

$$|U(\boldsymbol{f})| \leqslant N_q(\boldsymbol{u},\mu)\,N_p(\boldsymbol{f},\mu)\,.$$

The linear mapping $U:\boldsymbol{f}\to U(\boldsymbol{f})$ of $\mathscr{L}_E^p(\mu)$ into Z' is linear and continuous and we have

$$\langle U(\boldsymbol{f}), z\rangle = \int\langle\boldsymbol{u}\boldsymbol{f}, z\rangle\,\mathrm{d}\mu, \text{ for } \boldsymbol{f}\in\mathscr{L}_E^p(\mu) \text{ and } z\in Z\,.$$

Let us show that $|||U|||\leqslant N_p(\boldsymbol{u},\mu)$. Let $\boldsymbol{f}=\Sigma\,\varphi_{A_i}x_i$ be a Borel step function with A_i disjoint and $N_p(\boldsymbol{f},\mu)\leqslant 1$. Then (proposition 11.4)

$$\begin{aligned}\Sigma\,|U(\varphi_{A_i}x_i)| &\leqslant \Sigma\int^* |\boldsymbol{u}|\,\varphi_{A_i}|x_i|\,\mathrm{d}|\mu| = \int|\boldsymbol{u}|\,\Sigma\,\varphi_{A_i}|x_i|\,\mathrm{d}|\mu| = \\ &= \int^* |\boldsymbol{u}|\,|\Sigma\varphi_{A_i}x_i|\,\mathrm{d}|\mu| = \int^* |\boldsymbol{u}|\,|\boldsymbol{f}|\,\mathrm{d}|\mu| \leqslant N_q(\boldsymbol{u},\mu)\,,\end{aligned}$$

hence $|||U|||\leqslant N_p(\boldsymbol{u},\mu)$.

Consider now the cases where U has values in F. The case $F=Z'$ is trivial. Assume that $\boldsymbol{u}$ is simply μ-measurable. In this case, for every $\boldsymbol{f}\in\mathscr{L}_E^p(\mu)$, the function $\boldsymbol{u}\boldsymbol{f}$ is μ-integrable and we have

$$\langle U(\boldsymbol{f}), z\rangle = \int \langle \boldsymbol{u}\boldsymbol{f}, z\rangle \,\mathrm{d}\mu = \langle \int \boldsymbol{u}\boldsymbol{f}\,\mathrm{d}\mu, z\rangle$$

for every $\boldsymbol{f} \in \mathscr{L}_E^p(\mu)$ and $z \in Z$, hence

$$U(\boldsymbol{f}) = \int \boldsymbol{u}(t)\boldsymbol{f}(t)\,\mathrm{d}\mu \in F, \text{ for every } \boldsymbol{f} \in \mathscr{L}_E^p(\mu).$$

Assume now that condition (3) from (iii) is fulfilled. Let $x \in E$ and let $\mathscr{K} = (K_j)_{j\in J}$ be a family of disjoint compact sets with $T \setminus \cup K_j$ μ-negligible, such that the closure A_j of the convex balanced hull of the set $\{\boldsymbol{u}(t)x;\ t \in K_j\}$ is compact in F for the topology $\sigma(F, Z)$.

Let $\varphi \in \mathscr{K}(T)$ with $\|\varphi\| \leqslant 1$. Then $\boldsymbol{u}(t)\varphi(t)x \in A_j$ for every $t \in K_j$. Let (K_n) be the sequence of sets of the family $\mathscr{K}$ which intersects the support of φ.

For each K_n, the set A_n is compact and convex in the algebraic dual Z^* of Z, for the topology $\sigma(Z, Z^*)$. There exists then a family $(z_i)_{i\in I}$ of elements of Z such that

$$A_n = \cap_{i\in I} \{y;\ y \in Z^*, |\langle y, z_i\rangle| \leqslant 1\}.$$

Then

$$|\langle \boldsymbol{u}(t)\varphi(t)x, z_i\rangle| \leqslant 1, \text{ for } i \in I \text{ and } t \in K_n.$$

It follows that for every n we have

$$|\langle U(\varphi\varphi_{K_n}x), z_i\rangle| \leqslant \int_{K_n} |\langle \boldsymbol{u}(t)\varphi(t)x, z_i\rangle|\,\mathrm{d}|\mu| \leqslant |\mu|(K_n),$$

hence

$$U(\varphi\varphi_{K_n}x) \in |\mu|(K_n)\cdot A_n \subset F.$$

We deduce then that

$$U(\varphi x) = \Sigma_n U(\varphi\varphi_{K_n}x) \in F.$$

The relation $U(\varphi x) \in F$ is deduced then for every $\varphi \in \mathscr{K}(T)$ and every $x \in E$. If follows that we have $U(\boldsymbol{f}) \in F$ for every linear combination $\boldsymbol{f} = \Sigma\varphi_i x_i$ with $\varphi_i \in \mathscr{K}(T)$ and $x_i \in E$ and then, passing to the limit, for every $\boldsymbol{f} \in \mathscr{L}_E^p(\mu)$.

Let us prove now the equality $|||U||| = N_q(\boldsymbol{u}, \mu)$. Since the inequality $|||U||| \leqslant N_p(\boldsymbol{u}, \mu)$ was already proved, if remains to prove the converse inequality, in the cases of point (iii).

From theorem 18.42 we deduce that there exists a Z-weakly locally μ-integrable function $\boldsymbol{v} : T \to \mathscr{L}(E, Z')$ such that $|||U||| = N_q(\boldsymbol{v}, \mu)$ and

$$\langle U(\boldsymbol{f}), z\rangle = \int \langle \boldsymbol{v}\boldsymbol{f}, z\rangle\,\mathrm{d}\mu, \text{ for } \boldsymbol{f} \in \mathscr{L}_E^p(\mu) \text{ and } z \in Z.$$

In addition, for each lifting ρ of $\mathscr{L}^{\infty}(\mu)$ we can choose $\boldsymbol{v}$ such that $\rho[\boldsymbol{v}]=\boldsymbol{v}$. We have then

$$\int \langle \boldsymbol{u}\varphi x, z\rangle \, d\mu = \int \langle \boldsymbol{v}\varphi x, z\rangle \, d\mu, \text{ for } \varphi\in\mathscr{K}(T),\ x\in E \text{ and } z\in Z,$$

hence $\langle \boldsymbol{u}(t)x, z\rangle = \langle \boldsymbol{v}(t)x, z\rangle$ μ-almost everywhere, for each $x\in E$ and each $z\in Z$, that is $\boldsymbol{u}\equiv\boldsymbol{v}$.

We shall show that in each of the cases of point (iii) we have $|\boldsymbol{u}(t)|=|\boldsymbol{v}(t)|$ μ-almost everywhere, whence it will follow that $N_q(\boldsymbol{u}, \mu)=|||U|||$.

If there exists a lifting ρ of $\mathscr{L}^{\infty}(\mu)$ such that $\rho[\boldsymbol{u}]=\boldsymbol{u}$, then taking $\boldsymbol{v}$ such that we have also $\rho[\boldsymbol{v}]=\boldsymbol{v}$, we deduce $\boldsymbol{u}(t)=\boldsymbol{v}(t)$ μ-almost everywhere, hence $|\boldsymbol{u}(t)|=|\boldsymbol{v}(t)|$ μ-almost everywhere (see property 4 of the functions with the lifting property, §14, or the proof of proposition 17.14).

If condition (2) of point (iii) is fulfilled, then again it follows that $\boldsymbol{u}(t)=\boldsymbol{v}(t)$, μ-almost everywhere (see the proof of the proposition 17.15).

Assume now that condition (3) of point (iii) is fulfilled. Since $\boldsymbol{u}$ is simply μ-measurable, for each $x\in E$ the function $t\to\boldsymbol{u}(t)x$ is μ-measurable. For every $x\in E$ and every compact set $K\subset T$ we have $\varphi_K x\in\mathscr{L}^p_E(\mu)$ hence, by condition (i),

$$\int_K^* |\boldsymbol{u}(t)x| \, d|\mu| < \infty .$$

It follows that for every $x\in E$, the function $\boldsymbol{u}(t)x$ is locally μ-integrable (but $\boldsymbol{u}$ is not necessarily simply locally μ-integrable, since we can have $\int_K |\boldsymbol{u}| \, d|\mu| = +\infty$).

Choose $\boldsymbol{v}$ such that $\rho[\boldsymbol{v}]=\boldsymbol{v}$, for a certain lifting ρ of $\mathscr{L}^{\infty}(\mu)$. Then the function $t\to\boldsymbol{v}(t)x$ is also Z-weakly μ-measurable and $\rho[\boldsymbol{v}x]=\boldsymbol{v}x$ for each $x\in E$. It follows that the function $t\to|\boldsymbol{v}(t)x|$ is μ-measurable. Since $|\boldsymbol{v}|$ is locally μ-integrable, we have

$$\int_K^* |\boldsymbol{v}(t)x| \, d|\mu| \leqslant |x| \int_K^* |\boldsymbol{v}(t)| \, d|\mu| < \infty ,$$

for every compact set K, hence the function $t\to\boldsymbol{v}(t)x$ is Z-weakly locally μ-integrable and we have

$$\int \langle \boldsymbol{u}(t)x(t), z\rangle \, d\mu = \int \langle \boldsymbol{v}(t)x\varphi(t), z\rangle \, d\mu$$

for every $x\in E$, $z\in Z$ and $\varphi\in\mathscr{L}^p(\mu)$, in particular for $\varphi\in\mathscr{K}(T)$. It follows that $(\boldsymbol{u}x)\mu=(\boldsymbol{v}x)\mu$ for each $x\in E$.

From theorem 18.17 we deduce that $|(\boldsymbol{u}x)\mu|=|\boldsymbol{u}x|\,|\mu|$ and $|(\boldsymbol{v}x)\mu|=|\boldsymbol{v}x|\,|\mu|$, hence $|\boldsymbol{u}x|\,|\mu|=|\boldsymbol{v}x|\,|\mu|$; from proposition 17.5, we deduce that $|\boldsymbol{u}(t)x|=|\boldsymbol{v}(t)x|$, μ-almost everywhere for each $x\in E$. Letting x run over a countable set dense in E, we deduce that $|\boldsymbol{u}(t)| = |\boldsymbol{v}(t)|$, μ-almost everywhere.

Assume, finally, that $\boldsymbol{u}$ is locally μ-integrable and choose $\boldsymbol{v}$ such that $\rho[\boldsymbol{v}] = \boldsymbol{v}$ for a certain lifting ρ of $\mathscr{L}^{\infty}(\mu)$. From the equality $\boldsymbol{u} \equiv \boldsymbol{v}$ we deduce that $\boldsymbol{u}\mu = \boldsymbol{v}\mu$. By theorem 18.17 we have $|\boldsymbol{u}\mu| = |\boldsymbol{u}|\,|\mu|$ and $|\boldsymbol{v}\mu| = |\boldsymbol{v}|\,|\mu|$ hence $|\boldsymbol{u}|\,|\mu| = |\boldsymbol{v}|\,|\mu|$, whence, by proposition 17.5, $|\boldsymbol{u}(t)| = |\boldsymbol{v}(t)|$, μ-almost everywhere.

Thus the theorem is completely proved.

Remarks.

1. If we consider the condition

(i′) $\int^* |\boldsymbol{u}(t)\boldsymbol{f}(t)|\,\mathrm{d}|\mu| < \infty$, for every $\boldsymbol{f} \in \mathscr{L}^p_E(\mu)$,

instead of condition (i), we deduce as in the proof of the theorem, that there exists a linear mapping $U : \mathscr{L}^p_E(\mu) \to Z'$ such that

$$\langle U(\boldsymbol{f}), z\rangle = \int \langle \boldsymbol{u}\boldsymbol{f}, z\rangle\,\mathrm{d}\mu, \text{ for } \boldsymbol{f} \in \mathscr{L}^p_E(\mu) \text{ and } z \in Z,$$

and

$$|||U||| \leqslant N_q(\boldsymbol{u}, \mu) \leqslant \infty .$$

We remark that (i) implies (i′).

2. If condition (i′) and one of the conditions of point (iii) are fulfilled, then $|||U||| = N_q(\boldsymbol{u}, \mu)$.

In fact, if $|||U||| = \infty$, then we have $N_q(\boldsymbol{u}, \mu) = \infty$.

If $|||N||| < \infty$, then we reason as in the proof of the theorem.

18.45 Corollary. *Let $\boldsymbol{u} : T \to \mathscr{L}^*(E, F)$ be a Z-weakly locally μ-measurable and essentially bounded function. There exists then a linear continuous mapping $U : \mathscr{L}^1_E(\mu) \to Z'$ such that $\| U \| \leqslant N_\infty(\boldsymbol{u}, \mu)$ and*

$$\langle U(\boldsymbol{f}), z\rangle = \int \langle \boldsymbol{u}\boldsymbol{f}, z\rangle\,\mathrm{d}\mu, \textit{ for } \boldsymbol{f} \in \mathscr{L}^1_E(\mu) \textit{ and } z \in Z.$$

We have $\| U \| = N_\infty(\boldsymbol{u}, \mu)$ in each of the following cases:

(1) *there exists a lifting ρ of $\mathscr{L}^{\infty}(\mu)$ with $\rho(\boldsymbol{u}) = \boldsymbol{u}$;*

(2) *E is of countable type and there exists a countable set $S \subset Z$ norming for E;*

(3) *E is of countable type and $\boldsymbol{u}$ is simply μ-measurable;*

(4) *$\boldsymbol{u}$ is μ-measurable.*

We have $U(\boldsymbol{f}) \in F$ for every $\boldsymbol{f} \in \mathscr{L}^1_E(\mu)$ in each of the following cases:

(1) *$F = Z'$;*

(2) *$\boldsymbol{u}$ is simply μ-measurable; in particular F is of countable type;*

(3) *for each* $x \in E$ *there exists a locally countable family* (K_j) *of disjoint compact sets with* $T \setminus \cup K_j$ *μ-negligible such that, for each j, the convex balanced hull of the set* $\{\boldsymbol{u}(t)x\,;\, t \in K_j\}$ *is relatively compact in F for the topology* $\sigma(F, Z)$.

Now we are able to prove that there exists no lifting of the space $\mathscr{L}^p(\mu)$ with $1 \leqslant p < \infty$, if μ is not atomic.

We prove first the following proposition.

18.46 PROPOSITION. *Let* μ *be a positive measure and let* $\mathscr{L}^p(\mu)$ *with* $1 \leqslant p < \infty$. *Every positive linear functional on* $\mathscr{L}^p(\mu)$ *is continuous.*

Let U be a positive linear functional on $\mathscr{L}^p(\mu)$. Assume the contrary, that U is not continuous. Then for every $n \in N$ there exists a function $f_n \in \mathscr{L}^p(\mu)$ with $N_p(f_n) \leqslant 1$, $f_n \geqslant 0$ and $U(f_n) > n^3$. For the μ-measurable function

$$f = \Sigma n^{-2} f_n$$

we have

$$N_p(f) \leqslant \Sigma n^{-2} N_p(f_n) \leqslant \Sigma n^{-2} < \infty\,,$$

hence $f \in \mathscr{L}^p(\mu)$. But

$$U(f) \geqslant n^{-2}\, U(f_n) \geqslant n, \text{ for every } n\,,$$

that is, $U(f) = +\infty$, which is absurd.

Thus U is continuous and the proposition is proved.

We recall that a linear lifting of the space $\mathscr{L}^p(\mu)$ is a mapping $\rho : \mathscr{L}^p(\mu) \to \mathscr{L}^p(\mu)$ which has the following properties:

(1) $\rho(f) \equiv f$;

(2) $f \equiv g$ implies $\rho(f) = \rho(g)$;

(3) $f \geqslant 0$ implies $\rho(f) \geqslant 0$;

(4) $\rho(\alpha f + \beta g) = \alpha\rho(f) + \beta\rho(g)$.

$f \equiv g$ means $f(t) = g(t)$, μ-almost everywhere.

18.47 PROPOSITION. *Let* $\mu \neq 0$ *be a nonatomic measure. Then there is no linear lifting of the space* $\mathscr{L}^p(\mu)$ *with* $1 \leqslant p < \infty$.

Assume the contrary, that there exists a linear lifting ρ of $\mathscr{L}^p(\mu)$. For each $t_0 \in T$, the mapping $f \to \rho(f)(t_0)$ is a positive linear functional on $\mathscr{L}^p(\mu)$, therefore it is continuous. It follows that there exists a function $g(t_0) \in \mathscr{L}^q$, $p^{-1}+q^{-1}=1$, such that

$$\rho(f)(t_0) = \int fg(t_0)\,\mathrm{d}\mu, \text{ for } f \in \mathscr{L}^p(\mu).$$

In this way we define a mapping $t \to g(t)$ of T into $\mathscr{L}^q(\mu)$ such that

$$\rho(f)(t) = \int fg(t)\,\mathrm{d}\mu, \text{ for } f \in \mathscr{L}^p(\mu) \text{ and } t \in T.$$

Let K be a μ-measurable set with $0 < \mu(K) < \infty$ and denote $a = \mu(K)$ and $A_n = \{t; t \in K, \|g(t)\|_q \leqslant n\}$ for each n. Since $\|g(t)\|_q < \infty$ for every $t \in T$, we have $K = \cup A_n$, hence there exists a number $n' \in N$ with $\mu^*(A_{n'}) \neq 0$. Let r be an integer such that

$$r \geqslant a(n'+1)^p$$

and let $B_1, \ldots, B_r$ be disjoint μ-measurable subsets of K with union K and such that

$$\mu(B_1) = \ldots = \mu(B_r) = a/r.$$

Since the sets B_i cover K, there exists a set, for example, B_1, such that $\mu^*(A_{n'} \cap B_1) \neq 0$. Consider now the set $B_1^* = \{t; \rho(\varphi_{B_1})(t) = 1\}$. Then $B_1^* \equiv B_1$, consequently

$$\mu^*(A_{n'} \cap B_1^*) \neq 0.$$

For the function $f = (n'+1)\varphi_{B_1}$ we have

$$N_p(f) = (n'+1)\,N_p(\varphi_{B_1}) = (n'+1)(a/r)^{1/p} \leqslant 1.$$

On the other hand, since $\mu^*(A_{n'} \cap B_1^*) \neq 0$, we deduce $A_{n'} \cap B_1^* \neq \varnothing$. But if $t_0 \in A_{n'} \cap B_1 \neq \varnothing$, then

$$n'+1 = \rho((n'+1)\varphi_{B_1})(t_0) = \rho(f)(t_0) = \int fg(t_0)\,\mathrm{d}\mu \leqslant \|g(t_0)\|_q \leqslant n'$$

and we obtained a contradiction.

Thus the proposition is proved.

Chapter V

SUMS OF MEASURES. IMAGES OF MEASURES.

§19 Summable families of measures

1 *Summable families of positive measures*

Let $(\mu_i)_{i \in I}$ be a family of positive measures on T.

19.1 DEFINITION. *We say that the family $(\mu_i)_{i \in I}$ of positive measures is summable if for each function f of $\mathcal{K}(T)$, the family of numbers $(\mu_i(f))_{i \in I}$ is summable.*★

If the family (μ_i) is summable and if for each function $f \in \mathcal{K}(T)$ we write

$$\mu(f) = \Sigma_{i \in I} \mu_i(f)$$

the mapping $f \to \mu(f)$ is a *positive linear functional* on $\mathcal{K}(T)$, hence μ *is a positive measure* on T, called the *sum of the family* $(\mu_i)_{i \in I}$. We write $\mu = \Sigma_{i \in I} \mu_i$.

19.2 PROPOSITION. *If the family $(\mu_i)_{i \in I}$ is summable, then for each function $f \in \mathcal{K}(T)$ the family of numbers $(\mu_i(f))_{i \in I}$ is absolutely summable. The family $(\mu_i)_{i \in I}$ is summable if (and only if) for each positive function f of $\mathcal{K}(T)$ we have*

$$\Sigma_{i \in I} \mu_i(f) < +\infty\ .$$

★ A family $(x_i)_{i \in I}$ of elements of a Banach space X is summable if there exists $x \in X$ with the property that for each $\varepsilon > 0$, there exists a finite family $J_\varepsilon \subset I$ such that for every finite family $J \supset J_\varepsilon$ we have $|\Sigma_{i \in J} x_i - x| < \varepsilon$. We say that x is the sum of the family $(x_i)_{i \in I}$ and we write $x = \Sigma_{i \in I} x_i$. The family $(x_i)_{i \in I}$ is *absolutely summable* if the family of positive numbers $(|x_i|)_{i \in I}$ is summable. Every absolutely summable family is summable. For a summable family $(\alpha_i)_{i \in I}$ of *positive numbers* we have $\Sigma_{i \in I} \alpha_i = \sup_{J \subset I} \Sigma_{i \in J} \alpha_i < +\infty$, J finite. If the family is not summable, we have $\Sigma_{i \in I} \alpha_i = \sup_{J \subset I} \Sigma_{i \in J} \alpha_i = +\infty$

Assume that the family $(\mu_i)_{i\in I}$ is summable. If $f\in\mathscr{K}(T)$, then $|f|\in\mathscr{K}(T)$, therefore the family of positive numbers $(\mu_i(|f|))_{i\in I}$ is summable. Since for each i we have $|\mu_i(f)|\leqslant\mu_i(|f|)$, it follows that the family of positive numbers $(|\mu_i(f)|_{i\in I}$ is summable, that is, the family of numbers $(\mu_i(f))_{i\in I}$ is absolutely summable.

Assume now that for each positive function f of $\mathscr{K}(T)$ we have

$$\Sigma_{i\in I}\mu_i(f)<+\infty\,.$$

From the first part of the proof we deduce that for every function $g\in\mathscr{K}(T)$, the family $(\mu_i(g))$ is absolutely summable, therefore summable, and hence the family of measures $(\mu_i)_{i\in I}$ is summable.

19.3 PROPOSITION. *The family of measures $(\mu_i)_{i\in I}$ is summable if and only if for every compact set $K\subset T$ we have*

$$\Sigma_{i\in I}\mu_i(K)<+\infty\,.$$

Assume that the family $(\mu_i)_{i\in I}$ is summable and let $K\subset T$ be a compact set. There exists a continuous function with compact support $f\colon T\to[0,1]$ equal to 1 on K. We have $\varphi_K\leqslant f$, therefore $\mu_i(K)\leqslant\mu_i(f)$ for every $i\in I$. Then

$$\Sigma_{i\in I}\mu_i(K)\leqslant\Sigma_{i\in I}\mu_i(f)<+\infty\,.$$

Conversely, assume that for each compact set $K\subset T$ we have

$$\Sigma_{i\in I}\mu_i(K)<+\infty\,.$$

Let f be a positive function of $\mathscr{K}(T)$ and K be the support of f. We have $f\leqslant\|f\|\varphi_K$ therefore $\mu_i(f)\leqslant\|f\|\mu_i(K)$ for every $i\in I$. Then

$$\Sigma_{i\in I}\mu_i(f)\leqslant\Sigma_{i\in I}\|f\|\mu_i(K)=\|f\|\Sigma_{i\in I}\mu_i(K)<+\infty\,.$$

By proposition 19.2, the family $(\mu_i)_{i\in I}$ is summable.

Example. Every atomic measure $\mu\geqslant 0$ is the sum of a summable family of positive measures.

In fact, let α be a positive function defined on T, such that

$$\mu(f)=\Sigma_{t\in T}\alpha(t)f(t)<+\infty$$

for every $f\in\mathscr{K}(T)$. For every $t\in T$ denote by μ_t the measure defined by the mass $\alpha(t)$ at the point t:

$$\mu_t(f) = \alpha(t) f(t), \text{ for } f \in \mathscr{K}(T).$$

Then

$$\mu(f) = \Sigma_{t \in T} \mu_t(f), \text{ for } f \in \mathscr{K}(T),$$

therefore

$$\mu = \Sigma_{t \in T} \mu_t .$$

2 *Integration with respect to the sum of a family of measures*

Let $(\mu_i)_{i \in I}$ be a summable family of positive measures and

$$\mu = \Sigma_{i \in I} \mu_i .$$

19.4 PROPOSITION. *For every lower semicontinuous function $f \geqslant 0$ on T we have*

$$\mu^*(f) = \Sigma_{i \in I} \mu_i^*(f).$$

In fact, for $g \in \mathscr{K}_+$ and J finite we have

$$\begin{aligned} \mu^*(f) &= \sup_{g \leqslant f} \mu(g) = \sup_{g \leqslant f} \Sigma_{i \in I} \mu_i(g) = \\ &= \sup_{g \leqslant f} \sup_{J \subset I} \Sigma_{i \in J} \mu_i(g) = \sup_{J \subset I} \sup_{g \leqslant f} \Sigma_{i \in J} \mu_i(g) = \\ &= \sup_{J \subset I} \lim_{g \leqslant f} \Sigma_{i \in J} \mu_i(g) = \sup_{J \subset I} \Sigma_{i \in J} \mu_i^*(f) = \Sigma_{i \in I} \mu_i^*(f) . \end{aligned}$$

19.5 PROPOSITION. *For every function $f \geqslant 0$ defined on T we have*

$$\mu^*(f) \geqslant \Sigma_{i \in I} \mu_i^*(f).$$

In fact, for every compact set $K \subset T$ and every lower semicontinuous function $g \geqslant f\varphi_K$ we have

$$\mu^*(g) = \Sigma_{i \in I} \mu_i^*(g) \geqslant \Sigma_{i \in I} \mu_i^*(f\varphi_K),$$

whence

$$\mu^*(f\varphi_K) = \inf_{g \geqslant f\varphi_K} \mu^*(g) \geqslant \Sigma_{i \in I} \mu_i^*(f\varphi_K), \quad g \in \mathscr{I}_+ .$$

Then we have

$$\mu^*(f) = \sup_K \mu^*(f\varphi_K) \geqslant \sup_K \sup_{J\subset I} \Sigma_{i\in J}\, \mu_i^*(f\varphi_K) =$$
$$= \sup_{J\subset I} \sup_K (\Sigma_{i\in J}\, \mu_i)^*(f\varphi_K) =$$
$$= \sup_{J\subset I} (\Sigma_{i\in J}\, \mu_i)^*(f) = \sup_{J\subset I} \Sigma_{i\in J}\, \mu_i^*(f) = \Sigma_{i\in I}\, \mu_i^*(f)\,,$$

where K are compact sets in T and J are finite subsets of I.

Remarks. For a positive atomic measure defined by $\mu(f)=\Sigma_{t\in T}\,\alpha(t)f(t)$, $f\in\mathscr{K}(T)$, the above propositions have been proved in the following form

$$\mu^*(f)=\Sigma_{t\in T}\,\alpha(t)f(t) \text{ for } f\in\mathscr{I}_+ \quad\text{and}\quad \mu^*(f)\geqslant\Sigma_{t\in T}\,\alpha(t)f(t) \text{ for } f\geqslant 0\,.$$

19.6 Theorem. *If $\boldsymbol{f}$ is a μ-integrable function defined on T with values in a Banach space E or in $\overline{R}$, then for every $i\in I$, the function $\boldsymbol{f}$ is μ_i-integrable, the family $(\int \boldsymbol{f}\mathrm{d}\mu_i)_{i\in I}$ is absolutely summable and*

$$\int \boldsymbol{f}\mathrm{d}\mu = \Sigma_{i\in I} \int \boldsymbol{f}\mathrm{d}\mu_i\,.$$

The integrability of $\boldsymbol{f}$ with respect to the measures μ_i follows from the inequality $\mu_i\leqslant\mu$, for every $i\in I$.

From the inequalities (proposition 19.5).

$$\Sigma_{i\in I}\left|\int \boldsymbol{f}\mathrm{d}\mu_i\right| \leqslant \Sigma_{i\in I}\int |\boldsymbol{f}|\,\mathrm{d}\mu_i \leqslant \int |\boldsymbol{f}|\,\mathrm{d}\mu = \int^* |\boldsymbol{f}|\,\mathrm{d}\mu < +\infty\,,$$

it follows that the family $(\int \boldsymbol{f}\mathrm{d}\mu_i)_{i\in I}$ is absolutely summable.

For each function $\varphi\in\mathscr{K}(T)$ and each $x\in E$ we have

$$\int \varphi x\,\mathrm{d}\mu = \left(\int \varphi\,\mathrm{d}\mu\right)x = \left(\Sigma_{i\in I}\int \varphi\,\mathrm{d}\mu_i\right)x =$$
$$= \Sigma_{i\in I}\left(\int \varphi\,\mathrm{d}\mu_i\right)x = \Sigma_{i\in I}\int \varphi x\,\mathrm{d}\mu_i\,.$$

Then for every finite sum of the form $g=\Sigma\,\varphi_k x_k$ with $\varphi_k\in\mathscr{K}(T)$ and $x_k\in E$ we have

$$\int \boldsymbol{g}\,\mathrm{d}\mu=\Sigma_{i\in I}\int \boldsymbol{g}\,\mathrm{d}\mu_i\,.$$

Since the set $\mathscr{H}_E(T)$ of these finite sums is dense in $\mathscr{L}_E^1(\mu)$,there exists a sequence $(\boldsymbol{g}_n)$ of functions of $\mathscr{H}_E(T)$ covering in the mean to $\boldsymbol{f}$:

$$\lim \int |\boldsymbol{f}-\boldsymbol{g}_n|\,\mathrm{d}\mu = 0\,.$$

It follows that

$$\lim \int \boldsymbol{g}_n\mathrm{d}\mu = \int \boldsymbol{f}\mathrm{d}\mu\,.$$

On the other hand, from proposition 19.5 we deduce that

$$\lim \Sigma_{i\in I} \int |f - g_n| \, d\mu_i = 0 .$$

Then

$$|\Sigma_{i\in I} \int f d\mu_i - \Sigma_{i\in I} \int g_n d\mu_i| = |\Sigma_{i\in I} (\int f d\mu_i - \int g_n d\mu_i)| =$$
$$= |\Sigma_{i\in I} \int (f - g_n) d\mu_i| \leqslant \Sigma_{i\in I} | \int (f - g_n) d\mu_i| \leqslant$$
$$\leqslant \Sigma_{i\in I} \int |(f - g_n| \, d\mu_i ,$$

therefore

$$\lim \Sigma_{i\in I} \int g_n d\mu = \Sigma_{i\in I} \int f d\mu_i .$$

Since for each function g_n we have

$$\int g_n d\mu = \Sigma_{i\in I} \int g_n d\mu_i ,$$

passing to the limit we obtain

$$\int f d\mu = \Sigma_{i\in I} \int f d\mu_i$$

and the theorem is proved.

Remark. For an atomic measure, the equality in the statement of theorem 19.6 is written

$$\int f d\mu = \Sigma_{i\in I} \alpha(t) f(t) .$$

19.7 COROLLARY. *For every relatively compact Borel set $A \subset T$ we have*

$$\mu(A) = \Sigma_{i\in I} \mu_i(A) .$$

In fact, A is μ-integrable.

19.8 PROPOSITION. *A set $N \subset T$ is μ-negligible if and only if it is μ_i-negligible for every $i \in I$.*

From the inequalities $\mu_i \leqslant \mu$ it follows that if N is μ-negligible, then N is μ_i-negligible for every $i \in I$. Conversely, assume that $\mu_i^*(N) = 0$ for every $i \in I$. Let $K \subset T$ be a compact set and U a relatively compact neighbourhood of K. From the preceding corollary it follows

$$\Sigma_{i\in I} \mu_i(U) = \mu(U) < +\infty .$$

Let $\varepsilon > 0$; there exists a finite subset $J \subset I$ such that

$$\Sigma_{i \notin J} \mu_i(U) < \varepsilon/2 .$$

Let n be the number of elements of J. For each $i \in J$, we have $\mu_i^*(K \cap N) = 0$, therefore there exists an open neighbourhood $V_i \subset U$ of $K \cap N$ such that

$$\mu_i^*(V_i) < \varepsilon/2n .$$

Then $V = \cap_{1 \leqslant i \leqslant n} V_i$ is an open neighbourhood of $K \cap N$ and for every $i \in J$ we have $\mu_i(V) < \varepsilon/2n$. Then

$$\mu(V) = \Sigma_{i \in I} \mu_i(V) = \Sigma_{i \in J} \mu_i(V) + \Sigma_{i \notin J} \mu_i(V) \leqslant$$
$$\leqslant n\varepsilon/2n + \Sigma_{i \notin J} \mu_i(U) < (\varepsilon/2) + (\varepsilon/2) = \varepsilon ,$$

therefore $\mu^*(K \cap N) = 0$. It follows that N is μ-negligible.

Remarks. For an atomic measure defined by $\mu(f) = \Sigma_{t \in T} \alpha(t) f(t)$ for $f \in \mathscr{K}(T)$, a function f is μ-negligible if and only if $\alpha(t) f(t) \equiv 0$.

19.9 PROPOSITION. *A function f defined on T with values in a topological space X is μ-measurable if and only if f is μ_i-measurable for every $i \in I$.*

From the inequalities $\mu_i \leqslant \mu$ it follows that if f is μ-measurable then f is μ_i-measurable, for every $i \in I$.

Conversely, assume that for each $i \in I$, the function f is μ_i-measurable. Let $K \subset T$ be a compact set and $\varepsilon > 0$. Since

$$\Sigma_{i \in I} \mu_i(K) = \mu(K) < +\infty ,$$

there exists a finite subset $J \subset I$ such that

$$\Sigma_{i \notin J} \mu_i(K) < \varepsilon/2 .$$

Let n be the number of the elements of J. For each $i \in J$ there exists a compact set $K_i \subset K$ such that $\mu_i(K \setminus K_i) \leqslant \varepsilon/2n$ and the restriction of f to K_i is continuous. The set $K' = \cap_{i \in J} K_i$ is compact, the restriction of f to K' is continuous and $\mu_i(K \setminus K') \leqslant \mu_i(K \setminus K_i) < \varepsilon/2n$ for each $i \in J$. Then

$$\mu(K \setminus K') = \Sigma_{i \in I} \mu_i(K \setminus K') = \Sigma_{i \in J} \mu_i(K \setminus K') + \Sigma_{i \notin J} \mu_i(K \setminus K') \leqslant$$
$$\leqslant n \cdot \varepsilon/2n + \Sigma_{i \notin J} \mu_i(K) \leqslant (\varepsilon/2) + (\varepsilon/2) = \varepsilon .$$

It follows that f is μ-measurable.

Remark. For an atomic measure every function is measurable.

19.10 PROPOSITION. *If the function $f \geqslant 0$ is μ-measurable and vanishes outside the union of a sequence (A_n) of μ-integrable sets, then*

$$\int^* f \mathrm{d}\mu = \Sigma_{i \in I} \int^* f \mathrm{d}\mu_i \, .$$

We may assume that the sequence (A_n) is increasing. For each $n \in N$ the function $f_n = \inf(n, f\varphi_{A_n})$ is bounded, μ-measurable and vanishes outside the μ-integrable set A_n. It follows that f_n is μ-integrable. From theorem 19.6 we deduce that f_n is μ_i-integrable for each $i \in I$ and that

$$\int f_n \mathrm{d}\mu = \Sigma_{i \in I} \int f_n \mathrm{d}\mu_i \, .$$

But the sequence f_n is increasing and $\lim f_n(t) = \sup f_n(t) = f(t)$. Then

$$\int^* f \mathrm{d}\mu = \sup \int f_n \mathrm{d}\mu = \sup \Sigma_{i \in I} \int f_n \mathrm{d}\mu_i =$$

$$= \sup \sup_{J \subset I} \Sigma_{i \in J} \int f_n \mathrm{d}\mu_i = \sup_{J \subset I} \sup \int f_n \mathrm{d}(\Sigma_{i \in J} \mu_i) =$$

$$= \sup_{J \subset I} \int^* f \mathrm{d}(\Sigma_{i \in J} \mu_i) = \sup_{J \subset I} \Sigma_{i \in J} \int^* f \, \mathrm{d}\mu_i = \Sigma_{i \in I} \int^* f \mathrm{d}\mu_i .$$

19.11 PROPOSITION. *Assume that T is countable at infinity. A function $\boldsymbol{f}$ defined on T with values in a Banach space E or in $\overline{R}$ is μ-integrable if and only if for each $i \in I$, the function $\boldsymbol{f}$ is μ_i-integrable and $\Sigma_{i \in I} \int^* |\boldsymbol{f}| \mathrm{d}\mu_i < +\infty$.*

If $\boldsymbol{f}$ is μ-integrable, from theorem 19.6 it follows that $\boldsymbol{f}$ is μ_i-integrable for each $i \in I$. From proposition 19.5 it follows that

$$\Sigma_{i \in I} \int^* |\boldsymbol{f}| \mathrm{d}\mu \leqslant \int^* |\boldsymbol{f}| \mathrm{d}\mu < +\infty \, .$$

Conversely, assume that $\boldsymbol{f}$ is μ_i-integrable for every $i \in I$ and that $\Sigma_{i \in I} \int^* |\boldsymbol{f}| \mathrm{d}\mu_i < +\infty$. Then $\boldsymbol{f}$ is μ_i-measurable for each $i \in I$, therefore $\boldsymbol{f}$ is μ-measurable. From proposition 19.10 it follows that $\int^* |\boldsymbol{f}| \mathrm{d}\mu = \Sigma_{i \in I} |\boldsymbol{f}| \mathrm{d}\mu_i < +\infty$. By the integrability criterion, the function $\boldsymbol{f}$ is μ-integrable.

3 *Summable families of vector measures*

Let X be a Banach space and $(\boldsymbol{m}_i)_{i \in I}$ a family of dominated measures on T with values in X.

We say that the family of vector measures $(\boldsymbol{m}_i)_{i \in I}$ is summable, if the family of positive measures $(|\boldsymbol{m}_i|)_{i \in I}$ is summable.

It follows that if the family $(\boldsymbol{m}_i)_{i\in I}$ is summable, then for every function $f\in \mathscr{K}(T)$ the family $(\boldsymbol{m}_i(f))_{i\in I}$ of elements of X is absolutely summable.

In fact if $f\in\mathscr{K}(T)$, then

$$\Sigma_{i\in I}|\boldsymbol{m}_i(f)|\leqslant \Sigma_{i\in I}|\boldsymbol{m}_i|(f)< +\infty \ .$$

For each function $f\in\mathscr{K}(T)$ put

$$\boldsymbol{m}(f)=\Sigma_{i\in I}\,\boldsymbol{m}_i(f)\,.$$

The mapping $f\to\boldsymbol{m}(f)$ of $\mathscr{K}(T)$ into X is linear. It is dominated by the positive measure $\mu=\Sigma_{i\in I}|\boldsymbol{m}_i|$:

$$|\boldsymbol{m}(f)|=|\Sigma_{i\in I}\boldsymbol{m}_i(f)|\leqslant\Sigma_{i\in I}|\boldsymbol{m}_i(f)|\leqslant\Sigma_{i\in I}|\boldsymbol{m}_i|(|f|)=\mu(|f|)\,,$$

therefore $\boldsymbol{m}$ is a dominated measure on T and $|\boldsymbol{m}|\leqslant\mu$.

The measure $\boldsymbol{m}$ is called the sum of the family $(\boldsymbol{m}_i)_{i\in I}$ of vector measures and we write

$$\boldsymbol{m}=\Sigma_{i\in I}\,\boldsymbol{m}_i\ .$$

The inequality $|\boldsymbol{m}|\leqslant\mu$ is now written

$$|\Sigma_{i\in I}\,\boldsymbol{m}_i|\leqslant\Sigma_{i\in I}|\boldsymbol{m}_i|$$

and the inequality is generally strict, even for a finite family of measures.

19.12 THEOREM. *Let E and F be two Banach spaces such that $X\subset\mathscr{L}(E,F)$. If $\boldsymbol{f}:T\to E$ is a μ-integrable function then $\boldsymbol{f}$ is $\boldsymbol{m}$-integrable and $\boldsymbol{m}_i$-integrable, for every $i\in I$, and the family $(\int\boldsymbol{f}\,\mathrm{d}\boldsymbol{m}_i)_{i\in I}$ is absolutely summable and*

$$\int\boldsymbol{f}\,\mathrm{d}\boldsymbol{m}=\Sigma_{i\in I}\int\boldsymbol{f}\,\mathrm{d}\boldsymbol{m}_i\ .$$

From the inequality $|\boldsymbol{m}|\leqslant\mu$ and $|\boldsymbol{m}_i|\leqslant\mu$ it follows that the function $\boldsymbol{f}$ is $\boldsymbol{m}$-integrable and $\boldsymbol{m}_i$-integrable for each $i\in I$. The fact that the family $(\int\boldsymbol{f}\,\mathrm{d}\boldsymbol{m}_i)_{i\in I}$ is absolutely summable follows from the inequalities:

$$\Sigma_{i\in I}|\int\boldsymbol{f}\,\mathrm{d}\boldsymbol{m}_i|\leqslant\Sigma_{i\in I}\int|\boldsymbol{f}|\,\mathrm{d}|\boldsymbol{m}_i|=\int|\boldsymbol{f}|\,\mathrm{d}\mu<+\infty\ .$$

For each function $\varphi\in\mathscr{K}(T)$ and each $x\in E$ we have

$$\begin{aligned}(\Sigma_{i\in I}\int\varphi\,\mathrm{d}\boldsymbol{m}_i)x&=(\lim_{J\subset I}\Sigma_{i\in J}\int\varphi\,\mathrm{d}\boldsymbol{m}_i)x=\\&=\lim_{J\subset I}(\int\varphi\,\mathrm{d}(\Sigma_{i\in J}\boldsymbol{m}_i))x=\\&=\lim_{J\subset I}\int\varphi x\,\mathrm{d}(\Sigma_{i\in J}\boldsymbol{m}_i)=\\&=\lim_{J\subset I}\Sigma_{i\in J}\int\varphi x\,\mathrm{d}\boldsymbol{m}_i=\Sigma_{i\in I}\int\varphi x\,\mathrm{d}\boldsymbol{m}_i\ .\end{aligned}$$

Hence we deduce

$$\int \varphi x \, \mathrm{d}\boldsymbol{m} = (\int \varphi \, \mathrm{d}\boldsymbol{m})x = (\Sigma_{i\in I} \int \varphi \, \mathrm{d}\boldsymbol{m}_i)x = \Sigma_{i\in I} \int \varphi x \, \mathrm{d}\boldsymbol{m}_i \, .$$

Then for every finite sum of the form $\boldsymbol{g} = \Sigma \varphi_j x_j$ with $\varphi_j \in \mathscr{K}(T)$ and $x_j \in E$ we have

$$\int \boldsymbol{g} \, \mathrm{d}\boldsymbol{m} = \Sigma_{i\in I} \int \boldsymbol{g} \, \mathrm{d}\boldsymbol{m}_i \, .$$

Let $(\boldsymbol{g}_n)$ be a sequence of finite sums of the form $\Sigma \varphi_j x_j$ with $\varphi_j \in \mathscr{K}(T)$ and $x_j \in E$, converging to $\boldsymbol{f}$ in the mean with respect to μ:

$$\lim \int |\boldsymbol{f} - \boldsymbol{g}_n| \, \mathrm{d}\mu = 0 \, .$$

From the inequality $|\boldsymbol{m}| \leqslant \mu$ we deduce

$$\lim \int |\boldsymbol{f} - \boldsymbol{g}_n| \, \mathrm{d}|\boldsymbol{m}| = 0 \, ,$$

therefore

$$\lim \int \boldsymbol{g}_n \, \mathrm{d}\boldsymbol{m} = \int \boldsymbol{f} \mathrm{d}\boldsymbol{m} \, .$$

On the other hand we deduce that

$$\lim_n \Sigma_{i\in I} \int |\boldsymbol{f} - \boldsymbol{g}_n| \, \mathrm{d}|\boldsymbol{m}_i| = 0 \, .$$

Then

$$|\Sigma_{i\in I} \int \boldsymbol{f} \mathrm{d}\boldsymbol{m}_i - \Sigma_{i\in I} \int \boldsymbol{g}_n \mathrm{d}\boldsymbol{m}_i| = |\Sigma_{i\in I} \int (\boldsymbol{f} - \boldsymbol{g}_n) \mathrm{d}\boldsymbol{m}_i| \leqslant$$
$$\leqslant \Sigma_{i\in I} |\int (\boldsymbol{f} - \boldsymbol{g}_n) \mathrm{d}\boldsymbol{m}_i| \leqslant \Sigma_{i\in I} \int |\boldsymbol{f} - \boldsymbol{g}_n| \mathrm{d}|\boldsymbol{m}_i| \, ,$$

therefore

$$\lim_n \Sigma_{i\in I} \int \boldsymbol{g}_n \mathrm{d}\boldsymbol{m}_i = \Sigma_{i\in I} \int \boldsymbol{f} \mathrm{d}\boldsymbol{m}_i \, .$$

For the functions $\boldsymbol{g}_n$ we have

$$\int \boldsymbol{g}_n \mathrm{d}\boldsymbol{m} = \Sigma_{i\in I} \int \boldsymbol{g}_n \mathrm{d}\boldsymbol{m} \, .$$

Passing to the limit we obtain

$$\int \boldsymbol{f} \mathrm{d}\boldsymbol{m} = \Sigma_{i\in I} \int \boldsymbol{f} \, \mathrm{d}\boldsymbol{m}_i \, .$$

19.13 COROLLARY. *For every relatively compact Borel set A, the family $(\boldsymbol{m}_i(A))_{i\in I}$ is absolutely summable and*

$$\boldsymbol{m}(A) = \Sigma_{i\in I} \boldsymbol{m}_i(A) \, .$$

19.14 PROPOSITION. *If a set $N \subset T$ is $\boldsymbol{m}_i$-negligible for every $i \in I$, then N is $\boldsymbol{m}$-negligible.*

In fact, N is $|\boldsymbol{m}_i|$-negligible for every $i \in I$, hence (proposition 19.8) it is μ-negligible. From the inequality $|\boldsymbol{m}| \leqslant \mu$ it follows that N is $\boldsymbol{m}$-negligible.

19.15 PROPOSITION. *Let f be a function defined on T with values in a topological space E. If f is $\boldsymbol{m}_i$-measurable for every $i \in I$, then f is $\boldsymbol{m}$-measurable.*

In fact, f is μ-measurable, hence $\boldsymbol{m}$-measurable.

19.16 PROPOSITION. *Assume that T is countable at infinity and let $\boldsymbol{f}$ be a function defined on T with values in a Banach space E or in $\overline{R}$. If $\boldsymbol{f}$ is $\boldsymbol{m}_i$-integrable for every $i \in I$ and if $\Sigma_{i \in I} \int |\boldsymbol{f}| \mathrm{d} |\boldsymbol{m}_i| < +\infty$, then f is $\boldsymbol{m}$-integrable and*

$$\int \boldsymbol{f} \mathrm{d}\boldsymbol{m} = \Sigma_{i \in I} \int \boldsymbol{f} \mathrm{d}\boldsymbol{m}_i \, .$$

In fact, from proposition 19.11 it follows that $\boldsymbol{f}$ is μ-integrable and from the inequality $|\boldsymbol{m}| \leqslant \mu$ it follows that $\boldsymbol{f}$ is $\boldsymbol{m}$-integrable. By theorem 19.12 we deduce then that

$$\int \boldsymbol{f} \mathrm{d}\boldsymbol{m} = \Sigma_{i \in I} \int \boldsymbol{f} \mathrm{d}\boldsymbol{m}_i \, .$$

§20 Images of measures

1 *Definition of images of measures*

Let μ be a *positive* measure on T, S a locally compact space and $\alpha : T \to S$ a function. We shall define a measure ν on S, by means of the function α and the measure μ on T. It is natural to choose the measure ν such that a function f defined on S is integrable with respect to ν if and only if $f \circ \alpha$ is integrable with respect to μ.

In particular, for every function $f \in \mathscr{K}(S)$ the function $f \circ \alpha$ must be integrable with respect to μ.

20.1 DEFINITION. *We say that a function $\alpha : T \to S$ is proper for the measure μ, or μ-proper, if:*

(1) *α is μ-measurable;*

(2) *for every function $f \in \mathscr{K}(S)$, the function $f \circ \alpha : T \to R$ is μ-integrable.*

If $\boldsymbol{m}$ is a dominated measure on T, we say that α is $\boldsymbol{m}$-proper if α is $|\boldsymbol{m}|$-proper.

Examples.

1. If μ is a bounded positive measure (in particular, if μ has compact support), then every μ-measurable function $\alpha : T \to S$ is μ-proper. In fact, for every continuous function $f \in \mathscr{K}(S)$, the function $f \circ \alpha$ is μ-measurable and

$$\int^* |f(\alpha(t))| \mathrm{d}\mu(t) \leqslant \|f\| \mu^*(T) < +\infty ,$$

hence $f \circ \alpha$ is μ-integrable.

2. If $\alpha : T \to S$ is a *proper continuous* function (that is, for every compact set $K \subset S$, $\alpha^{-1}(K)$ is compact), then α is proper for every measure μ on T.

In fact, α is measurable for every measure μ on T; for every continuous function $f \in \mathscr{K}(S)$ with support K, the function $f \circ \alpha$ is continuous on T and has compact support contained in the compact set $\alpha^{-1}(K)$, consequently $f \circ \alpha$ is integrable for every measure μ on T.

Let X be a Banach space, $\boldsymbol{m}$ a *dominated* measure on T with values in X and $\alpha : T \to S$ an $\boldsymbol{m}$-proper function.

For every function $f \in \mathscr{K}(S)$, the real function $t \to f(\alpha(t))$ defined on T is $\boldsymbol{m}$-integrable. Put

$$\boldsymbol{n}(f) = \int f(\alpha(t)) \mathrm{d}\boldsymbol{m}(t)$$

and

$$v(f) = \int f(\alpha(t) \mathrm{d}|\boldsymbol{m}|(t) .$$

The mapping $f \to v(f)$ is a linear and positive functional on $\mathscr{K}(S)$, consequently it is a *positive measure* on S.

The mapping $f \to \boldsymbol{n}(f)$ of $\mathscr{K}(S)$ into X is *linear* and *dominated* by v:

$$|\boldsymbol{n}(f)| = |\int f(\alpha(t)) \mathrm{d}\boldsymbol{m}(t)| \leqslant \int |f(\alpha(t))| \mathrm{d}|\boldsymbol{m}| \leqslant$$
$$\leqslant \int |f| (\alpha(t)) \mathrm{d}|\boldsymbol{m}| = v(|f|) ,$$

hence $\boldsymbol{n}$ is a *dominated measure on S, with values in X.*

20.2 DEFINITION. *Let $\boldsymbol{m}$ be a dominated measure on T with values in X and $\alpha : T \to S$ an $\boldsymbol{m}$-proper function.*

The dominated measure $\boldsymbol{n}$ on S with values in X defined by the equality

$$\boldsymbol{n}(f) = \int f \circ \alpha \mathrm{d}\boldsymbol{m}, \; f \in \mathscr{K}(S)$$

is called the image of the measure $\boldsymbol{m}$ by the function α and it is denoted by $\alpha(\boldsymbol{m})$.

Therefore, we have

$$\int f \mathrm{d}\alpha(\boldsymbol{m}) = \int f \circ \alpha \mathrm{d}\boldsymbol{m}, \text{ for } f \in \mathscr{K}(S).$$

The measure $\boldsymbol{n} = \alpha(\boldsymbol{m})$ is dominated by the positive measure $\nu = \alpha(|\boldsymbol{m}|)$, therefore $|\boldsymbol{n}| \leqslant \nu$, that is,

$$|\alpha(\boldsymbol{m})| \leqslant \alpha(|\boldsymbol{m}|).$$

Remarks.

1. The inequality $|\alpha(\boldsymbol{m})| \leqslant \alpha(|\boldsymbol{m}|)$ may be strict.

Example. Consider $T = \{0, 1\}$, $S = \{0\}$, $\mu = \varepsilon_0 - \varepsilon_1$ and $\alpha : T \to S$ the only possible mapping of T into S. Evidently the function α is μ-proper. For every function $f \in \mathscr{K}(S)$, the function $f \circ \alpha$ is constant and

$$\int f \mathrm{d}\alpha(\mu) = \int (f \circ \alpha) \mathrm{d}\mu = \varepsilon_0(f \circ \alpha) - \varepsilon_1(f \circ \alpha) = f(\alpha(0)) - f(\alpha(1)) = 0,$$

therefore $\alpha(\mu) = 0$ and $|\alpha(\mu)| = 0$.

But $|\mu| = \varepsilon_0 + \varepsilon_1$ and for every function $f \neq 0$ of $\mathscr{K}(S)$ we have

$$\int f \mathrm{d}\alpha(|\mu|) = \int f \circ \alpha \mathrm{d}|\mu| = \varepsilon_0(f \circ \alpha) + \varepsilon_1(f \circ \alpha) = \\ = f(\alpha(0)) + f(\alpha(1)) = 2f(0),$$

therefore $\alpha(|\mu|) \neq 0$ and thus $|\alpha(\mu)| < \alpha(|\mu|)$.

2. The measure $\boldsymbol{m}$ and $\alpha(\boldsymbol{m})$ take on values in the same space X.

For instance, if μ is positive, then $\alpha(\mu)$ is positive; if μ is real, then $\alpha(\mu)$ is real, and if μ is complex, then $\alpha(\mu)$ is complex.

2 *The upper integral with respect to the image of a positive measure*

Let μ be *a positive* measure on T, $\alpha : T \to S$ a μ-proper function and $\nu = \mu(\alpha)$. We have $\nu \geqslant 0$. For every function $f \in \mathscr{K}(S)$ we have

$$\int f(s) \mathrm{d}\nu(s) = \int f(\alpha(t)) \mathrm{d}\mu(t).$$

We shall prove that this relation still remains valid for the upper integrals of all functions $f \geqslant 0$ defined on S.

20.3 Lemma. *If $f \geqslant 0$ is a lower semicontinuous function defined on S, then*

$$\int^* f(s) \mathrm{d}\nu(s) = \int^* f(\alpha(t)) \mathrm{d}\mu(t).$$

Let $(g_i)_{i\in I}$ be the directed family of all functions of $\mathscr{K}(S)$ satisfying the inequality $0\leqslant g_i\leqslant f$. We have

$$f(s)=\sup_{i\in I} g_i(s)\,,$$

hence

$$f(\alpha(t))=\sup_{i\in I} g_i(\alpha(t)) \text{ for } t\in T,$$

that is, $f\circ\alpha=\sup_{i\in I} g_i\circ\alpha$.

The mapping $t\to(g_i(\alpha(t)))_{i\in I}$ of T into $\bar{R}^I$is μ-measurable. In fact, let $K\subset T$ be a compact set and $\varepsilon>0$. There exists a compact set $K'\subset K$ such that $\mu(K\setminus K')<\varepsilon$ and such that the restriction of the function α to K' is continuous. Then the mapping $t\to g_i(\alpha(t))$ is continuous on K', consequently the function $t\to(g_i(\alpha(t)))_{i\in I}$ is also continuous on K'; it follows that this mapping is μ-measurable.

By proposition 16.1, we have

$$\int^* f\circ\alpha\,\mathrm{d}\mu=\sup_{i\in I}\int g_i\circ\alpha\,\mathrm{d}\mu\,.$$

But

$$\int^* f\mathrm{d}\nu=\sup_{i\in I}\int g_i\,\mathrm{d}\nu \text{ and } \int g_i\mathrm{d}\nu=\int g_i\circ\alpha\,\mathrm{d}\mu\,,$$

hence

$$\int^* f\mathrm{d}\nu=\int^* f\circ\alpha\,\mathrm{d}\mu\,.$$

20.4 Lemma. *For every function $f\geqslant 0$ defined on S with compact support we have*

$$\int^* f(s)\mathrm{d}\nu(s)\geqslant\int^* f(\alpha(t))\mathrm{d}\mu(t)\,.$$

Let $h\geqslant 0$ be a lower semicontinuous function on S such that $h\geqslant f$. From lemma 20.3 it follows that

$$\int^* h(s)\,\mathrm{d}\nu(s)=\int^* h(\alpha(t))\mathrm{d}\mu(t)\geqslant\int^* f(\alpha(t))\,\mathrm{d}\mu(t)\,.$$

Taking in the left-hand side the infimum for all functions $h\in\mathscr{I}_+(s)$ with $h\geqslant f$ we obtain

$$\int^* f(s)\mathrm{d}\nu(s)\geqslant\int^* f(\alpha(t))\,\mathrm{d}\mu(t)\,,$$

since f has compact support.

Remarks. From lemma 20.4 it follows that:

(1) If f is ν-negligible and has compact support, then $f \circ \alpha$ is μ-negligible.

(2) If $N \subset S$ is ν-negligible and relatively compact, then $\alpha^{-1}(N)$ is μ-negligible.

20.5 Lemma. *If $f \geqslant 0$ is a ν-integrable function with compact support defined on S, then the function $f \circ \alpha$ is μ-integrable and*

$$\int f(s)\,\mathrm{d}\nu(s) = \int f(\alpha(t))\,\mathrm{d}\mu(t)\,.$$

Let (f_n) be a sequence of functions of $\mathscr{K}(S)$ such that

$$\lim \int |f_n(s) - f(s)|\,\mathrm{d}\nu(s) = 0\,.$$

Since the functions $|f_n - f|$ have compact support, by lemma 20.4 we have

$$\lim \int^* |f_n(\alpha(t)) - f(\alpha(t))|\,\mathrm{d}\mu(t) = 0\,.$$

The functions $f_n \circ \alpha$ are μ-integrable, since $f_n \in \mathscr{K}(S)$. It follows that $f \circ \alpha$ is μ-integrable and

$$\int f \circ \alpha\,\mathrm{d}\mu = \lim \int f_n \circ \alpha\,\mathrm{d}\mu = \lim \int f_n\,\mathrm{d}\nu = \int f\,\mathrm{d}\nu\,.$$

Remark. By lemma 20.5 it follows that if $A \subset S$ is relatively compact and ν-integrable, then the set $\alpha^{-1}(A) \subset T$ is μ-integrable and

$$\nu(A) = \mu(\alpha^{-1}(A))\,.$$

In fact, $\varphi_A \circ \alpha = \varphi_{\alpha^{-1}(A)}$.

20.6 Lemma. *If $f \geqslant 0$ is a μ-measurable function with compact support defined on S, then*

$$\int^* f(s)\,\mathrm{d}\nu(s) = \int^* f(\alpha(t))\,\mathrm{d}\mu(t)\,.$$

For every natural number n put $f_n = \inf(n, f)$. The function f_n is ν-measurable, bounded, and with compact support, therefore it is ν-integrable. By lemma 20.5 we have

$$\int f_n\,\mathrm{d}\nu = \int f_n \circ \alpha\,\mathrm{d}\mu\,.$$

The sequence (f_n) is increasing, $f = \sup f_n$ and $f \circ \alpha = \sup f_n \circ \alpha$. Taking the supremum in both sides of the equality we obtain

$$\int^* f \mathrm{d}\nu = \int^* f \circ \alpha \, \mathrm{d}\mu .$$

20.7 LEMMA. *Let H and K be two compact spaces and α a continuous mapping of H onto K. If $h \geqslant 0$ is a lower semicontinuous function on H then the function $f'(s) = \inf_{\alpha(t)=s} h(t)$ is lower semicontinuous on K.*

We remark first that for every point $s \in K$ there exists a point $t_s \in H$ such that $\alpha(t_s) = s$ and $f'(s) = h(t_s)$. In fact, $\{s\}$ is a closed set in K and α is continuous on H, therefore $\alpha^{-1}(s)$ is closed in H and hence compact. Since h is lower semicontinuous and $\alpha^{-1}(s)$ is compact, there exists a point $t_s \in \alpha^{-1}(s)$ such that $h(t_s) = \inf_{\alpha(t)=s} h(t)$; hence $f'(s) = h(t_s)$.

Let now a be an arbitrary real number and prove that the set $A = \{s \mid s \in K, f'(s) > a\}$ is open; it will then follow that f' is lower semicontinuous. Let $s_0 \in A$. We have $f'(s_0) > a$, therefore $h(t) > a$ for every $t \in \alpha^{-1}(s_0)$. Since h is lower semicontinuous, there exists a neighbourhood V of $\alpha^{-1}(s_0)$ such that $h(t) > a$ for every $t \in V$. In addition, we can consider the set V saturated for the equivalence relation $\alpha(t) = \alpha(t')$. Then $\alpha(V)$ is a neighbourhood of s_0. For every $s \in \alpha(V)$ there exists $t \in \alpha^{-1}(s)$ such that $h(t) = f'(s)$. Therefore, we have $f'(s) > a$ for every $s \in \alpha(V)$, consequently s_0 is an interior point of A, hence A is open. It follows that f' is lower semicontinuous.

20.8 LEMMA. *If $f \geqslant 0$ is a function with compact support defined on S, then*

$$\int^* f(s) \, \mathrm{d}\nu(s) = \int^* f(\alpha(t)) \, \mathrm{d}\mu(t) .$$

By lemma 20.4 we have

$$\int^* f(s) \, \mathrm{d}\nu(s) \geqslant \int^* f(\alpha(t)) \, \mathrm{d}\mu(t) .$$

It remains to prove the converse inequality. Let $K \subset S$ be the support of f. Since K is compact and ν-integrable, the set $A = \alpha^{-1}(K)$ is μ-integrable. Since α is μ-measurable, A is the union of a μ-negligible set N and of a sequence (H_n) of mutually disjoint compact subsets of T, such that the restriction of α to each H_n is continuous.

For each n, the set $K_n = \alpha(H_n)$ is compact in S. Denote $B = \cup H_n$ and $B' = \cup K_n$. Since the sets H_n are mutually disjoint, we have $\alpha(B) = B'$.

Since

$$\int^* f \circ \alpha \, \mathrm{d}\mu = \int^* (f \circ \alpha) \varphi_B \mathrm{d}\mu ,$$

it will be sufficient to prove that for every lower semicontinuous function h defined on T such that $f(\alpha(t))\,\varphi_B(t) \leqslant h(t)$, we have

$$\int^* f(s)\,\mathrm{d}\nu(s) \leqslant \int^* h(t)\,\mathrm{d}\mu(t)\,,$$

since, taking the infimum in the right-hand side for all functions $h \geqslant (f\circ\alpha)\varphi_B$ of $\mathscr{I}_+(T)$ it will follow

$$\int^* f\mathrm{d}\nu \leqslant \int^* (f\circ\alpha)\,\varphi_B \mathrm{d}\mu = \int^* f\circ\alpha\,\mathrm{d}\mu\,.$$

therefore

$$\int^* f\mathrm{d}\nu = \int^* f\circ\alpha\,\mathrm{d}\mu\,.$$

For this purpose it is sufficient to prove that for every function $h \geqslant (f\circ\alpha)\varphi_B$ of $\mathscr{I}_+(T)$ we can find a function f' defined on S having the following properties:

(1) $f \leqslant f'$, therefore $\int^* f\mathrm{d}\nu \leqslant \int^* f'\,\mathrm{d}\nu$;

(2) f' is ν-measurable and with compact support, therefore $\int^* f'\,\mathrm{d}\nu = \int^* f'\circ\alpha\,\mathrm{d}\mu$;

(3) $f'\circ\alpha \leqslant h$, μ-almost everywhere, therefore $\int^* f'\circ\alpha\,\mathrm{d}\mu \leqslant \int^* h\,\mathrm{d}\mu$.

It will then follow that

$$\int^* f\mathrm{d}\nu \leqslant \int^* h\,\mathrm{d}\mu$$

and the lemma will be proved.

Let $h \geqslant (f\circ\alpha)\varphi_B$ be a function of $\mathscr{I}_+(T)$. Take a function $g \geqslant f$ of $\mathscr{I}_+(S)$ and put

$$f'(s) = g(s), \qquad \text{if } s \in K \setminus B'\,,$$

$$f'(s) = \inf\nolimits_{\alpha(t)=s} h(t), \text{ if } s \in B' = \cup K_n \setminus \cup\,\alpha(H_n)\,,$$

$$f'(s) = 0, \qquad \text{if } s \notin K\,.$$

It remains to verify that f' fulfills the above three conditions

(1) $f \leqslant f'$. In fact:

– for $s \in B'$ we have $f(s) = f(\alpha(t))$ for every $t \in \alpha^{-1}(s)$, therefore $f(s) = \inf_{\alpha(t)=s} f(\alpha(t)) \leqslant \inf_{\alpha(t)=s} h(t) = f'(s)$;

– for $s \in K \setminus B'$ we have $f(s) \leqslant g(s) = f'(s)$;

– for $s \notin K$ we have $f(s) = f'(s) = 0$.

(2) f' has compact support K, and is ν-measurable. In fact:

– f' is ν-measurable on $K \setminus B'$ since f' is equal on this set to the lower semicontinuous function g;

– f' is ν-measurable on each set K_n, by lemma 20.7, therefore f' is ν-measurable on B';

– f' is ν-measurable on $T \setminus K$ since it vanishes on this set.

Since the sets $K \setminus B'$, B and $T \setminus K$ are ν-measurable, it follows that f' is ν-measurable.

(3) $f'(\alpha(t)) \leqslant h(t)$ for $t \notin N$. In fact:

– for $t \notin A$ we have $\alpha(t) \notin K$, therefore $f'(\alpha(t)) = 0 \leqslant h(t)$;

– for $t \in B$, denoting $s = \alpha(t)$, we have $t \in \alpha^{-1}(s)$, therefore $f'(\alpha(t)) = f'(s) \leqslant h(t)$.

Thus the lemma is proved.

20.9 PROPOSITION. *For every function $f \geqslant 0$ defined on S we have*

$$\int^* f(s)\,d\nu(s) = \int^* f(\alpha(t))\,d\mu(t)\,.$$

For every compact set $K \subset S$ we have, by lemma 20.8,

$$\int^* f\varphi_K\,d\nu = \int^* (f \circ \alpha)\,\varphi_{\alpha^{-1}(K)}\,d\mu \leqslant \int^* f \circ \alpha\,d\mu\,,$$

therefore

$$\int^* f\,d\nu \leqslant \int^* f \circ \alpha\,d\mu\,.$$

It remains to prove the converse inequality. Let $H \subset T$ be a compact set. There exists a μ-negligible set $N \subset K$ and an increasing sequence (H_n) of compact sets with union equal to $H \setminus N$ such that the restriction of α to each H_n is continuous. Then the sets $K_n = \alpha(H_n)$ are compact, $\alpha(H \setminus N)$ is the union of the increasing sequence (K_n) and $H_n \subset \alpha^{-1}(H_n)$. We deduce that

$$\int^* (f \circ \alpha)\varphi_H\,d\mu = \int^* (f \circ \alpha)\varphi_{H \setminus N}\,d\mu =$$

$$= \sup \int^* (f \circ \alpha)\varphi_{H_n}\,d\mu \leqslant \sup \int^* (f \circ \alpha)\varphi_{\alpha^{-1}(K_n)}\,d\mu =$$

$$= \sup \int^* f\varphi_{K_n}\,d\nu \leqslant \int^* f\,d\nu\,.$$

Taking the supremum for all the compact sets $H \subset T$, we obtain

$$\int^* f \circ \alpha\,d\mu \leqslant \int^* f\,d\nu$$

and thus the proposition is proved.

20.10 COROLLARY. *For every set $A \subset S$ we have*

$$\nu^*(A) = \mu^*(\alpha^{-1}(A))\,.$$

3 *Integration with respect to the image of a positive measure*

Let μ be a positive measure on T and $\alpha : T \to S$ a μ-proper function.

20.11 PROPOSITION. *A function f defined on S with values in a Banach space E or in $\overline{R}$ is $\alpha(\mu)$-negligible, if and only if the function $f \circ \alpha$ defined on T is μ-negligible.*

In fact, setting $\nu = \alpha(\mu)$ we have

$$\int^* |f| \, d\nu = \int^* |f| \circ \alpha \, d\mu = \int^* |f \circ \alpha| \, d\mu ,$$

therefore $\nu^*(|f|) = 0$ if and only if $\mu^*(|f \circ \alpha|) = 0$.

20.12 COROLLARY. *A set $A \subset S$ is ν-negligible if and only if the set $\alpha^{-1}(A)$ is μ-negligible.*

20.13 COROLLARY. *If μ is concentrated on a set $M \subset T$, then $\alpha(\mu)$ is concentrated on the set $\alpha(M) \subset S$.*

Put $N = S \setminus \alpha(M)$. Then $\alpha^{-1}(N)$ is disjoint from M, therefore $\alpha^{-1}(N)$ is μ-negligible. It follows that N is $\alpha(\mu)$-negligible, therefore $\alpha(\mu)$ is concentrated on $\alpha(M)$.

20.14 COROLLARY. *If α is continuous and $S(\mu)$ is the support of μ, then the closure of $\alpha(S(\mu))$ is the support of $\alpha(\mu)$:*

$$S(\alpha(\mu)) = \overline{\alpha(S(\mu))} .$$

From corollary 20.13 it follows that $\alpha(\mu)$ is concentrated on $\alpha(S(\mu))$, therefore it is concentrated also on the closure of $\alpha(S(\mu))$. The support $S(\nu)$ of $\alpha(\mu)$ is the least *closed* set on which $\nu = \alpha(\mu)$ is concentrated, therefore $S(\nu) \subset \overline{\alpha(S(\mu))}$.

On the other hand, the set $\alpha^{-1}(S \setminus S(\nu))$ is an *open* set since α is continuous, and it is *μ-negligible* since $S \setminus S(\nu)$ is ν-negligible. It follows that $\alpha^{-1}(S \setminus S(\nu)) \subset T \setminus S(\mu)$, therefore $S(\mu) \subset \alpha^{-1}(S(\nu))$, whence $\alpha(S(\mu)) \subset S(\nu)$. Since $S(\nu)$ is a closed set, we have, also $\overline{\alpha(S(\mu))} \subset S(\nu)$ hence $S(\nu) = \overline{\alpha(S(\mu))}$.

20.15 PROPOSITION. *A mapping f of S into a topological space G is $\alpha(\mu)$-measurable if and only if the function $f \circ \alpha : T \to G$ is μ-measurable.*

Put $\alpha(\mu)=\nu$ and assume first that f is ν-measurable. Let $H \subset T$ be a compact set. Since α is μ-measurable, there exists a μ-negligible set $N \subset H$ and a sequence (H_n) of mutually disjoint compact sets with union $H \setminus N$ such that the restriction of α to each H_n is continuous. The sets $K_n=\alpha(H_n)$ are compact subsets of S. For each n there exists a ν-negligible set $M_n \subset K_n$ and a sequence $(C_{nm})_m$ of mutually disjoint compact sets with union $K_n \setminus M_n$ and such that the restriction of f to each set C_{nm} is continuous. The sets $\alpha^{-1}(M_n)$ are then μ-negligible, the sets $\alpha^{-1}(C_{nm})$ are μ-integrable and the restriction of the function $f \circ \alpha$ to each set $\alpha^{-1}(C_{nm}) \cap H_n$ is continuous.

Since H is the union of the μ-negligible set $N \cup (\cup \alpha^{-1}(M_n))$ and of the countable family $(\alpha^{-1}(C_{n,m}) \cap H_n)_{n,m}$, it follows that the function $f \circ \alpha$ is μ-measurable.

Conversely, assume that $f \circ \alpha$ is μ-measurable. Let $K \subset S$ be a compact set. The set $\alpha^{-1}(K)$ is μ-integrable, therefore there exists a μ-negligible set $N \subset \alpha^{-1}(K)$ and a sequence (H_n) of mutually disjoint compact sets with union $\alpha^{-1}(K) \setminus N$ such that the restriction of the functions $f \circ \alpha$ and α to each H_n are continuous. Then the sets $K_n=\alpha(H_n)$ are compact.

The restriction of the function $h=f \circ \alpha$ is continuous on H_n, therefore it reaches the infimum on each compact set $\alpha^{-1}(s) \cap H_n$ with $s \in K_n$. For each $s \in K_n$ there exists $t_s \in \alpha^{-1}(s) \cap H_n$ such that

$$f'(s)=\inf_{\alpha(t)=s,\, t \in H_n} h(t)=h(t_s)=f(\alpha(t_s))=f(s)\,.$$

But, by lemma 20.7, the function f' is lower semicontinuous, therefore ν-measurable on each compact set K_n; it follows that the restriction of f to each set K_n is ν-measurable, therefore it is ν-measurable on $\cup K_n$. But $H_n \subset \alpha^{-1}(K_n)$, therefore $\alpha^{-1}(K \setminus \cup K_n)=\alpha^{-1}(K) \setminus \cup \alpha^{-1}(K_n) \subset \alpha^{-1}(K) \setminus \cup H_n=N$, consequently $K \setminus \cup K_n$ is ν-negligible. It follows that f is ν-measurable on K, and hence f is ν-measurable.

20.16 COROLLARY. *A set $A \subset S$ is $\alpha(\mu)$-measurable if and only if $\alpha^{-1}(A)$ is μ-measurable.*

20.17 THEOREM. *A function $\boldsymbol{f}$ defined on S with values in a Banach space E or in $\overline{R}$ is $\alpha(\mu)$-integrable if and only if the function $\boldsymbol{f} \circ \alpha$ is μ-integrable. In this case we have*

$$\int \boldsymbol{f} \, d\alpha(\mu)=\int \boldsymbol{f} \circ \alpha \, d\mu\,.$$

In fact, f is $\alpha(\mu)$-measurable if and only if $f\circ\alpha$ is μ-measurable. From the equality

$$\int^* |f|\, d\alpha(\mu) = \int^* |f\circ\alpha|\, d\mu$$

it follows that $\int^* |f|\, d\alpha(\mu) < +\infty$ if and only if $\int^* |f\circ\alpha|\, d\mu < +\infty$. Thus f is $\alpha(\mu)$-integrable if and only if $f\circ\alpha$ is μ-integrable.

Assume now that f is $\alpha(\mu)$-integrable. There exists a sequence (f_n) of functions of $\mathscr{K}_E(S)$ (or of $\mathscr{K}(S)$) of the form $\Sigma\, \varphi_i x_i$ (finite sum) with $\varphi_i \in \mathscr{K}(S)$, $x_i \in E$, such that

$$\lim \int |f - f_n|\, d\alpha(\mu) = 0\,,$$

therefore

$$\lim \left|\int f\, d\alpha(\mu) - \int f_n\, d\alpha(\mu)\right| = 0\,,$$

whence

$$\lim \int f_n\, d\alpha(\mu) = \int f\, d\alpha(\mu)\,.$$

On the other hand

$$\left|\int f\circ\alpha\, d\mu - \int f_n\circ\alpha\, d\mu\right| \leqslant \int |f\circ\alpha - f_n\circ\alpha|\, d\mu = \int |f - f_n|\circ\alpha\, d\mu = \int |f - f_n|\, d\alpha(\mu)\,,$$

therefore

$$\lim \int f_n\circ\alpha\, d\mu = \int f\circ\alpha\, d\mu\,.$$

For $\varphi \in \mathscr{K}(S)$ and $x \in E$ we have

$$\int \varphi x\, d\alpha(\mu) = \left(\int \varphi\, d\alpha(\mu)\right) x = \left(\int \varphi\circ\alpha\, d\mu\right) x = \\ = \int (\varphi\circ\alpha) x\, d\mu = \int (\varphi x)\circ\alpha\, d\mu\,,$$

and the equality remains valid for finite sums of functions of the form φx. It follows that for the functions f_n we have

$$\int f_n\, d\alpha(\mu) = \int f_n\circ\alpha\, d\mu\,.$$

Passing to the limit we obtain

$$\int f\, d\alpha(\mu) = \int f\circ\alpha\, d\mu\,.$$

20.18 Corollary. *A set $A \subset S$ is $\alpha(\mu)$-integrable if and only if the set $\alpha^{-1}(A)$ is μ-integrable. In this case*

$$\alpha(\mu)(A) = \mu(\alpha^{-1}(A))\,.$$

20.19 PROPOSITION. *Let μ be a positive measure on T. A μ-measurable function $\alpha : T \to S$ is μ-proper if and only if for every compact set $K \subset S$ the set $\alpha^{-1}(K) \subset T$ is μ-integrable.*

If α is μ-proper, every compact set $K \subset S$ is $\alpha(\mu)$-integrable, therefore $\alpha^{-1}(K)$ is μ-integrable.

Conversely, assume that $\alpha^{-1}(K)$ is μ-integrable for every compact set $K \subset S$.

Let $f \in \mathscr{K}(S)$ and let K be the support of f. The set $A = \alpha^{-1}(K)$ is μ-integrable, therefore

$$\int^* |f(\alpha(t))| \, d\mu(t) = \int^* |f(\alpha(t))| \, \varphi_K(\alpha(t)) \, d\mu(t) \leqslant$$
$$\leqslant \|f\| \int^* \varphi_A(t) \, d\mu(t) < \infty .$$

The function $f \circ \alpha$ is also μ-measurable, therefore it is μ-integrable. It follows that α is μ-proper.

4 *Integration with respect to the image of a dominated measure*

Let X be a Banach space and $\boldsymbol{m}$ a dominated measure on T with values in X. Let E and F be two Banach spaces such that $X \subset \mathscr{L}(E, F)$. Let S be a locally compact space and $\alpha : T \to S$ an $\boldsymbol{m}$-proper function. Consider the measure $\alpha(\boldsymbol{m})$ on S with values in X, defined by the equality

$$\int f \, d\alpha(\boldsymbol{m}) = \int f \circ \alpha \, d\boldsymbol{m}, \quad f \in \mathscr{K}(S) .$$

20.20 THEOREM. *Let $\boldsymbol{f} : S \to E$ be a function.*

(1) *If the function $\boldsymbol{f} \circ \alpha$ is $\boldsymbol{m}$-negligible, then the function $\boldsymbol{f}$ is $\alpha(\boldsymbol{m})$-negligible;*

(2) *If $\boldsymbol{f} \circ \alpha$ is $\boldsymbol{m}$-measurable, then $\boldsymbol{f}$ is $\alpha(\boldsymbol{m})$-measurable;*

(3) *If $\boldsymbol{f} \circ \alpha$ is $\boldsymbol{m}$-integrable, then $\boldsymbol{f}$ is $\alpha(\boldsymbol{m})$-integrable and*

$$\int \boldsymbol{f} \, d\alpha(\boldsymbol{m}) = \int \boldsymbol{f} \circ \alpha \, d\boldsymbol{m} .$$

We have $|\alpha(\boldsymbol{m})| \leqslant \alpha(|\boldsymbol{m}|)$. If $\boldsymbol{f} \circ \alpha$ is $\boldsymbol{m}$-negligible, that is, $|\boldsymbol{m}|$-negligible, then $\boldsymbol{f}$ is $\alpha(|\boldsymbol{m}|)$-negligible; therefore, $\boldsymbol{f}$ is $|\alpha(\boldsymbol{m})|$-negligible, that is, $\alpha(\boldsymbol{m})$-negligible.

In the same way we prove that if $\boldsymbol{f} \circ \alpha$ is $\boldsymbol{m}$-measurable, then $\boldsymbol{f}$ is $\alpha(\boldsymbol{m})$-measurable.

From the inequality

$$\int^* |\boldsymbol{f}| \, d|\alpha(\boldsymbol{m})| \leqslant \int^* |\boldsymbol{f}| \, d\alpha(|\boldsymbol{m}|) = \int^* |\boldsymbol{f} \circ \alpha| \, d|\boldsymbol{m}|$$

it follows that if $\boldsymbol{f}\circ\alpha$ is $|\boldsymbol{m}|$-integrable (that is, $\boldsymbol{m}$-integrable), then the function $\boldsymbol{f}$ is $|\alpha(\boldsymbol{m})|$-integrable, that is, $\alpha(\boldsymbol{m})$-integrable.

Assume $\boldsymbol{f}\circ\alpha$ is $\boldsymbol{m}$-integrable, that is, $|\boldsymbol{m}|$-integrable. Then $\boldsymbol{f}$ is $\alpha(|\boldsymbol{m}|)$-integrable and $\alpha(\boldsymbol{m})$-integrable. There exists a sequence $(\boldsymbol{f}_n)$ of functions of $\mathscr{K}_E(S)$ of the form $\Sigma\,\varphi_i x_i$ (finite sum) with $\varphi_i\in\mathscr{K}(S)$ and $x_i\in E$, such that

$$\lim \int |\boldsymbol{f}-\boldsymbol{f}_n|\,\mathrm{d}\alpha(|\boldsymbol{m}|)=0\,.$$

Since $|\alpha(\boldsymbol{m})|\leqslant\alpha(|\boldsymbol{m}|)$, we have

$$|\int \boldsymbol{f}\,\mathrm{d}\alpha(\boldsymbol{m})-\int \boldsymbol{f}_n\mathrm{d}\alpha(\boldsymbol{m})|=|\int(\boldsymbol{f}-\boldsymbol{f}_n)\,\mathrm{d}\alpha(\boldsymbol{m})|\leqslant$$
$$\leqslant\int|\boldsymbol{f}-\boldsymbol{f}_n|\,\mathrm{d}|\alpha(\boldsymbol{m})|\leqslant\int \boldsymbol{f}-\boldsymbol{f}_n|\,\mathrm{d}\alpha(|\boldsymbol{m}|)\,,$$

therefore

$$\lim\int \boldsymbol{f}_n\,\mathrm{d}\alpha(\boldsymbol{m})=\int \boldsymbol{f}\,\mathrm{d}\alpha(\boldsymbol{m})\,.$$

On the other hnad

$$|\int \boldsymbol{f}\circ\alpha\,\mathrm{d}\boldsymbol{m}-\int \boldsymbol{f}_n\circ\alpha\,\mathrm{d}\boldsymbol{m}|=|\int(\boldsymbol{f}-\boldsymbol{f}_n)\circ\alpha\,\mathrm{d}\boldsymbol{m}|\leqslant$$
$$\leqslant\int|\boldsymbol{f}-\boldsymbol{f}_n|\circ\alpha\,\mathrm{d}|\boldsymbol{m}|=\int|\boldsymbol{f}-\boldsymbol{f}_n|\,\mathrm{d}\alpha(|\boldsymbol{m}|),$$

therefore

$$\lim\int \boldsymbol{f}_n\circ\alpha\,\mathrm{d}\boldsymbol{m}=\int \boldsymbol{f}\circ\alpha\,\mathrm{d}\boldsymbol{m}\,.$$

If $\varphi\in\mathscr{K}(S)$ and $x\in E$ we have

$$\int\varphi x\,\mathrm{d}\alpha(\boldsymbol{m})=(\int\varphi\,\mathrm{d}\alpha(\boldsymbol{m}))x=(\int\varphi\circ\alpha\,\mathrm{d}\boldsymbol{m})x=$$
$$=\int(\varphi\circ\alpha)x\,\mathrm{d}\boldsymbol{m}=\int(\varphi x)\circ\alpha\,\mathrm{d}\boldsymbol{m}$$

and the equality remains true also for finite sums of functions of the form φx. Thus, for the functions $\boldsymbol{f}_n$ we have

$$\int \boldsymbol{f}_n\,\mathrm{d}\alpha(\boldsymbol{m})=\int \boldsymbol{f}_n\circ\alpha\,\mathrm{d}\boldsymbol{m}\,.$$

Passing to the limit we obtain

$$\int \boldsymbol{f}\mathrm{d}\alpha(\boldsymbol{m})=\int \boldsymbol{f}\circ\alpha\,\mathrm{d}\boldsymbol{m}\,.$$

5 *Properties of images of measures*

20.21 Proposition. *Let μ and ν be two positive measures on T. If $\mu\leqslant\nu$ and $\alpha: T\rightarrow S$ is a ν-proper function, then α is μ-proper and $\alpha(\mu)\leqslant\alpha(\nu)$.*

In fact, α is ν-measurable, therefore it is μ-measurable; for every function $\varphi \in \mathscr{K}(S)$ the function $\varphi \circ \alpha$ is μ-integrable, therefore it is ν-integrable; if $\varphi \geqslant 0$, we have

$$\int \varphi \, d\alpha(\mu) = \int \varphi \circ \alpha \, d\mu \leqslant \int \varphi \circ \alpha \, d\nu = \int \varphi \, d\alpha(\nu)$$

that is $\alpha(\mu) \leqslant \alpha(\nu)$.

20.22 PROPOSITION. *Let μ and ν be two positive measures on T. A function $\alpha : T \to S$ is proper for the measure $\mu + \nu$ if and only if it is proper for μ and for ν. In this case we have*

$$\alpha(\mu + \nu) = \alpha(\mu) + \alpha(\nu).$$

In fact, α is $\mu + \nu$-measurable if and only if it is μ-measurable and ν-measurable. For every function $f \in \mathscr{K}(S)$, the function $f \circ \alpha$ is $\mu + \nu$-integrable if and only if $f \circ \alpha$ is μ-integrable and ν-integrable. It follows that α is $\mu + \nu$-proper if and only if it is μ-proper and ν-proper.

Let $f \in \mathscr{K}(S)$. We have

$$\int f \, d\alpha(\mu + \nu) = \int f \circ \alpha \, d(\mu + \nu) = \int f \circ \alpha \, d\mu + \int f \circ \alpha \, d\nu =$$

$$= \int f \, d\alpha(\mu) + \int f \, d\alpha(\nu) = \int f \, d(\alpha(\mu) + \alpha(\nu)),$$

therefore

$$\alpha(\mu + \nu) = \alpha(\mu) + \alpha(\nu).$$

20.23 PROPOSITION. *Let $\boldsymbol{m}$ and $\boldsymbol{n}$ be two dominated measures on T with values in a Banach space X. If $\alpha : T \to S$ is a proper function for $\boldsymbol{n}$ and for $\boldsymbol{m}$, then α is proper for $\boldsymbol{m} + \boldsymbol{n}$ and*

$$\alpha(\boldsymbol{m} + \boldsymbol{n}) = \alpha(\boldsymbol{m}) + \alpha(\boldsymbol{n}).$$

The proposition follows from the inequality $|\boldsymbol{m} + \boldsymbol{n}| \leqslant |\boldsymbol{m}| + |\boldsymbol{n}|$. The equality $\alpha(\boldsymbol{m} + \boldsymbol{n}) = \alpha(\boldsymbol{m}) + \alpha(\boldsymbol{n})$ is proved as in proposition 20.22.

20.24 PROPOSITION. *Let $\boldsymbol{m}$ be a dominated measure on T and a a scalar. If $\alpha : T \to S$ is $\boldsymbol{m}$-proper, then α is $a\boldsymbol{m}$-proper and*

$$\alpha(a\boldsymbol{m}) = a\alpha(\boldsymbol{m}).$$

The proof is immediate.

20.25 PROPOSITION. *Let μ be a real measure on T. A function $\alpha: T \to S$ is μ-proper if and only if it is proper for μ_+ and μ_-. In this case we have*

$$\alpha(\mu) = \alpha(\mu_+) - \alpha(\mu_-) .$$

In fact, if α is μ-proper, then it is $|\mu|$-proper, therefore it is proper for the measures

$$\mu_+ = \tfrac{1}{2}(|\mu| + \mu) \text{ and } \mu_- = \tfrac{1}{2}(|\mu| - \mu) .$$

Conversely, if α is proper for μ_+ and μ_- then it is proper for the measure $|\mu| = \mu_+ + \mu_-$, hence it is μ-proper. The equality

$$\alpha(\mu) = (\alpha\mu_+) - \alpha(\mu_-)$$

follows from the equality $\mu = \mu_+ - \mu_-$, using propositions 20.23 and 20.24.

Remark. The measures $\alpha(\mu_+)$ and $\alpha(\mu_-)$ are positive, but in general they are different from the positive and negative part of the measure $\alpha(\mu)$.

In fact, if we had

$$(\alpha(\mu))_+ = \alpha(\mu_+) \text{ and } (\alpha(\mu))_- = \alpha(\mu_-)$$

then it would follow $|\alpha(\mu)| = \alpha(|\mu|)$; but we proved that we may have $|\alpha(\mu)| < \alpha(|\mu|)$.

20.26 PROPOSITION. *Let $\mu = \mu_1 + i\mu_2$ be a complex measure on T, where μ_1 and μ_2 are real measures. A function $\alpha: T \to S$ is μ-proper, if and only if it is proper for μ_1 and μ_2. In this case we have*

$$\alpha(\mu) = \alpha(\mu_1) + i\alpha(\mu_2) .$$

The assertion follows from the inequalities $|\mu_1| \leqslant |\mu|$, $|\mu_2| \leqslant |\mu|$ and $|\mu| \leqslant |\mu_1| + |\mu_2|$. The equality follows from propositions 20.23 and 20.24.

20.27 PROPOSITION. *Let μ and ν be two real measures on T. The function $\alpha: T \to S$ is proper for μ and for ν if and only if it is proper for* $\sup(\mu, \nu)$ *and for* $\inf(\mu, \nu)$. *In this case we have*

$$\sup(\alpha(\mu), \alpha(\nu)) \leqslant \alpha(\sup(\mu, \nu))$$
$$\inf(\alpha(\mu), \alpha(\nu)) \geqslant \alpha(\inf(\mu, \nu)) .$$

The assertion follows from the inequalities

$$|\mu| \leqslant |\sup(\mu, \nu)|, \quad |\nu| \leqslant |\sup(\mu, \nu)| \leqslant |\mu| + |\nu|,$$

and from the similar inequalities for $\inf(\mu, \nu)$.

Then we have

$$\begin{aligned}\sup(\alpha(\mu), \alpha(\nu)) &= \tfrac{1}{2}(\alpha(\mu)+\alpha(\nu)+|\alpha(\mu)-\alpha(\nu)|) = \\ &= \tfrac{1}{2}(\alpha(\mu+\nu)+|\alpha(\mu-\nu)|) \leqslant \tfrac{1}{2}(\alpha(\mu+\nu)+\alpha(|\mu-\nu|) = \\ &= \alpha(\tfrac{1}{2}(\mu+\nu+|\mu-\nu|)) = \alpha(\sup(\mu, \nu))\end{aligned}$$

and also

$$\begin{aligned}\inf(\alpha(\mu), \alpha(\nu)) &= -\sup(-\alpha(\mu), -\alpha(\nu)) = \\ &= -\sup(\alpha(-\mu), \alpha(-\nu)) \geqslant -\alpha(\sup(-\mu, -\nu)) = \\ &= \alpha(-\sup(-\mu, -\nu)) = \alpha(\inf(\mu, \nu)).\end{aligned}$$

20.28 PROPOSITION. *Let T, S, R be three locally compact spaces, μ a measure on T and $\alpha : T \to S$ a μ-proper function. A function $\beta : S \to R$ is $\alpha(\mu)$-proper, if and only if the function $\beta \circ \alpha$ is μ-proper. In this case we have*

$$(\beta \circ \alpha)(\mu) = \beta(\alpha(\mu)).$$

In fact, by proposition 20.15 it follows that the function $\beta : S \to R$ is $\alpha(\mu)$-measurable if and only if $\beta \circ \alpha$ is μ-measurable.

For every function $f \in \mathscr{K}(R)$ we have $(f \circ \beta) \circ \alpha = f \circ (\beta \circ \alpha)$; therefore, $f \circ \beta$ is $\alpha(\mu)$-integrable if and only if $f \circ (\beta \circ \alpha)$ is μ-integrable. It follows that β is $\alpha(\mu)$-proper if and only if $\beta \circ \alpha$ is μ-proper.

Assume that $\beta \circ \alpha$ is μ-proper and let $f \in \mathscr{K}(R)$. Then

$$\begin{aligned}\int f \mathrm{d}(\beta \circ \alpha)(\mu) &= \int f \circ (\beta \circ \alpha) \mathrm{d}\mu = \int (f \circ \beta) \circ \alpha \mathrm{d}\mu = \\ &= \int f \circ \beta \mathrm{d}\alpha(\mu) = \int f \mathrm{d}\beta(\alpha(\mu)),\end{aligned}$$

whence

$$(\beta \circ \alpha)(\mu) = \beta(\alpha(\mu)).$$

20.29 COROLLARY. *Let $\boldsymbol{m}$ be a dominated measure on T, $\alpha : T \to S$ an $\boldsymbol{m}$ proper function and let $\beta : S \to R$. If the function $\beta \circ \alpha : T \to R$ is $\boldsymbol{m}$-proper, then β is $\alpha(\boldsymbol{m})$-proper and*

$$(\beta \circ \alpha)(\boldsymbol{m}) = \beta(\alpha(\boldsymbol{m})).$$

The assertion follows from the inequality $|\alpha(\boldsymbol{m})| \leqslant \alpha(|\boldsymbol{m}|)$. If $\beta \circ \alpha$ is $\boldsymbol{m}$-proper, that is, $|\boldsymbol{m}|$-proper, then β is $\alpha(|\boldsymbol{m}|)$-proper hence $|\alpha(\boldsymbol{m})|$-proper, that is, $\alpha(\boldsymbol{m})$-proper. The equality follows in the same way as in the preceding proposition.

20.30 COROLLARY. *Let T and S be two locally compact spaces, $\boldsymbol{m}$ a dominated measure on T and α a bijective mapping of T onto S. The function α is $\boldsymbol{m}$-proper if and only if the function α^{-1} is $\alpha(\boldsymbol{m})$-proper. In this case*

$$\alpha^{-1}(\alpha(\boldsymbol{m})) = \boldsymbol{m} .$$

In fact, if α is $\boldsymbol{m}$-proper, taking in corollary 20.29, $\beta = \alpha^{-1}$, the function $\beta \circ \alpha$ is the identity mapping of T onto T, therefore it is $\boldsymbol{m}$-proper; it follows then that $\alpha^{-1} = \beta$ is also $\alpha(\boldsymbol{m})$-proper and

$$\alpha^{-1}(\alpha(\boldsymbol{m})) = \beta(\alpha(\boldsymbol{m})) = (\beta \circ \alpha)(\boldsymbol{m}) = \boldsymbol{m} .$$

Conversely, if α^{-1} is $\alpha(\boldsymbol{m})$-proper, it follows as above that α is proper for the measure $\alpha^{-1}(\alpha(\boldsymbol{m})) = \boldsymbol{m}$.

20.31 COROLLARY. *If $\boldsymbol{m}$ is a dominated measure on T and if $\alpha : T \to S$ is a bijective mapping of T onto S, proper for $\boldsymbol{m}$, then*

$$|\alpha(\boldsymbol{m})| = \alpha(|\boldsymbol{m}|) .$$

In fact,

$$|\alpha(\boldsymbol{m})| \leqslant \alpha(|\boldsymbol{m}|) .$$

On the other hand, α^{-1} is proper for the measure $\alpha(\boldsymbol{m})$, that is, for $|\alpha(\boldsymbol{m})|$ and

$$|\boldsymbol{m}| = |\alpha^{-1}(\alpha(\boldsymbol{m}))| \leqslant \alpha^{-1}(|\alpha(\boldsymbol{m})|) .$$

Then α is proper for the measure $\alpha^{-1}(|\alpha(\boldsymbol{m})|)$, therefore it is also proper for $|\boldsymbol{m}|$ and

$$\alpha(|\boldsymbol{m}|) \leqslant \alpha(\alpha^{-1}(|\alpha(\boldsymbol{m})|)) = |\alpha(\boldsymbol{m})| .$$

It follows then that

$$|\alpha(\boldsymbol{m})| = \alpha(|\boldsymbol{m}|) .$$

20.32 COROLLARY. *Let $\boldsymbol{m}$ be a dominated measure on T with values in a*

Banach space X and $\alpha : T \to S$ a bijective mapping of T onto S, proper for the measure $\boldsymbol{m}$. Let E and F be two Banach spaces such that $X \subset \mathscr{L}(E, F)$ and let $\boldsymbol{f}$ be a function defined on S with values in E.

(1) *The function $\boldsymbol{f}$ is $\alpha(\boldsymbol{m})$-negligible if and only if $\boldsymbol{f} \circ \alpha$ is $\boldsymbol{m}$-negligible.*
(2) *The function $\boldsymbol{f}$ is $\alpha(\boldsymbol{m})$-measurable if and only if $\boldsymbol{f} \circ \alpha$ is $\boldsymbol{m}$-measurable.*
(3) *The function $\boldsymbol{f}$ is $\alpha(\boldsymbol{m})$-integrable if and only if $\boldsymbol{f} \circ \alpha$ is $\boldsymbol{m}$-integrable. In this case,*

$$\int \boldsymbol{f} \, \mathrm{d}\alpha(\boldsymbol{m}) = \int \boldsymbol{f} \circ \alpha \, \mathrm{d}\boldsymbol{m} .$$

The assertion follows from corollary 20.31 and theorem 20.20.

In particular, taking $E = C$ and $F = X$ the corollary remains valid for scalar functions f.

20.33 PROPOSITION. *Let μ be a positive measure on T, $\alpha : T \to S$ a μ-proper function and $g : S \to C$ a scalar function such that the scalar function $g \circ \alpha : T \to C$ is locally μ-integrable. The function g is locally integrable for $\alpha(\mu)$ if and only if α is proper for the measure $(g \circ \alpha)\mu$. In this case we have*

$$\alpha((g \circ \alpha)\mu) = g \cdot \alpha(\mu) .$$

We remark first that α is measurable for μ, therefore for every measure with bases μ, in particular for the measure $(g \circ \alpha)\mu$.

Since $g \circ \alpha$ is *scalar* and $|(g \circ \alpha)\mu| = |g \circ \alpha|\mu$, a function $\boldsymbol{h}$ defined on T with values in a Banach space E or in $\bar{R}$ is $(g \circ \alpha)\mu$-integrable if and only if the function $\boldsymbol{h}(g \circ \alpha)$ is μ-integrable.

The function g is locally $\alpha(\mu)$-integrable if and only if the function gf is $\alpha(\mu)$-integrable, for every function $f \in \mathscr{K}(S)$ (proposition 15.2); this last property is fulfilled if and only if the function $(gf) \circ \alpha = (g \circ \alpha)(f \circ \alpha)$ is μ-integrable (theorem 20.17); therefore (by the above remark) if and only if the function $f \circ \alpha$ is $(g \circ \alpha)\mu$-integrable.

Since α is $(g \circ \alpha)\mu$-measurable, it follows that g is locally $\alpha(\mu)$-integrable if and only if α is $(g \circ \alpha)\mu$-proper.

If g is locally $\alpha(\mu)$-integrable, for every function $f \in \mathscr{K}(T)$ we have

$$\int f \mathrm{d} g \alpha(\mu) = \int fg \, \mathrm{d}\alpha(\mu) = \int (fg) \circ \alpha \, \mathrm{d}\mu = \int (f \circ \alpha)(g \circ \alpha) \, \mathrm{d}\mu =$$

$$= \int f \circ \alpha \, \mathrm{d}(g \circ \alpha)\mu = \int f \, \mathrm{d}\alpha((g \circ \alpha)\mu) ,$$

therefore

$$\alpha((g \circ \alpha)\mu) = g \cdot \alpha(\mu) .$$

20.34 COROLLARY. *Let* $\boldsymbol{m}$ *be a dominated measure defined on* T *with values in a Banach space* X, $\alpha : T \to S$ *a bijective mapping of* T *onto* S, *proper for the measure* $\boldsymbol{m}$ *and* $g : S \to C$ *a scalar function such that the scalar function* $g \circ \alpha : T \to C$ *is locally* $\boldsymbol{m}$*-integrable, hence*

$$|(g \circ \alpha)\boldsymbol{m}| = |g \circ \alpha|\, |\boldsymbol{m}| .$$

The function g *is locally integrable for the measure* $\alpha(\boldsymbol{m})$ *if and only if* α *is proper for the measure* $(g \circ \alpha)\boldsymbol{m}$. *In this case*

$$\alpha((g \circ \alpha)\boldsymbol{m}) = g \cdot \alpha(\boldsymbol{m}) .$$

The proof is the same as for the proposition 20.33, using corollary 20.32 instead of theorem 20.17.

20.35 PROPOSITION. *Let* $\boldsymbol{m}$ *be a dominated measure defined on* T *with values in a Banach space* X, $\alpha : T \to S$ *an* $\boldsymbol{m}$*-proper function,* E *and* F *two Banach spaces such that* $X \subset \mathscr{L}(E, F)$ *and* $\boldsymbol{g} : S \to \mathscr{L}(E, F)$ *a function such that* $\boldsymbol{g} \circ \alpha : T \to E$ *is locally* $\boldsymbol{m}$*-integrable and*

$$|(\boldsymbol{g} \circ \alpha)\boldsymbol{m}| = |\boldsymbol{g} \circ \alpha|\, |\boldsymbol{m}| .$$

If α *is proper for the measure* $(\boldsymbol{g} \circ \alpha)\boldsymbol{m}$, *then the function* $\boldsymbol{g}$ *is locally* $\alpha(\boldsymbol{m})$*-integrable and*

$$\boldsymbol{g} \cdot \alpha(\boldsymbol{m}) = \alpha((\boldsymbol{g} \circ \alpha)\boldsymbol{m}) .$$

Let $f \in \mathscr{K}(S)$. Since α is proper for the measure $(\boldsymbol{g} \circ \alpha)\boldsymbol{m}$, the function $f \circ \alpha$ is $(\boldsymbol{g} \circ \alpha)\boldsymbol{m}$-integrable. From the hypotheses on $\boldsymbol{m}$ and $\boldsymbol{g} \circ \alpha$, by corollary 16.23, it follows that the function $(\boldsymbol{g} \circ \alpha)(f \circ \alpha) = (\boldsymbol{g}f) \circ \alpha$ is $\boldsymbol{m}$-integrable. Since α is $\boldsymbol{m}$-proper, by theorem 20.20 it follows that $\boldsymbol{g}f$ is $\alpha(\boldsymbol{m})$-integrable. Then we deduce that the function $\boldsymbol{g}$ is locally $\alpha(\boldsymbol{m})$-integrable. The equality $\boldsymbol{g}\alpha(\boldsymbol{m}) = \alpha((\boldsymbol{g} \circ \alpha)\boldsymbol{m})$ is proved as in proposition 20.33.

Remark. The propositions 20.34 and 20.35 are valid in particular if m is scalar, or if $g : S \to C$ is a scalar function such that the function $g \circ \alpha : T \to C$ is locally m-integrable.

In fact, in these cases we have the equality

$$|(\boldsymbol{g} \circ \alpha)\boldsymbol{m}| = |\boldsymbol{g} \circ \alpha|\, |\boldsymbol{m}| .$$

6 *Application. Lebesgue measure as image of diffuse measures*

In this section we shall prove that for every diffuse measure $\mu \geqslant 0$ on a *compact* space T, there exists a *continuous* mapping α of T onto $[0, 1]$ such that $\alpha(\mu)$ is the Lebesgue measure on $[0, 1]$.

To do this we need some preliminary notions.

Let T be a locally compact space and μ a positive measure on T.

We say that a set $A \subset T$ is *carrable* for the measure μ, or μ-carrable, if its boundary is μ-negligible.

Since the set of discontinuities of the characteristic function φ_A is the boundary set $\mathrm{Fr}A$, it follows that A is μ-carrable if and only if φ_A is μ-almost everywhere continuous.

Every μ-carrable set A is μ-measurable, since it is the union of the open set $\mathring{A}$ and a μ-negligible set

$$A \setminus \mathring{A} \subset \bar{A} \setminus \mathring{A} = \mathrm{Fr}A\,.$$

If A is μ-carrable, then

$$\mu^*(\mathring{A}) = \mu^*(\bar{A})$$

since $\bar{A} = \mathring{A} \cup \mathrm{Fr}A$ and $\mu^*(\mathrm{Fr}A) = 0$.

Conversely, if A is μ-integrable and if

$$\mu^*(\mathring{A}) = \mu^*(\bar{A})\,,$$

then A is μ-carrable since

$$\mu^*(\mathrm{Fr}A) = \mu^*(\bar{A}) - \mu^*(\mathring{A}) = 0\,.$$

If A is carrable, every set B such that $\mathring{A} \subset B \subset \bar{A}$ is carrable; in particular $\mathring{A}$ and $\bar{A}$ are carrable.

20.36 Proposition. *Every point of T possesses a fundamental system of carrable closed neighbourhoods.*

Let $t_0 \in T$ be an arbitrary point and U an open integrable neighbourhood of t_0. We must prove that there exists an open integrable set V such that

$$t_0 \in V \subset \bar{V} \subset U \text{ and } \mu(V) = \mu(\bar{V}).$$

To do this, let $f: T \to [0, 1]$ be a continuous function with $f(t_0) = 1$ and $f(t) = 0$ for $t \notin U$. For every number $\alpha > 0$, the set $F_\alpha = f^{-1}(\alpha)$ is closed and

$F_\alpha \subset U$. For every finite family $(\alpha_i)_{1 \leqslant i \leqslant n}$ of distinct numbers with $0 < \alpha_i < 1$, the sets F_{α_i} are mutually disjoint and

$$\Sigma \mu(F_{\alpha_i}) = \mu(\cup F_{\alpha_i}) \leqslant \mu(U) < +\infty .$$

It follows that

$$\Sigma_{0<\alpha<1} \mu(F_\alpha) \leqslant \mu(U) < +\infty ,$$

that is the family of positive numbers $(\mu(F_\alpha))_{0<\alpha<1}$ is summable. It follows that we have $\mu(F_\alpha) > 0$ at most for a countable set of indices α. There exists then a number β with $0 < \beta < 1$ and $\mu(F_\beta) = 0$. The set $G_\beta = \{t \mid f(t) > \beta\}$ is an open neighbourhood of t_0 contained in U, therefore it is integrable, and the set

$$G_\beta \cup F_\beta = \{t \mid f(t) \geqslant \beta\}$$

is closed. It follows that $\overline{G}_\beta \subset G_\beta \cup F_\beta \subset U$ and

$$\mu(G_\beta) \leqslant \mu(\overline{G}_\beta) \leqslant \mu(G_\beta) + \mu(F_\beta) \leqslant \mu(G_\beta) ,$$

whence $\mu(G_\beta) = \mu(\overline{G}_\beta)$, hence G_β is carrable. Taking $V = G_\beta$ we have $t_0 \in V \subset \overline{V} \subset U$ and the proposition is proved.

20.37 Corollary. *For every compact set $K \subset T$ and every open set $U \supset K$ there exists an open, integrable and carrable set V with*

$$K \subset V \subset \overline{V} \subset U .$$

In fact, for every point $t \in K$ there exists an open integrable and carrable set V_t with

$$t \in V_t \subset \overline{V}_t \subset U .$$

There exists then a finite family $(V_i)_{1 \leqslant i \leqslant n}$ of open integrable and carrable sets which cover K with $V_i \subset \overline{V}_i \subset U$ for every i. The set $V = \cup V_i$ is open integrable and carrable and

$$K \subset V \subset \overline{V} \subset U .$$

The diffuse measures have a property similar to the Darboux property: the set of values in an interval.

20.38 Proposition. *Let K be a compact set and $U \supset K$ an open set with*

$\mu^*(U) > \mu(K)$. *If μ is diffuse, for every number a with $\mu(K) < a \leqslant \mu^*(U)$ there exists an open μ-integrable set G_a with*

$$K \subset G_a \subset U \text{ and } \mu(G_a) = a .$$

Since

$$\mu(K) = \inf_{G \supset K} \mu(G), \ G \text{ open} ,$$

there exist open integrable sets G with $K \subset G \subset U$ and $\mu(G) < a$.

Denote by $\mathscr{G}$ the set of all open subsets G with $K \subset G \subset U$ and $\mu^*(G) \leqslant a$. The set $\mathscr{G}$ is inductively ordered by inclusion. In fact, if (G_α) is a totally ordered family of subsets of $\mathscr{G}$ then the set $G = \cup G_\alpha$ is open, $K \subset G \subset U$ and

$$\mu^*(G) = \mu^*(\cup G_\alpha) = \sup \mu^*(G_\alpha) \leqslant a ,$$

therefore $G \in \mathscr{G}$. Let G_0 be a maximal element of $\mathscr{G}$. We have $K \subset G_0 \subset U$, G_0 is open and $\mu^*(G_0) \leqslant a$. We shall prove that $\mu^*(G_0) = a$.

In fact, if we had $\mu^*(G_0) < a$, then $G_0 \neq U$; taking $t_0 \in U \setminus G_0$ we have $\mu(\{t_0\}) = 0$ (since μ is diffuse). There exists then an open set G_1 with $\{t_0\} \subset G_1 \subset U$ and

$$\mu(G_1) < a - \mu(G_0) .$$

Then $K \subset G_0 \cup G_1 \subset U$ and $\mu(G_0 \cup G_1) \leqslant \mu(G_0) + \mu(G_1) < a$, therefore $G_0 \cup G_1 \in \mathscr{G}$, hence G_0 is not an maximal element in $\mathscr{G}$. Thus, taking $G_a = G_0$, we have

$$K \subset G_a \subset U \text{ and } \mu(G_a) = a$$

and the proposition is proved.

20.39 Lemma. *Let V and W be two open, relatively compact, μ-carrable sets, and a and b two numbers such that*

$$V \subset W \text{ and } \mu(V) < a < b < \mu(W) .$$

If μ is diffuse, there exists an open μ-carrable set U such that

$$\bar{V} \subset U \subset \bar{U} \subset W \text{ and } a < \mu(U) < b .$$

Let $a < \alpha < b$. Since V is compact and μ is diffuse, by proposition 20.38, there exists an open set G with

$$\bar{V} \subset G \subset W \text{ and } \mu(G) = \alpha .$$

Since

$$\mu(G) = \sup_{K \subset G} \mu(K), \ K \text{ compact},$$

there exists a compact set K with

$$\bar{V} \subset K \subset G \text{ and } a < \mu(K) \leqslant \mu(G) = \alpha .$$

By corollary 20.37 there exists an open carrable set U with

$$K \subset U \subset \bar{U} \subset G .$$

It follows that

$$\bar{V} \subset U \subset \bar{U} \subset W$$

and

$$a < \mu(K) \leqslant \mu(U) \leqslant \mu(G) = \alpha < b .$$

20.40 PROPOSITION. *Let V and W be two open relatively compact μ-carrable sets, and a a number such that*

$$\bar{V} \subset W \text{ and } \mu(V) < a < \mu(W) .$$

If μ is diffuse, there exists an open μ-carrable set U such that,

$$\bar{V} \subset U \subset \bar{U} \subset W \text{ and } \mu(U) = a .$$

Let (a_n) be a strictly increasing sequence and (b_n) a strictly decreasing sequence both of them convergent to a, such that

$$\mu(V) < a_n < a < b_n < \mu(W), \text{ for every } n .$$

There exists an open μ-carrable set V_1 with

$$\bar{V} \subset V_1 \subset \bar{V}_1 \subset W \text{ and } a_1 < \mu(V_1) < a_2 .$$

There exists then an open μ-carrable set W_1 with

$$\bar{V} \subset W_1 \subset \bar{W}_1 \subset W \text{ and } b_2 < \mu(W_1) < b_1 .$$

Assume that we have found two open μ-carrable sets V_n and W_n with

$$\bar{V} \subset V_n \subset \bar{V}_n \subset W_n \subset \bar{W}_n \subset W \text{ and } a_n < \mu(V_n) < a_{n+1} < b_{n+1} < \mu(W_n) < b_n .$$

We can then find an open μ-carrable set V_{n+1} with

$$\bar{V}_n \subset V_{n+1} \subset \bar{V}_{n+1} \subset W_n \text{ and } a_{n+1} < \mu(V_{n+1}) < a_{n+2}$$

and then an open μ-carrable set W_{n+1} with

$$\overline{V}_{n+1} \subset W_{n+1} \subset \overline{W}_{n+1} \subset W_n \text{ and } b_{n+2} < \mu(W_{n+1}) < b_{n+1} \, .$$

Thus we have proved by recurrence that we can construct two sequences (V_n) and (W_n) of open μ-carrable sets such that for every n we have

$$\overline{V} \subset V_n \subset \overline{V}_n \subset \overline{V}_{n+1} \subset W_{n+1} \subset \overline{W}_{n+1} \subset W_n \subset \overline{W}_n \subset W$$

and

$$a_n < \mu(V_n) < a_{n+1} < b_{n+1} < \mu(W_n) < b_n \, .$$

Put $U = \cup V_n$. The set U is open. Since the sequence (V_n) is increasing, we have

$$\mu(U) = \mu(\cup V_n) = \sup \mu(V_n) = \lim \mu(V_n) = a.$$

On the other hand, we have

$$U = \cup V_n \subset \cap W_n \subset \cap \overline{W}_n$$

and the last set is closed, therefore

$$\overline{U} \subset \cap \overline{W}_n \, .$$

Thus

$$\mu(\overline{U}) \leqslant \mu(\cap \overline{W}_n) = \inf \mu(\overline{W}_n) = \lim \mu(\overline{W}_n) = a$$

and U is μ-integrable, therefore it is μ-carrable. The set U satisfies the inclusions

$$\overline{V} \subset U \subset \overline{U} \subset W$$

and the proposition is proved.

Now we can prove the result stated at the beginning of this section.

20.41 THEOREM. *If T is compact and if μ is a diffusive measure on T with $\mu(T)=1$, there exists a continuous mapping α of T onto the segment $[0, 1]$ such that the measure $\alpha(\mu)$ is the Lebesgue measure on $[0, 1]$.*

Take $G(0) = \varnothing$ and $G(1) = T$. These are carrable sets and

$$\overline{G(0)} \subset G(1), \quad 0 = \mu(G(0)) < \mu(G(1)) = 1 \, .$$

By proposition 20.40, there exists an open μ-carrable set $G(\frac{1}{2})$ with

$$\overline{G(\tfrac{1}{2})} \subset G(1) \text{ and } \mu(G(\tfrac{1}{2})) = \tfrac{1}{2} \, .$$

Assume that for every *diadic* number $k/2^n$, $k=0, 1, \ldots, 2^n$, we found an open μ-carrable set $G(k/2^n)$ such that for $k=0, 1, \ldots, 2^n-1$ we have

$$\overline{G(k/2^n)} \subset G[(k+1)/2^n] \text{ and } \mu[G(k/2^n)]=k/2^n .$$

For every diadic number of the form $(2k+1)/2^{n+1}$, $k=0, 1, \ldots, 2^n-1$, there exists then, by proposition 20.40, an open carrable set $G[(2k+1)/2^{n+1}]$ with

$$G(k/2^n) \subset G[(2k+1)/2^{n+1}] \subset \overline{G[(2k+1)/2^{n+1}]} \subset G[(k+1)/2^n]$$

and

$$\mu\{G[(2k+1)/2^{n+1}]\} = (2k+1)/2^{n+1} .$$

We proved thus by induction that for each diadic number $r \in [0, 1]$ there exists an open μ-carrable set $G(r)$ with $\mu(G(r))=r$ such that if $r<s$, then $\overline{G(r)} \subset G(s)$.

For every real number $x \in [0, 1]$ put

$$G(x) = \cup_{r \leqslant x} G(r) .$$

For $x=r$ diadic, we have $G(x)=G(r)$. If $0 \leqslant x < x' \leqslant 1$, then there exist two diadic numbers s and s' with $x<s<s'<x'$ and therefore

$$G(x) \subset G(s) \subset \overline{G(s)} \subset G(s') \subset G(x') ,$$

whence

$$\overline{G(x)} \subset G(x') .$$

The set $G(x)$ is open, μ-carrable and

$$\mu(G(x)) = x .$$

In fact,

$$\mu(G(x)) = \mu(\cup_{r \leqslant x} G(r)) = \sup_{r \leqslant x} \mu(G(r)) = \sup_{r \leqslant x} r = x .$$

We have then

$$\overline{G(x)} \subset \cap_{s>x} G(s) ,$$

therefore

$$\mu(\overline{G(x)}) \leqslant \mu(\cap_{s>x} G(s)) \leqslant \inf_{s>x} \mu(G(s)) = \inf_{s>x} s = x ,$$

hence

$$\mu(G(x)) = \mu(\overline{G(x)}) = x$$

and thus $G(x)$ is carrable.

For each point $t \in T$ define

$$\alpha(t) = \inf\{x; t \in G(x)\}.$$

Let us prove that the function α is continuous. Let $t_0 \in T$ and $\varepsilon > 0$. Take $a = \alpha(t_0)$. The set $V = G(a+\varepsilon) \setminus G(a-\varepsilon)$ is a neighbourhood of t_0, since it is open and $t_0 \in G(a+\varepsilon) \setminus G(a-\varepsilon)$. Here we have denoted $G(a-\varepsilon) = G(0)$ if $a - \varepsilon \leqslant 0$ and $G(a+\varepsilon) = G(1)$ if $a + \varepsilon \geqslant 1$.

For every $t \in G(a+\varepsilon) \setminus \overline{G(a-\varepsilon)}$ we have $a - \varepsilon \leqslant \alpha(t) \leqslant a + \varepsilon$ therefore $|\alpha(t) - \alpha(t_0)| \leqslant \varepsilon$. Consequently, α is continuous in t_0. As t_0 is arbitrary in T, it follows that α is continuous on T.

Let us prove that $\alpha(T) = [0, 1]$.

Let $x \in [0, 1]$. We have seen that if $t \in G(x+\varepsilon) \setminus \overline{G(x-\varepsilon)}$, then $|\alpha(t) - x| < \varepsilon$. But

$$\cap_{\varepsilon > 0} [G(x+\varepsilon) \setminus \overline{G(x-\varepsilon)}] \supset \cap_{\varepsilon > 0} [\overline{G(x+\varepsilon)} \setminus G(x-\varepsilon)] \neq \varnothing$$

since the sets $\overline{G(x+\varepsilon)} \setminus G(x-\varepsilon)$ are closed, and finite intersections of such sets are sets of the same type, hence they are non empty sets.

For $t \in \cap_{\varepsilon > 0} [G(t+\varepsilon) \setminus \overline{G(t-\varepsilon)}$ we have $|\alpha(t) - x| < \varepsilon$ for every $\varepsilon > 0$, therefore $\alpha(t) = x$.

Since α is continuous and T is compact, α is μ-proper. Let us prove that $\alpha(\mu)$ is the Lebesgue measure on $[0, 1]$. Put $\nu = \alpha(\mu)$. For every ν-integrable set $A \subset [0, 1]$ we have

$$\nu(A) = \mu(\alpha^{-1}(A)).$$

In particular, for $A = [0, a] \subset [0, 1]$ we have

$$\nu([0, a]) = \mu(\alpha^{-1}([0, a])).$$

But

$$\overline{G(a)} \subset \alpha^{-1}([0, a]) \subset \cap_{x > a} G(x)$$

and

$$\mu(\overline{G(a)}) = \mu(G(a)) = a$$

and

$$\mu(\cap_{r > a} G(r)) = \inf_{r > a} \mu(G(r)) = \inf_{r > a} r = a,$$

therefore

$$\mu[\alpha^{-1}([0, a])) = a,$$

whence

$$\nu([0, a]) = a .$$

It follows that ν is the Lebesgue measure on $[0, 1]$ and the theorem is proved.

§21 Induced measures

1 *Definition of induced measures*

Let $\boldsymbol{m}$ be a dominated measure on T with values in a Banach space X and $S \subset T$ a *locally compact* subspace.

By means of the measure $\boldsymbol{m}$ on T we shall define a measure $\boldsymbol{m}_S$ on S, which is in some sense the *restriction* of the measure $\boldsymbol{m}$ to S. It is natural to choose the measure $\boldsymbol{m}_S$ such that the functions f defined on S, integrable with respect to $\boldsymbol{m}_S$ are exactly the restrictions of functions defined on T and integrable with respect to $\boldsymbol{m}$, or equivalently, those which extended with 0 outside S are integrable with respect to $\boldsymbol{m}$. In particular, this condition must be imposed on continuous functions with compact support defined on S. For these functions the condition is automatically satisfied.

For each function $\boldsymbol{f}$ defined on S, with values in a Banach space or in $\overline{R}$, denote by $\boldsymbol{f}'$ the function defined on T by

$$\boldsymbol{f}'(t) = \boldsymbol{f}(t), \text{ if } t \in S ,$$
$$\boldsymbol{f}'(t) = 0, \quad \text{ if } t \notin S .$$

The restriction of the function $\boldsymbol{f}'$ to S is equal to $\boldsymbol{f}$.

It is easy to verify the following relations:

(1) $(\boldsymbol{f}+\boldsymbol{g})' = \boldsymbol{f}' + \boldsymbol{g}'$, $(\alpha \boldsymbol{f})' = \alpha \boldsymbol{f}'$, α scalar;

(2) $|\boldsymbol{f}|' = |\boldsymbol{f}'|$;

(3) if $f \geqslant 0$, then $f' \geqslant 0$; $f \leqslant g$ if and only if $f' \leqslant g'$;

(4) $\|\boldsymbol{f}'\| = \|\boldsymbol{f}\|$.

21.1 Proposition. *Let E be a Banach space. If $\boldsymbol{f} \in \mathscr{K}_E(S)$, then the function $\boldsymbol{f}'$ is integrable for every measure on T.*

Denote by $K \subset S$ the support of $\boldsymbol{f}$; K is a compact subspace of S, therefore it is also a compact subspace of T. Therefore, $\boldsymbol{f}'$ has compact support K;

from the equality $\|\boldsymbol{f}'\| = \|\boldsymbol{f}\|$ we deduce that $\boldsymbol{f}'$ is bounded. It remains to show that $\boldsymbol{f}'$ is measurable for every measure on T.

Every compact set $H \subset T$ is the union of the disjoint sets $H \cap K$ and $H \setminus K$. The restriction of $\boldsymbol{f}'$ to $H \cap K$ is equal to the restriction of $\boldsymbol{f}$ to $H \cap K$, therefore it is continuous; the restriction of $\boldsymbol{f}'$ to $H \setminus K$ is identically zero, therefore it is continuous. As $H \cap K$ is compact and $H \setminus K$ is the difference of two compact sets, the sets $H \cap K$ and $H \setminus K$ are integrable with respect to every measure, therefore the function $\boldsymbol{f}'$ is measurable with respect to every measure. Since for every measure $\mu \geqslant 0$ on T we have

$$\int^* |\boldsymbol{f}'| \mathrm{d}\mu = \int^* |\boldsymbol{f}'| \varphi_K \mathrm{d}\mu \leqslant \|\boldsymbol{f}'\| \, \mu(K) < +\infty$$

it follows that $\boldsymbol{f}'$ is integrable with respect to every measure, and thus the proposition is proved.

For every function $f \in \mathscr{K}(S)$ put

$$\boldsymbol{n}(f) = \int f' \mathrm{d}\boldsymbol{m}$$

and

$$\nu(f) = \int f' \mathrm{d}|\boldsymbol{m}| .$$

It is easy to see that ν is a linear positive functional on $\mathscr{K}(S)$, therefore ν is a *positive measure* on S. It is also easy to see that the mapping $\boldsymbol{n} : \mathscr{K}(S) \to X$ is linear and dominated by ν:

$$|\boldsymbol{n}(f)| = |\int f' \mathrm{d}\boldsymbol{m}| \leqslant \int |f'| \mathrm{d}|\boldsymbol{m}| = \int |f|' \mathrm{d}|\boldsymbol{m}| = \nu(|f|) ,$$

therefore $\boldsymbol{n}$ is a dominated measure on S with values in X and $|\boldsymbol{n}| \leqslant \nu$.

21.2 DEFINITION. *The measure* $\boldsymbol{n}$ *defined on* S *is called the restriction of the measure* $\boldsymbol{m}$ *to the locally compact subspace* S, *or the measure induced by* $\boldsymbol{m}$ *on* S, *and it is denoted by* $\boldsymbol{m}_S$:

$$\boldsymbol{m}_S(f) = \int f' \mathrm{d}\boldsymbol{m}, \; f \in \mathscr{K}(S) .$$

By this definition, ν is the restriction to S of the measure $|\boldsymbol{m}|$, therefore $\nu = |\boldsymbol{m}|_S$:

$$|\boldsymbol{m}|_S(f) = \int f' \mathrm{d}|\boldsymbol{m}|, \; f \in \mathscr{K}(S) .$$

The inequality $|\boldsymbol{n}| \leqslant \nu$ is now written

$$|\boldsymbol{m}_S| \leqslant |\boldsymbol{m}|_S .$$

We shall further show that we have always

$$|\boldsymbol{m}_S| = |\boldsymbol{m}|_S .$$

Remarks.

1. The locally compact subspace $S \subset T$ is the intersection of an open set with a closed set of T, therefore S is a measurable set with respect to every measure on T.

2. If G is an *open* subset of T, G is a locally compact subset of T.

For every continuous function $f: G \to R$ with compact support the function $f' : T \to R$ is continuous on T and has the same compact support. The space $\mathscr{K}(G)$ can be therefore identified with the subspace $\mathscr{K}(T, G)$ of $\mathscr{K}(T)$. In this case the measure $\boldsymbol{m}_G$ on G is the restriction of the measure $\boldsymbol{m}$ from $\mathscr{K}(T)$ to the subspace $\mathscr{K}(T, G)$.

3. According to our conventions, we shall denote by $\int_S f \mathrm{d}\boldsymbol{m}$ the integral $\int f' \mathrm{d}\boldsymbol{m} = \int f' \varphi_S \mathrm{d}\boldsymbol{m}$. In this case

$$\int f \mathrm{d}\boldsymbol{m}_S = \int_S f \mathrm{d}\boldsymbol{m} .$$

We shall also denote by $\int_S^* f \mathrm{d}\mu$ the upper integral $\int^* f' \mathrm{d}\mu = \int^* f' \varphi_S \mathrm{d}\mu$ of a function $f \geqslant 0$ defined on S with respect to a measure $\mu \geqslant 0$.

We shall show that for every function $f \geqslant 0$ defined on S and every measure $\mu \geqslant 0$ we have the equality

$$\int^* f \mathrm{d}\mu_S = \int_S^* f \mathrm{d}\mu .$$

4. The induced measure $\boldsymbol{m}_S$ has values in the same space X as the measure $\boldsymbol{m}$.

For instance, if μ is positive, μ_S is positive, and if μ is real or complex, then μ_S is real, respectively complex.

But it is possible, however, that μ is real without being positive and μ_S is positive (for example if S is such that $\mu_+ = \varphi_S \mu$).

2 *The upper integral with respect to a positive induced measure*

Let μ be a positive measure on T and $S \subset T$ a locally compact subset.

21.3 Lemma. *If $f \geqslant 0$ is a lower semicontinuous function on S, then the function f' is μ-measurable and*

$$\int^* f \mathrm{d}\mu_S = \int^* f' \mathrm{d}\mu .$$

Let $(g_\alpha)_{\alpha\in A}$ be the directed family of all functions $g\in\mathscr{K}(S)$ satisfying $0\leqslant g\leqslant f$.

Then $f' = \sup_{\alpha\in A} g'_\alpha$, the functions g'_α are μ-measurable (proposition 21.1) and the mapping $t\to(g'_\alpha(t))_{\alpha\in A}$ of T in $\bar{R}^A$ is μ-measurable.

In fact, let $K\subset T$ be a compact set and $\varepsilon>0$. The sets $K\cap S$ and $K\setminus S$ are relatively compact Borel sets, therefore they are μ-integrable. There exists a compact set $K_1\subset K\cap S$ and a compact set $K_2\subset K\setminus S$ such that $\mu((K\cap S)\setminus K_1)<\varepsilon/2$ and $\mu((K\setminus S)\setminus K_2)<\varepsilon/2$. Then $\mu(K\setminus(K_1\cup K_2))<\varepsilon$. The restriction of each function g'_α to K_1 is equal to the restriction of g_α to K_1 (since $K_1\subset K$), therefore it is continuous; the restriction of g'_α to K_2 vanishes identically (since $K_2\cap S=\emptyset$) therefore it is continuous. Since K_1 and K_2 are disjoint compact sets, the restriction of each function g'_α to $K_1\cup K_2$ is continuous. It follows that the restriction of the function $(g'_\alpha)_{\alpha\in A}$ to $K_1\cup K_2$ is also continuous, therefore the function $(g'_\alpha)_{\alpha\in A}$ is μ-measurable.

By proposition 16.1, f' is μ-measurable and

$$\int^* f'\mathrm{d}\mu = \sup_{\alpha\in A}\int g'_\alpha\mathrm{d}\mu\,.$$

But

$$\int g'_\alpha\,\mathrm{d}\mu=\int g_\alpha\,\mathrm{d}\mu_S \text{ and } \sup_{\alpha\in A}\int g_\alpha\mathrm{d}\mu_S=\int^* f\mathrm{d}\mu_S\,,$$

therefore

$$\int^* f\mathrm{d}\mu_S=\int^* f'\mathrm{d}\mu\,.$$

21.4 Lemma. *For every function $f\geqslant 0$ with compact support defined on S we have*

$$\int^* f\mathrm{d}\mu_S=\int^* f'\mathrm{d}\mu\,.$$

Let $h\geqslant 0$ be a lower semicontinuous function on S with $h\geqslant f$. By lemma 21.3 we deduce

$$\int^* h\,\mathrm{d}\mu_S=\int^* h'\,\mathrm{d}\mu\geqslant\int^* f'\,\mathrm{d}\mu\,\cdot$$

Taking the infimum for all functions $h\geqslant f$ from $\mathscr{I}_+(S)$ we obtain

$$\int^* f\mathrm{d}\mu_S\geqslant\int^* f'\mathrm{d}\mu\,.$$

In order to prove the converse inequality, let h_1 be a lower semicontinuous function defined on T such that $h_1\geqslant f'$. The restriction h of h_1 to S is lower semicontinuous on S and we have $h \geqslant f, h' \geqslant f'$ and $h_1 \geqslant h'$. By lemma 21.3 we deduce

$$\int^* f \mathrm{d}\mu_S \leqslant \int^* h \,\mathrm{d}\mu_S = \int^* h' \mathrm{d}\mu \leqslant \int^* h_1 \,\mathrm{d}\mu .$$

Taking in the right-hand side the infimum for all functions $h_1 \geqslant f'$ from $\mathscr{I}_+(T)$ and taking into account the fact that f' has compact support, we obtain

$$\int^* f \mathrm{d}\mu_S \leqslant \int^* f' \mathrm{d}\mu .$$

From the two opposite inequalities it follows the desired equality

21.5 PROPOSITION. *For every function $f \geqslant 0$ defined on S we have*

$$\int^* f \mathrm{d}\mu_S = \int^* f' \mathrm{d}\mu .$$

By lemma 21.4, for every compact set $K \subset S$ we have

$$\int^* f \varphi_K \mathrm{d}\mu_S = \int^* f' \varphi_K \mathrm{d}\mu \leqslant \int^* f' \mathrm{d}\mu ,$$

where φ_K is the characteristic function of K considered either in S or in T. Taking the supremum for all compact sets we obtain

$$\int^* f \mathrm{d}\mu_S \leqslant \int^* f' \mathrm{d}\mu .$$

In order to prove the converse inequality, let $K' \subset T$ be a compact set. Then $K' \cap S$ is μ-measurable and relatively compact, therefore it is μ-integrable. There exists a μ-negligible set $N \subset K' \cap S$ and an increasing sequence (K_n) of compact sets with union $(K' \cap S) \setminus N$. Since

$$f' \varphi_{K'} = f' \varphi_{K' \cap S}$$

we deduce

$$\int^* f' \varphi_{K'} \mathrm{d}\mu = \int^* f' \varphi_{K' \cap S} \mathrm{d}\mu = \int^* f' \varphi_{(K' \cap S) \setminus N} \mathrm{d}\mu =$$
$$= \sup \int^* f' \varphi_{K_n} \mathrm{d}\mu = \sup \int^* f \varphi_{K_n} \mathrm{d}\mu_S \leqslant \int^* f \mathrm{d}\mu_S .$$

Taking the supremum for all compact sets $K' \subset T$ we obtain

$$\int^* f' \mathrm{d}\mu \leqslant \int^* f \,\mathrm{d}\mu_S$$

and the proposition is proved.

21.6 COROLLARY. *For every set $A \subset S$ we have*

$$\mu_S^*(A) = \mu^*(A).$$

3 *Integration with respect to the restriction of a positive measure*

Let μ be a positive measure on T and $S \subset T$ a locally compact subspace.

21.7 PROPOSITION. *A function f defined on S with values in a Banach space E or in $\bar{R}$ is μ_S-negligible if and only if f' is μ-negligible.*

In fact,

$$\int^* |f| \mathrm{d}\mu_S = \int^* |f|' \mathrm{d}\mu = \int^* |f'| \mathrm{d}\mu .$$

21.8 COROLLARY. *A set $A \subset S$ is μ_S-negligible if and only if $A \subset T$ is μ-negligible.*

21.9 PROPOSITION. *Let G be a topological space. A function $f: S \to G$ is μ_S-measurable if and only if any extension f' of f to T, constant outside S, is μ-measurable.*

Assume first that f is μ_S-measurable. Let $K \subset T$ be a compact set. The sets $K \cap S$ and $K \setminus S$ are relatively compact Borel sets, therefore they are μ-integrable.

Since f is μ_S-measurable, there exists a μ_S-negligible set $N \subset K \cap S$ and a sequence (K_n) of mutually disjoint compact sets, with union $(K \cap S) \setminus N$, such that the restriction of f to each K_n is continuous. The restriction of f' to the μ-integrable set $K_0 = K \setminus S$ is constant, therefore continuous. Since the set N is also μ-negligible, it follows that we have decomposed the compact set $K \subset T$ into a μ-negligible set $N \subset K$ and a sequence $(K_n)_{0 \leqslant n < +\infty}$ of μ-integrable sets such that the restriction of f' to each K_n is continuous, hence f' is μ-measurable.

Conversely if f' is μ-measurable, each compact set $K \subset S$ is the union of a μ-negligible set $N \subset K$, and sequence (K_n) of mutually disjoint compact sets, such that the restriction of f' to each K_n is continuous.

But N is also μ_S-negligible, and the restriction of f to K_n is equal to the restriction of f' to K_n, hence it is continuous. It follows that f is μ_S-measurable.

21.10 COROLLARY. *A set $A \subset S$ is μ_S-measurable if and only if $A \subset T$ is μ-measurable.*

21.11 THEOREM. *A function f defined on S with values in a Banach space E or in $\bar{R}$ is μ_S-integrable, if and only if the function f' is μ-integrable. In this case we have*

$$\int f \, d\mu_S = \int f' \, d\mu .$$

The function f is μ_S-measurable if and only if the function f' is μ-measurable. From the equality

$$\int^* |f| \, d\mu_S = \int^* |f'| \, d\mu$$

it follows that $\int^* |f| \, d\mu_S < +\infty$ if and only if $\int^* |f'| \, d\mu < +\infty$. We deduce that f is μ_S-integrable if and only if f' is μ-integrable.

Assume now that f is μ_S-integrable. There exists a sequence (f_n) of functions which are finite linear combinations of the form $\Sigma \varphi_i x_i$ with $\varphi_i \in \mathscr{K}(S)$ and $x_i \in E$ (or $x_i \in R$) such that

$$\lim \int |(f - f_n| \, d\mu_S = 0 .$$

Then

$$|\int f \, d\mu_S - \int f_n \, d\mu_S| = |\int (f - f_n) \, d\mu_S| \leqslant \int |f - f_n| \, d\mu_S , \text{ therefore}$$

$$\lim \int f_n \, d\mu_S = \int f \, d\mu_S .$$

We have then

$$|\int f' \, d\mu - \int f_n' \, d\mu| = |\int (f' - f_n') \, d\mu| \leqslant \int |f' - f_n'| \, d\mu = \int |f - f_n| \, d\mu_S ,$$

whence

$$\lim \int f_n' \, d\mu = \int f' \, d\mu .$$

For $\varphi \in \mathscr{K}(S)$ and $x \in E$ we have

$$\int \varphi x \, d\mu_S = x \int \varphi \, d\mu_S = x \int \varphi' \, d\mu = \int \varphi' x \, d\mu = \int (\varphi x)' \, d\mu$$

and the equality holds also for a finite sum of functions of the form φx, therefore for each function f_n we have

$$\int f_n \, d\mu_S = \int f_n' \, d\mu .$$

Passing to the limit, we obtain

$$\int f \, d\mu_S = \int f' \, d\mu$$

and the theorem is proved.

21.12 COROLLARY. *A set $A \subset S$ is μ_S-integrable if and only if $A \subset T$ is μ-integrable. In this case we have*

$$\mu_S(A) = \mu(A).$$

4 Properties of the positive induced measures

Let μ be a positive measure on T and $S \subset T$ a locally compact subspace.

21.13 PROPOSITION. *The canonical mapping $j : S \to T$ (defined by the equality $j(t) = t$ for $t \in S$) is μ_S-proper and*

$$j(\mu_S) = \varphi_S \mu.$$

In fact, j is continuous, therefore μ_S-measurable. Let $f \in \mathscr{K}(T)$. The function $f \circ j$ is the restriction of f to S, therefore $(f \circ j)' = f\varphi_S$. The function $f\varphi_S$ is μ-integrable, that is, the function $(f \circ j)'$ is μ-integrable, therefore (theorem 21.11) $f \circ j$ is μ_S-integrable.

This means that j is μ_S-proper.

For $f \in \mathscr{K}(T)$ we have

$$\int f \mathrm{d}(\varphi_S \mu) = \int f \varphi_S \mathrm{d}\mu = \int (f \circ j)' \mathrm{d}\mu = \int f \circ j \mathrm{d}\mu_S = \int f \mathrm{d}j(\mu_S),$$

therefore

$$\varphi_S \mu = j(\mu_S).$$

21.14 COROLLARY. *Let T' be a locally compact space, $\alpha : T \to T'$ a μ-proper function and $\alpha_S : S \to T'$ the restriction of α to S. The function α_S is μ_S-proper and we have*

$$\alpha_S(\mu_S) = \alpha(\varphi_S \cdot \mu).$$

In fact, we have $\alpha_S = \alpha \circ j$. Since $j : S \to T$ is μ_S-proper and $\alpha : T \to T'$ is $j(\mu_S) = \varphi_S \mu$-proper, it follows that $\alpha \circ j = \alpha_S : S \to T'$ is μ_S-proper (proposition 20.28) and

$$\alpha_S(\mu_S) = (\alpha \circ j)(\mu_S) = \alpha(j(\mu_S)) = \alpha(\varphi_S \mu).$$

21.15 PROPOSITION. *If R is a locally compact subspace of T such that $R \subset S$, then the restriction of μ to R is equal to the restriction of μ_S to R:*

$$\mu_R = (\mu_S)_R.$$

Let $f \in \mathscr{K}(R)$ and f' the extension of f to S such that $f'(t)=0$ for $t \in S \setminus R$. We deduce

$$\int f \mathrm{d}(\mu_S)_R = \int f' \mathrm{d}\mu_S .$$

Since f' is μ_S-integrable, the extension f'' of f' to T such that $f''(t)=0$ for $t \in T \setminus S$ is μ-integrable and

$$\int f' \mathrm{d}\mu_S = \int f'' \mathrm{d}\mu .$$

But f'' is also the extension of f to T such that $f''(t)=0$ for $t \in T \setminus R$, therefore

$$\int f \mathrm{d}\mu_R = \int f'' \mathrm{d}\mu .$$

From the three equalities it follows

$$\int f \mathrm{d}\mu_R = \int f \mathrm{d}(\mu_S)_R ,$$

that is

$$\mu_R = (\mu_S)_R .$$

21.16 PROPOSITION. *If $g \geqslant 0$ is a locally μ-integrable function defined on T and g_S is the restriction of g to S, then the function g_S is locally μ_S-integrable and*

$$g_S \cdot \mu_S = (g\mu)_S .$$

Let $f \in \mathscr{K}(S)$. The function f' is integrable for every measure on T, in particular for the measure $g\mu$ and we have

$$\int f \mathrm{d}(g\mu)_S = \int f' \mathrm{d}(g\mu) .$$

Then the function $f'g$ is μ-integrable and we have

$$\int f' \mathrm{d}(g\mu) = \int f' g \mathrm{d}\mu .$$

But $f'g = f'g\varphi_S$ and $g\varphi_S = (g_S)'$, therefore $f'g = f'(g_S)' = (fg_S)'$. Thus $(fg_S)'$ is μ-integrable. It follows that fg_S is μ_S-integrable and

$$\int (fg_S) \mathrm{d}\mu_S = \int (fg_S)' \mathrm{d}\mu = \int f' g \mathrm{d}\mu .$$

It follows that g_S is locally μ_S-integrable and

$$\int f \mathrm{d}(g_S \mu_S) = \int f g_S \mathrm{d}\mu_S .$$

From the above equalities we deduce successively

$$\int f\mathrm{d}(g_S\mu_S)=\int fg_S\mathrm{d}\mu_S=\int f'g\mathrm{d}\mu=\int f'\mathrm{d}(g\mu)=\int f\mathrm{d}(g\mu)_S\,,$$

therefore

$$g_S\mu_S=(g\mu)_S\,.$$

21.17 COROLLARY. *We have* $\mu_S=(\varphi_S\mu)_S$.

In fact, we take $g=\varphi_S$ in proposition 21.16 and then $g_S=\varphi_S$ and $g_S\mu_S=\varphi_S\mu_S=\mu_S$.

21.18 PROPOSITION. *A positive measure λ on S is the restriction of a positive measure on T if and only if the set $K\cap S$ is λ-integrable for every compact set $K\subset T$.*

If there exists a measure $\nu\geqslant 0$ on T such that $\lambda=\nu_S$, then for every compact set $K\subset T$ the set $K\cap S$ is a relatively compact Borel set in T, consequently ν-integrable and therefore $K\cap S$ is $\nu_S=\lambda$-integrable.

Conversely, assume that for every compact set $K\subset T$ is the set $K\cap S$ is λ-integrable. Then the canonical mapping $j:S\to T$ is λ-proper (since $(j^{-1}(K)=K\cap S)$.

Then λ is the restriction of the measure $\nu=j(\lambda)$ on T. In fact, let $f\in\mathscr{K}(S)$. The function f' is $\nu=j(\lambda)$-integrable and $f'\circ j=f$, therefore

$$\int f'\mathrm{d}\nu=\int f'\mathrm{d}j(\lambda)=\int f'\circ j\mathrm{d}\lambda=\int f\mathrm{d}\lambda\,.$$

This means that $\lambda=\nu_S$.

21.19 COROLLARY. *Every bounded measures $\lambda\geqslant 0$ on S is the restriction of a positive measure on T.*

In fact, for every compact set $K\subset T$, the set $K\cap S$ is closed in S, therefore it is a Borel set in S. Since λ is bounded, $K\cap S$ is λ-integrable. From proposition 21.18 it follows that λ is the restriction of a positive measure on T.

21.20 COROLLARY. *If S is a closed set in T, then every measure $\lambda\geqslant 0$ on S is the restriction of a positive measure on T.*

In fact, for every compact set $K\subset T$ the set $K\cap S$ is closed in T therefore it is compact, hence it is λ-integrable. Then we can apply proposition 21.18.

21.21 PROPOSITION. *If the measure μ is concentrated on a set $M \subset T$, then the induced measure μ_S is concentrated on $M \cap S$.*

In fact, $S \setminus M = S \cap (T \setminus M)$, therefore

$$\mu_S^*(S \setminus M) = \mu_S^*(S \cap (T \setminus M)) = \mu^*(T \setminus M) = 0 ;$$

therefore, μ_S is concentrated on $S \setminus (S \setminus M) = S \cap M$.

21.22 PROPOSITION. *The measure μ_S is identically zero if and only if S is μ-negligible.*

In fact,

$$\mu_S^*(S) = \mu^*(S) .$$

21.23 COROLLARY. *If $S(\mu)$ is the support of μ and $S(\mu_S)$ is the support of μ_S, we have*

$$S(\mu_S) \subset S \cap S(\mu) .$$

In fact, μ is concentrated on the closed set $S(\mu)$, therefore μ_S is concentrated on the set $S \cap S(\mu)$ which is closed in S, and $S(\mu_S)$ is the least closed set in S on which μ_S is concentrated.

Remark. The inclusion $S(\mu_S) \subset S \cap S(\mu)$ may be strict. For example if $\mu \neq 0$ is diffuse and $S = \{a\}$, then $\mu_S = 0$, therefore it has empty support.

5 *Integration with respect to the restriction of a dominated measure*

Let $\boldsymbol{m}$ be a dominated measure on T with values in a Banach space X and $S \subset T$ a locally compact subspace.

21.24 LEMMA. *Let $f: T \to R$ be a real function. If the function $f\varphi_S$ is $\boldsymbol{m}$-integrable, then the restriction f_S of f to S is $|\boldsymbol{m}_S|$-integrable and*

$$\int f_S \mathrm{d}\boldsymbol{m}_S = \int f\varphi_S \mathrm{d}\boldsymbol{m} .$$

If $f\varphi_S$ is $\boldsymbol{m}$-integrable, that is, $|\boldsymbol{m}|$-integrable, then f_S is $|\boldsymbol{m}|_S$-integrable (theorem 21.11) since $(f_S)' = f\varphi_S$. From the inequality $|\boldsymbol{m}_S| \leqslant |\boldsymbol{m}|_S$, it follows that f_S is $|\boldsymbol{m}_S|$-integrable, that is, $\boldsymbol{m}_S$-integrable.

Let (f_n) be a sequence of functions of $\mathscr{K}(S)$ such that

$$\lim \int |f_n - f_S| \mathrm{d}|\boldsymbol{m}|_S = 0 .$$

Then

$$|\int f_n \mathrm{d}\boldsymbol{m}_S - \int f_S \mathrm{d}\boldsymbol{m}_S| = |\int (f_n - f_S) \mathrm{d}\boldsymbol{m}_S| \leqslant \int |f_n - f_S| \mathrm{d}|\boldsymbol{m}_S| \leqslant$$
$$\leqslant \int |f_n - f_S| \mathrm{d}|\boldsymbol{m}|_S ,$$

therefore

$$\lim \int f_n' \mathrm{d}\boldsymbol{m} = \int f\varphi_S \mathrm{d}\boldsymbol{m} .$$

Also

$$|\int f_n' \mathrm{d}\boldsymbol{m} - \int f\varphi_S \mathrm{d}\boldsymbol{m}| = |\int (f_n' - f\varphi_S) \mathrm{d}\boldsymbol{m}| \leqslant$$
$$\leqslant \int |f_n' - f_S'| \mathrm{d}|\boldsymbol{m}| = \int |f_n - f_S|' \mathrm{d}|\boldsymbol{m}| \leqslant \int |f_n - f_S| d|\boldsymbol{m}|_S ,$$

therefore

$$\lim \int f_n' \mathrm{d}\boldsymbol{m} = \int f\varphi_S \mathrm{d}\boldsymbol{m} .$$

For the functions f_n we have

$$\int f_n \mathrm{d}\boldsymbol{m}_S = \int f_n' \mathrm{d}\boldsymbol{m} .$$

Passing to the limit we obtain

$$\int f_S \mathrm{d}\boldsymbol{m}_S = \int f\varphi_S \mathrm{d}\boldsymbol{m} .$$

21.25 THEOREM. *For every dominated measure* **m** *on T with values in X we have*

$$|\boldsymbol{m}_S| = |\boldsymbol{m}|_S .$$

The inequality $|\boldsymbol{m}_S| \leqslant |\boldsymbol{m}|_S$ was already proved. In order to prove the converse inequality, let $K \subset T$ be a compact set. $K \cap S$ is a relatively compact Borel set, therefore it is $|\boldsymbol{m}|$-integrable and thus (corollary 21.12), $K \cap S$ is also $|\boldsymbol{m}|_S$-integrable, hence it is $|\boldsymbol{m}_S|$-integrable.

By proposition 21.18 it follows that the positive measure $|\boldsymbol{m}_S|$ is the restriction of a positive measure ν on T:

$$|\boldsymbol{m}_S| = \nu_S .$$

By corollary 21.17 we deduce

$$|\boldsymbol{m}|_S = (\varphi_S|\boldsymbol{m}|)_S \quad\text{and}\quad |\boldsymbol{m}_S| = \nu_S = (\varphi_S \nu)_S .$$

By lemma 21.24 we deduce that for every function $f \in \mathscr{K}(T)$ we have

$$\int f_S \mathrm{d}\boldsymbol{m}_S = \int f \varphi_S \mathrm{d}\boldsymbol{m} .$$

But

$$|\int f_S \mathrm{d}\boldsymbol{m}_S| \leqslant \int |f_S| \mathrm{d}|\boldsymbol{m}_S| = \int |f_S| \mathrm{d}\nu_S = \int |f_S'| \mathrm{d}\nu = \int |f| \varphi_S \mathrm{d}\nu$$

therefore

$$|\boldsymbol{m}(f)| = |\int f\varphi_S \mathrm{d}\boldsymbol{m} + \int f\varphi_{T\setminus S} \mathrm{d}\boldsymbol{m}| \leqslant |\int f_S \mathrm{d}\boldsymbol{m}_S| + |\int f\varphi_{T\setminus S}\mathrm{d}\boldsymbol{m}| \leqslant$$
$$\leqslant \int |f| \varphi_S \mathrm{d}\nu + \int |f| \varphi_{T\setminus S} \mathrm{d}|\boldsymbol{m}| = \int |f| \mathrm{d}(\varphi_S \nu + \varphi_{T\setminus S}|\boldsymbol{m}|) .$$

It follows that the positive measure $\varphi_S \nu + \varphi_{T\setminus S}|\boldsymbol{m}|$ dominates $\boldsymbol{m}$, therefore

$$|\boldsymbol{m}| \leqslant \varphi_S \nu + \varphi_{T\setminus S}|\boldsymbol{m}| .$$

We deduce that $\varphi_S|\boldsymbol{m}| \leqslant \varphi_S \nu$, therefore (corollary 21.17) $|\boldsymbol{m}|_S \leqslant \nu_S = |\boldsymbol{m}_S|$. Taking into account the converse inequality, we deduce $|\boldsymbol{m}|_S = |\boldsymbol{m}_S|$.

21.26 THEOREM. *Let E and F be two Banach spaces such that $X \subset \mathscr{L}(E, F)$ and $\boldsymbol{f}: S \to E$ a function*

(1) *The function $\boldsymbol{f}$ is $\mathbf{m}_S$-negligible if and only if $\boldsymbol{f}'$ is $\mathbf{m}$-negligible.*

(2) *The function $\boldsymbol{f}$ is $\mathbf{m}_S$-measurable if and only if $\boldsymbol{f}'$ is $\mathbf{m}$-measurable.*

(3) *The function $\boldsymbol{f}$ is $\mathbf{m}_S$-integrable if and only if $\boldsymbol{f}'$ is $\mathbf{m}$-integrable.*

In this case we have

$$\int \boldsymbol{f} \mathrm{d}\boldsymbol{m}_S = \int \boldsymbol{f}' \mathrm{d}\boldsymbol{m} .$$

The assertions follow from the equality $|\boldsymbol{m}_S| = |\boldsymbol{m}|_S$ and from theorem 21.11.

The equality $\int \boldsymbol{f} \mathrm{d}\boldsymbol{m}_S = \int \boldsymbol{f}' \mathrm{d}\boldsymbol{m}$ is proved as in lemma 21.24, taking a sequence $(\boldsymbol{f}_n)$ of finite linear combinations of the form $\Sigma \varphi_i x_i$ with $\varphi_i \in \mathscr{K}(S)$ and $x_i \in E$ and remarking that for the functions φx we have

$$\int \varphi x \,\mathrm{d}\boldsymbol{m}_S = (\int \varphi \mathrm{d}\boldsymbol{m}_S)x = (\int \varphi' \mathrm{d}\boldsymbol{m}_S)x = \int \varphi' x \,\mathrm{d}\boldsymbol{m} = \int (\varphi x)' \mathrm{d}\boldsymbol{m} ,$$

therefore for the functions $\boldsymbol{f}_n$ we have

$$\int \boldsymbol{f}_n \mathrm{d}\boldsymbol{m}_S = \int \boldsymbol{f}_n \mathrm{d}\boldsymbol{m} .$$

Remark. The assertion concerning the measurability remains valid for every function defined on S with values in an arbitrary topological space.

6 *Properties of restrictions of dominated measures*

Let X be a Banach space, $\boldsymbol{m}$ a dominated measure on T with values in X and $S \subset T$ a locally compact subspace.

All the properties of the restrictions of positive measures are still valid for restrictions of dominated measures.

21.27 PROPOSITION. *The canonical mapping $j: S \to T$ is $\boldsymbol{m}_S$-proper and*

$$j(\boldsymbol{m}_S) = \varphi_S \boldsymbol{m} .$$

In fact, j is proper for the positive measure $|\boldsymbol{m}|_S = |\boldsymbol{m}_S|$, therefore it is proper also for the measure $\boldsymbol{m}_S$ (proposition 21.13). For every function $f \in \mathscr{K}(T)$ we have

$$\int f \mathrm{d}(\varphi_S \boldsymbol{m}) = \int f \varphi_S \mathrm{d}\boldsymbol{m} = \int (f \circ j) \mathrm{d}\boldsymbol{m}_S = \int f \mathrm{d} j(\boldsymbol{m}_S),$$

therefore

$$j(\boldsymbol{m}_S) = \varphi_S \boldsymbol{m} .$$

21.28 COROLLARY *Let T' be a locally compact space, $\alpha : T \to T'$ an $\boldsymbol{m}$-proper function and $\alpha_S : S \to T'$ the restriction of α to S. The function α_S is $\boldsymbol{m}_S$-proper and*

$$\alpha_S(\boldsymbol{m}_S) = \alpha(\varphi_S \boldsymbol{m}) .$$

In fact, α is $|\boldsymbol{m}|$-proper therefore (corollary 21.14), α_S is $|\boldsymbol{m}|_S = |\boldsymbol{m}_S|$-proper, that is, $\boldsymbol{m}_S$-proper.

The equality $\alpha_S(\boldsymbol{m}_S) = \alpha(\varphi_S \boldsymbol{m})$ is proved as the similar one in corollary 21.14.

21.29 PROPOSITION. *If R and S are two locally compact subspaces of T such that $R \subset S$, then*

$$\boldsymbol{m}_R = (\boldsymbol{m}_S)_R .$$

The proof is the same as for proposition 21.15.

21.30 PROPOSITION. *Let E and F be two Banach spaces such that*

$X \subset \mathscr{L}(E, F)$. *If* $\boldsymbol{g} : T \to E$ *is (simply) locally* $\boldsymbol{m}$*-integrable, then the restriction* $\boldsymbol{g}_S : S \to E$ *is (simply) locally* $\boldsymbol{m}_S$*-integrable and*

$$\boldsymbol{g}_S \boldsymbol{m}_S = (\boldsymbol{g}\,\boldsymbol{m})_S\,.$$

The proof is the same as that of proposition 21.16, taking $f \in \mathscr{K}(S)$ if $\boldsymbol{g}$ is locally $\boldsymbol{m}$-integrable, or $f \in \mathscr{K}_E(S)$ if $\boldsymbol{g}$ is simply locally $\boldsymbol{m}$-integrable and taking into account the equality $|\boldsymbol{m}_S| = |\boldsymbol{m}|_S$.

21.31 COROLLARY. $\boldsymbol{m}_S = (\varphi_S \boldsymbol{m})_S$.

21.32 PROPOSITION. *A dominated measure* $\boldsymbol{n}$ *on* S *with values in* X *is the restriction of a measure on* T *if and only if the set* $K \cap S$, *is* $\boldsymbol{n}$*-integrable, for every compact set* $K \subset T$.

The proof is the same as that of the proposition 21.18.

In particular, the proposition is true if the measure $\boldsymbol{n}$ is bounded or if S is a closed subspace of T.

§22 Product of measures

1 *Definition of measures on a product space*

Let T and S be two locally compact spaces. The product space $T \times S$ is also locally compact.

Given a measure $\boldsymbol{m}$ on T and a measure $\boldsymbol{n}$ on S, we shall define in a natural way a measure on the product space $T \times S$ by means of the measures $\boldsymbol{m}$ and $\boldsymbol{n}$.

First we shall prove some lemmas.

22.1 LEMMA. *Let* U *be a relatively compact open set in* T, V *a relatively compact open set in* S *and* E *a Banach space.*

Every function $\boldsymbol{f} \in \mathscr{K}_E(T \times S, U \times V)$ *can be uniformly approximated on* $T \times S$ *by finite sums of the form* $\Sigma\, a_{ij} \varphi_i(t) \psi_j(s)$ *with* $a_{ij} \in E$, $\varphi_i \in \mathscr{K}_+(T, U)$ *and* $\psi_j \in \mathscr{K}_+(S, V)$.

Denote by K (respectively L) the projection of the support of $\boldsymbol{f}$ on T (respectively on S). We have $K \subset U$ and $L \subset V$.

The function $\boldsymbol{f}$ is continuous on the compact set $\bar{U} \times \bar{V}$, therefore it is uniformly continuous on this set. Let $\varepsilon > 0$. There exists a cover $(A_i)_{1 \leq i \leq n}$ of K consisting of open sets contained in U and a cover $(B_j)_{1 \leq j \leq m}$ of L consisting of open sets contained in V, such that the oscillation of the function $\boldsymbol{f}$ on each set $A_i \times B_j$ is $< \varepsilon$. Let $(\varphi_i)_{1 \leq i \leq n}$ (respectively $(\psi_j)_{1 \leq j \leq m}$) be a continuous partition of the unity subordinated to the cover (A_i) of K (respectively subordinated to the cover (B_j) of L).

For each i, the function $\varphi_i : T \to [0, 1]$ has the support contained in A_i; for each j, the function $\psi_j : S \to [0, 1]$ has the support contained in B_j and

$$\Sigma_{1 \leq i \leq n} \varphi_i(t) = 1 \text{ for } t \in K \text{ and } \Sigma_{1 \leq j \leq m} \psi_j(s) = \text{ for } s \in L.$$

We have then

$$\Sigma_{i,j} \varphi_i(t) \psi_j(s) = 1 \text{ for } (t, s) \in K \times L$$

and since $\boldsymbol{f}(t, s) = 0$ for $(t, s) \notin K \times L$ it follows that

$$\boldsymbol{f}(t, s) = \Sigma_{i,j} \boldsymbol{f}(t, s) \varphi_i(t) \psi_j(s), \text{ for } (t, s) \in T \times S.$$

For each i choose a point $t_i \in A_i$, for each j choose a point $s_j \in B_j$ and denote $a_{ij} = \boldsymbol{f}(t_i, s_j) \in E$. We have

$$|\boldsymbol{f}(t, s) - a_{ij}| < \varepsilon \text{ for } (t, s) \in A_i \times B_j,$$

therefore

$$|\boldsymbol{f}(t, s) - \Sigma_{i,j} a_{ij} \varphi_i(t) \psi_j(s)| < \varepsilon \text{ for every } (t, s) \in T \times S.$$

Thus the lemma is proved.

Remarks.

1. If the function f is ≥ 0, then $a_{ij} \geq 0$ and changing the notation, for example $\varphi_k = a_{ij} \varphi_j$, we deduce that:

Every positive function $f \in \mathscr{K}_+(T \times S, U \times V)$ can be uniformly approximated by finite sums of positive functions of the form

$$\Sigma \varphi_k(t) \psi_k(s) \text{ with } \varphi_k \in \mathscr{K}_+(T, U) \text{ and } \psi_k \in \mathscr{K}_+(S, V).$$

2. Using the notation $\boldsymbol{g}_k = a_{ij} \varphi_j$, lemma 22.1 can be stated as follows:

Every function $\boldsymbol{f} \in \mathscr{K}_E(T \times S, U \times V)$ can be uniformly approximated by finite sums of the form $\Sigma \boldsymbol{g}_k(t) \psi_k(s)$ with $\boldsymbol{g}_k \in \mathscr{K}_E(T, U)$ and $\varphi_k \in \mathscr{K}_+(S, V)$.

3. If E is a Banach algebra with unity, using the notations $\boldsymbol{g}_k = a_{ij}\varphi_j$ and $\boldsymbol{h}_k = e\psi_j$ we deduce that:

Every function $\boldsymbol{f} \in \mathscr{K}_E(T \times S, U \times V)$ can be uniformly approximated by finite sums of the form $\Sigma\, \boldsymbol{g}_k(t)\, \boldsymbol{h}_k(s)$ with $\boldsymbol{g}_k \in \mathscr{K}_E(T, U)$ and $\boldsymbol{h}_k \in \mathscr{K}_E(S, V)$.

22.2 LEMMA. *Let X, E and F be three Banach spaces, $(x, y) \to xy$ a bilinear mapping of $X \times E$ into F and $\boldsymbol{m}$ a dominated measure on T with values in X. Let $K \subset T$ and $L \subset S$ be two compact sets.*

For every function $\boldsymbol{f} \in \mathscr{K}_E(T \times S, K \times L)$, the function

$$\boldsymbol{h}(s) = \int \boldsymbol{f}(t, s)\,\mathrm{d}\boldsymbol{m}(t)$$

belongs to the space $\mathscr{K}_F(S, L)$.

We remind the reader first that since the measure $\boldsymbol{m} : \mathscr{K}(T) \to X$ is dominated and xy is a continuous bilinear mapping of $X \times E$ into F, $\boldsymbol{m}$ can be extended uniquely to a measure $\boldsymbol{m} : \mathscr{K}_E(T) \to F$ (theorem 8.61).

For each point $s \in S$ the partial function $\boldsymbol{f}_s$ defined on T by the equality $\boldsymbol{f}_s(t) = \boldsymbol{f}(s, t)$ is continuous with the support contained in K with values in E. The mapping $s \to \boldsymbol{f}_s$ of S into $\mathscr{K}_E(T, K)$ is continuous for the topology of the uniform convergence on $\mathscr{K}_E(T, K)$. Since $\boldsymbol{m}$ is continuous on $\mathscr{K}_E(T, K)$ for this topology, the composed function $s \to \boldsymbol{m}(\boldsymbol{f}_s)$ defined on T with values in F is continuous on S. But

$$\boldsymbol{m}(\boldsymbol{f}_s) = \int \boldsymbol{f}_s(t)\,\mathrm{d}\boldsymbol{m}(t) = \int \boldsymbol{f}(t, s)\,\mathrm{d}\boldsymbol{m}(t) = \boldsymbol{h}(s)\,,$$

therefore $\boldsymbol{h}$ is continuous on S. For $s \notin L$ we have $\boldsymbol{f}(t, s) = 0$ for every $t \in T$, therefore $\boldsymbol{h}(s) = 0$ and thus $\boldsymbol{h} \in \mathscr{K}_E(S, L)$.

Remarks

1. If we take $E = C$ (or $E = R$), $F = X$ and xy the multiplication by scalars in X, lemma 22.2 can be stated as follows:

If $f \in \mathscr{K}(T \times S, K \times L)$, then the function $\boldsymbol{h}(s) = \int f(t, s)\mathrm{d}\boldsymbol{m}(t)$ belongs to the space $\mathscr{K}_X(S, L)$.

2. If we take $E = X'$, $F = C$ and the bilinear application $(x, x') \to \langle x, x' \rangle$, lemma 22.2 can be stated as follows:

If $\boldsymbol{f} \in \mathscr{K}_E(T \times S, K \times L)$, then the scalar function $h(s) = \int \boldsymbol{f}(t, s)\,\mathrm{d}\boldsymbol{m}(t)$ is continuous on S with the support contained in L.

3. If X is a Banach algebra, taking $E=F=X$ and xy the multiplication in the algebra X, lemma 22.2 can be stated as follows:

If $\boldsymbol{f}\in\mathscr{K}_X(T\times S, K\times L)$, *then the function* $\boldsymbol{h}(s)=\int\boldsymbol{f}(t,s)\,\mathrm{d}\boldsymbol{m}(t)$ *belongs to the space* $\mathscr{K}_X(S, L)$.

4. If we are given from the beginning the measure $\boldsymbol{m}:\mathscr{K}_E(T)\to F$, lemma 22.2 remains valid even if $\boldsymbol{m}$ is not dominated. Thus, in remark 1, above, it is not necessary that the measure $\boldsymbol{m}$ should be dominated.

22.3 THEOREM. *Let X, Y and Z be three Banach spaces, $(x,y)\to xy$ a continuous bilinear mapping of $X\times Y$ into Z, $\boldsymbol{m}$ a dominated measure on T with values in X and $\boldsymbol{n}$ a dominated measure on S with values in Y. Then the linear mapping $\boldsymbol{p}:\mathscr{K}(T\times S)\to Z$ defined by the equality*

$$\boldsymbol{p}(f)=\int\left(\int f(t,s)\,\mathrm{d}\boldsymbol{m}(t)\right)\mathrm{d}\boldsymbol{n}(s),\ \text{for } f\in\mathscr{K}(T\times S)$$

is a dominated measure on $T\times S$ with values in Z, and it is the only measure on $T\times S$ satisfying the equality

$$\int g(t)h(s)\,\mathrm{d}\boldsymbol{p}(t,s)=\left(\int g(t)\,\mathrm{d}\boldsymbol{m}(t)\right)\left(\int h(s)\,\mathrm{d}\boldsymbol{n}(s)\right),$$

for every function $g\in\mathscr{K}(T)$ and every function $h\in\mathscr{K}(S)$.

We have also

$$\boldsymbol{p}(f)=\int\left(\int f(t,s)\,\mathrm{d}\boldsymbol{n}(s)\right)\mathrm{d}\boldsymbol{m}(t),\ \text{for } f\in\mathscr{K}(T\times S).$$

By lemma 22.2 (remark 1.) if $f\in\mathscr{K}(T\times S)$, then the function

$$s\to\int f(t,s)\,\mathrm{d}\boldsymbol{m}(t)$$

defined on S with values in X is continuous with compact support. Since the measure $\boldsymbol{n}:\mathscr{K}(T)\to Y$ is dominated and the bilinear mapping xy is continuous, we can extend the measure $\boldsymbol{n}$ to a measure $\boldsymbol{n}:\mathscr{K}_X(T)\to Z$, therefore it makes sense to set

$$\boldsymbol{p}(f)=\int\left(\int f(t,s)\,\mathrm{d}\boldsymbol{m}(t)\right)\mathrm{d}\boldsymbol{n}(s).$$

The mapping $f\to\boldsymbol{p}(f)$ of $\mathscr{K}(T\times S)$ into Z is linear.

It also makes sense to set

$$\pi(f)=\int\left(\int f(t,s)\,\mathrm{d}|\boldsymbol{m}|(t)\right)\mathrm{d}|\boldsymbol{n}|(s),\ \text{for } f\in\mathscr{K}(T\times S).$$

The mapping $f\to\pi(f)$ is a *positive* linear functional on $\mathscr{K}(T\times S)$, therefore it is a positive measure on $T\times S$.

Let $a>0$ be a number such that $|xy|\leqslant a|x|\,|y|$ for every $x\in X$ and $y\in Y$. Then

$$|\boldsymbol{p}(f)|=|\textstyle\int(\int f(t,s)\,\mathrm{d}\boldsymbol{m}(t))\,\mathrm{d}\boldsymbol{n}(s)|\leqslant$$

$$\leqslant a\textstyle\int|\int f(t,s)\,\mathrm{d}\boldsymbol{m}(t)|\,\mathrm{d}|\boldsymbol{n}|(s)\leqslant$$

$$\leqslant a\textstyle\int(\int|f(t,s)|\,\mathrm{d}|\boldsymbol{m}|(t))\,\mathrm{d}|\boldsymbol{n}|(s)=a\pi(|f|)\,,$$

therefore $\boldsymbol{p}$ is a measure dominated by the positive measure $a\,\pi$.

Now let $g\in\mathscr{K}(T)$ and $h\in\mathscr{K}(S)$ and put $f(t,s)=g(t)h(s)$.

For each $s\in S$ we have

$$\textstyle\int g(t)h(s)\,\mathrm{d}\boldsymbol{m}=(\int g(t)\,\mathrm{d}\boldsymbol{m}(t))h(s)=\boldsymbol{m}(g)h(s)$$

and

$$\textstyle\int \boldsymbol{m}(g)h(s)\,\mathrm{d}\boldsymbol{n}(s)=\boldsymbol{m}(g)\int h(s)\,\mathrm{d}\boldsymbol{m}(s)=\boldsymbol{m}(g)\,\boldsymbol{n}(h)\,,$$

therefore

$$\textstyle\int g(t)h(s)\,\mathrm{d}\boldsymbol{n}(t,s)=\boldsymbol{p}(hg)\;=\;\int(\int g(t)h(s)\,\mathrm{d}\boldsymbol{m}(t))\,\mathrm{d}\boldsymbol{n}(s)=$$

$$=\textstyle\int \boldsymbol{m}(g)h(s)\,\mathrm{d}\boldsymbol{n}(s)=\boldsymbol{m}(g)\,\boldsymbol{n}(h)=$$

$$=(\textstyle\int g(t)\,\mathrm{d}\boldsymbol{m}(t))\,(\int h(s)\,\mathrm{d}\boldsymbol{n}(s))\,.$$

If $\boldsymbol{p}'=\mathscr{K}(T\times S)\to Z$ is a measure which satisfies also the equality

$$\textstyle\int g(t)h(s)\,\mathrm{d}\boldsymbol{p}'(t,s)=(\int g(t)\,\mathrm{d}\boldsymbol{m}(t))\,(\int h(s)\,\mathrm{d}\boldsymbol{n}(s))$$

for $g\in\mathscr{K}(T)$ and $h\in\mathscr{K}(S)$, by lemma 22.1, we have then $\boldsymbol{p}(f)=\boldsymbol{p}'(f)$ for every function $f\in\mathscr{K}(T\times S)$, therefore $\boldsymbol{p}=\boldsymbol{p}'$ and thus the uniqueness of the measure $\boldsymbol{p}$ is proved.

If we denote by $\boldsymbol{q}$ the measure on $T\times S$ defined by the equality

$$\boldsymbol{q}(f)\;=\;\textstyle\int(\int f(t,s)\,\mathrm{d}\boldsymbol{n}(s))\,\mathrm{d}\boldsymbol{m}(t),\quad\text{for } f\in\mathscr{K}(T\times S)$$

then we can show as above, that

$$\textstyle\int g(t)h(s)\,\mathrm{d}\boldsymbol{q}(t,s)=(\int g(t)\,\mathrm{d}\boldsymbol{m}(t))\,(\int h(s)\,\mathrm{d}\boldsymbol{n}(s))$$

for every function $g\in\mathscr{K}(T)$ and every function $h\in\mathscr{K}(S)$.

From the uniqueness of the measure $\boldsymbol{p}$ it follows that $\boldsymbol{p}=\boldsymbol{q}$, therefore

$$\boldsymbol{p}(f)\;=\;\textstyle\int(\int f(t,s)\,\mathrm{d}\boldsymbol{n}(s))\,\mathrm{d}\boldsymbol{m}(t),\quad\text{for } f\in\mathscr{K}(T\times S)$$

and thus the theorem is proved.

Remark. If we are given from the beginning the measure $\boldsymbol{n}:\mathcal{K}_X(S)\to E$, we can define

$$\boldsymbol{p}(f)=\int(\int f(t,s)\,\mathrm{d}\boldsymbol{m}(t))\,\mathrm{d}\boldsymbol{n}(s),\ \text{for } f\in\mathcal{K}(T\times S)$$

without the assumption that $\boldsymbol{m}$ and $\boldsymbol{n}$ are dominated; in this case, $\boldsymbol{p}$ is *a measure* (not necessary dominated) on $T\times S$ with values in Z.

In fact, let $C\subset T\times S$ be a compact set, K the projection of C on T and L the projection of C on S. Let a_K and b_L be two numbers satisfying the inequalities

$$|\boldsymbol{m}(g)|\leqslant a_K\|g\|,\ \text{for } g\in\mathcal{K}(T,K),$$

$$|\boldsymbol{n}(\boldsymbol{h})|\leqslant b_L\|\boldsymbol{h}\|,\ \text{for } \boldsymbol{h}\in\mathcal{K}_X(S,L).$$

Then, for every function $f\in\mathcal{K}(T\times S,C)$, we have

$$\boldsymbol{p}(f)=|\int(\int f(t,s)\,\mathrm{d}\boldsymbol{m}(t))\,\mathrm{d}\boldsymbol{n}(s)|\leqslant b_L\sup_{s\in S}|\int f(t,s)\,\mathrm{d}\boldsymbol{m}(t)|\leqslant$$
$$\leqslant b_L a_K \sup_{s\in S,\,t\in T}|f(t,s)|=b_L a_K\|f\|,$$

therefore $\boldsymbol{p}$ is continuous on $\mathcal{K}(T\times S,C)$ for the topology of the uniform convergence, consequently $\boldsymbol{p}$ is a measure on $T\times S$.

It is also easily verified, as in the proof of the theorem, that $\boldsymbol{p}$ satisfies the equality

$$\int g(t)h(s)\,\mathrm{d}\boldsymbol{p}(t,s)=(\int g(t)\,\mathrm{d}\boldsymbol{m}(t))\ (\int h(s)\,\mathrm{d}\boldsymbol{n}(s))$$

for $f\in\mathcal{K}(T)$ and $h\in\mathcal{K}(S)$ and that $\boldsymbol{p}$ is the only measure on $T\times S$ satisfying this equality.

If, in addition, $\boldsymbol{m}$ is dominated or if $\boldsymbol{m}:\mathcal{K}_Y(T)\to Z$ and $X=\mathcal{L}(Y,Z)$, then $\boldsymbol{p}$ also satisfies the equality

$$\boldsymbol{p}(f)=\int(\int f(t,s)\,\mathrm{d}\boldsymbol{n}(s))\,\mathrm{d}\boldsymbol{m}(t)\ \text{for } f\in\mathcal{K}(T\times S).$$

22.4 Definition. *Let X, Y and Z be three Banach spaces and $(x,y)\to xy$ a bilinear continuous mapping of $X\times Y$ into Z.*

Given two dominated measures $\boldsymbol{m}:\mathcal{K}(T)\to X$ and $\boldsymbol{n}:\mathcal{K}(S)\to Y$, the dominated measure $\boldsymbol{p}:\mathcal{K}(T\times S)\to Z$ defined in the theorem 22.3 is called the product of the measures $\boldsymbol{m}$ and $\boldsymbol{n}$ and it is denoted by $\boldsymbol{m}\otimes\boldsymbol{n}$.

Instead of

$$\int(\int f(t,s)\,\mathrm{d}\boldsymbol{m}(t))\,\mathrm{d}\boldsymbol{n}(s)\ \text{and}\ \int(\int f(t,s)\,\mathrm{d}\boldsymbol{n}(s))\,\mathrm{d}\boldsymbol{m}(t),$$

we shall write, respectively

$$\int \mathrm{d}\boldsymbol{n}(s) \int f(t, s) \mathrm{d}\boldsymbol{m}(t) \text{ and } \int \mathrm{d}\boldsymbol{m}(t) \int f(t, s) \mathrm{d}\boldsymbol{n}(s)$$

and we shall call them the iterated integrals.

For the integral with respect to $\boldsymbol{m} \otimes \boldsymbol{n}$ of a function f defined on $T \times S$, we use the notations

$$\int\int f \, \mathrm{d}\boldsymbol{m} \mathrm{d}\boldsymbol{n} \quad \text{or} \quad \int\int f \, \mathrm{d}\boldsymbol{n} \mathrm{d}\boldsymbol{m} \quad \text{or}$$
$$\int\int f(t, s) \mathrm{d}\boldsymbol{m}(t) \mathrm{d}\boldsymbol{n}(s) \quad \text{or} \quad \int\int f(t, s) \mathrm{d}\boldsymbol{n}(s) \mathrm{d}\boldsymbol{m}(t)$$

and we call it *the double integral of f with respect to* **m** *and* **n**.

With these notations we have the equalities

$$\int\int f(t, s) \mathrm{d}\boldsymbol{m}(t) \mathrm{d}\boldsymbol{n}(s) = \int \mathrm{d}\boldsymbol{m}(t) \int f(t, s) \mathrm{d}\boldsymbol{n}(s) = \int \mathrm{d}\boldsymbol{n}(s) \int f(t, s) \mathrm{d}\boldsymbol{m}(t)$$

for every function $f \in \mathscr{K}(T \times S)$.

Remarks

1. The product $\boldsymbol{m} \otimes \boldsymbol{n}$ depends also on the continuous bilinear mapping xy. Every time we are talking about the product of two measures this bilinear mapping will be specified.

For example, if $\boldsymbol{m}$ and $\boldsymbol{n}$ are scalar measures, namely if $X = Y = Z = C$, xy is the usual product of numbers. If $X = C$ and $Y = Z$, xy is the product by scalars in Z.

2. The positive measure π defined by the equality

$$\pi(f) = \int \mathrm{d}|\boldsymbol{m}|(t) \int f(t, s) \mathrm{d}|\boldsymbol{n}|(s) \text{ for } f \in \mathscr{K}(T \times S)$$

is the product of the positive measures $|\boldsymbol{m}|$ and $|\boldsymbol{n}|$:

$$\pi = |\boldsymbol{m}| \otimes |\boldsymbol{n}| .$$

If $a > 0$ is a number satisfying the equality $|xy| \leqslant a|x|\,|y|$ for $x \in X$ and $y \in Y$, then from the proof of theorem 22.3 it follows that

$$|\boldsymbol{m} \otimes \boldsymbol{n}| \leqslant a|\boldsymbol{m}| \otimes |\boldsymbol{n}| .$$

If $|xy| \leqslant |x|\,|y|$ for every $x \in X$ and $y \in Y$, then

$$|\boldsymbol{m} \otimes \boldsymbol{n}| \leqslant |\boldsymbol{m}| \otimes |\boldsymbol{n}| .$$

3. It is easy to verify the following equalities

$$\alpha(\boldsymbol{m} \otimes \boldsymbol{n}) = (\alpha \boldsymbol{m}) \otimes \boldsymbol{n} = \boldsymbol{m} \otimes (\alpha \boldsymbol{n}), \ \alpha \text{ scalar} ,$$

$$\boldsymbol{m} \otimes (\boldsymbol{n}+\boldsymbol{n}') = \boldsymbol{m} \otimes \boldsymbol{n} + \boldsymbol{m} \otimes \boldsymbol{n}' ,$$

$$(\boldsymbol{m}+\boldsymbol{m}') \otimes \boldsymbol{n} = \boldsymbol{m} \otimes \boldsymbol{n} + \boldsymbol{m}' \otimes \boldsymbol{n} .$$

2 *Integration with respect to the product of two measures*

Let μ be a positive measure on T and ν a positive measure on S.

The double integral with respect to μ and ν is defined by the equalities

$$\int\int f(t, s)\mathrm{d}\mu(t)\mathrm{d}\nu(s) = \int \mathrm{d}\mu(t) \int f(t, s)\mathrm{d}\nu(s) = \int \mathrm{d}\nu(s) \int f(t, s)\mathrm{d}\mu(t)$$

for $f \in \mathscr{K}(T \times S)$.

We shall show that these equalities remain valid also for lower semicontinuous functions; however, for arbitrary functions $f \geqslant 0$ we have in general inequalities.

22.5 Lemma. *For every lower semicontinuous function $f \geqslant 0$ on $T \times S$, the function*

$$t \to \int^* f(t, s)\mathrm{d}\nu(s)$$

is lower semicontinuous on T and

$$\int\int^* f(t, s)\mathrm{d}\mu(t)\mathrm{d}\nu(s) = \int^* \mathrm{d}\mu(t) \int^* f(t, s)\mathrm{d}\nu(s).$$

Let A be the set of functions $h \in \mathscr{K}(T \times S)$ such that $0 \leqslant h \leqslant f$. We have

$$\begin{aligned}\int\int^* f(t, s)\mathrm{d}\mu(t)\mathrm{d}\nu(s) &= \sup_{h \in A} \int\int h(t, s)\mathrm{d}\mu(t)\mathrm{d}\nu(s) = \\ &= \sup_{h \in A} \int \mathrm{d}\mu(t) \int h(t, s)\mathrm{d}\nu(s) .\end{aligned}$$

By lemma 22.2, the functions

$$t \to \int h(t, s)\mathrm{d}\nu(s)$$

are continuous on T and with compact support, therefore the function

$$t \to \sup_{h \in A} \int h(t, s)\mathrm{d}\nu(s) = \int^* \sup_{h \in A} h(t, s)\mathrm{d}\nu(s) = \int^* f(t, s)\mathrm{d}\nu(s)$$

is also lower semicontinuous on T and

$$\begin{aligned}\int^* \mathrm{d}\mu(t) \int^* f(t, s)\mathrm{d}\nu(s) &= \int^* \mathrm{d}\mu(t) \sup_{h \in A} \int h(t, s)\mathrm{d}\nu(s) = \\ &= \sup_{h \in A} \int \mathrm{d}\mu(t) \int h(t, s)\mathrm{d}\nu(s) .\end{aligned}$$

It follows then that

$$\int\int^* f(t, s)\mathrm{d}\mu(t)\mathrm{d}\nu(s) = \int^* \mathrm{d}\mu(t) \int^* f(t, s)\mathrm{d}\nu(s) .$$

Remarks.

1. In the same way we can prove that if $f \geqslant 0$ is a lower semicontinuous function on $T \times S$, then the function

$$s \to \int^* f(t, s)\,\mathrm{d}\mu(t)$$

is lower semicontinuous on S and

$$\int \int^* f(t, s)\,\mathrm{d}\mu(t)\,\mathrm{d}\nu(s) = \int^* \mathrm{d}\nu(s) \int f(t, s)\,\mathrm{d}\mu(t)\,.$$

2. If $f \geqslant 0$ is lower semicontinuous on $T \times S$ and if

$$\int \int^* f(t, s)\,\mathrm{d}\mu(t)\,\mathrm{d}\nu(s) < \infty\,,$$

then the set

$$\{t;\, \int^* f(t, s)\,\mathrm{d}\nu(s) \neq 0\}$$

is equal to the union of a sequence of μ-integrable open sets, since the function $t \to \int^* f(t, s)\,\mathrm{d}\nu(s)$ is lower semicontinuous (see property 6, section 3, §5).

3. Lemma 22.5 is valid, in particular, if $f(t, s)$ is continuous on $T \times S$.

22.6 Lemma. *For every function $f \geqslant 0$ defined on $T \times S$ we have*

$$\inf_{h \geqslant f} \int \int^* h(t, s)\,\mathrm{d}\mu(t)\,\mathrm{d}\nu(s) \geqslant \int^* \mathrm{d}\mu(t) \int f(t, s)\,\mathrm{d}\nu(s), \quad h \in \mathscr{I}_+(T \times S)\,.$$

In fact, for every lower semicontinuous functions $h \geqslant f$ we have

$$\int\int^* h(t, s)\,\mathrm{d}\mu(t)\,\mathrm{d}\nu(s) = \int^* \mathrm{d}\mu(t) \int^* h(t, s)\,\mathrm{d}\nu(s) \geqslant \int^* \mathrm{d}\mu(t) \int^* f(t, s)\,\mathrm{d}\nu(s)\,.$$

Taking the infimum in the left-hand side for all functions $h \geqslant f$ of $\mathscr{I}_+(T \times S)$ we obtain the stated inequality.

22.7 Proposition. *For every function $f \geqslant 0$ defined on $T \times S$ which vanishes outside the union of a sequence of $\mu \otimes \nu$-integrable open sets we have*

$$\int \int^* f(t, s)\,\mathrm{d}\mu(t)\,\mathrm{d}\nu(s) \geqslant \int^* \mathrm{d}\mu(t) \int^* f(t, s)\,\mathrm{d}\nu(s)\,.$$

If in addition $\int \int^* f(t, s)\,\mathrm{d}\mu(t)\,\mathrm{d}\nu(s) < +\infty$, then the function $t \to \int^* f(t, s)\,\mathrm{d}\nu(s)$ vanishes outside the union of a sequence of μ-integrable open sets of T.

In fact, in this case we have (proposition 11.19)

$$\int \int^* f(t, s)\,\mathrm{d}\mu(t)\,\mathrm{d}\nu(s) = \inf_{f \geqslant h} \int \int^* h(t, s)\,\mathrm{d}\mu(t)\,\mathrm{d}\nu(s), \quad h \in \mathscr{I}_+(T \times S)\,,$$

and the inequality follows from lemma 22.6.

If, in addition, the integral of the left-hand side of the preceding equality is finite, there exists a function $h \geqslant f$ of $\mathscr{I}_+(T \times S)$ with

$$\int\int^* h(t, s) \mathrm{d}\mu(t) \mathrm{d}\nu(s) < +\infty .$$

For every $t \in T$ we have

$$\int^* f(t, s) \mathrm{d}\nu(s) \leqslant \int^* h(t, s) \mathrm{d}\nu(s) .$$

By remark 2. following lemma 22.5, the function $t \to \int^* h(t, s) \mathrm{d}\nu(s)$ vanishes outside the union of a sequence of μ-integrable open sets of T. From the last inequality we deduce that the function $t \to \int^* f(t, s) \mathrm{d}\nu(s)$ has the same property and the proposition is proved.

Remarks

1. In the conditions of the proposition we also have

$$\int\int^* f(t, s) \mathrm{d}\mu(t) \mathrm{d}\nu(s) \geqslant \int^* \mathrm{d}\nu(s) \int^* f(t, s) \mathrm{d}\mu(t) .$$

2. It will be proved that if in addition f is measurable, this inequality is in fact an equality.

3. If the set $\{(t, s) | f(t, s) > 0\}$ is not contained in the union of a sequence of open $\mu \otimes \nu$ -integrable sets, it is possible to have

$$0 = \int\int^* f(t, s) \mathrm{d}\mu(t) \mathrm{d}\nu(s) < \int^* \mathrm{d}\mu(t) \int^* f(t, s) \mathrm{d}\nu(s) <$$

$$< \inf_{h \geqslant f} \int\int^* h(t, s) \mathrm{d}\mu(t) \mathrm{d}\nu(s) = +\infty, \; h \in \mathscr{I}_+(T \times S) .$$

22.8 Corollary. *If f is a $\mu \otimes \nu$-negligible function defined on $T \times S$ with values in a Banach space E or in $\overline{\mathrm{R}}$ and vanishing outside the union of a sequence of $\mu \otimes \nu$-integrable open sets, then there exists a μ-negligible set $A \subset T$ such that for $t \notin A$, the function $s \to f(t, s)$ is ν-negligible.*

In fact,

$$0 = \int\int^* |f(t, s)| \mathrm{d}\mu(t) \mathrm{d}\nu(t) \geqslant \int^* \mathrm{d}\mu(t) \int^* |f(t, s)| \mathrm{d}\nu(s) .$$

From the equality

$$\int^* \mathrm{d}\mu(t) \int^* |f(t, s)| \mathrm{d}\nu(s) = 0$$

it follows that the positive function $t \to \int^* |f(t, s)| \mathrm{d}\nu(s)$ is μ-negligible, therefore there exists a μ-negligible set $A \subset T$, such that for $t \notin A$ we have

$\int^* |f(t, s)| \, d\nu(s) = 0$.

This means that for $t \notin A$, the function $s \to f(t, s)$ is ν-negligible.

22.9 COROLLARY. *If $N \subset T \times S$ is a $\mu \otimes \nu$-negligible set contained in the union of a sequence of $\mu \otimes \nu$-integrable open sets, there exists a μ-negligible set $A \subset T$, such that for $t \notin A$, its t-section $N(t)$ defined by*

$$N(t) = \{s \mid s \in S, (t, s) \in N\}$$

is ν-negligible.

22.10 PROPOSITION. *If f is a $\mu \otimes \nu$-measurable function defined on T with values in a topological space, constant outside the union of a sequence (G_n) of $\mu \otimes \nu$-integrable open sets, then there exists a μ-negligible set $A \subset T$ such that for $t \notin A$, the function $s \to f(t, s)$ is ν-measurable.*

Each set G_n is the union of a $\mu \otimes \nu$-negligible set and of a sequence of mutually disjoint compact sets on which the restriction of f is continuous.

There exists then a partition of $T \times S$ consisting of a $\mu \otimes \nu$-negligible set $N \subset \cup G_n$, a sequence (K_n) of mutually disjoint compact sets contained in $\cup G_n$ such that the restriction of f to each K_n is continuous, and the $\mu \otimes \nu$-measurable set $B = T \setminus \cup G_n$, on which f is constant.

There exists then a μ-negligible set $A \subset T$ such that for $t \notin A$, the t-section $N(t)$ is ν-negligible. The t-sections $K_n(t)$ of the sets K_n are compact subsets of S, and for each $t \in T$, the partial function $s \to f(t, s)$ is continuous on each $K_n(t)$ and constant on $B(t)$.

It follows that for $t \notin A$ the function $s \to f(t, s)$ is ν-measurable by the localization principle.

22.11 COROLLARY. *If $E \subset T \times S$ is a $\mu \otimes \nu$-measurable set contained in the union of a sequence (G_n) of $\mu \otimes \nu$-integrable open sets, then there exists a μ-negligible set $A \subset T$ such that for $t \notin A$, the section $E(t)$ of E is a ν-measurable subset of S.*

22.12 THEOREM (*Lebesgue–Fubini*). *If a function $\boldsymbol{f}$ defined on $T \times S$ with values in a Banach space E or in $\bar{R}$ is $\mu \otimes \nu$-integrable and vanishes outside the union of a sequence of $\mu \otimes \nu$-integrable open sets, then:*

(1) *there exists a μ-negligible set $A \subset T$ such that for $t \notin A$ the function $s \to \boldsymbol{f}(t, s)$ is ν-integrable;*

(2) *the function* $t \to \int \boldsymbol{f}(t, s)\mathrm{d}\nu(s)$ *defined* μ*-almost everywhere on* T *is* μ*-integrable and*

$$\iint \boldsymbol{f}(t, s)\mathrm{d}\mu(t)\mathrm{d}\nu(s) = \int \mathrm{d}\mu(t) \int \boldsymbol{f}(t, s)\mathrm{d}\nu(s).$$

There exists a sequence $(\boldsymbol{h}_n)$ of functions of $\mathscr{K}_E(T \times S)$ converging $\mu \otimes \nu$-almost everywhere and in the mean (in $\mathscr{L}_E^1(T \times S, \mu \otimes \nu)$) to $\boldsymbol{f}$, and a lower semicontinuous, $\mu \otimes \nu$-integrable function $g \geqslant 0$ defined on $T \times S$ such that $|\boldsymbol{h}_n(t, s)| \leqslant g(t, s)$ for every n and every $(t, s) \in T \times S$ (corollary 7.21). In addition, we can assume that each function $\boldsymbol{h}_n$ is a finite sum of the form $\Sigma x_i \varphi_i$ with $x_i \in E$ and $\varphi_i \in \mathscr{K}(T \times S)$. There exists therefore a $\mu \otimes \nu$-negligible set $N \subset T \times S$, such that for $(t, s) \notin N$ we have

$$\lim \boldsymbol{h}_n(t, s) = \boldsymbol{f}(t, s).$$

We have also

$$\lim \iint |\boldsymbol{h}_n - \boldsymbol{f}| \mathrm{d}\mu \mathrm{d}\nu = 0 \text{ and } \lim \iint \boldsymbol{h}_n \mathrm{d}\mu \mathrm{d}\nu = \iint \boldsymbol{f} \mathrm{d}\mu \mathrm{d}\nu.$$

From the hypothesis it follows that the set N is contained in the union of a sequence of $\mu \otimes \nu$-integrable open sets, therefore there exists a μ-negligible set $A_1 \subset T$, such that for each $t \notin A_1$ we have

$$\lim \boldsymbol{h}_n(t, s) = \boldsymbol{f}(t, s)$$

for ν-almost every $s \in S$ (corollary 22.9).

For each $t \in T$, the function $s \to g(t, s)$ is lower semicontinuous on S and

$$\int^* \mathrm{d}\mu(t) \int^* g(t, s)\mathrm{d}\nu(s) = \iint^* g(t, s)\mathrm{d}\mu(t)\mathrm{d}\nu(s) < +\infty.$$

There exists therefore a μ-negligible set $A_2 \subset T$ such that for $t \notin A_2$ we have

$$\int^* g(t, s)\mathrm{d}\nu(s) < +\infty.$$

If we put $A = A_1 \cup A_2$, the set A is μ-negligible and for $t \notin A$ we have

$$\lim \boldsymbol{h}_n(t, s) = \boldsymbol{f}(t, s), \text{ for } \nu\text{-almost every } s \in S,$$

$$\int^* g(t, s)\mathrm{d}\nu(s) < +\infty$$

and

$$|\boldsymbol{h}_n(t, s)| \leqslant g(t, s), \text{ for every } n \text{ and every } s \in S.$$

By Lebesgue's theorem, for $t \notin A$ the function $s \to \boldsymbol{f}(t, s)$ is ν-integrable. If for $t \in A$ we put $\int \boldsymbol{f}(t, s)\mathrm{d}\nu(s) = 0$, we have

$$\int^* d\mu(t)|\int \boldsymbol{h}_n(t, s)d\nu(s) - \int \boldsymbol{f}(t, s)d\nu(s)| =$$
$$= \int^* d\mu(t)|\int [\boldsymbol{h}_n(t, s) - \boldsymbol{f}(t, s)]d\nu(s)| \leqslant$$
$$= \int^* d\mu(t)\int |\boldsymbol{h}_n(t, s) - \boldsymbol{f}(t, s)|d\nu(s) \leqslant \int\int^* |\boldsymbol{h}_n(t, s) - \boldsymbol{f}(t, s)|d\mu(t)d\nu(s)$$

therefore

$$\lim \int^* d\mu(t)|\int \boldsymbol{h}_n(t, s)d\nu(s) - \int \boldsymbol{f}(t, s)d\nu(s)| = 0.$$

Since the functions $t \to \int \boldsymbol{h}_n(t, s)\,d\nu(s)$ are continuous on T and have compact support, it follows that the function $t \to \int \boldsymbol{f}(t, s)d\nu(s)$ is μ-integrable and

$$\int d\mu(t)\int \boldsymbol{f}(t, s)d\nu(s) = \lim \int d\mu(t)\int \boldsymbol{h}_n(t, s)d\nu(s).$$

For each $x \in E$ and $\varphi \in \mathscr{K}(T \times S)$ we have

$$\int\int x\varphi(t, s)d\mu(s)d\nu(t) = x\int\int \varphi(t, s)d\mu(t)d\nu(s) =$$
$$= x\int d\mu(t)\int \varphi(s, t)d\nu(s) =$$
$$= \int d\mu(t)x\int \varphi(t, s)d\nu(s) = \int d\mu(t)\int x\varphi(t, s)d\nu(s),$$

therefore for the function $\boldsymbol{h}_n$ we have

$$\int\int \boldsymbol{h}_n(t, s)d\mu(t)d\nu(s) = \int d\mu(t)\int \boldsymbol{h}_n(t, s)d\nu(s).$$

Passing to the limit in this equality we obtain

$$\int\int \boldsymbol{f}(t, s)d\mu(t)d\nu(s) = \int d\mu(t)\int \boldsymbol{f}(t, s)d\nu(s).$$

Remarks.

1. In the same way we can prove that *if $\boldsymbol{f}$ is $\mu\otimes\nu$-integrable and vanishes outside the union of a sequence of $\mu\otimes\nu$-integrable open sets, then*

(1′) *there exists a ν-negligible set $B \subset S$ such that for $s \notin B$ the function $t \to \boldsymbol{f}(t, s)$ is μ-integrable;*

(2′) *the function $s \to \int \boldsymbol{f}(t, s)d\nu(t)$ defined ν-almost everywhere on S is ν-integrable and*

$$\int\int \boldsymbol{f}(t, s)d\mu(t)d\nu(s) = \int d\nu(s)\int \boldsymbol{f}(t, s)d\mu(t).$$

2. In the conditions of the theorem, the function $t \to \int \boldsymbol{f}(t, s)d\nu(s)$ vanishes outside the union of a sequence of μ-integrable open sets of T (proposition 22.7).

3. Let $\boldsymbol{f}(t, s)$ be a function defined on $T \times S$ with values in E or $\overline{R}$. If the function $s \to \boldsymbol{f}(t, s)$ is ν-integrable for μ-almost every t, and if the function $t \to \int \boldsymbol{f}(t, s)d\nu(s)$ defined μ-almost everywhere is μ-integrable, we say that there exists the iterated integral $\int d\mu \int \boldsymbol{f}\,d\nu$.

In a similar way, if the function $t \to f(t, s)$ is μ-integrable for ν-almost every s, and if the function $s \to \int f(t, s) \mathrm{d}\mu(t)$ defined ν-almost everywhere is ν-integrable, we say that there exists the iterated integral $\int \mathrm{d}\nu \int f \mathrm{d}\mu$.

The Lebesgue–Fubini theorem is now stated as follows:

If f is $\mu \otimes \nu$-integrable and vanishes outside the union of a sequence of $\mu \otimes \nu$-integrable open sets, then the iterated integrals $\int \mathrm{d}\mu \int\int f \mathrm{d}\nu$ and $\int \mathrm{d}\nu \int f \mathrm{d}\mu$ exist and we have the equality

$$\int\int f \mathrm{d}\mu \mathrm{d}\nu = \int \mathrm{d}\mu \int f \mathrm{d}\nu = \int \mathrm{d}\nu \int f \mathrm{d}\mu .$$

But if one or both iterated integrals exist, it does not follow that f is $\mu \otimes \nu$-integrable, nor that the iterated integrals are equal (if f is $\mu \otimes \nu$-measurable, then see corollary 22.17).

22.13 Corollary. *If μ and ν are bounded and if $f(t, s)$ is $\mu \otimes \nu$-integrable, then the interated integrals exist, and*

$$\int\int f \mathrm{d}\mu \mathrm{d}\nu = \int \mathrm{d}\mu \int f \mathrm{d}\nu = \int \mathrm{d}\nu \int f \mathrm{d}\mu .$$

In fact, by lemma 22.5, $\mu \otimes \nu$ is bounded, therefore $T \times S$ is a $\mu \otimes \nu$-integrable open set.

22.14 Corollary. *If $f(t, s)$ is a lower semicontinuous positive function on $T \times S$ and if it is $\mu \otimes \nu$-integrable, then the iterated integrals exist and*

$$\int\int f \mathrm{d}\mu \mathrm{d}\nu = \int \mathrm{d}\mu \int f \mathrm{d}\nu = \int \mathrm{d}\nu \int f \mathrm{d}\mu .$$

In fact, an integrable lower semicontinuous function vanishes outside the union of a sequence of integrable open sets.

22.15 Corollary. *If $E \subset T \times S$ is a $\mu \otimes \nu$-integrable set contained in the union of a sequence of $\mu \otimes \nu$-integrable open sets then:*

there exists a μ-negligible set $A \subset T$ such that for $t \notin A$, $E(t) \subset S$ is ν-integrable;

the function $t \to \nu(E(t))$ defined μ-almost everwhere on T is μ-integrable and

$$(\mu \otimes \nu(E) = \int \nu(E(t)) \mathrm{d}\mu(t) .$$

Remark. If E is open or if μ and ν are bounded, the condition that E is contained in the union of a sequence of $\mu\otimes\nu$-integrable open sets is superfluous.

Proposition 22.7 and theorem 22.3 can now be stated more precisely:

22.16 PROPOSITION. *If $f \geqslant 0$ is a $\mu\otimes\nu$-measurable function vanishing outside the union of a sequence (G_n) of $\mu\otimes\nu$-integrable open sets, then the function $t \to \int^* f(t,s)\,\mathrm{d}\nu(s)$ is μ-measurable and*

$$\iint^* f(t,s)\,\mathrm{d}\mu(t)\,\mathrm{d}\nu(s) = \int^* \mathrm{d}\mu(t) \int^* f(t,s)\,\mathrm{d}\nu(s)\,.$$

We may assume that the sequence (G_n) is increasing. For each n set $f_n = \inf(n, f\varphi_{G_n})$. Each function f_n is $\mu\otimes\nu$-measurable, bounded, and vanishes outside the $\mu\otimes\nu$-integrable open set G_n, therefore f_n is $\mu\otimes\nu$-integrable and $f = \sup f_n$. For each $t \in T$ we have

$$\int^* f(t,s)\,\mathrm{d}\nu(s) = \int^* \sup f_n(t,s)\,\mathrm{d}\nu(s) = \sup \int^* f_n(t,s)\,\mathrm{d}\nu(s)\,.$$

By theorem 22.3, the function $t \to f_n(t,s)$ is integrable, except for the points t belonging to a μ-negligible set, and the function $t \to \int f_n(t,s)\,\mathrm{d}\nu(s)$, defined μ-almost everywhere, is μ-integrable. It follows that the function

$$t \to \int^* f_n(t,s)\,\mathrm{d}\nu(s)$$

is μ-almost everywhere equal to the μ-integrable function

$$t \to \int f_n(t,s)\,\mathrm{d}\nu(s)\,,$$

hence it is μ-measurable. Then the function

$$t \to \int^* f(t,s)\,\mathrm{d}\nu(s),$$

is μ-measurable as the upper envelope of a sequence of μ-measurable functions.

Since

$$\int^* f_n(t,s)\,\mathrm{d}\nu(s) = \int f_n(t,s)\,\mathrm{d}\nu(s)$$

μ-almost everywhere for each n, and the functions in each member of this equality form an increasing sequence of positive functions, we have

$$\begin{aligned}\iint^* f(t,s)\,\mathrm{d}\mu(t)\,\mathrm{d}\nu(s) &= \sup \iint f_n(t,s)\,\mathrm{d}\mu(t)\,\mathrm{d}\nu(s) = \sup \int \mathrm{d}\mu(t) \int f_n(t,s)\,\mathrm{d}\nu(s) = \\ &= \sup \int^* \mathrm{d}\mu(t) \int^* f_n(t,s)\,\mathrm{d}\nu(s) = \\ &= \int^* \mathrm{d}\mu(t) \sup \int^* f_n(t,s)\,\mathrm{d}\nu(s) = \int^* \mathrm{d}\mu(t) \int^* f(t,s)\,\mathrm{d}\nu(s)\end{aligned}$$

and the proposition is proved.

22.17 COROLLARY. *Let $\boldsymbol{f}$ be a function defined on $T \times S$ with values in a Banach space E or in $\bar{R}$, $\mu \otimes \nu$-measurable and vanishing outside the union of a sequence of $\mu \otimes \nu$-integrable open sets. Then the integrals*

$$\iint \boldsymbol{f}(t, s) \mathrm{d}\mu(t) \mathrm{d}\nu(s), \ \int \mathrm{d}\mu(t) \int \boldsymbol{f}(t, s) d\nu(s), \ \int \mathrm{d}\nu(s) \int \boldsymbol{f}(t, s) \mathrm{d}\mu(t)$$

exist (all of them) and are equal if and only if one of the integrals

$$\int^* \mathrm{d}\mu(t) \int^* |\boldsymbol{f}(t, s)| \mathrm{d}\nu(s), \ \int^* \mathrm{d}\nu(s) \int^* |\boldsymbol{f}(t, s)| \mathrm{d}\mu(t)$$

is finite.

If, for example,

$$\int^* \mathrm{d}\mu(t) \int^* |\boldsymbol{f}(t, s)| \mathrm{d}\nu(s) < +\infty ,$$

by the preceding proposition and by the integrability criterion, $\boldsymbol{f}$ is $\mu \otimes \nu$-integrable. From theorem 22.12 if follows that the two iterated integrals exist and are equal to the integral of $\boldsymbol{f}$ with respect to $\mu \otimes \nu$.

Conversely, if $\boldsymbol{f}$ is $\mu \otimes \nu$-integrable, then $|\boldsymbol{f}|$ is $\mu \otimes \nu$-integrable, therefore the iterated integrals

$$\int \mathrm{d}\mu(t) \int |f(t, s)| \mathrm{d}\nu(s) \ \text{ and } \ \int \mathrm{d}\nu(s) \int |f(t, s)| \mathrm{d}\mu(t)$$

exist, therefore the iterated upper integrals are finite.

Remark. A lower semicontinuous function $f \geqslant 0$ on $T \times S$ is $\mu \otimes \nu$-integrable if and only if one of the iterated integrals exists.

In fact, f is $\mu \otimes \nu$-measurable and from lemma 22.5 it follows that

$$\iint^* f \mathrm{d}\mu \mathrm{d}\nu < \infty .$$

3 *Integration of scalar functions with respect to the product of two vector measures*

We shall consider several cases.

(a) Let X, Y and Z be three Banach spaces and $(x, y) \to xy$ a continuous bilinear mapping of $X \times Y$ into Z.

Let $\boldsymbol{m}$ be a dominated measure defined on T with values in X, $\boldsymbol{n}$ a dominated measure defined on S with values in Y and $\boldsymbol{m} \otimes \boldsymbol{n}$ the product of the measures $\boldsymbol{m}$ and $\boldsymbol{n}$ with respect to the bilinear mapping xy.

22.18 THEOREM. *If f is a scalar function defined on $T \times S$, integrable with respect to the measure $|\boldsymbol{m}| \otimes |\boldsymbol{n}|$ and vanishing outside the union of a sequence of $\boldsymbol{m}\otimes\boldsymbol{n}$-integrable open sets, then*:

(1) *the function f is $\boldsymbol{m}\otimes\boldsymbol{n}$-integrable*;

(2) *there exists an $\boldsymbol{m}$-negligible set $A \subset T$ such that for $t \notin A$, the function $s \to f(t, s)$ is $\boldsymbol{n}$-integrable.*

(3) *the function $t \to \int f(t, s)\mathrm{d}\boldsymbol{n}(s)$ defined $\boldsymbol{m}$-almost everywhere on T is $\boldsymbol{m}$-integrable and*

$$\iint f(t, s)\mathrm{d}\boldsymbol{m}(t)\mathrm{d}\boldsymbol{n}(s) = \int \mathrm{d}\boldsymbol{m}(t) \int f(t, s)\mathrm{d}\boldsymbol{n}(s) .$$

The fact that f is $\boldsymbol{m}\otimes\boldsymbol{n}$-integrable follows from the inequality $|\boldsymbol{m}\otimes\boldsymbol{n}| \leqslant a|\boldsymbol{m}| \otimes |\boldsymbol{n}|$, where $a > 0$ is a number, which satisfies the inequality $|xy| \leqslant a|x|\,|y|$ for $x \in X$ and $y \in Y$.

From the Lebesgue–Fubini theorem it follows that there exists a $|\boldsymbol{m}|$-negligible, hence $\boldsymbol{m}$-negligible set $A \subset T$ such that for $t \notin A$ the function $s \to f(t, s)$ is $|\boldsymbol{n}|$-integrable, hence $\boldsymbol{n}$-integrable.

In order to prove the last point, let (f_n) be a sequence of scalar continuous functions with compact support defined on $T \times S$ converging $|\boldsymbol{m}| \otimes |\boldsymbol{n}|$-almost everywhere and in the mean in $\mathscr{L}^1_{\mathrm{C}}(T \times S, |\boldsymbol{m}| \otimes |\boldsymbol{n}|)$ to f.

We have then

$$\lim \int |f(t, s) - f_n(t, s)|\,\mathrm{d}|\boldsymbol{m}| \otimes |\boldsymbol{n}| = 0 .$$

Then

$$\lim \int |f(t, s) - f_n(t, s)|\,\mathrm{d}|\boldsymbol{m}\otimes\boldsymbol{n}| = 0 ,$$

therefore

$$\lim \int f_n(t, s)\mathrm{d}\boldsymbol{m}\otimes\boldsymbol{n} = \int f(t, s)\mathrm{d}\boldsymbol{m}\otimes\boldsymbol{n} ,$$

that is,

$$\lim \iint f_n(t, s)\mathrm{d}\boldsymbol{m}(t)\mathrm{d}\boldsymbol{n}(s) = \iint f(t, s)\mathrm{d}\boldsymbol{m}(t)\mathrm{d}\boldsymbol{n}(s) .$$

On the other hand

$$\begin{aligned}
&\int^* \mathrm{d}|\boldsymbol{m}|(t)\left| \int f_n(t, s)\mathrm{d}\boldsymbol{n}(s) - \int f(t, s)\mathrm{d}\boldsymbol{n}(s)\right| = \\
&\quad = \int^* \mathrm{d}|\boldsymbol{m}|(t)\left|\int [f_n(t, s) - f(t, s)]\mathrm{d}\boldsymbol{n}(s)\right| \leqslant \\
&\quad \leqslant \int^* \mathrm{d}|\boldsymbol{m}|(t)\int |f_n(t, s) - f(t)s|\,\mathrm{d}|\boldsymbol{n}|(s) = \\
&\quad = \int |f_n(t, s) - f(t, s)|\,\mathrm{d}|\boldsymbol{m}| \otimes |\boldsymbol{n}| ,
\end{aligned}$$

therefore

$$\lim \int^* \mathrm{d}|\boldsymbol{m}|(t)|\int f_n(t, s)\mathrm{d}\boldsymbol{n}(s) - \int f(t, s)\mathrm{d}\boldsymbol{n}(s)| = 0\,.$$

Since the functions $t \to \int f_n(t, s)\mathrm{d}\boldsymbol{n}(s)$ are $\boldsymbol{m}$-integrable, it follows that the function $t \to \int f(t, s)\mathrm{d}\boldsymbol{n}(s)$ defined $\boldsymbol{m}$-almost everywhere on T is $\boldsymbol{m}$-integrable and

$$\lim \int \mathrm{d}\boldsymbol{m}(t) \int f_n(t, s)\mathrm{d}\boldsymbol{n}(s) = \int \mathrm{d}\boldsymbol{m}(t) \int f(t, s)\mathrm{d}\boldsymbol{n}(s)\,.$$

For the functions f_n we have

$$\int\int f_n(t, s)\mathrm{d}\boldsymbol{m}(t)\mathrm{d}\boldsymbol{n}(s) = \int \mathrm{d}\boldsymbol{m}(t) \int f_n(t, s)\mathrm{d}\boldsymbol{n}(s)\,.$$

Passing to the limit in this equality, we obtain the equality stated in the theorem.

22.19 COROLLARY. *If $|\boldsymbol{m} \otimes \boldsymbol{n}| = |\boldsymbol{m}| \otimes |\boldsymbol{n}|$ and if f is a scalar $\boldsymbol{m}\otimes\boldsymbol{n}$-integrable function defined on $T \times S$, vanishing outside the union of a sequence of $\boldsymbol{m}\otimes\boldsymbol{n}$-integrable open sets, then*

(1) *there exists an $\boldsymbol{m}$-negligible set $A \subset T$ such that for $t \notin A$, the function $s \to f(t, s)$ is $\boldsymbol{n}$-integrable;*

(2) *the function $t \to \int f(t, s)\mathrm{d}\boldsymbol{n}(s)$ defined $\boldsymbol{m}$-almost everywhere on T is $\boldsymbol{m}$-integrable and*

$$\int\int f(t, s)\mathrm{d}\boldsymbol{m}(t)\mathrm{d}\boldsymbol{n}(s) = \int \mathrm{d}\boldsymbol{m}(t) \int f(t, s)\mathrm{d}\boldsymbol{n}(s)\,.$$

(b) Let E, Y and G be three Banach spaces and $(x, y) \to xy$ a continuous bilinear mapping of $E \times Y$ into G.

Let μ be a scalar measure on T, $\boldsymbol{n}$ a dominated measure on S with values in Y and $\mu\otimes\boldsymbol{n}$ the product of the measures μ and $\boldsymbol{n}$ with respect to the product by scalars in Y. The measure $\mu\otimes\boldsymbol{n}$ takes on values in Y as does the measure $\boldsymbol{n}$.

20.20 THEOREM. *If $\boldsymbol{f} : T \times S \to E$ is a $|\mu| \otimes |\boldsymbol{n}|$-integrable function vanishing outside the union of a sequence of $|\mu| \otimes |\boldsymbol{n}|$-integrable open sets, then:*

(1) *the function $\boldsymbol{f}$ is $\mu\otimes\boldsymbol{n}$-integrable;*

(2) *there exists a μ-negligible set $A \subset T$ such that for $t \notin A$ the function $s \to \boldsymbol{f}(t, s)$ is $\boldsymbol{n}$-integrable:*

(3) *the function $t \to \int \boldsymbol{f}(t, s)\mathrm{d}\boldsymbol{n}(s))$ defined μ-almost everywhere on T with values in G is μ-integrable and*

$$\int\int \boldsymbol{f}(t, s)\mathrm{d}\mu(t)\mathrm{d}\boldsymbol{n}(s) = \int \mathrm{d}\mu(t) \int \boldsymbol{f}(t, s)\mathrm{d}\boldsymbol{n}(s)\,.$$

The proof is the same as for theorem 22.18.

The points (2) and (3) of the theorem can be replaced by:

(2′) *there exists an* $\boldsymbol{n}$*-negligible set* $B \subset S$ *such that for* $s \notin B$, *the function* $t \to \boldsymbol{f}(t, s)$ *is* μ*-integrable*;

(3′) *the function* $s \to \int \boldsymbol{f}(t, s) \mathrm{d}\mu(t)$ *defined* $\boldsymbol{n}$*-almost everywhere on* S *with values in* E *is* $\boldsymbol{n}$*-integrable and*

$$\iint \boldsymbol{f}(t, s) \mathrm{d}\mu(t) \mathrm{d}\boldsymbol{n}(s) = \int \mathrm{d}\boldsymbol{n}(s) \int \boldsymbol{f}(t, s) \mathrm{d}\mu(t) .$$

Remark. The theorem is valid, in particular, if $Y = C$, $G = E$ and n is a scalar measure. In this case the theorem will be specified further.

(c) Let E be a Banach algebra, $\boldsymbol{m}$ a dominated measure on T with values in E, $\boldsymbol{n}$ a dominated measure on S with values in E and $\boldsymbol{m} \otimes \boldsymbol{n}$ the product of the measures $\boldsymbol{m}$ and $\boldsymbol{n}$ with respect to the product xy in the algebra E.

22.21 Theorem. *If* $\boldsymbol{f}: T \times S \to E$ *is a* $|\boldsymbol{m}| \otimes |\boldsymbol{n}|$*-integrable function and vanishes outside the union of a sequence of* $|\boldsymbol{m}| \otimes |\boldsymbol{n}|$*-integrable open sets, then*:

(1) *the function* $\boldsymbol{f}$ *is* $\boldsymbol{m} \otimes \boldsymbol{n}$*-integrable*;

(2) *there exists an* $\boldsymbol{m}$*-negligible set* $A \subset T$ *such that for* $t \notin A$ *the function* $s \to \boldsymbol{f}(t, s)$ *is* $\boldsymbol{n}$*-integrable*;

(3) *the function* $t \to \int \boldsymbol{f}(t, s) \mathrm{d}\boldsymbol{n}(s)$ *defined* $\boldsymbol{m}$*-almost everywhere on* T *with values in* E *is* $\boldsymbol{m}$*-integrable and*

$$\iint \boldsymbol{f}(t, s) \mathrm{d}\boldsymbol{m}(t) \mathrm{d}\boldsymbol{n}(s) = \int \mathrm{d}\boldsymbol{m}(t) \int \boldsymbol{f}(t, s) \mathrm{d}\boldsymbol{n}(s) .$$

The proof is similar to that of theorem 22.18.

4 *Properties of the product of measures*

a. *Properties of the upper integral*

Let μ be a positive measure on T and ν a positive measure on S.

22.22 Lemma. *For every function* $g \geqslant 0$ *defined on* T *and every function* $h \geqslant 0$ *defined on* S, *we have*

$$\int^* \mathrm{d}\mu(t) \int^* g(t) h(s) \mathrm{d}\nu(s) = \left(\int^* g(t) \mathrm{d}\mu(t)\right) \left(\int^* h(s) \mathrm{d}\nu(s)\right) .$$

In fact, for every $t \in T$ we have

$$\int^* g(t)h(s)\mathrm{d}\nu(s) = g(t)\int^* h(s)\mathrm{d}\nu(s)$$

even if $g(t) = +\infty$. Then we have

$$\int^* \mathrm{d}\mu(t)\int^* g(t)h(s)\mathrm{d}\nu(s) = \int^* \left(\int^* h(s)\mathrm{d}\nu(s)\right)g(t)\mathrm{d}\mu(t) =$$

$$= \left(\int^* h(s)\mathrm{d}\nu(s)\right)\left(\int^* g(t)\mathrm{d}\mu(t)\right)$$

even if $\int^* h(s)\mathrm{d}\nu(s) = +\infty$.

Remark. We have also

$$\int^* \mathrm{d}\nu(s)\int^* g(t)h(s)\mathrm{d}\mu(t) = \left(\int^* g(t)\mathrm{d}\mu(t)\right)\left(\int^* h(s)\mathrm{d}\nu(s)\right).$$

22.23 PROPOSITION. *For every function $g \geqslant 0$ defined on T and every function $h \geqslant 0$ defined on S we have*

$$\iint^* g(t)h(s)\mathrm{d}\mu(t)\mathrm{d}\nu(s) = \left(\int^* g(t)\mathrm{d}\mu(t)\right)\left(\int^* h(s)\mathrm{d}\nu(s)\right).$$

(a) Assume first that g and h have compact supports. Then the function $(t, s) \to g(t)h(s)$ has compact support in $T \times S$.

By proposition 22.7 we have then

$$\iint^* g(t)h(s)\mathrm{d}\mu(t)\mathrm{d}\nu(s) \geqslant \int^* \mathrm{d}\mu(t)\int^* g(t)h(s)\mathrm{d}\nu(s)),$$

therefore by the preceding lemma

$$\iint^* g(t)h(s)\mathrm{d}\mu(t)\mathrm{d}\nu(s) \geqslant \left(\int^* g(t)\mathrm{d}\mu(t)\right)\left(\int^* h(s)\mathrm{d}\nu(s)\right).$$

(b) In order to prove the converse inequality assume first that the right-hand side is not of the form $0 \cdot \infty$.

Let $g' \geqslant g$ be a lower semicontinuous function on T and $h' \geqslant h$ a lower semicontinuous function on S. The function $g'(t)h'(t)$ is lower semicontinuous on $T \times S$. Using lemmas 22.5 and 22.22 we deduce

$$\iint^* g(t)h(s)\mathrm{d}\mu(t)\mathrm{d}\nu(s) \leqslant \iint^* g'(t)h'(s)\mathrm{d}\mu(t)\mathrm{d}\nu(s) =$$

$$= \int^* \mathrm{d}\mu(t)\int^* g'(t)h'(s)\mathrm{d}\nu(s) =$$

$$= \left(\int^* g'(t)\mathrm{d}\mu(t)\right)\left(\int^* h'(s)\mathrm{d}\nu(s)\right).$$

Taking in the right-hand side the infimum for all the functions $g' \in \mathscr{I}_+(T)$ and $h' \in \mathscr{I}_+(S)$ with $g' \geqslant g$ and $h' \geqslant h$ we obtain the inequality

$$\iint^* g(t)h(s)\,d\mu(t)\,d\nu(s) \leqslant \left(\int^* g(t)\,d\mu(t)\right)\left(\int^* h(s)\,d\nu(s)\right),$$

whence follows the required equality in this case.

Before proceeding further observe that if $A \subset T$ and $B \subset S$ are relatively compact, then taking $g = \varphi_A$ and $h = \varphi_B$ we obtain

$$(\mu \otimes \nu)^*(A \times B) = \mu^*(A)\,\nu^*(B).$$

In particular, if $A \subset T$ is μ-negligible and $B \subset S$ is relatively compact, then $A \times B$ is $\mu \otimes \nu$-negligible.

(c) Assume now that the right-hand side is of the form $0 \cdot \infty$.

If for example $\int^* g(t)\,d\mu(t) = 0$, then the set $A = \{t \in T \mid g(t) \neq 0\}$ is μ-negligible, the set $B = \{s \in S \mid h(s) \neq 0\}$ is relatively compact, therefore $A \times B$ is $\mu \otimes \nu$-negligible. Since $\{(t, s) \in T \times S \mid g(t)h(s) \neq 0\} \subset A \times B$, it follows that the function $g(t)h(s)$ is $\mu \otimes \nu$-negligible, therefore

$$\iint^* g(t)h(s)\,d\mu(t)\,d\nu(s) = \left(\int^* g(t)\,d\mu(t)\right)\left(\int^* h(s)\,d\nu(s)\right) = 0.$$

(d) If g and h do not have compact supports, then for every compact sets $K \subset T$ and $L \subset S$ we have

$$\iint^* g(t)\varphi_K(t)h(s)\varphi_L(s)\,d\mu(t)\,d\nu(s) =$$
$$= \left(\int^* g(t)\varphi_K(t)\,d\mu(t)\right)\left(\int^* h(s)\varphi_L(s)\,d\nu(s)\right).$$

Since every compact set $C \subset T \times S$ is contained in a compact set of the form $K \times L$ with $K \subset T$ and $L \subset S$, taking the supremum for all the compact sets $K \subset T$ and $L \subset S$, we obtain the stated equality.

Thus the proposition is completely proved.

22.24 Corollary. *If μ and ν are bounded, then $\mu \otimes \nu$ is also bounded and* $\|\mu \otimes \nu\| = \|\mu\|\,\|\nu\|$.

22.25 Corollary. *For every set $A \subset T$ and every set $B \subset S$ we have*

$$(\mu \otimes \nu)^*(A \times B) = \mu^*(A)\,\nu^*(B).$$

22.26 Corollary. *If $A \subset T$ is μ-negligible, then $A \times B$ is $\mu \otimes \nu$-negligible for every set $B \subset S$.*

In particular, if A is μ-negligible, then $A \times S$ is $\mu \otimes \nu$-negligible, and conversely, if $\nu \neq 0$ and $A \times S$ is $\mu \otimes \nu$-negligible, then A is μ-negligible.

Also, if B is ν-negligible, then the set $A \times B$ is $\mu \otimes \nu$-negligible for every set $A \subset T$.

b. *Properties of measurable functions*

Let X, Y and Z be three Banach spaces, $(x, y) \to xy$ a continuous bilinear mapping of $X \times Y$ into Z and $a > 0$ a number such that $|xy| \leqslant a|x||y|$ for every $x \in X$ and $y \in Y$.

Let $\boldsymbol{m}$ be a dominated measure on T with values in X, $\boldsymbol{n}$ a dominated measure on S with values in Y and $\boldsymbol{m} \otimes \boldsymbol{n}$ the product of the measures $\boldsymbol{m}$ and $\boldsymbol{n}$ with respect to the bilinear mapping xy.

22.27 PROPOSITION. *If $A \subset T$ is an $\boldsymbol{m}$-negligible set, then $A \times B$ is $\boldsymbol{m} \otimes \boldsymbol{n}$-negligible for every $B \subset S$.*

In fact, A is $|\boldsymbol{m}|$-negligible, therefore $A \times B$ is $|\boldsymbol{m}| \otimes |\boldsymbol{n}|$-negligible. From the inequality $|\boldsymbol{m} \otimes \boldsymbol{n}| \leqslant |\boldsymbol{m}| \otimes |\boldsymbol{n}|$ it follows that $A \times B$ is $|\boldsymbol{m} \otimes \boldsymbol{n}|$-negligible, that is, $\boldsymbol{m} \otimes \boldsymbol{n}$-negligible.

Conversely, if $|\boldsymbol{m} \otimes \boldsymbol{n}| = |\boldsymbol{m}| \otimes |\boldsymbol{n}|$ and $\boldsymbol{n} \neq 0$ and if $A \times S$ is $\boldsymbol{m} \otimes \boldsymbol{n}$-negligible, then A is $\boldsymbol{m}$-negligible.

22.28 COROLLARY. *The supports of the measures $\boldsymbol{m}$, $\boldsymbol{n}$ and $\boldsymbol{m} \otimes \boldsymbol{n}$ satisfy the following relation*

$$S(\boldsymbol{m} \otimes \boldsymbol{n}) \subset S(\boldsymbol{m}) \times S(\boldsymbol{n}) .$$

If $|\boldsymbol{m} \otimes \boldsymbol{n}| = |\boldsymbol{m}| \otimes |\boldsymbol{n}|$, *then* $S(\boldsymbol{m} \otimes \boldsymbol{n}) = S(\boldsymbol{m}) \times S(\boldsymbol{n})$.

In fact, $T \setminus S(\boldsymbol{m})$ is open and $\boldsymbol{m}$-negligible and $S \setminus S(\boldsymbol{n})$ is open and $\boldsymbol{n}$-negligible, therefore $(T \setminus S(\boldsymbol{m})) \times S$ and $T \times (S \setminus S(\boldsymbol{n}))$ are open and $\boldsymbol{m} \otimes \boldsymbol{n}$-negligible. Since $T \times S \setminus (S(\boldsymbol{m}) \times S(\boldsymbol{n}))$ is the union of the sets $(T \setminus S(\boldsymbol{m})) \times S$ and $T \times (S \setminus S(\boldsymbol{n}))$, we deduce that $T \times S \setminus (S(\boldsymbol{m}) \times S(\boldsymbol{n}))$ is open and $\boldsymbol{m} \otimes \boldsymbol{n}$-negligible. It follows that $S(\boldsymbol{m} \otimes \boldsymbol{n}) \subset S(\boldsymbol{m}) \times S(\boldsymbol{n})$.

Assume now that $|\boldsymbol{m} \otimes \boldsymbol{n}| = |\boldsymbol{m}| \otimes |\boldsymbol{n}|$. Let $U \times V \subset T \times S$ be an $\boldsymbol{m} \otimes \boldsymbol{n}$-negligible open set, hence $|\boldsymbol{m}| \otimes |\boldsymbol{n}|$-negligible. From corollary 22.25 we deduce that one of the sets U or V is negligible, therefore

$$T \times S \setminus S(\boldsymbol{m} \otimes \boldsymbol{n}) \subset ((T \setminus S(\boldsymbol{m})) \times S) \cup (T \times (S \setminus S(\boldsymbol{n}))) = T \times S \setminus (S(\boldsymbol{m}) \times S(\boldsymbol{n})),$$

whence $S(\boldsymbol{m} \otimes \boldsymbol{n}) \supset S(\boldsymbol{m}) \times S(\boldsymbol{n})$ consequently

$$S(\boldsymbol{m} \otimes \boldsymbol{n}) = S(\boldsymbol{m}) \times S(\boldsymbol{n}) .$$

22.29 PROPOSITION. *Let E, F and G be three topological spaces and $u: E \times F \to G$ a continuous function. If $\boldsymbol{g}: T \to E$ is an $\boldsymbol{m}$-measurable function and $\boldsymbol{h}: S \to F$ is an $\boldsymbol{n}$-measurable function, then the function $(t, s) \to u(\boldsymbol{g}(t), \boldsymbol{h}(s))$ defined on $T \times S$ with values in G is $\boldsymbol{m} \otimes \boldsymbol{n}$-measurable.*

First prove that the mapping $(t, s) \to \boldsymbol{g}(t)$ of $T \times S$ into E is $\boldsymbol{m} \otimes \boldsymbol{n}$-measurable.

Let $C \subset T \times S$ be a compact set and let $K \subset T$ and $L \subset S$ be two compact sets such that $C \subset K \times L$. Since the function $\boldsymbol{g}: T \to E$ is $\boldsymbol{m}$-measurable, there exists a partition of K consisting of an $\boldsymbol{m}$-negligible set N and a sequence (K_n) of compact sets of T such that the restriction of $\boldsymbol{g}$ to each K_n is continuous. Then the set $N \times L$ is $\boldsymbol{m} \otimes \boldsymbol{n}$-negligible and the restriction of the function $(t, s) \to \boldsymbol{g}(t)$ to each compact set $K_n \times L$ is continuous. The sets $N \times L$ and $K_n \times L$ form a partition of $K \times L$, therefore the sets $C \cap (N \times L)$ and $C \cap (K_n \times L)$ form a partition of C. As the set $C \cap (N \times L)$ is $\boldsymbol{m} \otimes \boldsymbol{n}$-negligible, the sets $C \cap (K_n \times L)$ are compact and the restriction of the function $(t, s) \to \boldsymbol{g}(t)$ to each set $C \cap (K_n \times L)$ is continuous, the function $(t, s) \to \boldsymbol{g}(t)$ is $\boldsymbol{m} \otimes \boldsymbol{n}$-measurable.

Similarly we deduce that the mapping $(t, s) \to \boldsymbol{h}(s)$ of $T \times S$ into F is $\boldsymbol{m} \otimes \boldsymbol{n}$-measurable. Since the function $u: E \times F \to G$ is continuous, it follows that the mapping $(t, s) \to u(\boldsymbol{g}(t), \boldsymbol{h}(s))$ of $T \times S$ into G is $\mu \otimes \nu$-measurable.

In particular, the preceding proposition can be applied in the following cases:

(1) E is a Banach algebra, $u(x, y) = xy$ (the product in the algebra E), $\boldsymbol{g}: T \to E$ and $\boldsymbol{h}: S \to E$.

(2) E is a Banach space, $F = C$, $u(x, y) = xy$ (the product by scalars in E), $\boldsymbol{g}: T \to E$ and $\boldsymbol{h}: S \to C$.

(3) $F = E'$, $u(x, y) = \langle x, y \rangle$ for $x \in E$ and $y \in E'$, $\boldsymbol{g}: T \to E$ and $\boldsymbol{h}: S \to E'$.

(4) $E = F = C$, $u(x, y) = xy$ (the product of numbers), $\boldsymbol{g}: T \to C$ and $\boldsymbol{h}: S \to C$.

(5) $G = E \times F$. In this case the mapping $(t, s) \to (\boldsymbol{g}(t), \boldsymbol{h}(s))$ of $T \times S$ into $E \times F$ is $\boldsymbol{m} \otimes \boldsymbol{n}$-measurable.

22.30 COROLLARY. *If $A \subset T$ is an $\boldsymbol{m}$-measurable set and $B \subset S$ is an $\boldsymbol{n}$-measurable set, then the set $A \times B$ is $\boldsymbol{m} \times \boldsymbol{n}$-measurable.*

c. *Properties of integrable functions with respect to scalar measures*

22.31 PROPOSITION. *Let μ be a scalar measure on T and ν a scalar measure* *on S. Let E, F and G be three Banach spaces and $(x, y)\to xy$ a continuous bilinear mapping of $E\times F$ into G. If $\boldsymbol{g}: T\to E$ is a μ-integrable function and $\boldsymbol{h}: S\to F$ is a ν-integrable function, then the function $\boldsymbol{g}(t)\boldsymbol{h}(t)$ defined on $T\times S$ with values in G is $\mu\otimes\nu$-integrable and*

$$\int\int \boldsymbol{g}(t)\boldsymbol{h}(s)\mathrm{d}\mu(t)\mathrm{d}\nu(s) = \left(\int \boldsymbol{g}(t)\mathrm{d}\mu(t)\right)\left(\int \boldsymbol{h}(s)\mathrm{d}\nu(s)\right).$$

The function $\boldsymbol{g}(t)\boldsymbol{h}(s)$ is $\mu\otimes\nu$-measurable since $\boldsymbol{g}$ is μ-measurable and $\boldsymbol{h}$ is ν-measurable. Let $a>0$ be a number such that $|xy|\leqslant a|x||y|$ for every $x\in E$ and $y\in F$. We have

$$\begin{aligned}\int\int^* |\boldsymbol{g}(t)\boldsymbol{h}(s)\mathrm{d}|\mu|(t)\mathrm{d}|\nu|(s) &\leqslant a\int\int |\boldsymbol{g}(t)||\boldsymbol{h}(s)|\mathrm{d}|\mu|(t)\mathrm{d}|\nu|(s) = \\ &= a\left(\int^* |\boldsymbol{g}(t)|\mathrm{d}|\mu|(t)\right)\left(\int^* |\boldsymbol{h}(s)|\mathrm{d}|\nu|(s)\right) < +\infty .\end{aligned}$$

By the integrability criterion if follows that $\boldsymbol{g}(t)\boldsymbol{h}(t)$ is $|\mu|\otimes|\nu|$-integrable. From the inequality $|\mu\otimes\nu|\leqslant|\mu|\otimes|\nu|$ we deduce that $\boldsymbol{g}(t)\boldsymbol{h}(s)$ is $|\mu\otimes\nu|$-integrable, therefore $\mu\otimes\nu$-integrable. The stated equality follows then from the Lebesgue–Fubini theorem:

$$\begin{aligned}\int\int \boldsymbol{g}(t)\boldsymbol{h}(s)\mathrm{d}\mu(t)\mathrm{d}\nu(s) &= \int \mathrm{d}\mu(t)\int \boldsymbol{g}(t)\boldsymbol{h}(s)\mathrm{d}\nu(s) = \\ &= \int\left(\int \boldsymbol{g}(t)\boldsymbol{h}(s)\mathrm{d}\nu(s)\right)\mathrm{d}\mu(t) = \\ &= \int \boldsymbol{g}(t)\left(\int \boldsymbol{h}(s)\mathrm{d}\nu(s)\right)\mathrm{d}\mu(t) = \\ &= \left(\int \boldsymbol{g}(t)\mathrm{d}\mu(t)\right)\left(\int \boldsymbol{h}(s)\mathrm{d}\nu(s)\right)\end{aligned}$$

where we took into account the fact that for each $t\in T$, $\boldsymbol{g}(t)\in E$ is a continuous linear mapping of F into G and $\int \boldsymbol{h}(s)\mathrm{d}\nu(s)\in F$ is a continuous linear mapping of E into G.

22.32 COROLLARY. *If $A\subset T$ is a μ-integrable set, and $B\subset S$ is a ν-integrable set, then the set $A\times B\subset T\times S$ is ν-integrable and*

$$(\mu\otimes\nu)(A\times B) = \mu(A)\nu(B).$$

22.33 PROPOSITION. *Let μ be a scalar measure on T and ν a scalar measure on S. Let E, F and G be three Banach spaces and $(x, y)\to xy$ a bilinear continuous mapping of $E\times F$ into G.*

If $\boldsymbol{g}: T\to E$ is a locally μ-integrable function and $\boldsymbol{h}: S\to F$ a locally ν-integrable function, then the function $\boldsymbol{g}(t)\boldsymbol{h}(s)$ defined on $T\times S$ with values in G is locally $\mu\otimes\nu$-integrable and

$$(\boldsymbol{g}\mu)\otimes(\boldsymbol{h}\nu) = (\boldsymbol{g}\,\boldsymbol{h})(\mu\otimes\nu).$$

Since $\boldsymbol{g}$ is μ-measurable and $\boldsymbol{h}$ is ν-measurable, the function $\boldsymbol{g}(t)\,\boldsymbol{h}(s)$ is $\mu\otimes\nu$-measurable.

Let $C\subset T\times S$ be a compact set. Let $K\subset T$ and $L\subset S$ be two compact sets such that $C\subset K\times L$. Since the function $\boldsymbol{g}\varphi_K$ is μ-integrable, and the function $\boldsymbol{h}\varphi_L$ is ν-integrable, the function $\boldsymbol{g}(t)\,\varphi_K(t)\,\boldsymbol{h}(s)\,\varphi_L(s)$ is $\mu\otimes\nu$-integrable, therefore $\boldsymbol{g}(t)\,\boldsymbol{h}(s)\;\varphi_C(t,s)$ is also $\mu\otimes\nu$-integrable. It follows that the function $\boldsymbol{g}(t)\,\boldsymbol{h}(s)$ is locally $\mu\otimes\nu$-integrable.

For every function $\varphi\in\mathscr{K}(T)$ and every function $\psi\in\mathscr{K}(S)$ we have

$$\begin{aligned}\int\varphi(t)\psi(s)\mathrm{d}((\boldsymbol{g}\,\boldsymbol{h})(\mu\otimes\nu)) &= \int\varphi(t)\psi(s)\boldsymbol{g}(t)\,\boldsymbol{h}(s)\mathrm{d}\mu\otimes\nu=\\ &=\iint\varphi(t)\,\boldsymbol{g}(t)\psi(s)\,\boldsymbol{h}(s)\mathrm{d}\mu(t)\mathrm{d}\nu(s)=\\ &=\left(\int\varphi(t)\,\boldsymbol{g}(t)\mathrm{d}\mu(t)\right)\left(\int\psi(s)\,\boldsymbol{h}(s)\mathrm{d}\nu(s)\right)=\\ &=\left(\int\varphi\,\mathrm{d}(\boldsymbol{g}\,\mu)\right)\left(\int\psi\;\mathrm{d}(\boldsymbol{h}\nu)\right)\end{aligned}$$

whence

$$(\boldsymbol{g}\boldsymbol{h})(\mu\otimes\nu)=(\boldsymbol{g}\mu)\otimes(\boldsymbol{h}\nu)\,.$$

22.34 Corollary. *For every scalar measure μ on T and every scalar measure ν on S we have*

$$|\mu\otimes\nu|=|\mu|\otimes|\nu|\,.$$

There exists a locally $|\mu|$-integrable scalar function φ defined on T such that $|\varphi(t)|\equiv 1$ and $\mu=\varphi|\mu|$.

There exists also a locally $|\nu|$-integrable scalar function ψ defined on S such that $|\psi(s)|\equiv 1$ and $\nu=\psi|\nu|$. Then the function $\varphi(t)\psi(s)$ is locally $|\mu|\otimes|\nu|$-integrable and we have $|\varphi(t)\psi(s)|\equiv 1$, therefore

$$|(\varphi\psi)(|\mu|\otimes|\nu|)|=|\mu|\otimes|\nu|\,.$$

It follows that

$$|\mu\otimes\nu|=|(\varphi|\mu|)\otimes(\psi|\nu|)|=|(\varphi\psi(|\mu|\otimes|\nu|)|=|\mu|\otimes|\nu|\,.$$

22.35 Corollary. *We have $S(\mu\otimes\nu)=S(\mu)\times S(\nu)$.*

22.36 Corollary. *If μ is a real measure on T and ν is a real measure on S, then*

$$(\mu\otimes\nu)_+ = \mu_+\otimes\nu_+ + \mu_-\otimes\nu_-\ ,$$

$$(\mu\otimes\nu)_- = \mu_+\otimes\nu_- + \mu_-\otimes\nu_+\ .$$

In fact,

$$\mu\otimes\nu = (\mu_+ - \mu_-)\otimes(\nu_+ - \nu_-) = \\ = (\mu_+\otimes\nu_+ + \mu_-\otimes\nu_-) - (\mu_+\otimes\nu_- + \mu_-\otimes\nu_+)$$

and

$$|\mu\otimes\nu| = |\mu|\otimes|\nu| = (\mu_+ + \mu_-)\otimes(\nu_+ + \nu_-) = \\ = (\mu_+\otimes\nu_+ + \mu_-\otimes\nu_-) + (\mu_+\otimes\nu_- + \mu_-\otimes\nu_+)$$

whence

$$(\mu\otimes\nu)_+ = \tfrac{1}{2}(|\mu\otimes\nu| + \mu\otimes\nu) = \mu_+\otimes\nu_+ + \mu_-\otimes\nu_-$$

$$(\mu\otimes\nu)_- = \tfrac{1}{2}(|\mu\otimes\nu| - \mu\otimes\nu) = \mu_+\otimes\nu_- + \mu_-\otimes\nu_+\ .$$

22.37 PROPOSITION. *Let μ be a scalar measure on T, ν a scalar measure on S, E a Banach space, F and G two Banach spaces such that $E\subset\mathscr{L}(F, G)$ and $H\subset G'$ a vector subspace such that $\langle x, z\rangle = 0$ for every $z'\in H$ implies $x=0$.*

If $\boldsymbol{g}: T\to E$ is a H-weakly locally μ-integrable (respectively simply locally μ-integrable) function and h a locally ν-integrable scalar function defined on S, then the function $\boldsymbol{g}(t)h(s)$ defined on $T\times S$ with values in E is H-weakly locally $\mu\otimes\nu$-integrable (respectively simply locally $\mu\otimes\nu$-integrable) and

$$(\boldsymbol{g}\mu)\otimes(h\nu) = (\boldsymbol{g}h)(\mu\otimes\nu)\ .$$

In fact, $|\boldsymbol{g}|$ is locally μ-integrable and $|h|$ is locally ν-integrable, therefore $|\boldsymbol{g}h| = |\boldsymbol{g}|\,|h|$ is locally $\mu\otimes\nu$-integrable. On the other hand, if $\boldsymbol{g}$ is H-weakly locally μ-integrable, then for each $x\in F$ and $z'\in H$, the function $\langle\boldsymbol{g}(t)x, z'\rangle$ is μ-measurable, therefore, the function $\langle\boldsymbol{g}(t)h(s), z'\rangle = \langle\boldsymbol{g}(t)x, z'\rangle\ h(s)$ is $\mu\otimes\nu$-measurable, that is the function $\boldsymbol{g}(t)h(s)$ is H-weakly $\mu\otimes\nu$-measurable, consequently H-weakly locally $\mu\otimes\nu$-integrable. The assertion concerning the simple local integrability can be proved in the same way.

Assume that $\boldsymbol{g}$ is H-weakly locally μ-integrable. There exists then a dominated measure $\boldsymbol{g}\mu: \mathscr{K}_F(T)\to H'$ such that

$$\langle \textstyle\int \boldsymbol{f}(d)(\boldsymbol{g}\mu), z'\rangle = \int\langle\boldsymbol{g}(t)\boldsymbol{f}(t), z'\rangle d\mu(t)$$

for $\boldsymbol{f}\in\mathscr{K}_F(T)$ and $z'\in H$.

Since $g(t)h(s)$ is H-weakly locally $\mu\otimes\nu$-integrable, there exists a measure $(gh)(\mu\otimes\nu):\mathscr{K}_F(T\times S)\to H'$ such that

$$\langle\int f\,\mathrm{d}(gh)(\mu\otimes\nu)), z'\rangle = \int\langle f(t,s)g(t)h(s), z'\rangle\,\mathrm{d}\mu\otimes\nu$$

for $f\in\mathscr{K}_F(T\times S)$ and $z'\in H$.

For every functions $\varphi\in\mathscr{K}(T)$, $\psi\in\mathscr{K}(S)$, every $x\in F$ and every $z'\in H$, we have

$$\begin{aligned}\langle\int\varphi x\psi\,\mathrm{d}((gh)(\mu\otimes\nu)), z'\rangle &= \int\langle gh\varphi x\psi, z'\rangle\,\mathrm{d}\mu\otimes\nu =\\ &= \int\int\langle g(t)\varphi(t)x, z'\rangle h(s)\psi(s)\,\mathrm{d}\mu(t)\,\mathrm{d}\nu(s) =\\ &= \int\langle g(t)\varphi(t)x, z'\rangle\,\mathrm{d}\mu(t)(\int h(s)\psi(s)\,\mathrm{d}\nu(s)) =\\ &= (\langle\int\varphi(t)x\,\mathrm{d}(g\mu), z'\rangle)(\int\psi(s)\,\mathrm{d}(h\nu) =\\ &= \langle z', (\int\varphi(t)x(\mathrm{d}g\mu))(\int\psi(s)\mathrm{d}(h\nu)\rangle,\end{aligned}$$

whence

$$\int\varphi x\psi\,\mathrm{d}((gh)(\mu\otimes\nu)) = (\int\varphi x\,\mathrm{d}(g(\mu))(\int\psi\,\mathrm{d}(h\nu)$$

for every $x\in F$. Since

$$\int\varphi x\psi\,\mathrm{d}((gh)(\mu\otimes\nu)) = x\int\varphi\psi\,\mathrm{d}((gh)(\mu\otimes\nu))$$

and

$$\int\varphi x\,\mathrm{d}(g\mu) = x\int\varphi\,\mathrm{d}(g\mu),$$

it follows

$$\int\varphi\psi\,\mathrm{d}((gh)(\mu\otimes\nu)) = (\int\varphi\,\mathrm{d}(g\mu))(\int\psi\,\mathrm{d}(h\nu))$$

for every $\varphi\in\mathscr{K}(T)$ and $\psi\in\mathscr{K}(S)$, therefore

$$(gh)(\mu\otimes\nu) = (g\mu)\otimes(h\nu).$$

22.38 COROLLARY. *Let μ be a scalar measure on T, ν a scalar measure on S, E a Banach space, F and G two Banach spaces such that $E\subset\mathscr{L}(F,G)$ and $H\subset G'$ a norming subspace for G.*

Let $g:T\to E$ be a H-weakly locally μ-integrable function and $h:S\to C$ a locally ν-integrable scalar function. We have

$$|(g\mu)\otimes(h\nu)| = |g\mu|\otimes|h\nu|$$

in each of the following cases:

(1) *F is of countable type and there exists a countable set* $Z \subset H$ *norming for G.*
(2) *F is of countable type and* $\boldsymbol{g}$ *is simply locally* μ*-integrable.*
(3) $\boldsymbol{g}$ *is locally* μ*-integrable.*

In fact, in each of these cases we have

$$(\boldsymbol{g}\mu)\otimes(h\nu)=(\boldsymbol{g}h)(\mu\otimes\nu).$$

Using theorem 18.17 we deduce

$$|(\boldsymbol{g}\mu)\otimes(h\nu)|=|(\boldsymbol{g}h)(\mu\otimes\nu)|=|\boldsymbol{g}h|\,|\mu\otimes\nu|=$$
$$=(|\boldsymbol{g}|\,|h|)(|\mu|\otimes|\nu|)=(|\boldsymbol{g}|\,|\mu|)\otimes(|h|\,|\nu|)=|\boldsymbol{g}\mu|\otimes|h\nu|.$$

22.39 COROLLARY. *In each of the cases of corollary 22.38 we have*

$$S((\boldsymbol{g}\mu)\otimes(h\nu))=S(\boldsymbol{g}\mu)\times S(h\nu).$$

22.40 THEOREM. *Let* μ *be a scalar measure on T and* ν *a scalar measure on S. If* $\boldsymbol{f}$ *is a function defined on* $T\times S$ *with values in a Banach space E or in* $\overline{R}$, $\mu\otimes\nu$*-integrable and vanishing outside the union of a sequence of* $\mu\otimes\nu$*-integrable open sets, then:*

(1) *there exists a* μ*-negligible set* $A\subset T$ *such that for* $t\notin A$, *the function* $s\to\boldsymbol{f}(t,s)$ is ν*-integrable*;
(2) *the function* $t\to\int\boldsymbol{f}(t,s)\mathrm{d}\nu(s)$ *defined* μ*-almost everywhere on T is* μ*-integrable and*

$$\iint\boldsymbol{f}(t,s)\mathrm{d}\mu(t)\,\mathrm{d}\nu(s)=\int\mathrm{d}\mu(t)\int\boldsymbol{f}(t,s)\mathrm{d}\nu(s).$$

In fact, $|\mu\otimes\nu|=|\mu|\otimes|\nu|$; then theorem 22.20 is applied.

d. *Properties of integrable functions with respect to vector measures*

Let X, Y and Z be three Banach spaces and $(x,y)\to xy$ a continuous bilinear mapping of $X\times Y$ into Z.

Let $\boldsymbol{m}$ be a dominated measure on T with values in X, $\boldsymbol{n}$ a dominated measure on S with values in Y and $\boldsymbol{m}\otimes\boldsymbol{n}$ the product of the measures $\boldsymbol{m}$ and $\boldsymbol{n}$ with respect to the bilinear mapping xy.

22.41 PROPOSITION. *If g is an $\boldsymbol{m}$-integrable scalar function defined on T and h is an $\boldsymbol{n}$-integrable scalar function defined on S, then the scalar function $g(t)h(s)$ defined on $T \times S$ is $\boldsymbol{m}\otimes\boldsymbol{n}$-integrable and*

$$\iint g(t)h(s)\,\mathrm{d}\boldsymbol{m}(t)\,\mathrm{d}\boldsymbol{n}(s) = \iint g(t)\,\mathrm{d}\boldsymbol{m}(t))\left(\int h(s)\,\mathrm{d}\boldsymbol{n}(s)\right).$$

Since g is $|\boldsymbol{m}|$-integrable and h is $|\boldsymbol{n}|$-integrable, by proposition 22.31 we deduce that $g(t)h(s)$ is $|\boldsymbol{m}|\otimes|\boldsymbol{n}|$-integrable. It follows that $g(t)h(s)$ is $|\boldsymbol{m}\otimes\boldsymbol{n}|$-integrable, that is, $\boldsymbol{m}\otimes\boldsymbol{n}$-integrable.

The stated equality in proposition 22.41 follows immediately as in the proof of the proposition 22.31.

22.42 COROLLARY. *If $A \subset T$ is an $\boldsymbol{m}$-integrable set and $B \subset S$ is an $\boldsymbol{n}$-integrable set, then the set $A \times B$ is $\boldsymbol{m}\otimes\boldsymbol{n}$-integrable and*

$$(\boldsymbol{m}\otimes\boldsymbol{n})(A \times B) = \boldsymbol{m}(A)\,\boldsymbol{n}(B).$$

22.43 PROPOSITION. *If g is a locally $\boldsymbol{m}$-integrable scalar function defined on T and h is a locally $\boldsymbol{n}$-integrable scalar function defined on S, then the scalar function $g(t)h(s)$ defined on $T \times S$ is locally $\boldsymbol{m}\otimes\boldsymbol{n}$-integrable and*

$$(g\boldsymbol{m})\otimes(h\boldsymbol{n}) = (gh)(\boldsymbol{m}\otimes\boldsymbol{n}).$$

The proof is the same as that of proposition 22.33.

22.44 COROLLARY. *If $|\boldsymbol{m}\otimes\boldsymbol{n}| = |\boldsymbol{m}|\otimes|\boldsymbol{n}|$, then*

$$|(g\boldsymbol{m})\otimes(h\boldsymbol{n})| = |g\boldsymbol{m}|\otimes|h\boldsymbol{n}|.$$

In fact, by theorem 18.17, we have

$$|(gh)(\boldsymbol{m}\otimes\boldsymbol{n})| = |gh|\,|\boldsymbol{m}\otimes\boldsymbol{n}|$$

and also

$$|g\boldsymbol{m}| = |g|\,|\boldsymbol{m}| \text{ and } |h\boldsymbol{n}| = |h|\,|\boldsymbol{n}|.$$

Then

$$|(g\boldsymbol{m})\otimes(h\boldsymbol{n})| = |(gh)(\boldsymbol{m}\otimes\boldsymbol{n})| = |gh|\,|\boldsymbol{m}\otimes\boldsymbol{n}| =$$
$$= (|g|\,|h|(|\boldsymbol{m}|\otimes|\boldsymbol{n}|) = (|g|\,|\boldsymbol{m}|)\otimes(|h|\,|\boldsymbol{n}|) =$$
$$= |g\boldsymbol{m}|\otimes|h\boldsymbol{n}|.$$

e. *Properties of the product of a vector measure by a scalar measure*

Let X be a Banach space, $\boldsymbol{m}$ a dominated measure defined on T with values in X and ν a scalar measure defined on S.

Let F and G be two Banach spaces and $(x, y) \to xy$ a continuous bilinear mapping of $X \times F$ into G.

22.45 PROPOSITION. *If g is an $\boldsymbol{m}$-integrable scalar function defined on T and $\boldsymbol{h} : S \to F$ is a ν-integrable function, then the function $g(t)\boldsymbol{h}(s)$ defined on $T \times S$ with values in F is $\boldsymbol{m} \otimes \nu$-integrable and*

$$\iint g(t)\boldsymbol{h}(s)\,\mathrm{d}\boldsymbol{m}(t)\,\mathrm{d}\nu(s) = (\int(\int g(t)\,\mathrm{d}\boldsymbol{m}(t))\,(\int \boldsymbol{h}(s)\,\mathrm{d}\nu(s)) .$$

The proof is the same as that of proposition 22.31.

22.46 PROPOSITION. *If g is a locally $\boldsymbol{m}$-integrable scalar function defined on T and $\boldsymbol{h} : S \to F$ is a locally ν-integrable function, then the function $g(t)\boldsymbol{h}(s)$ defined on $T \times S$ with values in F is locally $\boldsymbol{m} \otimes \nu$-integrable and*

$$(g\boldsymbol{m}) \otimes (\boldsymbol{h}\nu) = (g\boldsymbol{h})(\boldsymbol{m} \otimes \nu) .$$

The proof is the same as that of proposition 22.33.

22.47 PROPOSITION. *Let X, F and G be three Banach spaces such that $X \subset \mathscr{L}(F, G)$, $\boldsymbol{m} : \mathscr{K}(T) \to X$ a dominated measure on T and ν a scalar measure on S.*

If F is of countable type and if there exists a countable set $Z \subset G'$, norming for G, then

$$|\boldsymbol{m} \otimes \nu| = |\boldsymbol{m}| \otimes |\nu| .$$

In fact, denoting $\mu = |\boldsymbol{m}|$, there exists a weakly locally μ-integrable function $\boldsymbol{g} : T \to \mathscr{L}(F, Z')$ such that $\boldsymbol{m} = \boldsymbol{g}\mu$ and $|\boldsymbol{m}| = |\boldsymbol{g}|\mu$ (theorem 18.15). Using corollary 22.38, we deduce then

$$|\boldsymbol{m} \otimes \nu| = |(\boldsymbol{g}\mu) \otimes \nu| = |\boldsymbol{g}\mu| \otimes |\nu| = |\boldsymbol{m}| \otimes |\nu| .$$

22.48 COROLLARY. *In the conditions of proposition 22.47 we have*

$$S(\boldsymbol{m} \otimes \nu) = S(\boldsymbol{m}) \times S(\nu) .$$

22.49 Corollary. *Let g be a locally $\boldsymbol{m}$-integrable scalar function and h a locally v-integrable scalar function. In the conditions of proposition 22.47 we have*

$$|(g\boldsymbol{m})\otimes(hv)| = |g\boldsymbol{m}|\otimes|hv|\,.$$

In fact, in this case we have $|\boldsymbol{m}\otimes v| = |\boldsymbol{m}|\otimes|v|$, $|g\boldsymbol{m}| = |g|\,|\boldsymbol{m}|$, $|hv| = |h|\,|v|$ and $|(gh)(\boldsymbol{m}\otimes v)| = |gh|\,|\boldsymbol{m}\otimes v|$, consequently

$$|(g\boldsymbol{m})\otimes(hv)| = |(gh)(\boldsymbol{m}\otimes v)| = |gh|\,|\boldsymbol{m}\otimes v| = (|g|\,|h|)(|\boldsymbol{m}|\otimes|v|) =$$
$$= (|g|\,|\boldsymbol{m}|)\otimes(|h|\,|v|) = |g\boldsymbol{m}|\otimes|hv|\,.$$

22.50 Theorem. *In the conditions of proposition 22.47, if $\boldsymbol{f}: T\times S\to F$ is an $\boldsymbol{m}\otimes v$-integrable function, vanishing outside the union of a sequence of $\boldsymbol{m}\otimes v$-integrable open sets, then:*

(1) *there exists an $\boldsymbol{m}$-negligible set $A\subset T$ such that for $t\notin A$, the function $s\to\boldsymbol{f}(t,s)$ is v-integrable;*
(2) *the function $t\to\int\boldsymbol{f}(t,s)\mathrm{d}v(t)$ defined $\boldsymbol{m}$-almost everywhere on T, with values in F, is $\boldsymbol{m}$-integrable and*

$$\iint\boldsymbol{f}(t,s)\,\mathrm{d}\boldsymbol{m}(t)\,\mathrm{d}v(s) = \int\,\mathrm{d}\boldsymbol{m}(t)\int\boldsymbol{f}(t,s)\,\mathrm{d}v(s)\,.$$

In fact,

$$|\boldsymbol{m}\otimes v| = |\boldsymbol{m}|\otimes|v|\,.$$

In the conditions of the theorem the following assertions are also true:

(1′) *there exists a v-negligible set $B\subset S$ such that for $s\notin B$, the function $t\to\boldsymbol{f}(t,s)$ is $\boldsymbol{m}$-integrable;*
(2′) *the functions $s\to\int\boldsymbol{f}(t,s)\mathrm{d}\boldsymbol{m}(t)$ defined v-almost everywhere on S with values in F is v-integrable and*

$$\iint\boldsymbol{f}(t,s)\,\mathrm{d}\boldsymbol{m}(t)\,\mathrm{d}v(s) = \int\mathrm{d}v(s)\int\boldsymbol{f}(t,s)\,\mathrm{d}\boldsymbol{m}(t)\,.$$

f. *Properties of measures with values in an algebra*

Let E be a Banach algebra, $\boldsymbol{m}$ a dominated measure on T with values in E, $\boldsymbol{n}$ a dominated measure on S with values in E and $\boldsymbol{m}\otimes\boldsymbol{n}$ the product of the measures $\boldsymbol{m}$ and $\boldsymbol{n}$ with respect to the product xy in the algebra E.

As before, the following propositions can be proved:

22.51 PROPOSITION. *If the function $\boldsymbol{g} : T \to E$ is $\boldsymbol{m}$-integrable and the function $\boldsymbol{h} : S \to E$ is $\boldsymbol{n}$-integrable, then the function $\boldsymbol{gh} : T \times S \to E$ is $\boldsymbol{m} \otimes \boldsymbol{n}$-integrable and*

$$\iint \boldsymbol{g}(t)\,\boldsymbol{h}(s)\,\mathrm{d}\boldsymbol{m}(t)\,\mathrm{d}\boldsymbol{n}(s) = \left(\int \boldsymbol{g}(t)\,\mathrm{d}\boldsymbol{m}(t)\right)\left(\int \boldsymbol{h}(s)\,\mathrm{d}\boldsymbol{n}(s)\right).$$

22.52 PROPOSITION. *If $\boldsymbol{g} : T \to E$ is locally $\boldsymbol{m}$-integrable and $\boldsymbol{h} : S \to E$ is locally $\boldsymbol{n}$-integrable, then $\boldsymbol{gh} : T \times S \to E$ is locally $\boldsymbol{m} \otimes \boldsymbol{n}$-integrable and*

$$(\boldsymbol{gm}) \otimes (\boldsymbol{hn}) = (\boldsymbol{gh})(\boldsymbol{m} \otimes \boldsymbol{n}).$$

22.53 THEOREM. *If E is of a countable type, if one of the measures $\boldsymbol{m}$ or $\boldsymbol{n}$ is scalar and if $\boldsymbol{f} : T \times S \to E$ is an $\boldsymbol{m} \otimes \boldsymbol{n}$-integrable function vanishing outside the union of a sequence of $\boldsymbol{m} \otimes \boldsymbol{n}$-integrable open sets, then:*

(1) *there exists an $\boldsymbol{m}$-integrable set $A \subset T$ such that for $t \notin A$, the function $s \to \boldsymbol{f}(t, s)$ is $\boldsymbol{n}$-integrable;*

(2) *the function $t \to \int \boldsymbol{f}(t, s)\,\mathrm{d}\boldsymbol{n}(t)$ defined $\boldsymbol{m}$-almost everywhere on T is $\boldsymbol{m}$-integrable and*

$$\iint \boldsymbol{f}(t, s)\,\mathrm{d}\boldsymbol{m}(t)\,\mathrm{d}\boldsymbol{n}(s) = \int \mathrm{d}\boldsymbol{m}(t) \int \boldsymbol{f}(t, s)\,\mathrm{d}\boldsymbol{n}(t).$$

g. *Properties of images and restrictions of measures*

Let X, Y and Z be three Banach spaces, $(x, y) \to xy$ a continuous bilinear mapping of $X \times Y$ into Z, $\boldsymbol{m}$ a dominated measure on T with values in X, $\boldsymbol{n}$ a dominated measure on S with values in Y and $\boldsymbol{m} \otimes \boldsymbol{n}$ the product of the measures $\boldsymbol{m}$ and $\boldsymbol{n}$ with respect to the bilinear mapping xy.

22.54 PROPOSITION. *Let T' and S' be two locally compact spaces. If $\alpha : T \to T'$ is an $\boldsymbol{m}$-proper function and $\beta : S \to S'$ is an $\boldsymbol{n}$-proper function, then the function $\alpha \times \beta : T \times S \to T' \times S'$ is $\boldsymbol{m} \otimes \boldsymbol{n}$-proper and*

$$(\alpha \times \beta)(\boldsymbol{m} \otimes \boldsymbol{n}) = \alpha(\boldsymbol{m}) \otimes \beta(\boldsymbol{n}).$$

Since α is $\boldsymbol{m}$-measurable and β is $\boldsymbol{n}$-measurable, the mapping $(t, s) \to$

$(\alpha(t), \beta(s))$ of $T\times S$ into $T'\times S'$ is $\boldsymbol{m}\otimes\boldsymbol{n}$-measurable, that is the mapping $(t, s)\to(\alpha\times\beta)(t, s)$ is $\boldsymbol{m}\otimes\boldsymbol{n}$-measurable.

Let $K'\subset T'$ and $L'\subset S'$ be two compact sets. Then $\alpha^{-1}(K')\subset T$ is $\boldsymbol{m}$-integrable and $\beta^{-1}(L')\subset S$ is $\boldsymbol{n}$-integrable, therefore the set $\alpha^{-1}(K')\times\beta^{-1}(D')=(\alpha\times\beta)^{-1}(K'\times L')$ is $\boldsymbol{m}\otimes\boldsymbol{n}$-integrable. It follows that for every compact set $C'\subset T'\times S'$, the set $(\alpha\times\beta)^{-1}(C')$ is $\boldsymbol{m}\otimes\boldsymbol{n}$-integrable, consequently $\alpha\times\beta$ is $\boldsymbol{m}\otimes\boldsymbol{n}$-proper.

For every functions $g'\in\mathscr{K}(T')$ and $h'\in\mathscr{K}(S')$ we have

$$\begin{aligned}\int f'(t')g'(s')\,\mathrm{d}(\alpha\times\beta)(\boldsymbol{m}\times\boldsymbol{n}) &= \int f'(\alpha(t))g'(\beta(s))\,\mathrm{d}\boldsymbol{m}\otimes\boldsymbol{n} =\\ &= \iint f'(\alpha(t))g'(\beta(s))\,\mathrm{d}\boldsymbol{m}(t)\,\mathrm{d}\boldsymbol{n}(s) =\\ &= \left(\int f'(\alpha(t))\,\mathrm{d}\boldsymbol{m}(t)\right)\left(\int g'(\beta(s))\,\mathrm{d}\boldsymbol{n}(s)\right) =\\ &= \left(\int f'\,\mathrm{d}\alpha(\boldsymbol{m})\right)\left(\int g'\,\mathrm{d}\beta(\boldsymbol{n})\right)\end{aligned}$$

whence

$$(\alpha\times\beta)(\boldsymbol{m}\otimes\boldsymbol{n})=\alpha(\boldsymbol{m})\otimes\beta(\boldsymbol{n})\,.$$

22.55 PROPOSITION. *If T' is a locally compact subspace of T and S' is a locally compact subspace of S, then*

$$(\boldsymbol{m}\otimes\boldsymbol{n})_{S'\times T'}=\boldsymbol{m}_{T'}\otimes\boldsymbol{n}_{S'}\,.$$

Let $g'\in\mathscr{K}(T')$ and $h'\in\mathscr{K}(S')$ and let g (respectively h) be the function obtained from g' (respectively h') extending it with 0 on $T\setminus T'$ (respectively $S\setminus S'$). The function g is $\boldsymbol{m}$-integrable, the function h is $\boldsymbol{n}$-integrable and

$$\int g'\,\mathrm{d}\boldsymbol{m}_{T'}=\int g\,\mathrm{d}\boldsymbol{m},\quad \int h'\,\mathrm{d}\boldsymbol{n}_{S'}=\int h\,\mathrm{d}\boldsymbol{n}\,.$$

The function obtained from $g'(t)h'(s)$ by extending it with 0 on $T\times S\setminus T'\times S'$ is equal to the function $g(t)h(t)$, therefore

$$\int g'(t)h'(s)\,\mathrm{d}(\boldsymbol{m}\otimes\boldsymbol{n})_{T'\times S'}=\int g(t)h(s)\,\mathrm{d}(\boldsymbol{m}\otimes n)\,.$$

Then

$$\begin{aligned}&\int g'(t)h'(s)\,\mathrm{d}(\boldsymbol{m}\otimes\boldsymbol{n})_{S'\times T'}=\iint g(t)h(s)\,\mathrm{d}\boldsymbol{m}(t)\,d\boldsymbol{n}(s)=\\ &=\left(\int g(t)\,\mathrm{d}\boldsymbol{m}(t)\right)\left(\int h(s)\,\mathrm{d}\boldsymbol{n}(s)\right)=\left(\int g'(t)\,\mathrm{d}\boldsymbol{m}_{T'}(t)\right)\left(\int h'(s)\,\mathrm{d}\boldsymbol{n}_{S'}(s)\right)\\ &=\iint g'(t)h'(s)\,\mathrm{d}\boldsymbol{m}_{T'}(t)\,\mathrm{d}\boldsymbol{n}_{S'}(s)=\int g'(t)h'(s)\,\mathrm{d}(\boldsymbol{m}_{T'}\otimes\boldsymbol{n}_{S'})\,,\end{aligned}$$

whence

$$(\boldsymbol{m}\otimes\boldsymbol{n})_{T'\times S'}=\boldsymbol{m}_{T'}\otimes\boldsymbol{n}_{S'}$$

5 Integration with respect to a finite product of measures

Let $(T_i)_{1\leqslant i\leqslant n}$ be a finite family of locally compact spaces and $T=\Pi_{1\leqslant i\leqslant n}T_i$ their product.

Using lemma 22.1 it follows by induction that the set of the linear combinations of functions of the form $(t_1, t_2, \ldots, t_n)\to f_1(t_1)f_2(t_2)\ldots f_n(t_n)$ with $f_i\in\mathscr{K}(T_i)$ for each i, is rich in $\mathscr{K}(T)$.

Let $(E_i)_{1\leqslant i\leqslant n}$ be a family of Banach spaces and $(u_1, u_2, \ldots, u_n)\to u_1u_2\ldots u_n$ a continuous and multilinear mapping on ΠE_i with values in a Banach space E.

For each i let $\boldsymbol{m}_i$ be a dominated measure on T_i with values in E_i.

There exists then a dominated measure $\boldsymbol{m}$ on T with values in E, and only one, satisfying the equality

$$\boldsymbol{m}(\Pi_{1\leqslant i\leqslant n}f_i)=\Pi_{1\leqslant i\leqslant n}\boldsymbol{m}(f_i),\ \text{for}\ f_i\in\mathscr{K}(T_i),\ 1\leqslant i\leqslant n\,.$$

The uniqueness of the measure follows from the fact that the vector space generated by the functions $\Pi_{1\leqslant i\leqslant n}f_i$ is rich in $\mathscr{K}(T)$. The existence of the measure $\boldsymbol{m}$ is proved by recurrence on n, taking

$$\boldsymbol{m}=(\boldsymbol{m}_1\otimes\boldsymbol{m}_2\otimes\ldots\otimes\boldsymbol{m}_{n-1})\otimes\boldsymbol{m}_n\,.$$

The measure $\boldsymbol{m}$ is called the product of the measures $\boldsymbol{m}_i$, $1\leqslant i\leqslant n$, and is denoted by $\boldsymbol{m}_1\otimes\boldsymbol{m}_2\otimes\ldots\otimes\boldsymbol{m}_n$ or $\otimes_{1\leqslant i\leqslant n}\boldsymbol{m}_i$. We have

$$|\otimes_{1\leqslant i\leqslant n}\boldsymbol{m}_i|\leqslant\otimes_{1\leqslant i\leqslant n}|\boldsymbol{m}_i|\,.$$

The equality

$$(\otimes_{i=1}\boldsymbol{m}_i)(\Pi_{1\leqslant i\leqslant n}f_i)=\Pi_{1\leqslant i\leqslant n}\boldsymbol{m}_i(f_i)$$

implies the following associativity formula of the product of measures:

$$\otimes_{1\leqslant k\leqslant p}(\otimes_{i\in I_k}\boldsymbol{m}_i)=\otimes_{1\leqslant i\leqslant n}\boldsymbol{m}_i$$

for every partition $(I_k)_{1\leqslant k\leqslant p}$ of the set $\{1, 2, \ldots, n\}$.

The integral of a function $f\in\mathscr{K}(T)$ with respect to the product measure $\otimes_{1\leqslant i\leqslant n}\boldsymbol{m}_i$ is denoted

$$\int f\,\mathrm{d}\boldsymbol{m}_1\,\mathrm{d}\boldsymbol{m}_2\ldots\mathrm{d}\boldsymbol{m}_n\ \text{or}\ \int\int\ldots\int f\,\mathrm{d}\boldsymbol{m}_1\,\mathrm{d}\boldsymbol{m}_2\ldots\mathrm{d}\boldsymbol{m}_n$$

or

$$\int\int \dots \int f(t_1, t_2, \dots, t_n)\,\mathrm{d}\boldsymbol{m}_1(t_1)\,\mathrm{d}\boldsymbol{m}_2(t_2)\dots \mathrm{d}\boldsymbol{m}_n(t_n)$$

and is called the *n-tuple integral.*

For every permutation σ of the set $\{1, 2, \dots, n\}$ and for every function $f\in\mathscr{K}(T)$ we have the equality

$$\int\int \dots \int f\,\mathrm{d}\boldsymbol{m}_1\,\mathrm{d}\boldsymbol{m}_2 \dots \mathrm{d}\boldsymbol{m}_n = \int \mathrm{d}\boldsymbol{m}_{\sigma(1)} \int \mathrm{d}\boldsymbol{m}_{\sigma(2)} \dots \int f\,\mathrm{d}\boldsymbol{m}_{\sigma(n)}\,.$$

This equality follows from the associativity formula and from the rule of change of order of integration.

The results concerning the product of two measures are easily extended to the product of a finite number of measures.

We shall state the Fubini theorem only for $n=3$ and for positive measures μ_i.

22.56 PROPOSITION. *If $f: T\to E$ is integrable for $\mu=\mu_1\otimes\mu_2\otimes\mu_3$ and vanishes outside the union of a sequence of μ-integrable open sets of T, then the integrated integrals of $\boldsymbol{f}$ exist and we have the equalities*

$$\begin{aligned}&\int\int\int \boldsymbol{f}(t_1, t_2, t_3)\,\mathrm{d}\mu_1(t_1)\,\mathrm{d}\mu_2(t_2)\,\mathrm{d}\mu_3(t_3) =\\ &\quad= \int\int \mathrm{d}\mu_1(t_1)\,\mathrm{d}\mu_2(t_2) \int f(t_1, t_2, t_3)\,\mathrm{d}\mu_3(t_3) =\\ &\quad= \int \mathrm{d}\mu_1(t_1) \int\int \boldsymbol{f}(t_1, t_2, t_3)\,\mathrm{d}\mu_2(t_2)\,\mathrm{d}\mu_3(t_3) =\\ &\quad= \int d\mu_1(t_1) \int \mathrm{d}\mu_2(t_2) \int \boldsymbol{f}(t_1, t_2, t_3)\,\mathrm{d}\mu_3(t_3)\,.\end{aligned}$$

In fact, applying the Fubini theorem we deduce that the function $(t_1, t_2)\to \int \boldsymbol{f}(t_1, t_2, t_3)\,\mathrm{d}\mu_3(t_3)$ is defined $\mu_1\otimes\mu_2$-almost everywhere on $T_1\times T_2$, vanishes on the complement of the union of a sequence of $\mu_1\otimes\mu_2$-integrable open sets of $T_1\times T_2$ (remark 2. after theorem 22.12), is $\mu_1\otimes\mu_2$-integrable and we have

$$\begin{aligned}&\int\int\int f(t_1, t_2, t_3)\,\mathrm{d}\mu_1(t_1)\,\mathrm{d}\mu_2(t_2)\,\mathrm{d}\mu_3(t_3) =\\ &\quad= \int\int \mathrm{d}\mu_1(t_1)\,\mathrm{d}\mu_2(t_2) \int f(t_1, t_2, t_3)\,\mathrm{d}\mu_3(t_3).\end{aligned}$$

Applying one more the Fubini theorem to the function $(t_1, t_2)\to \int \boldsymbol{f}(t_1, t_2, t_3)\,\mathrm{d}\mu_3(t_3)$ we deduce that the function $t_1\to \int \mathrm{d}\mu_2(t_2)$. $\int \boldsymbol{f}(t_1, t_2, t_3)\,\mathrm{d}\mu_3(t_3)$ is defined μ_1-almost everywhere, is μ_1-integrable and

$$\iint d\mu_1(t_1)\,d\mu_2(t_2)\int \boldsymbol{f}(t_1, t_2, t_3)\,d\mu_3(t_3) =$$
$$= \int d\mu_1(t_1)\int d\mu_2(t_2)\int \boldsymbol{f}(t_1, t_2, t_3)\,d\mu_3(t_3)\,.$$

In the same way it follows that the function

$$t_1 \to \int\int \boldsymbol{f}(t_1, t_2, t_3)\,d\mu_2(t_2)\,d\mu_3(t_3)$$

is defined μ_1-almost everywhere, is μ_1-integrable and

$$\iiint \boldsymbol{f}(t_1, t_2, t_3)\,d\mu_1(t_1)\,d\mu_2(t_2)\,d\mu_3(t_3) =$$
$$= \int d\mu_1(t_1)\iint \boldsymbol{f}(t_1, t_2, t_3)\,d\mu_2(t_2)\,d\mu_3(t_3)\,.$$

PART III

Chapter VI

MEASURES ON LOCALLY COMPACT GROUPS

§23 Haar measure

1 *Topological groups*

In the sequel, composition in a group G will be denoted multiplicatively, $x \cdot y$ or xy, and the unit element will be denoted by e.

If $A \subset G$, $B \subset G$ and $x \in G$, we use the following notations:

$$AB = \{xy \mid x \in A,\ y \in B\}, \quad xA = \{xy \mid y \in A\}, \quad Ax = \{yx \mid y \in A\},$$
$$A^{-1} = \{x^{-1} \mid x \in A\}.$$

23.1 Definition. *A topological group is a group G endowed with a topology satisfying the following two conditions*:

(1) *the mapping $(x, y) \to xy$ of $G \times G$ into G is continuous*;

(2) *the mapping $x \to x^{-1}$ of G into G is continuous.*

These two conditions are equivalent to the following condition:

the mapping $(x, y) \to xy^{-1}$ of $G \times G$ into G is continuous.

Conditions (1) and (2) express the fact that the group structure is consistent with the topological structure on G.

Let G be a topological group.

For every element $a \in G$ the translation to the left $x \to ax$, the translation to the right $x \to xa$, and the symmetry $x \to x^{-1}$ are homeomorphisms of G onto G.

It follows that if $a \in G$ and A is an open (respectively a closed) set in G, then aA, Aa and A^{-1} are open (respectively closed).

If A is an arbitrary set of G and B is open, then AB and BA are open. But if A and B are closed, then AB is not necessarily closed.

If G is separated, A is a compact set in G, and $a \in G$, then the sets aA, Aa and A^{-1} are compact. If, in addition, B is a closed set, then AB is closed; and if B is compact, then AB is compact.

If $\mathscr{V}$ is the filter of neighbourhoods of e, then for every element $a \in G$, the filter of neighbourhoods of a coincides with the family $a\mathscr{V} = \{aV \mid V \in \mathscr{V}\}$ and with the family $\mathscr{V}a = \{Va \mid V \in \mathscr{V}\}$.

The filter $\mathscr{V}$ of neighbourhoods of e is characterized by the following four properties;

(1) *for every* $U \in \mathscr{V}$ *there exists* $V \in \mathscr{V}$ with $V \cdot V \subset U$;

(2) *for every* $U \in \mathscr{V}$ we have $U^{-1} \in \mathscr{V}$.

These two properties are equivalent to the following property: *for every* $U \in \mathscr{V}$, *there exists* $V \in \mathscr{V}$ *with* $VV^{-1} \subset U$;

(3) *for every* $U \in \mathscr{V}$ *we have* $e \in U$;

(4) *for every* $a \in G$ *and every* $V \in \mathscr{V}$ we have $aVa^{-1} \in \mathscr{V}$.

Properties (1), (2) and (4) express the continuity of the mappings $(x, y) \to x \cdot y$, $x \to x^{-1}$ and $x \to axa^{-1}$. If G is commutative, we have $aVa^{-1} = V$, therefore property (4) is automatically satisfied even if the topological structure is not consistent with the group structure.

A set $\mathscr{U}$ of neighbourhoods of e is a *fundamental system* of neighbourhoods of e if and only if it fulfills the following conditions:

(1′) for every $U \in \mathscr{U}$ there exists $V \in \mathscr{U}$ with $V \cdot V \subset U$;

(2′) for every $U \in \mathscr{U}$ there exists $V \in \mathscr{U}$ with $V \subset U^{-1}$;

(3′) for every $U \in \mathscr{U}$ we have $e \in U$;

(4′) for every $a \in G$ and every $U \in \mathscr{U}$ there exists $V \in \mathscr{U}$ with $V \subset aUa^{-1}$.

If $\mathscr{U}$ is a fundamental system of neighbourhoods of e, then $a\mathscr{U}$ and $\mathscr{U}a$ are fundamental systems of neighbourhoods of a. A neighbourhood V of e is *symmetric* if $V = V^{-1}$. If $V \in \mathscr{V}$, then $V \cup V^{-1}$, $V \cap V^{-1}$, VV^{-1} are symmetric. The set of the symmetric neighbourhoods of e form a fundamental system of neighbourhoods of e.

If $\mathscr{U}$ is a fundamental system of neighbourhoods of e, then for each natural number n, the family $\{V^n \mid V \in \mathscr{U}\}$ is a fundamental system of neighbourhoods of e.

A topological group G is separated if and only if the set $\{e\}$ is closed.

We say that a function f defined on G with values in a Banach space E is left (respectively right) *uniformly continuous* on a set $A \subset G$ if:
for every $\varepsilon > 0$ *there exists a neighbourhood* V *of* e *such that* $|f(x) - f(y)| < \varepsilon$, *for every* $x \in G$ *and* $y \in A$ *with* $xy^{-1} \in V$ (*respectively, for every* $x \in A$ *and* $y \in G$ *with* $x^{-1}y \in V$).

It is easy to see that f *is left uniformly continuous on* A *if and only if for every* $\varepsilon > 0$, *there exists a neighbourhood* V *of* e *such that*

$$|f(sx) - f(x)| < \varepsilon, \text{ for every } x \in A \text{ and every } s \in V;$$

and that f *is right uniformly continuous on* A *if and only if for every* $\varepsilon > 0$, *there exists a neighbourhood* V *of* e *such that*

$$|f(xs) - f(x)| < \varepsilon, \text{ for every } x \in A \text{ and every } s \in V.$$

A (left, or right) uniformly continuous function on G will be called (left, respectively right) uniformly continuous function without other specifications.

If the group is commutative, the left uniform continuity is equivalent to the right uniform continuity. In this case we say simply, uniform continuity.

2 *Locally compact groups*

In the sequel we shall denote by G a locally compact topological group.

A topological group is locally compact if and only if the unit element e has a compact neighbourhood.

We shall show that the continuous functions with compact support defined on G are uniformly continuous. For this we prove first the following lemma.

23.2 LEMMA. *Let* S, T *and* E *be three topological spaces and* $f: S \times T \to E$ *a continuous function. If* W *is an open set in* E *and* K *is a compact set in* T, *then the set*

$$L = \{s \mid f(s, t) \in W \text{ for every } t \in K\}$$

is open in S.

Let $s_0 \in L$ be fixed. For each $t \in K$, the function f is continuous in (s_0, t) and we have $f(s_0, t) \in W$. There exists then an open set $U \times V$ in $S \times T$ which contains (s_0, t), such that

$$f(s, t) \in W \text{ for } (s, t) \in U \times V.$$

But K is compact, so there exists a finite family $(U_i \times V_i)$ of open sets, such that $s_0 \in U_i$ for $1 \leqslant i \leqslant n$, $K \subset \cup_{1 \leqslant i \leqslant n} V_i$ and for each i

$$f(s, t) \in W, \text{ for } (s, t) \in U_i \times V_i.$$

The set $V = \cap_{1 \leqslant i \leqslant n} V_i$ is a neighbourhood of s_0. For $(s, t) \in V \times K$ we have $f(s, t) \in W$, therefore $V \subset L$. It follows that s_0 is an interior point of L. As s_0 has been choosen arbitrarily, it follows that L is an open set.

23.3 PROPOSITION. *Let E be a Banach space. Every function* $f \in \mathscr{K}_E(G)$ *is left and right uniformly continuous.*

Let $f \in \mathscr{K}_E(G)$, K the support of f and U a compact symmetric neighbourhood of e. By the preceding lemma, the set

$$L = \{y \mid |f(yx) - f(x)| < \varepsilon \text{ for every } x \in UK\}$$

is open and contains e. On the other hand, if $y \in U$ and $x \notin UK$ we have $f(x) = 0$ and $f(yx) = 0$. Putting $V = L \cap U$, for every $y \in V$ and every $x \in G$ we have

$$|f(xy) - f(x)| < \varepsilon,$$

therefore f is left uniformly continuous.

Taking

$$L' = \{y \mid |f(xy) - f(x)| < \varepsilon \text{ for every } x \in KU\}$$

and denoting $V' = L' \cap U$, we deduce that for every $y \in V'$ and every $x \in G$ we have

$$|f(xy) - f(x)| < \varepsilon,$$

therefore f is right uniformly continuous and the proposition is proved.

If f is a function defined on G with values in a Banach space E, we shall denote by $\check{f}$ the function defined on G by the equality

$$\check{f}(x) = f(x^{-1}).$$

If $s \in G$ we shall denote by f_s and f^s the functions defined by the equalities

$$f_s(x) = f(sx), \quad f^s(x) = f(xs^{-1}).$$

We have $f_e = f^e = f$.

For every set $A \subset G$ we have $(\varphi_A)_s = \varphi_{s^{-1}A}$ and $(\varphi_A)^s = \varphi_{As}$. If G is commutative we have $f^s = f_{s^{-1}}$.

If $f \in \mathscr{K}_E(G)$, then the functions f_s, f^s and $\check{f}$ belong also to $\mathscr{K}_E(G)$; it follows that if μ is a positive measure on G and $1 \leqslant p \leqslant +\infty$ and if $f \in \mathscr{K}_E(G)$, then the functions f_s, f^s and $\check{f}$ belong to the space $\mathscr{L}_E^p(\mu)$.

23.4 PROPOSITION. *If $f \in \mathscr{K}_E(G)$, then the mappings $s \to f_s$ and $s \to f^s$ of G into $\mathscr{K}_E(G)$ are left uniformly continuous.*

Let $\varepsilon > 0$. Since f is left uniformly continuous, there exists a neighbourhood V of e such that

$$|f(y) - f(z)| < \varepsilon \text{ if } yz^{-1} \in V.$$

Let $x \in G$ and $st^{-1} \in V$. Taking $y = sx$ and $z = tx$ we have $yz^{-1} = st^{-1} \in V$, therefore

$$|f(sx) - f(tx)| < \varepsilon,$$

that is

$$|f_s(x) - f_t(x)| < \varepsilon.$$

It follows that

$$||f_s - f_t|| = \sup_{x \in G} |f_s(x) - f_t(x)| < \varepsilon \text{ if } st^{-1} \in V,$$

therefore the mapping $s \to f_s$ of G into $\mathscr{K}_E(G)$ is left uniformly continuous.

Since f is right uniformly continuous, there exists a neighbourhood W of e such that

$$|f(y) - f(z)| < \varepsilon, \text{ if } y^{-1}z \in W.$$

Let $x \in G$ and $st^{-1} \in V$. Taking $y = xs^{-1}$ and $z = xt^{-1}$ we have $y^{-1}z = st^{-1} \in V$, therefore

$$|f(xs^{-1}) - f(xt^{-1})| < \varepsilon$$

that is,

$$|f^s(x) - f^t(x)| < \varepsilon.$$

It follows that

$$||f^s - f^t|| < \varepsilon, \text{ if } st^{-1} \in W,$$

therefore the mapping $s \to f^s$ of G into $\mathscr{K}_E(G)$ is left uniformly continuous.

23.5 PROPOSITION. *Let E be a Banach space, μ a positive measure on G*

and $1 \leqslant p < +\infty$. *If* $\boldsymbol{f} \in \mathscr{K}_E(G)$, *then the mappings* $s \to \boldsymbol{f}_s$ *and* $s \to \boldsymbol{f}^s$ *of* G *into* $\mathscr{L}_E^p(\mu)$ *are continuous at* e.

Let K be the support of $\boldsymbol{f}$ and U a compact neighbourhood of e. The set $C = UKU$ is a compact neighbourhood of K. Let $\varepsilon > 0$ and V be a symmetric neighbourhood of e contained in U, such that

$$\|\boldsymbol{f}_s - \boldsymbol{f}_t\| \leqslant \varepsilon\mu(C)^{-1/p} \text{ and } \|\boldsymbol{f}^s - \boldsymbol{f}^t\| < \varepsilon\mu(C)^{-1/p} \text{ if } st^{-1} \in V .$$

In particular, taking $t = e$, we have

$$\|\boldsymbol{f}_s - \boldsymbol{f}\| < \varepsilon\mu(C)^{-1/p} \text{ and } \|\boldsymbol{f}^s - \boldsymbol{f}\| < \varepsilon\mu(C)^{-1/p} \text{ if } s \in V .$$

Let $s \in V$. If $s^{-1}x \in K$, then $x \in sK \subset VK \subset UKU$. Also, if $xs \in K$, then $x \in Ks^{-1} \subset KU \subset UKU$. Thus, if $x \notin UKU$, then $x \notin K$, $s^{-1}x \notin K$ and $xs \notin K$, therefore $\boldsymbol{f}(s^{-1}x) = \boldsymbol{f}(x\,s) = \boldsymbol{f}(x) = 0$.

It follows that the functions $\boldsymbol{f}_s$, $\boldsymbol{f}^s$ and $\boldsymbol{f}$ have their supports contained in the set $C = UKU$, therefore

$$N_p(\boldsymbol{f}_s - \boldsymbol{f}, \mu) = \left(\int |\boldsymbol{f}_s(x) - \boldsymbol{f}(x)|^p \mathrm{d}\mu\right)^{1/p} \leqslant \|\boldsymbol{f}_s - \boldsymbol{f}\| \mu(C)^{1/p} < \varepsilon$$

and

$$N_p(\boldsymbol{f}^s - \boldsymbol{f}, \mu) = \left(\int |\boldsymbol{f}^s(x) - \boldsymbol{f}(x)|^p \mathrm{d}\mu\right)^{1/p} \leqslant \|\boldsymbol{f}^s - \boldsymbol{f}\| \mu(C)^{1/p} < \varepsilon .$$

This means that the functions $s \to \boldsymbol{f}_s$ and $s \to \boldsymbol{f}^s$ are continuous at e.

3 *Invariant measures. Haar measure*

Let X be a Banach space and $\boldsymbol{m} : \mathscr{K}(G) \to X$ a measure.

We say that the measure $\boldsymbol{m}$ is *left invariant if*

$$\boldsymbol{m}(f_s) = \boldsymbol{m}(f), \text{ that is, } \int f(sx)\,\mathrm{d}\boldsymbol{m}(x) = \int f(x)\,\mathrm{d}\boldsymbol{m}(x)$$

for every $f \in \mathscr{K}(G)$ and $s \in G$.

The measure $\boldsymbol{m}$ is *right invariant* if

$$\boldsymbol{m}(f^s) = \boldsymbol{m}(f), \text{ that is, } \int f(xs)\,\mathrm{d}\boldsymbol{m}(x) = \int f(x)\,\mathrm{d}\boldsymbol{m}(x)$$

for every $f \in \mathscr{K}(G)$ and $s \in G$.

The positive invariant measures which *do not vanish identically* are called *Haar measures.*

If the group is commutative, the left invariant measures coincide with the right invariant measures. It will be proved that the compact groups possess this property too.

We shall prove first the existence of Haar measures. Then we shall prove that every invariant measure with finite variation is obtained from a Haar measure by multiplication by a (vector) constant.

23.6 THEOREM (*Haar*). *There exists a left (right) invariant positive Haar measure on G and, up to a multiplicative constant, only one.*

We shall prove the theorem in several steps, and only for the left invariant measure.

(a) Let f and $g \neq 0$ be two functions of $\mathscr{K}_+(G)$. There exists a finite family $(c_i)_{1 \leqslant i \leqslant n}$ of positive numbers and a finite family $(s_i)_{1 \leqslant i \leqslant n}$ of elements of G such that

$$f(x) \leqslant \Sigma_{1 \leqslant i \leqslant n} c_i g(s_i x), \text{ for } x \in G .$$

In fact, if $f(x) \equiv 0$, we take $c=0$ and we have $f(x) \leqslant cg(sx)$ for $x \in G$. If $f \neq 0$, there exists a relatively compact open set V such that the function f is strictly positive on $\bar{V}$. The function f is bounded on $\bar{V}$ and its infimum m on V belongs to $f(V)$. It follows that $m > 0$. Let K be the support of f. Since for every $s \in G$ the set sV is open and the family $(s^{-1} V)_{s \in G}$ covers G, therefore it covers K also, there exists a finite family $(s_i^{-1} V)_{1 \leqslant i \leqslant n}$ which covers K.

Let M be the supremum of the function f. If $f(x)=0$, then $f(x) \leqslant (M/m) \Sigma_{1 \leqslant i \leqslant n} g(s_i x)$. If $f(x) \neq 0$, then $x \in K$, therefore there exists i with $x \in s_i^{-1} V$, whence $s_i x \in V$, therefore $m \leqslant g(s_i x) \leqslant \Sigma_{1 \leqslant i \leqslant n} g(s_i x)$, therefore $f(x) \leqslant (M/m) \Sigma_{1 \leqslant i \leqslant n} g(s_i x)$. Thus, taking $c_i = M/m$ for $1 \leqslant i \leqslant n$, we have

$$f(x) \leqslant \Sigma_{1 \leqslant i \leqslant n} c_i g(s_i x), \text{ for every } x \in G .$$

Denote by $(f; g)$ the infimum of all the sums $\Sigma_{1 \leqslant i \leqslant n} c_i$ of finite families $(c_i)_{1 \leqslant i \leqslant n}$ of positive numbers, such that there exists a family $(s_i)_{1 \leqslant i \leqslant n}$ of elements of G satisfying the inequality

$$f(x) \leqslant \Sigma_{1 \leqslant i \leqslant n} c_i g(s_i x), \text{ for every } x \in G .$$

If $f(x) \equiv 0$, we evidently have, $(f; g) = 0$.

The function $(f; g)$ has the following properties:

(1) $(f_s; g) = (f, g)$ for every $s \in G$;

(2) $(cf; g) = c(f; g)$ for every number $c > 0$;

(3) $(f_1 + f_2; g) = (f_1; g) + (f_2; g)$;

(4) if $f_1 \leqslant f_2$, then $(f_1; g) \leqslant (f_2; g)$.

These properties are deduced immediately from the definition of the function $(f; g)$.

(5) $(f; h) \leqslant (f; g)(g; h)$ if $h \neq 0$ and $g \neq 0$.

In fact, if

$$f(x) \leqslant \Sigma_i c_i g(s_i x) \text{ and } g(x) \leqslant \Sigma_j d_j h(t_j x), \text{ then } f(x) \leqslant \Sigma_{ij} c_i d_j h(t_j s_i x),$$

whence

$$(f; h) \leqslant \inf \Sigma_{ij} c_i d_j = \inf \Sigma_i c_i \inf \Sigma_j d_j = (f; g)(g; h).$$

(6) If M_f and M_g are the suprema of f and g respectively, then

$$(f; g) \geqslant M_f / M_g .$$

In fact, there exists $x_0 \in G$ with $f(x_0) = M_f$. Then

$$M_f = f(x_0) \leqslant \Sigma_i c_i g(s_i x_0) \leqslant (\Sigma_i c_i) M_g,$$

whence

$$\Sigma_i c_i \geqslant M_f / M_g.$$

In particular, from property (6) it follows that

if $f \neq 0$ and $g \neq 0$, *then* $(f; g) > 0$.

(b) Choose a function $f_0 \neq 0$ from $\mathscr{K}_+(G)$ and set

$$I_\varphi(f) = (f; \varphi)/(f_0; \varphi), \text{ for } f, \varphi \in \mathscr{K}_+(G),\ \varphi \neq 0 .$$

From property (5) we deduce that if $f \neq 0$, then

$$0 < 1/(f_0; f) \leqslant I_\varphi(f) \leqslant (f; f_0) < +\infty,$$

therefore for each fixed function $f \neq 0$, the set $\{I_\varphi(f) \mid \varphi \in \mathscr{K}_+(G), \varphi \neq 0\}$ is bounded.

From the first three properties of the function $(f; g)$ we deduce that I_φ is an invariant functional:

$$I_\varphi(f_s) = I_\varphi(f),$$

homogeneous:

$$I_\varphi(cf) = c I_\varphi(f),$$

and subadditive:

$$I_\varphi(f_1+f_2)\leqslant I_\varphi(f_1)+I_\varphi(f_2)\,.$$

(c) We shall show that I_φ is almost additive for functions $\varphi\neq 0$ with sufficiently small support. More precisely:

For every functions f_1 and f_2 from $\mathscr{K}_+(G)$ and for every $\varepsilon>0$ there exists a neighbourhood V of the unit such that for every function $\varphi\in\mathscr{K}_+(G,V)$ we have

$$I_\varphi(f_1)+I_\varphi(f_2)\leqslant I_\varphi(f_1+f_2)+\varepsilon\,.$$

Let $f_1, f_2\in\mathscr{K}_+(G)$ and $\varepsilon>0$. If $f_1+f_2=0$, the inequality is satisfied. Assume then that $f_1+f_2\neq 0$. Choose a function $f'\in\mathscr{K}_+(G)$, equal to 1 on the support of the function f_1+f_2. Choose a number $\varepsilon'>0$ such that

$$2\varepsilon'(f_1+f_2;f_0)\leqslant\varepsilon/2$$

and a number $\delta>0$ such that

$$\delta(1+2\varepsilon')(f';f_0)<\varepsilon\,.$$

Denote $f=f_1+f_2+\delta f'$ and for $i=1,2$, put $h_i(t)=f_i(t)/f(t)$ if $f(t)\neq 0$, and $h_i(t)=0$ if $f(t)=0$.

The functions h_1 and h_2 belong to $\mathscr{K}_+(G)$. Since the functions h_i are uniformly continuous, there exists a neighbourhood V of e such that

$$|h_i(x)-h_i(y)|<\varepsilon' \text{ for } y^{-1}x\in V\,.$$

Let $\varphi\neq 0$ be a function of $\mathscr{K}_+(G,V)$ and $f(x)\leqslant\Sigma_j c_j\varphi(s_jx)$. If $\varphi(s_jx)\neq 0$, then $s_jx\in V$, therefore $|g_i(x)-h_i(s_j^{-1})|<\varepsilon'$, whence

$$f_i(x)=f(x)h_i(x)\leqslant\Sigma_j c_j\varphi(s_jx)h_i(x)\leqslant\Sigma_j c_j\varphi(s_jx)[h_i(s_j^{-1})+\varepsilon'],$$

therefore

$$(f_i;\varphi)\leqslant\Sigma_j c_j[h_i(s_j^{-1})+\varepsilon']\,,$$

consequently

$$(f_1;\varphi)+(f_2;\varphi)\leqslant\Sigma_j c_j(1+2\varepsilon')\,,$$

since $h_1+h_2\leqslant 1$. As $\Sigma_j c_j$ can be choosen arbitrarily close to $(f;\varphi)$, we have

$$\begin{aligned}I_\varphi(f_1)+I_\varphi(f_2)&\leqslant I_\varphi(f)(1+2\varepsilon')\leqslant[I_\varphi(f_1+f_2)+\delta I_\varphi(f')](1+2\varepsilon')=\\&=I_\varphi(f_1+f_2)+2\varepsilon' I_\varphi(f_1+f_2)+\delta(1+2\varepsilon')I_\varphi(f')\leqslant\\&\leqslant I_\varphi(f_1+f_2)+\varepsilon.\end{aligned}$$

(d) For every function $f \neq 0$ of $\mathscr{K}_+(G)$ denote by S_f the closed interval $[1/(f_0; f), (f; f_0)]$ and by S the product space of the compact spaces S_f, when f runs over $\mathscr{K}_+(G)$. For each function $\varphi \neq 0$ of $\mathscr{K}_+(G)$ we have $I_\varphi(f) \in S_f$, therefore I_φ can be identified with the point of the product space S, with projection $I_\varphi(f)$ on S_f.

For each neighbourhood V of the unit in G denote by C_V the closure of the set $\{I_\varphi \mid \varphi \in \mathscr{K}_+(G, V)\}$. The set C_V is compact in S. If $V_1, V_2, \ldots, V_n$ are neighbourhoods of the unit in G, then

$$\cap_{1 \leqslant i \leqslant n} C_{V_i} = C_{\cap_{1 \leqslant i \leqslant n} V_i} \neq \varnothing .$$

It follows that the intersection of the sets C_V when V runs over the filter of neighbourhoods of e is nonempty. Let μ be a point belonging to all the sets C_V. For every V, every finite family of functions $f_1, \ldots, f_n$ of $\mathscr{K}_+(G)$ and every $\varepsilon > 0$, there exists a function $\varphi \neq 0$ of $\mathscr{K}_+(G)$ such that $|\mu(f_i) - I_\varphi(f_i)| < \varepsilon$ for $1 \leqslant i \leqslant n$. Using the properties of I_φ, we deduce easily that μ is a positive, additive, homogeneous and invariant functional on $\mathscr{K}_+(G)$ and that

$$1/(f_0; f) \leqslant \mu(f) \leqslant (f; f_0),$$

therefore μ does not vanish identically. Moreover, if $f \neq 0$, then $\mu(f) > 0$.

Extending μ to $\mathscr{K}(G)$ by the equality

$$\mu(f_1 - f_2) = \mu(f_1) - \mu(f_2),$$

μ is a positive, left invariant, non-zero measure on G.

The uniqueness of μ follows from the following more general proposition.

Remark. The measure μ constructed in the proof of this theorem has the property that $\mu(f) > 0$ for every function $f \neq 0$ of $\mathscr{K}_+(G)$.

23.7 PROPOSITION. *Let ν be a positive left invariant Haar measure. Every left invariant measure is of the form.*

$$\boldsymbol{m} = a\nu .$$

Assume first that m is a left invariant scalar measure. Consider the measure μ constructed in the proof of the theorem, with the property that $\mu(f) > 0$ for every function $f \neq 0$ of $\mathscr{K}_+(G)$.

Let $f \geqslant 0$ be a function of $\mathscr{K}_+(G)$, let C be the support of f and U the interior of C. Let $f' : G \to [0, 1]$ be a continuous function with compact support, equal to 1 on C. Let $\varepsilon > 0$. Choose a symmetric neighbourhood V of of e such that

$$|f(yx) - f(xz^{-1})| < \varepsilon \text{ for } y, z \in V$$

and such that $CV \cup VC \subset U$. From this inclusion we deduce

$$f(xy) = f(xy)f'(x) \text{ and } f(yx) = f(yx)f'(x)$$

for $y \in V$ and $x \in G$; taking in the preceding inequality $z = y^{-1}$, we get

$$|f(xy) - f(yx)| < \varepsilon f'(x) \text{ for every } y \in V \text{ and } x \in G .$$

Let $h \neq 0$ be a function of $\mathscr{K}_+(G, V)$ with $h(x) = h(x^{-1})$ for every $x \in G$. We deduce that

$$\mu(h)m(f) = \int h(y)\,\mathrm{d}\mu(y) \cdot \int f(x)\,\mathrm{d}m(x) = \int \mathrm{d}\mu(y) \int h(y)f(x)\,\mathrm{d}m(x) =$$
$$= \int \mathrm{d}\mu(y) \int h(y) f(yx)\,\mathrm{d}m(x)$$

and

$$m(h)\mu(f) = \int h(x)\,\mathrm{d}m(x) \int f(y)\,\mathrm{d}\mu(y) = \int \mathrm{d}\mu(y) \int h(x)f(y)\,\mathrm{d}m(x) =$$
$$= \int \mathrm{d}\mu(y) \int h(y^{-1}x) f(y)\,\mathrm{d}m(x) =$$
$$= \int \mathrm{d}m(x) \int h(x^{-1}y) f(y)\,\mathrm{d}\mu(y) =$$
$$= \int \mathrm{d}m(x) \int h(y) f(xy)\,\mathrm{d}\mu(y) = \int \mathrm{d}\mu(y) h(y) f(xy)\,\mathrm{d}m(x) ,$$

therefore

$$|\mu(h)m(f) - m(h)\mu(f)| \leqslant \int \mathrm{d}\mu(y) \int h(y)\,|f(xy) - f(yx)|\,\mathrm{d}m(x) \leqslant$$
$$\leqslant \varepsilon \int \mathrm{d}\mu(y) \int h(y) f'(x)\,\mathrm{d}|m|(x) = \varepsilon\mu(h)\,|m|(f') .$$

Similarly, taking a function $g \neq 0$ of $\mathscr{K}_+(G)$ and a continuous function $g' : G \to [0, 1]$ with compact support, equal to 1 on the support of g, we deduce

$$|\mu(h)m(g) - m(h)\mu(g)| \leqslant \varepsilon\mu(h)|m|(g') .$$

Since $\mu(f) > 0$, $\mu(h) > 0$ and $\mu(g) > 0$, we have

$$|m(f)/\mu(f) - m(h)/\mu(h)| \leqslant \varepsilon\,|m|(f')/\mu(f)$$

and

$$|m(g)/\mu(g) - m(h)/\mu(h)| \leqslant \varepsilon\,|m|(g')/\mu(g)$$

whence

$$|m(f)/\mu(f) - m(g)/\mu(g)| \leqslant \varepsilon(|m|(f')/\mu(f) + |m|(g')/\mu(g)),$$

and f' and g' are independent on ε. Since ε is arbitrary, we deduce

$$m(f)/\mu(f) - m(g)/\mu(g) = 0 .$$

If we fix a function $g_0 \neq 0$ in $\mathscr{K}_+(G)$ and set $b = m(g_0)/\mu(g_0)$, we obtain

$$m(f) = b\mu(f)$$

for every function $f \in \mathscr{K}^+(G)$, therefore for every function $f \in \mathscr{K}(G)$. We deduce that $m(f) \neq 0$ for every function $f \neq 0$ of $\mathscr{K}_+(G)$.

It follows that for the positive left invariant Haar measure ν we have $\nu = \alpha\mu$ with $\alpha > 0$. Then $\mu = \alpha^{-1}\nu$ consequently putting $a = b/\alpha$ we have

$$m = a\nu .$$

Let now $\boldsymbol{m} : \mathscr{K}(G) \to X$ be a left invariant vector measure. For every $x' \in X'$, the functional $\boldsymbol{m}_{x'}$ defined on $\mathscr{K}(G)$ by the equality

$$\boldsymbol{m}_{x'}(f) = \langle \boldsymbol{m}(f), x' \rangle$$

is a left invariant scalar measure, therefore by the first part of the proof, it is of the form

$$\boldsymbol{m}_{x'}(f) = a(x')\nu(f), \text{ for } f \in \mathscr{K}(G),$$

where $a(x')$ is a number which does not depend on f. It is easy to verify that a is a linear functional on X'.

Taking $f \in \mathscr{K}(G)$ with $\nu(f) \neq 0$, we deduce

$$|a(x')| = |\boldsymbol{m}_{x'}(f)|/|\nu(f)| \leqslant |x'|\, |\boldsymbol{m}(f)|/|\nu(f)|,$$

therefore a is continuous, $a \in X''$. For each $f \in \mathscr{K}(G)$ and every $x' \in X'$ we have

$$\langle \boldsymbol{m}(f), x' \rangle = a(x')\nu(f)$$

therefore, for every $f \in \mathscr{K}(G)$, we have

$$\boldsymbol{m}(f) = a\nu(f) .$$

It follows that $a \in X$ and $\boldsymbol{m} = a\nu$. Thus the proposition is proved.

23.8 Corollary. *Every left invariant measure $\boldsymbol{m} : \mathscr{K}(G) \to X$ is dominated and $|\boldsymbol{m}|$ is a left invariant positive measure.*

In fact, if ν is a left invariant Haar measure, there exists $a \in X$ with $\boldsymbol{m} = a\nu$. Then $|\boldsymbol{m}| = |a|\nu$.

Remarks.

1. If the measure μ, constructed in the proof of the Haar theorem (23.6) is also right invariant (for example, if G is commutative), the proof of the uniqueness of the complex measure m is proposition 23.7 is simpler:

Let f and h be two functions of $\mathscr{K}_+(G)$ with $h \neq 0$. We have

$$\begin{aligned} m(h)\mu(f) &= \int \mathrm{d}m(y) \int h(y) f(x) \mathrm{d}\mu(x) = \int \mathrm{d}m(y) \int h(y) f(xy) \mathrm{d}\mu(x) = \\ &= \int \mathrm{d}\mu(x) \int h(y) f(xy) \mathrm{d}m(y) = \int \mathrm{d}\mu(x) \int h(x^{-1}y) f(y) \mathrm{d}m(y) = \\ &= \int \mathrm{d}m(y) \int \check{h}(y^{-1}x) f(y) \mathrm{d}\mu(x) = \\ &= \int \mathrm{d}m(y) \int \check{h}(x) f(y) \mathrm{d}\mu(x) = \mu(\check{h}) m(f), \end{aligned}$$

whence, denoting $b = m(h)/\mu(\check{h})$, we obtain

$$m(f) = b\mu(f)$$

and then as before, $m = av$.

2. If $\boldsymbol{m}$ is a left invariant non zero measure, then we have $\boldsymbol{m}(f) \neq 0$ for every function $f \neq 0$ of $\mathscr{K}_+(G)$.

3. Since all the left invariant measures are obtained from a left invariant positive Haar measure μ by multiplication by a (vector or scalar) constant, we shall study further only the measure μ which will be referred to as the left invariant Haar measure, without other specification.

4 *Properties of Haar measure*

Let μ be the left invariant Haar measure.

23.9 Proposition. *For every function $f \geqslant 0$ defined on G and every $s \in G$ we have*

$$\mu^*(f_s) = \mu^*(f) .$$

The function $\alpha(x) = sx$ is a homeomorphism of G onto G and $f \circ \alpha = f_s$. If $f \in \mathscr{K}_+$, then $f \circ \alpha \in \mathscr{K}(G)$, therefore α is μ-proper and $\alpha(\mu) = \mu$. It follows then that for every function $f \geqslant 0$ we have

$$\int^* f_s \mathrm{d}\mu = \int^* f \circ \alpha \mathrm{d}\mu = \int^* f \mathrm{d}\alpha(\mu) = \int^* f \mathrm{d}\mu ,$$

that is

$$\mu^*(f_s) = \mu^*(f) .$$

23.10 Corollary. *For every set $A \subset G$ and every $s \in G$ we have*

$$\mu^*(sA) = \mu^*(A) .$$

23.11 Corollary. *For every nonempty open set $U \subset G$ we have*

$$\mu^*(U) > 0 .$$

In fact, if $V \subset G$ were a nonempty open set with $\mu^*(V)=0$, then every compact set $K \subset G$ could be covered by a finite family of open sets of the form sV; and, since $\mu^*(sV) = \mu^*(V)=0$, it would follow that $\mu^*(K)=0$, whence it would follow that μ vanishes identically.

23.12 Corollary. *The Haar measure μ has support equal to G.*

23.13 Corollary. *If G is discrete, then there exists a number $a>0$ such that $\mu(\{x\})=a$ for every $x \in G$.*

In fact, $\{e\}$ is an open set, therefore $a=\mu(\{e\})>0$. Then

$$\mu(\{x\}) = \mu(x\{e\}) = \mu(\{e\}) = a .$$

Remark. If G is discrete, usually the Haar measure is chosen such that $\mu(\{x\})=1$ for every $x \in G$.

23.14 Proposition. *The group G is compact if and only if the Haar measure μ is bounded.*

If G is compact, every positive measure on G is bounded; in particular, the Haar measure μ is bounded.

Conversely, assume that G is not compact, and prove that μ is unbounded. Let V be a relatively compact neighbourhood of e. The family $(xV)_{x \in G}$ covers G; but since G is not compact, no finite subfamily covers G. Then we can find by recurrence a sequence (x_n) of points of G such that for each n we have $x_n \notin \cup_{i<n} x_i V$. Let U be a *symmetric* open neighbourhood of e with $U^2 \subset V$. The open sets $x_n U$ are mutually disjoint. In fact, assume the contrary, that there exist two indices $n<m$ with $x_n U \cap x_m U \neq \varnothing$. Let $x \in x_n U \cap x_m U$. Then $x \in x_n U$ and $x \in x_m U$, therefore $x_m \in x U^{-1} = xU \subset x_n U^2 \subset x_n V$ which contradicts the way in which the sequence (x_n) was chosen.

For each n we have $\mu^*(x_n U) = \mu^*(U) > 0$. It follows that

$$\mu^*(G) \geqslant \mu^*(\cup_{1 \leqslant n < \infty} x_n U) = \Sigma_{1 \leqslant n < \infty} \mu^*(x_n U) = +\infty$$

therefore μ is unbounded.

Thus if μ is bounded, then G is compact.

Remark. In case G is compact, usually we choose the Haar measure μ such that $\mu(G) = 1$.

In the sequel, when a measure will be mentioned without being specified, we shall understand that this is the left invariant Haar measure. We shall denote this measure by dx. We shall denote by $\mathscr{L}_E^p$ the spaces constructed by means of this measure.

23.15 PROPOSITION. *Let E be a Banach space, $s \in G$ and $f : G \to E$ a function.*

(1) *the function f is negligible if and only if f_s is negligible;*

(2) *the function f is measurable if and only if f_s is measurable;*

(3) *the function f is integrable if and only if f_s is integrable.*

In this case we have

$$\int f(sx)\,dx = \int f(x)\,dx .$$

The first two properties follow from proposition 23.9.

If $f \in \mathscr{K}_E(G)$, then $f_s \in \mathscr{K}_E(G)$ and for every $z \in E'$, the functions $\langle f, z\rangle$ and $\langle f_s, z\rangle = (\langle f, z\rangle)_s$ belong to $\mathscr{K}(G)$, therefore

$$\langle \int f(sx)\,dx, z\rangle = \int \langle f(sx), z\rangle\,dx =$$

$$= \int \langle f(x), z\rangle\,dx = \langle \int f(x)\,dx, z\rangle ,$$

whence

$$\int f(sx)\,dx = \int f(x)\,dx .$$

Now let $f \in \mathscr{L}_E^1$ be an arbitrary function. There exists a sequence (f_n) of functions of $\mathscr{K}_E(G)$, converging in the mean to f. Then

$$\int^* |f_n(sx) - f(sx)|\,dx = \int^* |f_n(x) - f(x)|\,dx ,$$

therefore the sequence $(f_n)_s$ of functions of $\mathscr{K}_E(G)$ converges in the mean to f_s. It follows that $f_s \in \mathscr{L}_E^1$ and

$$\int f(sx)\,dx = \lim \int f_n(sx)\,dx = \lim \int f_n(x)\,dx = \int f(x)\,dx .$$

If $f_s \in \mathscr{L}_E^1$, then $f = (f_s)_{s^{-1}} \in \mathscr{L}_E^1$ and thus the proposition is proved.

23.16 COROLLARY. *If $f: G \to E$ is locally integrable and $s \in G$ then f_s is locally integrable and for every $\varphi \in \mathcal{K}(G)$ we have*

$$\int \varphi(x) f_s(x) dx = \int \varphi(s^{-1}x) f(x) dx .$$

In fact, if $\varphi \in \mathcal{K}(G)$, the function $x \to \varphi(s^{-1}x)$ belongs to $\mathcal{K}(G)$, therefore the function $x \to \varphi(s^{-1}x) f(x)$ is integrable. It follows then that the function $x \to \varphi(x) f(sx)$ is integrable, therefore the function f_s is locally integrable and

$$\int \varphi(s^{-1}x) f(x) dx = \int \varphi(x) f(sx) dx = \int \varphi(x) f_s(x) dx .$$

23.17 COROLLARY. *If $f \in \mathcal{L}^p_E$, $1 \leqslant p \leqslant +\infty$ and $s \in G$, then $f_s \in \mathcal{L}^p_E$ and*

$$N_p(f_s) = N_p(f) .$$

In fact, if $f \in \mathcal{L}^p_E$, then f is measurable and $N_p(f) < +\infty$. It follows that f_s is measurable.

If $p = +\infty$, from part (1) of proposition 23.15 we deduce that $N_\infty(f_s) = N_\infty(f) < +\infty$, therefore $f_s \in \mathcal{L}^\infty_E$.

If $p < +\infty$, we have

$$N_p(f_s) = (\int^* |f(sx)|^p dx)^{1/p} = (\int^* |f(x)|^p dx)^{1/p} = N_p(f) < +\infty ,$$

therefore $f_s \in \mathcal{L}^p_E$.

23.18 PROPOSITION. *Let E be a Banach space and $1 \leqslant p < +\infty$. For every function $f \in \mathcal{L}^p_E$, the mapping $s \to f_s$ of G into $\mathcal{L}^p_E$ is left uniformly continuous.*

Let $f \in \mathcal{L}^p_E$ and $\varepsilon > 0$. There exists a function $g \in \mathcal{K}_E(G)$ with $N_p(f-g) < \varepsilon/3$. For every $s \in G$ we have

$$N_p(f_s - g_s) = N_p(f-g) < \varepsilon/3.$$

Since $g_s \in \mathcal{K}_E(G)$ we deduce that $f_s \in \mathcal{L}^p_E$ for every $s \in G$.

By proposition 23.5 we deduce that there exists a neighbourhood V of e such that

$$N_p(g_s - g) < \varepsilon/3, \text{ if } s \in V .$$

If $st^{-1} \in V$, then

$$N_p(g_s - g_t) = N_p(g_{st^{-1}} - g) < \varepsilon/3 ,$$

therefore

$$N_p(f_s - f_t) \leqslant N_p(f_s - g_s) + N_p(g_s - g_t) + N_p(f_t - g_t) < \varepsilon\,;$$

therefore, the mapping $s \to f_s$ of G into $\mathscr{L}_E^p$ is left uniformly continuous.

23.19 COROLLARY. *If for each $s \in G$ we put*

$$U_s f = f_{s^{-1}}, \text{ for } f \in \mathscr{L}_E^p, \quad 1 \leqslant p < +\infty\,,$$

then the mapping $U : s \to U_s$ is a simply continuous representation of G into the multiplicative group of linear operators of the space $\mathscr{L}_E^p$, and $\|U_s\| \equiv 1$.

The fact that U_s is linear is immediate. Then we have

$$N_p(U_s f) = N_p(f_{s^{-1}}) = N_p(f)\,,$$

therefore $\|U_s\| = 1$. For $s, t \in G$ we have

$$U_{st} f = f_{(st)^{-1}} = f_{t^{-1}s^{-1}} = U_s f_{t^{-1}} = U_s U_t f\,,$$

therefore $U_{st} = U_s U_t$. We have then $U_e f = f_e = f$, therefore $U_e = I$, consequently U is a representation. For each $f \in \mathscr{L}_E^p$, the mapping $s \to f_{s^{-1}}$ of G into $\mathscr{L}_E^p$ is continuous (proposition 23.18), therefore U is simply continuous.

Remark. The mapping $U_s f = f_{s^{-1}}$ is called the left regular representation of the group G into $\mathscr{L}(\mathscr{L}_E^p)$.

23.20 PROPOSITION. *For every function $f(x, y) \geqslant 0$ defined on $G \times G$ we have*

$$\int\int^* f(x, y)\,dx\,dy = \int\int^* f(x, x^{-1}y)\,dx\,dy\,.$$

The mapping $\alpha(x, y) = (x, x^{-1}y)$ of $G \times G$ onto $G \times G$ is proper with respect to the product measure $dx\,dy = dx \otimes dy$.

Let $f \in \mathscr{K}(G \times G)$. We have

$$\int\int f(x, y)\,dx\,dy = \int dx \int f(x, y)\,dy = \int dx \int f(x, x^{-1}y)\,dy =$$
$$= \int\int f(x, x^{-1}y)\,dx\,dy = \int\int f(\alpha(x, y))\,dx\,dy\,,$$

therefore $\alpha(dx\,dy) = dx\,dy$. Then, for every function $f(x, y) \geqslant 0$ we have (proposition 20.9).

$$\int\int^* f(x, y)\mathrm{d}x\,\mathrm{d}y = \int\int^* f(x, y)\,\alpha(\mathrm{d}x\,\mathrm{d}y) = \int\int^* f(\alpha(x, y))\mathrm{d}x\,\mathrm{d}y =$$
$$= \int\int^* f(x, x^{-1}y)\mathrm{d}x\,\mathrm{d}y\,.$$

23.21 PROPOSITION. *Let $\boldsymbol{f}(x, y)$ be a function defined on G with values in a Banach space or in $\overline{R}$.*

(1) *$\boldsymbol{f}(x, y)$ is negligible (for $\mathrm{d}x\,\mathrm{d}y$) if and only if $\boldsymbol{f}(x, x^{-1}y)$ is negligible;*

(2) *$\boldsymbol{f}(x, y)$ is measurable (for $\mathrm{d}x\,\mathrm{d}y$) if and only if $\boldsymbol{f}(x, x^{-1}y)$ is measurable;*

(3) *$\boldsymbol{f}(x, y)$ is integrable (for $\mathrm{d}x\,\mathrm{d}y$) if ahd only if $\boldsymbol{f}(x, x^{-1}y)$ is integrable. In this case*

$$\int\int \boldsymbol{f}(x, y)\mathrm{d}x\,\mathrm{d}y = \int\int \boldsymbol{f}(x, x^{-1}y)\mathrm{d}x\,\mathrm{d}y\,.$$

We use propositions 20.11, 20.15 and theorem 20.17.

23.22 COROLLARY. *Let X, Y and Z be three Banach spaces, $(u, v) \to uv$ a continuous bilinear mapping of $X \times Y$ into Z and two functions $\boldsymbol{f}: G \to X$ and $\boldsymbol{g}: G \to Y$.*

(1) *If $\boldsymbol{f}$ and $\boldsymbol{g}$ are negligible, then the function $\boldsymbol{f}(x)\boldsymbol{g}(x^{-1}y)$ is negligible (for $\mathrm{d}x\,\mathrm{d}y$);*

(2) *If $\boldsymbol{f}$ and $\boldsymbol{g}$ are measurable, then the function $\boldsymbol{f}(x)\boldsymbol{g}(x^{-1}y)$ is measurable (for $\mathrm{d}x\,\mathrm{d}y$);*

(3) *If $\boldsymbol{f}$ and $\boldsymbol{g}$ are integrable, then the function $\boldsymbol{f}(x)\boldsymbol{g}(x^{-1}y)$ is integrable (for $\mathrm{d}x\,\mathrm{d}y$) and*

$$\int\int \boldsymbol{f}(x)\boldsymbol{g}(y)\mathrm{d}x\,\mathrm{d}y = \int\int \boldsymbol{f}(x)\boldsymbol{g}(x^{-1}y)\,\mathrm{d}x\,\mathrm{d}y\,.$$

Remark. The mappings $(x, y) \to (y^{-1}x, y)$, $(x, y) \to (x, xy)$ and $(x, y) \to (yx, y)$ of $G \times G$ onto $G \times G$ are proper. Reasoning as before, we deduce that for every function $f(x, y) \geqslant 0$ we have

$$\int\int^* f(x, y)\mathrm{d}x\,\mathrm{d}y = \int\int^* f(y^{-1}x, y)\,\mathrm{d}x\,\mathrm{d}y = \int\int^* f(x, xy)\,\mathrm{d}x\,\mathrm{d}y =$$
$$= \int\int^* f(yx, y)\,\mathrm{d}x\,\mathrm{d}y\,.$$

Let $\boldsymbol{f}(x, y)$ be a function defined on $G \times G$ with values in a Banach space E or in $\overline{R}$;

(1) *If $\boldsymbol{f}(x, y)$ is negligible (for $\mathrm{d}x\,\mathrm{d}y$), then the functions $\boldsymbol{f}(y^{-1}x, y)$ $\boldsymbol{f}(x, xy)$ and $\boldsymbol{f}(yx, y)$ are negligible.*

(2) *If $\boldsymbol{f}(x, y)$ is measurable (for $\mathrm{d}x\,\mathrm{d}y$), then the functions $\boldsymbol{f}(y^{-1}x, y)$, $\boldsymbol{f}(x, xy)$ and $\boldsymbol{f}(yx, y)$ are measurable;*

(3) *If* $f(x, y)$ *is integrable* (*for* $dx\,dy$), *then the functions* $f(y^{-1}x, y)$, $f(x, yx)$ *and* $f(yx, y)$ *are integrable and we have*

$$\iint f(x, y)\,dx\,dy = \iint f(y^{-1}x, y)\,dx\,dy = \iint f(x, xy)\,dx\,dy =$$
$$= \iint f(yx, y)\,dx\,dy .$$

5 *The modular function*

Let μ be a left invariant Haar measure. In general, this measure is not right invariant. For each point $t \in G$ define the functional μ^t on $\mathcal{K}(G)$ by the equality

$$\mu^t(f) = \mu(f^t), \text{ for } f \in \mathcal{K}(G) .$$

It is easy to see that μ^t is a left invariant positive measure on G:

$$\mu^t(f_s) = \mu((f_s)^t) = \mu((f^t)_s) = \mu(f^t) = \mu^t(f) ,$$

therefore μ^t can be obtained from μ by multiplication with a positive number $\Delta(t)$:

$$\mu^t(f) = \Delta(t)\,\mu(f) .$$

This equality is also written

$$\mu(f^t) = \Delta(t)\,\mu(f) ,$$

that is,

$$\int f(xt^{-1})\,dx = \Delta(t) \int f(x)\,dx ,$$

or

$$\int f(xt)\,dx = \Delta(t^{-1}) \int f(x)\,dx .$$

The function $t \to \Delta(t)$ defined on G is called the *modular function* of the group G.

We say that the group G is *unimodular* if $\Delta(t) \equiv 1$, that is if the left invariant Haar measure is also right invariant.

23.23 PROPOSITION. *The commutative groups, the compact groups and the discrete groups are unimodular.*

The assertion concerning the commutative groups is immediate.

Assume that G is compact. Then the function $f(x) \equiv 1$ is integrable with respect to the measure μ and $f^s = f$, therefore

$$\mu(f) = \mu(f^s) = \Delta(s)\mu(f).$$

Since $\mu(f) > 0$, we deduce $\Delta(s) = 1$ for every $s \in G$.

Assume now that G is discrete. Let $f \neq 0$ be a positive function from $\mathscr{K}(G)$. The function f has a finite support $\{x_1, x_2, \ldots, x_n\}$ on which it takes on the values $\{a_1, a_2, \ldots, a_n\}$ different from zero. For every $t \in G$, the function f^t has the support $\{x_1 t, x_2 t, \ldots, x_n t\}$ on which it takes on the values $\{a_1, a_2, \ldots, a_n\}$. Then

$$\mu(f^t) = \Sigma a_i = \mu(f) > 0$$

and therefore

$$\mu(f^t) = \Delta(t)\mu(f) = \Delta(t)\mu(f^t) > 0$$

whence

$$\Delta(t) = 1.$$

23.24 PROPOSITION. *The function $\Delta(t)$ is a continuous homomorphism of G into the multiplicative group R_+ of the strictly positive numbers.*

In fact, let $f \neq 0$ be a function of $\mathscr{K}_+(G)$. For every $t \in G$ the function f^t belongs to $\mathscr{K}_+(G)$ and does not vanish identically, therefore $\mu(f) > 0$ and $\mu(f^t) > 0$. It follows that

(1) $\Delta(t) = \mu(f^t)/\mu(f) > 0.$

Then we have

$$\Delta(st)\mu(f) = \mu(f^{st}) = \mu((f^s)^t) = \Delta(t)\mu(f^s) = \Delta(t)\Delta(s)\mu(f),$$

whence

(2) $\Delta(st) = \Delta(s)\Delta(t).$

Taking $s = e$ we obtain $\Delta(t) = \Delta(e)\Delta(t)$, therefore

(3) $\Delta(e) = 1.$

We deduce then

$$1 = \Delta(e) = \Delta(s\,s^{-1}) = \Delta(s)\Delta(s^{-1})$$

whence

$$\Delta(s^{-1}) = \Delta(s)^{-1}.$$

Let us show that $\Delta(s)$ is continuous on G. Let $f \in \mathscr{K}_+(G)$ with $\mu(f) > 0$. Then

$$\Delta(t) = \mu(f^t)/\mu(f).$$

By proposition 23.5, the mapping $t \to f^t$ of G into $\mathscr{L}^1$ is continuous at e. Since μ is continuous on $\mathscr{L}^1$, it follows that the mapping $t \to \mu(f^t)$ of G into $\mathscr{L}^1$ is continuous at e, therefore $\Delta(t)$ is continuous at e.

Now let $s \in G$ be an arbitrary point and $\varepsilon > 0$.

Since Δ is continuous at e, there exists a neighbourhood V of e such that

$$|\Delta(z) - \Delta(e)| < \varepsilon/\Delta(s), \text{ for } z \in V.$$

The set sV is a neighbourhood of s and for $t \in sV$ we have $s^{-1}t \in V$, therefore

$$|\Delta(s^{-1}t) - \Delta(e)| < \varepsilon/\Delta(s),$$

whence, multiplying by $\Delta(s)$, we deduce

$$|\Delta(t) - \Delta(s)| < \varepsilon.$$

Thus, Δ is continuous at s; and as s is arbitrary, it follows that Δ is continuous on G.

23.25 PROPOSITION. *For every function $f \geqslant 0$ defined on G and every $s \in G$ we have*

$$\int^* f(xs^{-1})\mathrm{d}x = \Delta(s) \int^* f(x)\mathrm{d}x.$$

In fact, the function $\alpha(x) = xs^{-1}$ is a homeomorphism of G onto G, therefore it is proper with respect to the Haar measure $\mathrm{d}x = \mathrm{d}\mu$. We have then $f \circ \alpha = f^s$ and $\alpha(\mu) = \mu^s = \Delta(s)\mu$. By proposition 20.9, we have

$$\begin{aligned}\int^* f(xs^{-1})\mathrm{d}x &= \int^* f^s(x)\mathrm{d}x = \int^* f \circ \alpha \, \mathrm{d}\mu = \int^* f \, \mathrm{d}\alpha(\mu) = \\ &= \int^* f(x)\mathrm{d}\mu^s(x) = \Delta(s) \int^* f(x)\mathrm{d}\mu(x) = \\ &= \Delta(s) \int^* f(x)\mathrm{d}x.\end{aligned}$$

23.26 COROLLARY. *For every set $A \subset G$ and every $s \in G$ we have*

$$\mu^*(As) = \Delta(s)\mu^*(A).$$

23.27 Corollary. *For every function $f \geqslant 0$ defined on G and every $s \in G$ we have*

$$\int^* f(xs)\Delta(x^{-1})dx = \int^* f(x)\Delta(x^{-1})dx .$$

In fact, denoting $g(x) = f(x)\Delta(x^{-1})$ we have

$$f(xs)\Delta(x^{-1}) = f(xs)\Delta(s^{-1}x^{-1})\Delta(s) = g(xs)\Delta(s) ;$$

therefore

$$\int^* f(xs)\Delta(x^{-1})dx = \Delta(s)\int^* g(xs)dx = $$
$$= \Delta(s)\Delta(s^{-1})\int^* g(x)dx = \int^* f(x)\Delta(x^{-1})dx .$$

Remark. Corollary 23.27 shows that the measure $\Delta(x^{-1})dx$ is right invariant on G.

23.28 Corollary. *Every left invariant Haar measure is equivalent to every right invariant Haar measure.*

In fact, since the function $\Delta(x)$ does not vanish, the measures dx and $\Delta(x^{-1})dx$ are equivalent. Every left invariant measure is equivalent to dx and every right invariant Haar measure is equivalent to $\Delta(x^{-1})dx$, whence the corollary follows.

It follows that the negligible functions and the measurable functions are the same for the left invariant Haar measures and for the right invariant ones.

23.29 Proposition. *Let $\boldsymbol{f}$ be a function defined on G with values in a Banach space E or in $\overline{\mathrm{R}}$.*

(1) *The function $\boldsymbol{f}^s$ is negligible if and only if $\boldsymbol{f}$ is negligible;*

(2) *The function $\boldsymbol{f}^s$ is measurable if and only if $\boldsymbol{f}$ is measurable;*

(3) *The function $\boldsymbol{f}^s$ is integrable if and only if $\boldsymbol{f}$ is integrable. In this case we have*

$$\int \boldsymbol{f}(xs^{-1})dx = \Delta(s)\int \boldsymbol{f}(x)dx .$$

We use propositions 23.25, 20.11, 20.15 and theorem 20.17.

23.30 Corollary. *If $\boldsymbol{f} \in \mathscr{L}_E^p$, $1 \leqslant p \leqslant +\infty$ and $s \in G$, then $\boldsymbol{f}^s \in \mathscr{L}_E^p$ and*

$$N_p(\boldsymbol{f}^s) = \Delta(s)^{1/p} N_p(\boldsymbol{f}) .$$

In fact, if $f \in \mathscr{L}_E^p$, then f is measurable and $N_p(f) < +\infty$. It follows that f^s is measurable. If $p < +\infty$ we have

$$N_p(f^s) = (\int^* |f(xs^{-1})|^p \mathrm{d}x)^{1/p} =$$
$$= (\Delta(s) \int^* |f(x)|^p \mathrm{d}x)^{1/p} = \Delta(s)^{1/p} N_p(f),$$

therefore $f^s \in \mathscr{L}_E^p$.

If $p = +\infty$, there exists a negligible set $N \subset G$ such that for $x \notin N$ we have $|f(x)| \leqslant N_\infty(f)$. Then Ns is negligible and for $x \notin Ns$ we have $xs^{-1} \notin N$ therefore $|f(xs^{-1})| \leqslant N_\infty(f)$, that is $|f^s(x)| \leqslant N_\infty(f)$. It follows that $N_\infty(f^s) \leqslant N_\infty(f) < +\infty$, therefore $f^s \in \mathscr{L}_E^\infty$. As we have also $f = (f^s)^{s^{-1}}$, it follows that $N_\infty(f) \leqslant N_\infty(f^s)$, therefore $N_\infty(f^s) = N_\infty(f)$.

Since, in this case, $\Delta(s)^{1/p} \equiv 1$, it follows that the equality $N_p(f^s) = \Delta(s)^{1/p} \cdot N_p(f)$ is satisfied for every p.

23.31 COROLLARY. *If $f \in \mathscr{L}_E^\infty$ and $s \in G$, then $f^s \in \mathscr{L}_E^\infty$ and $N_\infty(f^s) = N_\infty(f)$.*

23.32 COROLLARY. *If G is unimodular, for every function $f \in \mathscr{L}_E^p$, $1 \leqslant p \leqslant +\infty$ and $s \in G$ we have $f^s \in \mathscr{L}_E^p$ and*

$$N_p(f^s) = N_p(f).$$

If $f \in \mathscr{L}_E^1$, then

$$\int f(xs)\mathrm{d}x = \int f(x)\mathrm{d}x.$$

23.33 PROPOSITION. *Let E be a Banach space and $1 \leqslant p < +\infty$. If $f \in \mathscr{L}_E^p$, the mapping $s \to f^s$ of G into $\mathscr{L}_E^p$ is continuous.*

Let $f \in \mathscr{L}_E^p$ and $\varepsilon > 0$. Choose $g \in \mathscr{K}_E(G)$ such that

$$N_p(f - g) < \varepsilon/6.$$

Since by proposition 23.5, the mapping $s \to g^s$ of G into $\mathscr{L}_E^p$ is continuous at e, and the modular function Δ is also continuous at e, there exists a neighbourhood V of e such that

$$N_p(g^s - g) < \varepsilon/3 \text{ and } |\Delta(s) - \Delta(e)| < 1 \text{ for } s \in V.$$

It follows that for $s \in V$ we have

$$\Delta(s) < \Delta(e) + 1 = 2,$$

therefore

$$N_p(f^s - g^s) = \Delta(s)^{1/p} N_p(f-g) < \varepsilon/3$$

consequently

$$N_p(f^s - f) \leqslant N_p(f^s - g^s) + N_p(g^s - g) + N_p(g - f) < \varepsilon .$$

We deduce that the mapping $s \to f^s$ of G into $\mathscr{L}_E^p$ is continuous at e.

Let now $a \in G$ and show that the mapping $s \to f^s$ is continuous at a. Let $\varepsilon > 0$ and let V be a neighbourhood of e such that

$$N_p(f^s - f) < \varepsilon \Delta(a)^{-1/p}, \text{ for } s \in V .$$

For every point $t \in Va$ we have then $ta^{-1} \in V$, therefore

$$N_p(f^t - f^a) = N_p(f^{ta^{-1}a} - f^a) = \Delta(a)^{1/p} N_p(f^{ta^{-1}} - f) < \varepsilon$$

consequently the mapping $t \to f^t$ of G into $\mathscr{L}_E^p$ is continuous at a. Since a is arbitrary in G, we deduce that this mapping is continuous on G and thus the proposition is completely proved.

23.34 Corollary. *If for each $s \in G$ we put*

$$V_s f = f^{s^{-1}}, \text{ for } t \in \mathscr{L}_E^p, \quad 1 \leqslant p < +\infty ,$$

then $V : s \to V_s$ is a simply continuous representation of G into the multiplicative group of the linear continuous operators of the space $\mathscr{L}_E^p$.

It is immediately verified that $V_{st} = V_s V_t$ and $V_e = I$. We have then

$$N_p(V_s f) = N_p(f^{s^{-1}}) = \Delta(s^{-1})^{1/p} N_p(f) ,$$

therefore $\|V_s\| = \Delta(s^{-1})^{1/p} < +\infty$, hence V_s is continuous. The simple continuity of the representation V follows from the continuity of the mapping $s \to f^{s^{-1}}$.

Remark. The representation V is called the *regular right representation* of the group G.

23.35 Proposition. *Let E be a Banach space. If $f \in \mathscr{K}_E(G)$, then $\check{f}\,\check{\Delta} \in \mathscr{K}_E(G)$ and*

$$\int f(x^{-1}) \Delta(x^{-1}) \mathrm{d}x = \int f(x) \mathrm{d}x .$$

For every function $f \in \mathscr{K}_E(G)$ the mapping $x \to f(x^{-1})\Delta(x^{-1})$ of G into E belongs to $\mathscr{K}_E(G)$. Consider the functional ν defined on $\mathscr{K}(G)$ by the equality

$$\nu(f) = \int \check{f}(x)\check{\Delta}(x)\mathrm{d}x = \int f(x^{-1})\Delta(x^{-1})\mathrm{d}x, \text{ for } f \in \mathscr{K}(G).$$

It is easy to see that ν is a positive linear functional, therefore it is a positive measure on G. This measure is left invariant,

$$\nu(f_s) = \int f_s(x^{-1})\Delta(x^{-1})\mathrm{d}x = \int f(sx^{-1})\Delta(x^{-1})\mathrm{d}x =$$
$$= \int \check{f}(xs^{-1})\check{\Delta}(xs^{-1})\Delta(s^{-1})\mathrm{d}x = \int \check{f}(x)\check{\Delta}(x)\mathrm{d}x.$$

It follows that there exists a number $c \geqslant 0$ such that

$$\nu(f) = c \int f(x)\mathrm{d}x, \text{ for } f \in \mathscr{K}(G).$$

Let us show that $c = 1$. Let $\varepsilon > 0$. Since Δ is continuous at e and $\Delta(e) = 1$, there exists a *symmetric* neighbourhood V of an e such that

$$|1 - \Delta(x)| < \varepsilon, \text{ for } x \in V.$$

Choose a *symmetric* function $f \in \mathscr{K}_E(G)$, $f(x^{-1}) = f(x)$, with the support contained in V, and

$$\int f(x)\mathrm{d}x = 1.$$

Then

$$|1 - c| = |(1 - c)\int f(x)\mathrm{d}x| = |\int f(x)\mathrm{d}x - \nu(f)| =$$
$$= |\int (1 - \Delta(x^{-1}))f(x)\mathrm{d}x| < \varepsilon \int f(x)\mathrm{d}x = \varepsilon.$$

Since $\varepsilon > 0$ is arbitrary, we deduce that $c = 1$, therefore

$$\nu(f) = \int f(x)\mathrm{d}x$$

for every function $f \in \mathscr{K}(G)$.

If $f \in \mathscr{K}_E(G)$, then for every $z \in E'$ we have $\langle f, z\rangle \in \mathscr{K}(G)$, therefore

$$\langle \int f(x^{-1})\Delta(x^{-1})\mathrm{d}x, z\rangle = \int \langle f(x^{-1})\Delta(x^{-1}), z\rangle \mathrm{d}x =$$
$$= \int \langle f(x^{-1}), z\rangle \Delta(x^{-1})\mathrm{d}x = \int \langle f(x), z\rangle \mathrm{d}x =$$
$$= \langle \int f(x)\mathrm{d}x, z\rangle,$$

whence

$$\int f(x^{-1})\Delta(x^{-1})\mathrm{d}x = \int f(x)\mathrm{d}x$$

and the proposition is proved.

We shall show further that the proposition remains valid also for functions of $\mathscr{L}^1_E$.

23.36 PROPOSITION. *For every function $f \geqslant 0$ defined on G we have*

$$\int^* f(x^{-1})\Delta(x^{-1})dx = \int^* f(x)dx .$$

Denote $\nu = \Delta(x^{-1})dx$ and $\alpha(x) = x^{-1}$. Since α is a homeomorphism of G onto G, it is proper with respect to the measure ν. By proposition 23.35 we have $\alpha(\nu) = dx$. Then (proposition 20.9)

$$\int^* f(x^{-1})\Delta(x^{-1})dx = \int^* f(\alpha(x))d\nu(x) = \int^* f(x)d\alpha(\nu) = \int^* f(x)dx .$$

23.37 COROLLARY. *If G is unimodular (in particular, if G is commutative, discrete or compact), for every function $f \geqslant 0$ defined on G we have*

$$\int^* f(x^{-1})dx = \int^* f(x)dx$$

and for every set $A \subset G$ we have

$$\mu^*(A^{-1}) = \mu^*(A)$$

where $d\mu = dx$.

23.38 PROPOSITION. *Let $\boldsymbol{f}$ be a function defined on G with values in a Banach space E or in $\overline{R}$.*

(1) *The function $\boldsymbol{f}$ is negligible if and only if the function $\check{\boldsymbol{f}}$ is negligible.*

(2) *The function $\boldsymbol{f}$ is measurable if and only if the function $\check{\boldsymbol{f}}$ is measurable.*

(3) *The function $\boldsymbol{f}$ is integrable if and only if the function $\check{\boldsymbol{f}}\check{\Delta}$ is integrable. In this case we have*

$$\int \boldsymbol{f}(x^{-1})\Delta(x^{-1})dx = \int \boldsymbol{f}(x)dx .$$

In fact, since $\Delta(x^{-1})$ is continuous and does not vanish, the measure $\Delta(x^{-1})dx$ is equivalent to the measure dx (proposition 18.10). On the other hand, denoting $\alpha(x) = x^{-1}$, we have $\alpha(\check{\Delta}\,dx) = dx$. Then $\boldsymbol{f}$ is negligible for $dx = \alpha(\check{\Delta}\,dx)$ if and only if $\check{\boldsymbol{f}} = \boldsymbol{f} \circ \alpha$ is negligible for $\check{\Delta}\,dx$ (proposition 20.11), and the function $\check{\boldsymbol{f}}$ is negligible for $\check{\Delta}\,dx$ if and only if $\check{\boldsymbol{f}}$ is negligible for dx (definition 18.1). The point (2) of the proposition can be proved in the same way. The point (3) is proved using the results of §§16 and 20.

23.39 COROLLARY. *If* $f \in \mathcal{L}_E^p$, $1 \leqslant p \leqslant +\infty$, *then* $\check{f}\check{\Delta}^{1/p} \in \mathcal{L}_E^p$ *and*

$$N_p(f) = N_p(\check{f}\check{\Delta}^{1/p}).$$

If $p = +\infty$ and if $f \in \mathcal{L}_E^\infty$, then, by the points (1) and (2) of proposition 23.38, we deduce that $\check{f} \in \mathcal{L}_E^\infty$ and $N_\infty(f) = N_\infty(\check{f})$ and the proposition is proved in this case.

Assume that $1 \leqslant p < +\infty$ and let $f \in \mathcal{L}_E^p$. Then $\check{f}$ is measurable and $\check{\Delta}^{1/p}$ is continuous, therefore $\check{f}\check{\Delta}^{1/p}$ is measurable. We have then

$$N_p(\check{f}\check{\Delta}^{1/p}) = (\int^* |\check{f}\check{\Delta}^{1/p}|^p \mathrm{d}x)^{1/p} = (\int^* |f(x^{-1})|^p \Delta(x^{-1})\mathrm{d}x)^{1/p} =$$
$$= (\int^* |f(x)|^p \mathrm{d}x)^{1/p} = N_p(f) < +\infty ,$$

therefore $\check{f}\check{\Delta}^{1/p} \in \mathcal{L}_E^p$ and the proposition is completely proved.

23.40 COROLLARY. *If* $f \in \mathcal{L}_E^\infty$, *then* $\check{f} \in \mathcal{L}_E^\infty$ *and* $N_\infty(f) = N_\infty(\check{f})$.

23.41 COROLLARY. *Assume that G is unimodular. If* $f \in \mathcal{L}_E^p$, $1 \leqslant p \leqslant +\infty$, *then* $\check{f} \in \mathcal{L}_E^p$ *and*

$$N_p(f) = N_p(\check{f}).$$

If $f \in \mathcal{L}_E^1$, *then*

$$\int f(x^{-1})\mathrm{d}x = \int f(x)\mathrm{d}x .$$

§24 Convolution

1 *The convolution of two measures*

Let X, Y, Z be three Banach spaces and $(u, v) \to uv$ a continuous bilinear mapping of $X \times Y$ into Z with $|uv| \leqslant |u|\,|v|$ for every $u \in X$ and $v \in Y$.

Let $\boldsymbol{m} : \mathcal{K}(G) \to X$ and $\boldsymbol{n} : \mathcal{K}(G) \to Y$ be two dominated measures.

Assume that for every function $f \in \mathcal{K}(G)$, the function $f(xy)$ of two variables defined on $G \times G$ is integrable with respect to the positive measure $|\boldsymbol{m}| \otimes |\boldsymbol{n}|$. Since $|\boldsymbol{m} \otimes \boldsymbol{n}| \leqslant |\boldsymbol{m}| \otimes |\boldsymbol{n}|$, it follows that the function $f(xy)$ is integrable with respect to $\boldsymbol{m} \otimes \boldsymbol{n}$. Denote then

$$\boldsymbol{I}(f) = \int\int f(xy)\,\mathrm{d}\boldsymbol{m}(x)\mathrm{d}\boldsymbol{n}(y) \tag{1}$$

It is easy to see that $\boldsymbol{I}$ is a linear mapping of $\mathcal{K}(G)$ into Z.

Denote also

$$\lambda(f) = \iint f(xy)\mathrm{d}|\boldsymbol{m}|(x)\mathrm{d}|\boldsymbol{n}|(y)\cdot \tag{1'}$$

It is easy to verify that λ is a positive linear functional on $\mathscr{K}(G)$, therefore it is a positive measure on G. The linear mapping $\boldsymbol{I}: \mathscr{K}(G) \to Z$ is dominated by λ:

$$|\boldsymbol{I}(f)| = |\iint f(xy)\mathrm{d}\boldsymbol{m}\otimes\boldsymbol{n}(x, y)| \leqslant \iint |f(xy)|\mathrm{d}|\boldsymbol{m}\otimes\boldsymbol{n}|(x, y) \leqslant$$
$$\leqslant \iint |f(x, y)|\,\mathrm{d}|\boldsymbol{m}|\otimes|\boldsymbol{n}|(x, y) = \lambda(|f|).$$

It follows that $\boldsymbol{I}$ is a dominated measure on G with values in Z.

24.1 Definition. *Assume that for every $f \in \mathscr{K}(G)$ the function $f(xy)$ is $|\boldsymbol{m}|\otimes|\boldsymbol{n}|$-integrable. The measure* **I** *defined on G by the equality* (1) *is called the convolution of the measures* **m** *and* **n** *and is denoted by* $\boldsymbol{m}*\boldsymbol{n}$:

$$(\boldsymbol{m}*\boldsymbol{n})(f) = \iint f(xy)\mathrm{d}\boldsymbol{m}(x)\mathrm{d}\boldsymbol{n}(y), \text{ for } f \in \mathscr{K}(G).$$

By this definition, the measure λ defined by the equality (1′) is the convolution of the positive measures $|\boldsymbol{m}|$ and $|\boldsymbol{n}|$. The inequality $|\boldsymbol{I}| \leqslant \lambda$ is written now

$$|\boldsymbol{m}*\boldsymbol{n}| \leqslant |\boldsymbol{m}|*|\boldsymbol{n}|.$$

Remarks.

1. If $f \in \mathscr{K}(G)$, then the function $f(xy)$ of two variables is continuous and bounded on $G \times G$ but it is not necessarily integrable with respect to the measure $|\boldsymbol{m}|\otimes|\boldsymbol{n}|$ if $\boldsymbol{m}$ and $\boldsymbol{n}$ are arbitrary. Thus, for two arbitrary measures $\boldsymbol{m}$ and $\boldsymbol{n}$, their convolution $\boldsymbol{m}*\boldsymbol{n}$ is not necesarily defined.

2. By definition 24.1, to say that $\boldsymbol{m}\otimes\boldsymbol{n}$ makes sense is equivalent to the fact that $|\boldsymbol{m}|*|\boldsymbol{n}|$ makes sense.

24.2 Proposition. *If one of the measures* **m** *and* **n** *has compact support, then* $\boldsymbol{m}*\boldsymbol{n}$ *exists.*

Assume that $\boldsymbol{m}$ has compact support. Denote $\mu = |\boldsymbol{m}|$ and $\nu = |\boldsymbol{n}|$. Let C be the compact support of $\boldsymbol{m}$ and μ.

Let $f \in \mathscr{K}(G)$ and let K be the support of the function f. Let C' be the support of the measure $\mu\otimes\nu$ and K' the support of the function $f(xy)$ of two variables. Denote $I=C'\cap K\subset G\times G$. We shall show that

$$I\subset C\times C^{-1}K\,.$$

Let $(x, y)\in I$. Then $(x, y)\in C'\subset C\times G$, therefore $x\in C$. We have also $(x, y)\in K'$, therefore $xy\in K$. It follows that $y\in x^{-1}K$ and, since $x\in C$, we deduce that $y\in C^{-1}K$. It follows that $(x, y)\in C\times C^{-1}K$ and the above inclusion is proved. We deduce then that I is a relatively compact set. The function $f(xy)$ is continuous and bounded on $G\times G$ and

$$\int\int^* |f(xy)|\,\mathrm{d}\mu\otimes\nu\leqslant\|f\|\,\mu\otimes\nu(I)<+\infty\,.$$

It follows that $f(xy)$ is $\mu\otimes\nu$-integrable, therefore $\mu*\nu$ and $\boldsymbol{m}*\boldsymbol{n}$ exist.

24.3 PROPOSITION. *For every dominated measure $\boldsymbol{m}:\mathscr{K}(G)\to X$ we have*

$$\boldsymbol{m}*\varepsilon_e=\varepsilon_e*\boldsymbol{m}=\boldsymbol{m}\,.$$

In fact, ε_e has compact support, therefore the convolutions $\boldsymbol{m}*\varepsilon_e$ and $\varepsilon_e*\boldsymbol{m}$ make sense. For every function $f\in\mathscr{K}(G)$ we have

$$\begin{aligned}\int f\,\mathrm{d}(\boldsymbol{m}*\varepsilon_e)&=\int\int f(xy)\,\mathrm{d}\boldsymbol{m}(x)\,\mathrm{d}\varepsilon_e(y)=\\&=\int \mathrm{d}\boldsymbol{m}(x)\int f(xy)\,\mathrm{d}\varepsilon_e(y)=\int f(x)\,\mathrm{d}\boldsymbol{m}(x)\end{aligned}$$

and

$$\begin{aligned}\int f\mathrm{d}(\varepsilon_e*\boldsymbol{m})&=\int\int f(xy)\,\mathrm{d}\varepsilon_e(x)\,\mathrm{d}\boldsymbol{m}(y)=\\&=\int \mathrm{d}\boldsymbol{m}(y)\int f(xy)\,\mathrm{d}\varepsilon_e(x)=\int f(y)\,\mathrm{d}\boldsymbol{m}(y)\,,\end{aligned}$$

therefore $\boldsymbol{m}*\varepsilon_e=\boldsymbol{m}$ and $\varepsilon_e*\boldsymbol{m}=\boldsymbol{m}$.

24.4 PROPOSITION. *If $\boldsymbol{m}$ and $\boldsymbol{n}$ are bounded, then $\boldsymbol{m}*\boldsymbol{n}$ exists. The measure $\boldsymbol{m}*\boldsymbol{n}$ is bounded and*

$$\|\boldsymbol{m}*\boldsymbol{n}\|\leqslant\|\boldsymbol{m}\|\,\|\boldsymbol{n}\|\,.$$

Denote $\mu=|\boldsymbol{m}|$ and $\nu=|\boldsymbol{n}|$. By hypothesis it follows that μ and ν are bounded, therefore $\mu\otimes\nu$ is bounded. If $f: G\to R$ is continuous and bounded, then the function $f(xy)$ is continuous and bounded on $G\times G$, therefore it is $\mu\otimes\nu$-integrable. In particular, if $f\in\mathscr{K}(G)$, then $f(xy)$ is $\mu\otimes\nu$-integrable; therefore, $\mu*\nu$ makes sense and consequently $\boldsymbol{m}*\boldsymbol{n}$ makes sense too.

For every $f\in\mathscr{K}(G)$ we have

$$(\mu*\nu)(f)=\int\int f(xy)\,\mathrm{d}\mu\otimes\nu\,,$$

therefore

$$|(\mu*\nu)(f)| \leqslant ||f|| \, ||\mu \otimes \nu||,$$

$$|(\mu * v)(f)| \leq \|f\| \, \|\mu \otimes v\|,$$

whence

$$||\mu*\nu|| \leqslant ||\mu \otimes \nu|| < +\infty .$$

We deduce that $\mu*\nu$ is bounded. From the inequality

$$|\boldsymbol{m}*\boldsymbol{n}| \leqslant |\boldsymbol{m}|*|\boldsymbol{n}| = \mu*\nu$$

we deduce that $|\boldsymbol{m}*\boldsymbol{n}|$ is bounded, that is, $\boldsymbol{m}*\boldsymbol{n}$ is bounded.

From the equality

$$(\boldsymbol{m}*\boldsymbol{n})(f) = \int\int f(xy)\, \mathrm{d}\boldsymbol{m}\, \mathrm{d}\boldsymbol{n}, \quad \text{for } f \in \mathscr{K}(G)$$

we deduce, as before, that

$$||\boldsymbol{m}*\boldsymbol{n}|| \leqslant ||\boldsymbol{m} \otimes \boldsymbol{n}|| .$$

But $||\boldsymbol{m} \otimes \boldsymbol{n}|| \leqslant ||\boldsymbol{m}|| \, ||\boldsymbol{n}||$, therefore

$$||\boldsymbol{m}*\boldsymbol{n}|| \leqslant ||\boldsymbol{m}|| \, ||\boldsymbol{n}|| .$$

It is easy to prove that if $\boldsymbol{m}*\boldsymbol{n}$ and $\boldsymbol{m}'*\boldsymbol{n}$ both exist and α and β are scalars, then $(\alpha \boldsymbol{m} + \beta \boldsymbol{m}')*\boldsymbol{n}$ exists and

$$(\alpha \boldsymbol{m} + \beta \boldsymbol{m}')*\boldsymbol{n} = \alpha(\boldsymbol{m}*\boldsymbol{n}) + \beta(\boldsymbol{m}'*\boldsymbol{n}),$$

therefore the set of dominated measures $\boldsymbol{m}$ for which $\boldsymbol{m}*\boldsymbol{n}$ exists is a vector space and the convolution $\boldsymbol{m}*\boldsymbol{n}$ is linear with respect to $\boldsymbol{m}$.

Also for a given measure $\boldsymbol{m}$ the set of the measures $\boldsymbol{n}$ for which the convolution $\boldsymbol{m}*\boldsymbol{n}$ exists, is a vector space and $\boldsymbol{m}*\boldsymbol{n}$ is linear in $\boldsymbol{n}$.

In particular, $\boldsymbol{m}*\boldsymbol{n}$ is a bilinear mapping of the product $\mathscr{M}_X^1(G) \times \mathscr{M}_Y^1(G)$ into $\mathscr{M}_Z^1(G)$.

24.5 Proposition. *The convolution is commutative if and only if the group G is commutative.*

The measures $\boldsymbol{m}*\boldsymbol{n}$ and $\boldsymbol{n}*\boldsymbol{m}$ are defined by the equalities:

$$(\boldsymbol{m}*\boldsymbol{n})(f) = \int\int f(xy)\, \mathrm{d}\boldsymbol{m}(x)\, \mathrm{d}\boldsymbol{n}(y),$$

$$(\boldsymbol{n}*\boldsymbol{m})(f) = \int\int f(xy)\, \mathrm{d}\boldsymbol{n}(x)\, \mathrm{d}\boldsymbol{m}(y) = \int\int f(xy)\, \mathrm{d}\boldsymbol{m}(y)\, \mathrm{d}\boldsymbol{n}(x) =$$

$$= \int\int f(yx)\, \mathrm{d}\boldsymbol{m}(x)\, \mathrm{d}\boldsymbol{n}(y)$$

for $f \in \mathscr{K}(G)$.

If G is commutative, then $f(xy)=f(yx)$, therefore $\boldsymbol{m}*\boldsymbol{n}=\boldsymbol{n}*\boldsymbol{m}$.

Conversely, assume that $\boldsymbol{m}*\boldsymbol{n}=\boldsymbol{n}*\boldsymbol{m}$ for every measures $\boldsymbol{m}$ and $\boldsymbol{n}$ for which $\boldsymbol{m}*\boldsymbol{n}$ and $\boldsymbol{n}*\boldsymbol{m}$ exist, and show that G is commutative. For every functions $g, h\in\mathcal{K}(G)$ we consider the real measures μ and ν defined by the equalities

$$\mathrm{d}\mu = g\,\mathrm{d}x, \quad \mathrm{d}\nu = h\,\mathrm{d}x\,.$$

For every function $f\in\mathcal{K}(G)$ we have

$$(\mu*\nu)(f)=\int\int f(xy)g(x)h(y)\,\mathrm{d}x\,\mathrm{d}y\,,$$

$$(\nu*\mu)(f)=\int\int f(xy)h(x)g(y)\,\mathrm{d}x\,\mathrm{d}y=\int\int f(yx)\,\mathrm{d}\boldsymbol{m}(x)\,\mathrm{d}\boldsymbol{n}(y)$$

therefore

$$\int\int[f(xy)-f(yx)]g(x)h(y)\,\mathrm{d}x\,\mathrm{d}y=(\mu*\nu)(f)-(\nu*\mu)(f)=0\,.$$

Since the finite linear combinations of the form $\Sigma g_i(x)h_i(y)$ with g_i, $h_i\in\mathcal{K}(G)$ are dense in $\mathcal{K}(G\times G)$, for the topology of the uniform convergence (lemma 22.1), we deduce

$$\int\int \Phi(x, y)[f(xy)-f(yx)]\,\mathrm{d}x\,\mathrm{d}y=0$$

for every $\Phi\in\mathcal{K}(G\times G)$ and $f\in\mathcal{K}(G)$.

This means that for each function $f\in\mathcal{K}(G)$ the measure $[f(xy)-f(yx)]\,\mathrm{d}x\,\mathrm{d}y$ vanishes identically, therefore $f(xy)-f(yx)=0$ almost everywhere with respect to the measure $\mathrm{d}x\,\mathrm{d}y$. Since the function $f(xy)-f(yx)$ is continuous and the support of the measure $\mathrm{d}x\,\mathrm{d}y$ is $G\times G$, we deduce that

$$f(xy)-f(yx)=0$$

for every $f\in\mathcal{K}(G)$.

If there existed two points $x,y\in G$ with $xy\neq yx$, then choosing an open neighbourhood V of xy which does not contain yx, we can select a function $f\in\mathcal{K}(G)$ with $f(xy)=1$ and $f(z)=0$ for $z\notin V$; in particular $f(yx)=0$, therefore $f(xy)\neq f(yx)$ which contradicts the above conclusion.

Thus, we have $xy=yx$ for every $x, y\in G$, consequently G is commutative.

24.6 PROPOSITION. *For every pair of points $s, t\in G$ we have*

$$\varepsilon_s*\varepsilon_t=\varepsilon_{st}\,.$$

If $s\neq t$ then $\|\varepsilon_s-\varepsilon_t\|=2$.

For $f \in \mathscr{K}(G)$ we have

$$(\varepsilon_s * \varepsilon_t)(f) = \int\int f(xy)\,\mathrm{d}\varepsilon_s(x)\,\mathrm{d}\varepsilon_t(y) = \int \mathrm{d}\varepsilon_s(x) \int f(xy)\,\mathrm{d}\varepsilon_t(y) =$$
$$= \int f(xt)\,\mathrm{d}\varepsilon_s(x) = f(st) = \varepsilon_{st}(f),$$

therefore $\varepsilon_s * \varepsilon_t = \varepsilon_{st}$.

If $s \neq t$, there exists a function $f \in \mathscr{K}(G)$ with $f(s) = 1$, $f(t) = -1$ and $\|f\| = 1$. We have then

$$\|\varepsilon_s - \varepsilon_t\| \geqslant |\varepsilon_s(f) - \varepsilon_t(f)| = 1 + 1 = 2.$$

On the other hand

$$\|\varepsilon_s - \varepsilon_t\| \leqslant \|\varepsilon_s\| + \|\varepsilon_t\| = 2,$$

therefore

$$\|\varepsilon_s - \varepsilon_t\| = 2.$$

2 *Integration with respect to the convolution of two measures*

Let μ and ν be two *positive* measures on G such that $\mu * \nu$ exists. This means that for every function $\varphi \in \mathscr{K}(G)$, the function $\varphi(xy)$ is $\mu \otimes \nu$-integrable and

$$(\mu * \nu)(\varphi) = \int\int \varphi(xy)\,\mathrm{d}\mu(x)\,\mathrm{d}\nu(y).$$

Denote by $\alpha : G \times G \to G$ the product in G, $\alpha(x, y) = xy$. Observe that α is continuous on $G \times G$, therefore it is $\mu \otimes \nu$-measurable; and the condition that φ is $\mu * \nu$-integrable means that the composition $\varphi \circ \alpha$ is $\mu \otimes \nu$-integrable.

Thus, to say that $\mu * \nu$ exist is equivalent to saying that the function α is $\mu \otimes \nu$-proper; and then the convolution $\mu * \nu$ is the image by α of the measure $\mu \otimes \nu$:

$$\mu * \nu = \alpha(\mu \otimes \nu).$$

From the results concerning images of measure in §20 we deduce the following properties of the convolution $\mu * \nu$.

24.7 PROPOSITION. *For every function $f \geqslant 0$ defined on G we have*

$$(\mu * \nu)^*(f) = \int \int^* f(xy)\,\mathrm{d}\mu(x)\,\mathrm{d}\nu(y).$$

We use proposition 20.9.

24.8 COROLLARY. *For every set $A \subset G$ we have*

$$(\mu * \nu)^*(A) = (\mu \otimes \nu)^* \{(x, y) \mid xy \in A\}.$$

24.9 Corollary. *We have* $||\mu * \nu|| = ||\mu \otimes \nu|| = ||\mu|| \, ||\nu||$.

In fact, in corollary 24.8 we take $A = G$.

24.10 Proposition. *A function* f *defined on G with values in a Banach space E or in* $\bar{R}$ *is* $\mu * \nu$*-negligible if and only if the function* $f(xy)$ *is* $\mu \otimes \nu$*-negligible.*

We use proposition 20.11

24.11 Corollary. *A set* $A \subset G$ *is* $\mu * \nu$*-negligible if and only if the set* $\{(x, y) \mid xy \in A\} \subset G \times G$ *is* $\mu \otimes \nu$*-negligible.*

24.12 Corollary. *If A is the support of the measure* μ *and B is the support of the measure* ν, *then* $\overline{AB}$ *is the support of the measure* $\mu * \nu$.

We use corollary 22.28 and corollary 20.14.

24.13 Corollary. *If* μ *and* ν *have compact support, then* $\mu * \nu$ *has compact support.*

24.14 Proposition. *A mapping* f *of G into a topological space is* $\mu * \nu$*-measurable, if and only if the function* $f(xy)$ *is* $\mu \otimes \nu$*-measurable.*

We use proposition 20.15.

24.15 Corollary. *A set* $A \subset G$ *is* $\mu * \nu$*-measurable if and only if the set* $\{(x, y) \mid xy \in A\}$ *is* $\mu \otimes \nu$*-measurable.*

24.16 Proposition. *A function* f *defined on G with values in a Banach space E or in* $\bar{R}$ *is* $\mu * \nu$*-integrable if and only if* $f(xy)$ *is* $\mu \otimes \nu$*-integrable. In this case we have*

$$\int f \, \mathrm{d}(\mu * \nu) = \int \int f(xy) \, \mathrm{d}\mu(x) \, \mathrm{d}\nu(y) .$$

We use theorem 20.17.

24.17 Proposition. *The convolution* $\mu * \nu$ *exists if and only if for every compact set* $K \subset G$, *the set* $\{(x, y) \mid xy \in K\} \subset G \times G$ *is* $\mu \otimes \nu$*-integrable.*

We use proposition 20.19.

Let $\boldsymbol{m} : \mathscr{K}(G) \to X$ and $\boldsymbol{n} : \mathscr{K}(G) \to Y$ be two dominated measures such that $|\boldsymbol{m}| * |\boldsymbol{n}|$ exists. Then the convolution $\boldsymbol{m} * \boldsymbol{n} = \alpha(\boldsymbol{m} \otimes \boldsymbol{n})$ also exists.

The inequalities $|\alpha(\boldsymbol{m} \otimes \boldsymbol{n}) \leqslant \alpha(|\boldsymbol{m} \otimes \boldsymbol{n}|) \leqslant \alpha(|\boldsymbol{m}| \otimes |\boldsymbol{n}|)$ are written

$$|\boldsymbol{m} * \boldsymbol{n}| \leqslant \alpha(|\boldsymbol{m} \otimes \boldsymbol{n}|) \leqslant |\boldsymbol{m}| * |\boldsymbol{n}|,$$

whence, for every function $\varphi \geqslant 0$,

$$\int^* \varphi \, \mathrm{d} |\boldsymbol{m} * \boldsymbol{n}| \leqslant \iint^* \varphi(xy) \, \mathrm{d} |\boldsymbol{m} \otimes \boldsymbol{n}| \leqslant \int^* \varphi \, \mathrm{d}(|\boldsymbol{m}| * |\boldsymbol{n}|).$$

It follows that

$$\|\boldsymbol{m} * \boldsymbol{n}\| \leqslant \| \, |\boldsymbol{m} * \boldsymbol{n}| \, \| \leqslant \| \, |\boldsymbol{m} \otimes \boldsymbol{n}| \, \| \leqslant \| \, |\boldsymbol{m}| * |\boldsymbol{n}| \, \| = \| \, |\boldsymbol{m}| \, \| \cdot \| \, |\boldsymbol{n}| \, \|.$$

24.18 PROPOSITION. *Let f be a scalar function defined on G.*

(1) *If $f(xy)$ is $\boldsymbol{m} \otimes \boldsymbol{n}$-negligible, then f is $\boldsymbol{m} * \boldsymbol{n}$-negligible;*

(2) *If $f(xy)$ is $\boldsymbol{m} \otimes \boldsymbol{n}$-measurable, then f is $\boldsymbol{m} * \boldsymbol{n}$-measurable;*

(3) *If $f(xy)$ is $\boldsymbol{m} \otimes \boldsymbol{n}$-integrable, then f is $\boldsymbol{m} * \boldsymbol{n}$-integrable and*

$$\int f \, \mathrm{d}(\boldsymbol{m} * \boldsymbol{n}) = \iint f(xy) \, \mathrm{d}\boldsymbol{m}(x) \, \mathrm{d}\boldsymbol{n}(y).$$

(*see theorem* 20.20).

Let μ be a scalar measure on G and $\boldsymbol{n} : \mathscr{K}(G) \to Y$ a dominated measure such that the convolution $|\mu| * \boldsymbol{n}$ exists; then the convolution $\mu * \boldsymbol{n} = \alpha(\mu \otimes \boldsymbol{n}) : \mathscr{K}(G) \to Y$ also exists.

24.19 PROPOSITION. *Let $\boldsymbol{f} : G \to X$ be a function*

(1) *If $\boldsymbol{f}(xy)$ is $\mu \otimes \boldsymbol{n}$-negligible, then $\boldsymbol{f}$ is $\mu * \boldsymbol{n}$-negligible;*

(2) *If $\boldsymbol{f}(xy)$ is $\mu \otimes \boldsymbol{n}$-measurable, then $\boldsymbol{f}$ is $\mu * \boldsymbol{n}$-measurable;*

(3) *If $\boldsymbol{f}(xy)$ is $\mu \otimes \boldsymbol{n}$-integrable, then $\boldsymbol{f}$ is $\mu * \boldsymbol{n}$-integrable and*

$$\int \boldsymbol{f} \, \mathrm{d}(\mu * \boldsymbol{n}) = \iint \boldsymbol{f}(xy) \, \mathrm{d}\mu(x) \, \mathrm{d}\boldsymbol{n}(y).$$

We can state similar proposition for $\boldsymbol{n} * \mu$ and $\boldsymbol{n} \otimes \mu$.

3 *Convolution of a measure with a function*

Let $\boldsymbol{m} : \mathscr{K}(G) \to X$ be a dominated measure and $\boldsymbol{f}$ a function defined almost everywhere on G with values in Y.

If $\boldsymbol{f}$ is locally integrable, then denoting $\boldsymbol{n} = \boldsymbol{f} \mathrm{d}y$, the problem of the convolutions $\boldsymbol{m} * \boldsymbol{n}$ and $\boldsymbol{n} * \boldsymbol{m}$ arises. We shall show that, under certain conditions, the measure $\boldsymbol{m}*\boldsymbol{f}\,\mathrm{d}y$ exists and is of bases $\mathrm{d}y$, that is, of the form $\boldsymbol{m}*\boldsymbol{f}\,\mathrm{d}y=\boldsymbol{g}\,\mathrm{d}y$, where the function $\boldsymbol{g}$ is defined by the equality

$$\boldsymbol{g}(y)=\int \boldsymbol{f}(x^{-1}y)\mathrm{d}\boldsymbol{m}(x),$$

and that the measure $\boldsymbol{f}\,\mathrm{d}y*\boldsymbol{m}$ exists and is of bases $\mathrm{d}y$, that is of the form $\boldsymbol{f}\,\mathrm{d}y*\boldsymbol{m}=\boldsymbol{h}\,\mathrm{d}y$, where the function $\boldsymbol{h}$ is defined by the equality

$$\boldsymbol{h}(y)=\int \boldsymbol{f}(yx^{-1})\Delta(x^{-1})\mathrm{d}\boldsymbol{m}(x).$$

Observe that if $\boldsymbol{f}$ is defined almost everywhere, then, for each $y\in G$, the functions $x\to\boldsymbol{f}(x^{-1}y)$ and $x\to\boldsymbol{f}(yx^{-1})$ are defined also almost everywhere.

24.20 Definition. (a) *If the set A of points $y\in G$ such that the function $x\to\boldsymbol{f}(x^{-1}y)$ is $\boldsymbol{m}$-integrable, is not empty, then the function $\boldsymbol{m}*\boldsymbol{f}: A\to Z$ defined by the equality.*

$$(\boldsymbol{m}*\boldsymbol{f})(y)=\int \boldsymbol{f}(x^{-1}y)\mathrm{d}\boldsymbol{m}(x),\ \textit{for}\ \ y\in A,$$

is called the convolution of the measure $\boldsymbol{m}$ *with the function* $\boldsymbol{f}$.

(b) *If the set B of the points $y\in G$ such that the function $x\to\boldsymbol{f}(yx^{-1})\Delta(x^{-1})$ is $\boldsymbol{m}$-integrable, is not empty, then the function $\boldsymbol{f}*\boldsymbol{m}: B\to Z$ defined by the equality*

$$(\boldsymbol{f}*\boldsymbol{m})(y)=\int \boldsymbol{f}(yx^{-1})\Delta(x^{-1})\mathrm{d}\boldsymbol{m}(x),\ \textit{for}\ \ y\in B,$$

is called the convolution of the function $\boldsymbol{f}$ *with the measure* $\boldsymbol{m}$.

Remarks.

1. To say that the function $x\to\boldsymbol{f}(x^{-1}y)$ is $\boldsymbol{m}$-integrable means that this function is $|\boldsymbol{m}|$-integrable. Thus the functions $|\boldsymbol{m}|*\boldsymbol{f}$ and $\boldsymbol{m}*\boldsymbol{f}$ are defined on the same set A.

Also, the functions $\boldsymbol{f}*\boldsymbol{m}$ and $\boldsymbol{f}*|\boldsymbol{m}|$ are defined on the same set B.

2. If the function $x\to\boldsymbol{f}(x^{-1}y)$ is $\boldsymbol{m}$-measurable for every $y\in G$, in particular if $\boldsymbol{f}$ is a Borel function, to say that the function $x\to\boldsymbol{f}(x^{-1}y)$ is $|\boldsymbol{m}|$-integrable, means that the function $x\to|\boldsymbol{f}(x^{-1}y)|$ is $|\boldsymbol{m}|$-integrable. Thus, in this case, the functions $\boldsymbol{m}*\boldsymbol{f}$, $|\boldsymbol{m}|*\boldsymbol{f}$ and $|\boldsymbol{m}|*|\boldsymbol{f}|$ are defined on the same set A.

Similarly, if the function $x\to\boldsymbol{f}(yx^{-1})$ is measurable for every $y\in G$, the functions $\boldsymbol{f}*\boldsymbol{m}$, $\boldsymbol{f}*|\boldsymbol{m}|$ and $|\boldsymbol{f}|*|\boldsymbol{m}|$ are defined on the same set B.

3. If $\boldsymbol{m}$ is *bounded* and $\boldsymbol{f}$ is a *bounded Borel* function defined everywhere,

then the convolution $\boldsymbol{m}*\boldsymbol{f}$ is defined on the whole group G, because for each $y\in G$ the function $x\to\boldsymbol{f}(x^{-1}y)$ is a bounded Borel function, therefore it is $\boldsymbol{m}$-integrable.

If, in addition, G is unimodular, then the function $\boldsymbol{f}*\boldsymbol{m}$ is also defined on G.

The function $\boldsymbol{f}$ is locally integrable, but it is possible that the convolutions of measures $\boldsymbol{m}*\boldsymbol{f}\,\mathrm{d}y$ and $\boldsymbol{f}\,\mathrm{d}y*\boldsymbol{m}$ make no sense.

4. If $\boldsymbol{m}$ has compact support and $\boldsymbol{f}$ is a bounded (or more general bounded on the support of $\boldsymbol{m}$) Borel function, then $\boldsymbol{m}*\boldsymbol{f}$ and $\boldsymbol{f}*\boldsymbol{m}$ are defined everywhere on G (even if G is not unimodular), and the measures $\boldsymbol{m}*\boldsymbol{f}\,\mathrm{d}y$ and $\boldsymbol{f}\,\mathrm{d}y*\boldsymbol{m}$ exist.

The following proposition gives the connection between the measures $\boldsymbol{m}*\boldsymbol{f}\,\mathrm{d}y$ and $\boldsymbol{f}\,\mathrm{d}y*\boldsymbol{m}$ and the functions $\boldsymbol{m}*\boldsymbol{f}$ and $\boldsymbol{f}*\boldsymbol{m}$.

24.21 PROPOSITION. *Assume that* $\boldsymbol{m}$ *is bounded and* $\boldsymbol{f}$ *is locally integrable.*

(1) *If the measure* $\boldsymbol{m}*\boldsymbol{f}\,\mathrm{d}y$ *exists and if the function* $(x, y)\to\boldsymbol{f}(x^{-1}y)$ *is* $|\boldsymbol{m}|\otimes\mathrm{d}y$*-measurable (in particular if* $\boldsymbol{f}$ *is a Borel function) then the function* $x\to\boldsymbol{f}(x^{-1}y)$ *is* $\boldsymbol{m}$*-integrable for almost every* $y\in G$*, the function*

$$(\boldsymbol{m}*\boldsymbol{f})(y)=\int\boldsymbol{f}(x^{-1}y)\,\mathrm{d}\boldsymbol{m}(x)$$

defined almost everywhere is locally integrable and

$$\boldsymbol{m}*(\boldsymbol{f}\,\mathrm{d}y)=(\boldsymbol{m}*\boldsymbol{f})\,\mathrm{d}y\,.$$

(2) *If the measure* $\boldsymbol{f}\,\mathrm{d}y*\boldsymbol{m}$ *exists and if the function* $(x, y)\to\boldsymbol{f}(yx^{-1})$ *is* $\mathrm{d}y\otimes|\boldsymbol{m}|$*-measurable (in particular if* $\boldsymbol{f}$ *is a Borel function), then the function* $x\to\boldsymbol{f}(yx^{-1})\Delta(x^{-1})$ *is* $\boldsymbol{m}$*-integrable for almost every* $y\in G$*, the function*

$$(\boldsymbol{f}*\boldsymbol{m})(y)=\int\boldsymbol{f}(yx^{-1})\,\Delta(x^{-1})\,\mathrm{d}\boldsymbol{m}(x)$$

defined almost everywhere is locally integrable and

$$(\boldsymbol{f}\,\mathrm{d}y)*\boldsymbol{m}=(\boldsymbol{f}*\boldsymbol{m})\,\mathrm{d}y\,.$$

Denote $\boldsymbol{n}=\boldsymbol{f}\,\mathrm{d}y$, $\mu=|\boldsymbol{m}|$ and $\nu=|\boldsymbol{n}|=|\boldsymbol{f}|\,\mathrm{d}y$. Assume first that $\boldsymbol{m}*\boldsymbol{n}$ exists. This means that $\mu*\nu$ exists.

Let $\varphi\geqslant 0$ be a function of $\mathscr{K}(G)$. The function $\varphi(xy)$ is $\mu\otimes\nu$-integrable. By the Fubini theorem and the left invariance of the Haar measure, we deduce

$$\int \varphi \, d(\mu * \nu) = \int\int \varphi(xy) d\mu(x) |f(y)| dy = \int d\mu(x) \int \varphi(xy) |f|(y) dy =$$
$$= \int d\mu(x) \int \varphi(y) |f(x^{-1}y)| dy =$$
$$= \int d\mu(x) \int |f(x^{-1}y)| (\varphi(y) dy) .$$

Since, by hypothesis (1), the function $|f(x^{-1}y)|$ is $\mu \otimes dy$-measurable, the function $|f(x^{-1}y)| \varphi(y)$ is also $\mu \otimes dy$-measurable, therefore $|f(x^{-1}y)|$ is measurable with respect to the bounded measure $\varphi(y)(\mu \otimes dy) = \mu \otimes \varphi \, dy$. Since the iterated integral of the function $|f(x^{-1}y)|$ with respect to μ and $\varphi \, dy$ is finite, it follows that $f(x^{-1}y)$ is $\mu \otimes \varphi \, dy$-integrable and

$$\int d\mu(x) \int |f(x^{-1}y)| (\varphi(y) dy) = \int \varphi(y) dy \int |f(x^{-1}y)| d\mu(x) .$$

This means that for each function $\varphi \in \mathscr{K}(G)$ the function $x \to f(x^{-1}y)$ is μ-integrable, except for the points y of a $\varphi \, dy$-negligible set, and the function

$$y \to \int |f(x^{-1}y)| d\mu(x)$$

defined $\varphi \, dy$-almost everywhere is $\varphi \, dy$-integrable.

The set N of the points y for which the preceding integral does not exist, is independent of φ and is $\varphi \, dy$-negligible for every $\varphi \in \mathscr{K}(G)$. It follows that N is negligible for the Haar measure dy. Thus the function $(\mu * |f|)(y) = \int |f(x^{-1}y)| dy \, \mu(x)$ is defined dy-almost everywhere and it is $\varphi \, dy$-integrable for every $\varphi \in \mathscr{K}(G)$, therefore $(\mu * f)\varphi$ is dy-integrable for every $\varphi \in \mathscr{K}(G)$, that is $\mu * |f|$ is locally integrable for the Haar measure. By the Fubini theorem and the above results, we deduce that for every function $\varphi \in \mathscr{K}(G)$ we have

$$\int \varphi \, d(\boldsymbol{m} * \boldsymbol{n}) = \int \int \varphi(xy) d\boldsymbol{m}(x) |f(y)| dy = \int d\boldsymbol{m}(x) \int \varphi(xy) |f(y)| dy =$$
$$= \int d\boldsymbol{m}(x) \int \varphi(y) f(x^{-1}y) dy = \int \varphi(y) dy \int f(x^{-1}y) d\boldsymbol{m}(x) =$$
$$= \int \varphi(y) (\boldsymbol{m} * f)(y) dy ,$$

therefore $\boldsymbol{m} * \boldsymbol{n} = (\boldsymbol{m} * f) dy$ that is $\boldsymbol{m} * (f \, dy) = (\boldsymbol{m} * f) dy$.

Assume now that $\boldsymbol{n} * \boldsymbol{m}$ exists, that is, $\nu * \mu$ exists. Let $\varphi \geqslant 0$ be a function of $\mathscr{K}(G)$. The function $\varphi(xy)$ is $\mu \otimes \nu$-integrable and as before,

$$\int \varphi \, d(\nu * \mu) = \int \int \varphi(yx) d\mu(x) |f(y)| dy = \int d\mu(x) \int \varphi(yx) |f(y)| dy =$$
$$= \int d\mu(x) \int \varphi(y) |f(yx^{-1})| \Delta(x^{-1}) dy =$$
$$= \int d\mu(x) \int |f(yx^{-1})| \Delta(x^{-1}) (\varphi(y) dy) .$$

The function $|f(yx^{-1})| \Delta(x^{-1})$ is measurable with respect to the bounded measure $\mu\otimes\varphi\,dy$ and the iterated integral is finite. It follows that the function $|f(yx^{-1})|\Delta(x^{-1})$ is integrable with respect to $m\otimes\varphi\,dy$ and

$$\int dm(x)\int f(yx^{-1})\Delta(x^{-1})(\varphi(y)dy)=\int\varphi(y)\,dy\int f(yx^{-1})\Delta(x^{-1})dm(x)\,.$$

As before, we deduce that the function $x\to f(yx^{-1})\Delta(x^{-1})$ is m-integrable, except for the points y of a negligible set and the function

$$y\to\int f(yx^{-1})\Delta(x^{-1})\,dm(x)$$

defined almost everywhere is locally integrable; therefore, the function $f*m$, defined almost everywhere by the equality $(f*m)(y)=\int f(yx^{-1})\Delta(x^{-1})dx$ is locally integrable and

$$\int\varphi\,d(n*m)=\int\int\varphi(yx)\,dm(x)\,f(y)dy=\int dm(x)\int\varphi(yx)\,f(y)dy=$$

$$=\int dm(x)\int\varphi(y)f(yx^{-1})\Delta(x^{-1})dy=$$

$$=\int\varphi(y)dy\int f(yx^{-1})\Delta(x^{-1})\,dm(x)=\int\varphi(y)(f*m)(y)dy$$

therefore $n*m=(f*m)dy$, that is $(f\,dy)*m=(f\;*\;m)dy$.

Remarks.

1. The proposition remains true if m is not bounded provided that G is the union of a sequence of open m-integrable sets, since in this case corollary 22.17 can still be applied.

2. Let $f':G\to X$ be a function almost everywhere equal to f.

If f and f' fulfill the conditions of point (1) in the preceding proposition, then

$$m*f=m*f',\ \text{almost everywhere}\,,$$

If f and f' fulfill the conditions of point (2) in the preceding proposition, then

$$f*m=f'*m,\ \text{almost everywhere.}$$

In fact, in the first case, $m*f\,dy=m*f'dy$, $m*f\,dy=(m*f)dy$ and $m*f'dy=(m*f')dy$ therefore $(m*f)dy=(m*f')dy$, hence $m*f=m*f'$ almost everywhere (proposition 17.5). The second case is proved in a similar way.

3. If m has compact support, or if f has compact support, then the convolutions $m*f\,dy$ and $f\,dy*m$ exist.

If m is bounded and $f\in\mathscr{L}^1_Y$, then the measure $f\,dy$ is bounded, therefore $m*f\,dy$ and $f\,dy*m$ exist. We have a more general result:

24.22 PROPOSITION. *If* $\boldsymbol{m}$ *is bounded and if* $f \in \mathscr{L}_Y^p$, $1 \leqslant p \leqslant +\infty$, *then the measures* $\boldsymbol{m} * f \mathrm{d}y$ *and* $f \mathrm{d}y * \boldsymbol{m}$ *exist and we have*

$$\|\boldsymbol{m} * f \mathrm{d}y\| \leqslant \|\boldsymbol{m}\| N_p(f), \quad \|f \mathrm{d}y * \boldsymbol{m}\| \leqslant \|\boldsymbol{m}\| N_p(f).$$

If, in addition, the function $f(x^{-1}y)$ *(respectively* $f(yx^{-1})$*) is* $|\boldsymbol{m}| \otimes \mathrm{d}y$*-measurable, in particular, if* f *is a Borel function, then the function* $\boldsymbol{m} * f$ *(respectively* $f * \boldsymbol{m}$*) is defined almost everywhere, we have* $\boldsymbol{m} * f \in \mathscr{L}_Z^p$ *(respectively* $f * \boldsymbol{m} \in \mathscr{L}_Z^p$*) and*

$$N_p(\boldsymbol{m} * f) \leqslant \|\boldsymbol{m}\| N_p(f), \quad \boldsymbol{m} * (f \mathrm{d}y) = (\boldsymbol{m} * f) \mathrm{d}y$$

(respectively $N_p(f * \boldsymbol{m}) \leqslant \|\boldsymbol{m}\| N_p(f)$, $(f \mathrm{d}y) * \boldsymbol{m} = (f * \boldsymbol{m}) \mathrm{d}y$.*)*

If g *is equal almost everywhere to* f *and fulfills the above conditions, then* $\boldsymbol{m} * f = \boldsymbol{m} * g$ *(respectively* $f * \boldsymbol{m} = g * \boldsymbol{m}$*) almost everywhere.*

Denote $\mu = |\boldsymbol{m}|$ and show that $\mu * |f| \mathrm{d}y$ exists. Let $\varphi \geqslant 0$ be a function of $\mathscr{K}(G)$. Denote $K_1 = \mathrm{pr}_1 C$, $K_2 = \mathrm{pr}_2 C$ and $K = K_1 K_2$. The set $K \subset G$ is compact and for $(x, y) \in C$ we have $x \in K_1$ and $y \in K_2$, therefore $xy \in K_1 K_2$, hence $\varphi_C(x, y) \leqslant \varphi_K(xy)$.

The function $\varphi(xy) \varphi_{K \times K}(x, y)$ has compact support contained in $K \times K$, and its restriction to $K \times K$ is continuous, therefore it is integrable for the measure $\mu \otimes |f| \mathrm{d}y$.

Denote

$$I_K = \int\int_{K \times K} \varphi(xy) \mathrm{d}\mu(x) |f(y)| \mathrm{d}y.$$

The Fubini theorem can be applied and we have

$$I_K = \int_K \mathrm{d}\mu(x) \int_K \varphi(xy) |f(y)| \mathrm{d}y.$$

Since $f \in \mathscr{L}_Y^p$ for each $x \in G$, the function $y \to \varphi(xy) |f(y)|$ is integrable with respect to the Haar measure $\mathrm{d}y$ and

$$\int \varphi(xy) |f(y)| \mathrm{d}y \leqslant N_q(\varphi) N_p(f).$$

It follows that

$$I_K \leqslant \int_K^* \mathrm{d}\mu(x) \int \varphi(xy) |f(y)| \mathrm{d}y \leqslant N_q(\varphi) N_p(f) \|\mu\|,$$

whence

$$\int\int^* \varphi(xy) \mathrm{d}\mu(x) |f(y)| \mathrm{d}y = \sup_C \int\int_C \varphi(xy) \mathrm{d}\mu(x) |f(y)| \mathrm{d}y \leqslant$$

$$\leqslant \sup_K I_K \leqslant N_p(f) N_q(\varphi) \|\mu\| < +\infty.$$

The function $\varphi(xy)$ is continuous and from the preceding inequality we deduce that it is $\mu\otimes|\boldsymbol{f}|\mathrm{d}y$-integrable. It follows that $\mu*|\boldsymbol{f}|\mathrm{d}y$ exists, therefore $\boldsymbol{m}*\boldsymbol{f}\mathrm{d}y$ exists.

In order to prove that $\boldsymbol{f}\mathrm{d}y*\boldsymbol{m}$ exists, we replace in the above proof $\varphi(xy)$ by $\varphi(yx)$.

Assume now that the function $\boldsymbol{f}(x^{-1}y)$ is $|\boldsymbol{m}|\otimes\mathrm{d}y$-measurable. By proposition 24.21 we deduce that $\boldsymbol{m}*\boldsymbol{f}$ is defined almost everywhere, it is locally integrable—therefore measurable—and that $\boldsymbol{m}*(\boldsymbol{f}\mathrm{d}y)=(\boldsymbol{m}*\boldsymbol{f})\mathrm{d}y$. On the other hand, for $\varphi\in\mathscr{K}(G)$, we have

$$\int|\varphi|\,|\boldsymbol{m}*\boldsymbol{f}|\,\mathrm{d}y=\int|\varphi(y)|\,\mathrm{d}y\,|\int\boldsymbol{f}(x^{-1}y)\mathrm{d}\boldsymbol{m}(x)|\leqslant$$

$$\leqslant\int\int|\varphi(xy)|\,\mathrm{d}\mu(x)|\boldsymbol{f}(y)|\,\mathrm{d}y\leqslant N_q(\varphi)N_p(\boldsymbol{f})\,||\mu||\,,$$

whence

$$N_p(\boldsymbol{m}*\boldsymbol{f})\leqslant N_p(\boldsymbol{f})\,||\mu||<+\infty\,.$$

It follows that $\boldsymbol{m}*\boldsymbol{f}\in\mathscr{L}_Z^p$.

If the function $\boldsymbol{f}(yx^{-1})$ is $|\boldsymbol{m}|\otimes\mathrm{d}y$-measurable, we deduce that $\boldsymbol{f}*\boldsymbol{m}$ is defined almost everywhere, is measurable and for $\varphi\in\mathscr{K}(G)$ we have

$$\int|\varphi|\,|\boldsymbol{f}*\boldsymbol{m}|\,\mathrm{d}y=\int|\varphi(y)|\,\mathrm{d}y|\,\int\boldsymbol{f}(yx^{-1})\Delta(x^{-1})\mathrm{d}\boldsymbol{m}(x)|\leqslant$$

$$\leqslant\int|\varphi(y)|\,\mathrm{d}y\int|\boldsymbol{f}(yx^{-1})|\,\Delta(x^{-1})\mathrm{d}\mu(x)=$$

$$=\int\mathrm{d}\mu(x)\int|\varphi(y)|\,|\boldsymbol{f}(yx^{-1})|\,\Delta(x^{-1})\mathrm{d}y=$$

$$=\int\mathrm{d}\mu(x)\int|\varphi(yx)|\,|\boldsymbol{f}(y)|\,\mathrm{d}y\leqslant N_q(\varphi)\,N_p(\boldsymbol{f})\,||\mu||$$

whence

$$N_p(\boldsymbol{f}*\boldsymbol{m})\leqslant N_p(\boldsymbol{f})\,||\mu||<+\infty$$

consequently $\boldsymbol{f}*\boldsymbol{m}\in\mathscr{L}_Z^p$. Thus the proposition is proved.

24.23 Proposition. *For every $s\in G$ we have*

$$\varepsilon_s*\boldsymbol{f}=\boldsymbol{f}_{s^{-1}}\ \text{ and }\ \boldsymbol{f}*\varepsilon_s=\boldsymbol{f}^s\Delta(s^{-1})\,.$$

If $\boldsymbol{f}$ is locally integrable, then the functions $\varepsilon_s*\boldsymbol{f}$ and $\boldsymbol{f}*\varepsilon_s$ are locally integrable and we have

$$\varepsilon_s*(\boldsymbol{f}\,\mathrm{d}y)=(\varepsilon_s*\boldsymbol{f})\,\mathrm{d}y,\ \ (\boldsymbol{f}\,\mathrm{d}y)*\varepsilon_s=(\boldsymbol{f}*\varepsilon_s)\,\mathrm{d}y\,.$$

For every $y\in G$ the functions $x\to\boldsymbol{f}(x^{-1}y)$ and $x\to\boldsymbol{f}(yx^{-1})\Delta(x^{-1})$ are ε_s-integrable and

$$(\varepsilon_s * f)(y) = \int f(x^{-1}y)\,d\varepsilon_s(x) = f(s^{-1}y) = f_{s^{-1}}(y),$$
$$(f*\varepsilon_s)(y) = \int f(yx^{-1})\Delta(x^{-1})\,d\varepsilon_s(x) = f(yx^{-1})\Delta(s^{-1}) = f^s(y)\Delta(s^{-1})\cdot$$

therefore $\varepsilon_s * f = f_{s^{-1}}$ and $f*\varepsilon_s = f^s\Delta(s^{-1})$.

Assume now that f is locally integrable. Since the measure ε_s has compact support, the convolutions $\varepsilon_s * f\,dy$ and $f\,dy*\varepsilon_s$ exist. Since the function $f(x)$ is measurable, the function $(x, y)\to f(x^{-1}y)$ is $\varepsilon_s * dy$-measurable and the function $(x, y)\to f(yx^{-1})$ is $dy*\varepsilon_s$-measurable. By proposition 24.21 the functions $\varepsilon_s * f$ and $f*\varepsilon_s$ are locally integrable and

$$\varepsilon_s * (f\,dy) = (\varepsilon_s * f)\,dy, \quad (f\,dy) * \varepsilon_s = (f*\varepsilon_s)\,dy\,.$$

24.24 Proposition. *For every* $s\in G$ *we have*

$$(f*m)_s = f_s * m \text{ and } (m*f)^s = m*f^s\,.$$

Let $y\in G$ be such that $(f_s * m)(y)$ exists; this means that the function $x\to f_s(yx^{-1})\Delta x^{-1}$ is m-integrable, that is, the function $x\to f(syx^{-1})\Delta(x^{-1})$ is m-integrable; this means that $(f*m)(sy)$ exists, that is $(f*m)_s(y)$ exists. Thus, the functions $f_s * m$ and $(f*m)_s$ are defined on the same set. If $(f*m)_s$ exists, then

$$(f*m)_s(y) = (f*m)(sy) = \int f(syx^{-1})\Delta(x^{-1})\,dm(s) =$$
$$= \int f_s(yx^{-1})\Delta(x^{-1})\,dm(x) = f_s * m(y)\,,$$

therefore

$$(f*m)_s = f_s * m\,.$$

The other equality is proved similarly.

24.25 Proposition. *If* A *is the support of the measure* m *and* B *is the support of the function* f*, then the support of the function* $m*f$ *is contained in* $\overline{AB}$ *and the support of the function* $f*m$ *is contained in* $\overline{BA}$.

In fact, let y be a point for which there exists the integral

$$(m*f)(y) = \int f(x^{-1}y)\,dm(x)\,.$$

If $y\notin AB$, then for every $x\in A$ we have $x^{-1}y\notin B$ therefore $f(x^{-1}y)=0$, hence $\int f(x^{-1}y)\,dm(x) = 0$, that is, $(m*f)(y)=0$. It follows that the support of the function $m*f$ is contained in $\overline{AB}$. Now let y be a point for which there exists the integral

$$(f*m)(y) = \int f(yx^{-1})\Delta(x^{-1})\mathrm{d}m(x)\,.$$

If $y \notin BA$, then for every $x \in A$ we have $yx^{-1} \notin B$, therefore $f(yx^{-1}) = 0$, hence $\int f(yx^{-1})\Delta(x^{-1})\mathrm{d}m(x) = 0$, that is, $(f*m)(y) = 0$. It follows that the support of the function $f*m$ is contained in $\overline{BA}$.

24.26 Corollary. *If m and f have compact supports, then the functions $m*f$ and $f*m$ have compact supports.*

4 Convolution of bounded measures with bounded functions

Let $m : \mathscr{K}(G) \to X$ be a *bounded* measure and $f : G \to Y$ a *bounded Borel* function.

As we have remarked, the function

$$(m*f)(y) = \int f(x^{-1}y)\mathrm{d}m(x)$$

is defined everywhere on G and if G is unimodular, the function

$$(f*m)(y) = \int f(yx^{-1})\Delta(x^{-1})\mathrm{d}m(x)$$

is also defined everywhere on G.

If m has *compact support*, or if f has *compact support*, then the function

$$(f*m)(y) = \int f(yx^{-1})\Delta(x^{-1})\mathrm{d}m(x)$$

is defined everywhere on G (even if G is not unimodular).

We impose now additional conditions on the function f and we shall then show that the functions $m*f$ and $f*m$ have additional properties.

24.27 Lemma. *If m is bounded and f is a bounded Borel function then for every $\varepsilon > 0$ there exists a compact set $K \subset G$ such that for every $y \in G$ we have*

$$|\int f(x^{-1}y)\mathrm{d}m(x) - \int_K f(x^{-1}y)\mathrm{d}m(x)| < \varepsilon$$

and

$$|\int f(yx^{-1})\mathrm{d}m(x) - \int_K f(yx^{-1})\mathrm{d}m(x)| < \varepsilon\,.$$

In fact, denoting $\mu = |m|$, then, since μ is bounded, we have

$$\mu(G) = \sup_K \mu(K) < +\infty\,,\ K \text{ compact}\,.$$

There exists, therefore, a compact set $K \subset G$ such that

$$\mu(G \setminus K) < \varepsilon/\|f\|.$$

For every $y \in G$ we have

$$|\int f(x^{-1}y)\mathrm{d}\boldsymbol{m}(x) - \int_K f(x^{-1}y)\mathrm{d}\boldsymbol{m}(x)| = |\int f(x^{-1}y)\varphi_{G\setminus K}(x)\mathrm{d}\boldsymbol{m}(x)| \leqslant$$

$$\leqslant \int |f(x^{-1}y)|\varphi_{G\setminus K}(x)\mathrm{d}\mu(x) \leqslant \|f\|\mu(G\setminus K) < \varepsilon$$

and also

$$|\int f(yx^{-1})\mathrm{d}\boldsymbol{m}(x) - \int_K f(yx^{-1})\mathrm{d}\boldsymbol{m}(x)| \leqslant \|f\|\mu(G\setminus K) < \varepsilon.$$

24.28 Corollary. *If* $\boldsymbol{m}$ *has compact support and if* f *is a bounded Borel function, then for every* $\varepsilon > 0$ *there exists a compact set* $K \subset G$ *such that for every* $y \in G$ *we have*

$$|\int f(yx^{-1})\Delta(x^{-1})\mathrm{d}\boldsymbol{m}(x) - \int_K f(yx^{-1})\Delta(x^{-1})\mathrm{d}\boldsymbol{m}(x)| < \varepsilon.$$

In fact, the measure $\mathrm{d}\boldsymbol{n}(x) = \Delta(x^{-1})\mathrm{d}\boldsymbol{m}(x)$ is bounded, therefore, by lemma 24.27, for every $\varepsilon > 0$ there exists a compact set $K \subset G$ such that for every $y \in G$ we have

$$|\int f(yx^{-1})\mathrm{d}\boldsymbol{n}(x) - \int_K f(yx^{-1})\mathrm{d}\boldsymbol{n}(x)| < \varepsilon,$$

that is,

$$|\int f(yx^{-1})\Delta(x^{-1})\mathrm{d}\boldsymbol{m}(x) - \int_K f(yx^{-1})\Delta(x^{-1})\mathrm{d}\boldsymbol{m}(x)| < \varepsilon.$$

24.29 Proposition. *If* $\boldsymbol{m}$ *is bounded and if* f *is continuous and bounded then the function* $\boldsymbol{m} * f$ *is continuous and bounded.*

If, in addition, G is unimodular, or if $\boldsymbol{m}$ *has compact support, then* $f * \boldsymbol{m}$ *is continuous and bounded.*

Let $x_0 \in G$ and $\varepsilon > 0$. We shall show that there exists a neighbourhood V of x_0 such that, denoting $g = \boldsymbol{m} * f$, we have $|g(x) - g(x_0)| < \varepsilon$ for $x \in V$.

By lemma 24.27 there exists a compact set $K \subset G$ such that for every $y \in G$ we have

$$|g(y) - \int_K f(x^{-1}y)\mathrm{d}\boldsymbol{m}(x)| < \varepsilon/4.$$

Let W be a compact neighbourhood of e. The set $H = K^{-1}Wx_0$ is compact, therefore f is right uniformly continuous on H. There exists therefore a neighbourhood U of e (which can be chosen contained in W), such that for $x \in K$ and $s \in U$ we have

$$|f(x^{-1}x_0) - f(x^{-1}x_0 s)| \leqslant \varepsilon/\mu(K),$$

since $x^{-1}x_0 = x^{-1}e\,x_0 \in K^{-1}Wx_0 = H$. Then, for every $s \in U$, we have

$$|\textstyle\int_K [f(x^{-1}x_0) - f(x^{-1}x_0 s)]\,\mathrm{d}\boldsymbol{m}(x)| \leqslant$$
$$\leqslant \textstyle\int_K |f(x^{-1}x_0) - f(x^{-1}x_0 s)|\,\mathrm{d}\mu(x) \leqslant \varepsilon,$$

therefore

$$|g(x_0) - g(x_0 s)| \leqslant |g(x_0) - g(x_0 s) - \textstyle\int_K [f(x^{-1}x_0) - f(x^{-1}x_0 s)]\,\mathrm{d}\boldsymbol{m}(x)|$$
$$+ |\textstyle\int_K [f(x^{-1}x_0) - f(x^{-1}x_0 s)]\,\mathrm{d}\boldsymbol{m}(x)| \leqslant 2\varepsilon/4 + \varepsilon/2 = \varepsilon.$$

If we denote $V = x_0 U$, every $y \in V$ is of the form $y = x_0 s$ with $s \in V$, therefore for every $y \in V$ we have $|g(x_0) - g(y)| < \varepsilon$, that is, $g = \boldsymbol{m} * f$ is continuous at x_0. As x_0 is arbitrary, it follows that $\boldsymbol{m} * f$ is continuous on G. Since $\boldsymbol{m} * f \in \mathscr{L}_Z^\infty$ and it is continuous, it follows that $\boldsymbol{m} * f$ is bounded.

If G is unimodular, we show in the same way that $f * \boldsymbol{m}$ is continuous, taking $g = f * \boldsymbol{m}$ and $H = x_0 WK^{-1}$ and using the other inequality of lemma 24.27. If $\boldsymbol{m}$ has compact support, we proceed in the same way, using corollary 24.28.

24.30 PROPOSITION. *If $\boldsymbol{m}$ is bounded and f is right uniformly continuous and bounded, then $\boldsymbol{m} * f$ is bounded and right uniformly continuous.*

If $\boldsymbol{m}$ is bounded and f is bounded and left uniformly continuous and if G is unimodular or if $\boldsymbol{m}$ has compact support, then $f * \boldsymbol{m}$ is bounded and left uniformly continuous.

In fact, in the condition of the proposition, the function $g = f * \boldsymbol{m}$ or the function $h = f * \boldsymbol{m}$ is continuous and bounded by proposition 24.29.

Assume that f is right uniformly continuous and let $\varepsilon > 0$. There exists a neighbourhood V of e such that

$$|f(x^{-1}ys) - f(x^{-1}y)| < \varepsilon/||\mu|| \text{ for } s \in V \text{ and } x, y \in G.$$

It follows that for $s \in V$ and $y \in G$ we have

$$|g(ys) - g(y)| = |\textstyle\int f(x^{-1}ys)\,\mathrm{d}\boldsymbol{m}(x) - \int f(x^{-1}y)\,\mathrm{d}\boldsymbol{m}(x)| \leqslant$$
$$\leqslant \textstyle\int |f(x^{-1}ys) - f(x^{-1}y)|\,\mathrm{d}\mu(x) < \varepsilon,$$

therefore $\boldsymbol{g}=\boldsymbol{m}*\boldsymbol{f}$ is right uniformly continuous. In the same way we prove the second part of the proposition, taking V such that

$$|\boldsymbol{f}(syx^{-1})-\boldsymbol{f}(yx^{-1})|<\varepsilon/||\mu||, \text{ for } s\in V \text{ and } x, y\in G.$$

24.31 LEMMA. *Let $S\subset G$ be a closed set and c a number such that $|\boldsymbol{f}(x)|\leqslant c$ for $x\notin S$. If $\boldsymbol{m}$ is bounded and $\boldsymbol{f}$ is a bounded Borel function, then for every closed set $K\subset G$ we have*

$$|\textstyle\int_K \boldsymbol{f}(x^{-1}y)\,\mathrm{d}\boldsymbol{m}(x)|\leqslant c||\,|\boldsymbol{m}|\,||, \textit{ for } y\notin KS$$

and

$$|\textstyle\int_K \boldsymbol{f}(yx^{-1})\mathrm{d}\boldsymbol{m}(x)|\leqslant c||\,|\boldsymbol{m}|\,||, \textit{ for } y\notin SK.$$

We remark first that the integrals exist for every $y\in G$. Let $K\subset G$ be a closed set and let $y\in G$ be such that for every $x\in K$ we have $x^{-1}y\notin S$, that is such that $K^{-1}y\subset G\setminus S$. Then, denoting $\mu=|\boldsymbol{m}|$, we have

$$|\textstyle\int_K \boldsymbol{f}(x^{-1}y)\mathrm{d}\boldsymbol{m}(x)|\leqslant \int_K |\boldsymbol{f}(x^{-1}y)|\mathrm{d}\mu(x)\leqslant c||\mu||.$$

In particular, this inequality is true for $y\notin KS$, since in this case, for every $x\in K$, we have $x^{-1}y\notin S$. The second inequality is proved similarly.

24.32 PROPOSITION. *If $\boldsymbol{m}$ is bounded and if $\boldsymbol{f}\in\overline{\mathscr{K}_Y(G)}$, then $\boldsymbol{m}*\boldsymbol{f}\in\overline{\mathscr{K}_Z(G)}$. If, in addition, G is unimodular or if $\boldsymbol{m}$ has compact support, then $\boldsymbol{f}*\boldsymbol{m}\in\overline{\mathscr{K}_Z(G)}$.*

Let $\varepsilon>0$. Denote $\mu=|\boldsymbol{m}|$. By lemma 24.27, there exists a compact set $K\subset G$ such that for every $y\in G$ we have

$$|(\boldsymbol{m}*\boldsymbol{f})(y)-\textstyle\int_K \boldsymbol{f}(x^{-1}y)\,\mathrm{d}\boldsymbol{m}(x)|<\varepsilon/2.$$

Since $\boldsymbol{f}\in\mathscr{K}_Y(G)$, there exists a compact set $S\subset G$ such that for $x\notin S$ we have $|\boldsymbol{f}(x)|\leqslant\varepsilon/2||\mu||$. By lemma 24.21 we have then

$$|\textstyle\int_K f(x^{-1}y)\mathrm{d}\boldsymbol{m}(x)|<\varepsilon/2 \text{ for } y\notin KS,$$

therefore, for $y\notin KS$, we have

$$|(\boldsymbol{m}*\boldsymbol{f})(y)|\leqslant|(\boldsymbol{m}*\boldsymbol{f})(y)-\textstyle\int_K \boldsymbol{f}(x^{-1}y)\mathrm{d}\boldsymbol{m}(x)|+|\int_K \boldsymbol{f}(x^{-1})\mathrm{d}\boldsymbol{m}(x)|\leqslant\varepsilon,$$

that is, $\boldsymbol{m}*\boldsymbol{f}$ vanishes at infinity. Since $\boldsymbol{m}*\boldsymbol{f}$ is continuous, it follows that $\boldsymbol{m}*\boldsymbol{f}\in\overline{\mathscr{K}_Z(G)}$.

If G is unimodular, we prove in the same way that $\boldsymbol{m}*\boldsymbol{f}\in\overline{\mathscr{K}_Z(G)}$.

In case $\boldsymbol{m}$ has compact support, we use corollary 24.28 and lemma 24.31 for the bounded measure $\boldsymbol{n}=\check{\Delta}\boldsymbol{m}$ to prove that $\boldsymbol{f}*\boldsymbol{m}\in\overline{\mathscr{K}_Z(G)}$.

24.33 PROPOSITION. *If $\boldsymbol{m}$ has compact support and if $\boldsymbol{f}\in\mathscr{K}_Y(G)$, then $\boldsymbol{m}*\boldsymbol{f}\in\mathscr{K}_Z(G)$ and $\boldsymbol{f}*\boldsymbol{m}\in\mathscr{K}_Z(G)$.*

In fact, $\boldsymbol{m}*\boldsymbol{f}$ is continuous. If A is the support of $\boldsymbol{m}$ and B is the support of $\boldsymbol{f}$, then the support of the function $\boldsymbol{m}*\boldsymbol{f}$ is contained in the compact set AB, and the support of the function $\boldsymbol{f}*\boldsymbol{m}$ is contained in the compact set BA.

24.34 PROPOSITION. *If $\boldsymbol{m}$ is bounded and if $\boldsymbol{f}$ is a bounded Borel function, then*

$$\boldsymbol{m}(\boldsymbol{f})=(\boldsymbol{m}*\check{\boldsymbol{f}})(e)=(\check{\boldsymbol{f}}\,\check{\Delta}*\boldsymbol{m})(e)\,.$$

In fact, the function $\check{\boldsymbol{f}}(x)=\boldsymbol{f}(x^{-1})$ is also a bounded Borel function, therefore the function $x\to\check{\boldsymbol{f}}(x)$ is $\boldsymbol{m}$-integrable and

$$(\boldsymbol{m}*\check{\boldsymbol{f}})(e)=\int\check{\boldsymbol{f}}(x^{-1}e)\mathrm{d}\boldsymbol{m}(x)=\int\boldsymbol{f}(x)\mathrm{d}\boldsymbol{m}(x)=\boldsymbol{m}(\boldsymbol{f})\,,$$
$$(\check{\boldsymbol{f}}\,\check{\Delta}*\boldsymbol{m})(e)=\int\check{\boldsymbol{f}}(ex^{-1})\check{\Delta}(ex^{-1})\Delta(x^{-1})\mathrm{d}\boldsymbol{m}(x)=\int\boldsymbol{f}(x)\mathrm{d}\boldsymbol{m}(x)=\boldsymbol{m}(\boldsymbol{f})\,.$$

24.35 COROLLARY. *If $\boldsymbol{m}$ is bounded and if $\boldsymbol{m}*\boldsymbol{f}=0$ (respectively if $\boldsymbol{f}*\boldsymbol{m}=0$) for every function $\boldsymbol{f}\in\mathscr{K}_Y(G)$, then $\boldsymbol{m}=0$.*

In fact, for every $\boldsymbol{f}\in\mathscr{K}_Y$ we have $\check{\boldsymbol{f}}\in\mathscr{K}_Y$ and $\check{\boldsymbol{f}}\,\check{\Delta}\in\mathscr{K}_Y$, therefore

$$\boldsymbol{m}(\boldsymbol{f})=(\boldsymbol{m}*\check{\boldsymbol{f}})(e)=0\,,$$

respectively

$$\boldsymbol{m}(\boldsymbol{f})=(\check{\boldsymbol{f}}\,\check{\Delta}*\boldsymbol{m})(e)=0\,.$$

5 *Convolution of two functions*

Let $\boldsymbol{f}:G\setminus M\to X$ and $\boldsymbol{g}:G\setminus N\to Y$, where M and N are negligible sets.

If $\boldsymbol{f}$ is locally integrable and we set $\boldsymbol{m}=\boldsymbol{f}\,\mathrm{d}x$, we can define the convolution $\boldsymbol{m}*\boldsymbol{g}$:

$$(\boldsymbol{m}*\boldsymbol{g})(y)=\int\boldsymbol{g}(x^{-1}y)\mathrm{d}\boldsymbol{m}(x)=\int\boldsymbol{f}(x)\boldsymbol{g}(x^{-1}y)\mathrm{d}x$$

for those values of y for which the integral exists. By the left invariance of the Haar measure, we deduce that we have also

$$(\boldsymbol{m}*\boldsymbol{g})(y)=\int \boldsymbol{f}(yx)\boldsymbol{g}(x^{-1})\mathrm{d}x\,.$$

If $\boldsymbol{g}$ is locally integrable and if we set $\boldsymbol{n}=\boldsymbol{g}\,\mathrm{d}x$, we can define the convolution $\boldsymbol{f}*\boldsymbol{n}$:

$$(\boldsymbol{f}*\boldsymbol{n})(y)=\int \boldsymbol{f}(yx^{-1})\Delta(x^{-1})\mathrm{d}\boldsymbol{n}(x)=\int \boldsymbol{f}(yx^{-1})\boldsymbol{g}(x)\Delta(x^{-1})\mathrm{d}x\,.$$

By proposition 23.38, we have

$$(\boldsymbol{f}*\boldsymbol{n})(y)=\int \boldsymbol{f}(yx)\boldsymbol{g}(x^{-1})\mathrm{d}x\,;$$

therefore, if $\boldsymbol{f}$ and $\boldsymbol{g}$ are locally integrable, we have $\boldsymbol{m}*\boldsymbol{g}=\boldsymbol{f}*\boldsymbol{n}$. We are led, in this way, to the following definition:

24.36 Definition. *If the set A of points $y\in G$ for which the function $x\to \boldsymbol{f}(x)\boldsymbol{g}(x^{-1}y)$ is integrable, is nonempty, the function $\boldsymbol{f}*\boldsymbol{g}:A\to Z$ defined by the equality*

$$(\boldsymbol{f}*\boldsymbol{g})(y)=\int \boldsymbol{f}(x)\boldsymbol{g}(x^{-1}y)\mathrm{d}x=\int \boldsymbol{f}(yx)\boldsymbol{g}(x^{-1})\mathrm{d}x$$

is called the convolution of the function $\boldsymbol{f}$ with the function $\boldsymbol{g}$.

By propositions 23.29 and 23.28, we have also

$$(\boldsymbol{f}*\boldsymbol{g})(y)=\int \boldsymbol{f}(x^{-1})\boldsymbol{g}(xy)\Delta(x^{-1})\mathrm{d}x=\int \boldsymbol{f}(yx^{-1})\boldsymbol{g}(x)\Delta(x^{-1})\mathrm{d}x\,.$$

From this definition we deduce that if $\boldsymbol{f}$ is locally integrable, we have $(\boldsymbol{f}\mathrm{d}x)*\boldsymbol{g}=\boldsymbol{f}*\boldsymbol{g}$; if $\boldsymbol{g}$ is locally integrable, we have $\boldsymbol{f}*(\boldsymbol{g}\,\mathrm{d}x)=\boldsymbol{f}*\boldsymbol{g}$; and if $\boldsymbol{f}$ and $\boldsymbol{g}$ are both integrable, we have $(\boldsymbol{f}\mathrm{d}x)*\boldsymbol{g}=\boldsymbol{f}*(\boldsymbol{g}\,\mathrm{d}x)=\boldsymbol{f}*\boldsymbol{g}$.

We shall see that in some cases the function $\boldsymbol{f}*\boldsymbol{g}$ is defined everywhere or almost everywhere, $\boldsymbol{f}*\boldsymbol{g}$ is locally integrable, and $(\boldsymbol{f}\mathrm{d}x)*(\boldsymbol{g}\,\mathrm{d}y)=(\boldsymbol{f}*\boldsymbol{g})\mathrm{d}y$.

24.37 Proposition. *If A is the support of $\boldsymbol{f}$ and B is the support of $\boldsymbol{g}$, then the support of $\boldsymbol{f}*\boldsymbol{g}$ is contained in $\overline{AB}$.*

Let y be a point for which there exists

$$(\boldsymbol{f}*\boldsymbol{g})(y)=\int \boldsymbol{f}(x)\boldsymbol{g}(x^{-1}y)\mathrm{d}x\,.$$

If $y\notin AB$, then for every $x\in A$ we have $x^{-1}y\notin B$ therefore $\boldsymbol{g}(x^{-1}y)=0$ hence $\int \boldsymbol{f}(x)\boldsymbol{g}(x^{-1}y)\mathrm{d}x=0$.

24.38 Corollary. *If $\boldsymbol{f}$ and $\boldsymbol{g}$ have compact supports, then $\boldsymbol{f}*\boldsymbol{g}$ has compact support.*

It is easy to prove that for every $s \in G$ we have:

$$(\boldsymbol{f}*\boldsymbol{g})_s = \boldsymbol{f}_s*\boldsymbol{g} \text{ and } (\boldsymbol{f}*\boldsymbol{g})^s = \boldsymbol{f}*\boldsymbol{g}^s .$$

24.39 PROPOSITION. *If $\boldsymbol{f} \in \mathscr{L}^1_X$, if $\boldsymbol{g} : G \to Y$ is locally integrable and the measure $\boldsymbol{f}\,\mathrm{d}x*\boldsymbol{g}\,\mathrm{d}y$ (respectively the measure $\boldsymbol{g}\,\mathrm{d}y*\boldsymbol{f}\,\mathrm{d}x$) exists, then the function $\boldsymbol{f}*\boldsymbol{g}$ (respectively the function $\boldsymbol{g}*\boldsymbol{f}$) is defined almost everywhere, is locally integrable and*

$$\boldsymbol{f}\,\mathrm{d}x*\boldsymbol{g}\,\mathrm{d}y = (\boldsymbol{f}*\boldsymbol{g})\,\mathrm{d}y \text{ (respectively } \boldsymbol{g}\,\mathrm{d}y*\boldsymbol{f}\,\mathrm{d}x = (\boldsymbol{g}*\boldsymbol{f})\,\mathrm{d}x) .$$

If $\boldsymbol{f}'$ is equal almost everywhere to $\boldsymbol{f}$ and $\boldsymbol{g}'$ is equal almost everywhere to $\boldsymbol{g}$, then $\boldsymbol{f}'*\boldsymbol{g}' = \boldsymbol{f}*\boldsymbol{g}$ almost everywhere.

We shall show that the conditions of proposition 24.21 are satisfied. The measure $\boldsymbol{m} = \boldsymbol{f}\,\mathrm{d}x$ is bounded and $\boldsymbol{m}*(\boldsymbol{g}\,\mathrm{d}y)$ exists. The function $|\boldsymbol{f}(x)|\,\boldsymbol{g}(x^{-1}y)$ is measurable for the measure $\mathrm{d}x\,\mathrm{d}y = \mathrm{d}x \otimes \mathrm{d}y$, therefore the function $\boldsymbol{g}(x^{-1}y)$ is measurable for the measure $|\boldsymbol{f}(x)|\,\mathrm{d}x \otimes \mathrm{d}y = |\boldsymbol{m}| \otimes \mathrm{d}y$.

By proposition 24.21, the function $\boldsymbol{f}*\boldsymbol{g} = (\boldsymbol{f}\,\mathrm{d}x)*\boldsymbol{g} = \boldsymbol{m}*\boldsymbol{g}$ is defined almost everywhere, is locally integrable and $\boldsymbol{m}*(\boldsymbol{g}\,\mathrm{d}y) = (\boldsymbol{m}*\boldsymbol{g})\,\mathrm{d}y$, that is

$$(\boldsymbol{f}\,\mathrm{d}x)*(\boldsymbol{g}\,\mathrm{d}y) = (\boldsymbol{f}*\boldsymbol{g})\,\mathrm{d}y .$$

The assertion concerning the function $\boldsymbol{g}*\boldsymbol{f}$ is proved similarly.

We have then $\boldsymbol{m} = \boldsymbol{f}'\,\mathrm{d}x$ therefore $\boldsymbol{m}*\boldsymbol{g}' = \boldsymbol{m}*\boldsymbol{g}$ almost everywhere (proposition 24.22) whence $\boldsymbol{f}'*\boldsymbol{g}' = \boldsymbol{f}*\boldsymbol{g}$ almost everywhere.

24.40 PROPOSITION. *If $\boldsymbol{f} \in \mathscr{L}^1_X$ and $\boldsymbol{g} \in \mathscr{L}^p_Y$, $(1 \leqslant p \leqslant +\infty)$, then the measures $\boldsymbol{f}\,\mathrm{d}x*\boldsymbol{g}\,\mathrm{d}y$ and $\boldsymbol{g}\,\mathrm{d}y*\boldsymbol{f}\,\mathrm{d}x$ exist, the functions $\boldsymbol{f}*\boldsymbol{g}$ and $\boldsymbol{g}*\boldsymbol{f}$ are defined almost everywhere, are locally integrable and we have*

$$N_p(\boldsymbol{f}*\boldsymbol{g}) \leqslant N_1(\boldsymbol{f})\,N_p(\boldsymbol{g}),\; N_p(\boldsymbol{g}*\boldsymbol{f}) \leqslant N_p(\boldsymbol{g})\,N_1(\boldsymbol{f})$$

and

$$\boldsymbol{f}\,\mathrm{d}x*\boldsymbol{g}\,\mathrm{d}y = (\boldsymbol{f}*\boldsymbol{g})\,\mathrm{d}y,\; \boldsymbol{g}\,\mathrm{d}y*\boldsymbol{f}\,\mathrm{d}x = (\boldsymbol{g}*\boldsymbol{f})\,\mathrm{d}x .$$

If $\boldsymbol{f}'$ is equal almost everywhere to $\boldsymbol{f}$ and $\boldsymbol{g}'$ is equal almost everywhere to $\boldsymbol{g}$, then

$$\boldsymbol{f}'*\boldsymbol{g}' = \boldsymbol{f}*\boldsymbol{g} \text{ and } \boldsymbol{g}'*\boldsymbol{f}' = \boldsymbol{g}*\boldsymbol{f}, \text{ almost everywhere} .$$

We use the proposition 24.24, taking $\boldsymbol{m} = \boldsymbol{f}\,\mathrm{d}x$, and the preceding proposition, noticing that $\|\boldsymbol{m}\| = N_1(\boldsymbol{f})$.

24.41 COROLLARY. *If $f \in \mathscr{L}^1_X$ and $g \in \mathscr{L}^1_{Y^{-1}}$ then the function $f*g$ is defined almost everywhere, is integrable and we have*

$$N_1(f*g) \leqslant N_1(f)N_1(g) \text{ and } f\mathrm{d}x*g\mathrm{d}y = (f*g)\mathrm{d}y\,.$$

For the functions $g \in \mathscr{L}^\infty_Y$ we have additional properties.

24.42 PROPOSITION. *If $f \in \mathscr{L}^1_X$ and $g \in \mathscr{L}^\infty_Y$, then the function $f*g$ is defined everywhere, is continuous, bounded and*

$$\|f*g\| \leqslant N_1(f)N_\infty(g)\,.$$

If G is unimodular or if f has compact support, then $g*f$ is defined everywhere, is continuous, bounded and

$$\|g*f\| \leqslant N_1(f)N_\infty(g)\,.$$

In fact, for every $y \in G$ the function $x \to f(yx)$ belongs to the space $\mathscr{L}^1_X$ and the function $x \to g(x^{-1})$ belongs to the space $\mathscr{L}^\infty_Y$. Then the function $x \to f(yx)g(x^{-1})$ is integrable, therefore there exists

$$(f*g)(y) = \int f(yx)g(x^{-1})\mathrm{d}x = \int f(x)g(x^{-1}y)\mathrm{d}x$$

for every $y \in G$. We have

$$|(f*g)(y)| \leqslant \int |f(yx)|\,|g(x^{-1})|\,\mathrm{d}x \leqslant N_1(f)N_\infty(g)\,,$$

therefore $f*g$ is bounded and

$$\|f*g\| \leqslant N_1(f)N_\infty(g)\,.$$

We show now that $f*g$ is continuous. Let $x_0 \in G$ and $\varepsilon > 0$. If $g = 0$, then $f*g = 0$. Assume that $N_\infty(g) \neq 0$. Denoting $h = f*g$, for every $s \in G$ we have

$$|h(x_0) - h(x_0 s)| = \left|\int [f(x_0 x) - f(s x_0 x)]\, g(x^{-1})\mathrm{d}x\right| \leqslant$$
$$\leqslant N_\infty(g) \int |f(x_0 x) - f(s x_0 x)|\,\mathrm{d}x\,.$$

Let f' be a function from $\mathscr{K}_X(G)$ with $N_1(f - f') < \varepsilon/3N_\infty(g)$. We have $\int |f(zx) - f'(zx)|\,\mathrm{d}x = \int |f(x) - f'(x)|\,\mathrm{d}x = N_1(f - f') < \varepsilon/3N_\infty(g)$ for every $z \in G$.

Let K be the compact support of f and V a compact neighbourhood of e. We have $K \subset KV$, therefore if $z \notin KV$, then $f'(z) = 0$. The function $x \to f'(x_0 x)$ is left uniformly continuous. There exists therefore a neighbourhood $W \subset V$

of e such that $|f(z)-f(sz)|<\varepsilon/3N_\infty(g)\mu(x_0^{-1}K)$ for $s\in W$ and $z\in G$. If $s\in W$ and $x\notin x_0^{-1}K$, then $x_0x\notin K$ and $xs_0x\notin WK$ therefore $f'(x_0x)=f'(sx_0x)=0$; if $s\in W$ and $x\in x_0^{-1}K$, then $x_0x\in K$ therefore $|f(x_0x)-f(sx_0x)|<\varepsilon/3N_\infty(g)\mu(x_0^{-1}K)$.

It follows that for $s\in W$ we have

$$N_\infty(g)\int|f(x_0x)-f(sx_0x)|\,\mathrm{d}x<\varepsilon/3\,.$$

Then, for $s\in W$,

$$|h(x_0)-h(sx_0)|\leqslant N_\infty(g)\int|f(x_0x)-f(x_0x)|\,\mathrm{d}x+$$
$$+N_\infty(g)\int|f'(x_0x)-f'(sx_0x)|\,\mathrm{d}x+N_\infty(g)\int|f'(sx_0x)-f(sx_0x)|\,\mathrm{d}x<\varepsilon,$$

therefore $h=f*g$ is continuous at x_0. As x_0 is arbitrary, $f*g$ is continuous on G.

If G is unimodular, we show in the same way that the integral

$$(g*f)(y)=\int f(xy)g(x^{-1})\mathrm{d}x$$

exists for every $y\in X$ and that for every $x_0\in G$ and every $\varepsilon>0$ there exists a neighbourhood W of e such that, denoting $h=g*f$, for every $s\in W$ we have $|h(x_0)-h(x_0s)|<\varepsilon$; that is, $g*f$ is continuous on G.

Assume now that f has compact support C. Since the function $\Delta(x^{-1})$ is continuous on G, it is bounded on C, therefore the function $f(x)\Delta(x^{-1})$ is integrable. The function $x\to g(yx^{-1})$ belongs to the space $\mathscr{L}_Y^\infty$ for every $y\in G$, therefore the integral

$$(g*f)(y)=\int g(yx^{-1})f(x)\Delta(x^{-1})\mathrm{d}x$$

exists for every $y\in G$. Thus, in this case too, $g*f$ is defined everywhere and

$$|(g*f)(y)|\leqslant N_\infty(g)\int|f(x)|\Delta(x^{-1})\mathrm{d}x=$$
$$=N_\infty(g)\int|f(x^{-1})|\,\mathrm{d}x=N_\infty(g)N_1(\check{f}\,)\,.$$

The continuity of the function $g*f$ is proved as before, replacing $g(x)$ by $g(x)\Delta(x^{-1})\varphi_C(x)$.

24.43 COROLLARY. *Let $f\in\mathscr{L}_X^1$ and $g\in\mathscr{L}_Y^\infty$.*

(1) *If g is right uniformly continuous and bounded, then $f*g$ is bounded and right uniformly continuous;*

(1′) *If g is bounded and left uniformly continuous and G is unimodular or f has compact support, then $g*f$ is bounded and left uniformly continuous;*

(2) *If* $g \in \overline{\mathcal{K}_Y(G)}$, *then* $f*g \in \overline{\mathcal{K}_Z(G)}$;

(2′) *If* $g \in \overline{\mathcal{K}_Y(G)}$, *and G is unimodular or f has compact support, then* $g*f \in \overline{\mathcal{K}_Z(G)}$;

(3) *If* $f \in \mathcal{K}_X(G)$ *and* $g \in \mathcal{K}_Y(G)$, *then* $f*g \in \mathcal{K}_Z(G)$ *and* $g*f \in \mathcal{K}_Z(G)$.

We use Propositions 24.30, 24.32. and 24.33.

24.44 PROPOSITION. *Assume that G is unimodular. If* $f \in \mathcal{L}_X^p$ *and* $g \in \mathcal{L}_Y^q$, $1 < p, q < +\infty$, $p^{-1}+q^{-1}=1$, *then the function* $f*g$ *is defined everywhere on G and belongs to the space* $\overline{\mathcal{K}_Z(G)}$, *the measure* $f\,\mathrm{d}x * g\,\mathrm{d}y$ *exists and we have*

$$\|f*g\| \leqslant N_p(f)\,N_q(g) \quad \text{and} \quad f\,\mathrm{d}x * g\,\mathrm{d}y = (f*g)\,\mathrm{d}y\,.$$

Since G is unimodular, the function $x \to g(x^{-1})$ belongs to $\mathcal{L}_Y^q$. For every $y \in G$, the function $x \to g(x^{-1}y)$ belongs to $\mathcal{L}_Y^q$ therefore the function $x \to f(x)g(x^{-1}y)$ is integrable, consequently the function $(f*g)(y) = \int f(x)g(x^{-1}y)\mathrm{d}x$, exists for every $y \in G$. We deduce

$$|(f*g)(y)| \leqslant \int |f(x)|\,|g(x^{-1}y)|\,\mathrm{d}x \leqslant N_p(f)\,N_q(g)$$

therefore $f*g$ is bounded and

$$\|f*g\| \leqslant N_p(f)\,N_q(g)\,.$$

Let us show that $f*g$ is continuous. Let (f_n) be a sequence from $\mathcal{K}_X(G)$ and (g_n) a sequence from $\mathcal{K}_Y(G)$ such that $N_p(f-f_n) \to 0$ and $N_q(g-g_n) \to 0$.

Then

$$\|f*g - f_n*g_n\| = \|(f-f_n)*g + f_n*(g-g_n)\| \leqslant$$
$$\leqslant \|(f-f_n)*g\| + \|f_n*(g-g_n)\| \leqslant$$
$$\leqslant N_p(f-f_n)\,N_q(g) + N_p(f_n)\,N_q(g-g_n)\,.$$

But $N_p(f_n) \to N_p(f)$ therefore $\|f*g - f_n g_n\| \to 0$, that is, the sequence f_n*g_n converges uniformly on G to $f*g$. As the functions f_n*g_n are continuous and with compact support, it follows that $f*g \in \overline{\mathcal{K}_Z(G)}$.

Let us show now that $f\,\mathrm{d}x * g\,\mathrm{d}y$ exists.

Let $\varphi \geqslant 0$ be a function from $\mathcal{K}(G)$, $C \subset G \times G$ a compact set, $K_1 = \mathrm{pr}_1 C$, $K_2 = \mathrm{pr}_2 C$ and $K = K_1 K_2$.

The function $\varphi(xy)\varphi_{K \times K}(x, y)$ is integrable for the measure

$|f|\,dx \otimes |g|\,dy$ and the function $y \to \varphi(xy)|g(y)|$ is integrable for every $x \in G$. Then

$$I_K = \iint_{K\times K} \varphi(xy)|f(x)|dx|g(y)|dy = \int_K |f(x)|dx \int_K \varphi(xy)|g(y)|dy \leqslant$$
$$\leqslant \int_K^* |f(x)|dx \int \varphi(xy)|g(y)|dy = \int_K^* |f(x)|dx \int \varphi(y)|g(x^{-1}y)|dy\,.$$

We have then

$$\int_K \varphi(xy)|g(y)|\,dy = \int \varphi_x(y)|g(y)|\,dy \leqslant N_p(\varphi_x)N_q(g) = N_p(\varphi)N_q(g)\,,$$

therefore

$$\int_K^* |f(x)|dx \int |g(x^{-1}y)|\,\varphi(y)dy = \int_K^* |f(x)|dx \int \varphi(xy)\,|g(y)|\,dy \leqslant$$
$$\leqslant N_p(\varphi)\,N_q(g) \int_K |f(x)|dx < +\infty\,.$$

Since $g(x^{-1}y)$ is measurable for $dx\,dy$, it is measurable with respect to the bounded measure $\varphi_K|f|\,dx \otimes \varphi\,dy$. Since the iterated integral exists, by corollary 22.17, the function $g(x^{-1}y)$ is integrable with respect to the measure $\varphi_K f\,dx \otimes \varphi\,dy$ and

$$I_K \leqslant \int_K |f(x)|dx \int \varphi(y)|g(x^{-1}y)|dy = \int \varphi(y)dy \int_K |f(x)|\,|g(x^{-1}y)|dx \leqslant$$
$$\leqslant N_p(f)N_q(g) \int \varphi\,dy\,.$$

It follows that

$$\iint^* \varphi(xy)|f(x)|dx|g(y)|dy = \sup_K I_K \leqslant N_p(f)N_q(g) \int \varphi\,dy < +\infty\,,$$

therefore $\varphi(xy)$ is integrable with respect to $|f|\,dx \otimes |g|\,dy$.

We deduce that the convolution $|f|\,dx \otimes |g|\,dy$ exists, therefore the convolution $f\,dx \otimes g\,dy$ exists as well. For every function $\varphi \in \mathcal{K}(G)$ we have then

$$\int \varphi f\,dx \otimes g\,dy = \iint \varphi(xy)f(x)dx\,g(y)dy = \int f(x)dx \int \varphi(xy)g(y)dy =$$
$$= \int f(x)dx \int \varphi(y)g(x^{-1}y)dx = \int \varphi(y)dy \int f(x)g(x^{-1}y)dy =$$
$$= \int \varphi(y)(f * g)(y)\,dy\,,$$

therefore

$$f\,dx * g\,dy = (f * g)dy$$

and the proposition is proved.

24.45 Corollary. *If G is unimodular and if $f \in \mathcal{L}_X^2$ and $g \in \mathcal{L}_Y^2$, then $f * g \in \mathcal{K}_Z(G)$ and $\|f * g\| \leqslant N_2(f)N_2(g)$.*

6 *The approximating unit*

For each neighbourhood V of e choose a positive function u_V defined on G to be symmetric, $(u_V(x^{-1}) = u_V(x))$, continuous, with the support contained in V, and such that $\int u_V(x)\mathrm{d}x = 1$. Such a function u_V can be obtained from an arbitrary function $\varphi \in \mathscr{K}(G, V)$, taking $u_V = (\varphi - \check{\varphi})/\int(\varphi - \check{\varphi})\mathrm{d}x$. The family (u_V) will be called the *approximating unit* for the convolution. The reason will follow from the following propositions.

24.46 PROPOSITION. *Let* f *be a function defined on G with values in a Banach space E.*

It f *is left uniformly continuous, then for every* $\varepsilon > 0$, *there exists a neighbourhood V of e such that*

$$|(u_V * f)(y) - f(y)| < \varepsilon, \text{ for every } y \in G .$$

If f *is right uniformly continuous, then for every* $\varepsilon > 0$ *there exists a neighbourhood V of e such that*

$$|(f * u_V)(y) - f(y)| < \varepsilon, \text{ for every } y \in G .$$

Assume that f is left uniformly continuous and let $\varepsilon > 0$' There exists a neighbourhood V of e, such that

$$|f(z) - f(y)| < \varepsilon \text{ for } yz^{-1} \in V.$$

For every $x \in V$ and every $y \in G$ we have $y(x^{-1}y)^{-1} = x \in V$, therefore $|f(x^{-1}y) - f(y)| < \varepsilon$. The function $u_V * f$ is defined everywhere and we have

$$|(u_V * f)(y) - f(y)| = |\int u_V(x) f(x^{-1}y)\mathrm{d}x - \int u_V(x) f(y)\mathrm{d}x| \leqslant$$
$$\leqslant \int u_V(x) |f(x^{-1}y) - f(y)| \mathrm{d}x < \varepsilon .$$

If f is right uniformly continuous, we choose the neighbourhood V such that we have

$$|f(z) - f(y)| < \varepsilon, \text{ for } y^{-1}z \in V .$$

For every $x \in V$ and every $y \in G$ we have $y^{-1}(yx) = x \in V$, therefore

$$|f(yx) - f(y)| < \varepsilon ,$$

whence, u_V being symmetric,

$$\begin{aligned}|(f*u_V)(y)-f(y)| &= |\int f(yx)u_V(x^{-1})\mathrm{d}x - \int f(y)u_V(x)\mathrm{d}x| = \\ &= |\int f(yx)u_V(x)\mathrm{d}x - \int f(y)u_V(x)\mathrm{d}x| \leqslant \\ &\leqslant \int u_V(x)|f(yx)-f(y)|\mathrm{d}x < \varepsilon\,.\end{aligned}$$

Remark. The preceding proposition states the fact that if f is left uniformly continuous, the family of functions (u_V*f) is uniformly converging to f on G, the limit being taken following the order defined by the inclusion on the set of neighbourhoods of e. Also, if f is right uniformly continuous then $(f*u_V)$ converges uniformly to f.

In particular, if $f\in\mathscr{K}_E(G)$, then both u_V*f and $f*u_V$ converge uniformly to f.

24.47 PROPOSITION. *If $f\in\mathscr{L}_E^p$, $1\leqslant p<+\infty$, then for every $\varepsilon>0$, there exists a neighbourhood V of e such that*

$$N_p(u_V*f-f)<\varepsilon \text{ and } N_p(f*u_V-f)<\varepsilon\,.$$

Let $f\in\mathscr{L}_E^p$ and $\varepsilon>0$. The mapping $s\to f_s$ of G into $\mathscr{L}_E^p$ is left uniformly continuous (proposition 23.18). There exists, therefore a symmetric neighbourhood V of e such that

$$N_p(f_s-f)<\varepsilon \text{ for } s\in V\,.$$

Let $h\in\mathscr{K}(G)$. Since $|u_V*f|$ has p-integrable power, the function $|h|\,|u_V*f|$ is integrable and

$$\int |h(y)|\,\mathrm{d}y\int u_V(x)|f(x^{-1}y)|\,\mathrm{d}x = \int |h(y)|\,|u_V*f(y)|\,\mathrm{d}y\,.$$

Since the function $f(x^{-1}y)$ is measurable with respect to the bounded measure $h\,\mathrm{d}y\otimes u_V\,\mathrm{d}x$, we deduce that $f(x^{-1}y)$ is integrable with respect to this measure. Applying the Fubini theorem we deduce

$$\begin{aligned}|\int[u_V*f(y)-f(y)]\,h(y)\mathrm{d}y &= |\int(u_V*f)(y)h(y)\mathrm{d}y-\int f(y)h(y)\mathrm{d}y| = \\ &= |\int h(y)\mathrm{d}y\int u_V(x)f(x^{-1}y)\mathrm{d}x-\int h(y)f(y)\mathrm{d}y| \\ &= |\int u_V(x)\mathrm{d}x\int h(y)f(x^{-1}y)\mathrm{d}y - \\ &\quad -\int u_V(x)\mathrm{d}x\int h(y)f(y)\mathrm{d}y| = \\ &= |\int u_V(x)\mathrm{d}x\int h(y)[f_{x^{-1}}(y)-f(y)]\mathrm{d}y| \leqslant \\ &\leqslant \int u_V(x)\mathrm{d}x\int|h(y)|\,|f_{x^{-1}}(y)-f(y)|\mathrm{d}y \leqslant \\ &\leqslant N_q(h)\,N_p(f_{x^{-1}}-f)<\varepsilon N_q(h)\,,\end{aligned}$$

whence

$$N_p(u_V * f - f) < \varepsilon .$$

In order to prove the other inequality we choose the symmetric neighbourhood V such that

$$N_p(f^{s^{-1}} - f) < \varepsilon, \text{ for } s \in V ,$$

and, remarking that $u_V(x) = u_V(x^{-1})$, we deduce, as before, for $h \in \mathcal{K}(G)$,

$$|\int [(f * u_V)(y) - f(y)]\, h(y)\,dy| = |\int h(y)\,dy \int f(xy)\, u_V(x^{-1})\,dx$$

$$- \int f(y) h(y)\,dy| = |\int u_V(x^{-1})\,dx \int f(yx) h(y)\,dy - \int u_V(x^{-1})\,dx \int f(y) h(y)\,dy| =$$

$$= |\int u_V(x^{-1})\,dx \int [f^{x^{-1}}(y) - f(y)]\, h(y)\,dy| \leqslant$$

$$\leqslant \int u_V(x^{-1})\,dx \int |f^{x^{-1}}(y) - f(y)|\, |h(y)|\,dy \leqslant$$

$$\leqslant \int u_V(x^{-1})\,dx\, N_p(f^{x^{-1}} - f)\, N_q(h) \leqslant \varepsilon N_q(h) ,$$

whence $N_p(f * u_V - f) < \varepsilon$ and the proposition is proved.

Remark. The proposition states the fact that $u_V * f$ and $f * u_V$ tend to f in the topology of the space $\mathcal{L}^p_E$.

§25 The group algebra

1 *The algebra* $\mathcal{M}^1$

In the sequel we shall denote by A a (complex) Banach algebra with unit element. In particular A can be the field C of the complex numbers.

We can embed C isometrically in A by identifying a scalar $\alpha \in C$ with the vector $\alpha e \in A$ and we can consider $C \subset A$.

In this case, we can consider the convolution of two measures $\boldsymbol{m}$ and $\boldsymbol{n}$ with values in A, or of a measure $\boldsymbol{m} : \mathcal{K}(G) \to A$ with a function $f : G \to A$, or even, the convolution of two functions $f, g : G \to A$. In each case, the convolution is a measure or a function with values still in A.

We can define the product $a\boldsymbol{m}$ and $\boldsymbol{m}a$ of a measure $\boldsymbol{m} : \mathcal{K}(G) \to A$ by an element $a \in A$ by the equality $(a\boldsymbol{m})(\varphi) = a \cdot \boldsymbol{m}(\varphi)$ and $(\boldsymbol{m}a)(\varphi) = \boldsymbol{m}(\varphi) \cdot a$, for every $\varphi \in \mathcal{K}(G)$.

25.1 PROPOSITION. *The space $\mathcal{M}_A^1$ of bounded measures $\boldsymbol{m}:\mathcal{K}(G)\to A$ is an algebra with unit element, for the convolution $\boldsymbol{m}*\boldsymbol{n}$.*

By proposition 24.4, if $\boldsymbol{m}, \boldsymbol{n}\in\mathcal{M}_A^1$, then $\boldsymbol{m}*\boldsymbol{n}$ exists and $\boldsymbol{m}*\boldsymbol{n}\in\mathcal{M}_A^1$.

It was already noticed that the mapping $(\boldsymbol{m}, \boldsymbol{n})\to\boldsymbol{m}*\boldsymbol{n}$ of $\mathcal{M}_A^1\times\mathcal{M}_A^1$ into $\mathcal{M}_A^1$ is bilinear, hence

(1) $(\boldsymbol{m}_1+\boldsymbol{m}_2)*\boldsymbol{n}=\boldsymbol{m}_1*\boldsymbol{n}+\boldsymbol{m}_2*\boldsymbol{n}$,

(2) $\boldsymbol{m}*(\boldsymbol{n}_1+\boldsymbol{n}_2)=\boldsymbol{m}*\boldsymbol{n}_1+\boldsymbol{m}*\boldsymbol{n}_2$

and that ε_e is a unit element (proposition 24.3):

(3) $\varepsilon_e*\boldsymbol{m}=\boldsymbol{m}*\varepsilon_e=\boldsymbol{m}$.

It is immediately seen that

(4) $\alpha(\boldsymbol{m}*\boldsymbol{n})=(\alpha\boldsymbol{m})*\boldsymbol{n}=\boldsymbol{m}*(\alpha\boldsymbol{n})$, α scalar,

(4') $a(\boldsymbol{m}*\boldsymbol{n})=(a\boldsymbol{m})*\boldsymbol{n}=\boldsymbol{m}*(a\boldsymbol{n})$,
$(\boldsymbol{m}*\boldsymbol{n})a=(\boldsymbol{m}a)*\boldsymbol{n}=\boldsymbol{m}*(\boldsymbol{n}a)$ for every $a\in A$.

It remains to see that the convolution is associative:

$$\boldsymbol{l}*(\boldsymbol{m}*\boldsymbol{n})=(\boldsymbol{l}*\boldsymbol{m})*\boldsymbol{n}$$

for every $\boldsymbol{l}, \boldsymbol{m}, \boldsymbol{n}\in\mathcal{M}_A^1$.

Let $\varphi\in\mathcal{K}(G)$. Using the Fubini theorem we have

$$(\boldsymbol{l}*(\boldsymbol{m}*\boldsymbol{n}))(\varphi)=\int\int\varphi(xy)\,\mathrm{d}\boldsymbol{l}(x)\,\mathrm{d}(\boldsymbol{m}*\boldsymbol{n})(y)=\int\mathrm{d}\boldsymbol{l}(x)\int\varphi(xy)\,\mathrm{d}(\boldsymbol{m}*\boldsymbol{n})(y).$$

But, for each $x\in G$, we have

$$\int\varphi(xy)\,\mathrm{d}(\boldsymbol{m}*\boldsymbol{n})(y)=\int\int\varphi(xuv)\,\mathrm{d}\boldsymbol{m}(u)\,\mathrm{d}\boldsymbol{n}(v)\,,$$

therefore

$$(\boldsymbol{l}*(\boldsymbol{m}*\boldsymbol{n}))(\varphi)=\int\mathrm{d}\boldsymbol{l}(x)\int\int\varphi(xuv)\,\mathrm{d}\boldsymbol{m}(u)\,\mathrm{d}\boldsymbol{n}(v)\,.$$

The measure $\boldsymbol{l}\otimes\boldsymbol{m}\otimes\boldsymbol{n}$ is bounded on $G\times G\times G$ and the function φ is continuous and bounded, therefore it is integrable with respect to this measure, and by the Fubini theorem we have

$$\int\int\int\varphi(xuv)\,\mathrm{d}\boldsymbol{l}(x)\,\mathrm{d}\boldsymbol{m}(u)\,\mathrm{d}\boldsymbol{n}(v)=\int\mathrm{d}\boldsymbol{l}(x)\int\varphi(xuv)\,\mathrm{d}\boldsymbol{m}(u)\,\mathrm{d}\boldsymbol{n}(v)\,,$$

therefore

$$(\boldsymbol{l}*(\boldsymbol{m}*\boldsymbol{n}))(\varphi)=\int\int\int\varphi(xuv)\,\mathrm{d}\boldsymbol{l}(x)\,\mathrm{d}\boldsymbol{m}(u)\,\mathrm{d}\boldsymbol{n}(v)\,.$$

On the other hand

$$((\boldsymbol{l}*\boldsymbol{m})*\boldsymbol{n})(\varphi)=\int\int\varphi(yv)\mathrm{d}(\boldsymbol{l}*\boldsymbol{m})(y)\mathrm{d}\boldsymbol{n}(v)=\int\mathrm{d}\boldsymbol{n}(v)\int\varphi(yv)\mathrm{d}(\boldsymbol{l}*\boldsymbol{m})(y)=$$
$$=\int\mathrm{d}\boldsymbol{n}(v)\int\varphi(xuv)\mathrm{d}\boldsymbol{l}(x)\mathrm{d}\boldsymbol{m}(u)=$$
$$\int\int\int\varphi(xuv)\mathrm{d}\boldsymbol{l}(x)\mathrm{d}\boldsymbol{m}(u)\mathrm{d}\boldsymbol{n}(v)$$

and then $\boldsymbol{l}*(\boldsymbol{m}*\boldsymbol{n})=(\boldsymbol{l}*\boldsymbol{m})*\boldsymbol{n}$.

Thus the proposition is proved.

By proposition 24.5, the algebra $\mathscr{M}_A^1$ is commutative if and only if the group G is commutative.

We shall denote by $\mathscr{M}^1$ instead of $\mathscr{M}_C^1$, the space of the bounded complex measures. We deduce that the space $\mathscr{M}^1$ is an algebra.

Identifying a scalar measure $\mu\in\mathscr{M}^1$ with the measure $e\mu\in\mathscr{M}_A^1$ we deduce that $\mathscr{M}^1$ is a subalgebra of the algebra $\mathscr{M}_A^1$.

By corollary 24.13, the set $\mathscr{M}_A^c$ of the measures of $\mathscr{M}_A^1$ with compact support is an algebra, and the set $\mathscr{M}^c$ of the complex measures with compact support is an algebra.

The algebra $\mathscr{M}_A^1$ is a subspace of the space $\mathscr{L}(\mathscr{K}(G), A)$ of the linear continuous mappings of $\mathscr{K}(G)$ into A. Such linear mappings are measures, but they might not be dominated, and even if they are dominated, their modulus might not be a bounded measure, therefore they might not belong to $\mathscr{M}_A^1$. We shall consider on this algebra the norm

$$|||\boldsymbol{m}|||=||\,|\boldsymbol{m}|\,||.$$

For the scalar measures $\mu\in\mathscr{M}_A^1$ we have $|||\mu|||=||\mu||$, since $||\mu||=||\,|\mu|\,||$.

Since $|\boldsymbol{m}*\boldsymbol{n}|\leqslant|\boldsymbol{m}|*|\boldsymbol{n}|$ and $||\,|\boldsymbol{m}|*|\boldsymbol{n}|\,||\leqslant||\,|\boldsymbol{m}|\,||\cdot||\,|\boldsymbol{n}|\,||$, we deduce $|||\boldsymbol{m}*\boldsymbol{n}|||\leqslant|||\boldsymbol{m}|||\;|||\boldsymbol{n}|||$, therefore $\mathscr{M}_A^1$ is a normed algebra.

With this norm, $\mathscr{M}^1$ is a Banach algebra, being equal to the dual of the normed space $\mathscr{K}(G)$ with the norm $||f||=\sup_{x\in G}|f(x)|$.

The set of the measures $(\varepsilon_s)_{s\in G}$ is a discrete subgroup of the multiplicative subgroup of the algebra $\mathscr{M}^1$, since $|||\varepsilon_s-\varepsilon_t|||=2$ (proposition 24.6).

2 *The algebra* L_A^1

If $\boldsymbol{f}$ and $\boldsymbol{g}$ are two functions from $\mathscr{L}_A^1$, their convolution $\boldsymbol{f}*\boldsymbol{g}$ is an integrable function *defined almost everywhere on* G, therefore it does not necessarily belong to the space $\mathscr{L}_A^1$. Thus, in general, $\mathscr{L}_A^1$ is not an algebra for the convolution. As we have noticed, if $\boldsymbol{f}'$ and $\boldsymbol{g}'$ are two functions equivalent respectively to $\boldsymbol{f}$ and $\boldsymbol{g}$, then the functions $\boldsymbol{f}*\boldsymbol{g}$ and $\boldsymbol{f}'*\boldsymbol{g}'$ are equal almost everywhere. That is why we shall define the convolution $\dot{f}*\dot{g}$ of two equivalence classes $\dot{f}$ and $\dot{g}$ from $\mathscr{L}_A^1$ by the equality

$$\dot{f}*\dot{g}=\widehat{f*g}^{\bullet}$$

The class $\widehat{f*g}^{\bullet}$ does not depend on the functions f and g but only on the classes $\dot{f}$ and $\dot{g}$. We shall show that L_A^1 is an algebra for this product.

We shall embed the space L_A^1 isometrically in the space $\mathscr{M}_A^1$, by identifying an equivalence class $\dot{f}\in L_A^1$ with the bounded measure $\boldsymbol{m}=f\,\mathrm{d}x$. The measure $\boldsymbol{m}$ does not depend on the function f but only on the equivalence class $\dot{f}$ and we have

$$|||\boldsymbol{m}|||=\|\,|\boldsymbol{m}|\,\|=\|\,|f\mathrm{d}x|\,\|=\|\,|f|\mathrm{d}x\|=N_1(f)=\|\dot{f}\|_1\,.$$

By this identification, the convolution $\boldsymbol{m}*\boldsymbol{n}$ on $\mathscr{M}_A^1$ is an extension of the convolution $\dot{f}*\dot{g}$ on L_A^1, since

$$f\,\mathrm{d}x*g\,\mathrm{d}y=(f*g)\,\mathrm{d}y\,.$$

It follows that the product $\dot{f}*\dot{g}$ is associative, therefore L_A^1 is an algebra.

By this identification, L_A^1 consists of the bounded measures with basis the Haar measure and density from $\mathscr{L}_A^1$.

25.2 PROPOSITION. *L_A^1 is a two-sided ideal in the algebra $\mathscr{M}_A^1$.*

Let $\boldsymbol{m}\in\mathscr{M}_A^1$ and $\dot{f}\in L_A^1$. Choose a *Borel* function $f\in\dot{f}$. By proposition 24.22, the functions $f*\boldsymbol{m}$ and $\boldsymbol{m}*f$ are defined almost everywhere, are integrable and we have

$$(f\,\mathrm{d}x)*\boldsymbol{m}=(f*\boldsymbol{m})\mathrm{d}x \quad\text{and}\quad \boldsymbol{m}*(f\,\mathrm{d}x)=(\boldsymbol{m}*f)\mathrm{d}x\,,$$

that is,

$$\dot{f}*\boldsymbol{m}=\widehat{f*\boldsymbol{m}}^{\bullet}\in L_A^1 \quad\text{and}\quad \boldsymbol{m}*\dot{f}=\widehat{\boldsymbol{m}*f}^{\bullet}\in L_A^1\,.$$

Hence L_A^1 is a two-sided ideal in $\mathscr{M}_A^1$.

25.3 COROLLARY. *L^1 is a two-sided ideal in the algebra $\mathscr{M}^1$.*

The algebra L^1 is called the *group algebra*. Some authors call $\mathscr{M}^1$ the group algebra.

Identifying a function $f\in\mathscr{L}^1$ with the function $ef\in\mathscr{L}_A^1$, we can embed the space $\mathscr{L}^1$ isometrically in $\mathscr{L}_A^1$ and the space L^1 in the space L_A^1.

By this identification, L^1 is a subalgebra of the algebra L^1_A and of the algebra $\mathcal{M}^1_A$.

25.4 PROPOSITION. *The space $L^1_A \cap L^p_A$, $1 \leqslant p \leqslant \infty$, is a two-sided ideal in the algebra $\mathcal{M}^1_A$.*

In fact, let $\boldsymbol{m} \in \mathcal{M}^1_A$ and $\dot{\boldsymbol{f}} \in L^1_A \cap L^p_A$. Choose a Borel function $\boldsymbol{f} \in \dot{\boldsymbol{f}}$. We have $\boldsymbol{f} \in \mathcal{L}^1_A \cap \mathcal{L}^p_A$. By proposition 24.22, we deduce that $\boldsymbol{m} * \boldsymbol{f}$ and $\boldsymbol{f} * \boldsymbol{m}$ are defined almost everywhere, are equal almost everywhere to functions from $\mathcal{L}^1_A \cap \mathcal{L}^p_A$ and

$$(\boldsymbol{f}\,\mathrm{d}x) * \boldsymbol{m} = (\boldsymbol{f} * \boldsymbol{m})\,\mathrm{d}x, \quad \boldsymbol{m} * (\boldsymbol{f}\,\mathrm{d}x) = (\boldsymbol{m} * \boldsymbol{f})\,\mathrm{d}x\,.$$

We deduce

$$\dot{\boldsymbol{f}} * \boldsymbol{m} = \dot{\overbrace{\boldsymbol{f} * \boldsymbol{m}}} \in L^1_A \cap L^p_A \text{ and } \boldsymbol{m} * \dot{\boldsymbol{f}} = \dot{\overbrace{\boldsymbol{m} * \boldsymbol{f}}} \in L^1_A \cap L^p_A$$

and the proposition is proved.

25.5 COROLLARY. *The space $L^1 \cap L^p$, $1 \leqslant p \leqslant +\infty$, is a two-sided ideal in the algebra $\mathcal{M}^1$.*

If two *continuous* functions $\boldsymbol{f}, \boldsymbol{g} : G \to A$ are equal almost everywhere, then they are equal everywhere. Thus an equivalence class contains at most one continuous function. That is why we shall identify a continuous function $\boldsymbol{f} : G \to A$ whit its equivalence class $\dot{\boldsymbol{f}}$ and we shall consider that the space of the bounded continuous functions $\mathscr{C}^\infty_A$ with the norm $\|\boldsymbol{f}\| = \sup_{x \in G} |\boldsymbol{f}(x)|$ is a subspace of the space L^∞_A, and that the space $\overline{\mathscr{K}_A(G)}$ of the continuous functions vanishing at infinity is a subspace of L^∞_A.

Also, we shall consider the space $\mathscr{K}_A(G)$ with the norm N_p as a subspace of the space L^p_A, $1 \leqslant p \leqslant +\infty$.

25.6 PROPOSITION. *The spaces $L^1_A \cap \mathscr{C}^\infty_A$ and $L^1_A \cap \overline{\mathscr{K}_A(G)}$ are left ideals in the algebra $\mathcal{M}^1_A$.*

If G is unimodular, $L^1_A \cap \mathscr{C}^\infty_A$ and $L^1_A \cap \overline{\mathscr{K}_A(G)}$ are two-sided ideals in the algebra $\mathcal{M}^1_A$.

In fact, if $\boldsymbol{m} \in \mathcal{M}^1_A$ and $\boldsymbol{f} \in L^1_A \cap \mathscr{C}^\infty_A$, we have, by proposition 24.29, $\boldsymbol{m} * \boldsymbol{f} \in \mathscr{C}^\infty_A$ and by proposition 25.4, $\boldsymbol{m} * \boldsymbol{f} \in L^1_A$; therefore, $\boldsymbol{m} * \boldsymbol{f} \in L^1_A \cap \mathscr{C}^\infty_A$, hence $L^1_A \cap \mathscr{C}^\infty_A$ is a left ideal.

Using proposition 24.32, we deduce similarly that $L_A^1 \cap \overline{\mathscr{K}_A(G)}$ is a left ideal.

If G is unimodular, we deduce from the same propositions that $L_A^1 \cap \mathscr{C}_A^\infty$ and $L_A^1 \cap \overline{\mathscr{K}_A(G)}$ are right ideals.

25.7 PROPOSITION. *The space $\mathscr{K}_A(G)$ is an algebra for the convolution $f * g$ and a two-sided ideal in the algebra $\mathscr{M}_A^c$ of the measures with compact support.*

We use proposition 24.33.

Remark. To simplify the notations, we shall identify, in the sequel, two functions which are equal almost everywhere, and we shall denote an equivalence class $\dot{f}$ still by f. It will be easy to realize whether f is a function (when writing, for example $f \in \mathscr{L}_A^1$) or f is an equivalence class (when writing $f \in L_A^1$).

25.8 PROPOSITION. *If G is discrete then $L_A^1 = \mathscr{M}_A^1$ and L_A^1 has a unit element. Conversely, if L^1 has a unit element, then G is discrete.*

Assume that G is discrete. Then the points are open sets and have the same measure. We may assume that each point has measure 1. Then the measure ε_e has basis the Haar measure dx and density the function u defined on G by : $u(e) = 1$, and $u(x) = 0$ if $x \neq e$. In fact, for every function $\varphi \in \mathscr{K}(G)$ we have

$$\varepsilon_e(\varphi) = \varphi(e) = \Sigma_{x \in G}\, \varphi(x)u(x) = \int \varphi(x)u(x)dx\,,$$

therefore $d\varepsilon_e = u(x)dx$, consequently $\varepsilon_e \in L_A^1$. Since L_A^1 is an ideal in $\mathscr{M}_A^1$ and contains the unit element, we deduce that $L_A^1 = \mathscr{M}_A^1$.

Conversely assume that L^1 has a unit element for the convolution.

This means that there exists a function $u \in \mathscr{L}^1$ such that for every function $f \in \mathscr{L}^1$ we have

$$(u * f)(x) = f(x) \text{ and } (f * u)(x) = f(x) \text{ almost everywhere.}$$

Let us show that there exists a number $a > 0$ such that the measure of every open set is $\geqslant a$. Assume, on the contrary, that there is no such number $a > 0$. This means that for every number $a > 0$ there exists an open neighbourhood V of e with measure $< a$.

Let $\varepsilon > 0$. We may find an open neighbourhood V of e such that $\int_V |u(x)|\,dx < \varepsilon$. In fact, for every n we may find an open neighbourhood V_n of e, with measure $< n^{-1}$, such that $V_{n+1} \subset V_n$ for each n. Then the set $\cap V_n$ is negligible, therefore the sequence of characteristic functions φ_{V_n} is decreasing and converges to 0 almost everywhere. It follows that the sequence $(|u|\varphi_{V_n})$ is decreasing and converges to 0 almost everywhere, therefore

$$\int_{V_n} |u(x)|\,dx \to 0\,.$$

There exists then an n such that for $V = V_n$ we have

$$\int_V |u(x)|\,dx < \varepsilon\,.$$

Take a symmetric neighbourhood U of e with $U^2 \subset V$. For its characteristic function φ_U we have $\varphi_U(z) = \varphi_U(z^{-1})$ and

$$\varphi_U(y) = (u*\varphi_U)(y) = \int u(x)\,\varphi_U(x^{-1}y)\,dx = \int u(x)\,\varphi_U(y^{-1}x)\,dx =$$
$$= \int u(x)\,\varphi_{yU}(x)\,dx \leqslant \int_V |u|(x)\,dx < \varepsilon$$

for almost every y, which contradicts the fact that $\varphi_U(y) = 1$ for $y \in U$.

Thus there exists a number $a > 0$ such that the measure of every open set is $\geqslant a$.

It follows that every integrable open set U contains only a finite number of points. In fact, if U contains a sequence (x_n) then for each n we can separate the points $x_1, \ldots, x_n$ by disjoint open sets $V_1, \ldots, V_n$ contained in U. As each V_n has measure $\geqslant a$, it would follow that U has measure $\geqslant na$ for every n, that is, U has measure $+\infty$ and thus we have obtained a contradiction.

In particular, every relatively compact open set is of the form $U = \{x_1, x_2, \ldots, x_n\}$. Since G is separated, each point is a closed set. The set $G\setminus\{x_1\} = \{x_2, \ldots, x_n\} \cup (G\setminus U)$ is closed, therefore $\{x_1\}$ is an open set. Similarly, we show that the points $x_2, \ldots, x_n$ are open. As each point $x \in G$ is contained in a relatively compact open set, it follows that each point is an open set, therefore G is discrete.

3 *Involution algebras*

Here we assume that on the algebra A there is defined an involution $x \to \bar{x}$ with $|\bar{x}| = |x|$.

If for each measure $\boldsymbol{m} \in \mathscr{M}_A^1$ we consider the measure $\bar{\boldsymbol{m}}$ defined by the equalities

$$\bar{\boldsymbol{m}}(\varphi)=\overline{\boldsymbol{m}(\varphi)}, \text{ for } \varphi\in\mathscr{K}(G),$$

the mapping $\boldsymbol{m}\to\bar{\boldsymbol{m}}$ has the following properties:

(1) $\overline{\boldsymbol{m}+\boldsymbol{n}}=\bar{\boldsymbol{m}}+\bar{\boldsymbol{n}}$;

(2) $\overline{\alpha\boldsymbol{m}}=\bar{\alpha}\,\bar{\boldsymbol{m}}$, α scalar;

(3) $\bar{\bar{\boldsymbol{m}}}=\boldsymbol{m}$;

(4) $\overline{\boldsymbol{m}*\boldsymbol{n}}=\bar{\boldsymbol{m}}*\bar{\boldsymbol{n}}$;

(5) $|\bar{\boldsymbol{m}}|=|\boldsymbol{m}|$;

(6) $\overline{\boldsymbol{m}(\boldsymbol{f})}=\bar{\boldsymbol{m}}(\bar{\boldsymbol{f}})$ for $\boldsymbol{f}\in\mathscr{L}^1_A(\boldsymbol{m})$.

Let us prove property (4); for $\varphi\in\mathscr{K}(G)$ we have

$$\overline{\boldsymbol{m}*\boldsymbol{n}}(\varphi)=\overline{(\boldsymbol{m}*\boldsymbol{n})(\varphi)}=\overline{\int\int\varphi(xy)\,\mathrm{d}\boldsymbol{m}(x)\,\mathrm{d}\boldsymbol{n}(y)}=$$

$$=\overline{\int\mathrm{d}\boldsymbol{m}(x)\int\varphi(xy)\,\mathrm{d}\boldsymbol{n}(y)}=\int\mathrm{d}\bar{\boldsymbol{m}}(x)\,\overline{\int\varphi(xy)\,\mathrm{d}\boldsymbol{n}(y)}=$$

$$=\int\mathrm{d}\bar{\boldsymbol{m}}(x)\int\varphi(xy)\,\mathrm{d}\bar{\boldsymbol{n}}(y)=\int\int\varphi(xy)\,\mathrm{d}\bar{\boldsymbol{m}}(x)\,\mathrm{d}\bar{\boldsymbol{n}}(y)=(\bar{\boldsymbol{m}}*\bar{\boldsymbol{n}})(\varphi).$$

It follows that if G is commutative, then the algebra $\mathscr{M}^1_A$ is commutative, therefore the mapping $\boldsymbol{m}\to\bar{\boldsymbol{m}}$ is an involution on $\mathscr{M}^1_A$.

But if G is not commutative, then the equality $\overline{\boldsymbol{m}*\boldsymbol{n}}=\bar{\boldsymbol{n}}*\bar{\boldsymbol{m}}$ is not necessarily true, therefore the mapping $\boldsymbol{m}\to\bar{\boldsymbol{m}}$ is not an involution on $\mathscr{M}^1_A$.

We shall define now another mapping on $\mathscr{M}^1_A$ which is an involution.

Let μ be a *positive* measure on G. We define the mapping $\mu^\sim:\mathscr{K}(G)\to R$ by the equality

$$\mu^\sim(\varphi)=\mu(\check{\varphi}), \text{ for } \varphi\in\mathscr{K}(G).$$

It is easy to verify that $\mu^\sim$ is *additive* and *positive*, therefore $\mu^\sim$ is a positive measure on G.

If $\mu\leqslant\nu$, then $\mu^\sim\leqslant\nu^\sim$.

25.9 PROPOSITION. *For every function $f\geqslant 0$ defined on G we have*

$$\int^* f(x)\,\mathrm{d}\mu^\sim(x)=\int^* f(x^{-1})\,\mathrm{d}\mu(x).$$

In fact, the function $\alpha(x)=x^{-1}$ is a homeomorphism of G onto G, therefore α is a μ-proper function. We have $\check{h}(x)=h(\alpha(x))$ for every function h defined on G, therefore

$$\int\varphi(x)\,\mathrm{d}\mu^\sim(x)=\int\varphi(x^{-1})\,\mathrm{d}\mu(x)=\int\varphi\circ\alpha\,\mathrm{d}\mu=\int\varphi\,\mathrm{d}\alpha(\mu),$$

for every function $\varphi\in\mathscr{K}(G)$. It follows that $\mu^\sim=\alpha(\mu)$. Then we apply proposition 20.9.

25.10 PROPOSITION. *Let E be a Banach space and $\boldsymbol{f}: G \to E$ a function:*

(1) *the function $\boldsymbol{f}$ is $\mu^\sim$-negligible if and only if the function $\check{\boldsymbol{f}}$ is μ-negligible;*

(2) *the function $\boldsymbol{f}$ is $\mu^\sim$-measurable if and only if the function $\check{\boldsymbol{f}}$ is μ-measurable;*

(3) *the function $\boldsymbol{f}$ is $\mu^\sim$-integrable if and only if the function $\check{\boldsymbol{f}}$ is μ-integrable. In this case we have*

$$\int \boldsymbol{f}(x) \mathrm{d}\mu^\sim(x) = \int \check{\boldsymbol{f}}(x) \mathrm{d}\mu(x) = \int \boldsymbol{f}(x^{-1}) \mathrm{d}\mu(x).$$

We apply proposition 20.11, proposition 20.15 and theorem 20.17, for the μ-proper function $\alpha(x) = x^{-1}$.

25.11 COROLLARY. *We have $\boldsymbol{f} \in \mathscr{L}_E^p(\mu^\sim)$, $1 \leqslant p \leqslant +\infty$, if and only if $\check{\boldsymbol{f}} \in \mathscr{L}_E^p(\mu)$. In this case we have*

$$N_p(\boldsymbol{f}, \mu^\sim) = N_p(\check{\boldsymbol{f}}, \mu).$$

For every dominated measure $\boldsymbol{m} : \mathscr{K}(G) \to A$ we define the mapping $\boldsymbol{m}^\sim : \mathscr{K}(G) \to A$ by the equality

$$\boldsymbol{m}^\sim(\varphi) = \bar{\boldsymbol{m}}(\check{\varphi}) = \overline{\boldsymbol{m}(\check{\varphi})}, \text{ for } \check{\varphi} \in \mathscr{K}(G).$$

It is easy to see that $\boldsymbol{m}^\sim$ is linear. We have

$$|\boldsymbol{m}^\sim|(\varphi) = |\overline{\boldsymbol{m}(\check{\varphi})}| = |\boldsymbol{m}(\check{\varphi})| \leqslant |\boldsymbol{m}|(\check{\varphi}) = |\boldsymbol{m}|^\sim(\varphi),$$

therefore $\boldsymbol{m}^\sim$ is a measure dominated by the positive measure $|\boldsymbol{m}|^\sim$. Thus $|\boldsymbol{m}^\sim| \leqslant |\boldsymbol{m}|^\sim$.

If $\boldsymbol{m}$ is bounded, then $\boldsymbol{m}^\sim$ is bounded.

25.12 PROPOSITION. *The mapping $\boldsymbol{m} \to \boldsymbol{m}^\sim$ is an involution on $\mathscr{M}_A^1$ and $|\boldsymbol{m}^\sim| = |\boldsymbol{m}|^\sim$.*

In fact, the following properties are immediately verified:

(1) $(\boldsymbol{m}+\boldsymbol{n})^\sim = \boldsymbol{m}^\sim + \boldsymbol{n}^\sim$;

(2) $(\alpha \boldsymbol{m})^\sim = \bar{\alpha} \boldsymbol{m}^\sim$;

(3) $(\boldsymbol{m}^\sim)^\sim = \boldsymbol{m}$.

It remains to prove that

(4) $(\boldsymbol{m} * \boldsymbol{n})^\sim = \boldsymbol{n}^\sim * \boldsymbol{m}^\sim$;

(5) $|\boldsymbol{m}^\sim| = |\boldsymbol{m}|^\sim$.

The last equality follows in the following way: we have $|\boldsymbol{m}^{\sim}| \leqslant |\boldsymbol{m}|^{\sim}$ and $|\boldsymbol{m}| = |\boldsymbol{m}^{\sim\sim}| \leqslant |\boldsymbol{m}^{\sim}|^{\sim}$ whence $|\boldsymbol{m}^{\sim}| \leqslant |\boldsymbol{m}^{\sim}|^{\sim\sim} = |\boldsymbol{m}^{\sim}|$, hence $|\boldsymbol{m}^{\sim}| = |\boldsymbol{m}|^{\sim}$.

Let $\varphi \in \mathscr{K}(G)$. We have

$$\begin{aligned}(\boldsymbol{m}*\boldsymbol{n})^{\sim}(\varphi) &= \overline{\boldsymbol{m}*\boldsymbol{n}(\overline{\check{\varphi}})} = \overline{\int\int \overline{\check{\varphi}}(xy)\,\mathrm{d}\boldsymbol{m}(x)\,\mathrm{d}\boldsymbol{n}(y)} = \\ &= \overline{\int \mathrm{d}\boldsymbol{m}(x) \int \overline{\varphi(y^{-1}x^{-1})}\,\mathrm{d}\boldsymbol{n}(y)} = \\ &= \int \mathrm{d}\bar{\boldsymbol{m}}(x)\, \overline{\int \overline{\varphi(y^{-1}x^{-1})}\,\mathrm{d}\boldsymbol{n}(y)} = \int \mathrm{d}\bar{\boldsymbol{m}}(x) \int \varphi(yx^{-1})\,\mathrm{d}\boldsymbol{n}^{\sim}(y) = \\ &= \int \mathrm{d}\boldsymbol{m}^{\sim}(x) \int \varphi(yx)\,\mathrm{d}\boldsymbol{n}^{\sim}(y) = \int \varphi(yx)\,\mathrm{d}\boldsymbol{n}^{\sim}(y)\,\mathrm{d}\boldsymbol{m}^{\sim}(x) = \\ &= (\boldsymbol{n}^{\sim}*\boldsymbol{m}^{\sim})(\varphi),\end{aligned}$$

whence $(\boldsymbol{m}*\boldsymbol{n})^{\sim} = \boldsymbol{n}^{\sim}*\boldsymbol{m}^{\sim}$ and the proposition is proved.

Remark. If $\boldsymbol{m}: \mathscr{K}(G) \to A$ is a not necessary dominated measure, then $\boldsymbol{m}^{\sim}$ is also a measure. In fact, let $K \subset G$ be a compact set and let $a > 0$ be such that

$$|\boldsymbol{m}(h)| \leqslant a\,\|h\|, \quad \text{for } h \in \mathscr{K}(G, K^{-1}).$$

For every function $\varphi \in \mathscr{K}(G, K)$ we have then $\check{\varphi} \in \mathscr{K}(G, K^{-1})$, therefore

$$|\boldsymbol{m}^{\sim}(\varphi)| = |\overline{\boldsymbol{m}(\overline{\check{\varphi}})}| = |\boldsymbol{m}(\check{\varphi})| \leqslant a\,\|\check{\varphi}\| = a\,\|\varphi\|,$$

that is, $\boldsymbol{m}^{\sim}$ is a measure on G.

Properties (1), (2) and (3) of proposition 25.12 remain valid for every measure (not necessarily dominated) and property (5) remains valid for every dominated measure $\boldsymbol{m}$, (not necessary bounded).

25.13 PROPOSITION. *Let* $\boldsymbol{m}: \mathscr{K}(G) \to A$ *be a dominated measure and* $\boldsymbol{f}: G \to A$ *a function.*

(1) *the function* $\boldsymbol{f}$ *is* $\boldsymbol{m}^{\sim}$*-negligible if and only if* $\check{\boldsymbol{f}}$ *is* $\boldsymbol{m}$*-negligible;*

(2) *the function* $\boldsymbol{f}$ *is* $\boldsymbol{m}^{\sim}$*-measurable if and only if* $\check{\boldsymbol{f}}$ *is* $\boldsymbol{m}$*-measurable;*

(3) *the function* $\boldsymbol{f}$ *is* $\boldsymbol{m}^{\sim}$*-integrable if and only if* $\check{\boldsymbol{f}}$ *is* $\boldsymbol{m}$*-integrable. In this case we have*

$$\int \boldsymbol{f}(x)\,\mathrm{d}\boldsymbol{m}^{\sim}(x) = \overline{\int \overline{\check{\boldsymbol{f}}}(x)\,\mathrm{d}\boldsymbol{m}(x)} = \overline{\int \overline{\boldsymbol{f}(x^{-1})}\,\mathrm{d}\boldsymbol{m}(x)}.$$

In fact, since $|\boldsymbol{m}^{\sim}| = |\boldsymbol{m}|^{\sim}$, if $\boldsymbol{f}$ is $\boldsymbol{m}^{\sim}$-negligible, that is $|\boldsymbol{m}^{\sim}|$-negligible, then $\boldsymbol{f}$ is $|\boldsymbol{m}|^{\sim}$-negligible and, by proposition 25.10, $\check{\boldsymbol{f}}$ is $|\boldsymbol{m}|$-negligible, that is $\boldsymbol{m}$-negligible. Conversely, if $\check{\boldsymbol{f}}$ is $\boldsymbol{m}$-negligible, it follows that $\boldsymbol{f}$ is $\boldsymbol{m}^{\sim}$-negligible.

The other two properties are proved similarly.

The equality $\boldsymbol{m}^{\sim}(\boldsymbol{f})=\overline{\boldsymbol{m}(\check{\boldsymbol{f}})}$ for $\boldsymbol{f}\in\mathscr{L}_A^1(\boldsymbol{m}^{\sim})$ is deduced from corollary 20.32 and theorem 22.40, remarking that α is bijective and $\boldsymbol{m}^{\sim}=\alpha(\bar{\boldsymbol{m}})$.

For every function $\boldsymbol{f}: G\to A$ put

$$\boldsymbol{f}^{\sim}=\bar{\check{\boldsymbol{f}}}\,\check{\Delta}$$

Remark. We have to be careful not to mix up the function $\boldsymbol{f}^{\sim}=\bar{\check{\boldsymbol{f}}}\,\check{\Delta}$ with the equivalence class of the function f, which in §6 was denoted also by $\boldsymbol{f}^{\sim}$. We shall often use also the notation $\boldsymbol{f}^{*}=\bar{\check{\boldsymbol{f}}}\,\check{\Delta}$.

25.14 Proposition *If* $\boldsymbol{f}$ *is locally integrable and* $\boldsymbol{m}=\boldsymbol{f}\,\mathrm{d}x$, then $\boldsymbol{f}^{\sim}$ *is locally integrable and* $\boldsymbol{m}^{\sim}=\boldsymbol{f}^{\sim}\mathrm{d}x$.

In fact, if $\varphi\in\mathscr{K}(G)$, then $\check{\varphi}\in\mathscr{K}(G)$, therefore $\check{\varphi}\boldsymbol{f}\in\mathscr{L}_A^1$ hence (proposition 23.38) $\varphi\check{\boldsymbol{f}}\,\check{\Delta}\in\mathscr{L}_A^1$, therefore (theorem 8.62) $\varphi\boldsymbol{f}^{\sim}=\varphi\bar{\check{\boldsymbol{f}}}\,\check{\Delta}=\overline{\varphi\check{\boldsymbol{f}}\,\check{\Delta}}\in\mathscr{L}_A^1$. It follows that $\boldsymbol{f}^{\sim}$ is locally integrable. We have then

$$\begin{aligned}\boldsymbol{m}^{\sim}(\varphi)&=\overline{\boldsymbol{m}(\check{\bar{\varphi}})}=\overline{\textstyle\int\varphi(x^{-1})\mathrm{d}\boldsymbol{m}(x)}=\textstyle\int\varphi(x^{-1})\overline{\boldsymbol{f}(x)}\mathrm{d}x=\\&=\textstyle\int\varphi(x)\overline{\boldsymbol{f}(x^{-1})}\Delta(x^{-1})\mathrm{d}x=\int\varphi(x)\boldsymbol{f}^{\sim}(x)\,\mathrm{d}x\,,\end{aligned}$$

therefore $\boldsymbol{m}^{\sim}=\boldsymbol{f}^{\sim}\mathrm{d}x$ and the proposition is proved.

25.15 Proposition. *For every function* $f\geqslant 0$ *we have*

$$\textstyle\int^{*} f^{\sim}(x)\mathrm{d}x=\int^{*} f(x)\,\mathrm{d}x\,.$$

In fact, this equality is another form of the equality (proposition 23.35)

$$\textstyle\int^{*} f(x^{-1})\Delta(x^{-1})\,\mathrm{d}x=\int^{*} f(x)\,\mathrm{d}x\,.$$

25.16 Proposition. *Let* $\boldsymbol{f}: G\to A$.

(1) *the function* $\boldsymbol{f}$ *is negligible if and only if* $\boldsymbol{f}^{*}$ *is negligible*;

(2) *the function* $\boldsymbol{f}$ *is measurable if and only if* $\boldsymbol{f}^{*}$ *is measurable*;

(3) *the function* $\boldsymbol{f}$ *is integrable if and only if* $\boldsymbol{f}^{*}$ *is integrable. In this case we have*

$$\textstyle\int \boldsymbol{f}^{*}(x)\mathrm{d}x=\overline{\int \boldsymbol{f}(x)\,\mathrm{d}x}\,.$$

We use proposition 23.36, the equality $\boldsymbol{f}^{*}=\bar{\check{\boldsymbol{f}}}\,\check{\Delta}$ and theorem 8.62.

25.17 COROLLARY. *We have* $f \in \mathscr{L}_A^p$, $1 \leqslant p \leqslant +\infty$, *if and only if* $f^* \in \mathscr{L}_A^p$. *If* $f^* \in \mathscr{L}_A^p$, *we have* $N_p(f^*) = N_p(f)$.

It is easy to see that the mapping $f \to f^*$ is an involution on each *vector space* $\mathscr{L}_A^p$:

(1) $(f+g)^* = f^* + g^*$;
(2) $(\alpha f)^* = \bar{\alpha} f^*$;
(3) $f^{**} = f$;
(4) $N_p(f^*) = N_p(f)$.

For each function $f: G \to A$ put

$$\dot{f}^* = \dot{\widehat{f^*}}.$$

By proposition 25.16 we deduce that $\dot{f}^*$ is independent on the function f and depends only on the equivalence class $\dot{f}$.

25.18 PROPOSITION. *The mapping* $\dot{f} \to \dot{f}^*$ *is an involution on the algebra* L_A^1.

In fact (proposition 25.13), if $f \in L_A^1$ and if $\boldsymbol{m} = f\,\mathrm{d}x$, then $\boldsymbol{m}^\sim = f^\sim \mathrm{d}x = f^* \mathrm{d}x$, therefore the mapping $\dot{f} \to \dot{f}^*$ is the restriction on the subalgebra L_A^1 of the involution $\boldsymbol{m} \to \boldsymbol{m}^\sim$ of the algebra $\mathscr{M}_A^1$.

Remark. By our convention, we shall write f instead of $\dot{f}$ and f^* instead of $\dot{f}^*$. By proposition 25.14, the involution $f \to f^*$ is the restriction on L_A^1 of the involution $\boldsymbol{m} \to \boldsymbol{m}^\sim$ of $\mathscr{M}_A^1$.

§26 Representations

1 *Representations of the group G*

We call a *representation* of the group G into an algebra A with unit, any function $U: G \to A$ such that

$$U_{st} = U_s U_t \text{ for } s, t \in G, \text{ and } U_e = 1,$$

where 1 denotes the unit element in A.

If, in addition, A is an algebra of operators on a Hilbert space and if U_s are *unitary operators*, we say that U is a *unitary representation*.

We shall assume in the sequel that A is a Banach algebra with unit.

26.1 PROPOSITION. *If U is a representation of the group G into A, then U_s is invertible for every $s \in G$ and*

$$U_s^{-1} = U_{s^{-1}}.$$

In fact, for every $s \in G$ we have

$$1 = U_e = U_{ss^{-1}} = U_s U_{s^{-1}} \text{ and } 1 = U_e = U_{s^{-1}s} = U_{s^{-1}} U_s,$$

therefore U_s is invertible and $U_s^{-1} = U_{s^{-1}}$.

26.2 PROPOSITION. *Let $U : G \to A$ be a mapping such that $U_{st} = U_s U_t$ for every s and t in G. If there exists $s_0 \in G$ such that U_s is invertible, then U is a representation.*

It is sufficient to prove that $U_e = 1$. We have

$$U_{s_0} U_{s_0 e} = U_{s_0} U_e$$

and multiplying on the left by $U_{s_0}^{-1}$ we obtain $1 = U_e$.

26.3 PROPOSITION. *If U is a bounded representation of G into A, then*

$$|U_s| \geqslant 1 \text{ for every } s \in G.$$

Let $\alpha = \|U\| = \sup_{s \in G} |U_s|$. Since $|U_e| = |1| = 1$, we have $\alpha \geqslant 1$.

Assume that there exists $a \in G$ with $|U_a| < 1$. Then $|U_a|^n \to 0$, therefore there exists a number N such that

$$|U_a|^N < \alpha^{-1}.$$

But

$$|U_{a^N}| = |U_a^N| \leqslant |U_a|^N < \alpha^{-1}$$

and

$$|U_{a^{-N}}|\,|U_{a^N}| \geqslant |U_{a^{-N}} U_{a^N} = |U_e| = 1,$$

therefore

$$|U_{a^{-N}}| > |U_{a^N}|^{-1} > \alpha,$$

which contradicts the inequality $|U_s| \leqslant \alpha$ for every $s \in G$. Thus, $|U_s| \geqslant 1$ for every $s \in G$.

Remarks.

1. The conclusion of the proposition remains true if U is bounded, $U_{st} = U_s U_t$ for every $s, t \in G$ and $U_e \neq 0$ (not necessarily $U_e = 1$). We replace in the proof α by $\alpha |U_e|$.

2. If, in addition, $|U_s^n| = |U_s|^n$ for every $s \in G$ and every natural number n, then $|U_s| \equiv 1$.

In fact, if there exists $a \in G$ with $|U_a| > 1$, then $|U_{a^n}| = |U_a|^n \to +\infty$ which contradicts the fact that U is bounded.

3. We shall show that if U is continuous and bounded and if G is commutative, then we have $|U_s| \equiv 1$.

Examples. Let $1 \leqslant p \leqslant +\infty$ and $f \in L_E^p$.

The left regular representation

$$U_s f = f_{s^{-1}}$$

and the right regular representation

$$V_s f = f^{s^{-1}}$$

are representations of the group G into the algebra $A = \mathscr{L}(L_E^p)$ and we have

$$|U_s| \equiv 1 \text{ and } |V_s| = \Delta(s^{-1})^{1/p}.$$

If $1 \leqslant p < +\infty$, these representations are *simply continuous* (corollary 23.19 and 23.34), that is, for each $f \in L_E^p$, the mappings $s \to U_s f$ and $s \to V_s f$ of G into L_E^p are continuous and we have $\|U_s f\|_p = \|f\|_p$ and $\|V_s f\|_p = \Delta(s^{-1})^{1/p}\|f\|_p$.

If H is a Hilbert space, then L_H^2 is a Hilbert space for the scalar product

$$\langle f, g \rangle = \int \langle f(x), g(x) \rangle \, dx .$$

In this case the left regular representation U is a *unitary representation.* In fact, for each $s \in G$ we have

$$\langle U_s g, U_s h \rangle = \int \langle g(s^{-1}x), h(s^{-1}x) \rangle dx = \int \langle g(x), h(x) \rangle dx = \langle g, h \rangle$$

for $g, h \in L_H^2$, therefore U_s is a unitary operator.

2 *Representations of the algebra L^1*

We shall call a *representation* of an algebra A_1 into an algebra A_2 every linear and multiplicative mapping $T : A_1 \to A_2$:

$$T_{a+b} = T_a + T_b,\ T_{\alpha a} = \alpha T_a,\ T_{ab} = T_a T_b\,.$$

If A_1 and A_2 are involution algebras, we say that the representation $T : A_1 \to A_2$ is symmetric if

$$T_{a^*} = (T_a)^* \text{ for every } a \in A_1\,.$$

Let A be a Banach algebra with unit element.

26.4 Proposition. *If A is commutative and $U : G \to A$ is a measurable bounded representation, then the mapping $T : L^1_A \to A$ defined by the equality*

$$Tf = \int U_s f(s)\,\mathrm{d}s, \text{ for } f \in L^1_A\,,$$

is a continuous representation and we have

$$\|T\| = \|U\|_\infty \text{ and } T(L^1_A) = A\,.$$

We remark that $U \in L^\infty_A$, therefore the integral in the statement exists for every $f \in L^1_A$. The linearity of the mapping $f \to T_f$ can be easily verified. Moreover, we have

$$T_{af} = a\,T_f \text{ for } f \in L^1_A \text{ and } a \in A\,.$$

This mapping is continuous:

$$|T_f| = |\int U_s f(s)\mathrm{d}s| \leqslant \int |U_s|\,|f(s)|\,\mathrm{d}s \leqslant \|U\|_\infty\,\|f\|_1\,.$$

The equality $\|T\| = \|U\|_\infty$ follows from theorem 18.44, since U is locally integrable.

Let now f and g be two functions from $\mathscr{L}^1_A$. Modifying these functions on a negligible set, which does not modify the operators T_f and T_g, we may assume that f and g vanish outside the union of a sequence of compact subsets of G.

The function $f(t)g(s)$ is integrable for $\mathrm{d}t\,\mathrm{d}s$, therefore the function $f(t)g(t^{-1}s)$ is integrable for $\mathrm{d}t\,\mathrm{d}s$, consequently the function $U_s f(t)g(t^{-1}s)$ is integrable for $\mathrm{d}t\,\mathrm{d}s$. Replacing s by ts we deduce that the function $U_{ts}f(t)g(s)$ is integrable and

$$\iint U_s f(t)g(t^{-1}s)\,\mathrm{d}t\,\mathrm{d}s = \iint U_{ts} f(t)g(s)\,\mathrm{d}t\,\mathrm{d}s\,.$$

Applying the Fubini theorem—since the functions in these integrals vanish outside the union of a sequence of compact subsets of $G \times G$—we obtain

$$\begin{aligned} T_{f*g} &= \int U_s f*g(s)\,ds = \int U_s ds \int f(t)\,g(t^{-1}s)\,dt = \\ &= \int\int U_s f(t)\,g(t^{-1}s)\,dt\,ds = \\ &= \int\int U_{st} f(t)\,g(s)\,dt\,ds = \int\int U_s U_t f(t)\,g(s)\,dt\,ds = \\ &= \int U_t f(t)\,dt \int U_s g(s)\,ds = T_f T_g\,. \end{aligned}$$

It remains to show that $T(L_A^1) = A$. Let $a \in A$. There exists a function $\varphi \in \mathscr{K}(G)$ such that $\int \varphi(s)\,ds = 1$. The function $s \to U_s^{-1}$ is measurable and bounded, hence the function $s \to f(s) = a\varphi(s)U_s^{-1}$ is integrable. We have

$$T_f = \int U_s f(s)\,ds = a\int e U_s \varphi(s)\,U_s^{-1}\,ds = a\int \varphi(s)\,ds = a\,.$$

Thus $T(L_A^1) = A$ and the proposition is proved.

Remarks.

1. There exists a function $g \in \mathscr{K}_A(G)$ such that T_g^{-1} exists.

In fact, since $T(L_A^1) = A$, there exists a function $f \in L_A^1$ such that T_f^{-1} exists. There exists then a neighbourhood V of T_f such that for every $a \in V$, a^{-1} exists. Since T is continuous, $U = T^{-1}(V)$ is a neighbourhood of f, therefore there exists $g \in U \cap \mathscr{K}_A(G)$. Then $T_g \in V$, hence T_g^{-1} exists.

2. If G is *commutative*, we shall show that every continuous representation $T: L_A^1 \to A$ such that $T(L_A^1) = A$ can be obtained as in proposition 26.4, from a continuous bounded representation $U: G \to A$, and T can be uniquely extended to a representation of $\mathscr{M}_A^1$.

26.5 Proposition. *If $U: G \to A$ is a continuous bounded representation, then the mapping $T: \mathscr{M}_A^1 \to A$ defined by the equality*

$$T_m = \int U_s\,dm(s), \text{ for } m \in \mathscr{M}_A^1\,,$$

is a continuous representation and $\|T\| = \|U\|_\infty$.

Let us remark that the continuous bounded function U_s is integrable with respect to each bounded measure $m \in \mathscr{M}_A^1$. It is easy to see that T is linear. We have

$$|T_m| = |\int U_s\,dm(s)| \leqslant \int |U_s|\,d|m| \leqslant \|U\|_\infty\,\|\,|m|\,\|\,,$$

hence T is continuous and $\|T\| \leqslant \|U\|_\infty$. For the restriction T_0 of T to L_A^1 we have $\|T_0\| \leqslant \|T\|$; on the other hand, by proposition 26.4, $\|T_0\| = \|U\|_\infty$. It follows that $\|T\| = \|U\|_\infty$.

If $\boldsymbol{m}, \boldsymbol{n} \in \mathscr{M}_A^1$, then applying theorem 22.21 we obtain

$$T_{\boldsymbol{m}*\boldsymbol{n}} = \int U_s \, \mathrm{d}\boldsymbol{m}*\boldsymbol{n} = \int\int U_{st} \, \mathrm{d}\boldsymbol{m}(s)\mathrm{d}\boldsymbol{n}(t) = \int\int U_s U_t \, \mathrm{d}\boldsymbol{m}(s)\mathrm{d}\boldsymbol{n}(t) = \\ = \int U_s \, \mathrm{d}\boldsymbol{m}(s) \int U_t \, \mathrm{d}\boldsymbol{n}(t) = T_{\boldsymbol{m}} T_{\boldsymbol{n}},$$

therefore T is a representation.

Remark. The proposition remains valid if U is measurable with respect to each bounded measure, in particular, if U is a Borel function.

3 *Weakly measurable representations*

Let E be a Banach space. We can consider representations with values in the algebra $\mathscr{L}(E)$. In this case we can study the simply measurable or weakly measurable representations.

We shall state first a previous result (corollary 18.45).

26.6 Proposition. *Let $U: G \to \mathscr{L}(E)$ be a bounded weakly measurable function. There exists a continuous linear mapping $T: L^1 \to \mathscr{L}(E, E'')$ such that*

$$\langle T_f x, x' \rangle = \int f(s) \langle U_s x, x' \rangle \mathrm{d}s, \text{ for } f \in L^1, x \in E \text{ and } x' \in E'$$

and

$$\|T\| \leqslant \|U\|_\infty$$

We have the equality $\|T\| = \|U\|_\infty$ in each of the following cases:

(1) *there exists a lifting ρ of $\mathscr{L}^\infty(G)$ such that $\rho(U) = U$;*

(2) *E is of countable type. In this case U is simply measurable and we have $T_f \in \mathscr{L}(E)$ and*

$$T_f x = \int f(s) U_s x \, \mathrm{d}s, \text{ for } f \in L^1 \text{ and } x \in E;$$

(3) *U is measurable. In this case we have $T_f \in \mathscr{L}(E)$ and*

$$T_f = \int f(s) U_s \, \mathrm{d}s, \text{ for } f \in L^1.$$

26.7 Proposition. *Let $U: G \to \mathscr{L}(E)$ be a bounded weakly measurable representation. Assume that the corresponding linear mapping $T: f \to T_f$ of L^1 into $\mathscr{L}(E)$ defined by the equality,*

$$\langle T_f x, x' \rangle = \int f(s) \langle U_s x, x' \rangle \, \mathrm{d}s, \text{ for } f \in \mathscr{L}^1, x \in E \text{ and } x' \in E',$$

takes on values in $\mathscr{L}(E)$. Then T is a continuous representation of the algebra L^1 into the algebra $\mathscr{L}(E)$.

If E is a Hilbert space and U is a unitary representation, then T is symmetric:

$$T_{f^*} = T_f^* \text{ for } f \in L^1 .$$

Let $f, g \in L^1$. We shall show that $T_{f*g} = T_f T_g$. Modifying the functions f and g on a negligible set we may assume that f and g vanish outside the union of a sequence of compact sets, which does not modify T_f, T_g and T_{f*g}.

The function $f(t)g(s)$ is integrable for $\mathrm{d}t\,\mathrm{d}s$, therefore the function $f(t)g(t^{-1}s)$ is integrable for $\mathrm{d}t\,\mathrm{d}s$, therefore $\langle U_s x, x' \rangle f(t) g(t^{-1}s)$ is integrable. Replacing t by ts we deduce that the function $\langle U_{ts} x, x' \rangle f(t) g(s)$ is integrable and

$$\int\int \langle U_s x, x' \rangle f(t) g(t^{-1}s) \mathrm{d}t\,\mathrm{d}s = \int\int \langle U_{ts} x, x' \rangle f(t) g(s) \mathrm{d}t\,\mathrm{d}s .$$

Remarking that $\langle U_{ts} x, x' \rangle = \langle U_t U_s x, x' \rangle = \langle U_s x, {}^t U_t x' \rangle$ and that the Fubini theorem can be applied for the functions in the above integrals, since they vanish outside the union of a sequence of compact subsets of $G \times G$, we obtain

$$\begin{aligned}\langle T_{f*g} x, x' \rangle &= \int \langle U_s x, x' \rangle f*g(s) \mathrm{d}s = \\ &= \int \langle U_s x, x' \rangle \mathrm{d}s \int f(t) g(t^{-1}s) \mathrm{d}t = \\ &= \int\int \langle U_s x, x' \rangle f(t) g(t^{-1}s) \mathrm{d}t\,\mathrm{d}s = \\ &= \int\int \langle U_{st} x, x' \rangle f(t) g(s) \mathrm{d}t\,\mathrm{d}s = \\ &= \int f(t) \mathrm{d}t \int \langle U_s x, {}^t U_t x' \rangle g(s) \mathrm{d}s = \int f(t) \langle U_g x, {}^t U_t x' \rangle \mathrm{d}t = \\ &= \int f(t) \langle U_t U_g x, x' \rangle \mathrm{d}t = \langle T_f T_g x, x' \rangle\end{aligned}$$

whence $T_{f*g} = T_f T_g$.

Assume now that E is a Hilbert space and that all the operators U_s are unitary: $U_{s^{-1}} = U_s^{-1} = U_s^*$. Then for every $f \in L^1$ and $x, x' \in E$ we have

$$\begin{aligned}\langle T_{f^*} x, x' \rangle &= \int f^*(s) \langle U_s x, x' \rangle \mathrm{d}s = \\ &= \int \overline{f(s^{-1})} \Delta(s^{-1}) \langle U_s x, x' \rangle \mathrm{d}s = \\ &= \int \overline{f(s)} \langle U_{s^{-1}} x, x' \rangle \mathrm{d}s = \int \overline{f(s)} \langle U_s^* x, x' \rangle \mathrm{d}s = \\ &= \int \overline{f(s)} \langle x, U_s x' \rangle \mathrm{d}s = \int \overline{f(s)} \overline{\langle U_s x', x \rangle} \mathrm{d}s = \\ &= \overline{\langle T_f x', x \rangle} = \langle x, T_f x' \rangle .\end{aligned}$$

whence $T_{f^*} = T_f^*$.

Remark. In case U is weakly continuous we can not deduce that the mapping $T : \mathcal{M}^1 \to \mathcal{L}(E)$ is a representation.

But if U is continuous we can deduce that $T_{\mu * \nu} = T_\mu T_\nu$ and moreover, that $T_{\boldsymbol{m} * \boldsymbol{n}} = T_{\boldsymbol{m}} T_{\boldsymbol{n}}$ for every bounded measures $\boldsymbol{m}$ and $\boldsymbol{n}$ with values in $\mathcal{L}(E)$ (proposition 26.5).

26.8 PROPOSITION. *Let $T : f \to T_f$ be a continuous representation of the algebra L^1 onto $\mathcal{L}(E)$ such that the set $F = \{T_f x \mid f \in L^1, x \in E\}$ is dense in E. There exists then a simply continuous representation U of the group G into $\mathcal{L}(E)$ such that*

$$T_f x = \int f(s) U_s x \,ds, \text{ for } f \in L^1 \text{ and } x \in E .$$

If E is a Hilbert space and if $T_{f^*} = T_f^*$ for every $f \in L^1$, then U is a unitary representation of the group G.

Let $a \in G$. Denote by (u) the approximating unit in L^1. For every function $f \in L^1$ we have $u_a * f = (u * f)_a$ and $u * f \to f$ therefore $(u * f)_a \to f_a$, whence $u_a * f \to f_a$. Since the representation $T : f \to T_f$ is continuous, we deduce $\|T_{u_a} T_f - T_{f_a}\| \to 0$.

It follows that for every $x \in E$ we have

$$\lim_u T_{u_a} T_f x = T_{f_a} x ,$$

the convergence taking place in E. Thus, for every $y \in F$ there exists the limit

$$V_a y = \lim_u T_{u_a} y ,$$

and if $y = T_f x$ with $f \in \mathcal{L}^1$ and $x \in E$, then

$$V_a T_f x = T_{f_a} x \in F .$$

We deduce that V_a is a linear mapping of F into F (the linearity of V_a follows from the linearity of the limit $\lim_u T_{u_a} y$). This mapping is also continuous, since

$$\|T_{u_a}\| \leqslant \|T\| N_1(u_a) \leqslant \|T\|,$$

therefore

$$|V_a y| < \|T\| \, |y| .$$

It follows that V_a can be extended uniquely to a linear continuous mapping of E into E, also denoted by V_a, and we have $|V_a| \leqslant \|T\|$.

If $a, b \in G$ and $f \in \mathscr{L}^1$ and $x \in E$, we have

$$V_{ab} T_f x = T_{f_{ab}} x = T_{(f_a)_b} x = V_b T_{f_a} x = V_b V_a T_f x.$$

Since the elements of the form $T_f x$ with $f \in \mathscr{L}^1$ and $x \in E$ form a dense set in E, we deduce

$$V_{ab} = V_b V_a .$$

Similarly, we have $V_e T_f x = T_{f_e} x = T_f x$ for $f \in \mathscr{L}^1$ and $x \in E$, therefore $V_e x = x$ for every $x \in E$, hence $V_e = I$.

Let $f \in \mathscr{L}^1$. The mapping $a \to f_a$ of G into $\mathscr{L}^1$ is continuous, therefore the mapping $a \to T_{f_a}$ of G into $\mathscr{L}(E)$ is continuous. In particular, for each $x \in E$ the mapping $a \to V_a T_f x = T_{f_a} x$ of G into E is continuous. Thus, for each $y \in F$ the mapping $a \to V_a y$ of G into E is continuous.

Let us show that for every $x \in E$ the mapping $a \to V_a x$ is continuous on G.

Let $x \in E$ and $\varepsilon > 0$. There exists $y \in F$ with $|y - x| < \varepsilon/3 \|T\|$. Then $|V_s x - V_s y| = |V_s(x - y)| < \varepsilon/3$ for every $s \in G$. Let $a_0 \in G$. Since the mapping $a \to V_a y$ is continuous at a_0, there exists a neighbourhood V of a_0 such that for $a \in V$ we have $|V_a y - V_{a_0} y| < \varepsilon/3$. Then, for $a \in V$ we have

$$|V_a x - V_{a_0} x| \leqslant |V_a x - V_a y| + |V_a y - V_{a_0} y| + |V_{a_0} y - V_{a_0} x| < \varepsilon ,$$

therefore the mapping $a \to V_a x$ is continuous at a_0. Since a_0 is arbitrary in G, it follows that the mapping $a \to V_a x$ is continuous on G.

If for each $s \in G$ we denote

$$U_s = V_{s^{-1}}$$

we have $U_e = I$ and $U_{st} = U_s U_t$ for $s, t \in G$, therefore the mapping U is a representation of G into $\mathscr{L}(E)$ and this representation is simply continuous and bounded. For every $f \in \mathscr{L}^1$ and $s \in G$ we have

$$U_s T_f = T_{f_{s^{-1}}} .$$

It remains to show that

$$T_f x = \int f(s) U_s x \, \mathrm{d}x, \text{ for } f \in \mathscr{L}^1 \text{ and } x \in E .$$

Let $x \in E$ and $x' \in E'$. If for every function $f \in \mathscr{K}(G)$ we put $\mu(f) = \langle T_f x, x' \rangle$, then μ is a scalar measure on G, absolutely continuous with respect to the Haar measure and

$$|\mu(f)| \leqslant |x| \, |x'| \, \|T\| \int |f(s)| \mathrm{d}s .$$

It follows that there exists a bounded complex function $\chi(t)$, measurable with respect to the Haar measure, such that $\mathrm{d}\mu = \chi \mathrm{d}t$ and

$$\sup_{t\in G} |\chi(t)| \leqslant |x|\,|x'|\,||T||\,.$$

For $f, g \in \mathscr{K}(G)$, we have

$$\begin{aligned}\langle T_f T_g x, x'\rangle &= \langle T_{f*g} x, x'\rangle = \mu(f*g) = \int \chi(t) f*g(t)\,\mathrm{d}t = \\ &= \int \chi(t)\mathrm{d}t \int f(s) h(s^{-1}t)\,\mathrm{d}s = \int f(s)\mathrm{d}s \int \chi(t) g(s^{-1}t)\mathrm{d}t = \\ &= \int f(s)\mathrm{d}s \int \chi(t) g_{s^{-1}}(t)\,\mathrm{d}t = \\ &= \int f(s) \langle T_{g_{s^{-1}}}\, x, x'\rangle\,\mathrm{d}s = \int f(s)\langle U_s T_g x, x'\rangle\,\mathrm{d}s\,.\end{aligned}$$

Since the function $s \to U_s T_g x$ is continuous and bounded, the function $s \to f(s)\, U_s T_g x$ is integrable and

$$T_f T_g x = \int f(s)\, U_s T_g x\,\mathrm{d}s\,.$$

Thus for $y \in F$ and $f \in \mathscr{K}(G)$ we have

$$T_f y = \int f(s)\, U_s y\,\mathrm{d}s\,.$$

Then, for $x \in E$ we choose a sequence $y_n \in E$ converging to x; applying the Lebesgue theorem we deduce

$$T_f x = \int f(s)\, U_s x\,\mathrm{d}s, \text{ for } f \in \mathscr{K}(G) \text{ and } x \in E.$$

Finally this equality remains valid for every function $f \in \mathscr{L}^1$ and every $x \in E$, as we can see by choosing a sequence (f_n) of functions from $\mathscr{K}(G)$ converging in the mean and almost everywhere to f.

Assume now that E is a Hilbert space and that $T_{f*} = T_f^*$, for every $f \in L^1$. Show that each U_s is unitary, that is, $U_s^* = U_s^{-1} = U_{s^{-1}}$. If (u) is the approximating unit in $\mathscr{L}^1$, then for every function $g \in \mathscr{L}^1$ we have $g^s * u \to g^s$. Then $g * u_{s^{-1}} = \Delta(s^{-1}) g^s * u \to \Delta(s^{-1}) g^s = ((g^*)_s)^*$. It follows that for every function $f \in \mathscr{L}^1$ we have (denoting $g = f^*$, therefore $g^* = f$), $f^* * u_{s^{-1}} \to (f_s)^*$, whence $(u_{s^{-1}})^* * f \to f_s$; since T is continuous, we deduce

$$T_{u_{s^{-1}}}\, T_f = T_{(u_{s^{-1}})^*} T_f \to T_{f_s} = U_{s^{-1}} T_f\,.$$

Since every $y \in F$ is of the form $T_f x$ with $f \in \mathscr{L}^1$ and $x \in E$ we deduce $T_{u_{s^{-1}}}\, y \to U_{s^{-1}}\, y$, therefore for every $z \in F$ we have

$$\langle T^*_{u_{s^{-1}}}\, y, z\rangle = \langle U_{s^{-1}} y, z\rangle\,.$$

On the other hand, for $x \in E$ and $f \in \mathscr{L}^1$ we have

$$T_{u_{s^{-1}}} T_f x = U_s T_f x ,$$

therefore for every $z \in F$ we have

$$T_{u_{s-1}} z \to U_s z$$

and thus for every $y \in F$ we have

$$\langle T_{u_{s-1}} z, y \rangle \to \langle U_s z, y \rangle ,$$

whence

$$\langle z, T^*_{u_{s-1}} y \rangle \to \langle z, U^*_s y \rangle ,$$

that is

$$\langle T^*_{u_{s-1}} y, z \rangle \to \langle U^*_s y, z \rangle .$$

It follows that $\langle U_s y, z \rangle = \langle U_{s^{-1}} y, z \rangle$, for every $y, z \in F$, whence $U^*_s y = U_{s^{-1}} y$ for $y \in F$ and hence $U_s = U_{s^{-1}}$. Thus the proposition is proved.

26.9 COROLLARY. *Let U be a bounded weakly measurable representation of G into $\mathscr{L}(E)$ such that the corresponding representation T of the algebra L^1 takes on values in $\mathscr{L}(E)$. If the set $F = \{T_f x \mid f \in L^1, x \in E\}$ is dense in E, then there exists a bounded simply continuous representation V of G into $\mathscr{L}(E)$ such that for $x \in E$ and $x' \in E'$ we have*

$$\langle U_s x, x' \rangle = \langle V_s x, x' \rangle \text{ for almost every } s \in G .$$

In fact, by proposition 26.7, T is continuous, and by proposition 26.8 there exists a simply continuous representation V of G into $\mathscr{L}(E)$ such that

$$T_f x = \int f(s) V_s x \, ds, \text{ for } f \in L^1 \text{ and } x \in E .$$

For every $x' \in E'$ we have then

$$\langle T_f x, x' \rangle = \int f(s) \langle V_s x, x' \rangle ds$$

therefore

$$\int f(s) \langle U_s x, x' \rangle ds = \int f(s) \langle V_s x, x' \rangle ds$$

for every $f \in L^1$. It follows that

$$\langle U_s x, x' \rangle = \langle V_s x, x' \rangle, \text{ for almost every } s \in G .$$

26.10 Corollary. *If U is bounded and weakly continuous and if the set $F=\{Tx \mid f\in L^1, x\in E\}$ is dense in E, then U is simply continuous.*

In fact, from the proof of the preceding corollary we deduce that

$$\langle U_s x, x'\rangle = \langle V_s x, x'\rangle$$

for *every* $s\in G$, $x\in E$ and $x'\in E'$, whence $U_s = V_s$ for every $s\in G$, hence U is simply continuous.

26.11 Corollary. *Let U be a bounded weakly measurable representation of G into $\mathscr{L}(E)$. If E is of countable type and if the set $F=\{Tx \mid f\in L^1, x\in E\}$ is dense in E, then U is equal almost everywhere to a simply continuous representation V of G into $\mathscr{L}(E)$.*

By corollary 26.9, there exists a simply continuous representation V of G into $\mathscr{L}(E)$, such that for every $x\in E$ and every $x'\in E'$ we have

$$\langle U_s x, x'\rangle = \langle V_s x, x'\rangle, \text{ almost everywhere}.$$

Since E is of countable type, following the proof of the proposition 17.15, we deduce that $U_s = V_s$ almost everywhere.

26.12 Corollary. *If E is a Hilbert space of countable type, every weakly measurable unitary representation U of G into $\mathscr{L}(E)$ is simply continuous.*

Let us show first that the set $F=\{T_f x \mid f\in\mathscr{K}(G), x\in E\}$ is dense in E. In fact, in the contrary case there would exist an element $y\neq 0$ of E such that for every $f\in\mathscr{K}(G)$ and every $x\in E$ we have $\langle T_f x, y\rangle = 0$, that is

$$\int f(s)\langle U_s x, y\rangle\, ds = 0.$$

It follows then that for each $x\in E$ there exists a negligible set $A(x)\subset G$ such that $\langle U_s x, y\rangle = 0$ for $s\notin A(x)$. Let (x_n) be a sequence dense in E and $A=\cup A(x_n)$. The set A is negligible and $\langle x_n, U_s^{-1} y\rangle = \langle U_s x_n, y\rangle = 0$, for every n and every $s\notin A$.

Since the sequence (x_n) is dense in E, it follows that $U_s^{-1} y = 0$ for every $s\notin A$. Since U_s^{-1} is a unitary operator, it is isometric, hence $y=0$, which is a contradiction.

Thus, the set F is dense in E. Let now $f\in\mathscr{K}(G)$, and $x\neq 0$, $y\neq 0$ be two elements in E. Show that the function $t\to\langle U_t T_f x, y\rangle$ is continuous.

Let $a \in G$. We have

$$\langle U_a T_f x, y \rangle = \langle T_f x, U_{a^{-1}} y \rangle = \int f(s) \langle U_s x, U_{a^{-1}} y \rangle \mathrm{d}s =$$
$$= \int f(s) \langle U_{as} x, y \rangle \mathrm{d}s = \int f(a^{-1} s) \langle U_s x, y \rangle \mathrm{d}s \,.$$

Let $\varepsilon > 0$. Since f is left uniformly continuous, there exists a neighbourhood V of e such that

$$|f(u) - f(v)| < \varepsilon / |x| \, |y|, \text{ if } uv^{-1} \in V \,.$$

For every $s \in G$ and $t \in aV$ we have $a^{-1}s(t^{-1}s^{-1}) = a^{-1}t \in V$, therefore

$$|(fa^{-1}s) - f(t^{-1}s)| < \varepsilon / |x| \, |y| \,,$$

consequently

$$| < U_a T_f x, y > - < U_t T_f x, y > | < \varepsilon \,.$$

It follows that the mapping $t \to \langle U_t T_f x, y \rangle$ is continuous at a. As a is arbitrary in G, this mapping is continuous on G. Thus, the mapping $t \to \langle U_t T_f x, y \rangle$ is continuous for every $x \in F$.

Now let $x \in E$. Since F is dense in E, there exists a sequence (x_n) of elements of F convergent to x. Then

$$|\langle U_t x, y \rangle - \langle U_t x_n, y \rangle| = |\langle U_t \, (x - x_n), y \rangle| \leqslant |x - x_n| \, |y| \,,$$

therefore the sequence of continuous functions $t \to \langle U_t x_n, y \rangle$ converges uniformly to the function $t \to \langle U_t x, y \rangle$, therefore this function is also continuous. Thus the unitary representation U is weakly continuous. By corollary 26.10, U is simply continuous.

4 *Regular representations*

Let $1 \leqslant p \leqslant +\infty$. For each measure $\boldsymbol{m} \in \mathscr{M}_A^1$ define the operator $T_{\boldsymbol{m}}$ on the algebra $B = \mathscr{L}(L_A^p)$ by the equality

$$T_{\boldsymbol{m}} \boldsymbol{g} = \boldsymbol{m} * \boldsymbol{g}, \text{ for } \boldsymbol{g} \in L_A^p \,.$$

It is immediately seen that the mapping $T : \boldsymbol{m} \to T_{\boldsymbol{m}}$ of $\mathscr{M}_A^1$ into the algebra B is linear and multiplicative:

$$T_{\boldsymbol{m}*\boldsymbol{n}} \boldsymbol{g} = (\boldsymbol{m} * \boldsymbol{n}) * \boldsymbol{g} = \boldsymbol{m} * (\boldsymbol{n} * \boldsymbol{g}) = T_{\boldsymbol{m}} (\boldsymbol{n} * \boldsymbol{g}) = T_{\boldsymbol{m}} T_{\boldsymbol{n}} \boldsymbol{g} \,,$$

therefore $T : \mathscr{M}_A^1 \to \mathscr{L}(L_A^p)$ is a representation. This representation is continuous:

$$\|T_{\boldsymbol{m}}\boldsymbol{g}\|_p = \|\boldsymbol{m}*\boldsymbol{g}\|_p \leqslant |||\boldsymbol{m}|||\;\|\boldsymbol{g}\|_p$$

therefore $\|T_{\boldsymbol{m}}\| \leqslant |||\boldsymbol{m}|||$, consequently $\|T\| \leqslant 1$.

The representation T defined above is called the *regular representation* of the algebra $\mathscr{M}_A^1$ into the algebra $\mathscr{L}(L_A^p)$.

The connection between the representation T of the algebra L^1 and the left regular representation U_s of the group G is given by the following proposition:

26.13 Proposition. *If* $1 \leqslant p < +\infty$, *for every function* $f \in \mathscr{L}^1$ *and* $\boldsymbol{g} \in L_E^p$, *the mapping* $s \to f(s)\,U_s\boldsymbol{g}$ *of* G *into* $\mathscr{L}(L_E^p)$ *is integrable and we have*

$$T_f\boldsymbol{g} = \int f(s)\,U_s\boldsymbol{g}\,\mathrm{d}s\,.$$

Let E' be the dual of E and $\boldsymbol{h} \in \mathscr{K}_{E'}(G)$. The function $|f(s)|\,|\boldsymbol{h}(t)|\,|\boldsymbol{g}(s^{-1}t)|$ is $\mathrm{d}s \otimes \mathrm{d}t$-measurable, therefore the function $\boldsymbol{g}(s^{-1}t)$ is measurable with respect to the bounded measure $|f|\,\mathrm{d}s \otimes |\boldsymbol{h}|\,\mathrm{d}t$. Since

$$\int^* |\boldsymbol{h}|(t)|\,\mathrm{d}t \int^* |f(s)|\,|\boldsymbol{g}(s^{-1}t)|\,\mathrm{d}s = \int^* |\boldsymbol{h}(t)|\,|f*\boldsymbol{g}(t)|\,\mathrm{d}t < +\infty\,,$$

it follows that the function $\boldsymbol{g}(s^{-1}t)$ is integrable with respect to the measure $|f|\,\mathrm{d}s \otimes |\boldsymbol{h}|\,\mathrm{d}t$, therefore also with respect to the measure $f\,\mathrm{d}s \otimes \boldsymbol{h}\,\mathrm{d}t$. Applying the Fubini theorem we deduce:

$$\begin{aligned}\langle T_f\boldsymbol{g}, \boldsymbol{h}\rangle &= \langle f*\boldsymbol{g}, \boldsymbol{h}\rangle = \int \langle f*\boldsymbol{g}(t), \boldsymbol{h}(t)\rangle\,\mathrm{d}t = \\ &= \int \langle \int f(s)\boldsymbol{g}(s^{-1}t)\mathrm{d}s, \boldsymbol{h}(t)\rangle\,\mathrm{d}t = \int \mathrm{d}t \int f(s)\langle \boldsymbol{g}(s^{-1}t), \boldsymbol{h}(t)\rangle\,\mathrm{d}s = \\ &= \int f(s)\mathrm{d}s \int \langle \boldsymbol{g}(s^{-1}t), \boldsymbol{h}(t)\rangle\,\mathrm{d}t = \int f(s)\mathrm{d}s \int \langle \boldsymbol{g}_{s^{-1}}(t), \boldsymbol{h}(t)\rangle\,\mathrm{d}t = \\ &= \int f(s)\langle \boldsymbol{g}_{s^{-1}}, \boldsymbol{h}\rangle\,\mathrm{d}s = \int \langle f(s)\,U_s\boldsymbol{g}, \boldsymbol{h}\rangle\,\mathrm{d}s\,.\end{aligned}$$

Since the function $s \to U_s\boldsymbol{g}$ is continuous, the function $s \to f(s)\,U_s\boldsymbol{g}$ is measurable and the function $s \to \langle f(s)\,U_s\boldsymbol{g}, \boldsymbol{h}\rangle$ is integrable for every $\boldsymbol{h} \in \mathscr{K}_{E'}(G)$. It follows that the function $s \to f(s)\,U_s\boldsymbol{g}$ is integrable and

$$\int \langle f(s)\,U\boldsymbol{g}, \boldsymbol{h}\rangle\,\mathrm{d}s = \langle \int f(s)\,U\boldsymbol{g}\,\mathrm{d}s, \boldsymbol{h}\rangle\,,$$

whence

$$\langle T_f\boldsymbol{g}, \boldsymbol{h}\rangle = \langle \int f(s)\,U_s\boldsymbol{g}\,\mathrm{d}s, \boldsymbol{h}\rangle\,,$$

consequently

$$T_f\boldsymbol{g} = \int f(s)\,U_s\boldsymbol{g}\,\mathrm{d}s\,.$$

Remark. If the two members of this equality are functions from $\mathscr{L}_E^p$, this equality is written

$$\int f(s)g(s^{-1}x)ds = [\int f(s)g_{s^{-1}}ds](x)$$

for almost every $x \in G$.

5 Representations of commutative groups

Here we shall assume that G is a locally compact *commutative* group and that A is a *commutative Banach* algebra with unit.

We have seen (proposition 26.4) that to every measurable bounded representation $U: G \to A$ there corresponds a continuous representation $T: L_A^1 \to A$ by the equality

$$T(f) = \int f(s)U(s)ds, \text{ for } f \in L_A^1,$$

and we have $T(L_A^1) = A$ and $\|T\| = \|U\|_\infty$.

This result is true even if G is not commutative.

If G is commutative, we also have a converse proposition.

Let $T: L_A^1 \to A$ be a *multiplicative* mapping:

$$T(f*g) = T(f)T(g) \text{ for } f, g \in L_A^1.$$

For every $s \in G$ and $f, g \in L_A^1$ we have

$$f*g^s = f^s*g = (f*g)^s$$

consequently

$$T(f)T(g^s) = T(f^s)T(g).$$

We deduce that if T_f and T_g are invertible, we have $T(f^s)T(f)^{-1} = T(g^s)T(g)^{-1}$.

26.14 Lemma. *Let $T: L_A^1 \to A$ be a multiplicative mapping. Assume that there exists $h \in L_A^1$ such that $T(h)$ is invertible. Then the mapping $U: G \to A$. defined by the equality*

$$U(s) = T(h^s)T(h)^{-1},$$

is a representation which for every $s \in G$ and $f \in L_A^1$ satisfies the equality

$$T(f_s) = U(s^{-1})T(f).$$

We remark first that $U(s)$ does not depend on the function $\boldsymbol{h}\in L_A^1$ for which $T(\boldsymbol{h})^{-1}$ exists, but depends only on s. We have

$$U(e)=T(\boldsymbol{h}^e)\,T(\boldsymbol{h})^{-1}=T(\boldsymbol{h})\,T(\boldsymbol{h})^{-1}=1\,.$$

If $s, t\in G$, then

$$T(\boldsymbol{h}^{st}*\boldsymbol{h})=T(\boldsymbol{h}^s*\boldsymbol{h}^t)=T(\boldsymbol{h}^s)\,T(\boldsymbol{h}^t)$$

therefore

$$\begin{aligned}U(st)&=T(\boldsymbol{h}^{st})\,T(\boldsymbol{h})^{-1}=T(\boldsymbol{h}^{st})\,T(\boldsymbol{h})\,T(\boldsymbol{h})^{-1}\,T(\boldsymbol{h}^{-1})=\\&=T(\boldsymbol{h}^{st}*\boldsymbol{h})\,T(\boldsymbol{h})^{-1}\,T(\boldsymbol{h}^{-1})=T(\boldsymbol{h}^s)\,T(\boldsymbol{h})^{-1}\,T(\boldsymbol{h}^t)\,T(\boldsymbol{h})^{-1}=\\&=U(s)\,U(t)\end{aligned}$$

consequently U is a representation

For every $s\in G$ and $\boldsymbol{f}\in L_A^1$ we have then

$$\begin{aligned}T(\boldsymbol{f}_s)=T(\boldsymbol{f}^{s^{-1}})\,T(\boldsymbol{h})\,T(\boldsymbol{h})^{-1}&=T(\boldsymbol{f})\,T(\boldsymbol{h}^{s^{-1}})\,T(\boldsymbol{h})^{-1}=\\&=T(\boldsymbol{f})\,U(s^{-1})=U(s^{-1})\,T(\boldsymbol{f})\,.\end{aligned}$$

26.15 LEMMA. *Let $T: L_A^1\to A$ be a multiplicative mapping such that $T(\boldsymbol{h})^{-1}$ exists at least for a function $\boldsymbol{h}\in L_A^1$ and let $U: G\to A$ be the representation which satisfies the equality*

$$T(\boldsymbol{f}_s)=U(s^{-1})\,T(\boldsymbol{f})\ \textit{for}\ s\in G\ \textit{and}\ \boldsymbol{f}\in L_A^1\,.$$

If T is bounded, then U is bounded. If T is continuous, then U is continuous and $T(\boldsymbol{g})^{-1}$ exists at least for a function $\boldsymbol{g}$ from $\mathscr{K}_A(G)$.

In fact, if T is bounded:

$$|T(\boldsymbol{f})|\leqslant\alpha\int|\boldsymbol{f}|\,\mathrm{d}t,\ \text{for every}\ \boldsymbol{f}\in L_A^1$$

then

$$\begin{aligned}|U(s)|&=|T(\boldsymbol{h}_{s^{-1}})\,T(\boldsymbol{h})^{-1}|\leqslant|T(\boldsymbol{h}_{s^{-1}})|\,|T(\boldsymbol{h})^{-1}|\leqslant\\&\leqslant\alpha|\,T(\boldsymbol{h})^{-1}|\int|\boldsymbol{h}_{s^{-1}}|\,\mathrm{d}t=\alpha|T(\boldsymbol{h})^{-1}\int|\boldsymbol{h}|\,\mathrm{d}t\,,\end{aligned}$$

for every $s\in G$, consequently U is bounded.

Assume that T is continuous. Since the mapping $s\to\boldsymbol{h}_{s^{-1}}$ of G into L_A^1 is continuous, the composition $s\to\boldsymbol{h}_{s^{-1}}\to T(\boldsymbol{h}_{s^{-1}})$ is continuous and then the function $s\to T(\boldsymbol{h}_{s^{-1}})\,T(\boldsymbol{h})^{-1}=U(s)$ is also continuous.

Since $T(\boldsymbol{h})^{-1}$ exists, there exists an open neighbourhood V of $T(\boldsymbol{h})$

such that a^{-1} exists for every $a \in V$. Since T is continuous, the set $V' = T^{-1}(V)$ is an open neighbourhood of $\boldsymbol{h}$, therefore there exists $g \in V' \cap \mathscr{K}_A(G)$. For $g \in V' \cap \mathscr{K}_A$ we have $T(\boldsymbol{g}) \in V$, consequently $T(\boldsymbol{g})^{-1}$ exists.

26.16 PROPOSITION. *For every continuous representation $T: L_A^1 \to A$ such that $T(\boldsymbol{h})^{-1}$ exists for at least one function $\boldsymbol{h} \in L_A^1$, there exists a continuous bounded representation $U: G \to A$ such that*

$$T(\boldsymbol{f}) = \int U(s)\boldsymbol{f}(s)\mathrm{d}s, \textit{ for every } \boldsymbol{f} \in L_A^1 .$$

Let $\boldsymbol{g} \in \mathscr{K}_A(G)$ be a function such that $T(\boldsymbol{g})^{-1}$ exists (lemma 26.15). Consider the representation $U: G \to A$ defined by the equality

$$U(s) = T(\boldsymbol{g}_{s^{-1}})T(\boldsymbol{g})^{-1}, \text{ for } s \in G .$$

Since T is linear and continuous, it is bounded. By lemma 26.15, U is continuous and bounded. It remains to prove the equality

$$T(\boldsymbol{f}) = \int U(s)\boldsymbol{f}(s)\mathrm{d}s, \text{ for } \boldsymbol{f} \in L_A^1 .$$

Let $\boldsymbol{h} \in \mathscr{K}_A(G)$. Denote by S the support of $\boldsymbol{h}$ and by K the support of $\boldsymbol{g}$. The set SK is compact. For each $s \in G$ the function $t \to \boldsymbol{h}(s)\boldsymbol{g}_{s^{-1}}(t)$ is continuous, with support contained in SK. In fact, if $s \notin S$, then $\boldsymbol{h}(s) = 0$, therefore $\boldsymbol{h}(s)\boldsymbol{g}_s(t) = 0$. If $s \in S$ and $t \notin SK$, then $t \notin sK$, therefore $s^{-1}t \notin K$, hence $\boldsymbol{g}(s^{-1}t) = 0$, consequently $\boldsymbol{h}(s)\boldsymbol{g}_{s^{-1}}(t) = \boldsymbol{h}(s)\boldsymbol{g}(s^{-1}t) = 0$. The function $(s, t) \to \boldsymbol{h}(s)\boldsymbol{g}(s^{-1}t)$ is thus continuous on $G \times G$, with compact support contained in $S \times SK$. Since $T: L_A^1 \to A$ is a continuous linear mapping, for every function $\boldsymbol{f} \in \mathscr{K}_A(G)$ we have

$$|T(\boldsymbol{f})| \leqslant \alpha \int |\boldsymbol{f}(t)| \mathrm{d}t = \int |\boldsymbol{f}(t)| \mathrm{d}(\alpha t) .$$

Thus, the restriction $\boldsymbol{m}$ of T to $\mathscr{K}_A(G)$ is a dominated measure.

We have

$$T(\boldsymbol{f}) = \int \boldsymbol{f} \, \mathrm{d}\boldsymbol{m} \text{ for } \boldsymbol{f} \in \mathscr{K}_A(G) .$$

The function $(s, t) \to \boldsymbol{h}(s)\boldsymbol{g}(s^{-1}t)$ is integrable with respect to the product measure $\mathrm{d}s \otimes \mathrm{d}\boldsymbol{m}(t)$ and we have

$$\begin{aligned} \iint \boldsymbol{h}(s)\boldsymbol{g}(s^{-1}t)\mathrm{d}s\,\mathrm{d}\boldsymbol{m}(t) &= \int \boldsymbol{h}(s)\mathrm{d}s \int \boldsymbol{g}_{s^{-1}}(t)\mathrm{d}\boldsymbol{m}(t) = \\ &= \int \boldsymbol{h}(s)T(\boldsymbol{g}_{s^{-1}})\mathrm{d}s = \int \boldsymbol{h}(s)U(s)T(\boldsymbol{g})\mathrm{d}s = \\ &= T(\boldsymbol{g})\int \boldsymbol{h}(s)U(s)\mathrm{d}s . \end{aligned}$$

On the other hand

$$\int\int \boldsymbol{h}(s)\boldsymbol{g}(s^{-1}t)\,ds\,d\boldsymbol{m}(t)=\int d\boldsymbol{m}(t)\int \boldsymbol{h}(s)\boldsymbol{g}(s^{-1}t)ds=$$
$$=\int(\boldsymbol{h}*\boldsymbol{g})(t)\,d\boldsymbol{m}(t)=T(\boldsymbol{h}*\boldsymbol{g})=T(\boldsymbol{h})\,T(\boldsymbol{g})\,.$$

Thus

$$T(\boldsymbol{h})\,T(\boldsymbol{g})=T(\boldsymbol{g})\int \boldsymbol{h}(s)\,U(s)\,ds\,.$$

Multiplying both sides by $T(\boldsymbol{g})^{-1}$ we deduce

$$T(\boldsymbol{h})=\int \boldsymbol{h}(s)\,U(s)ds,\ \text{for}\ \boldsymbol{h}\in\mathscr{K}_A(G)\,.$$

Since U is continuous and bounded, the mapping $T':L_A^1\to A$ defined by the equality

$$T'(\boldsymbol{f})=\int \boldsymbol{f}(s)\,U(s)ds\ \text{ for }\ \boldsymbol{f}\in L_A^1$$

is a continuous representation and we have

$$T(\boldsymbol{h})=T'(\boldsymbol{h}),\ \text{for}\ \boldsymbol{h}\in\mathscr{K}_A(G)\,.$$

Since $\mathscr{K}_A(G)$ is dense in $\mathscr{L}_A^1$, we deduce $T(\boldsymbol{f})=T'(\boldsymbol{f})$ for every $\boldsymbol{f}\in L_A^1$ therefore

$$T(\boldsymbol{f})=\int \boldsymbol{f}(s)\,U(s)\,ds,\ \text{for}\ \boldsymbol{f}\in L_A^1\,.$$

26.17 Corollary. *Let $T:L_A^1\to A$ be a continuous representation. We have $T(L_A^1)=A$ if and only if $T(\boldsymbol{h})^{-1}$ exists for at least one function $\boldsymbol{h}\in L_A^1$.*

We use propositions 26.4 and 26.16.

26.18 Corollary. *Every measurable bounded representation $V:G\to A$ is equal almost everywhere to a bounded continuous representation.*

By proposition 26.5, to the representation V there corresponds a continuous representation $T:L_A^1\to A$ by the equality

$$T(\boldsymbol{f})=\int \boldsymbol{f}(s)\,V(s)\,ds,\ \text{for}\ \boldsymbol{f}\in L_A^1$$

and we have $T(L_A^1)=A$. By proposition 26.16, there exists a continuous bounded representation $U:G\to A$ such that

$$T(\boldsymbol{f})=\int \boldsymbol{f}(s)\,U(s)ds,\ \text{for}\ \boldsymbol{f}\in L_A^1\,.$$

It follows that

$$\int f(s)V(s)\mathrm{d}s = \int f(s)U(s)\mathrm{d}s, \text{ for } f \in L_A^1$$

whence $V(s) = U(s)$ almost everywhere.

26.19 COROLLARY. *There exists a bijective correspondence between the set of continuous representations $T: L_A^1 \to A$ such that $T(L_A^1) = A$ and the set of continuous bounded representations $U: G \to A$, given by the equality*

$$T(f) = \int f(s)U(s)\mathrm{d}s, \textit{ for } f \in L_A^1,$$

and we have

$$||T|| = ||U||_\infty .$$

The one-to-one correspondence follows from propositions 26.5 and 26.16 and the equality $||T|| = ||U||_\infty$ from proposition 26.6.

26.20 COROLLARY. *Let $T: L_A^1 \to A$ be a continuous representation with $T(L_A^1) = A$ and $U: G \to A$ the corresponding bounded continuous representation. Then $||T|| \geqslant 1$. We have $||T|| = 1$ if and only if $|U(s)| \equiv 1$.*

By corollary 26.19 we deduce that

$$||T|| = ||U||_\infty.$$

But U is bounded, therefore $|U(s)| \geqslant 1$ for every $s \in G$ (proposition 26.3). It follows that $||U||_\infty \geqslant 1$, therefore $||T|| \geqslant 1$. If $|U(s) \equiv 1$, then $||T|| = ||U||_\infty = 1$. Conversely, if $||T|| = 1$, then $||U||_\infty = 1$, therefore $|U(s)| \leqslant 1$ almost everywhere, hence everywhere, since U is continuous and the support of the Haar measure is the whole space G. Since we also have $|\bar{U}(s)| \geqslant 1$, it follows that $|U(s)| \equiv 1$.

26.21 COROLLARY. *There exists a bijective and isometric correspondence between the set of continuous representations $T: L_A^1 \to A$ with $T(L_A^1) = A$ and the set of continuous representation $T': \mathscr{M}_A^1 \to A$ with $T'(L_A^1) = A$, given by the equality*

$$T(f) = T'(f) \textit{ for } f \in L_A^1 .$$

If $T': \mathscr{M}_A^1 \to A$ is a continuous representation with $T'(L_A^1) = A$, then the restriction $T : L_A^1 \to A$ of T' is a continuous representation with $T(L_A^1) = A$. The representation T' is uniquely determined by its restriction T. In fact, for every $\boldsymbol{m} \in \mathscr{M}_A^1$ and every $\boldsymbol{f} \in L_A^1$ we have $\boldsymbol{m} * \boldsymbol{f} \in L_A^1$ and

$$T(\boldsymbol{m} * \boldsymbol{f}) = T'(\boldsymbol{m} * \boldsymbol{f}) = T'(\boldsymbol{m}) T'(\boldsymbol{f}) = T'(\boldsymbol{m}) T(\boldsymbol{f})$$

If we choose $\boldsymbol{f} \in L_A^1$ such that $T(\boldsymbol{f})^{-1}$ exists, then

$$T'(\boldsymbol{m}) = T(\boldsymbol{m} * \boldsymbol{f}) T(\boldsymbol{f})^{-1}, \text{ for every } \boldsymbol{m} \in \mathscr{M}_A^1.$$

Thus, the correspondence $T' \to T$ is bijective.

Now let $T: L_A^1 \to A$ be a continuous bounded representation with $T(L_A^1) = A$. There exists a continuous bounded representation $U : G \to A$ such that

$$T(\boldsymbol{f}) = \int \boldsymbol{f}(s) U(s) \mathrm{d}s, \text{ for } \boldsymbol{f} \in L_A^1.$$

The function U is integrable for every measure $\boldsymbol{m} \in \mathscr{M}_A^1$. We put

$$T'(\boldsymbol{m}) = \int U(s) \mathrm{d}\boldsymbol{m}(s), \text{ for } \boldsymbol{m} \in \mathscr{M}_A^1.$$

The mapping $T' : \mathscr{M}_A^1 \to A$ is a continuous representation (proposition 26.5) and $\|T'\| = \|U\|_\infty$, and the restriction of T' to L_A^1 coincides with T. We have

$$\|T\| = \|U\|_\infty = \|T'\|.$$

26.22 PROPOSITION. *If $a \to \bar{a}$ is an involution on A such that for every invertible element $a \in A$ we have $a^{-1} = \bar{a}$, then every representation $T: \mathscr{M}_A^1 \to A$ with $T(L_A^1) \neq A$ is a hermitian representation, that is*

$$T(\boldsymbol{m}^\sim) = \overline{T(\boldsymbol{m})}, \textit{ for every } \boldsymbol{m} \in \mathscr{M}_A^1.$$

In fact, there exists a continuous bounded representation $U : G \to A$ such that

$$T(\boldsymbol{m}) = \int U(s) \mathrm{d}\boldsymbol{m}(s), \text{ for } \boldsymbol{m} \in \mathscr{M}_A^1$$

Then

$$T(\boldsymbol{m}^\sim) = \int U(s) \mathrm{d}\boldsymbol{m}^\sim(s) = \overline{\int \overline{U(s^{-1})} \mathrm{d}\boldsymbol{m}(s)} =$$
$$= \overline{\int \overline{U(s)^{-1}} \mathrm{d}\boldsymbol{m}(s)} = \overline{\int U(s) \mathrm{d}\boldsymbol{m}(s)} = \overline{T(\boldsymbol{m})}.$$

6 The group of representations

Assume that A is a *commutative* Banach algebra with unit.

Denote by $\hat{G}_A$ the set of continuous representations $U: G \to A$ with $|U(s)| \equiv 1$. By corollary 26.19, the set $\hat{G}_A$ is in bijective correspondence with the set $\mathscr{H}$ of continuous representations $T: L_A^1 \to A$ with $T(L_A^1) = A$ and $||T|| = 1$. The correspondence is given by the equality

$$T(\boldsymbol{f}) = \int \boldsymbol{f}(s) U(s) \mathrm{d}s, \text{ for } \boldsymbol{f} \in L_A^1.$$

We shall write $T = \Phi(U)$. The set $\mathscr{H}$ is a subset in the unit sphere of the space $\mathscr{L}(L_A^1, A)$.

We consider on $\mathscr{H}$ the topology of simple convergence.

The sets of the form

$$V(T_0; f_1, \ldots, f_p; \varepsilon) = \{T \mid T \in \mathscr{H}, |T(f_i) - T_0(f_i)| < \varepsilon, i = 1, 2, \ldots, p\},$$

where $\varepsilon > 0$ and $\boldsymbol{f}_1, \ldots, \boldsymbol{f}_p$ are elements from L_A^1, form a fundamental system of neighbourhoods of $T_0 \in \mathscr{H}$ in this topology.

Consider now on $\hat{G}_A$ the topology τ obtained carrying the preceding topology on $\mathscr{H}$ by the bijective correspondence Φ. The neighbourhoods of an element $U_0 \in \hat{G}_A$ are of the form

$$V(U_0; f_1, \ldots, f_p; \varepsilon) = \{U \mid U \in \hat{G}_A \mid \int [U(s) - U_0(s)] f_i(s) \mathrm{d}s| < \varepsilon, i = 1, 2, \ldots, p\},$$

where $\varepsilon > 0$ and $\boldsymbol{f}_1, \ldots, \boldsymbol{f}_p$ are elements from L_A^1. With these topologies, $\hat{G}_A$ and $\mathscr{H}$ are homeomorphic and Φ is a homeomorphism.

26.23 PROPOSITION. *The mapping $(U, s) \to U(s)$ of the product $\hat{G}_A \times G$ into A is continuous.*

Let $U_0 \in \hat{G}_A$ and $s_0 \in G$. Let $T_0 = \Phi(U_0)$ and let $\boldsymbol{g} \in L_A^1$ be such that $T_0(\boldsymbol{g})^{-1}$ exists. Then

$$|T(\boldsymbol{g}_0) - T_0(\boldsymbol{g}_{s_0})| \leqslant |T(\boldsymbol{g}_s) - T(\boldsymbol{g}_{s_0})| + |T(\boldsymbol{g}_{s_0}) - T_0(\boldsymbol{g}_{s_0})|.$$

Since the mapping $s \to \boldsymbol{g}_s$ of G into L_A^1 is continuous, for every $\varepsilon > 0$ there exists a neighbourhood V_0 of s_0 such that

$$||\boldsymbol{g}_s - \boldsymbol{g}_{s_0}|| < \varepsilon/2, \text{ for } s \in V_0.$$

If $T \in V(T_0; \boldsymbol{g}_{s_0}; \varepsilon/2) = V$, then

$$|T(\boldsymbol{g}_{s_0}) - T_0(\boldsymbol{g}_{s_0})| < \varepsilon/2.$$

Thus, if $(T, s) \in V \times V_0$, we have

$$|T(g_s) - T_0(g_{s_0})| < \varepsilon\,,$$

therefore the mapping $(T, s) \to T(g_s)$ is continuous on $\mathscr{H} \times G$. But $T(g_s) = U(s)\,T(g)$, hence the mapping $(T, s) \to U(s) = T(g_s)\,T(g)^{-1}$ is continuous on $\mathscr{H} \times G$.

Then the composed mapping $(U, s) \to (T, s) \to U(s)$ is continuous on $\hat{G}_A \times G$, since the mapping $(U, s) \to (T, s)$ is a homeomorphism, and the proposition is proved.

Consider now on $\hat{G}_A$ the topology τ' of the uniform convergence on the compact subsets of G. The sets of the form

$$V(U_0\,; K\,; \varepsilon) = \{U \mid U \in \hat{G}_A, |U(s) - U_0(s)| < \varepsilon \text{ for } s \in K\}\,,$$

where $\varepsilon > 0$ and K are compact sets in G, form a fundamental system of neighbourhoods of U_0 in the topology τ'.

26.24 Proposition. *The topologies τ and τ' are equivalent on $\hat{G}_A$.*

Let $U_0 \in \hat{G}_A$ and let $V(U_0\,; f_1, \ldots, f_p\,; \varepsilon)$ be a neighbourhood of U_0 in the topology τ. For every $U \in V(U_0\,; f_1, \ldots, f_p\,; \varepsilon)$ we have then

$$|\textstyle\int [U(s) - U_0(s)]\, f_i(s)\, ds| < \varepsilon, \text{ for } i = 1, 2, \ldots, p\,.$$

Denote

$$\alpha = \max_{1 \leqslant i \leqslant p} \textstyle\int |f_i(s)|\, ds\,.$$

Since for each i we have

$$\sup_K \textstyle\int_K |f_i(s)| ds = \int |f_i(s)|\, ds, \ K \text{ compact}\,,$$

there exists a compact set $K \subset G$ such that

$$\textstyle\int_{G \setminus K} |f_i(s)|\, ds < \varepsilon/(\alpha + 2), \text{ for } i = 1, 2, \ldots, p\,.$$

If $U \in V(U_0\,; K\,; \varepsilon/(\alpha+2)$ we have for each i

$$\begin{aligned} |\textstyle\int (U(s) - U_0(s))\, f_i(s) ds| &\leqslant |\textstyle\int_K (U(s) - U(s_0))\, f_i(s)\, ds| + \\ &+ |\textstyle\int_{G\setminus K} (U(s) - U_0(s))\, f_i(s)\, ds| \leqslant \varepsilon/(\alpha+2) \textstyle\int_K |f_i(s)|\, ds + \\ &+ \textstyle\int_{G\setminus K} (|U(s)| + |U_0(s)|)\, |f_i(s)|\, ds \leqslant \alpha\varepsilon/(\alpha+2) + 2\varepsilon/(\alpha+2) = \varepsilon\,, \end{aligned}$$

therefore $U \in V(U_0\,; f_1,, \ldots, f_p\,; \varepsilon)$, consequently

$$V(U_0; K; \varepsilon/(\alpha+2)) \subset V(U_0; f_1, \ldots, f_p; \varepsilon),$$

that is the topology τ' is finer than the topology τ.

Conversely, let $V(U_0; K; \varepsilon)$ be an arbitrary neighbourhood of U_0 in the topology τ'. Show that this neighbourhood is an open set in the topology τ.

The set $W = \{a \in A \mid |a| < \varepsilon\}$ is open in A and

$$V(U_0; K; \varepsilon) = \{U \mid U \in \hat{G}_A, |U(s) - U_0(s)| < \varepsilon, \text{ for } s \in K\} =$$
$$= \{U \mid U \in \hat{G}_A, U(s) - U_0(s) \in W, \text{ for } s \in K\}$$

and the mapping $(U, s) \to U(s) - U_0(s)$ of $\hat{G}_A \times G$ into A is continuous for the topology τ on $\hat{G}_A$. By lemma 23.2 the set $V(U_0; K; \varepsilon)$ is open for the topology τ. It follows that the topology τ is finer than the topolbgy τ', therefore the two topologies are equivalent.

We can define on the set $\hat{G}_A$ in a natural way a group structure: for every representations $U', U'' \in \hat{G}_A$ we define the product $U = U'U''$ by the equality

$$U(s) = U'(s)\,U''(s), \text{ for } s \in G.$$

It is immediately seen that

$$U(st) = U(s)\,U(t), \text{ for } s, t \in G$$

and

$$U(e) = 1,$$

where 1 is the unit element in A. Since U' and U'' are continuous, U is a continuous representation. We have then

$$|U(s)| \leqslant |U'(s)|\,|U''(s)| \leqslant 1, \text{ for every } s \in G,$$

therefore U is bounded. Then (proposition 26.3), we have also $|U(s)| \geqslant 1$ for every $s \in G$, consequently $|U(s)| \equiv 1$, that is $U \in \hat{G}_A$.

26.25 PROPOSITION. *The product $U'U''$ defines on $\hat{G}_A$ a structure of a commutative group consistent with the topology τ on $\hat{G}_A$.*

The product is associative because the product in the algebra A is so.

The representation $U_0(s) \equiv 1$ belongs to $\hat{G}_A$ and is the unit element for this product. If $U \in \hat{G}_A$, then the function U^{-1} defined on G by the equality

$$U^{-1}(s) = U(s)^{-1} = U(s^{-1})$$

is a representation in $\hat{G}_A$ and it is the inverse of the representation U. We have then $U'(s)\,U''(s) = U''(s)\,U'(s)$, consequently $U'U'' = U''U'$. Thus $\hat{G}_A$ is a commutative group.

Denote by U_0 the unit element of $\hat{G}_A (U_0(s) \equiv 1$, where 1 is the unit element in A) and consider the fundamental system of neighbourhoods of U_0, consisting of sets of the form

$$V(K;\varepsilon) = |U \in \hat{G}_A, |U(s) - 1| < \varepsilon, \text{ for } s \in K\},$$

where $\varepsilon > 0$ and K are compact sets. We have

$$V(K;\varepsilon/2)\,V(K;\varepsilon/2) \subset V(K;\varepsilon).$$

In fact, if $U' \in V(K;\varepsilon/2)$ and $U'' \in V(K;\varepsilon/2)$, then

$$|U'(s) - 1| < \varepsilon/2 \text{ and } |U''(s) - 1| < \varepsilon/2, \text{ for } s \in K,$$

whence

$$|U'(s)\,U''(s) - 1| = |U'(s)\,U''(s) - U'(s) + U'(s) - 1| \leqslant$$
$$\leqslant |U'(s)|\,|U''(s) - 1| + |U'(s) - 1| < \varepsilon,$$

therefore $U'U'' \in V(K;\varepsilon)$.

Thus, the mapping $(U', U'') \to U'U''$ is continuous. We have then

$$[V(K^{-1};\varepsilon)]^{-1} \subset V(K;\varepsilon).$$

In fact, if $U \in [V(K^{-1};\varepsilon)]^{-1}$, then $U^{-1} \in V(K^{-1}, \varepsilon)$, therefore

$$|U^{-1}(s) - 1| < \varepsilon \text{ for } s \in K^{-1}.$$

Then, for every $s \in K$ we have $s^{-1} \in K^{-1}$, therefore $|U^{-1}(s^{-1}) - 1| < \varepsilon$. But $U^{-1}(s^{-1}) = U(s)$, consequently

$$|U(s) - 1| < \varepsilon, \text{ for } s \in K$$

that is $U \in V(K;\varepsilon)$.

Thus, the mapping $U \to U^{-1}$ is continuous, and the proposition is proved.

§27 Harmonic analysis on locally compact commutative groups

1 *The groups of characters*

In this paragraph we shall assume that G is a *commutative* locally compact group. We shall denote by $\mathscr{L}^1$ the space of integrable *complex* functions, defined on G.

The *continuous* bounded representations of G into C will be called *characters* of the group G. We denote by $\hat{G}$ (instead of $\hat{G}_C$) the set of the characters of G. The group $\hat{G}$ with the topology τ will be called the *dual* of G. By remark 20 after proposition 26.3, for every character $\chi \in \hat{G}$ we have $|\chi(s)| \equiv 1$. Thus a character χ of the group G is a continuous complex function defined on G such that

$$\chi(st) = \chi(s)\chi(t) \text{ and } |\chi(s)| \equiv 1 .$$

The set $\hat{G}$ of characters of G is in bijective correspondence with the set $\mathscr{H}$ of the multiplicative, linear continuous functionals T defined on $L^1(G)$. The correspondence is given by the equality

$$T(f) = \int f(s)\chi(s)\mathrm{d}s, \text{ for } f \in L^1 .$$

Since $|\chi(s)| \equiv 1$, by corollary 26.20, we have $\|T\| = 1$ for every $T \in \mathscr{H}$, therefore $\mathscr{H}$ is a subset in the unit sphere of the dual of $L^1(G)$.

27.1 PROPOSITION. *The dual $\hat{G}$ of the group G is a locally compact commutative group.*

The fact that $\hat{G}$ is a commutative topological group has been proved in proposition 26.25. It remains to show that τ is a locally compact topology on G.

To do this it is sufficient to show that $\mathscr{H}$ is a locally compact space with the topology induced by the weak star topology of $(L^1)'$. The unit sphere S of $(L^1)'$ is compact for the weak star topology and $\mathscr{H} \subset S$.

It follows that $\overline{\mathscr{H}}$ is a weakly compact set. Let then $T \neq 0$ be a functional in $\overline{\mathscr{H}}$. This means that for each $\varepsilon > 0$ and $f, g \in L^1$ there exists a representation $T' \in \mathscr{H}$ such that

$$|T(f) - T'(f)| < \varepsilon/3, \, |T(g) - T'(g)| < \varepsilon/3 \text{ and } |T(f*g) - T'(f*g)| < \varepsilon/3.$$

But $T'(f*g) = T'(f)T'(g)$, therefore

$$|T(f*g) - T(f)T(g)| \leqslant T(f*g) - T'(f*g)| +$$
$$+ |T'(f)T'(g) - T'(f)T(g)| + |T'(f)T(g) - T(f)T(g)| < \varepsilon .$$

As ε is arbitrary, it follows that $T(f*g) = T(f)T(g)$ therefore $T \in \mathscr{H}$.

If $0 \notin \overline{\mathscr{H}}$, then $\mathscr{H} = \overline{\mathscr{H}}$, hence $\mathscr{H}$ is compact. If $0 \in \overline{\mathscr{H}}$, then $\overline{\mathscr{H}} = \mathscr{H} \cup \{0\}$, and consequently $\mathscr{H}$ is locally compact. Thus in both cases $\mathscr{H}$ is locally compact. It follows that $\hat{G}$, with the topology τ, is locally compact and the proposition is proved.

2 *The Fourier transform*

In the sequence we shall denote by $\hat{x}$ the characters of the group G and by $\langle x, \hat{x}\rangle$ the value of the character $\hat{x}$ at a point x of G. By proposition 26.23, the mapping $(x, \hat{x}) \to \langle x, \hat{x}\rangle$ of the product $G \times \hat{G}$ into C is continuous. We have

$$\langle xy, \hat{x}\rangle = \langle x, \hat{x}\rangle \langle y, \hat{x}\rangle ,$$

$$\langle x, \hat{x}\hat{y}\rangle = \langle x, \hat{x}\rangle \langle x, \hat{y}\rangle$$

and

$$|\langle x, \hat{x}\rangle| = 1 .$$

The Fourier transform of a *complex measure* $\mu \in \mathcal{M}^1(G)$ is the complex *function* denoted by $\hat{\mu}$ defined on G by the equality

$$\hat{\mu}(\hat{x}) = \int \langle x, \hat{x}\rangle \, \mathrm{d}\mu(x), \text{ for } \hat{x} \in \hat{G} .$$

We remark that the mapping $x \to \langle x, \hat{x}\rangle$ is continuous and bounded, therefore the integral exists.

In particular, if μ has basis $\mathrm{d}x$ and density $f \in L^1(G)$, then the Fourier transform of μ is denoted by $\hat{f}$ and is also called the Fourier transform of f:

$$\hat{f}(\hat{x}) = \int \langle x, \hat{x}\rangle f(x) \, \mathrm{d}x, \text{ for } \hat{x} \in \hat{G} .$$

Remark. If we denote by $T_{\hat{x}} : L^1 \to C$ the representation corresponding to the character $\hat{x}$:

$$T_{\hat{x}}(f) = \int \langle x, \hat{x}\rangle f(x) \mathrm{d}x, \text{ for } f \in L^1 ,$$

we deduce that the Fourier transform $\hat{f}$ is defined by the equality

$$\hat{f}(\hat{x}) = T_{\hat{x}}(f) .$$

27.2 Proposition. *The Fourier transform $\hat{\mu}$ of any measure $\mu \in \mathcal{M}^1(G)$ is a bounded uniformly continuous function on $\hat{G}$ and we have*

$$|\hat{\mu}(\hat{x})| \leqslant \|\mu\|, \textit{ for } \hat{x} \in \hat{G} .$$

In fact, we have first

$$|\hat{\mu}(\hat{x})| = |\int \langle x, \hat{x}\rangle \mathrm{d}\mu(x)| \leqslant \int |\langle x, \hat{x}\rangle| \mathrm{d}|\mu|(x) = \int \mathrm{d}|\mu| = \|\mu\|$$

for every $\hat{x} \in \hat{G}$.

Let now $\varepsilon > 0$. There exists a compact set $K \subset G$ such that

$$\int_{G \setminus K} \mathrm{d}|\mu| < \varepsilon/4 .$$

Then

$$\begin{aligned} |\hat{\mu}(\hat{x}) - \hat{\mu}(\hat{y})| &\leqslant \int_K |\langle x, \hat{x}\rangle - \langle x, \hat{y}\rangle| \mathrm{d}|\mu|(x) + \\ &\quad + \int_{G\setminus K} |\langle x, \hat{x}\rangle - \langle x, \hat{y}\rangle| \mathrm{d}|\mu|(x) \leqslant \\ &\leqslant \int_K |1 - \langle x, \hat{y}\hat{x}^{-1}\rangle| \mathrm{d}|\mu|(x) + \varepsilon/2 . \end{aligned}$$

But for $\hat{y}\hat{x}^{-1} \in V = V(\hat{e}, K, \varepsilon/2\|\mu\|)$, we have

$$|\langle x, \hat{y}\hat{x}^{-1}\rangle - 1| < \varepsilon/2\|\mu\|, \text{ for } x \in K ,$$

therefore

$$|\hat{\mu}(\hat{x}) - \hat{\mu}(\hat{y})| \leqslant \varepsilon \text{ for } \hat{y}\hat{x}^{-1} \in V ,$$

that is, $\hat{\mu}$ is uniformly continuous on $\hat{G}$ and the proposition is proved.

For functions in L^1, proposition 27.2 can *be made more precise*:

27.3 PROPOSITION. *If $f \in L^1(G)$ then $\hat{f} \in \overline{\mathscr{K}(\hat{G})}$ and $\|\hat{f}\|_\infty \leqslant \|f\|_1$.*

By proposition 27.2 it follows that $\hat{f}$ is uniformly continuous on $\hat{G}$ and $\|\hat{f}\|_\infty \leqslant \|f\|_1$. It remains to show that $\hat{f}$ vanishes at infinity, if $\hat{G}$ is not compact.

Let $\varepsilon > 0$. Denoting by S the unit sphere in $(L^1)'$, the set $\{T \in S \mid T(f) \geqslant \varepsilon\}$ is weak star compact. Denote by $\hat{K}$ the corresponding set of characters $\hat{x} \in \hat{G}$:

$$\hat{K} = \{\hat{x} \in \hat{G} \mid |\int \langle x, \hat{x}\rangle f(x) \mathrm{d}x| \geqslant \varepsilon\} .$$

The set $\hat{K}$ is compact and for $\hat{x} \notin \hat{K}$ we have

$$|\hat{f}(\hat{x})| = |\int \langle x, \hat{x}\rangle f(x) \mathrm{d}x| < \varepsilon ,$$

therefore $\hat{f}$ vanishes at infinity and the proposition is proved.

Remark. We can define by the same formulae, the Fourier transform of the measures $\boldsymbol{m} \in \mathscr{M}_E^1$ and of the functions $\boldsymbol{f} \in \mathscr{L}_E^1$. The extension to this case of the results concerning the scalar measures and functions is left to the reader.

27.4 PROPOSITION. *If A is an involutive algebra dense in L^1, then the set $\hat{A} = \{\hat{f} \mid f \in A\}$ is a self adjoint algebra, dense in $\mathscr{K}(\hat{G})$.*

By the equality

$$\hat{f}(\hat{x}) = \int \langle x, \hat{x} \rangle f(x)\,\mathrm{d}x$$

we deduce

$$\widehat{(f+g)} = \hat{f} + \hat{g},\ \widehat{(\alpha f)} = \alpha \hat{f} \text{ and } \widehat{(f * g)} = \hat{f}\hat{g}$$

for $f, g \in A$ and $\alpha \in C$, therefore $\hat{A}$ is an algebra. Also, if $f \in A$, then $\tilde{f} \in A$, therefore if $\hat{f} \in \hat{A}$ then $\bar{\hat{f}} = \hat{\tilde{f}} \in \hat{A}$, consequently $\hat{A}$ is self adjoint.

The algebra $\hat{A}$ separates the points of $\hat{G}$, since, if $\hat{x}_1 \neq \hat{x}_2$, then there exists $f \in A$ such that

$$\int \langle x, \hat{x}_1 \rangle f(x)\,\mathrm{d}x \neq \int \langle x, \hat{x}_2 \rangle f(x)\,\mathrm{d}x$$

(in the contrary case we would have $\int \langle x, \hat{x}_1 \rangle f(x)\,\mathrm{d}x = \int \langle x, \hat{x}_2 \rangle f(x)\,\mathrm{d}x$ for every $f \in A$, therefore for every $f \in L^1$, whence it would follow that $\langle x, \hat{x}_1 \rangle = \langle x, \hat{x}_2 \rangle$ almost everywhere, hence everywhere since $\hat{x}_1$ and $\hat{x}_2$ are continuous).

Thus $\hat{f}(\hat{x}_1) \neq \hat{f}(\hat{x}_2)$ and $\hat{f} \in \hat{A}$, therefore $\hat{A}$ separates the points of $\hat{G}$.

If $\hat{G}$ is compact, then $\mathscr{C}(\hat{G}) = \overline{\mathscr{K}(\hat{G})}$ and by the Weierstrass-Stone theorem, $\hat{A}$ is dense in $\mathscr{C}(\hat{G})$.

If $\hat{G}$ is not compact, adding the point ω at infinity, the space $K = \hat{G} \cup \{\omega\}$ is compact and the algebra generated by $\hat{A}$ and by the function identically equal to 1 on K, is dense in $\mathscr{C}(K)$ by the Weierstrass-Stone theorem. It follows that $\hat{A}$ is dense in the algebra $\overline{\mathscr{K}(\hat{G})}$ of the functions vanishing at ω. The proposition is thus proved.

Remark. In general we have $\hat{A} \neq \overline{\mathscr{K}(\hat{G})}$.

27.5 PROPOSITION. *For every function $f \in L^1(G)$ we have*

$$\|\hat{f}\|_\infty = \lim_\infty (\|f^n\|_1)^{1/n}$$

where $f^n = f * f * \ldots * f$ (n times).

For every n we have

$$|\hat{f}(\hat{x})|^n = |\hat{f}^n(\hat{x})| = |\widehat{f^n}(\hat{x})| = \int \langle x, \hat{x} \rangle f^n(x)\,\mathrm{d}x \leqslant$$

$$\leqslant \int |\langle x, \hat{x} \rangle|\,|f^n(x)|\,\mathrm{d}x \leqslant \|f^n\|_1$$

therefore $|\hat{f}(\hat{x})| \leqslant (\|f^n\|_1)^{1/n}$, consequently $\|\hat{f}\|_\infty \leqslant (\|f^n\|_1)^{1/n}$, whence

$$\|\hat{f}\|_\infty \leqslant \liminf (\|f^n\|_1)^{1/n}.$$

It remains to prove the inequality

$$\limsup (\|f^n\|_1)^{1/n} \leqslant \|\hat{f}\|_\infty .$$

From the theory of the commutative Banach algebras, it is known that for every $|\lambda| < \|f\|_\infty^{-1}$, λf has quasi-inverse (an element x of a commutative algebra A is the quasi-inverse of an element $y \in A$, if $x+y+xy=0$. The quasi-inverse of an element $z \in A$ is denoted by z'). The quasi-inverse of λf can be expressed as a series $(\lambda f)' = \Sigma_{1 \leqslant n < \infty} (\lambda f)^n$, which is an analytic function of λ in the disc $|\lambda| < \|f\|_\infty^{-1}$. Since the series $\Sigma_{1 \leqslant n < \infty} (\lambda f)^n$ is convergent in this disc, it follows that $\lambda^n f^n \to 0$, therefore the sequence $(\lambda^n f^n)$ is bounded. There exists $B > 0$ with $|\lambda|^n \|f^n\|_1 \leqslant B$, consequently $(\|f^n\|_1)^{1/n} \leqslant B^{1/n}/|\lambda| \leqslant (B)^{1/n} \|\hat{f}\|_\infty$, whence $\limsup (\|f^n\|_1)^{1/n} \leqslant \|f\|_\infty$. It follows that the sequence $(\|f^n\|_1)^{1/n}$ is convergent and

$$\lim (\|f^n\|_1)^{1/n} = \|\check{f}\|_\infty .$$

3 *Functions of positive type. The Bochner theorem*

A linear funcional Φ defined on an algebra A with involution $x \to x^*$ is of *positive type* if $\Phi(xx^*) \geqslant 0$ for every $x \in A$.

If Φ is a functional of positive type on A then,

$$|\Phi(xy^*)| \leqslant [\Phi(xx^*)]^{\frac{1}{2}}, \text{ for every } x \text{ and } y \text{ in } A .$$

In fact, put

$$\langle x, y \rangle = \Phi(xy^*) .$$

The functional $\langle x, y \rangle$ is linear in x and antilinear in y. On the other hand

$$\Phi((x+\lambda y)(x \times \lambda y)^*) \geqslant 0 ,$$

therefore $\lambda \Phi(yx^*) + \bar{\lambda} \Phi(xy^*)$ is real for every λ, consequently

$$\Phi(yx^*) = \overline{\Phi(xy^*)} ,$$

that is

$$\langle y, x \rangle = \langle x, y \rangle .$$

Thus $\langle x, y\rangle$ has all the properties of the scalar product, except that we may have $\langle x, x\rangle = 0$ for $x \neq 0$. The Schwarz inequality

$$|\langle x, y\rangle| \leqslant (\langle x, x\rangle)^{\frac{1}{2}}(\langle y, y\rangle)^{\frac{1}{2}}$$

is written then

$$|\Phi(xy^*)| \leqslant [\Phi(xx^*)]^{\frac{1}{2}}\,[\Phi(yy^*)]^{\frac{1}{2}}\,.$$

If A has a unit element e, putting $y = e$ in the preceding inequality, we deduce

$$|\Phi(x)|^2 \leqslant k\Phi(xx^*), \text{ where } k = \Phi(e)\,.$$

Also, if in the equality $\Phi(yx^*) = \overline{\Phi\,(xy^*)}$ we put $y = e$, we deduce

$$\Phi(x^*) = \overline{\Phi\,(x)}\,.$$

We say that a functional of positive type Φ on A is *extendable* if

$$|\Phi(x)|^2 \leqslant k\Phi(xx^*) \text{ and } \Phi(x^*) = \overline{\Phi\,(x)}\,.$$

The reason for this follows from the following lemma:

27.6 LEMMA. *Let A be an involution algebra without unit, A_e the algebra obtained by adding the unit. A functional of positive type Φ on A can be extended to a functional of positive type Φ' in A_e if and only if Φ is extendable.*

If Φ can be extended to a functional of positive type Φ' on A_e then

$$|\Phi'(x)|^2 \leqslant k\Phi'(xx^*) \text{ and } \Phi'(x^*) = \overline{\Phi'(x)}, \text{ for } x \in A_e,$$

therefore

$$|\Phi(x)|^2 \leqslant k\Phi(xx^*) \text{ and } \Phi(x^*) = \overline{\Phi\,(x)}, \text{ for } x \in A\,,$$

hence Φ is extendable.

Conversely, assume that Φ is extendable. Define

$$\Phi'(x + \lambda e) = \Phi(x) + \lambda k, \text{ for } x \in A \text{ and } \lambda \in C\,.$$

Φ' is a linear functional on A_e which coincide with Φ on A. We have then

$$\begin{aligned}\Phi'((x+\lambda e)(x+\lambda e)^*) &= \Phi(xx^*) + 2\,\mathrm{Re}\,\bar{\lambda}\Phi(x) + |\lambda|^2 k \geqslant \\ &\geqslant \Phi(xx^*) - 2|\lambda|k^{\frac{1}{2}}[\Phi(xx^*)]^{\frac{1}{2}} + |\lambda|^2 k = \\ &= ([\Phi(xx^*)]^{\frac{1}{2}} - |\lambda|k^{\frac{1}{2}})^2 \geqslant 0\,,\end{aligned}$$

consequently Φ' is of positive type on A_e.

27.7 Lemma. *Let Φ be a functional of positive type. If A is a Banach algebra with continuous involution and if Φ is extendable, then Φ is continuous.*

Consider the algebra A_e obtained from A by adding the unit. A_e is also a Banach algebra, and the involution on A is extended to a continuous involution on A_e. The functional Φ is extended to a functional of positive type on A_e, still denoted by Φ.

Let $x \in A_e$ with $|x| < 1$. Then $e - x$ possesses a square root $(e-x)^{\frac{1}{2}}$. If x is symmetric ($x^* = x$), then $y = (e-x)^{\frac{1}{2}}$ is symmetric. Then $\Phi(e-x) = \Phi(yy^*) \geqslant 0$, whence $\Phi(x) \leqslant \Phi(e)$. Similarly $\Phi(-x) \leqslant \Phi(e)$, therefore $|\Phi(x)| \leqslant \Phi(e)$, if $|x| < 1$ and $x^* = x$.

Now let x be an arbitrary element in A. We have

$$x = \tfrac{1}{2}\left[(x+x^*) - i(i(x-x^*))\right]$$

and $x + x^*$ and $i(x-x^*)$ are symmetric. If B is such that $|x^*| \leqslant B|x|$ for every $x \in A_e$, then

$$|\Phi(x)| \leqslant 2^{\frac{1}{2}}\Phi(e) \text{ for } |x| < 2/(B+1)$$

therefore Φ is continuous.

In the sequel we shall consider functionals of positive type on the involution algebras $\mathscr{K}(G)$ and $L^1(G)$, with the involution $f \to \tilde{f} = \bar{\check{f}}$.

A complex (not necessarily bounded) measure μ of positive type is written $\mu \gg 0$ and it is characterized by the inequality

$$\mu(f * \tilde{f}) \geqslant 0, \text{ for every } f \in \mathscr{K}(G).$$

By proposition 24.34, we have

$$\mu(f * \tilde{f}) = (\mu * (f * \tilde{f})^{\vee})(e) = (\mu * \check{f} * \check{\tilde{f}})(e)$$

therefore $\mu \gg 0$ if and only if

$$(\mu * f * \tilde{f})(e) \geqslant 0, \text{ for every } f \in \mathscr{K}(G).$$

Examples (1). *The Haar measure* d*x is of positive type.*
In fact

$$\begin{aligned}\int f * \tilde{f}(x)\,dx &= \int dx \int f(y)\tilde{f}(y^{-1}x)\,dy = \int f(y)\,dy \int \overline{f(x^{-1}y)}\,dx = \\ &= \int f(y)\,dy \int \overline{f(x)}\,dx = \int f(y)\,dy\, \overline{\int f(x)\,dx} = \\ &= \left|\int f(x)\,dx\right|^2 \geqslant 0.\end{aligned}$$

(2) The measure ε_e is of positive type.
In fact,

$$\varepsilon(f*\tilde{f}) = (f*\tilde{f})(e) = \int f(y)\tilde{f}(ey^{-1})\mathrm{d}y = \\ = \int f(y)\overline{f(y)}\mathrm{d}y = \int |f(y)|^2 \mathrm{d}y \geqslant 0\,.$$

We say that a complex function $\varphi : G \to C$ is of *positive type* and we write $\varphi \gg 0$, if φ is locally integrable and if the measure $\varphi\,\mathrm{d}x$ is of positive type. We say also that the equivalence class of φ is a function of positive type.

Thus, a locally integrable function $\varphi : G \to C$ is of positive type if and only if

$$\int (f*\tilde{f})(x)\varphi(x)\mathrm{d}x \geqslant 0, \text{ for } f \in \mathscr{K}(G)\,.$$

But

$$\begin{aligned} \int (f*\tilde{f})(x)\varphi(x)\mathrm{d}x &= \int \varphi(x)\mathrm{d}x \int \tilde{f}(y) f(y^{-1}x)\mathrm{d}y = \\ &= \overline{\int f(y^{-1})}\mathrm{d}y \int f(x)\varphi(xy)\mathrm{d}x = \\ &= \int\int \varphi(xy) f(x)\overline{f(y^{-1})}\mathrm{d}x\,\mathrm{d}y\,. \end{aligned}$$

It follows that a locally integrable function $\varphi : G \to C$ is of positive type if and only if

$$\int\int \varphi(xy) f(x)\overline{f(y^{-1})}\mathrm{d}x\,\mathrm{d}y \geqslant 0, \text{ for } f \in \mathscr{K}(G)\,.$$

Examples (3). *Every character $\hat{x} \in \hat{G}$ is a function of positive type on G.*

In fact, $\hat{x}$ is a continuous bounded function on G, therefore it is locally integrable. For every function $f \in \mathscr{K}(X)$ we have

$$\int\int \langle xy, \hat{x}\rangle f(x)\overline{f(y^{-1})}\mathrm{d}x\,\mathrm{d}y = \int \langle x, \hat{x}\rangle f(x)\mathrm{d}x \int \overline{\langle y^{-1}, x\rangle}\,\overline{f(y^{-1})}\mathrm{d}y = \\ = \int \langle x, \hat{x}\rangle f(x)\mathrm{d}x \overline{\int \langle y, \hat{x}\rangle f(y)\mathrm{d}y} \geqslant 0$$

therefore $\hat{x}$ is of positive type.

(4). *For every function $g \in \mathscr{L}^1(G)$, the function $g*\tilde{g}$ is of positive type.*

In fact, $g*\tilde{g} \in \mathscr{L}^1(G)$, therefore $g*\tilde{g}$ is locally integrable.

Putting $\mathrm{d}\mu = (g*\tilde{g})\mathrm{d}x$, for every function $f \in \mathscr{K}(G)$ we have

$$(\mu*f*\tilde{f})(e) = (g*\tilde{g}*f*\tilde{f})(e) = ((f*g)*(f*g)^{\sim})(e) = (h*\tilde{h})(e)\,,$$

where we set $h = f*g$. Furthermore,

$$(h*\tilde{h})(e) = \int h(y^{-1})\overline{h(y^{-1})}\mathrm{d}y = \int h(y)\overline{h(y)}\mathrm{d}x \geqslant 0\,,$$

hence $\mu(f*\tilde{f}) \geqslant 0$.

It follows that μ, hence also $g*\tilde{g}$, is of positive type.

A linear functional Φ defined on the algebra L^1 is of positive type if

$$\Phi(f*\tilde{f}) \geqslant 0, \text{ for } f \in L^1 .$$

27.8 PROPOSITION. *A linear functional of positive type Φ on $L^1(G)$ is continuous, if and only if it is extendable.*

If Φ is extendable, we already know that Φ is continuous (lemma 27.7).

Assume now that Φ is continuous. Let (u) be the approximating unit in L^1. Then

$$|\Phi(f)|^2 = \lim_u |\Phi(f*u)|^2 \leqslant \Phi(f*\tilde{f}) \limsup_u \Phi(u*\tilde{u}) \leqslant ||\Phi|| \Phi(f*\tilde{f}),$$

since $\Phi(u*\tilde{u}) \leqslant ||\Phi|| \, ||u*\tilde{u}||_1$, and

$$||u*\tilde{u}||_1 = \int u*\tilde{u}(x)\mathrm{d}x = \int \mathrm{d}x \int u(xy)\tilde{u}(y^{-1})\mathrm{d}y =$$
$$= \int \overline{u(y)}\,\mathrm{d}y \int u(xy)\mathrm{d}x = \overline{\int u(y)\mathrm{d}y} \int u(x)\mathrm{d}x = 1 .$$

We also have

$$\Phi(\tilde{f}) = \lim_u \Phi(\tilde{f}*u) = \lim_u \overline{\Phi(\tilde{u}*f)} = \overline{\Phi(f)}$$

consequently Φ is extendable and the proposition is proved.

27.9 PROPOSITION. *There exists a bijective correspondence between the set of continuous linear functionals of positive type $\Phi : L^1(G) \to C$ and the set of the functions of positive type φ of $L^\infty(G)$, given by the equality*

$$\Phi(f) = \int f(x)\varphi(x)\mathrm{d}x, \textit{ for } f \in L^1(G) .$$

First let $\varphi \in L^\infty(G)$ be a function of positive type. This means that the measure $\mathrm{d}\nu = \varphi\,\mathrm{d}x$ is of positive type. A function $f : G \to C$ is ν-integrable if and only if $f\varphi$ is $\mathrm{d}x$-integrable.

In particular, for every function $f \in L^1(G)$, the function $f\varphi$ is $\mathrm{d}x$-integrable, therefore f is ν-integrable and

$$\nu(f) = \int f(x)\varphi(x)\mathrm{d}x .$$

Denote by Φ the restriction of the functional $f \to \nu(f)$ to $L^1(G)$. Thus,

$$\Phi(f) = \int f(x)\varphi(x)\mathrm{d}x, \text{ for } f \in L^1$$

and Φ is continuous and $||\Phi|| = ||\varphi||_\infty$.

Let $f \in L^1$ and (f_n) be a sequence of functions of $\mathscr{K}(G)$ converging in the mean to f:

$$||f - f_n||_1 \to 0 .$$

Then $||\tilde{f} - \tilde{f}_n||_1 \to 0$ and $||f * \tilde{f} - f_n * \tilde{f}_n||_1 \to 0$. For each n we have

$$\Phi(f_n * \tilde{f}_n) = \nu(f_n * \tilde{f}_n) \geqslant 0 .$$

Since Φ is continuous, we deduce that

$$\Phi(f_n * \tilde{f}_n) \to \Phi(f * \tilde{f}) ,$$

consequently

$$\Phi(f * \tilde{f}) \geqslant 0 .$$

Conversely, let $\Phi : L^1 \to C$ be a continuous linear functional of positive type. This means that

$$\Phi(f * \tilde{f}) \geqslant 0, \text{ for } f \in L^1$$

and there exists a function $\varphi \in L^\infty$ such that

$$\Phi(f) = \int f(x) \varphi(x) \mathrm{d}x, \text{ for } f \in L^1$$

and $||\Phi|| = ||\varphi||_\infty$.

If we denote by ν the restriction of Φ to $\mathscr{K}(G)$:

$$\nu(f) = \int f(x) \varphi(x) \mathrm{d}x, \text{ for } f \in \mathscr{K}(G) ,$$

we deduce that ν is of positive type:

$$\nu(f * \tilde{f}) = \Phi(f * \tilde{f}) \geqslant 0 \text{ for } f \in \mathscr{K}(G) ;$$

on the other hand $\mathrm{d}\nu = \varphi \, \mathrm{d}x$, therefore φ is of positive type.

27.10 Corollary. *For every function $\varphi \in L^\infty$ of positive type we have $\tilde{\varphi} = \varphi$.*

In fact, let Φ be the continuous linear functional of positive type, defined on L^1 by the equality

$$\Phi(f) = \int f(x) \varphi(x) \mathrm{d}x, \text{ for } f \in L^1(G) .$$

By proposition 27.8, Φ is extendable, therefore

$$\Phi(\tilde{f}) = \overline{\Phi(f)}, \text{ for } f \in L^1 ,$$

consequently

$$\int f(x)\varphi(x)\,dx = \Phi(f) = \overline{\Phi(\tilde{f})} = \overline{\int \tilde{f}(x)\varphi(x)\,dx} =$$
$$= \int f(x^{-1})\overline{\varphi(x)}\,dx = \int f(x)\overline{\varphi(x^{-1})}\,dx =$$
$$= \int f(x)\tilde{\varphi}(x)\,dx$$

for every $f \in L^1$. It follows that $\varphi(x) = \tilde{\varphi}(x)$ almost everywhere.

27.11 THEOREM (*Bochner*). *There exists a bijective correspondence* $\varphi \leftrightarrow \mu$ *between the set of continuous functions* φ *of positive type of* $\mathscr{L}^{\infty}(G)$ *and the set of the positive bounded measures* μ *on* $\hat{G}$, *given by the equality*

$$\varphi(x) = \int \langle x, \hat{x} \rangle \, d\mu(\hat{x}).$$

First let μ be a positive bounded measure on $\hat{G}$. Put

$$\varphi(x) = \int \langle x, \hat{x} \rangle \, d\mu(\hat{x}), \text{ for } x \in G.$$

The function φ is *continuous* on G. Also put

$$v(f) = \int \hat{f}(\hat{x})\,d\mu(\hat{x}), \text{ for } f \in \mathscr{K}(G).$$

(the integral exists since $\hat{f} \in \overline{\mathscr{K}(\hat{G})}$ and μ is bounded).
Since

$$\hat{f}(\hat{x}) = \int \langle x, \hat{x} \rangle f(x)\,dx, \text{ for } f \in \mathscr{K}(G) \text{ and } \hat{x} \in \hat{G},$$

we have

$$v(f) = \int d\mu(\hat{x}) \int \langle x, \hat{x} \rangle f(x)\,dx = \int f(x)\,dx \int \langle x, \hat{x} \rangle \, d\mu(\hat{x}) =$$
$$= \int f(x)\varphi(x)\,dx, \text{ for } f \in \mathscr{K}(G).$$

We deduce that v is a measure of basis dx on G and $dv = \varphi\,dx$. Since the set $\hat{A} = \{\hat{f} \mid f \in L^1\}$ is a selfadjoint algebra, we have

$$(\widehat{f * \tilde{f}}) = \hat{f}(\hat{\tilde{f}}) = \hat{f}\bar{\hat{f}} = |\hat{f}|^2$$

therefore

$$v(f * \tilde{f}) = \int |\hat{f}(\hat{x})|^2 \, d\mu(\hat{x}) \geqslant 0,$$

consequently v is a measure of positive type on G. It follows that φ is a function of positive type on G. We have then

$$|\varphi(x)| \leqslant \int |\langle x, \hat{x} \rangle| \, d\mu(\hat{x}) \leqslant ||\mu||,$$

consequently $\varphi \in \mathscr{L}^{\infty}(G)$.

Conversely, let $\varphi \in \mathscr{L}^{\infty}(G)$ be a function of positive type. By proposition 27.9, the continuous functional

$$\Phi(f) = \int f(x) \varphi(x) dx, \text{ for } f \in L^1(G)$$

is of positive type and $||\Phi|| = ||\varphi||_{\infty}$. By proposition 27.8, for $f \in L^1$ we have

$$|\Phi(f)|^2 \leqslant ||\Phi|| \, \Phi(f * \tilde{f}) \leqslant ||\Phi||^{1+\frac{1}{2}} [\Phi(f * f)^2]^{\frac{1}{2}} \leqslant \ldots$$

$$\ldots \leqslant ||\Phi||^{1+2^{-1}+\ldots+2^{-n}} [\Phi((f * \tilde{f})^{2^n})]^{2^{-n}} \leqslant$$

$$\leqslant ||\Phi||^{1+2^{-1}+\ldots+2^{-n}} ||\Phi||^{2^{-n}} ||(f * \tilde{f})^{2^n}||_1^{2^{-n}}.$$

Passing to the limit as $n \to \infty$, we deduce

$$|\Phi(f)|^2 \leqslant ||\Phi||^2 ||\hat{f} \, \hat{\tilde{f}}||_{\infty} = ||\Phi||^2 ||\hat{f}||_{\infty}^2$$

therefore

$$|\Phi(f)| \leqslant ||\Phi|| \, ||\hat{f}||_{\infty}.$$

For every function $f \in L^1(G)$ put

$$\mu(\hat{f}) = \Phi(f).$$

It is easy to see that μ is linear on the algebra $\hat{A} = \{\hat{f} \mid f \in L^1(G)\}$. The preceding equality is now written

$$|\mu(\hat{f}) \leqslant ||\Phi|| \, ||\hat{f}||_{\infty},$$

therefore μ is continuous on $\hat{A}$ for the topology of the uniform convergence. Since $\hat{A}$ is dense in $\overline{\mathscr{K}(G)}$, μ can be extended to a continuous linear functional on $\overline{\mathscr{K}(\hat{G})}$, therefore its restriction to $\mathscr{K}(\hat{G})$ is a *bounded measure*, which will still be denoted by μ.

Let $h \geqslant 0$ be a function in $\mathscr{K}(\hat{G})$. Then $h^{\frac{1}{2}} \in \mathscr{K}(G)$ and $h^{\frac{1}{2}}$ can be uniformly approximated by a sequence $(\hat{f}_n)$ of functions from $\hat{A}$. Then h is uniformly approximated on $\hat{G}$ by the sequence $(|\hat{f}_n|^2)$, therefore

$$\mu(|f_n|^2) \to \mu(h).$$

But $|\hat{f}_n|^2 = \hat{f}_n \bar{\hat{f}}_n = \hat{f}_n \hat{\tilde{f}}_n = (\widehat{f_n * \tilde{f}_n})$ and

$$\mu(|\hat{f}_n|^2)=\mu((\widehat{f_n*\tilde{f}_n}))=\Phi(f_n*\tilde{f}_n)\geqslant 0\,,$$

therefore $\mu(h)\geqslant 0$. Thus μ is a *positive* bounded measure on $\hat{G}$. The equality $\Phi(f)=\mu(\hat{f})$ is written for $f\in\mathscr{K}(G)$,

$$\int f(x)\varphi(x)\mathrm{d}x=\int\hat{f}(\hat{x})\mathrm{d}\mu(\hat{x})=\int\mathrm{d}\mu(\hat{x})\int\langle x,\hat{x}\rangle f(x)\mathrm{d}x=$$
$$=\int f(x)\mathrm{d}x\int\langle x,\hat{x}\rangle\,\mathrm{d}\mu(\hat{x})\,.$$

It follows that

$$\varphi(x)=\int\langle x,\hat{x}\rangle\,\mathrm{d}\mu(\hat{x})\,,$$

almost everywhere on G; thus the proposition is proved.

Since the mapping $\hat{x}\to\int\langle x,\hat{x}\rangle\,\mathrm{d}\mu(\hat{x})$ is continuous, we deduce:

27.12 COROLLARY. *Every function $\varphi\in\mathscr{L}^\infty$ of positive type is equal almost everywhere to a continuous function of positive type of $\mathscr{L}^\infty$.*

By the Bochner theorem and proposition 27.9, we have

27.13 COROLLARY. *There exists a bijective correspondence $\Phi\leftrightarrow\mu$ between the set of continuous linear functionals of positive type Φ on $L^1(G)$ and the set of positive bounded measures μ on $\hat{G}$, given by the equality*

$$\Phi(f)=\int\hat{f}(\hat{x})\mathrm{d}\mu(\hat{x}),\ \textit{for}\ f\in L^1(G)\,.$$

In fact, from the correspondence

$$\Phi(f)=\int f(x)\varphi(x)\mathrm{d}x,\ f\in L^1\,,$$

where $\varphi\in L^\infty$ are functions of positive type, and from the correspondence

$$\varphi(x)=\int\langle x,\hat{x}\rangle\,\mathrm{d}\mu(\hat{x})\,,$$

we deduce

$$\Phi(f)=\int f(x)\mathrm{d}x\int\langle x,\hat{x}\rangle\,\mathrm{d}\mu(\hat{x})=$$
$$=\int\mathrm{d}\mu(\hat{x})\int\langle x,\hat{x}\rangle f(x)\mathrm{d}x=\int\hat{f}(\hat{x})\mathrm{d}\mu(\hat{x})\,.$$

Remark. In the form stated in the corollary, the Bochner theorem can be generalized to the case of an involution Banach algebra.

4 *Inversion formula. The Plancherel theorem*

Denote by P the set of functions of positive type of L^∞ and by $[L^1 \cap P]$ the vector space generated by $L^1 \cap P$.

27.14 PROPOSITION. *For every functions $f, g \in L^2$ we have $f*g \in [L^1 \cap P]$.*

In fact

$$f*g = \tfrac{1}{4}[(f+\tilde{g})*(f+\tilde{g})^{\sim} - (f-\tilde{g})*(f-\tilde{g})^{\sim} + \\ + i(f+i\tilde{g})*(f+i\tilde{g})^{\sim} - i(f-i\tilde{g})*(f-i\tilde{g})^{\sim}],$$

therefore $f*g$ is a linear combination of functions of the form $h*\tilde{h}$ with $h \in L^1(G)$. As each function $h*\tilde{h}$ is of positive type (example 4) in $L^1 \cap P$, it follows that $f*g \in [L^1 \cap P]$.

27.15 PROPOSITION. *The space $[L^1 \cap P]$ is dense in L^1 and in L^2.*

Let $g \in \mathscr{K}(G)$ and let (u) be an approximating unit. We have $\lim_u u*g = g$, in the norm of the space L^1 as well as in the norm of the space L^2. Since $\mathscr{K}(G)$ is dense in L^1 and L^2, it follows that the set of the functions of the form $f*g$ with $f, g \in \mathscr{K}(G)$ is dense in L^1 and L^2. Since these functions are contained in $[L^1 \cap P]$ by proposition 27.14, we deduce that $[L^1 \cap P]$ is dense in L^1 and L^2.

27.16 COROLLARY. *The set $\{\hat{f} \mid f \in [L^1 \cap P]\}$ is dense in $\overline{\mathscr{K}(\hat{G})}$.*

Since $\tilde{\varphi} = \varphi$ for $\varphi \in [L^1 \cap P]$ (corollary 27.10), $[L^1 \cap P]$ is a selfadjoint algebra dense in L^1.

By proposition 27.4, the functions $\hat{f}$ with $f \in [L^1 \cap P]$ form an algebra dense in $\overline{\mathscr{K}(\hat{G})}$.

For every continuous, bounded, integrable function $f \in L^1 \cap \mathscr{C}^\infty$, put

$$\Phi(f) = f(e).$$

The set $L^1 \cap \mathscr{C}^\infty$ is an ideal in L^1 (proposition 25.6) since $\mathscr{C}^\infty \subset L^\infty$. If $f \in L^1 \cap \mathscr{C}^\infty$ then $\tilde{f} \in L^1 \cap \mathscr{C}^\infty$ and $f*\tilde{f} \in L^1 \cap \mathscr{C}^\infty$ and we have

$$\Phi(f*\tilde{f}) = f*\tilde{f}(e) = \int f(x)\tilde{f}(x^{-1})\mathrm{d}x = \int |f(x)|^2 \mathrm{d}x \geqslant 0.$$

Thus Φ is a linear functional of positive type on $L^1 \cap \mathscr{C}^\infty$.

Denote by $\mathscr{I}$ the set of functions $p \in L^1 \cap \mathscr{C}^\infty$ which have the property that the linear functional Φ_p defined on L^1 by the equality

$$\Phi_p(f) = \Phi(p*f), \text{ for } f \in L^1 ,$$

is of positive type and continuous ($p*f \in L^1 \cap \mathscr{C}^\infty$, since $L^1 \cap \mathscr{C}^\infty$ is an ideal).

For $p \in \mathscr{I}$ and $f \in L^1$ we have

$$\Phi_p(f) = (p*f)(e) = \int f(x)p(x^{-1})\mathrm{d}x = \int f(x^{-1})p(x)\mathrm{d}x .$$

It is immediately seen that if $p, q \in \mathscr{I}$ and $\alpha > 0$, then $p+q \in \mathscr{I}$ and $\alpha p \in \mathscr{I}$.

27.17 LEMMA. *We have $L^1 \cap P \subset \mathscr{I}$.*

In fact, let $p \in L^1 \cap P$ be a function of positive type. By proposition 27.9, the mapping $f \to \int f(x)p(x)\mathrm{d}x$ is a linear functional of positive type on L^1, therefore

$$\int (f*\tilde{f})(x)p(x)\mathrm{d}x \geqslant 0, \text{ for } f \in L^1 .$$

Then

$$\Phi_p(f*\tilde{f}) = \int (f*\tilde{f})(x^{-1})p(x)\mathrm{d}x = \int (f*\tilde{f})(x)p(x)\mathrm{d}x \geqslant 0 ,$$

since $(f*f)(x^{-1}) = (f*\tilde{f})(x) = (\tilde{f}*f)(x)$, therefore $p \in \mathscr{I}$.

In particular, for every function $h \in L^1 \cap L^\infty$ we have $h*\tilde{h} \in \mathscr{I}$. In fact, $h \in L^1$ and $\tilde{h} \in L^\infty$, therefore (proposition 24.42) the function $h*\tilde{h}$ is continuous, bounded, and belongs to $\mathscr{C}^\infty$; on the other hand, $h*\tilde{h} \in L^1$, hence $h*\tilde{h} \in L^1 \cap \mathscr{C}^\infty$. The function $h*\tilde{h}$ is of positive type (example[4]), consequently, by lemma 27.17, we have $h*\tilde{h} \in \mathscr{I}$.

By corollary 27.13, for every function $p \in \mathscr{I}$ there exists a positive bounded measure μ_p on $\hat{G}$ such that

$$(p*f)(e) = \Phi_p(f) = \int \hat{f}(\hat{x})\mathrm{d}\mu_p(\hat{x}), \text{ for } f \in L^1(G) .$$

If $p, q \in \mathscr{I}$, then

$$\mu_q(\hat{f}\hat{p}) = (p*q*f)(e) = \mu_p(\hat{f}\hat{q}), \text{ for } f \in L^1(G) .$$

Since the set $\{\hat{f} \mid f \in L^1\}$ is dense in $\mathscr{K}(\hat{G})$ for the topology of the uniform convergence, we deduce

$$\mu_q(g\hat{p}) = \mu_p(g\hat{q}), \text{ for } g \in \mathscr{K}(\hat{G}) .$$

27.18 LEMMA. *For each function $h \in \mathscr{K}(\hat{G})$ there exists a function $p \in \mathscr{I}$ such that $\hat{p}$ has strictly positive infimum on the support of h.*

Let S_h be the support of h. For each $\hat{x} \in S_h$ there exists a function $f \in \mathscr{K}(G)$ such that $\hat{f}(\hat{x}) \neq 0$, since the set $\{\hat{f} \mid f \in \mathscr{K}(G)\}$ is dense in $\overline{\mathscr{K}(\hat{G})}$. Then $|\hat{f}|^2$ does not vanish on a neighbourhood of $\hat{x}$, and $|\hat{f}|^2 = (f * \tilde{f})\hat{}$.

Since S_h is compact, there exists a finite number of functions $f_i \in \mathscr{K}(G)$ such that the function

$$\Sigma_i (f_i * \tilde{f}_i)\hat{} = \Sigma_i |\hat{f}_i|^2$$

does not vanish on a neighbourhood of S_h, therefore it has strictly positive infimum on S_h. For each i we have $f_i * \tilde{f}_i \in \mathscr{I}$, therefore putting $p = \Sigma_i (f_i * \tilde{f}_i)$, we have $p \in \mathscr{I}$ and $\hat{p}$ is positive and has strictly positive infimum on S_h.

Let now $h \in \mathscr{K}(\hat{G})$, and let p and q be two functions in $\mathscr{I}$ such that $\hat{p}$ and $\hat{q}$ are positive and have strictly positive infimum on the support of h. Then

$$g = h/\hat{p}\hat{q} \in \mathscr{K}(\hat{G}) .$$

The equality

$$\mu_q(g\hat{p}) = \mu_p(g\hat{q})$$

is now written

$$\mu_q(h/\hat{q}) = \mu_p(h/\hat{p}) .$$

For each function $h \in \mathscr{K}(\hat{G})$ put

$$\mu(h) = \mu_p(h/\hat{p})$$

where $\hat{p}$ is choosen as above, such that $h/\hat{p} \in \mathscr{K}(\hat{G})$. The definition of $\mu(h)$ does not depend on p but only on h. It is easy to see that μ is linear on $\mathscr{K}(\hat{G})$ and if $h \geqslant 0$, then taking the function p as above, we have $\hat{p} \geqslant 0$, therefore $h/\hat{p} \geqslant 0$, consequently $\mu_p(h/\hat{p}) \geqslant 0$. Thus $\mu(h) \geqslant 0$, that is μ is a positive measure on $\hat{G}$. We shall show that this measure satisfies the equality

$$\mu(g\hat{p}) = \mu_p(g), \text{ for every } p \in \mathscr{I} \text{ and } g \in \overline{\mathscr{K}(\hat{G})} .$$

Denote by S the set of points $\hat{x} \in \hat{G}$ with the property that for every neighbourhood V of $\hat{x}$ there exists $p \in \mathscr{I}$ such that μ_p does not vanish identically in V. The set S is closed and contains the support $S(\mu_p)$ of each measure μ_p with $p \in \mathscr{I}$.

27.19 LEMMA. *If $q \in \mathscr{I}$ and if $\hat{q}(\hat{x}_1) > 0$ at a point $x_1 \in S$, then $\hat{x}_1 \in S(\mu_q)$.*

In fact, assume that there exists a neighbourhood U of $\hat{x}_1$ such that the restriction of μ_q to $\mathscr{K}(\hat{G}, U)$ vanishes. Since $\hat{q}$ is continuous we can take U contained in the open set.

$$\{\hat{x} \in \hat{G} \mid \hat{q}(\hat{x}) > \tfrac{1}{2}\hat{q}(\hat{x}_1)\}\,.$$

Then, for every $p \in \mathscr{I}$ and every $h \geqslant 0$ in $\mathscr{K}(\hat{G}, U)$, we have

$$0 = \int h\hat{p}\, d\mu_q = \int h\hat{q}\, d\mu_p \geqslant \tfrac{1}{2}\hat{q}(\hat{x}_1) \int h\, d\mu_p \geqslant 0$$

therefore $\mu_p(h) = 0$ for $h \in \mathscr{K}(\hat{G}, U)$ and $p \in \mathscr{I}$ which contradicts the hypothesis that $\hat{x}_1 \in S$.

27.20 LEMMA. *If $p \in \mathscr{I}$, then $\hat{p}(\hat{x}) \geqslant 0$ for every $\hat{x} \in S$.*

Assume that there exists $\hat{x}_1 \in S$ with $\hat{p}(\hat{x}_1) < 0$. Since $\hat{p}$ is continuous, there exists a neighbourhood U of $\hat{x}$, such that we have $\hat{p}(\hat{x}) \leqslant \frac{1}{2}\hat{p}(\hat{x}_1) < 0$, for every $\hat{x} \in U$. Choose $f_1 \in L^2 \cup \mathscr{C}^\infty$ such that $\hat{f}_1(\hat{x}_1) \neq 0$. Such a function f_1 exists since the set $\{\hat{f} \mid f \in \mathscr{K}(G)\}$ is dense in $\overline{\mathscr{K}(\hat{G})}$. Then $q = \tilde{f}_1 * f_1 \in \mathscr{I}$ and $\hat{q}(\hat{x}_1) > 0$. We can choose the neighbourhood U of $\hat{x}_1$ such that we have $\hat{q}(\hat{x}) > 0$ for $\hat{x} \in U$. Then, for every positive function $h \in \mathscr{K}(\hat{G}, U)$ we have

$$0 \leqslant \int h\hat{q}\, d\mu_p = \int h\hat{p}\, d\mu_q < \tfrac{1}{2}\hat{p}(\hat{x}_1) \int h\, d\mu_q \leqslant 0\,,$$

therefore $\mu_q(h) = 0$ for every $h \in \mathscr{K}_+(\hat{G}, U)$ consequently $U \cap S(\mu_q) = \varnothing$. But by lemma 27.19, we have $\hat{x}_1 \in S(\mu_q) \cap U$, since $\hat{q}(x_1) > 0$ and we have got a contradiction.

Remark that, since all the measures μ_p have the support contained in S, the measure μ also has its support contained in S.

Let $p \in \mathscr{I}$. Put,

$$F = \{\hat{x} \in S \mid \hat{p}(\hat{x}) = 0\}, \quad O = \{\hat{x} \in S \mid \hat{p}(\hat{x}) > 0\}\,,$$
$$C_n = \{\hat{x} \in S \mid \hat{p}(\hat{x}) \geqslant n^{-1}\}, \quad D_n = \{\hat{x} \in S \mid \hat{p}(\hat{x}) \leqslant (n+1)^{-1}\}\,.$$

Since $\hat{p}$ vanishes at infinity, all the sets C_n are compact. For each n there exists a function $h_n \in \mathscr{K}(\hat{G})$ with values in $[0, 1]$ with $h_n(\hat{x}) = 0$ for $\hat{x} \in C_n$ and $h_n(\hat{x}) = 1$ for $\hat{x} \in D_n$.

27.21 LEMMA. *We have $\mu_p(\varphi_F g) = 0$ for every $g \in \overline{\mathscr{K}(\hat{G})}$.*

Remark that on S we have $h_n \searrow \varphi_F$ and $h_n\hat{p} \searrow 0$, therefore $\mu_p(h_n\hat{p}) \searrow 0$ and $\mu_p(h_n\hat{q}) \searrow \mu_p(\varphi_F\hat{q})$. Thus $\mu_p(\varphi_F\hat{q}) = 0$ for every $q \in \mathscr{I}$. The *algebra* generated by the set $\{\hat{q} \mid q \in \mathscr{I}\}$ is dense in $\overline{\mathscr{K}(\hat{G})}$, since $L^1 \cap P \subset \mathscr{I}$ and $L^1 \cap P$ is dense in L^1 (proposition 27.4). Passing to the limit in the equality $\mu_p(\varphi_F\hat{q}) = 0$ it follows that

$$\mu_p(\varphi_F g) = 0, \text{ for every } g \in \overline{\mathscr{K}(\hat{G})}.$$

27.22 Lemma. *We have* $\mu(g\hat{p}) = \mu_p(g)$ *for every* $g \in \overline{\mathscr{K}(\hat{G})}$.

Let $g \geqslant 0$ in $\overline{\mathscr{K}(\hat{G})}$. Since $\hat{p}(\hat{x}) > (n+1)^{-1} > 0$ for every $\hat{x}$ in S for which $g(1-h_n)\hat{p} \neq 0$, we have

$$\mu(g(1-h_n)\hat{p}) = \mu_p(g(1-h_n)).$$

But on the set S we have

$$g(1-h_n)\hat{p} \nearrow g\varphi_0\hat{p} = g\hat{p} \text{ and } g(1-h_n) \nearrow g\varphi_0$$

consequently passing to the limit in the preceding equality,

$$\mu(g\hat{p}) = \mu_p(g\varphi_0) = \mu_p(g) - \mu_p(g\varphi_F) = \mu_p(g).$$

The equality remains valid for every $g \in \overline{\mathscr{K}(\hat{G})}$.

27.23 Lemma. *For every* $p \in \mathscr{I}$ *we have* $\hat{p} \in L^1(\hat{G}, \mu) \cap L^2(\hat{G}, \mu)$.

In fact, if $\hat{p} \geqslant 0$ on S and $S(\mu) \subset S$, we deduce that $\mu(|\hat{p}|) = \mu(\hat{p}) = \mu_p(1) < +\infty$.

On the other hand $\hat{p} \in \overline{\mathscr{K}(\hat{G})}$, therefore $|\hat{p}|^2 = \hat{p}\bar{\hat{p}}$ is μ-integrable and thus $\hat{p} \in L^2(\hat{G}, \mu)$.

27.24 Lemma. *For every* p *and* q *in* $\mathscr{I}$ *we have*

$$\int p\bar{q}\,\mathrm{d}x = \int \hat{p}\bar{\hat{q}}\,\mathrm{d}\mu.$$

In fact

$$\Phi(p*\tilde{q}) = \mu_p(\bar{\hat{q}}) = \mu(\hat{p}\bar{\hat{q}})$$

and on the other hand

$$\Phi(p*\tilde{q}) = (p*\tilde{q})(e) = \int p(x)\tilde{q}(x^{-1})\mathrm{d}x = \int p(x)\overline{q(x)}\,\mathrm{d}x.$$

27.25 THEOREM. *If* $f \in [L^1 \cap P]$, *then* $\hat{f} \in L^1(\hat{G})$ *and*

$$f(x) = \int \langle x, \hat{x} \rangle \hat{f}(\hat{x}) \, \mathrm{d}\hat{x} \,,$$

for almost every x, where $\mathrm{d}\hat{x}$ *is a convenient Haar measure on* $\hat{G}$.

The stated equality is called the *Fourier inversion formula.*

Consider the measure μ on $\hat{G}$ defined above. Let $p \in L^1 \cap P$. Since $p \in \mathscr{I}$, by lemma 27.23 we have $\hat{p} \in L^1(\hat{G}, \mu)$. On the other hand, for every $f \in L^1$ we have $\hat{f} \in \mathscr{K}(\hat{G})$, therefore by lemma 27.22 $\mu_p(\hat{f}) = \mu(\hat{f}\hat{p})$, that is

$$(p * f)(e) = \int \hat{f}(\hat{x}) \hat{p}(\hat{x}) \, \mathrm{d}\mu(\hat{x}) \,,$$

for $f \in L^1$.

Then, for $f \in L^1$ we have

$$\begin{aligned} \int p(x) \overline{f(x)} \mathrm{d}x = (p * f)(e) &= \int \hat{p}(\hat{x}) \overline{\hat{f}(\hat{x})} \mathrm{d}\mu(\hat{x}) = \\ &= \int \hat{p}(\hat{x}) \mathrm{d}\mu(\hat{x}) \int \overline{\langle x, \hat{x} \rangle f(x)} \mathrm{d}x = \\ &= \int \overline{f(x)} \mathrm{d}x \int \overline{\langle x, \hat{x} \rangle} \hat{p}(\hat{x}) \, \mathrm{d}\mu(\hat{x}) \,, \end{aligned}$$

whence

$$p(x) = \int \overline{\langle x, \hat{x} \rangle} \hat{p}(\hat{x}) \, \mathrm{d}\mu(\hat{x})$$

for almost every $x \in G$.

This equality remains then true for every function $f \in [L^1 \cap P]$.

It remains to show that μ is a Haar measure. If $p \in L^1 \cap P$ and $\hat{x}_0 \in \hat{G}$, then the function $x \to p(x) \langle x, \hat{x}_0 \rangle$ belongs to $L^1 \cap P$. In fact, the mapping $\alpha(\hat{x}) = \hat{x}_0 \hat{x}$ is a homeomorphism of $\hat{G}$, therefore it is proper with respect to the positive bounded measure μ_p on $\hat{G}$, which satisfies the equality

$$p(x) = \int \langle x, \hat{x} \rangle \, \mathrm{d}\mu_p(\hat{x}) \,.$$

The measure $\alpha(\mu_p)$ is positive and bounded, therefore, by Bochner's theorem, the function

$$q(x) = \int \langle x, \hat{x} \rangle \, \mathrm{d}\alpha(\mu_p(\hat{x}))$$

is of positive type. But

$$\begin{aligned} q(x) &= \int \langle x, \hat{x} \rangle \, \mathrm{d}\alpha(\mu_p(\hat{x})) = \int \langle x, \alpha(\hat{x}) \rangle \, \mathrm{d}\mu_p(\hat{x}) = \\ &= \int \langle x, \hat{x}_0 \, \hat{x} \rangle \, \mathrm{d}\mu_p(\hat{x}) = \langle x, \hat{x}_0 \rangle \int \langle x, \hat{x} \rangle \, \mathrm{d}\mu_p(\hat{x}) = \\ &= \langle x, \hat{x}_0 \rangle \, p(x) \end{aligned}$$

hence $\langle x, x_0 \rangle p(x)$ is of positive type. Since $p \in L^1 \cap P$ and $\hat{x}_0 \in P$, we deduce that $p\hat{x}_0 \in L^1 \cap P$. Remark that

$$\widehat{p\hat{x}_0}(\hat{x}) = \int \langle x, \hat{x} \rangle p(x) \langle x, \hat{x}_0 \rangle \mathrm{d}x =$$
$$= \int \langle x, \hat{x}_0 \hat{x} \rangle p(x) \mathrm{d}x = \hat{p}(\hat{x}_0 \hat{x}) = \hat{p}_{\hat{x}_0}(\hat{x})$$

therefore

$$\int \hat{p}(\hat{x}) \mathrm{d}\mu(\hat{x}) = p(e) = p(e) \langle e, \hat{x}_0 \rangle = (p\hat{x}_0)(e) =$$
$$= \int \widehat{p\hat{x}_0}(\hat{x}) \mathrm{d}\mu(\hat{x}) = \int \hat{p}(\hat{x}_0 \hat{x}) \mathrm{d}\mu(\hat{x})$$

for every $p \in L^1 \cap P$. Since $L^1 \cap P$ is dense in L^1, the set $\{\hat{p} \mid p \in L^1 \cap P\}$ is dense in $\overline{\mathscr{K}(\hat{G})}$. From the last equality we obtain, passing to the limit,

$$\int f(\hat{x}) \mathrm{d}\mu(\hat{x}) = \int f(\hat{x}_0 \hat{x}) \mathrm{d}\mu(\hat{x}), \text{ for } f \in \mathscr{K}(\hat{G})$$

consequently μ is a Haar measure. If we denote $\mathrm{d}\mu = \mathrm{d}\hat{x}$, then

$$p(x) = \int \overline{\langle x, \hat{x} \rangle} \hat{p}(\hat{x}) \mathrm{d}\hat{x}, \text{ for every } p \in [L^1 \cap P]$$

and the theorem is proved.

27.26 Corollary. *If $p \in L^1 \cap P$ then $\hat{p} \geqslant 0$.*

In fact, $p \in L^1(\hat{G})$ and

$$p(x) = \int \overline{\langle x, \hat{x} \rangle} \hat{p}(\hat{x}) \mathrm{d}\hat{x} = \int \langle x, \hat{x} \rangle \hat{p}(\hat{x}^{-1}) \mathrm{d}\hat{x}.$$

For every function $f \in L^1$ we have

$$\int f(x) p(x) \mathrm{d}x = \int \overline{\langle x, \hat{x} \rangle} \hat{p}(\hat{x}^{-1}) \mathrm{d}\hat{x} \int f(x) \mathrm{d}x =$$
$$= \int \hat{p}(\hat{x}^{-1}) \mathrm{d}\hat{x} \int \langle x, \hat{x} \rangle f(x) \mathrm{d}x = \int \hat{f}(\hat{x}) \hat{p}(\hat{x}^{-1}) \mathrm{d}\hat{x}.$$

Taking now $f * \tilde{f}$ instead of f, we obtain

$$\int f * \tilde{f}(x) p(x) \mathrm{d}x \geqslant 0,$$

since p is of positive type, therefore

$$\int |\hat{f}(x)|^2 \hat{p}(\hat{x}^{-1}) \mathrm{d}\hat{x} \geqslant 0.$$

The set $\{\hat{f} \mid f \in L^1\}$ is dense in $\overline{\mathscr{K}(\hat{G})}$ therefore passing to the limit, we deduce that for every function $h \geqslant 0$ in $\mathscr{K}(\hat{G})$ we have (taking $h^{\frac{1}{2}}$ instead of f in the preceding inequality)

$$\int h(\hat{x})\,\hat{p}(\hat{x}^{-1})\,d\hat{x} \geqslant 0\,.$$

Thus it follows that $\hat{p}(\hat{x}^{-1}) \geqslant 0$ for every $\hat{x} \in \hat{G}$, that is $\hat{p} \geqslant 0$.

27.27 COROLLARY. *If $h \geqslant 0$ is a function of $\mathscr{L}^1(\hat{G})$ and if the function*

$$f(x) = \int \overline{\langle x, \hat{x}\rangle}\, h(\hat{x})\,d\hat{x}$$

belongs to the space $\mathscr{L}^1(G)$, then

$$h(\hat{x}) = \int \langle x, \hat{x}\rangle f(x)\,dx = \hat{f}(\hat{x})$$

almost everywhere.

The measure $d\mu = h(\hat{x})\,d\hat{x}$ is positive and bounded. By the Bochner theorem, the function $f(x)$ is continuous, bounded and of positive type. Using also the hypothesis, we deduce that $f \in L^1 \cap P$.

By the inversion formula, we have

$$f(x) = \int \overline{\langle x, \hat{x}\rangle}\, \hat{f}(\hat{x})\,d\hat{x}, \quad \text{for } x \in G\,,$$

consequently

$$\int \langle x, \hat{x}\rangle \hat{f}(\hat{x})\,d\hat{x} = \int \langle x, \hat{x}\rangle\, h(\hat{x})\,dx, \quad \text{for } x \in G\,.$$

For every function $g \in L^1$ we have

$$\begin{aligned}\int \hat{g}(\hat{x}^{-1})\, h(\hat{x})\,d\hat{x} &= \int h(\hat{x})\,d\hat{x} \int \overline{\langle x, \hat{x}\rangle}\, g(x)\,dx = \\ &= \int g(x)\,dx \int \overline{\langle x, \hat{x}\rangle}\, h(\hat{x})\,d\hat{x} = \\ &= \int g(x)\,dx \int \overline{\langle x, \hat{x}\rangle}\, \hat{f}(\hat{x})\,d\hat{x} = \int \hat{g}(\hat{x}^{-1})\,\hat{f}(\hat{x})\,d\hat{x}\,,\end{aligned}$$

whence it follows that $h(\hat{x}) = \hat{f}(\hat{x})$ almost everywhere on G, since the set $\{\hat{g} \mid g \in L^1\}$ is dense in $\mathscr{K}(\hat{G})$.

27.28 THEOREM (*Plancherel*). *The Fourier transform $f \to Tf = \hat{f}$ maps the set $[L^1 \cap P]$ dense in $L^2(G)$ onto a dense set in $L^2(\hat{G})$ and preserves the scalar product,*

$$\int f(x)\,\overline{g(x)}\,dx = \int \hat{f}(\hat{x})\,\overline{\hat{g}(\hat{x})}\,dx$$

and consequently it can be extended to a linear isometric mapping of $L^2(G)$ onto $L^2(\hat{G})$.

By lemma 27.24, for every pair of functions $p, q \in L^1 \cap P \subset \mathscr{I}$ we have

$$\int p\bar{q}\,\mathrm{d}x = \int \hat{p}\bar{\hat{q}}\,\mathrm{d}\mu$$

and in the preceding theorem we have proved that μ is a Haar measure on $\hat{G}$, which was denoted by $\mathrm{d}\hat{x}$. Then for $f, g \in [L^1 \cap P]$, we deduce also that

$$\int f\bar{g}\,\mathrm{d}x = \int \hat{f}\bar{\hat{g}}\,\mathrm{d}\hat{x}\ ,$$

consequently the Fourier transform, considered on the space $[L^1 \cap P]$, preserves the scalar product. Since $[L^1 \cap P]$ is dense in $L^2(G)$, this transform is uniquely extended to a linear isometric mapping T of $L^2(G)$ into $L^2(\hat{G})$. It remains to show that the image of $L^2(G)$ by T is dense in $L^2(\hat{G})$, whence it will follow that this image is even equal to $L^2(\hat{G})$.

Consider the adjoint operator $T^* : L^2(\hat{G}) \to L^2(G)$. Let $f \in [L^1(G) \cap P]$ and $g \in L^1(\hat{G}) \cap L^2(\hat{G})$ and denote $g' = T^* g$. We have $\langle \hat{f}, g\rangle = \langle Tf, g\rangle = \langle f, T^* g\rangle = \langle f, g'\rangle$, that is

$$\int\int \langle x, \hat{x}\rangle f(x) g(\hat{x})\,\mathrm{d}x\,\mathrm{d}\hat{x} = \int f(x)\overline{g'(x)}\mathrm{d}x\ ,$$

whence, $[L^1(G) \cap P]$ being dense in $\mathscr{K}(G)$,

$$g'(x) = (T^* g)(x) = \int \overline{\langle x, \hat{x}\rangle} g(\hat{x})\ \mathrm{d}\hat{x} \ \text{ almost everywhere}\ .$$

Let now, $g, \varphi \in L^1(\hat{G}) \cap L^2(\hat{G})$, $g \geqslant 0$, $\varphi \geqslant 0$. Then $g * \varphi \in L^1(\hat{G}) \cap L^2(\hat{G})$. On the other hand the functions $g(\hat{y})\langle x, \hat{y}\rangle$ and $\varphi(\hat{x})\langle x, \hat{x}\rangle$ are integrable, therefore their product is integrable for the measure $\mathrm{d}\hat{x}\,\mathrm{d}\hat{y}$. Using the Fubini theorem and the above equality, we deduce

$$\begin{aligned} T^*(g*\varphi) &= \int (g*\varphi)(\hat{x})\overline{\langle x, \hat{x}\rangle}\,\mathrm{d}\hat{x} = \\ &= \int \overline{\langle x, \hat{x}\rangle}\,\mathrm{d}\hat{x} \int g(\hat{y})\varphi(\hat{y}^{-1}\hat{x})\mathrm{d}\hat{y} = \\ &= \int g(\hat{y})\mathrm{d}\hat{y} \int \overline{\langle x, \hat{x}\rangle}\, \varphi(\hat{y}^{-1}\hat{x})\,\mathrm{d}\hat{x} = \\ &= \int g(\hat{y})\mathrm{d}\hat{y} \int \overline{\langle x, \hat{y}\hat{x}\rangle}\, \varphi(\hat{x})\,\mathrm{d}\hat{x} = \\ &= \int g(\hat{y})\overline{\langle x, \hat{y}\rangle}\,\mathrm{d}\hat{y} \int \varphi(\hat{x})\,\langle x, \hat{x}\rangle\,\mathrm{d}\hat{x} = \\ &= T^* g(x)\, T^* \varphi(x) = (g'\,\varphi')(x)\ . \end{aligned}$$

Since $g', \varphi' \in L^2(G)$, we deduce that $g'\,\varphi' \in L^1(G)$. On the other hand $g'\,\varphi' = T^*(g*\varphi) \in L^2(G)$, therefore

$$T^*(g*\varphi) = g'\,\varphi' \in L^1(G) \cap L^2(G)\ .$$

Since $g \geqslant 0$ and $\varphi \geqslant 0$, we have $g*\varphi \geqslant 0$. Since, in addition, $g*\varphi \in L^1(\hat{G})$ and the function $(g*\varphi)'(x) = \int \langle x, \hat{x}\rangle (g*\varphi)(\hat{x})\,\mathrm{d}\hat{x}$ belongs to the space $L^1(G)$, by corollary 27.27, we deduce that we have

$$(g*\varphi)(\hat{x}) = \int \langle x, \hat{x} \rangle (g*\varphi)'(x)\,dx = T(g*\varphi)'(x)$$

almost everywhere, that is

$$g*\varphi = T(g*\varphi)' \in TL^2(G).$$

If g and φ belong to $L^1(\hat{G}) \cap L^2(\hat{G})$ but are not positive, writing them as linear combinations of positive functions, we deduce that $g*\varphi \in TL^2(G)$. But the functions of the form $g*\varphi$ with $g, \varphi \in L^1(\hat{G}) \cap L^2(\hat{G})$ form a dense set in $L^2(\hat{G})$, since among them there are all the functions of the form $g*\tilde{g}$ which are of positive type (proposition 27.14). Thus $TL^2(G)$ is dense in $L^2(\hat{G})$, therefore $TL^2(G) = L^2(\hat{G})$.

Thus the theorem is proved.

5 *The Pontryagin theorem*

Let $x \in G$. Consider the mapping X defined on $\hat{G}$ by the equality

$$X(\hat{x}) = \hat{x}(x), \text{ for } \hat{x} \in G.$$

We have

$$|X(\hat{x})| = |\hat{x}(x)| = 1$$

and

$$X(\hat{x}\hat{y}) = (\hat{x}\hat{y})(x) = \hat{x}(x)\hat{y}(x) = X(\hat{x})X(\hat{y}),$$

therefore $Z = XY$. Thus the mapping $x \to X$ is an algebraic homomorphism of plex numbers with modulus 1. Since the mapping $(x, \hat{x}) \to \hat{x}(x)$ of $G \times \hat{G}$ into C is continuous (proposition 26.23) it follows that the mapping $\hat{x} \to \hat{x}(x) = X(\hat{x})$ is continuous on $\hat{G}$, therefore X is a character of the group $\hat{G}$.

Remark that if $x, y \in G$ and $z = xy$, then

$$Z(\hat{x}) = \hat{x}(z) = \hat{x}(xy) = \hat{x}(x)\hat{x}(y) = X(\hat{x})\,Y(\hat{x}) = (XY)(\hat{x}),$$

therefore $Z = XY$. Thus the mapping $x \to X$ is an algebraic homeomorphism of the group G into the group $\hat{\hat{G}}$, the dual of the group $\hat{G}$.

This mapping is an isomorphism. In fact, if $x_1 \neq x_2$ are two elements of G, there exists a character $\hat{x}_0 \in \hat{G}$ such that $\hat{x}_0(x_1) \neq \hat{x}_0(x_2)$. Then $X_1(\hat{x}_0) \neq X_2(\hat{x}_0)$ therefore $X_1 \neq X_2$.

27.29 THEOREM (*Pontryagin*). *The mapping $x \to X$ is an isomorphism and and a homeomorphism of the group G onto the group $\hat{\hat{G}}$ of the characters of $\hat{G}$.*

Since $x \to X$ is a bijective mapping of G into $\hat{G}$ we shall identify x with X and we shall write $G \subset \hat{\hat{G}}$. We have to show that $G = \hat{\hat{G}}$ and that the topologies on G and $\hat{\hat{G}}$ are equivalent.

Denote by A the set of finite linear combinations $\Sigma_{1 \leqslant i \leqslant n} \alpha_i f_i * \varphi_i$ with $f_i, \varphi_i \in \mathscr{K}(G)$. The set A is an algebra dense in $L^1(G)$ and in $\mathscr{K}(G)$ (for the norm $\|f\|_\infty$).

The image $\hat{A} = TA$ of the algebra A by the Fourier transform T is dense in $L^1(\hat{G})$ since the set $\{\hat{f} = Tf \mid f \in \mathscr{K}(G)\}$ is dense in $L^2(\hat{G})$, hence the finite linear combinations $\Sigma \alpha_i \hat{f}_i \hat{g}_i = T\Sigma \alpha_i f_i * g_i$ form a dense set in $L^1(\hat{G})$.

We have $A \subset [L^1 \cap P]$, therefore for every $f \in A$ we have

$$f(x) = \int \overline{\langle x, \hat{x} \rangle} \hat{f}(\hat{x}) \, \mathrm{d}\hat{x}$$

for almost every $x \in G$, and $\hat{f} \in \hat{A} \cap L^1(\hat{G})$. On the other hand, the functions of $\mathscr{K}(G)$ separate the points of G and for every point $x_0 \in G$ there exists a function $f \in \mathscr{K}(G)$ such that $f(x_0) \neq 0$. Since A is dense in $\mathscr{K}(G)$, it follows that the functions of A separate the points of G, and for every $x_0 \in G$ there exists $f \in A$ with $f(x_0) \neq 0$. Since $A \subset \mathscr{K}(G)$, the functions of A vanish at infinity; it follows that the weak topology on G defined by the functions of A coincides with the topology of G.

The sets of G of the form

$$U(x_0; f_1, \ldots, f_n; \varepsilon) = \{x \in G \mid |\int [\overline{\langle x, \hat{x} \rangle} - \overline{\langle x_0, \hat{x} \rangle}] \hat{f}_i(\hat{x}) \, \mathrm{d}\hat{x}| < \varepsilon, \quad i = 1, 2, \ldots, n\},$$

where $\varepsilon > 0$ and $f_i \in A$, form a basis of neighbourhoods of x_0 in G. But the inequalities

$$|\int [\overline{\langle x, \hat{x} \rangle} - \overline{\langle x_0, \hat{x} \rangle}] \hat{f}_i(\hat{x}) \, \mathrm{d}\hat{x}| < \varepsilon$$

can be written

$$|\int [\overline{X(\hat{x})} - \overline{X_0(\hat{x})}] \hat{f}_i(\hat{x}) \, \mathrm{d}\hat{x}| < \varepsilon,$$

consequently the sets $U(x_0; \hat{f}_1, \ldots, \hat{f}_n; \varepsilon)$ form a fundamental system of neighbourhoods of x_0 in G, considered as a subspace of $\hat{\hat{G}}$.

Thus, G is a locally compact subspace of $\hat{\hat{G}}$ and G is closed in $\hat{\hat{G}}$. In fact, let G_∞ and $\hat{\hat{G}}_\infty$ be the compact spaces obtained, from G and $\hat{\hat{G}}$ by adding a point at infinity. Then G_∞ is a closed subspace of $\hat{\hat{G}}_\infty$ therefore G is closed in $\hat{\hat{G}}$.

It remains to show that G is dense in $\hat{\hat{G}}$, whence it will follow that $G = \hat{\hat{G}}$. If G were not dense in $\hat{\hat{G}}$, there would exist a relatively compact open set in $\hat{G}$, disjoint from G, therefore there would exist positive functions $f, g \in L^1(\hat{\hat{G}}) \cap L^2(\hat{\hat{G}})$, vanishing on G but not vanishing identically on $\hat{\hat{G}}$. Putting $h = f * g$, we deduce

$$h(\hat{\hat{x}}) = \int \langle \hat{\hat{x}}, \hat{x} \rangle \hat{h}(\hat{x})\, \mathrm{d}\hat{x}\,,$$

where $\hat{h} \in L^1(\hat{G}) \cap L^2(\hat{G})$. But $h(x) = 0$ for $x \in G$, hence

$$\overline{h(x)} = \int \overline{\langle x, \hat{x} \rangle}\, \overline{\hat{h}(\hat{x})}\, \mathrm{d}x = 0$$

for $x \in G$, that is $T^* \bar{\hat{h}} = 0$. It follows that $\hat{h} = 0$, therefore $h(\hat{\hat{x}}) = 0$, and we obtain a contradiction. Thus G is dense in $\hat{\hat{G}}$, therefore $G = \hat{\hat{G}}$ and the theorem is proved.

Due to this theorem, we may identify G with $\hat{\hat{G}}$ and we say that in turn, G is the dual of $\hat{G}$.

27.30 PROPOSITION. *The group G is compact if and only if $\hat{G}$ is discrete.*

If G is discrete, then the algebra $L^1(\hat{G})$ has unit element, therefore the group $G = \hat{\hat{G}}$ is compact, as space of continuous, linear and multiplicative functionals on $L^1(\hat{G})$.

Conversely, if G is compact, the set of characters $\{x \mid \|x - 1\|_\infty < \frac{1}{2}\}$ is an open neighbourhood of the character 1 and it consists of this character only. It follows that the topology on $\hat{G}$ is discrete and the proposition is proved.

6 *Examples*

(1) *The additive group R of the real numbers.*

The Haar measure $\mathrm{d}x$ on R is the Lebesgue measure.

The convolution is defined in this case by

$$(f*g)(x) = \int_{-\infty}^{+\infty} f(x-y)g(x)\,\mathrm{d}y = \int_{-\infty}^{+\infty} f(y)g(x-y)\,\mathrm{d}y\,.$$

We have

$$\int_{-\infty}^{+\infty} f(x)\,\mathrm{d}x = \int_{-\infty}^{+\infty} f(-x)\,\mathrm{d}x\,.$$

The involution $\tilde{f}$ is defined by the equality

$$\tilde{f}(x) = \overline{f(-x)}\,.$$

For each $y \in R$, the function $\chi_y(x) = e^{2\pi i x y}$ is a continuous homeomorphism of R into the multiplicative group of the complex numbers of modulus 1, therefore it is a character of R.

Conversely, it can be proved that the only continuous functions which fulfill the conditions;

$$|\chi(x)| \equiv 1 \text{ and } \chi(x+y) = \chi(x) \cdot \chi(y)$$

are of the form $\chi(x) = e^{2\pi ixy}$ with $y \in R$.

Thus, there exists a bijective correspondence between the set $\hat{R}$ of the characters of R and the set R.

This correspondence is an algebraic isomorphism and a homeomorphism, therefore the dual of R is R itself.

The Fourier transform of a function $f \in L^1(R)$ is written

$$\hat{f}(\xi) = \int_{-\infty}^{+\infty} f(x) e^{2\pi ix\xi} \mathrm{d}x .$$

The inversion formula is written

$$f(x) = \int_{-\infty}^{+\infty} \hat{f}(\xi) e^{-2\pi ix\xi} \mathrm{d}\xi$$

for the functions $f \in [L^1 \cap P]$.

The Plancherel formula:

$$\int_{-\infty}^{+\infty} f(x) \overline{g(x)} \mathrm{d}x = \int_{-\infty}^{+\infty} \hat{f}(\xi) \overline{\hat{g}(\xi)} \mathrm{d}\xi .$$

The functions φ of positive type of $L^\infty(R)$ are of the form

$$\varphi(x) = \int_{-\infty}^{+\infty} e^{2\pi ix\xi} \mathrm{d}\mu(\xi) ,$$

where μ are positive measures on R, that is of the form

$$\varphi(x) = \int_{-\infty}^{+\infty} e^{2\pi ix\xi} \mathrm{d}g(\xi) ,$$

where g are increasing left continuous functions on R, vanishing at the origin.

(2) *The additive group Z of the integers*

This group is discrete. We choose the Haar measure $\mathrm{d}y$ with mass 1 in each point.

A function $f: Z \to C$ is integrable for $\mathrm{d}x$ if and only if

$$\Sigma_{-\infty \leqslant n < +\infty} |f(n)| < +\infty$$

The convolution of two functions f and g is

$$(f*g)(n) = \Sigma_{-\infty \leqslant n < +\infty} f(n-m) g(m) = \Sigma_{-\infty \leqslant n < +\infty} f(n) g(n-m) .$$

The function h defined by $h(0)=1$ and $h(n)=0$ for $n\neq 0$ is the unit for the convolution.

For each complex number of modulus 1, $\xi=e^{2\pi i\theta}$ (with $\theta\in R$, θ depending on ξ), the function $\chi_\xi(n)=e^{2\pi in\theta}$ is a character of the group Z and it can be proved that these are the only characters of Z. The dual $\hat{Z}$ of the group Z is the multiplicative group U of the complex numbers of modulus 1.

The Fourier transform of a function $f\in L^1(Z)$ is

$$\hat{f}(\xi)=\hat{f}(e^{2\pi i\theta})=\Sigma_{-\infty\leqslant n<+\infty}f(n)e^{2\pi in\theta}\,.$$

The inversion formula is

$$f(n)=\int_0^1\hat{f}(e^{2\pi i\theta})e^{-2\pi in\theta}\,\mathrm{d}\theta=\int_U f(\xi)e^{2\pi in\theta(\xi)}\mathrm{d}\xi\,,$$

where $\mathrm{d}\xi$ is the Haar measure on U, the image of the measure $\mathrm{d}\theta$ on $[0, 1]$ by the mapping $\theta\to e^{2\pi i\theta}$.

The Plancherel formula:

$$\Sigma_{-\infty\leqslant n<+\infty}f(n)\overline{g(n)}=\int_0^1\hat{f}(e^{2\pi i\theta})\overline{\hat{g}(e^{2\pi i\theta})}\mathrm{d}\theta=\int_U\hat{f}(\xi)\overline{\hat{g}(\xi)}\,\mathrm{d}\xi\,.$$

The functions φ of positive type of $L^\infty(Z)$ are of the form

$$\varphi(n)=\int_0^1 e^{2\pi in\theta}\,\mathrm{d}\mu(\theta)=\int_U e^{2\pi in\theta(\xi)}\,\mathrm{d}\mu'(\xi)\,,$$

where μ are positive measures on $[0, 1]$ and μ' the images of the measures μ by the mapping $\theta\to e^{2\pi i\theta}$ of $[0, 1]$ onto U.

The functions φ of positive type can be written in the form

$$\varphi(n)=\int_0^1 e^{2\pi in\theta}\,\mathrm{d}g(\theta)\,,$$

where g are increasing functions on $[0, 1]$.

(3) *The multiplicative group U of the complex numbers of modulus* 1

This group is compact. The Haar measure $\mathrm{d}\xi$ is chosen such that the measure of U is equal to 1. The measure $\mathrm{d}\xi$ is the image of the Lebesgue measure $\mathrm{d}\theta$ on $[0, 1]$ by the mapping $\theta\to e^{2\pi i\theta}$.

The convolution:

$$(f*g)(\zeta)=\int_U f(\zeta\xi^{-1})g(\xi)\,\mathrm{d}\xi=\int_U f(\xi)g(\zeta\xi^{-1})\,\mathrm{d}\xi.$$

Since $\hat{Z}=U$, it follows that $\hat{U}=Z$. Thus, the characters of U are of the form $\chi_n(\theta)=e^{2\pi in\theta}$ with $n\in Z$.

The Fourier transform:

$$\hat{f}(n)=\int_0^1 e^{2\pi in\theta}f(e^{2\pi i\theta})\mathrm{d}\theta=\int_U e^{2\pi in\theta(\xi)}f(\xi)\,\mathrm{d}\xi\,.$$

The inversion formula:

$$f(\xi) = f(e^{2\pi i\theta}) = \Sigma_{-\infty \leqslant n < +\infty} \hat{f}(n) e^{-2\pi i n\theta} .$$

The functions of positive type φ of $L^{\infty}(U)$ are of the form

$$\varphi(\xi) = \varphi(e^{2\pi i\theta}) = \Sigma_{-\infty \leqslant n < +\infty} e^{2\pi i n\theta} g(n) ,$$

where g is a positive function on Z.

(4) *The commutative group G with n elements*

We choose the Haar measured dx with the mass $n^{-\frac{1}{2}}$ in each point. The group G is discrete and compact.

The convolution

$$(f*g)(s) = \Sigma_{t\in G} f(t) g(t^{-1}s) = \Sigma_{t\in G} f(st^{-1}) g(t) .$$

The function h defined by $h(e)=1$ and $h(s)=0$ if $s \neq e$, is the unit for the convolution.

The characters of the group G are the *bounded* representations of G into U. It can be proved that G has just n characters, therefore the dual of G is G again, and the characters χ of G satisfy the following relations:

$$\Sigma_{s\in G} \chi(s) = \begin{cases} n \text{ if } \chi = \chi_0 \\ 0 \text{ if } \chi \neq \chi_0 \end{cases}$$

where χ_0 is the character equal identically to 1, and

$$\Sigma_{s\in G} \chi(s) \overline{\chi'(s)} = \begin{cases} n \text{ if } \chi = \chi' , \\ 0 \text{ if } \chi \neq \chi' . \end{cases}$$

The Fourier transform of a function $f: G \to C$ is

$$\hat{f}(\chi) = n^{-\frac{1}{2}} \Sigma_{s\in G} f(s) \chi(s)$$

and the inversion formula is

$$f(s) = n^{-\frac{1}{2}} \Sigma_{\chi\in \hat{G}} \hat{f}(\chi) \overline{\chi(s)} .$$

The Plancherel formula is written

$$\Sigma_{s\in G} f(s) \overline{g(s)} = \Sigma_{\chi\in \hat{G}} \hat{f}(\chi) \overline{\hat{g}(\chi)} .$$

Chapter VII

SPACES OF VECTOR FIELDS

§28 $\mathscr{L}^p_{\mathscr{A}}$ spaces

1. *Fundamental families*

Let $\mathscr{E}=(E(t))_{t\in T}$ be a family of Banach spaces. For each $t\in T$, $E(t)$ is called the *space tangent to* T *at the point* t.

We call a *vector field* any function $\boldsymbol{x}$ defined on a subset $A\subset T$, such that for each $t\in A$ we have $\boldsymbol{x}(t)\in E(t)$. If $\boldsymbol{x}$ is a vector field we shall denote by $|\boldsymbol{x}|$ the positive function $t\rightarrow|\boldsymbol{x}(t)|$ and $\|\boldsymbol{x}\|_A=\sup_{t\in A}|\boldsymbol{x}(t)|$.

The set of vector fields defined on T will be denoted by $\mathscr{C}(\mathscr{E})$. The set $\mathscr{C}(\mathscr{E})$ is a vector space for the usual operations of addition and multiplication by scalars, and a module over the algebra of scalar functions defined on T, if we define the product $f\boldsymbol{x}$ by the equality $(f\boldsymbol{x})(t)=f(t)\boldsymbol{x}(t)$. Instead of $\|\boldsymbol{x}\|_T$ we shall write $\|\boldsymbol{x}\|$.

28.1 Definition. *We shall call a fundamental family of continuous vector fields any vector subspace* $\mathscr{A}\subset\mathscr{C}(\mathscr{E})$ *satisfying the following axioms*:

(1) *for every* $\boldsymbol{x}\in\mathscr{A}$, *the positive function* $|\boldsymbol{x}|$ *is continuous on* T;

(2) *for every* $t\in T$, *the set* $\{\boldsymbol{x}(t);\ \boldsymbol{x}\in\mathscr{A}\}$ *is dense in* $E(t)$.

Sometimes we shall consider fundamental families $\mathscr{A}$ which satisfy in addition the countability axiom of Godement:

(G): *there exists a countable subset* $\mathscr{A}_0\subset\mathscr{A}$ *such that for every* $t\in T$, *the set* $\{\boldsymbol{x}(t);\ \boldsymbol{x}\in\mathscr{A}_0\}$ *is dense in* $E(t)$.

If there exists a fundamental family $\mathscr{A}\subset\mathscr{C}(\mathscr{E})$ which satisfies the axiom (G), then all the spaces $E(t)$ are of countable type.

Example. If all the tangent spaces are equal to a Banach space E, a vector field is a usual function defined on T with values in E. In this case, the following sets are fundamental families:

(1) The set $\mathscr{C}_E(T)$ of continuous functions $f : T \to E$;

(2) The set $\mathscr{K}_E(T)$ of continuous functions $f : T \to E$ with compact support;

(3) The set of constant mappings of T into E. This fundamental family will later be identified with E.

Remark. If the tangent spaces are equal to the same space E, we shall always choose as fundamental family $\mathscr{A}$ the set of the constant mappings of T into E. In this case $\mathscr{A}$ will be identified with E.

2. *Continuous vector fields*

Let $\mathscr{A} \subset \mathscr{C}(\mathscr{E})$ be a fundamental family of continuous vector fields on T.

28.2 Definition. *We say that a vector field* $\boldsymbol{x}$ *defined on a set* A *is continuous* (*with respect to the fundamental family* $\mathscr{A}$) *at a point* $t_0 \in T$, *if for every number* $\varepsilon > 0$, *there exists a neighbourhood* V *of* t_0 *and a vector field* $\boldsymbol{y} \in \mathscr{A}$ *such that*

$$|\boldsymbol{x}(t) - \boldsymbol{y}(t)| < \varepsilon, \textit{ for every } t \in V \cap A.$$

A vector field $\boldsymbol{x}$ is continuous on a set if it is continuous at each point of this set.

By this definition, every field $\boldsymbol{x}$ of $\mathscr{A}$ is continuous on the whole space, since in definition 28.2 we can take $\boldsymbol{y} = \boldsymbol{x}$.

It is easy to see that the set of vector fields defined on A and continuous at a point $t_0 \in A$ (or on A) is a vector space and a module over the algebra of scalar functions defined on A and continuous at t_0 (respectively on A).

Example. If $E(t) = E$ for every $t \in T$ and if we take as fundamental family $\mathscr{A}$ the set of constant mappings of T into E, then a vector field $\boldsymbol{x}$ is continuous at t_0 if and only if the function $\boldsymbol{x} : T \to E$ is continuous at t_0 in the usual sense.

We get the same result if we take as fundamental family the space $\mathscr{C}_E(T)$ or the space $\mathscr{K}_E(T)$.

Remark. The following results are generalizations of the corresponding results established for usual functions in case $E(t) = E$ for every $t \in T$.

28.3 PROPOSITION. *If $\boldsymbol{x}$ is a vector field continuous at a point $t_0 \in A$, then the function $|\boldsymbol{x}|$ is continuous at t_0.*

Let $\varepsilon > 0$. By definition 28.2, there exists a neighbourhood V' of t_0 and a vector field $\boldsymbol{y} \in \mathscr{A}$ such that

$$|\boldsymbol{x}(t) - \boldsymbol{y}(t)| < \varepsilon/3, \text{ for } t \in V' \cap A .$$

It follows that:

$$||\boldsymbol{x}(t)| - |\boldsymbol{y}(t)|| < \varepsilon/3, \text{ for } t \in V' \cap A .$$

But the function $|\boldsymbol{y}|$ is continuous on T, therefore at t_0. There exists then a neighbourhood V'' of t_0 such that

$$||\boldsymbol{y}(t)| - |\boldsymbol{y}(t_0)|| < \varepsilon/3, \text{ for } t \in V'' \cap A .$$

If we take $V = V' \cap V''$, then for $t \in V \cap A$ we have

$$||\boldsymbol{x}(t)| - |\boldsymbol{x}(t_0)|| \leqslant ||\boldsymbol{x}(t)| - |\boldsymbol{y}(t)|| + \\ + ||\boldsymbol{y}(t)| - |\boldsymbol{y}(t_0)|| + ||\boldsymbol{y}(t_0)| - |\boldsymbol{x}(t_0)|| < \varepsilon,$$

therefore the function $t \to |\boldsymbol{x}(t)|$ is continuous at t_0.

28.4 COROLLARY. *If $\boldsymbol{x}$ is a vector field continuous on a set $A \subset T$, then the function $|\boldsymbol{x}|$ is continuous on A.*

28.5 COROLLARY. *The set of vector fields $\mathscr{C}_{\mathscr{A}}(T)$ defined on T and continuous (with respect to $\mathscr{A}$) on T, is a fundamental family.*

In fact $\mathscr{C}_{\mathscr{A}}(T)$ is a vector space; since $\mathscr{A} \subset \mathscr{C}_{\mathscr{A}}(T)$, the set $\{\boldsymbol{x}(t); \boldsymbol{x} \in \mathscr{C}_{\mathscr{A}}(T)\}$ is dense in $E(t)$ for every $t \in T$, since it contains the dense set $\{\boldsymbol{x}(t); \boldsymbol{x} \in \mathscr{A}\}$.

By corollary 28.4, for every $\boldsymbol{x} \in \mathscr{C}_{\mathscr{A}}(T)$, the function $|\boldsymbol{x}|$ is continuous on T, therefore all conditions of definition 28.1 are satisfied by $\mathscr{C}_{\mathscr{A}}(T)$.

28.6 DEFINITION. *We say that two fundamental families $\mathscr{A}_1$ and $\mathscr{A}_2$ are equivalent if the set of vector fields on T continuous with respect to $\mathscr{A}_1$ is equal to the set of the vector fields on T continuous with respect to $\mathscr{A}_2$.*

By corollary 28.5 it follows that every fundamental family $\mathscr{A}$ is equivalent to $\mathscr{C}_{\mathscr{A}}(T)$. It follows that if $\mathscr{A}_1$ is a vector space, obtained from $\mathscr{A}$ by adding continuous vector fields of $\mathscr{C}_{\mathscr{A}}(T)$ then $\mathscr{A}_1$ is a fundamental family equivalent to $\mathscr{A}$.

Example. If $E(t)=E$ for every $t\in T$, the fundamental families $\mathscr{C}_E(T)$ and $\mathscr{K}_E(T)$ are equivalent to the fundamental family of constant mappings of T into E.

3 *Properties of continuous vector fields*

We shall assume in the sequel that there exists a fundamental family $\mathscr{A}\subset\mathscr{C}(T)$, which will be kept fixed.

28.7 PROPOSITION. *A vector field $\boldsymbol{x}$ is continuous on a set $A\subset T$ if and only if for every $\boldsymbol{y}\in\mathscr{A}$, the function $|\boldsymbol{x}-\boldsymbol{y}|$ is continuous on A.*

If $\boldsymbol{x}$ is continuous on A, then for every $\boldsymbol{y}\in\mathscr{A}$, the field $\boldsymbol{x}-\boldsymbol{y}$ is continuous on A therefore the function $|\boldsymbol{x}-\boldsymbol{y}|$ is continuous on A.

Conversely, assume that the function $|\boldsymbol{x}-\boldsymbol{y}|$ is continuous on A for every field $\boldsymbol{y}\in\mathscr{A}$. Let $t_0\in A$ and $\varepsilon>0$.

Since the set $\{\boldsymbol{y}(t_0);\ \boldsymbol{y}\in\mathscr{A}\}$ is dense in $E(t_0)$, there exists a field $\boldsymbol{y}\in\mathscr{A}$ such that we have $|\boldsymbol{y}(t_0)-\boldsymbol{x}(t_0)|<\varepsilon$.

Since the function $|\boldsymbol{x}-\boldsymbol{y}|$ is continuous at t_0, there exists a neighbourhood V of t_0 such that

$$|\boldsymbol{y}(t)-\boldsymbol{x}(t)|<\varepsilon, \text{ for } t\in V\cap A,$$

therefore $\boldsymbol{x}$ is continuous at t_0. Since $t_0\in A$ is arbitrary it follows that $\boldsymbol{x}$ is continuous on A.

28.8 PROPOSITION. *A vector field $\boldsymbol{x}\in\mathscr{C}(\mathscr{E})$ is continuous on T if and only if it is the uniform limit, on each compact set, of vector fields of the form $\Sigma_{1\leqslant i\leqslant n}f_i\boldsymbol{x}_i$, where $\boldsymbol{x}_i\in\mathscr{A}$ and f_i are continuous scalar functions on T.*

Assume that $\boldsymbol{x}$ is the uniform limit, on each compact set, of fields of the form $\Sigma f_i\boldsymbol{x}_i$.

Let $t_0\in T$ be a point, K a compact neighbourhood of t_0 and $\varepsilon>0$. There exists a field $\boldsymbol{z}=\Sigma_{1\leqslant i\leqslant n}f_i\boldsymbol{x}_i$ such that $|\boldsymbol{x}(t)-\boldsymbol{z}(t)|<\varepsilon/2$ for $t\in K$. Since $\boldsymbol{z}$ is a continuous vector field, there exists a neighboorhood V of t_0 and a vector field $\boldsymbol{y}\in\mathscr{A}$ such that $|\boldsymbol{z}(t)-\boldsymbol{y}(t)|<\varepsilon/2$ for $t\in V$. Then $K\cap V$ is a neighbourhood of t_0 and for $t\in K\cap V$ we have

$$|\boldsymbol{x}(t)-\boldsymbol{y}(t)|\leqslant|\boldsymbol{x}(t)-\boldsymbol{z}(t)|+|\boldsymbol{z}(t)-\boldsymbol{y}(t)|<\varepsilon,$$

hence $\boldsymbol{x}$ is continuous at t_0. It follows that $\boldsymbol{x}$ is continuous on T.

Conversely, assume that $\boldsymbol{x}$ is continuous on T. Let $K \subset T$ be a compact set and $\varepsilon > 0$. For every point $t_0 \in K$ there exists (definition 28.2) a field $\boldsymbol{x}_0 \in \mathscr{A}$ such that we have $|\boldsymbol{x}(t_0) - \boldsymbol{x}_0(t_0)| < \varepsilon$.

Since $\boldsymbol{x} - \boldsymbol{x}_0$ is continuous on T, there exists a neighbourhood V_0 of t_0 such that

$$|\boldsymbol{x}(t) - \boldsymbol{x}_0(t)| < \varepsilon, \text{ for } t \in V_0 .$$

Since K is compact, we can find a finite cover $(V_i)_{1 \leqslant i \leqslant n}$ of K consisting of open sets, and a finite family $(\boldsymbol{x}_i)_{1 \leqslant i \leqslant n}$ of vector fields of $\mathscr{A}$, such that for each i we have

$$|\boldsymbol{x}(t) - \boldsymbol{x}_i(t)| < \varepsilon, \text{ for } t \in V_i .$$

Let $(f_i)_{1 \leqslant i \leqslant n}$ be a continuous partition of the unity on K, subordinated to the cover $(V_i)_{1 \leqslant i \leqslant n}$. We have then

$$|\boldsymbol{x}(t) - \Sigma_i f_i(t)\, \boldsymbol{x}_i(t)| < \varepsilon \text{ for } t \in V_i ,$$

and thus the proposition is proved.

Remarks

1. From the first part of the proof we deduce that if $(\boldsymbol{x}_n)$ is a sequence of vector fields uniformly convergent on a set $A \subset T$ to a vector field $\boldsymbol{x}$ and if the restriction of each field $\boldsymbol{x}_n$ to A is continuous, then the restriction of the limit $\boldsymbol{x}$ to A is also continuous.

2. From the second part of the proof we deduce that if $\mathscr{A}_0$ is a set of continuous vector fields, having only the property that for every $t \in T$ the set $\{\boldsymbol{x}(t);\ \boldsymbol{x} \in A_0\}$ is dense in $E(t)$, then every vector field $\boldsymbol{x}$ continuous on T, is the uniform limit on each compact set of fields of the form $\Sigma f_i \boldsymbol{x}_i$ with $\boldsymbol{x}_i \in \mathscr{A}_0$ and f_i continuous scalar functions.

28.9 LEMMA. *Let $K \subset T$ be a compact set, $\boldsymbol{x}$ a continuous vector field defined on K, $\eta > 0$ and $\boldsymbol{y} \in \mathscr{C}(\mathscr{E})$ a continuous vector field on T such that*

$$|\boldsymbol{y}(t) - \boldsymbol{x}(t)| < \eta, \textit{ for } t \in K .$$

For every $\varepsilon > 0$ there exists then a continuous vector field z on T such that

$$|z(t) - \boldsymbol{x}(t)| < \varepsilon \textit{ for } t \in K \textit{ and } |z(t) - \boldsymbol{y}(t)| < \varepsilon + \eta, \textit{ for } t \in T .$$

Let $t_0 \in K$. Since the set $\{\boldsymbol{x}(t_0);\ \boldsymbol{x} \in \mathscr{A}\}$ is dense in $E(t_0)$, there exists a field $z_0 \in \mathscr{A}$ such that

$$|z_0(t_0)-x(t_0)|<\varepsilon\,.$$

Then

$$|z_0(t_0)-y(t_0)|\leqslant|z_0(t_0)-x(t_0)|+|x(t_0)-y(t_0)|<\varepsilon+\eta\,.$$

Since the functions z_0-x and z_0-y are continuous at t_0, there exists an open neighbourhood V_0 of t_0 such that $|z_0(t)-x(t)|<\varepsilon$ for $t_0\in V_0\cap K$ and $|z_0(t)-y(t)|<\varepsilon+\eta$ for $t\in V_0$. Since K is compact, there exists a finite cover $(V_i)_{1\leqslant i\leqslant n}$ of K consisting of open sets and a finite family $(z_i)_{1\leqslant i\leqslant n}$ of fields of $\mathscr{A}$ such that for each i we have $|z_i(t)-x(t)|<\varepsilon$ for $t\in V_i\cap K$ and $|z_i(t)-y(t)|<\varepsilon+\eta$ for $t\in V_i$.

Let $(f_i)_{1\leqslant i\leqslant n}$ be a continuous partition of the unity on K, subordinated to the cover $(V_i)_{1\leqslant i\leqslant n}$. If we denote $z'=\Sigma f_i z_i$, then z' is continuous on T; denoting $U=\cup V_i$, we have

$$|z'(t)-x(t)|<\varepsilon \text{ for } t\in K \text{ and } |z'(t)-y(t)|<\varepsilon+\eta \text{ for } t\in U\,.$$

Since U is an open neighbourhood of K, there exists a continuous function $f\colon T\to[0,1]$, such that $f(t)=1$ for $t\in K$ and $f(t)=0$ for $t\notin U$. If we denote

$$z(t)=[1-f(t)]\,y(t)+f(t)\,z'(t)$$

then z is a continuous vector field on T and

$$|z(t)-x(t)|<\varepsilon \text{ for } t\in K \text{ and } |z(t)-y(t)|<\varepsilon+\eta, \text{ for } t\in T\,.$$

Thus the lemma is proved.

28.10 PROPOSITION. *Let $K\subset T$ be a compact set and $\varepsilon>0$. Every vector field $\mathbf{x}$ defined and continuous on K can be extended to a vector field $\mathbf{y}$ defined and continuous on T such that*

$$\|y\|\leqslant\|x\|_K+\varepsilon\,.$$

From the second part of the proof of proposition 28.8, it follows that x can be approximated uniformly on K by vector fields on the form

$$y=\Sigma f_i x_i\,,$$

with $x_i\in\mathscr{A}$ and f_i scalar functions defined and continuous on T. There exists then a vector field y_1 continuous on T such that we have

$$|\boldsymbol{y}_1(t)-\boldsymbol{x}(t)|<\tfrac{1}{2} \text{ for } t\in K .$$

By lemma 28.9, for $\varepsilon=2^{-2}$, there exists a vector field $\boldsymbol{y}_2$, continuous on T such that

$$|\boldsymbol{y}_2(t)-\boldsymbol{x}(t)|<2^{-2}, \text{ for } t\in K$$

and

$$|\boldsymbol{y}_2(t)-\boldsymbol{y}_1(t)|<2^{-1}, \text{ for } t\in T .$$

By recurrence, we deduce that there exists a sequence $(\boldsymbol{y}_n)$ of continuous fields on T, satisfying the inequalities

$$|\boldsymbol{y}_n(t)-\boldsymbol{x}(t)|<2^{-n}, \text{ for } t\in K ,$$
$$|\boldsymbol{y}_n(t)-\boldsymbol{y}_{n-1}(t)|<2^{-n+2}, \text{ for } t\in T .$$

The sequence $(\boldsymbol{y}_n)$ converges uniformly on T to a vector field $\boldsymbol{y}'$ which is continuous on T. Passing to the limit in the first of the preceding inequalities, we deduce that $\boldsymbol{y}'(t)=\boldsymbol{x}(t)$ for $t\in K$, therefore $\boldsymbol{y}'$ is an extension of $\boldsymbol{x}$.

For every point $t_0\in K$ we have

$$|\boldsymbol{y}'(t_0)\leqslant \|\boldsymbol{y}'\|_K=\|\boldsymbol{y}\|_K<\|\boldsymbol{x}\|_K+\varepsilon .$$

Since $|\boldsymbol{y}'|$ is continuous, the inequality $|\boldsymbol{y}'(t)|<\|\boldsymbol{x}\|_K+\varepsilon$ remains valid on an open neighbourhood V of t_0.

Since K is compact, we can find a cover $(V_i)_{1\leqslant i\leqslant n}$ of K consisting of open sets, such that, denoting $U=\cup V_i$, we have

$$|\boldsymbol{y}'(t)<\|\boldsymbol{x}\|_K+\varepsilon, \text{ for } t\in U .$$

Taking now a continuous function $f: T\to[0, 1]$, equal to 1 on K and vanishing on $T\setminus U$, the field $\boldsymbol{y}=f\boldsymbol{y}'$ is continuous on T, $\boldsymbol{y}(t)=\boldsymbol{x}(t)$ for $t\in K$ and

$$|\boldsymbol{y}(t)|\leqslant\|\boldsymbol{x}\|_K+\varepsilon, \text{ for every } t\in T .$$

28.11 Corollary. *Let $(t_i)_{1\leqslant i\leqslant n}$ be a finite family of points of T and for each i, let $a_i\in E(t_i)$. There exists a vector field $\boldsymbol{x}$ continuous on T such that*

$$\boldsymbol{x}(t_i)=a_i, \text{ for } i=1,2,\ldots,n .$$

In fact, the set K consisting of the points t_i is compact and the vector field $\boldsymbol{y}$

defined by $\boldsymbol{y}(t)=a_i$ for $t=t_i$, $i=1,2,\ldots,n$, and $\boldsymbol{y}(t)=0$ for $t\notin K$, is continuous on K.

We shall denote by $\mathscr{K}_{\mathscr{A}}(T)$ the set of vector fields continuous on T (with respect to the fundamental family $\mathscr{A}$) with compact support.

For every set $A\subset T$, $\mathscr{K}_{\mathscr{A}}(T, A)$ is the subspace of $\mathscr{K}_{\mathscr{A}}(T)$ consisting of the fields with support contained in A.

The sets $\mathscr{K}_{\mathscr{A}}(T)$ and $\mathscr{K}_{\mathscr{A}}(T, A)$ are vector spaces and modules over the space of continuous scalar functions on T.

In case $E(t)=E$ for every $t\in T$, we have

$$\mathscr{K}_{\mathscr{A}}(T)=\mathscr{K}_E(T) \text{ and } \mathscr{K}_{\mathscr{A}}(T, A)=\mathscr{K}_E(T, A).$$

28.12 Definition. *We say that a subspace $\mathscr{V}\subset\mathscr{K}_{\mathscr{A}}(T)$ is rich if for every compact set $K\subset T$ there exists a relatively compact neighbourhood $U\supset K$ such that every field $\boldsymbol{x}\in\mathscr{K}_{\mathscr{A}}(T, K)$ can be approximated uniformly by fields of $\mathscr{V}$ with support contained in U.*

28.13 Proposition. *The space $\mathscr{H}_{\mathscr{A}}(T)$ of linear combinations of the form*

$$\Sigma_{1\leqslant i\leqslant n} f_i\boldsymbol{x}_i \text{ with } \boldsymbol{x}_i\in\mathscr{A} \text{ and } f_i\in\mathscr{K}(T)$$

is rich.

Let K be a compact set and U a relatively compact neighbourhood of K. Let $\boldsymbol{x}\in\mathscr{K}_{\mathscr{A}}(T, K)$. By proposition 28.8 for each $\varepsilon>0$, there exists a field of the form $\boldsymbol{y}_\varepsilon=\Sigma g_i\boldsymbol{x}_i$ with $\boldsymbol{x}_i\in\mathscr{A}$ and g_i continuous scalar functions such that

$$|\boldsymbol{x}(t)-\boldsymbol{y}_\varepsilon(t)|<\varepsilon \text{ for } t\in\overline{U}.$$

Let $\varphi: T\to[0, 1]$ be a continuous function, equal to 1 on K and vanishing on $T\setminus U$. Each function $f_i=g_i\varphi$ is continuous, with support in U and for the field $\boldsymbol{x}_\varepsilon=\Sigma f_i\boldsymbol{x}_i$ we have

$$|\boldsymbol{x}(t)-\boldsymbol{x}_\varepsilon(t)|<\varepsilon, \text{ for } t\in T.$$

Remark. Let $\mathscr{A}_0$ be a set of continuous vector fields having only the property that for each $t\in T$ the set $\{\boldsymbol{x}(t);\ \boldsymbol{x}\in\mathscr{A}_0\}$ is dense in $E(t)$. Taking into account remark 2 following proposition 28.8, we deduce that the space of the linear combinations of the form $\Sigma f_i\boldsymbol{x}_i$, with $\boldsymbol{x}_i\in\mathscr{A}_0$ and $f_i\in\mathscr{K}_+(T)$ is rich.

4 p-integrable vector fields

Let μ be a positive measure on T.

For each vector field $\boldsymbol{x} \in \mathscr{C}(\mathscr{E})$ denote

$$N_p(\boldsymbol{x}, \mu) = (\int^* |\boldsymbol{x}(t)|^p \mathrm{d}\mu)^{1/p}, \qquad (1 \leqslant p < +\infty).$$

Instead of $N_p(\boldsymbol{x}, \mu)$ we shall write $N_p(\boldsymbol{x})$ if there is no confusion concerning the measure μ. The following two properties are deduced, as in the case of the usual functions:

(1) $N_p(\Sigma_{1 \leqslant n < \infty} \boldsymbol{x}_n) \leqslant \Sigma_{1 \leqslant n < \infty} N_p(\boldsymbol{x}_n)$, (theorem 7.6).

(2) *If* $N_p(\boldsymbol{x}) < +\infty$, *then* $\boldsymbol{x}(t) = 0$, *μ-almost everywhere outside the union of a sequence of compact sets* (corollary 11.7).

For each number p which satisfies the inequalities $1 \leqslant p < +\infty$, we denote by $\mathscr{F}^p_{\mathscr{E}}(T, \mu)$ or $\mathscr{F}^p_{\mathscr{E}}(\mu)$ or $\mathscr{F}^p_{\mathscr{E}}$, the set of vector fields $\boldsymbol{x} \in \mathscr{C}(\mathscr{E})$ with $N_p(\boldsymbol{x}) < +\infty$.

In case $E(t) = E$ for each $t \in T$, we have $\mathscr{F}^p_{\mathscr{E}} = \mathscr{F}^p_{\mathrm{E}}$ (see section 3 of §7).

$\mathscr{F}^p_{\mathscr{E}}$ is a vector space and the mapping $\boldsymbol{x} \to N_p(\boldsymbol{x})$ is a seminorm which defines on $\mathscr{F}^p_{\mathscr{E}}$ the topology of the convergence in the mean of order p. The adherence of the zero field, in this topology, is the subspace $\mathscr{N}_{\mathscr{E}}$ of the μ-negligible fields (vanishing μ-almost everywhere).

The quotient space $\mathscr{F}^p_{\mathscr{E}} / \mathscr{N}_{\mathscr{E}}$ is denoted by $F^p_{\mathscr{E}}$ and the mapping $\tilde{\boldsymbol{x}} \to \|\tilde{\boldsymbol{x}}\|_p = N_p(\boldsymbol{x})$ is a norm on this space.

All the properties of the spaces $\mathscr{F}^p_{\mathrm{E}}$ remain valid, with the same proofs, for the spaces $\mathscr{F}^p_{\mathscr{E}}$:

(1) *If* $\Sigma_{1 \leqslant n < \infty} N_p(\boldsymbol{x}_n) < +\infty$, *then the series* $\Sigma_{1 \leqslant n < \infty} \boldsymbol{x}_n(t)$ *is absolutely convergent almost everywhere. Every field* $\boldsymbol{x} \in \mathscr{C}(\mathscr{E})$ *equal almost everywhere to the sum of the series belongs to the space* $\mathscr{F}^p_{\mathscr{E}}$ *and* $N_p(\boldsymbol{x} - \Sigma_{k \leqslant n} \boldsymbol{x}_k) \leqslant \Sigma_{n > k} N_p(\boldsymbol{x}_k)$ *(proposition 7.9).*

(2) *From every Cauchy sequence* $(\boldsymbol{x}_n)$ *of* $\mathscr{F}^p_{\mathscr{E}}$ *we can extract a subsequence* $(\boldsymbol{x}_{n_k})$ *converging almost everywhere and in the mean of order p to a field* $\boldsymbol{x} \in \mathscr{F}^p_{\mathscr{E}}$, *and there exist a function* $g \geqslant 0$ *with* $N_p(g) < +\infty$ *and* $|\boldsymbol{x}_{n_k}(t)| \leqslant g(t)$, *for every* $t \in T$ *and* $k \in N$ *(proposition 7.12).*

(3) *The space* $\mathscr{F}^p_{\mathscr{E}}$ *is complete (theorem 7.11).*

If $\boldsymbol{x} \in \mathscr{K}_{\mathscr{A}}(T)$, then $|\boldsymbol{x}| \in \mathscr{K}(T)$, therefore $N_p(\boldsymbol{x}) < +\infty$. It follows that the space $\mathscr{K}_{\mathscr{A}}(T)$ is contained in each space $\mathscr{F}^p_{\mathscr{E}}$.

28.14 DEFINITION. *For each p, $1 \leqslant p < +\infty$, denote by $\mathscr{L}^p_{\mathscr{A}}(T, \mu)$ or $\mathscr{L}^p_{\mathscr{A}}(\mu)$ or $\mathscr{L}^p_{\mathscr{A}}$, the adherence of the space $\mathscr{K}_{\mathscr{A}}(T)$ in $\mathscr{F}^p_{\mathscr{E}}$. The vector fields of $\mathscr{L}^p_{\mathscr{A}}$ are called vector fields of p-integrable power, or p-integrable fields.*

In case $E(t) = E$ for every $t \in T$, we have $\mathscr{L}^p_{\mathscr{A}} = \mathscr{L}^p_{\mathrm{E}}$ (see section 4 of §7).

By definition 28.14 it follows first that $\mathscr{L}^p_{\mathscr{A}}$ is complete for the topology defined by the seminorm N_p, and that $\mathscr{K}_{\mathscr{A}}(T)$ is dense in $\mathscr{L}^p_{\mathscr{A}}$ for this topology.

All the properties of the spaces $\mathscr{L}^p_{\mathscr{A}}$ of usual functions are preserved, with the same proofs, for the spaces $\mathscr{L}^p_{\mathscr{A}}$:

(1) *If $(\boldsymbol{x}_n)$ is a Cauchy sequence of fields of $\mathscr{L}^p_{\mathscr{A}}$ converging almost everywhere to a field $\boldsymbol{x} \in \mathscr{C}(\mathscr{E})$, then $\boldsymbol{x} \in \mathscr{L}^p_{\mathscr{A}}$ and $(\boldsymbol{x}_n)$ converges to $\boldsymbol{x}$ also in the mean of order p (corollary 7.19).*

(2) *If $\mathscr{L}$ is a dense set in $\mathscr{L}^p_{\mathscr{A}}$, then for every field $\boldsymbol{x} \in \mathscr{L}^p_{\mathscr{A}}$ there exists a sequence $(\boldsymbol{x}_n)$ of fields of $\mathscr{L}$ converging to $\boldsymbol{x}$ in the mean of order p and almost everywhere, and a function $g \geqslant 0$ with $N_p(g) < +\infty$ and $|\boldsymbol{x}_n(t)| \leqslant g(t)$, for every $t \in T$ and every $n \in N$ (proposition 7.20).*

In particular we can take $\mathscr{L} = \mathscr{K}_{\mathscr{A}}(T)$.

(3) *If $\boldsymbol{x} \in \mathscr{L}^p_{\mathscr{A}}$ then $|\boldsymbol{x}| \in \mathscr{L}^p$ (proposition 7.26).*

(4) *Lebesgue's theorem: Let $(\boldsymbol{x}_n)$ be a sequence of vector fields of $\mathscr{L}^p_{\mathscr{A}}$. If $(\boldsymbol{x}_n)$ converges almost everywhere to a field $\boldsymbol{x} \in \mathscr{C}(\mathscr{E})$ and if there exists a function $f \geqslant 0$ with $N_p(f) < +\infty$ and $|\boldsymbol{x}_n(t)| \leqslant f(t)$ almost everywhere, for each n, then $\boldsymbol{x} \in \mathscr{L}^p_{\mathscr{A}}$ and $\lim N_p(\boldsymbol{x}_n - \boldsymbol{x}) = 0$.*

(5) *$\boldsymbol{x} \in \mathscr{L}^p_{\mathscr{A}}$ if and only if $|\boldsymbol{x}|^{p-1}\boldsymbol{x} \in \mathscr{L}^1_{\mathscr{A}}$ (corollary 7.44)*

5 *Measurable vector fields*

Let μ be a positive measure on T.

28.15 DEFINITION. *We say that a vector field $\boldsymbol{x} \in \mathscr{C}(\mathscr{E})$ is measurable (with respect to $\mathscr{A}$ and μ) if for every compact set $K \subset T$ and every $\varepsilon > 0$, there exists a compact set $K' \subset K$ such that $\mu(K \setminus K') < \varepsilon$, and the restriction of $\boldsymbol{x}$ to K' is continuous.*

In case $E(t) = E$ for every $t \in T$, definition 28.15 reduces to the definition of the usual measurable functions (see §9).

The set $\mathscr{M}_{\mathscr{A}}(T, \mu)$ of the fields $\boldsymbol{x} \in \mathscr{C}(\mathscr{E})$ measurable with respect to $\mathscr{A}$

and μ is a vector space and a module over the algebra of μ-measurable scalar functions defined on T.

As for the usual measurable functions, one can prove the following properties:

(1) *A field* $\boldsymbol{x} \in \mathscr{C}(\mathscr{E})$ *is measurable if and only if for every compact set* $K \subset T$ *there exists a partition of* K *consisting of a negligible set* N *and a sequence* (K_n) *of compact sets, such that the restriction of* $\boldsymbol{x}$ *to each* K_n *is continuous* (proposition 9.5).

In this property as well as in definition 28.15, the compact sets can be replaced by integrable sets.

(2) *If* $\boldsymbol{x}$ *is a measurable field, then* $|\boldsymbol{x}|$ *is a measurable function* (*see proposition* 28.3).

(3) *If* $\boldsymbol{x}$ *has the property that for every compact set* $K \subset T$, *the field* $\varphi_K \boldsymbol{x}$ *is measurable, then* $\boldsymbol{x}$ *is measurable* (*corollary* 9.41).

(4) *Egoroff's theorem. If* $(\boldsymbol{x}_n)$ *is a sequence of measurable fields and if the* $\lim \boldsymbol{x}_n(t) = \boldsymbol{x}(t)$ *exists almost everywhere, then for every compact set* $K \subset T$ *and every* $\varepsilon > 0$ *there exists a compact set* $K_1 \subset K$ *with* $\mu(K \setminus K_1) < \varepsilon$ *such that the restrictions of the fields* $\boldsymbol{x}_n$ *to* K_1 *are continuous and uniformly convergent to* $\boldsymbol{x}$ (*theorem* 10.1).

In order to prove this last result, we remark that since the fields $\boldsymbol{x}_n$ are measurable, we can find a compact set $K_0 \subset K$ with $\mu(K \setminus K_0) < \varepsilon/2$, such that the restrictions of all the fields $\boldsymbol{x}_n$ to K_0 are continuous. The proof goes on as in the Egoroff theorem for the usual functions, replacing the distance $d(f_p(t), f_q(t))$ by the norm $|\boldsymbol{x}_p(t) - \boldsymbol{x}_q(t)|$.

(5) *If* $(\boldsymbol{x}_n)$ *is a sequence of measurable fields converging almost everywhere to a field* $\boldsymbol{x} \in \mathscr{C}(\mathscr{E})$, *then* $\boldsymbol{x}$ *is measurable* (*corollary* 10.2).

(6) *Integrability criterion. A vector field* $\boldsymbol{x} \in \mathscr{C}(\mathscr{E})$ *belongs to the space* $\mathscr{L}^p_{\mathscr{A}}$, $1 \leqslant p < +\infty$, *if and only if* $\boldsymbol{x}$ *is measurable and* $N_p(\boldsymbol{x}) < +\infty$ (*theorem* 11.10).

(7) *If* $\boldsymbol{x} \in \mathscr{L}^p_{\mathscr{A}}$ *and* A *is a measurable set, then* $\varphi_A \boldsymbol{x} \in \mathscr{L}^p_{\mathscr{A}}$, (*corollary* 11.14).

(8) *If* $\boldsymbol{x}$ *is a continuous vector field on* T *and* A *is a relatively compact Borel set, then* $\varphi_A \boldsymbol{x} \in \mathscr{L}^p_{\mathscr{A}}$.

In fact, $\|\boldsymbol{x}\|_A = \sup_{t \in A} |\boldsymbol{x}(t)| < +\infty$, therefore

$$N_p(\varphi_A x) = (\int \varphi_A |x|^p \mathrm{d}\mu)^{1/p} \leqslant \|x\|_A \mu(A)^{1/p} < +\infty$$

and $\varphi_A x$ is measurable, hence $\varphi_A x \in \mathscr{L}^p_{\mathscr{A}}$.

Remarks. We can take A a relatively compact measurable set or an integrable set on which x is bounded. It follows that every step vector field of $\mathscr{E}_{\mathscr{A}}(\mathscr{B})$ of the form

$$x = \Sigma \varphi_{A_i} x_i, \quad A_i \in \mathscr{B}, x_i \in \mathscr{A}, \quad 1 \leqslant i \leqslant n,$$

belongs to each space $\mathscr{L}^p_{\mathscr{A}}$. Moreover:

(9) *The space $\mathscr{E}_{\mathscr{A}}(\mathscr{B})$ of step vector fields is dense in each space $\mathscr{L}^p_{\mathscr{A}}$, $1 \leqslant p < +\infty$.*

This property follows from the fact that every field x of $\mathscr{K}_{\mathscr{A}}(T)$ can be approximated uniformly by step vector fields of $\mathscr{E}_{\mathscr{A}}(\mathscr{B})$.

In fact, let $x \in \mathscr{K}_{\mathscr{A}}(T)$, and let K be the support of x and $\varepsilon > 0$. Let $t_0 \in K$. Since x is continuous at t_0, there exists a field $x_0 \in \mathscr{A}$ and a neighbourhood V_0 of t_0 such that we have $|x(t) - x_0(t)| < \varepsilon$ for $t \in V_0$.

Since K is compact, there exists a finite family $(V_i)_{1 \leqslant i \leqslant n}$ of relatively compact open sets covering K, and a finite family $(x_i)_{1 \leqslant i \leqslant n}$ of fields of $\mathscr{A}$ such that for each i we have

$$|x(t) - x_i(t)| < \varepsilon \text{ for } t \in V_i .$$

There exists a finite family $(A_i)_{1 \leqslant i \leqslant n}$ of Borel sets of $\mathscr{B}$ with $A_i \subset V_i$ for each i and $\cup_{1 \leqslant i \leqslant n} = \cup_{1 \leqslant i \leqslant n} V_i$. Then $\Sigma \varphi_{A_i} x_i$ is a step field and

$$|x(t) - \Sigma \varphi_{A_i}(t) x_i(t)| < \varepsilon \text{ for every } t \in T .$$

Thus $\mathscr{E}_{\mathscr{A}}(\mathscr{B})$ is dense in $\mathscr{K}_{\mathscr{A}}(T)$ for the topology of the uniform convergence, therefore also for the topology of the convergence in the mean of order p, which is coarser. As $\mathscr{K}_{\mathscr{A}}(T)$ is dense in $\mathscr{L}^p_{\mathscr{A}}$ for this topology it follows that $\mathscr{E}_{\mathscr{A}}(\mathscr{B})$ is dense in $\mathscr{L}^p_{\mathscr{A}}$.

(10) *There are measurable fields $x \in \mathscr{C}(\mathscr{E})$ different from* 0 *at every point of T.*

Let $K \subset T$ be a compact set and $t_0 \in K$. Since the set $\{x(t_0) | x \in \mathscr{A}\}$ is dense in $E(t_0)$, there exists a field $x_0 \in \mathscr{A}$ with $x_0(t_0) \neq 0$.

Since x_0 is continuous, there exists an open neighbourhood V_0 of t_0 such that $x_0(t) \neq 0$ for $t \in V$. Since K is compact, there exists a finite family $(V_i)_{1 \leqslant i \leqslant n}$ of open sets which cover K and a finite family $(x_i)_{1 \leqslant i \leqslant n}$ of fields of $\mathscr{A}$, such that for each i we have

$x_i(t) \neq 0$ for $t \in V_i$.

Let $(A_i)_{1 \leqslant i \leqslant n}$ be a finite family of disjoint Borel sets such that $\cup A_i = K$ and $A_i \subset V_i$ for each i. Then the field $x = \Sigma \varphi_{Ai} x_i$ is measurable and $x(t) \neq 0$ for $t \in K$.

Let $\mathscr{K}$ be a class of mutually disjoint compact sets such that the set $N = T \setminus \cup_{K \in \mathscr{K}} K$ is negligible and such that every point $t \in T$ possesses a neighbourhood which intersects at most countable many sets of $\mathscr{K}$ (proposition 11.22).

For each set $K \in \mathscr{K}$ there exists a measurable field $x_K \in \mathscr{C}(\mathscr{E})$ with $x_K(t) \neq 0$ for $t \in T$. Define the field x on T by the equality

$$x(t) = \Sigma_{K \in \mathscr{K}} \varphi_K(t) x_K(t), \quad t \in T.$$

The field x vanishes only on the negligible set N. Let us show that x is measurable. Let $K \subset T$ be an arbitrary compact set. For each point $t \in K$ choose a neighbourhood V_t of t which intersects at most countably many sets of $\mathscr{K}$. As K can be covered by a finite family of such neighbourhoods, it follows that the set K intersects only a sequence (K_n) of sets of $\mathscr{K}$.

It follows that the restriction of x to K is equal almost everywhere to the restriction to K of the measurable field $\Sigma_{1 \leqslant n < \infty} \varphi_{K_n} x_n$, therefore x is measurable. Modifying now x on the negligible set N, the field x remains measurable and we can manage to be different from 0 at every point of T.

(11) *For every measurable scalar function f, there exists a measurable vector field $x \in \mathscr{C}(\mathscr{E})$ such that we have $|x(t)| = |f(t)|$ for every $t \in T$.*

In fact, there exists a measurable field y not vanishing on T. Then the field $z(t) = y(t)/|y(t)|$, for $t \in T$, is measurable and $|z(t)| = 1$ for $t \in T$. The field $x = fz$ is measurable and for every $t \in T$ we have $|x(t)| = |f(t)|$.

Let us denote by $\mathscr{E}'$ the family $(E'(t))_{t \in T}$ of the duals of the spaces $E(t)$ and by $\mathscr{C}(\mathscr{E}')$ the set of functional fields x' defined on T such that $x'(t) \in E'(t)$ for every $t \in T$.

Assume that there exists a fundamental family $\mathscr{A}' \subset \mathscr{C}(\mathscr{E}')$ of continuous functional fields such that the scalar function $t \to \langle x(t), x'(t) \rangle$ is continuous for every $x \in \mathscr{A}$ and $x' \in \mathscr{A}'$.

28.16 PROPOSITION. *Let $x \in \mathscr{C}(\mathscr{E})$ be a vector field measurable with respect to $\mathscr{A}$ and μ and $0 < a < 1$. There exists a functional field $x' \in \mathscr{A}'$ measurable with respect to $\mathscr{A}'$ and μ such that*

$$a|\boldsymbol{x}(t)| < |\langle \boldsymbol{x}(t), \boldsymbol{x}'(t) \rangle| \text{ and } a < |\boldsymbol{x}'(t)| < 2-a$$

μ-almost everywhere.

Let $K \subset T$ be a compact set. Assume first that the restriction of $\boldsymbol{x}$ to K is continuous and let $\varepsilon > 0$ be such that $a+\varepsilon < 1$. Let $t_0 \in K$. There exists an element $x'_0 \in E'(t_0)$ with $|x'_0| = 1$ and

$$|\langle \boldsymbol{x}(t_0), x'_0 \rangle| > |\boldsymbol{x}(t_0)|(a+\varepsilon).$$

There exists then a functional field $\boldsymbol{x}' \in \mathscr{A}'$ such that

$$|x'_0 - \boldsymbol{x}'(t_0)| < \varepsilon.$$

Then

$$\begin{aligned}|\langle \boldsymbol{x}(t_0), \boldsymbol{x}'(t_0) \rangle| &= |\langle \boldsymbol{x}(t_0), \boldsymbol{x}'(t_0) - x'_0 \rangle + \langle \boldsymbol{x}(t_0), x'_0 \rangle| \geqslant \\ &\geqslant |\langle \boldsymbol{x}(t_0), x'_0 \rangle| - |\langle \boldsymbol{x}(t_0), \boldsymbol{x}'(t_0) - x'_0 \rangle| \geqslant \\ &\geqslant |\boldsymbol{x}(t_0)|(a+\varepsilon) - |\boldsymbol{x}(t_0)|\,|\boldsymbol{x}'(t_0) - x'_0| \geqslant |\boldsymbol{x}(t_0)|(a+\varepsilon) - \\ &\quad - |\boldsymbol{x}(t_0)|\varepsilon = a|\boldsymbol{x}(t_0)|\end{aligned}$$

and

$$a < 1-\varepsilon = |x'_0| - \varepsilon < |\boldsymbol{x}'(t_0)| < |x'_0| + \varepsilon = 1+\varepsilon < 2-a.$$

Since the restrictions to K of the functions in this inequality are continuous, there exists a neighbourhood V_0 of t_0 such that for $t \in V_0 \cap K$ we have

$$|\langle \boldsymbol{x}(t), \boldsymbol{x}'(t) \rangle| > a|\boldsymbol{x}(t)| \text{ and } a < |\boldsymbol{x}'(t)| < 2-a.$$

Since K is compact, there exists a finite family $V_1, \ldots, V_n$ of open sets covering K and a finite family of functional fields $\boldsymbol{x}'_1, \ldots, \boldsymbol{x}'_n$ of $\mathscr{A}'$ such that for each i we have

$$|\langle \boldsymbol{x}(t), \boldsymbol{x}'_i(t) \rangle| > a|\boldsymbol{x}(t)| \text{ and } a < |\boldsymbol{x}'_i(t)| < 2-a, \ t \in V_i \cap K.$$

Let $A_1, \ldots, A_n$ be disjoint measurable sets such that $A_i \subset V_i$ for each i and $K = \cup_{1 \leqslant i \leqslant n} A_i$. The functional field $\boldsymbol{x}'_K = \Sigma_{1 \leqslant i \leqslant n} \varphi_{A_i} \boldsymbol{x}'_i$ is measurable with respect to $\mathscr{A}'$ and μ and for every $t \in K$ we have

$$|\langle \boldsymbol{x}(t), \boldsymbol{x}'_K(t) \rangle| > a|\boldsymbol{x}(t)| \text{ and } a|\boldsymbol{x}'_K(t)| < 2-a.$$

For $t \notin K$ we have $\boldsymbol{x}'_K(t) = 0$.

Assume now that $\boldsymbol{x}$ is measurable. There exists a negligible set $A \subset K$ and a partition (K_n) of $K \setminus A$ consisting of a sequence of compact sets such that the restriction of $\boldsymbol{x}$ to each K_n is continuous.

For each n there exists a functional field $\boldsymbol{x}'_n \in \mathscr{C}(\mathscr{E}')$ measurable with respect to $\mathscr{A}'$ and μ, such that for $t \in K_n$ we have

$$a|\boldsymbol{x}(t)| < |\langle \boldsymbol{x}(t), \boldsymbol{x}'_n(t)\rangle| \text{ and } a < |\boldsymbol{x}'_n(t)| < 2-a\,.$$

Then the functional field $\boldsymbol{x}'_K = \Sigma_{1 \leqslant n < \infty} \boldsymbol{x}'_n$ is measurable and for $t \in K \setminus A$ we have

$$a|\boldsymbol{x}(t)| < |\langle \boldsymbol{x}(t), \boldsymbol{x}'_K(t)\rangle| \text{ and } a < |\boldsymbol{x}'_K(t)| < 2-a\,.$$

Now let $\mathscr{K}$ be a family of mutually disjoint nonempty compact sets (proposition 11.22) whose union differs from T by a negligible set.

The field $\boldsymbol{x}'(t) = \Sigma_{K \in \mathscr{K}} \boldsymbol{x}'_K(t)$ is measurable with respect to $\mathscr{A}'$ and μ and fulfills the required conditions.

In the same way we can prove the following proposition:

28.17 PROPOSITION. *Let $\boldsymbol{x}' \in \mathscr{C}(\mathscr{E}')$ be a functional field, measurable with respect to $\mathscr{A}'$ and μ and $0 < a < 1$. There exists a vector field $\boldsymbol{x} \in \mathscr{C}(\mathscr{E})$, measurable with respect to $\mathscr{A}$ and μ such that*

$$a|\boldsymbol{x}'(t)| < |\langle \boldsymbol{x}(t), \boldsymbol{x}'(t)\rangle| \text{ and } a < |\boldsymbol{x}(t)| < 2-a\,,$$

μ-almost everywhere.

6 The space $\mathscr{L}^\infty_{\mathscr{A}}$

Let μ be a positive measure on T.

For each field $\boldsymbol{x} \in \mathscr{C}(\mathscr{E})$ we shall denote by $N_\infty(\boldsymbol{x}, \mu)$ or $N_\infty(\boldsymbol{x})$, the (finite or $+\infty$) positive number, defined by the equality

$$N_\infty(\boldsymbol{x}) = N_\infty(|\boldsymbol{x}|)\,.$$

We shall denote by $\mathscr{F}^\infty_{\mathscr{E}}(\mu)$ or $\mathscr{F}^\infty_{\mathscr{E}}$, the set of vector fields, $\boldsymbol{x} \in \mathscr{C}(\mathscr{E})$ with $N_\infty(\boldsymbol{x}) < +\infty$. The set $\mathscr{F}^\infty_{\mathscr{E}}$ is a vector space and a module over the algebra $\mathscr{F}^\infty$ of scalar functions.

This space is complete for the topology of the uniform convergence almost everywhere, defined by the seminorm N_∞.

Every bounded field $\boldsymbol{x} \in \mathscr{C}(\mathscr{E})$ belongs to the space $\mathscr{F}^\infty_{\mathscr{E}}$. In particular, $\mathscr{K}_{\mathscr{A}}(T) \subset \mathscr{F}^\infty_{\mathscr{E}}$.

We shall denote by $\mathscr{L}^\infty_{\mathscr{A}}$ the set of measurable vector fields of $\mathscr{F}^\infty_{\mathscr{E}}$. The set $\mathscr{L}^\infty_{\mathscr{A}}$ is a vector space, complete for the topology defined by the seminorm

N_∞. The adherence of the origin in this topology is the subspace of the negligible fields.

Evidently, we have $\mathscr{K}_{\mathscr{A}}(T) \subset \mathscr{L}^\infty_{\mathscr{A}}$ and also $\mathscr{E}_{\mathscr{A}}(\mathscr{B}) \subset \mathscr{L}^\infty_{\mathscr{A}}$. If $E(t) = E$ for every $t \in T$, we have $\mathscr{F}^\infty_{\mathscr{E}} = \mathscr{F}^\infty_E$ and $\mathscr{L}^\infty_{\mathscr{A}} = \mathscr{L}^\infty_E$.

7 *Measures defined on vector fields*

Let F be a Banach space.

28.18 DEFINITION. *We call an* $(\mathscr{A}, F)$ *vector measure on* T *any linear mapping* $\boldsymbol{m} : \mathscr{K}_{\mathscr{A}}(T) \to F$ *having the property that for every compact set* $K \subset T$, *the restriction of* $\boldsymbol{m}$ *to the subspace* $\mathscr{K}_{\mathscr{A}}(T, K)$ *is continuous for the topology of uniform convergence.*

To say that the linear mapping $\boldsymbol{m} : \mathscr{K}_{\mathscr{A}}(T) \to F$ is a measure means that for each compact set $K \subset T$ there exists a number $a_K > 0$ such that

$$|\boldsymbol{m}(\boldsymbol{x})| \leqslant a_K \|\boldsymbol{x}\| \quad \text{for every} \quad \boldsymbol{x} \in \mathscr{K}_{\mathscr{A}}(T, K).$$

The value $\boldsymbol{m}(\boldsymbol{x})$ of $\boldsymbol{m}$ at $\boldsymbol{x}$ is called the integral of $\boldsymbol{x}$ with respect to $\boldsymbol{m}$ and is denoted by $\int \boldsymbol{x} \, \mathrm{d}\boldsymbol{m}$ or $\int \boldsymbol{x}(t) \, \mathrm{d}\boldsymbol{m}(t)$.

If $E(t) = E$ for each $t \in T$, this is the definition of the usual measures $\boldsymbol{m} : \mathscr{K}_E(T) \to F$ (definition 1.9).

Denote $\|\boldsymbol{m}\| = \sup |\boldsymbol{m}(\boldsymbol{x})|$ for $\|\boldsymbol{x}\| \leqslant 1$ in $\mathscr{K}_{\mathscr{A}}(T)$. If $\boldsymbol{m} : \mathscr{K}_{\mathscr{A}}(T) \to F$ is a measure, then for each $z' \in F'$ the mapping $z' \circ \boldsymbol{m} : \mathscr{K}_{\mathscr{A}}(T) \to C$ defined by the equality

$$(z' \circ \boldsymbol{m})(\boldsymbol{x}) = \langle \boldsymbol{m}(\boldsymbol{x}), z' \rangle, \quad \text{for} \quad \boldsymbol{x} \in \mathscr{K}_{\mathscr{A}}(T)$$

is a measure.

28.19 DEFINITION. *We say that a measure* $\boldsymbol{m} : \mathscr{K}_{\mathscr{A}}(T) \to F$ *is dominated if there exists a positive measure* v *on* T *such that*

$$|\boldsymbol{m}(\boldsymbol{x})| \leqslant v(|\boldsymbol{x}|) \quad \textit{for every} \quad \boldsymbol{x} \in \mathscr{K}_{\mathscr{A}}(T).$$

One can prove as in the case of the usual measures that if $\boldsymbol{m} : \mathscr{K}_{\mathscr{A}}(T) \to F$ is a *dominated linear mapping*, then $\boldsymbol{m}$ is a dominated measure (proposition 3.8).

We shall show that for every dominated measure there exists a least positive measure which dominates it. To do that we shall prove first the following.

18.20 LEMMA. *Let $\boldsymbol{x}\in\mathscr{K}_{\mathscr{A}}(T)$ and $g\in\mathscr{K}_+(T)$ with $0\leqslant g\leqslant|\boldsymbol{x}|$ and define the vector field $\sigma_{\boldsymbol{x}}$* by $\sigma_{\boldsymbol{x}}(t)=\boldsymbol{x}(t)/|\boldsymbol{x}(t)|$ if $\boldsymbol{x}(t)\neq 0$ *and* $\sigma_{\boldsymbol{x}}(t)=0$ *if* $\boldsymbol{x}(t)=0$.

Then $\sigma_{\boldsymbol{x}}|\boldsymbol{x}|=\boldsymbol{x}$ and $g\sigma_{\boldsymbol{x}}\in\mathscr{K}_{\mathscr{A}}(T)$.

The equality $\sigma_{\boldsymbol{x}}|\boldsymbol{x}|=\boldsymbol{x}$ is evident. Let us show that $g\sigma_{\boldsymbol{x}}\in\mathscr{K}_{\mathscr{A}}(T)$. Let $t_0\in T$. If $\boldsymbol{x}(t_0)\neq 0$, there exists a neighbourhood V of t_0 such that $\boldsymbol{x}(t)\neq 0$ for $t\in V$. The function h defined on T by $h(t)=1/|\boldsymbol{x}(t)|$ for $t\in V$ and $h(t)=1$ for $t\notin V$ is continuous at t_0. Then $\sigma_{\boldsymbol{x}}(t)=h(t)\boldsymbol{x}(t)$, therefore $\sigma_{\boldsymbol{x}}$ is continuous at t_0. It follows that $g\sigma_{\boldsymbol{x}}$ is also continuous at t_0.

If $\boldsymbol{x}(t_0)=0$, then $g(t_0)=0$. Let $\varepsilon>0$; there exists a vector field $\boldsymbol{y}\in\mathscr{A}$ and a neighbourhood V of t_0 such that for $t\in V$ we have

$$|\boldsymbol{x}(t)-\boldsymbol{y}(t)|<\varepsilon/4,\quad g(t)<\varepsilon/2 \text{ and } |\boldsymbol{x}(t)|<\varepsilon/2\,.$$

Then, for $t\in V$ we have $|\boldsymbol{y}(t)|<\varepsilon/2$, therefore

$$|g(t)\sigma_{\boldsymbol{x}}(t)-\boldsymbol{y}(t)|\leqslant g(t)|\sigma_{\boldsymbol{x}}(t)|+|\boldsymbol{y}(t)|<\varepsilon/2+\varepsilon/2=\varepsilon$$

hence $g\sigma_{\boldsymbol{x}}$ is continuous at t_0.

Let $\boldsymbol{m}:\mathscr{K}_{\mathscr{A}}(T)\to C$ be a scalar valued measure. For every positive function $\varphi\in\mathscr{K}_+(T)$ put

$$\mu(\varphi)=\sup\nolimits_{|\boldsymbol{x}|\leqslant\varphi}|\boldsymbol{m}(\boldsymbol{x})|\,,\quad \boldsymbol{x}\in\mathscr{K}_{\mathscr{A}}(T)\,.$$

Using the preceding lemma, one can prove, as for the usual measures, that μ is *additive* and *positive* on $\mathscr{K}_+(T)$ (proposition 2.26), therefore it can be extended to $\mathscr{K}(T)$ as a positive measure denoted still by μ. This positive measure satisfies the inequality

$$|\boldsymbol{m}(\boldsymbol{x})|\leqslant\mu(|\boldsymbol{x}|) \text{ for } \boldsymbol{x}\in\mathscr{K}_{\mathscr{A}}(T)$$

and it is the least positive measure satisfying this inequality (proposition 2.27). The measure μ so defined is denoted by $|\boldsymbol{m}|$ and is called the modulus of $\boldsymbol{m}$.

Thus we have proved that every scalar valued measure $\boldsymbol{m}:\mathscr{K}_{\mathscr{A}}(T)\to C$ is dominated.

If $\boldsymbol{m}:\mathscr{K}_{\mathscr{A}}(T)\to F$ is a dominated measure, then there exists a least positive measure which dominates it, denoted by $|\boldsymbol{m}|$ and called, again, the modulus of $\boldsymbol{m}$.

The proof is done as for usual measures (proposition 3.10).

We say that a dominated measure $\boldsymbol{m}$ is *bounded* if its modulus $|\boldsymbol{m}|$ is a bounded measure.

8 *Integration of vector fields*

Let $\boldsymbol{m} : \mathscr{K}_{\mathscr{A}}(T) \to F$ be a dominated measure.

We say that a vector field $\boldsymbol{x} \in \mathscr{C}(\mathscr{E})$ is $\boldsymbol{m}$-integrable if $\boldsymbol{x}$ is $|\boldsymbol{m}|$-integrable.

The set of $\boldsymbol{m}$-integrable fields $\boldsymbol{x} \in \mathscr{C}(\mathscr{E})$ will be denoted by $\mathscr{L}^1_{\mathscr{A}}(\boldsymbol{m})$. We have therefore $\mathscr{L}^1_{\mathscr{A}}(\boldsymbol{m}) = \mathscr{L}^1_{\mathscr{A}}(|\boldsymbol{m}|)$.

We shall consider on the space $\mathscr{L}^1_{\mathscr{A}}(\boldsymbol{m})$ the topology of the convergence in the mean, defined by the seminorm

$$N_1(\boldsymbol{x}, \boldsymbol{m}) = N_1(\boldsymbol{x}, |\boldsymbol{m}|) = \int |\boldsymbol{x}(t)| \, \mathrm{d}|\boldsymbol{m}|(t) .$$

For every field $\boldsymbol{x}$ of $\mathscr{K}_{\mathscr{A}}(T)$ we have

$$|\int \boldsymbol{x} \, \mathrm{d}\boldsymbol{m}| \leqslant \int |\boldsymbol{x}| \, \mathrm{d}|\boldsymbol{m}| = N_1(\boldsymbol{x}) .$$

This inequality shows that the mapping $\boldsymbol{x} \to \int \boldsymbol{x} \, \mathrm{d}\boldsymbol{m}$ of $\mathscr{K}_{\mathscr{A}}(T)$ into F is continuous for the topology of the convergence in the mean, hence it can be extended uniquely to a continuous linear mapping $\boldsymbol{x} \to \int \boldsymbol{x} \, \mathrm{d}\boldsymbol{m}$ of $\mathscr{L}^1_{\mathscr{A}}(\boldsymbol{m})$ into F.

For each field $\boldsymbol{x} \in \mathscr{L}^1_{\mathscr{A}}(\boldsymbol{m})$, $\int \boldsymbol{x} \, \mathrm{d}\boldsymbol{m}$ is called the integral of $\boldsymbol{x}$ with respect to $\boldsymbol{m}$ and is also denoted by $\boldsymbol{m}(\boldsymbol{x})$. It also satisfies the inequality

$$|\int \boldsymbol{x} \, \mathrm{d}\boldsymbol{m}| \leqslant \int |\boldsymbol{x}| \, \mathrm{d}|\boldsymbol{m}| \quad \text{for} \quad \boldsymbol{x} \in \mathscr{L}^1_{\mathscr{A}}(\boldsymbol{m}) .$$

It follows that if a sequence $(\boldsymbol{x}_n)$ converges in the mean to $\boldsymbol{x}$ then

$$\lim \int \boldsymbol{x}_n \mathrm{d}\boldsymbol{m} = \int \boldsymbol{x} \, \mathrm{d}\boldsymbol{m} .$$

In fact,

$$|\int \boldsymbol{x}_n \mathrm{d}\boldsymbol{m} - \int \boldsymbol{x} \, \mathrm{d}\boldsymbol{m}| = |\int (\boldsymbol{x}_n - \boldsymbol{x}) \mathrm{d}\boldsymbol{m}| \leqslant \int |\boldsymbol{x}_n - \boldsymbol{x}| \, \mathrm{d}|\boldsymbol{m}| = N_1(\boldsymbol{x}_n - \boldsymbol{x}) .$$

28.21 Theorem. *Let $(\boldsymbol{x}_n)$ be a sequence of $\boldsymbol{m}$-integrable fields of $\mathscr{L}^1_{\mathscr{A}}(\boldsymbol{m})$. If the sequence $(\boldsymbol{x}_n)$ converges $\boldsymbol{m}$-almost everywhere to a field $\boldsymbol{x}_0 \in \mathscr{C}(\mathscr{E})$ and if there exists a function $\varphi \geqslant 0$ defined on T such that $\int^* \varphi \, \mathrm{d}|\boldsymbol{m}| < +\infty$ and $|\boldsymbol{x}_n(t)| \leqslant \varphi(t)$, $\boldsymbol{m}$-almost everywhere, for each $n \in N$, then the field $\boldsymbol{x}_0$ is $\boldsymbol{m}$-integrable and*

$$\lim \int \boldsymbol{x}_n \mathrm{d}\boldsymbol{m} = \int \boldsymbol{x} \, \mathrm{d}\boldsymbol{m} .$$

In fact, from the Lebesgue theorem it follows that the sequence $(\boldsymbol{x}_n)$ converges in the mean to $\boldsymbol{x}_0$.

28.22 COROLLARY. *Let $(\boldsymbol{x}_n)$ be a sequence of $\boldsymbol{m}$-integrable fields, of $\mathscr{L}^1_{\mathscr{A}}(\boldsymbol{m})$. If the series $\Sigma_{1 \leqslant n < \infty} \boldsymbol{x}_n(t)$ converges $\boldsymbol{m}$-almost everywhere to a field $\boldsymbol{x}_0 \in \mathscr{C}(\mathscr{E})$ and if there exists a function $\varphi \geqslant 0$ defined on T with $\int^* \varphi \mathrm{d}|\boldsymbol{m}| < +\infty$ and $|\Sigma_{k \leqslant n} \boldsymbol{x}_k(t)| \leqslant \varphi(t)$, μ-almost everywhere, for each $n \in N$, then the field $\boldsymbol{x}_0$ is $\boldsymbol{m}$-integrable and*

$$\int \boldsymbol{x}_0 \mathrm{d}\boldsymbol{m} = \Sigma_{1 \leqslant n < \infty} \int \boldsymbol{x}_n \mathrm{d}\boldsymbol{m} .$$

9 *Weakly measurable and weakly locally integrable operation fields*

For each $t \in T$ write $G(t) = \mathscr{L}(E(t)), F)$ and denote by $\mathscr{G}_F$ the family $(G(t))_{t \in T}$ and by $\mathscr{C}(\mathscr{G}_F)$ the set of operation fields U defined on T such that $U(t) \in G(t) = \mathscr{L}(E(t), F)$ for each $t \in T$. Let $Z \subset F'$ be a norming subspace for F, that is, such that

$$|\boldsymbol{y}| = \sup_{z \in Z} |\langle \boldsymbol{y}, z \rangle| / |z|, \text{ for every } y \in F .$$

If $F = C$, then $G(t) = E'(t)$ for each $t \in T$. In this case $\mathscr{E}' = (E'(t))_{t \in T}$ and we denoted by $\mathscr{C}(\mathscr{E}')$ the family of functional fields $\boldsymbol{x}'$ defined on T such that $\boldsymbol{x}'(t) \in E'(t)$ for each $t \in T$.

Let μ be a positive measure on T.

28.23 DEFINITION. *We say that an operation field $U \in \mathscr{C}(\mathscr{G}_F)$ is Z-weakly μ-measurable if for every field $\boldsymbol{x} \in \mathscr{A}$ and every $z \in Z$ the function $t \to \langle U(t)\boldsymbol{x}(t), z \rangle$ is μ-measurable.*

We say that U is simply μ-measurable, if for every $\boldsymbol{x} \in \mathscr{A}$, the function $t \to U(t)\boldsymbol{x}(t)$ with values in F is μ-measurable.

28.24 PROPOSITION. *An operation field $U \in \mathscr{C}(\mathscr{G}_F)$ is Z-weakly μ-measurable (respectively simply μ-measurable), if and only if for every field $\boldsymbol{x} \in \mathscr{C}(\mathscr{E})$, measurable with respect to $\mathscr{A}$ and μ and every $z \in Z$, the function $t \to \langle U(t)\boldsymbol{x}(t), Z \rangle$ is μ-measurable (respectively the function $t \to U(t)\boldsymbol{x}(t)$ is μ-measurable).*

Assume that U is Z-weakly μ-measurable. If $\boldsymbol{x}$ is a vector field, continuous with respect to $\mathscr{A}$, then $\boldsymbol{x}$ is a uniform limit, on every compact set $K \subset T$, of fields of the form $\Sigma \varphi_i \boldsymbol{x}_i$ with $\boldsymbol{x}_i \in \mathscr{A}$ and φ_i complex functions continuous on T. It follows that for each $z \in Z$, the function $t \to \langle U(t)\boldsymbol{x}(t), z \rangle$ is the uniform limit, on each compact set, of μ-measurable functions of the form

$\Sigma_i \varphi_i(t)\langle U(t)\boldsymbol{x}_i(t), z\rangle$ with $\varphi_i \in \mathscr{K}_C(T)$, $\boldsymbol{x}_i \in \mathscr{A}$ and $z \in Z$, therefore the function $t \to \langle U(t)\boldsymbol{x}(t), z\rangle$ is also measurable.

Let now $\boldsymbol{x} \in \mathscr{C}(\mathscr{E})$ be an arbitrary field, measurable with respect to $\mathscr{A}$ and μ. Let $z \in Z$ and let $K \subset T$ be compact. There exists a compact set $K' \subset K$ such that $\mu(K \setminus K') < \varepsilon/2$ and the restriction of $\boldsymbol{x}$ to K' is continuous. The restriction of $\boldsymbol{x}$ to K' can be extended to a vector field $\boldsymbol{y}$ continuous on the whole space T. It follows that the function

$$\langle U(t)\boldsymbol{x}(t), z\rangle \varphi_{K'}(t) = \langle U(t)\boldsymbol{y}(t), z\rangle \varphi_{K'}(t)$$

is μ-measurable. There exists then a compact set $K_1 \subset K'$ such that $\mu(K' \setminus K_1) < \varepsilon/2$, therefore $\mu(K \setminus K_1) < \varepsilon$, and the restriction of the function $\langle U(t)\boldsymbol{x}(t), z\rangle$ to K_1 is continuous. It follows that this function is μ-measurable.

The converse implication is evident, and thus the assertion with respect to the weak measurability is proved.

The assertion concerning the simple measurability is proved in the same way.

Remarks

1. Every simply measurable operation field is weakly measurable. Conversely, if F is of countable type, every weakly measurable operation field is simply measurable.

2. If U is a simply measurable operation field, then for every vector field $\boldsymbol{x}$, measurable with respect to $\mathscr{A}$ and μ, the function $t \to |U(t)\boldsymbol{x}(t)|$ is measurable.

In some conditions, this function is measurable even if U is weakly measurable.

Let U be a *Z-weakly measurable* operation field and $\boldsymbol{x}$ a vector field, measurable with respect to $\mathscr{A}$ and μ.

(1) *If there exists a countable set $S \subset Z$ norming for F, then the function $t \to |U(t)\boldsymbol{x}(t)|$ is measurable.*

The proof is the same as for usual functions (proposition 15.28). In particular:

(2) *If F is the dual of a Banach space S of countable type, then the function $t \to |U(t)\boldsymbol{x}(t)|$ is measurable.*

(3) *If the function $t \to |U(t)\boldsymbol{x}(t)|$ is measurable for every vector field $\boldsymbol{x} \in \mathscr{A}$*

and if the family $\mathscr{A}$ satisfies the countability axiom (G), *then the function* $t \to |U(t)|$ *is measurable.*

In fact, let $(\boldsymbol{x}_n)$ be a sequence of $\mathscr{A}$, such that for each $t \in T$, the sequence $(\boldsymbol{x}_n(t))$ is dense in $E(t)$.

We have

$$|U(t)| = \sup_n |U(t)\boldsymbol{x}_n(t)|/|\boldsymbol{x}_n(t)|$$

where we agree to put $|U(t)\boldsymbol{x}_n(t)|/|\boldsymbol{x}_n(t)| = 0$ if $\boldsymbol{x}_n(t) = 0$.

As each function $|U(t)\boldsymbol{x}_n(t)|/|\boldsymbol{x}_n(t)|$ is μ-measurable, if follows that the function $t \to |U(t)|$ is measurable.

In particular:

If the family $\mathscr{A}$ satisfies axiom (G) *and F is of countable type or the dual of a Banach space of a countable type, then for every weakly measurable operation field U, the function $t \to |U(t)|$ is measurable.*

(4) *Assume that there exists a fundamental family $\mathscr{D} \subset \mathscr{C}(\mathscr{G}_F)$ of continuous operation fields which satisfy the following axiom:*

(*T*) *The function $t \to V(t)\boldsymbol{x}(t)$ is μ-measurable for every $V \in \mathscr{D}$ and $\boldsymbol{x} \in \mathscr{A}$.*

If $\mathscr{A}$ and $\mathscr{D}$ satisfy axiom (G) *and if U is weakly μ-measurable, then U is measurable with respect to $\mathscr{D}$ and μ.*

In fact, let (V_n) be a sequence of elements of $\mathscr{D}$ such that for each $t \in T$, the sequence $(V_n(t))$ is dense in the space $G(t) = \mathscr{L}(E(t), F)$. We deduce that $G(t)$ is of countable type, therefore F is of countable type. Then for every field $\boldsymbol{x} \in \mathscr{A}$ the function $t \to U(t)\boldsymbol{x}(t)$ is μ-measurable. It follows that the function $t \to |V_n(t) - U(t)|$ is μ-measurable.

Let $\varepsilon > 0$ and $K \subset T$ be a compact set. For each n, the set

$$B_n = \{t \mid |V_n(t) - U(t)| \leqslant \varepsilon\} \cap K$$

is μ-measurable and $\cup B_n = K$. Choose a sequence (A_n) of disjoint μ-measurable sets, with $A_n \subset B_n$ for each n and $\cup A_n = K$. Put

$$U_\varepsilon = \Sigma \varphi_{A_n} V_n .$$

The operation field U_ε is measurable with respect to μ and $\mathscr{D}$ and for every $t \in K$ we have $|U_\varepsilon(t) - U(t)| \leqslant \varepsilon$.

It follows that U is the uniform limit, on every compact set, of operation fields measurable with respect to $\mathscr{D}$ and μ, therefore U is measurable with respect to $\mathscr{D}$ and μ.

28.25 DEFINITION. *We say that an operation field* $U \in \mathscr{C}(\mathscr{G}_F)$ *is Z-weakly locally* μ*-integrable (respectively simply locally* μ*-integrable), if U is Z-weakly* μ*-measurable (respectively simply* μ*-measurable) and if the function* $t \to |U(t)|$ *is locally* μ*-integrable.*

Every simply locally integrable operation field is weakly locally integrable.

28.26 PROPOSITION. *If U is a Z-weakly locally* μ*-integrable operation field, then for every field* $\boldsymbol{x} \in \mathscr{K}_{\mathscr{A}}(T)$ *and every* $z \in Z$*, the function* $\langle U\boldsymbol{x}, z\rangle$ *is* μ*-integrable.*

In fact, the function $\langle U\boldsymbol{x}, z\rangle$ is μ-measurable and

$$\int^* |\langle U(t)\boldsymbol{x}(t), z\rangle| \, d\mu(t) \leqslant |z| \int^* |U(t)| \, |\boldsymbol{x}(t)| \, d\mu(t) < \infty ,$$

since the function $|U|$ is locally μ-integrable and $|\boldsymbol{x}| \in \mathscr{K}(T)$.

28.27 PROPOSITION. *Let U and V be two fields such that for every* $\boldsymbol{x} \in \mathscr{K}_{\mathscr{A}}(T)$ *and* $z \in Z$*, the functions* $\langle U\boldsymbol{x}, z\rangle$ *and* $\langle V\boldsymbol{x}, z\rangle$ *are* μ*-integrable and*

$$\int \langle U\boldsymbol{x}, z\rangle \, d\mu = \int \langle V\boldsymbol{x}, z\rangle \, d\mu .$$

Then for every continuous field $\boldsymbol{x}$ *and every* $z \in Z$ *we have*

$$|\langle U(t)\boldsymbol{x}(t), z\rangle| = \langle V(t)\boldsymbol{x}(t), z\rangle, \ \mu \text{ almost everywhere.}$$

In each of the following cases we have

$U(t) = V(t)$, μ*-almost everywhere:*

(1) *There exists a lifting* ρ *of* $\mathscr{L}^\infty(\mu)$ *and a countable set* $\mathscr{B} \subset \mathscr{A}$ *such that the set* $\{\boldsymbol{x}(t) : \boldsymbol{x} \in \mathscr{B}\}$ *is dense in* $E(t)$ *for every* $t \in T$*, and such that*

$$\rho[U\boldsymbol{x}] = U\boldsymbol{x} \ \text{ and } \ \rho[V\boldsymbol{x}] = V\boldsymbol{x}, \text{ for every } \boldsymbol{x} \in \mathscr{B} .$$

(2) $\mathscr{A}$ *satisfies axiom (G) and there exists a countable set* $S \subset Z$ *with the property that if* $y \in F$ *and* $\langle y, s\rangle = 0$ *for every* $s \in S$*, then* $y = 0$.

(3) $\mathscr{A}$ *satisfies axiom (G) and* $U\boldsymbol{x}$ *and* $V\boldsymbol{x}$ *are* μ*-integrable for every* $\boldsymbol{x} \in \mathscr{K}_{\mathscr{A}}(T)$.

Let $\boldsymbol{x}$ be a continuous field and $z \in Z$. For every function $\varphi \in \mathscr{K}(T)$ we have $\boldsymbol{x}\varphi \in \mathscr{K}_{\mathscr{A}}(T)$ therefore

$$\int \langle U\boldsymbol{x}, z\rangle \varphi \, d\mu = \int \langle V\boldsymbol{x}, z\rangle \varphi \, d\mu ,$$

whence

$$\langle U(t)\boldsymbol{x}(t), z\rangle = \langle V(t)\boldsymbol{x}(t), z\rangle, \ \mu\text{-almost everywhere.}$$

We remark that in all three cases, $\mathscr{A}$ satisfies axiom (G). Let $\mathscr{B} \subset \mathscr{A}$ be a countable set such that the set $\{\boldsymbol{x}(t); \boldsymbol{x} \in \mathscr{B}\}$ is dense in $E(t)$ for every $t \in T$.

Denote $W = U - V$. For each continuous field $\boldsymbol{x}$ and each $z \in Z$ we have

$$\langle W(t)\boldsymbol{x}(t), z\rangle = 0, \ \mu\text{-almost everywhere}.$$

We shall show that in all three cases we have $W(t)\boldsymbol{x}(t) = 0$, μ-almost everywhere, for every $\boldsymbol{x} \in \mathscr{B}$.

In fact, in case 1 we deduce that $\rho[W\boldsymbol{x}] = W\boldsymbol{x}$ for each $\boldsymbol{x} \in \mathscr{B}$, therefore $W(t)\boldsymbol{x}(t) = 0$, μ-almost everywhere.

In case 2 we deduce that for every $\boldsymbol{x} \in \mathscr{B}$, and even for every continuous field $\boldsymbol{x}$ and for every $s \in S$ there exists a μ-negligible set $N(\boldsymbol{x}, s)$ such that for $t \notin N(\boldsymbol{x}, s)$ we have $\langle U(t)\boldsymbol{x}(t), s\rangle = 0$. The set $N(\boldsymbol{x}) = \cup_{s \in S} N(\boldsymbol{x}, s)$ is μ-negligible and for $t \notin N(\boldsymbol{x})$ we have $U(t)\boldsymbol{x}(t) = 0$.

If $U\boldsymbol{x}$ and $V\boldsymbol{x}$ are μ-integrable for every $\boldsymbol{x} \in \mathscr{K}_{\mathscr{A}}(T)$, then $W\boldsymbol{x}$ is μ-integrable for every $\boldsymbol{x} \in \mathscr{K}_{\mathscr{A}}(T)$ and we have

$$\langle \int W\boldsymbol{x}\,\mathrm{d}\mu, z\rangle = \int \langle W\boldsymbol{x}, z\rangle\,\mathrm{d}\mu = 0$$

for every $\boldsymbol{x} \in \mathscr{K}_{\mathscr{A}}(T)$ and every $z \in Z$, therefore

$$\int W\boldsymbol{x}\,\mathrm{d}\mu = 0, \text{ for every } \boldsymbol{x} \in \mathscr{K}_{\mathscr{A}}(T).$$

If $\boldsymbol{x}$ is a continuous field, then for every $\varphi \in \mathscr{K}(T)$ we have $\boldsymbol{x}\varphi \in \mathscr{K}_{\mathscr{A}}(T)$ therefore

$$\int W\boldsymbol{x}\varphi\,\mathrm{d}\mu = 0,$$

whence $W(t)\boldsymbol{x}(t) = 0$, μ-almost everywhere. In particular, this relation is true for $\boldsymbol{x} \in \mathscr{B}$.

For each $\boldsymbol{x} \in \mathscr{B}$ let $A(\boldsymbol{x})$ be a μ-negligible set such that

$$W(t)\boldsymbol{x}(t) = 0, \text{ for } t \notin A(\boldsymbol{x}).$$

The union $A = \cup A(\boldsymbol{x})$, when $\boldsymbol{x}$ runs over the countable set $\mathscr{B}$, is μ-negligible and we have

$$W(t)\boldsymbol{x}(t) = 0, \text{ for every } t \notin A \text{ and every } \boldsymbol{x} \in \mathscr{B}.$$

For $t \notin A$ we have, therefore,

$$|W(t)| = \sup_{x \in \mathscr{B}} |W(t)x(t)|/|x(t)| = 0\,,$$

where we agree to set $W(t)x(t)/|x(t)| = 0$ if $x(t) = 0$.

It follows that $W(t) = 0$, μ-almost everywhere, that is

$U(t) = V(t)$, μ-almost everywhere.

10 *Measures defined by densities*

28.28 THEOREM. *Let $Z \subset F'$ be a norming subspace for F, μ a scalar measure on T and $U \in \mathscr{C}(\mathscr{G}_F)$ a Z-weakly locally μ-integrable operation field. There exists then a dominated measure $\boldsymbol{m}$: $\mathscr{K}_{\mathscr{A}}(T) \to Z'$ such that*

$$\langle \boldsymbol{m}(x), z\rangle = \int \langle U(t)x(t), z\rangle \mathrm{d}\mu \text{ for } x \in \mathscr{K}_{\mathscr{A}}(T) \text{ and } z \in Z$$

and

$$|\boldsymbol{m}| \leqslant |U|\,|\mu|\,.$$

The measure $\boldsymbol{m}$ has values in F in each of the following cases:

(1) $F = Z'$;

(2) *U is simply μ-measurable*; *in particular F is of countable type*;

(3) *for every $x \in \mathscr{A}$ there exists a locally countable μ-negligible family $(K_j)_{j\in J}$ of disjoint compact subsets with $T \setminus \cup K_j$ negligible, such that the convex balanced hull of the set $\{U(t)x(t);\, t \in K_j\}$ is relatively compact in F for the topology $\sigma(F, Z)$.*

For every $x \in \mathscr{K}_{\mathscr{A}}(T)$ and $z \in Z$, the function $\langle Ux, z\rangle$ is μ-integrable (proposition 28.26). Put

$$M(x, z) = \int \langle U(t)x(t), z\rangle \mathrm{d}\mu\,.$$

We have

$$|M(x, z)| \leqslant |z| \int |U|\,|x|\,\mathrm{d}\mu < \infty\,.$$

The mapping $\boldsymbol{m}(x) : z \to M(x, z)$ is a linear continuous functional on Z and we have

$$|\boldsymbol{m}(x)| \leqslant \int |U|\,|x|\,\mathrm{d}|\mu|\,,$$

therefore $\boldsymbol{m}(x) \in Z'$ for every $x \in \mathscr{K}_{\mathscr{A}}(T)$. The mapping $\boldsymbol{m} : x \to \boldsymbol{m}(x)$ of $\mathscr{K}_{\mathscr{A}}(T)$ into Z' is linear and dominated by the positive measure $|U|\,|\mu|$, therefore $\boldsymbol{m}$ is a dominated measure and we have $|\boldsymbol{m}| \leqslant |U|\,|\mu|$.

For $\boldsymbol{x} \in \mathscr{K}_{\mathscr{A}}(T)$ and every $z \in Z$ we have

$$\langle \boldsymbol{m}(\boldsymbol{x}), z \rangle = M(\boldsymbol{x}, z) = \int \langle U\boldsymbol{x}, z \rangle \, \mathrm{d}\mu .$$

Let $\boldsymbol{x} \in \mathscr{K}_{\mathscr{A}}(T)$, $z \in Z$ and K a compact set, and show that we have

$$\langle \int_K \boldsymbol{x} \, \mathrm{d}\boldsymbol{m}, z \rangle = \int_K \langle U\boldsymbol{x}, z \rangle \, \mathrm{d}\mu .$$

Since $\boldsymbol{x}\varphi_K$ is an integrable field with respect to every measure, there exists a sequence $(\boldsymbol{x}_n)$ of fields of $\mathscr{K}_{\mathscr{A}}(T)$ such that

$$\lim \int |\boldsymbol{x}_n - \boldsymbol{x} \, \varphi_K| \, \mathrm{d}(|U| \, |\mu|) = 0 .$$

Since $|\boldsymbol{m}| \leqslant |U| \, |\mu|$, we deduce that

$$\lim \int |\boldsymbol{x}_n - \boldsymbol{x}\varphi_K| \, \mathrm{d}|\boldsymbol{m}| = 0 ,$$

therefore

$$\lim \int \boldsymbol{x}_n \mathrm{d}\boldsymbol{m} = \int \boldsymbol{x}\varphi_K \mathrm{d}\boldsymbol{m} ,$$

consequently

$$\lim \langle \int \boldsymbol{x}_n \mathrm{d}\boldsymbol{m}, z \rangle = \langle \int \boldsymbol{x}\varphi_K \mathrm{d}\boldsymbol{m}, z \rangle .$$

Since the functions $\langle U\boldsymbol{x}_n, z \rangle$ and $\langle U\boldsymbol{x}_n\varphi_K, z \rangle$ are μ-integrable, we have

$$|\int \langle U\boldsymbol{x}_n, z \rangle \, \mathrm{d}\mu - \int \langle U\boldsymbol{x}_n\varphi_K, z \rangle \, \mathrm{d}\mu| \leqslant |z| \int |U| \, |\boldsymbol{x}_n - \boldsymbol{x}| \varphi_K \mathrm{d}|\mu| =$$

$$= |z| \int |\boldsymbol{x}_n - \boldsymbol{x}\varphi_K| \, \mathrm{d}(|U| \, |\mu|) ,$$

therefore

$$\lim \int \langle U\boldsymbol{x}_n, z \rangle \, \mathrm{d}\mu = \int \langle U\boldsymbol{x}\varphi_K, z \rangle \, \mathrm{d}\mu .$$

As for each n we have

$$\langle \int \boldsymbol{x}_n \mathrm{d}\boldsymbol{m}, z \rangle = \int \langle U\boldsymbol{x}_n, z \rangle \, \mathrm{d}\mu ,$$

passing to the limit we obtain

$$\langle \int \boldsymbol{x}\varphi_K \mathrm{d}\boldsymbol{m}, z \rangle = \int \langle U\boldsymbol{x}\varphi_K, z \rangle \mathrm{d}\mu .$$

Consider now the case where $\boldsymbol{m}$ has values in F. The case $F = Z'$ is evident. Assume that U is simply μ-measurable.

Then for every $\boldsymbol{x} \in \mathscr{K}_{\mathscr{A}}(T)$ the function $t \to U(t)\boldsymbol{x}(t)$ is μ-measurable, therefore μ-integrable, and we have

$$\langle \boldsymbol{m}(\boldsymbol{x}), z \rangle = \int \langle U\boldsymbol{x}, z \rangle \, \mathrm{d}\mu = \langle \int U \boldsymbol{x} \mathrm{d}\mu, z \rangle$$

for every $z \in Z$, therefore

$$\boldsymbol{m}(\boldsymbol{x}) = \int U\boldsymbol{x}\,\mathrm{d}\mu \in F .$$

Finally, assume condition 3 fulfilled. Let $\boldsymbol{x} \in \mathscr{A}$ and let $\mathscr{K} = (K_j)_{j\in J}$ be a family of disjoint compact subsets such that for each $j \in J$, the convex closed balanced hull A_j of the set $\{U(t)\boldsymbol{x}(t);\ t \in K_j\}$ is compact for the topology $\sigma(F, Z)$. Let $\varphi \in \mathscr{K}(T)$ with $\|\varphi\| \leqslant 1$. Then $U(t)\boldsymbol{x}(t)\varphi(t) \in A_j$ for $t \in K_j$. Let (K_n) be the sequence of sets of $\mathscr{K}$ which intersect the support of φ. For each K_n, the set A_n is compact in Z^* for the topology $\sigma(Z^*, Z)$, therefore there exists a family $(Z_i)_{i\in I}$ of elements of Z such that

$$A_n = \cap_{i\in I} \{y \in Z^*;\ |\langle y, z_i\rangle| \leqslant 1\} .$$

Then

$$|\langle U(t)\boldsymbol{x}(t)\varphi(t), z_i\rangle| \leqslant 1 \text{ for } t \in K_n \text{ and } i \in I .$$

It follows that for every n we have

$$|\langle \textstyle\int_{K_n} \varphi\boldsymbol{x}\,\mathrm{d}\boldsymbol{m}, z_i\rangle| \leqslant \int_{K_n} |\langle U(t)\varphi(t)\boldsymbol{x}(t), z_i\rangle|\,\mathrm{d}|\mu| \leqslant |\mu|(K_n),$$

therefore

$$\textstyle\int_{K_n} \varphi\boldsymbol{x}\,\mathrm{d}\boldsymbol{m} \in |\mu|(K_n)\,A_n \subset F .$$

The relation $\boldsymbol{m}(\varphi\varphi_K\boldsymbol{x}) \in F$ remains valid for every $\varphi \in \mathscr{K}(T)$ and every $\boldsymbol{x} \in \mathscr{A}$, therefore we have $\boldsymbol{m}(\boldsymbol{x}) \in F$ for every $\boldsymbol{x} = \Sigma\varphi_i\boldsymbol{x}_i$ with $\varphi_i \in \mathscr{K}(T)$ and $\boldsymbol{x}_i \in \mathscr{A}$ and then, passing to the limit, for every $\boldsymbol{x} \in \mathscr{K}_{\mathscr{A}}(T)$.

Thus the theorem is proved.

Remarks.

1. The measure $\boldsymbol{m}$ depends on the space Z.
2. If U is simply locally μ-integrable, we have

$$\boldsymbol{m}(\boldsymbol{x}) = \int U(t)\boldsymbol{x}(t)\,\mathrm{d}\mu, \text{ for } \boldsymbol{x} \in \mathscr{K}_{\mathscr{A}}(T) .$$

In this case, the proof of the existence of the measure $\boldsymbol{m}$ is immediate.

28.29 Definition. *Let μ be a scalar measure, $U \in \mathscr{C}(\mathscr{G}_F)$ a Z-weakly locally μ-integrable operation field and the measure $\boldsymbol{m} : \mathscr{K}_{\mathscr{A}}(T) \to Z'$ defined by the equality*

$$\langle \boldsymbol{m}(\boldsymbol{x}), z\rangle = \int \langle U(t)\boldsymbol{x}(t), z\rangle\,\mathrm{d}\mu \textit{ for } \boldsymbol{x} \in \mathscr{K}_{\mathscr{A}}(T) \textit{ and } z \in Z .$$

We say that the measure $\boldsymbol{m}$ is of density U and of basis μ, or that $\boldsymbol{m}$ is the product of μ by U and we write

$$\boldsymbol{m} = U\mu .$$

We have then

$$\langle \int \boldsymbol{x} \mathrm{d}(U\mu), z \rangle = \int \langle U\boldsymbol{x}, z \rangle \mathrm{d}\mu, \text{ for } \boldsymbol{x} \in \mathcal{K}_{\mathcal{A}}(T) \text{ and } z \in Z .$$

We shall show that the equality remains valid for every field $\boldsymbol{x} \in \mathcal{L}^1_{\mathcal{A}}(\boldsymbol{m})$. The inequality $|\boldsymbol{m}| \leqslant |U| \, |\mu|$ is written now

$$|U\mu| \leqslant |U| \, |\mu| .$$

We shall show that in certain cases we even have the equality $|U\mu| = |U| \, |\mu|$.

We shall show also that every dominated measure is the product of a positive measure μ by a weakly locally μ-integrable operation field.

If U is a simply locally μ-integrable field, then we have

$$\int \boldsymbol{x} \mathrm{d}(U\mu) = \int U\boldsymbol{x} \mathrm{d}\mu, \text{ for } \boldsymbol{x} \in \mathcal{K}_{\mathcal{A}}(T) .$$

If $E(t) = E$ for every $t \in T$, the product $U\mu$ coincides with the one defined in §15.

28.30 THEOREM. *Let $U \in \mathcal{C}(\mathcal{G}_F)$ be a Z-weakly locally μ-integrable operation field and $\boldsymbol{x} \in \mathcal{C}(\mathcal{E})$ a vector field.*

(1) *If $\boldsymbol{x}$ is $|U| \, |\mu|$-negligible, then $\boldsymbol{x}$ is $U\mu$-negligible and the function $U\boldsymbol{x}$ is μ-negligible.*

(2) *If $\boldsymbol{x}$ is $|U| \, |\mu|$-measurable, then $\boldsymbol{x}$ is $U\mu$-measurable and the function $\langle U\boldsymbol{x}, z \rangle$ is μ-measurable for every $z \in Z$.*

(3) *If $\boldsymbol{x}$ is $|U| \, |\mu|$-integrable, then $\boldsymbol{x}$ is $U\mu$-integrable and the function $\langle U\boldsymbol{x}, z \rangle$ is μ-integrable for every $z \in Z$ and we have*

$$\langle \int \boldsymbol{x} \mathrm{d}(U\mu), z \rangle = \int \langle U\boldsymbol{x}, z \rangle \mathrm{d}\mu .$$

The proof is the same as for theorem 16.16.

28.31 COROLLARY. *Let $U \in \mathcal{C}(\mathcal{G}_F)$ be a Z-weakly locally μ-integrable operation field such that*

$$|U\mu| = |U| \, |\mu|$$

and let $\boldsymbol{x} \in \mathcal{C}(\mathcal{E})$ be a vector field.

(1) *If $\boldsymbol{x}$ is $U\mu$-negligible, then the function $U\boldsymbol{x}$ is μ-negligible;*

(2) *If $\boldsymbol{x}$ is $U\mu$-measurable, then the function $\langle U\boldsymbol{x}, z\rangle$ is μ-measurable for every $z\in Z$;*

(3) *If $\boldsymbol{x}$ is $U\mu$-integrable, then the function $\langle U\boldsymbol{x}, z\rangle$ is μ-integrable for for every $z\in Z$ and we have*

$$\langle \textstyle\int \boldsymbol{x}\,\mathrm{d}(U\mu), z\rangle = \int \langle U\boldsymbol{x}, z\rangle\,\mathrm{d}\mu\,.$$

11 *Absolutely continuous measures*

We say that a dominated measure $\boldsymbol{m} : \mathscr{K}_{\mathscr{A}}(T)\to F$ is absolutely continuous with respect to a positive measure μ if the modulus $|\boldsymbol{m}|$ is absolutely continuous with respect to μ.

If $\boldsymbol{m}$ is of the form $\boldsymbol{m}=U\mu$, where U is a weakly locally μ-integrable function, then $\boldsymbol{m}$ is absolutely continuous with respect to μ.

Conversely, under some countability conditions on the family $\mathscr{A}$, the Lebesgue–Nikodym theorem can be generalized.

28.32 Theorem. *Let $Z\subset F'$ be a norming subspace for F, μ a scalar measure and $\boldsymbol{m} : \mathscr{K}_{\mathscr{A}}(T)\to F$ a measure absolutely continuous with respect to μ. If the family $\mathscr{A}$ satisfies the countability axiom* (G), *then there exists an operation field $U_{\boldsymbol{m}}\in\mathscr{C}(\mathscr{G}_{Z'})$ with the following properties:*

(1) *The function $|U_{\boldsymbol{m}}|$ is locally μ-integrable and we have $|\boldsymbol{m}| = |U_{\boldsymbol{m}}|\,|\mu|$, that is*

$$\textstyle\int \varphi\,\mathrm{d}|\boldsymbol{m}| = \int |U_{\boldsymbol{m}}|\varphi\,\mathrm{d}|\mu|, \textit{ for } \varphi\in\mathscr{L}^1(|\boldsymbol{m}|)\,.$$

(2) *The field $U_{\boldsymbol{m}}$ is Z-weakly locally μ-integrable and we have $\boldsymbol{m}=U_{\boldsymbol{m}}\mu$, that is*

$$\langle \textstyle\int \boldsymbol{x}\,\mathrm{d}\boldsymbol{m}, z\rangle = \int \langle U_{\boldsymbol{m}}(t)\boldsymbol{x}(t), z\rangle\,\mathrm{d}\mu, \textit{ for } \boldsymbol{x}\in\mathscr{L}^1_{\mathscr{A}}(\boldsymbol{m}) \textit{ and } z\in Z\,.$$

(3) *If ρ is a lifting of $\mathscr{L}^\infty(\mu)$ and $\mathscr{B}\subset\mathscr{A}$ is a countable subset such that the set $\{\boldsymbol{x}(t)\,;\,\boldsymbol{x}\in\mathscr{B}\}$ is dense in $E(t)$ for each $t\in T$, then we can choose $U_{\boldsymbol{m}}$ such that*

$$\rho[U_{\boldsymbol{m}}\boldsymbol{x}] = U_{\boldsymbol{m}}\boldsymbol{x}, \textit{ for } \boldsymbol{x}\in\mathscr{B}\,.$$

(4) *We can choose $U_{\boldsymbol{m}}\in\mathscr{C}(\mathscr{G}_F)$ in each of the following cases:*

(*a*) $F=Z'$,

(*b*) *for each $\boldsymbol{x}\in\mathscr{A}$ the convex balanced hull of the set $A_{\boldsymbol{x}}=\{\boldsymbol{m}(\varphi\boldsymbol{x});\ \varphi\in\mathscr{K}(T), \int|\varphi|\,\mathrm{d}|\boldsymbol{m}|\leqslant 1\}$ is relatively compact in F for the topology $\sigma(F, Z)$.*

Let ρ be a lifting of $\mathscr{L}^\infty(\mu)$. If $\boldsymbol{x}\in A$ and $\varphi\in\mathscr{K}(T)$, then $\varphi\boldsymbol{x}\in\mathscr{K}_{\mathscr{A}}(T)$. For each $\boldsymbol{x}\in\mathscr{A}$, the mapping $\boldsymbol{m}_{\boldsymbol{x}}:\mathscr{K}(T)\to F=\mathscr{L}(R,F)$ defined by the equality

$$\boldsymbol{m}_{\boldsymbol{x}}(\varphi)=\boldsymbol{m}(\varphi\boldsymbol{x}),\ \text{for}\ \varphi\in\mathscr{K}(T),$$

is a dominated measure and we have

$$|\boldsymbol{m}_{\boldsymbol{x}}(\varphi)|\leqslant|\boldsymbol{m}|(|\varphi|\,|\boldsymbol{x}|)=(|\boldsymbol{x}|\,|\boldsymbol{m}|)(|\varphi|)$$

consequently

$$|\boldsymbol{m}_{\boldsymbol{x}}|\leqslant|\boldsymbol{x}|\,|\boldsymbol{m}|.$$

Since $|\boldsymbol{m}|$ is absolutely continuous with respect to μ, the measure $\boldsymbol{m}_{\boldsymbol{x}}$ is absolutely continuous with respect to μ, hence, by theorem 18.16, there exists a weakly locally μ-integrable function $\boldsymbol{g}_{\boldsymbol{x}}:T\to Z'$ such that $\boldsymbol{m}_{\boldsymbol{x}}=\boldsymbol{g}_{\boldsymbol{x}}\mu$ $|\boldsymbol{m}_{\boldsymbol{x}}|=|\boldsymbol{g}_{\boldsymbol{x}}|\,|\mu|$ and $\rho[\boldsymbol{g}_{\boldsymbol{x}}]=\boldsymbol{g}_{\boldsymbol{x}}$.

We deduce then

$$\boldsymbol{g}_{\alpha\boldsymbol{x}+\beta\boldsymbol{y}}=\alpha\boldsymbol{g}_{\boldsymbol{x}}+\beta\boldsymbol{g}_{\boldsymbol{y}}, \tag{1}$$

for every $\boldsymbol{x},\boldsymbol{y}\in\mathscr{A}$ and every scalars α, β.

Since $|\boldsymbol{m}|$ is absolutely continuous with respect to μ, there exists a locally μ-integrable scalar function g with $|\boldsymbol{m}|=g\mu=|g|\,|\mu|$ and $\rho[g]=g$.

From the inequality $|\boldsymbol{m}_{\boldsymbol{x}}|\leqslant|\boldsymbol{x}|\,|\boldsymbol{m}|$ we deduce $|\boldsymbol{g}_{\boldsymbol{x}}|\,|\mu|\leqslant|\boldsymbol{x}|\,|g|\,|\mu|$ therefore

$$|\boldsymbol{g}_{\boldsymbol{x}}(t)|\leqslant|\boldsymbol{x}(t)|\,|g(t)|,\ \mu\text{-almost everywhere} \tag{2}$$

for each $\boldsymbol{x}\in\mathscr{A}$.

From relations (1) and (2) we deduce

$$|\boldsymbol{g}_{\boldsymbol{x}}(t)-\boldsymbol{g}_{\boldsymbol{y}}(t)|\leqslant|\boldsymbol{x}(t)-\boldsymbol{y}(t)|\,|g(t)|,\ \mu\text{-almost everywhere} \tag{3}$$

for each $\boldsymbol{x},\boldsymbol{y}\in\mathscr{A}$.

Since the family $\mathscr{A}$ satisfies axiom (G), we can choose a sequence $(\boldsymbol{x}_n)$ of fields of $\mathscr{A}$, such that for each $t\in T$ the sequence $(\boldsymbol{x}_n(t))$ is dense in $E(t)$. Let $\mathscr{A}_0$ be the set of finite linear combinations of terms of the sequence $(\boldsymbol{x}_n)$ with rational coefficients (real or complex, according to whether the spaces $E(t)$ are real or complex). For each $t\in T$, the set $E_0(t)=\{\boldsymbol{x}(t);\boldsymbol{x}\in\mathscr{A}_0\}$ is a vector space (with respect to the field of the rational numbers) dense in $E(t)$.

Since $\mathscr{A}_0$ is countable, there exists a μ-negligible set $N\subset T$ such that for $t\notin N$ the relations (1), (2) and (3) are satisfied for every fields $\boldsymbol{x},\boldsymbol{y}\in\mathscr{A}_0$ and every rational numbers α and β.

For each $\boldsymbol{x} \in \mathscr{A}_0$ we modify the function $\boldsymbol{g}_{\boldsymbol{x}}$ at the points $t \in N$ putting $\boldsymbol{g}_{\boldsymbol{x}}(t) = 0$. Then we still have $\rho[\boldsymbol{g}_{\boldsymbol{x}}] = \boldsymbol{g}_{\boldsymbol{x}}$ and the relations (1), (2) and (3) are valid now for every $t \in T$, every $\boldsymbol{x}, \boldsymbol{y} \in \mathscr{A}_0$ and every α, β rational. From relation (3) we deduce that for each $t \in T$ we have

$$\boldsymbol{g}_{\boldsymbol{x}}(t) = \boldsymbol{g}_{\boldsymbol{y}}(t), \tag{4}$$

for every fields $\boldsymbol{x}, \boldsymbol{y} \in \mathscr{A}_0$ with $\boldsymbol{x}(t) = \boldsymbol{y}(t)$.

Let $t \in T$. For each $a \in E_0(t)$ put

$$\boldsymbol{g}_a(t) = \boldsymbol{g}_{\boldsymbol{x}}(t),$$

where $\boldsymbol{x} \in \mathscr{A}$ is choosen such that $\boldsymbol{x}(t) = a$. From equality (4) it follows that $\boldsymbol{g}_a(t)$ does not depend on $\boldsymbol{x}$ but only on a.

From the inequality (2) we deduce

$$|\boldsymbol{g}_a(t)| \leqslant |a|\,|\boldsymbol{g}(t)|.$$

The mapping $a \to \boldsymbol{g}_a(t)$ of $E_0(t)$ into Z' is linear with respect to the field of the rational numbers and it is continuous, therefore it can be extended to a continuous linear mapping $U_{\boldsymbol{m}}(t): E(t) \to Z'$. It follows that $U_{\boldsymbol{m}}(t) \in \mathscr{L}(E(t), Z') = G(t)$ and that

$$|U_{\boldsymbol{m}}(t)| \leqslant |\boldsymbol{g}(t)|.$$

For every $\boldsymbol{x} \in \mathscr{A}_0$ and every $t \in T$ we have

$$U_{\boldsymbol{m}}(t)\,\boldsymbol{x}(t) = \boldsymbol{g}_{\boldsymbol{x}(t)}(t) = \boldsymbol{g}_{\boldsymbol{x}}(t),$$

therefore $\rho[U_{\boldsymbol{m}}\boldsymbol{x}] = \rho[\boldsymbol{g}_{\boldsymbol{x}}] = U_{\boldsymbol{m}}\boldsymbol{x} = \boldsymbol{g}_{\boldsymbol{x}}$ and for every $\varphi \in \mathscr{K}(T)$ and every $z \in Z$ we have

$$\begin{aligned}\langle \boldsymbol{m}(\varphi\boldsymbol{x}), z\rangle &= \langle \boldsymbol{m}_{\boldsymbol{x}}(\varphi), z\rangle = \int \langle \boldsymbol{g}_{\boldsymbol{x}}(t)\,\varphi(t), z\rangle\,\mathrm{d}\mu = \\ &= \int \langle U_{\boldsymbol{x}}(t)\,\boldsymbol{x}(t)\,\varphi(t), z\rangle\,\mathrm{d}\mu\,.\end{aligned}$$

Let $\boldsymbol{x} \in \mathscr{K}_{\mathscr{A}}(T)$ and let K be the support of $\boldsymbol{x}$. The field $\boldsymbol{x}$ is the uniform limit on K of fields of the form

$$\Sigma_{1 \leqslant i \leqslant n} \boldsymbol{x}_i \varphi_i, \text{ with } \boldsymbol{x}_i \in \mathscr{A}_0 \text{ and } \varphi_i \in \mathscr{K}(T, K).$$

As for these fields we have

$$\langle \boldsymbol{m}(\Sigma\,\boldsymbol{x}_i\varphi_i), z\rangle = \int \langle U_{\boldsymbol{m}} \Sigma\,\boldsymbol{x}_i \varphi_i, z\rangle\,\mathrm{d}\mu\,,$$

it follows that

$$\langle \boldsymbol{m}(\boldsymbol{x}), z\rangle = \int \langle U_{\boldsymbol{m}}\boldsymbol{x}, z\rangle\,\mathrm{d}\mu, \text{ for } \boldsymbol{x} \in \mathscr{K}_{\mathscr{A}}(T) \text{ and } z \in Z\,.$$

Since $\boldsymbol{g}_x$ is weakly locally μ-integrable, the function $|\boldsymbol{g}_x|$ is μ-measurable. From the inequality

$$|U_{\boldsymbol{m}}(t)| = \sup_n |U_{\boldsymbol{m}}(t)\boldsymbol{x}_n(t)|/|\boldsymbol{x}_n(t)| = \sup_n |g\boldsymbol{x}_n(t)|/|\boldsymbol{x}_n(t)|$$

we deduce that the function $|U_{\boldsymbol{m}}|$ is μ-measurable. We put

$$U_{\boldsymbol{m}}(t)\boldsymbol{x}_n(t)/|\boldsymbol{x}_n(t)| = 0 \text{ if } \boldsymbol{x}_n(t) = 0 .$$

It follows that the function $U_{\boldsymbol{m}}$ is weakly locally μ-integrable, hence $\boldsymbol{m} = U_{\boldsymbol{m}}\mu$.

From the relation $\boldsymbol{m} = U_{\boldsymbol{m}}\mu$ we deduce

$$|g|\,|\mu| = |\boldsymbol{m}| \leqslant |U_{\boldsymbol{m}}|\,|\mu| ,$$

consequently $|U_{\boldsymbol{m}}(t)| \geqslant |g(t)|$, μ-almost everywhere. Since we have also $|U_{\boldsymbol{m}}(t)| \leqslant |g(t)|$ for every $t \in T$, if follows that $|U_{\boldsymbol{m}}(t)| = |g(t)|$, μ-almost everywhere, consequently $|\boldsymbol{m}| = |U_{\boldsymbol{m}}|\,|\mu|$. The case $U_{\boldsymbol{m}} \in \mathscr{C}(\mathscr{G}_F)$ is proved as in theorem 18.15.

Thus the theorem is completely proved.

Remarks.

1. If $\mathscr{A}$ satisfies axiom (*G*) and if *F* is of countable type, then the operation field $U_{\boldsymbol{m}}$ in the preceding theorem is *simply locally* μ-*integrable, therefore*

$$\int \boldsymbol{x}\,\mathrm{d}\boldsymbol{m} = U_{\boldsymbol{m}}(t)\boldsymbol{x}(t)\,\mathrm{d}\mu(t) \text{ for } \boldsymbol{x} \in \mathscr{L}^1_{\mathscr{A}}(\boldsymbol{m}) .$$

In fact, in this case the field $U_{\boldsymbol{m}}$ is simply measurable and the function $|U|$ is locally μ-integrable.

2. Assume that there exists a fundamental family $\mathscr{D} \subset \mathscr{C}(\mathscr{G}_{Z'})$ of continuous operation fields, which fulfills the following condition:

(*T*) *The function* $t \to V(t)\boldsymbol{x}(t)$ *is* $\boldsymbol{m}$-*measurable, for every* $V \in \mathscr{D}$ *and* $\boldsymbol{x} \in \mathscr{A}$.

Then the operation field $U_{\boldsymbol{m}}$ is measurable with respect to $\mathscr{D}$ and μ, consequently $U_{\boldsymbol{m}} \in \mathscr{L}^{\infty}_{\mathscr{D}}(\mu)$.

28.33 THEOREM. *Let* μ *be a scalar measure,* $Z \subset F'$ *a norming subspace for* F *and* $U \in \mathscr{C}(\mathscr{G}_F)$ *a* Z-*weakly locally* μ-*integrable operation field. We have*

$$|U\mu| = |U|\,|\mu|$$

in each of the following cases:

(1) *$\mathscr{A}$ satisfies axiom (G) and there exists a lifting ρ of $\mathscr{L}^\infty(\mu)$ and a countable set $\mathscr{B} \subset \mathscr{A}$ such that for every $t \in T$ the set $\{\boldsymbol{x}(t); \boldsymbol{x} \in \mathscr{B}\}$ is dense in $E(t)$ and $\rho[U\boldsymbol{x}] = U\boldsymbol{x}$ for every $\boldsymbol{x} \in \mathscr{B}$;*

(2) *$\mathscr{A}$ satisfies axiom (G) and there exists a countable set $S \subset Z$, norming for F;*

(3) *$\mathscr{A}$ satisfies the axiom (G) and U is simply locally μ-integrable;*

(4) *There exists a fundamental family $\mathscr{A}' \subset \mathscr{C}(\mathscr{E}')$ of continuous functional fields such that the function $t \to \langle \boldsymbol{x}(t), \boldsymbol{x}'(t) \rangle$ is continuous for every $\boldsymbol{x} \in \mathscr{A}$ and every $\boldsymbol{x}' \in \mathscr{A}'$, and $U \in \mathscr{C}(\mathscr{E}')$ is measurable with respect to $\mathscr{A}'$ and μ.*

Set $\boldsymbol{m} = U\mu$. By theorem 28.32, there exists a weakly locally μ-integrable operation field $U_{\boldsymbol{m}} \in \mathscr{C}(\mathscr{G}_{Z'})$ such that we have $|\boldsymbol{m}| = |U_{\boldsymbol{m}}|\,|\mu|$ and $\boldsymbol{m} = U_{\boldsymbol{m}}\mu$.

We have then

$$\int \langle U\boldsymbol{x}, z \rangle \, d\mu = \int \langle U_{\boldsymbol{m}}\boldsymbol{x}, z \rangle \, d\mu, \quad \text{for } \boldsymbol{x} \in \mathscr{L}^1_{\mathscr{A}}(\boldsymbol{m}) \text{ and } z \in Z,$$

whence $\langle U(t)\boldsymbol{x}(t), z \rangle = \langle U_{\boldsymbol{m}}(t)\boldsymbol{x}(t), z \rangle$, μ-almost everywhere for every $\boldsymbol{x} \in \mathscr{A}$ and $z \in Z$, that is $U\boldsymbol{x} = U_{\boldsymbol{m}}\boldsymbol{x}$ for every $\boldsymbol{x} \in \mathscr{A}$.

In order to prove that $|\boldsymbol{m}| = |U|\,|\mu|$, it is sufficient to show that $|U(t)| = |U_{\boldsymbol{m}}(t)|$, μ-almost everywhere.

If condition (2) is satisfied, then by proposition 28.27, we deduce that $U(t) = U_{\boldsymbol{m}}(t)$ μ-almost everywhere, therefore $|U\mu| = |\boldsymbol{m}| = |U_{\boldsymbol{m}}|\,|\mu| = |U|\,|\mu|$.

Assume that condition (1) is fulfilled. Then we can choose $U_{\boldsymbol{m}}$ such that we also have

$$\rho[U_{\boldsymbol{m}}\boldsymbol{x}] = U_{\boldsymbol{m}}\boldsymbol{x}, \quad \text{for } \boldsymbol{x} \in \mathscr{B}.$$

By proposition 28.27, we deduce still in this case, that we have $U(t) = U_{\boldsymbol{m}}(t)$, μ-almost everywhere, consequently $|\boldsymbol{m}| = |U_{\boldsymbol{m}}\mu| = |U_{\boldsymbol{m}}|\,|\mu| = |U|\,|\mu|$.

Assume now condition (3) is satisfied. Let ρ be a lifting of $\mathscr{L}^\infty(\mu)$ and $\mathscr{B} \subset \mathscr{A}$ a countable set, such that for each $t \in T$ the set $\{\boldsymbol{x}(t); \boldsymbol{x} \in \mathscr{B}\}$ is dense in $E(t)$. Choose $U_{\boldsymbol{m}}$ such that we have $\rho[U_{\boldsymbol{m}}\boldsymbol{x}] = U_{\boldsymbol{m}}\boldsymbol{x}$ for $\boldsymbol{x} \in \mathscr{B}$. Then, for each $\boldsymbol{x} \in \mathscr{B}$, the function $|U_{\boldsymbol{m}}\boldsymbol{x}|$ is μ-measurable, consequently $U_{\boldsymbol{m}}\boldsymbol{x}$ is weakly locally μ-integrable, and we have

$$|(U_{\boldsymbol{m}}\boldsymbol{x})\mu| = |U_{\boldsymbol{m}}\boldsymbol{x}|\,|\mu|.$$

For each $\boldsymbol{x} \in \mathscr{B}$, the function $U_{\boldsymbol{x}}$ is locally μ-integrable, therefore we have

$$|(U\boldsymbol{x})\mu| = |U\boldsymbol{x}|\,|\mu|.$$

But for every $\boldsymbol{x} \in \mathscr{A}$, $\varphi \in \mathscr{K}(T)$ and $z \in Z$ we have

$$\langle \int \boldsymbol{x}\varphi \,\mathrm{d}\boldsymbol{m}, z\rangle = \int \langle U_{\boldsymbol{m}} \boldsymbol{x}\varphi, z\rangle \,\mathrm{d}\mu = \int \langle U\boldsymbol{x}\varphi, z\rangle \,\mathrm{d}\mu\,,$$

therefore

$$\langle \int \varphi \,\mathrm{d}(U_{\boldsymbol{m}}\boldsymbol{x}), z\rangle = \langle \int \varphi \,\mathrm{d}(U\boldsymbol{x}), z\rangle\,, \text{ that is, } (U_{\boldsymbol{m}}\boldsymbol{x})\mu = (U\boldsymbol{x})\mu\,.$$

Then

$$|U\boldsymbol{x}|\,|\mu| = |(U\boldsymbol{x})\mu| = |(U_{\boldsymbol{m}}\boldsymbol{x})\mu| = |U_{\boldsymbol{m}}\boldsymbol{x}|\,|\mu|\,,$$

whence

$$|U(t)\,\boldsymbol{x}(t)| = |U_{\boldsymbol{m}}(t)\,\boldsymbol{x}(t)|,\ \mu\text{-almost everywhere, for each } \boldsymbol{x} \in \mathscr{B}\,,$$

consequently

$$|U(t)\boldsymbol{x}(t)|/|\boldsymbol{x}(t)| = |U_{\boldsymbol{m}}(t)\boldsymbol{x}(t)|/|\boldsymbol{x}(t)|,\ \mu\text{-almost everywhere}$$

for each $\boldsymbol{x} \in \mathscr{B}$, where, by convention, the two members of the equality vanish if $\boldsymbol{x}(t) = 0$. Taking the supremum for $\boldsymbol{x} \in \mathscr{B}$, we deduce that $|U(t)| = |U_{\boldsymbol{m}}(t)|$ μ-almost everywhere, consequently $|U\mu| = |U|\,|\mu|$.

Assume now that condition (4) is fulfilled.

Assume first that $|U(t)| \equiv 1$ and $\mu \geqslant 0$, and show that $|\boldsymbol{m}| = \mu$. Let $\varphi \in \mathscr{K}_+(T)$ and let K be the support of φ. Let $0 < a < 1$. There exists a field $\boldsymbol{y} \in \mathscr{C}(\mathscr{E})$ measurable with respect to $\mathscr{A}$ and μ such that

$$a\,|U(t)| \leqslant \langle \boldsymbol{y}(t), U(t)\rangle \quad\text{and}\quad a \leqslant |\boldsymbol{y}(t)| \leqslant 2 - a$$

μ-almost everywhere. Let $\varepsilon > 0$; there exists a compact set $K' \subset K$ such that $\mu(K \setminus K') < \varepsilon/(2 - a + \varepsilon)\|\varphi\|$ and such that $\boldsymbol{y}$ and U are continuous on K', therefore the function $t \to \langle \boldsymbol{y}(t), U(t)\rangle$ is continuous on K'. There exists a field $\boldsymbol{y}_0 \in \mathscr{K}_{\mathscr{A}}(T)$ with $\boldsymbol{y}(t) = \boldsymbol{y}_0(t)$ for $t \in K'$ and $\|\boldsymbol{y}_0\| \leqslant 2 - a + \varepsilon$.

We have

$$\begin{aligned}
a\mu(\varphi) &= a\int |U(t)|\,\varphi(t)\,\mathrm{d}\mu \leqslant \int_K \langle \boldsymbol{y}(t), U(t)\rangle\,\varphi(t)\,\mathrm{d}\mu = \\
&= \int_{K'} \langle \boldsymbol{y}(t), U(t)\rangle\,\varphi(t)\,\mathrm{d}\mu + \int_{K\setminus K'} \langle \boldsymbol{y}(t)\, U(t)\rangle\,\varphi(t)\,\mathrm{d}\mu \leqslant \\
&\leqslant \Big|\int_K \langle \boldsymbol{y}_0(t)\,\varphi(t), U(t)\rangle\,\mathrm{d}\mu - \\
&\quad - \int_{K\setminus K'} \langle \boldsymbol{y}_0(t), U(t)\rangle\,\varphi(t)\,\mathrm{d}\mu\Big| + \int_{K\setminus K'} \langle \boldsymbol{y}(t), U(t)\rangle\,\varphi(t)\,\mathrm{d}\mu| \leqslant \\
&\leqslant |\boldsymbol{m}(\varphi\boldsymbol{y}_0)| + \Big|\int_{K\setminus K'} |\boldsymbol{y}_0(t)|\,\varphi(t)\,\mathrm{d}\mu + \int_{K\setminus K'} |\boldsymbol{y}(t)|\,\varphi(t)\,\mathrm{d}\mu\Big| \leqslant \\
&\leqslant |\boldsymbol{m}|(\varphi|\boldsymbol{y}_0|) + (2 - a + \varepsilon)\mu(K\setminus K')\|\varphi\| + (2 - a)\mu(K\setminus K')\|\varphi\| \\
&\leqslant (2 - a + \varepsilon)\,|\boldsymbol{m}|\,(\varphi) + 2\varepsilon\,.
\end{aligned}$$

Since a and ε are arbitrary, it follows that $\mu(\varphi) \leqslant |\boldsymbol{m}|(\varphi)$ for every $\varphi \in \mathscr{K}_+(T)$, consequently $\mu \leqslant |\boldsymbol{m}|$. On the other hand we have $|\boldsymbol{m}| \leqslant |U|\mu = \mu$, therefore $|\boldsymbol{m}| = \mu$.

Consider now the general case. Denote $\lambda = |U|\,|\mu|$ and write $\mu = g|\mu|$, where $|g(t)| \equiv 1$. Put $V(t) = g(t)/|U(t)|$, if $U(t) \neq 0$ and $V(t) = a'_t \in E'(t)$ with $|a'_t| = 1$, if $U(t) = 0$.

We have $|V(t)| \equiv 1$ and V is measurable with respect to $\mathscr{A}'$ and λ, since $V(t)|U(t)| = U(t)g(t)$, and Ug is measurable with respect to $\mathscr{A}'$ and μ. Then

$$\begin{aligned}\boldsymbol{m}(\boldsymbol{x}) &= \int \langle \boldsymbol{x}(t), U(t)\rangle \,\mathrm{d}\mu = \int \langle \boldsymbol{x}(t), V(t)\rangle |U(t)|\, g(t)^{-1} \,\mathrm{d}\mu(t) = \\ &= \int \langle \boldsymbol{x}, V\rangle |U| \,\mathrm{d}|\mu| = \int \langle \boldsymbol{x}, V\rangle \,\mathrm{d}\lambda\,.\end{aligned}$$

From the first part of the proof it follows that $|\boldsymbol{m}| = \lambda$ therefore $|\boldsymbol{m}| = |U|\,|\mu|$.

Two dominated measured $\boldsymbol{m}, \boldsymbol{n} : \mathscr{K}_{\mathscr{A}}(T) \to F$ are singular if the positive measures $|\boldsymbol{m}|$ and $|\boldsymbol{n}|$ are singular.

The next theorem is a generalization of the Lebesgue theorem concerning the decomposition of a measure.

28.34 THEOREM. *Let ν be a positive measure. If $\mathscr{A}$ satisfies axiom (G), every dominated measure $\boldsymbol{m} : \mathscr{K}_{\mathscr{A}}(T) \to F'$ is written in the form*

$$\boldsymbol{m} = \boldsymbol{n}' + \boldsymbol{m}'\,,$$

where $\boldsymbol{n}'$ and $\boldsymbol{m}'$ are dominated measures such that $\boldsymbol{n}'$ is absolutely continuous with respect to ν, and $\boldsymbol{m}'$ is singular with respect to ν.

Let μ be the modulus of $\boldsymbol{m}$. By the Lebesgue theorem, μ is written in the form

$$\mu = \nu' + \mu'\,,$$

where ν' and μ' are positive measures, such that ν' is absolutely continuous with respect to ν and μ' is singular with respect to ν.

On the other hand, $\boldsymbol{m}$ can be written

$$\langle \boldsymbol{m}(\boldsymbol{x}), z\rangle = \int \langle U(t)\boldsymbol{x}(t), z\rangle \,\mathrm{d}\mu(t)$$

for $z \in F'$ and $\boldsymbol{x} \in \mathscr{K}_{\mathscr{A}}(T)$, where U is weakly locally μ-integrable and $|U(t)| \equiv 1$. Then

$$\langle \boldsymbol{m}(\boldsymbol{x}), z\rangle = \int \langle U(t)\boldsymbol{x}(t), z\rangle \,\mathrm{d}\nu'(t) + \int \langle U(t)\boldsymbol{x}(t), z\rangle \,\mathrm{d}\mu'(t)$$

and U is weakly locally ν-integrable and weakly locally μ'-integrable.

If for $\boldsymbol{x} \in \mathscr{K}_{\mathscr{A}}(T)$ and $z \in F$ we put

$$\langle \boldsymbol{n}'(\boldsymbol{x}), z\rangle = \int \langle U(t)\boldsymbol{x}(t), z\rangle \, \mathrm{d}\nu'(t),$$

$$\langle \boldsymbol{m}'(\boldsymbol{x}), z\rangle = \int \langle U(t)\boldsymbol{x}(t), z\rangle \, \mathrm{d}\mu'(t),$$

then $\boldsymbol{n}'$ and $\boldsymbol{m}'$ are dominated measures, $\boldsymbol{m} = \boldsymbol{n}' + \boldsymbol{m}'$, $|\boldsymbol{n}'| \leqslant \nu'$, and $|\boldsymbol{m}'| \leqslant \mu'$, therefore $\boldsymbol{n}'$ is absolutely continuous with respect to ν and $\boldsymbol{m}'$ is singular with respect to μ. The uniqueness of the decomposition is proved as in theorem 18.29.

12 *Linear operations on the space $\mathscr{L}^p_{\mathscr{A}}$.*

Let μ be a positive measure on T and $1 \leqslant p < \infty$. Consider the space $\mathscr{L}^p_{\mathscr{A}}(\mu)$ and a linear mapping $U : \mathscr{L}^p_{\mathscr{A}}(\mu) \to F$. Set

$$|||U||| = \sup \Sigma_i |U(\varphi_{A_i}\boldsymbol{x}_i)|,$$

$$||U|| = \sup |\Sigma_i U(\varphi_{A_i}\boldsymbol{x}_i| \; ||U|| = \sup |\Sigma_i U(\varphi_{A_i}\boldsymbol{x}_i)|$$

the supremum being considered for all step vector fields $\boldsymbol{x} = \Sigma \varphi_{A_i}\boldsymbol{x}_i$ with $A_i \in \mathscr{B}$, $\boldsymbol{x}_i \in \mathscr{A}$ and $1 \leqslant i \leqslant n$ such that the sets A_i are disjoint and $N_p(\boldsymbol{x}, \mu) \leqslant 1$.

We have $||U|| \leqslant |||U||| < \infty$. In fact, for every step vector field $\boldsymbol{x} = \Sigma \varphi_i \boldsymbol{x}_i$ with A_i disjoint and $N_p(\boldsymbol{x}, \mu) \leqslant 1$ we have

$$|U(\boldsymbol{x})| = |U(\Sigma \varphi_{A_i}\boldsymbol{x}_i) = |\Sigma U(\varphi_{A_i}\boldsymbol{x}_i)| \leqslant \Sigma |U(\varphi_{A_i}\boldsymbol{x}_i)| \leqslant |||U|||\,;$$

taking the supremum we obtain $||U|| \leqslant |||U|||$.

28.35 PROPOSITION. *For every linear mapping $U : \mathscr{L}^1_{\mathscr{A}}(\mu) \to F$ we have $||U|| = |||U|||$.*

In fact, if $||U|| = +\infty$, then we have also $|||U||| = +\infty$.

Assume therefore that $||U|| < \infty$. For every step vector field $\boldsymbol{x} = \Sigma \varphi_{A_i}\boldsymbol{x}_i$ with A_i *disjoint* and $N_1(\boldsymbol{x}, \mu) \leqslant 1$ we have

$$\Sigma |U(\varphi_{A_i}\boldsymbol{x}_i)| \leqslant \Sigma ||U|| N_1(\varphi_{A_i}\boldsymbol{x}_i, \mu) = ||U|| N_1(\Sigma \varphi_{A_i}\boldsymbol{x}_i, \mu) = ||U||,$$

whence $|||U||| \leqslant ||U||$, consequently $||U|| = |||U|||$.

28.36 PROPOSITION. *For every linear functional $U : \mathscr{L}^p_{\mathscr{A}}(\mu) \to C$, with $1 < p < \infty$, we have $||U|| = |||U|||$.*

If $||U|| = +\infty$, then we also have $|||U||| = +\infty$. Assume that $||U|| < \infty$

and let $x = \Sigma \varphi_{A_i} x_i$ be a step vector field with A_i *disjoint* and $N_p(x, \mu) \leqslant 1$.

For every i, there exists a complex number θ_i with $|\theta_i| = 1$ and

$$|U(\varphi_{A_i} x_i)| = \theta_i U(\varphi_{A_i} x_i) = U(\varphi_{A_i} \theta_i x_i).$$

Then

$$\Sigma |U(\varphi_{A_i} x_i)| = U(\Sigma \varphi_{A_i} \theta_i x_i) \leqslant \|U\| N_p(\Sigma \varphi_{A_i} \theta_i x_i, \mu) =$$

$$= \|U\| N_p(\Sigma \varphi_{A_i} x_i, \mu) \leqslant \|U\|,$$

whence $|||U||| \leqslant \|U\|$, consequently $\|U\| = |||U|||$.

Remarks

1. If $1 < p < \infty$ and $F \neq C$ we may have $\|U\| < \|U\||$.

2. The set of the continuous linear mappings $U : \mathscr{L}^p_{\mathscr{A}} \to F$ with $|||U||| < \infty$ is a vector space contained in the Banach space $\mathscr{L}(\mathscr{L}^p_{\mathscr{A}}(\mu), F)$, and $|||U|||$ is a norm.

28.37 THEOREM. *For every continuous linear mapping* $U : \mathscr{L}^p_{\mathscr{A}}(\mu) \to F$, $1 \leqslant p < \infty$ *with* $|||U||| < \infty$, *its restriction* m *to* $\mathscr{K}_{\mathscr{A}}(T)$ *is a dominated measure absolutely continuous with respect to* μ, *such that*

$$\int |\varphi| \mathrm{d}|m| \leqslant |||U||| N_p(\varphi, \mu), \textit{ for } \varphi \in \mathscr{K}(T).$$

Conversely, every dominated measure $m : \mathscr{K}_{\mathscr{A}}(T) \to F$ *absolutely continuous with respect to* μ, *having the property that there exists* $\alpha > 0$ *such that*

$$\int |\varphi| \mathrm{d}|m| \leqslant \alpha N_p(\varphi, \mu), \textit{ for every } \varphi \in \mathscr{K}(T),$$

can be extended uniquely to a continuous linear operation $U : \mathscr{L}^p_{\mathscr{A}}(\mu) \to F$ *with* $|||U||| < \infty$. *We have then* $|m| = g\mu$ *with*

$$N_q(g, \mu) = |||U|||, \ p^{-1} + q^{-1} = 1,$$

$$\mathscr{L}^p_{\mathscr{A}}(\mu) \subset \mathscr{L}^1_{\mathscr{A}}(m),$$

$$U(x) = \int x \mathrm{d}m, \textit{ for } x \in \mathscr{L}^p_{\mathscr{A}}(\mu),$$

and

$$\int |x| \mathrm{d}|m| \leqslant |||U||| N_p(x, \mu) \textit{ for } x \in \mathscr{L}^p_{\mathscr{A}}(\mu).$$

Let $U : \mathscr{L}^p_{\mathscr{A}}(\mu) \to F$ with $|||U||| < \infty$. For every relatively compact μ-integrable set $A \subset T$ and every field $x \in \mathscr{A}$ we have $\varphi_A x \in \mathscr{L}^p_{\mathscr{A}}$. Put

$$m(A)x = U(\varphi_A x).$$

If for $\boldsymbol{x}, \boldsymbol{y} \in \mathscr{A}$ we have $\varphi_A \boldsymbol{x} = \varphi_A \boldsymbol{y}$, then

$$\boldsymbol{m}(A)\boldsymbol{x} = \boldsymbol{m}(A)\boldsymbol{y} .$$

For each A, $\boldsymbol{m}(A)$ is a linear mapping of $\mathscr{A}$ into F.

If for every $\boldsymbol{x} \in \mathscr{A}$ and every A we put

$$||\boldsymbol{x}||_A = \sup_{t \in A} |\boldsymbol{x}(t)|$$

and if $||\boldsymbol{x}||_A \leqslant 1$, then

$$|\boldsymbol{m}(A)\boldsymbol{x}| = |U(\varphi_A \boldsymbol{x})| \leqslant |||U||| \, N_p(\varphi_A \boldsymbol{x}, \mu) \leqslant |||U||| \, N_p(\varphi_A, \mu) .$$

If for $A \subset B$ we put

$$||\boldsymbol{m}(A)|| = \sup_{||\boldsymbol{x}||_B \leqslant 1} |\boldsymbol{m}(A)\boldsymbol{x}| ,$$

then $||\boldsymbol{m}(A)||$ does not depend on B, but only on A and we have

$$||\boldsymbol{m}(A)|| \leqslant |||U||| \, N_p(\varphi_A, \mu) .$$

The set function $\boldsymbol{m}(A)$ is additive. We shall show that if we put

$$v(A) = \sup \Sigma \, ||\boldsymbol{m}(A_i)|| ,$$

where the supremum is taken for all the finite partitions $(A_i)_{1 \leqslant i \leqslant n}$ of A in μ-integrable sets, then

$$v(A) \leqslant |||U||| \, N_p(\varphi_A, \mu) < \infty .$$

In fact, let $(c_i)_{1 \leqslant i < n}$ be a finite family of numbers and let $\varepsilon > 0$. For each i there exists a field $x_i \in \mathscr{A}$ with $||\boldsymbol{x}_i||_A \leqslant 1$ such that

$$||\boldsymbol{m}(A_i)|| \, |c_i| \leqslant |\boldsymbol{m}(A_i)\boldsymbol{x}_i| + \varepsilon/n .$$

Then

$$\Sigma_{1 \leqslant i \leqslant n} ||\boldsymbol{m}(A_i)|| \, |c_i| \leqslant \Sigma |\boldsymbol{m}(A_i)\boldsymbol{x}_i| + \varepsilon = \Sigma |U(\varphi_{A_i} c_i \boldsymbol{x}_i)| + \varepsilon \leqslant$$
$$\leqslant |||U||| \, N_p(\Sigma \varphi_{A_i} c_i \boldsymbol{x}_i, \mu) + \varepsilon \leqslant |||U||| \, N_p(\Sigma \varphi_{A_i} c_i, \mu) + \varepsilon .$$

As ε is arbitrary, we deduce

$$\Sigma ||\boldsymbol{m}(A_i)|| \, |c_i| \leqslant |||U||| \, N_p(\Sigma \varphi_{A_i} c_i, \mu) .$$

In particular, taking $c_i = 1$ we obtain

$$\Sigma \, ||\boldsymbol{m}(A_i)|| \leqslant |||U||| \, N_p(\varphi_A, \mu),$$

therefore

$$\nu(A) \leqslant |||U|||\, N_p(\varphi_A, \mu)\,.$$

It is easy to see that ν is also additive. One can prove as in theorem 18.40 that we have

$$\Sigma\, \nu(A_i)\,|c_i| \leqslant |||U|||\, N_p(\Sigma \varphi_{A_i} c_i, \mu)\,.$$

Setting

$$v(\varphi) = \Sigma\, \nu(A_i) c_i, \text{ for } \varphi = \Sigma\, \varphi_{A_i} c_i\,,$$

one can prove as in theorem 18.40, that ν can be extended continuously to $\mathscr{L}^p(\mu)$ and that we have

$$|U(\boldsymbol{x})| \leqslant v(|\boldsymbol{x}|) \text{ for } \boldsymbol{x} \in \mathscr{L}^p_{\mathscr{A}}(\mu)\,.$$

The rest of the proof is similar to that of theorem 18.40.

28.38 THEOREM. *Let* $Z \subset F'$ *be a norming subspace for* F, μ *a scalar measure and* $U: \mathscr{L}^p{}_A(\mu) \to F$ *a linear mapping with* $|||U||| < \infty$. *If the family* $\mathscr{A}$ *satisfies axiom* (G), *then there exists an operation field* $\boldsymbol{u} \in \mathscr{C}(\mathscr{G}_{Z'})$ *with the following properties*:

(1) *The function* $|\boldsymbol{u}|$ *belongs to* $\mathscr{L}^q(\mu)$ *and*

$N_q(\boldsymbol{u}, \mu) = |||U|||$, $p^{-1} + q^{-1}$;

(2) *The field* $\boldsymbol{u}$ *is Z-weakly locally* μ*-integrable and we have*

$\langle U(\boldsymbol{x}, z\rangle = \int \langle \boldsymbol{u}(t)\boldsymbol{x}(t), z\rangle \mathrm{d}\mu$, *for* $\boldsymbol{x} \in \mathscr{L}^p_{\mathscr{A}}(\mu)$ *and* $z \in Z$;

(3) *If* ρ *is a lifting of* $\mathscr{L}^\infty(\mu)$ *and* $\mathscr{B} \in \mathscr{A}$ *is a countable set such that for each* $t \in T$, *the set* $\{\boldsymbol{x}(t); \boldsymbol{x} \in \mathscr{B}\}$ *is dense in* $E(t)$, *then we can choose* $\boldsymbol{u}$ *such that*

$\rho[\boldsymbol{ux}] = \boldsymbol{ux}$, *for* $\boldsymbol{x} \in \mathscr{B}$;

(4) *We have* $\boldsymbol{x} \in \mathscr{C}(\mathscr{G}_F)$ *in each of the following cases*;

(*a*) $F = Z'$;

(*b*) *for each* $\boldsymbol{x} \in \mathscr{A}$, *the convex balanced hull of the set* $\{U(\varphi\boldsymbol{x}); \varphi \in \mathscr{K}(T), \int |\varphi|\mathrm{d}|\mu| \leqslant 1\}$ *is relatively compact in* F *for the topology* $\sigma(F, Z)$.

The proof is the same as for theorem 18.42, using theorems 28.32 and 28.37, instead of theorem 18.16 and 18.40.

Remarks.

1. If, in addition, F is of countable type, then $\boldsymbol{u}$ is simply measurable and

$$U(\boldsymbol{x}) = \int \boldsymbol{u}(t)\boldsymbol{x}(t)\mathrm{d}\mu, \text{ for } \boldsymbol{x} \in \mathscr{L}^p_{\mathscr{A}}(\mu).$$

2. If, in addition, there exists a fundamental family $\mathscr{D} \subset \mathscr{C}(\mathscr{G}_F)$ such that the function $t \to V(t)\boldsymbol{x}(t)$ is μ-measurable, for every $V \in \mathscr{D}$ and $\boldsymbol{x} \in \mathscr{A}$, then $\boldsymbol{u} \in \mathscr{L}^q_{\mathscr{D}}(\mu)$.

28.39 COROLLARY. *If $\mathscr{A}$ satisfies axiom (G), then every continuous linear mapping $U : \mathscr{L}^1_{\mathscr{A}}(\mu) \to F$ is represented in the form*

$$\langle U(\boldsymbol{x}), z\rangle = \int \langle \boldsymbol{u}(t)\boldsymbol{x}(t), z\rangle \mathrm{d}\mu, \textit{ for } \boldsymbol{x} \in \mathscr{L}^1_{\mathscr{A}}(\mu) \textit{ and } z \in Z,$$

where $\boldsymbol{u}$ is a Z-weakly locally μ-integrable operation field with

$$\|U\| = N_\infty(\boldsymbol{u}, \mu).$$

28.40 COROLLARY. *If $\mathscr{A}$ satisfies axiom (G), every linear continuous functional $U : \mathscr{L}^p_{\mathscr{A}}(\mu) \to C$ with $1 \leqslant p < \infty$ is represented in the form*

$$U(\boldsymbol{x}) = \int \langle \boldsymbol{x}(t), \boldsymbol{u}(t)\rangle \mathrm{d}\mu, \text{ for } \boldsymbol{x} \in \mathscr{L}^p_{\mathscr{A}}(\mu),$$

where $\boldsymbol{u}$ is a weakly locally integrable functional field with

$$\|U\| = N_q(\boldsymbol{u}, \mu), \; p^{-1} + q^{-1} = 1$$

Without any other restriction on the family $\mathscr{A}$, we have the following theorem:

28.41 THEOREM. *Let $Z \subset F'$ be a norming subspace for F and $\boldsymbol{u} \in \mathscr{C}(\mathscr{G}_F)$ a Z-weakly locally μ-integrable operation field. If*

(i) $N_q(\boldsymbol{u}, \mu) < \infty \quad p^{-1} + q^{-1} = 1,$

then there exists a mapping $U : \mathscr{L}^p_{\mathscr{A}}(\mu) \to Z'$ such that we have

$$\langle U(\boldsymbol{x}), z\rangle = \int \langle \boldsymbol{u}\boldsymbol{x}, z\rangle \mathrm{d}\mu, \textit{ for } \boldsymbol{x} \in \mathscr{L}^p_{\mathscr{A}}(\mu) \textit{ and } z \in Z.$$

In this case we have

$$|||U||| \leqslant N_q(\boldsymbol{u}, \mu).$$

(ii) *We have $U(\boldsymbol{x}) \in F$ for every $\boldsymbol{x} \in \mathscr{L}^p_{\mathscr{A}}(\mu)$ in each of the following cases:*
1. *$F = Z'$;*
2. *$\boldsymbol{u}$ is simply μ-measurable; in particular F is of countable type;*
3. *for every $\boldsymbol{x} \in \mathscr{A}$ there exists a locally countable family $(K_j)_{j \in J}$* of dis-

joint compact subsets with $T\setminus\cup K_j$ μ-negligible, such that for each $j\in J$, the convex balanced hull of the set $\{\boldsymbol{u}(t)\boldsymbol{x}(t);\ t\in K_j\}$ is relatively compact in F for the topology $\sigma(F, Z)$.

(iii) *In each of the following cases the function $|\boldsymbol{u}|$ is μ-measurable and we have*

$$|||U||| = N_q(\boldsymbol{u}, \mu).$$

1. *A satisfies axiom (G) and there exists a lifting ρ of $\mathscr{L}^\infty(\mu)$ and a countable set $\mathscr{B}\subset\mathscr{A}$ such that for every $t\in T$ the set $\{\boldsymbol{x}(t);\ \boldsymbol{x}\in\mathscr{B}\}$ is dense in $E(t)$ and*

$$\rho[\boldsymbol{ux}] = \boldsymbol{ux},\ \textit{for}\ \ \boldsymbol{x}\in\mathscr{B}\ .$$

2. *$\mathscr{A}$ satisfies axiom (G) and there exists a countable set $S\subset Z$ norming for F.*

3. *$\mathscr{A}$ satisfies axiom (G) and $\boldsymbol{u}$ is simply μ-measurable.*

4. *There exists a fundamental family $\mathscr{A}'\in\mathscr{C}(\mathscr{E}')$ such that the function $t\to\langle\boldsymbol{x}(t), \boldsymbol{x}'(t)\rangle$ is continuous for every $\boldsymbol{x}\in\mathscr{A}$ and every $\boldsymbol{x}'\in\mathscr{A}'$, and $\boldsymbol{u}\in\mathscr{C}(\mathscr{E}')$ is a functional field, measurable with respect to $\mathscr{A}'$ and μ.*

The proof is the same as that of theorem 18.44, using theorems 28.32 and 28.37.

§29 Orlicz spaces

1 *The $\mathcal{O}^{\Phi}_{\mathscr{A}}$ spaces*

Let $\Phi(u)$ be a positive real (finite or infinite) function defined on $[0, +\infty]$ which is *increasing, left continuous* and

$$\lim_{u\to 0}\Phi(u) = 0 \ \text{and}\ \lim_{u\to\infty}\Phi(u) = +\infty\ .$$

It follows that Φ is continuous at 0 and $\Phi(0)=0$. The function Φ is Lebesgue measurable.

29.1 DEFINITION. *A measurable vector field $\boldsymbol{x}\in\mathscr{C}(\mathscr{E})$ is Φ-integrable if the scalar function $t\to\Phi(|\boldsymbol{x}(t)|)$ is μ-integrable.*

We shall denote by $\mathcal{O}_{\mathcal{A}}^{\Phi}$ the set of Φ-integrable fields and we shall write

$$|\boldsymbol{x}|_{\Phi} = \int \Phi(|\boldsymbol{x}(t)|)\, \mathrm{d}\mu(t) .$$

We shall say that a sequence $(\boldsymbol{x}_n)$ of vector fields of $\mathcal{O}_{\mathcal{A}}^{\Phi}$ is Φ-convergent to a vector field $\boldsymbol{x} \in \mathcal{O}_{\mathcal{A}}^{\Phi}$ if $\lim |\boldsymbol{x}_n - \boldsymbol{x}|_{\Phi} = 0$. We shall also say that $\boldsymbol{x}_n \to \boldsymbol{x}$ in $\mathcal{O}_{\mathcal{A}}^{\Phi}$, or with respect to Φ.

29.2 Proposition. *If $\Phi(u) > 0$ for $u > 0$, then $|\boldsymbol{x}|_{\Phi} = 0$ if and only if $\boldsymbol{x}$ is μ-negligible.*

The proof is immediate.

29.3 Proposition. *If α_n and α are complex numbers and if $\alpha_n \to \alpha$ then $\alpha_n \boldsymbol{x} \to \alpha \boldsymbol{x}$ with respect to Φ for every $\boldsymbol{x} \in \mathcal{O}_{\mathcal{A}}^{\Phi}$.*

We can assume that $|\alpha_n - \alpha| < 1$ for every n. Since Φ is continuous at 0, we have $\lim_n \Phi(|\alpha_n - \alpha|\,|\boldsymbol{x}(t)|) = 0$ for every $t \in T$. Remarking that for each n we have

$$\Phi(|\alpha_n - \alpha|\,|\boldsymbol{x}(t)|) \leqslant \Phi(|\boldsymbol{x}(t)|)$$

and applying the Lebesgue theorem we deduce

$$\lim_{n \to \infty} \int \Phi(|\alpha_n - \alpha|\,|\boldsymbol{x}(t)|)\, \mathrm{d}\mu(t) = 0 ,$$

that is,

$$\lim_{n \to \infty} |\alpha_n \boldsymbol{x} - \alpha \boldsymbol{x}|_{\Phi} = 0 .$$

The space $\mathcal{O}_{\mathcal{A}}^{\Phi}$ is not linear, in general, if the function Φ does not fulfill some additional conditions.

In the sequel we shall assume that the following condition is satisfied:

(Δ) *there exists a number $M > 1$ such that $\Phi(2u) \leqslant M\Phi(u)$ for $u \geqslant 0$.*

From condition (Δ) we deduce the following properties for the function Φ:

(1) $0 < \Phi(u) < +\infty$ *if* $0 < u < +\infty$;

(2) $\Phi(u+v) \leqslant M(\Phi(u) + \Phi(v))$, *for* $u \geqslant 0$ *and* $v \geqslant 0$;

(3) $\Phi(\alpha u) \leqslant M^n \Phi(u)$, *if* $0 \leqslant \alpha \leqslant 2^n$;

(4) $\Phi(\Sigma_{1 \leqslant n < \infty} a_n) \leqslant \Sigma_{1 \leqslant n < \infty} M^n \Phi(a_n)$, *for every sequence* (a_n) *of positive numbers*;

(5) $|\Phi(u)-\Phi(v)| \leqslant M\Phi(|u-v|)$, for $u \geqslant 0$ and $v \geqslant 0$.

Proof.

1. If there exists $u_0 > 0$ with $\Phi(u_0)=0$; we have $\Phi(2^n u_0) \leqslant M^n(u_0)$ for every n, consequently $\Phi(u) \equiv 0$, which contradicts the condition $\lim_{u\to\infty} \Phi(\mu) = +\infty$.

If there exists $u_1 < +\infty$ with $\Phi(u_1) = +\infty$, we have $\Phi(u_1) \leqslant M^n \Phi(u_1 2^{-n})$, consequently $\Phi(u) \equiv +\infty$, which contradicts the condition $\lim_{u\to 0} \Phi(u) = 0$. Thus, if $0 < u < \infty$, then $0 < \Phi(u) < +\infty$.

2. Assume, for example, that $u \geqslant v \geqslant 0$; then

$$\Phi(u+v) \leqslant \Phi(2u) \leqslant M\Phi(u) \leqslant M(\Phi(u)+\Phi(v)).$$

Property 3. is evident.

4. We have $\Phi(a_1) \leqslant M\Phi(a_1)$. Assume that inequality 4. is true for every family of n numbers:

$$\Phi(\Sigma_{k\leqslant n} a_k) \leqslant \Sigma_{k\leqslant n} M^k \Phi(a_k).$$

Let $(a_k)_{1\leqslant k\leqslant n+1}$ be a family of $n+1$ numbers. We have:

$$\begin{aligned}\Phi(\Sigma_{k\leqslant n+1}) &\leqslant M\Phi(a_1) + M\Phi(\Sigma_{2\leqslant k\leqslant n+1} a_k) \leqslant \\ &\leqslant M\Phi(a_1) + M\,\Sigma_{k\leqslant 2\leqslant n+1} M^{k-1} \Phi(a_n) = \\ &= \Sigma_{k\leqslant n+1} M^k \Phi(a_k).\end{aligned}$$

It follows, by induction, that the inequality (4) is true for every finite family of numbers.

Let now (a_n) be a sequence of positive numbers. Since Φ is left continuous, we have

$$\begin{aligned}\Phi(\Sigma_{1\leqslant k<\infty} a_k) = \Phi(\lim_{n\to\infty} \Sigma_{k\leqslant n} a_k) = \lim_{n\to\infty} \Phi(\Sigma_{k\leqslant n} a_k) \leqslant \\ \leqslant \lim_{n\to\infty} \Sigma_{k\leqslant n} M^k \Phi(a_k) = \Sigma_{1\leqslant k<\infty} M^k \Phi(a_k).\end{aligned}$$

5. Assume first that $u \geqslant v \geqslant 0$. Denote $a=u-v$ and $b=v$; we have $a \geqslant 0$ and $b \geqslant 0$. Assuming, for example, that $a \geqslant b$, we have

$$\Phi(a+b) \leqslant \Phi(2a) \geqslant M\Phi(a) \geqslant M\Phi(a) + \Phi(b)$$

hence

$$0 \leqslant \Phi(a+b) - \Phi(b) \leqslant M\Phi(a),$$

whence

$$0 \leqslant \Phi(u) - \Phi(v) \leqslant M\Phi(u-v) = M\Phi(|u-v|).$$

If we have $v < u$, then

$$0 \leqslant \Phi(v) - \Phi(u) \leqslant M\Phi(v-u) = M\Phi(|u-v|).$$

It follows that

$$|\Phi(u) - \Phi(v)| \leqslant M\Phi(|u-v|).$$

29.4 PROPOSITION. *$\mathcal{O}^{\Phi}_{\mathcal{A}}$ is a vector space.*

This follows from properties (2) and (3).

29.5 PROPOSITION. *The function $x \to |x|_{\Phi}$ defined on $\mathcal{O}^{\Phi}_{\mathcal{A}}$ has the following properties:*

(i) $|\boldsymbol{x}|_{\Phi} = 0$, *if and only if $\boldsymbol{x}$ is negligible;*
(ii) $|\boldsymbol{x}+\boldsymbol{y}|_{\Phi} \leqslant M(|\boldsymbol{x}|_{\Phi} + |\boldsymbol{y}|_{\Phi})$;
(iii) $|\alpha\boldsymbol{x}|_{\Phi} \leqslant M^n |\boldsymbol{x}|_{\Phi}$, if $|\alpha| \leqslant 2^n$;
(iv) *if $\boldsymbol{x}_n \to \boldsymbol{x}$ and $\boldsymbol{y}_n \to \boldsymbol{y}$ with respect to Φ, then $\boldsymbol{x}_n + \boldsymbol{y}_n \to \boldsymbol{x}+\boldsymbol{y}$ with respect to Φ;*
(v) *if $\alpha_n \to \alpha$ then $\alpha_n \boldsymbol{x} \to \alpha\boldsymbol{x}$ with respect to Φ, for every $\boldsymbol{x} \in \mathcal{O}^{\Phi}_{\mathcal{A}}$;*
(vi) *if $\boldsymbol{x}_n \to \boldsymbol{x}$ with respect to Φ, then $\alpha\boldsymbol{x}_n \to \alpha\boldsymbol{x}$ with respect to Φ, for every scalar α;*
(vii) *if $\boldsymbol{x}_n \to \boldsymbol{x}$ with respect to Φ and $\alpha_n \to \alpha$ then $\alpha_n \boldsymbol{x}_n \to \alpha\boldsymbol{x}$ with respect to Φ;*
(viii) *if $\boldsymbol{x}_n \to \boldsymbol{x}$ with respect to Φ, then $|\boldsymbol{x}_n|_{\Phi} \to |\boldsymbol{x}|_{\Phi}$;*
(ix) $||\boldsymbol{x}|_{\Phi} - |\boldsymbol{y}|_{\Phi}| \leqslant M|\boldsymbol{x}-\boldsymbol{y}|_{\Phi}$.

Proof. Properties (i) and (v) have already been proved (propositions 29.2 and 29.3). Properties (ii) and (iii) follow from properties (2) and (3) of the function Φ, and property (vi) follows from (iii).

Property (iv) is proved in the following way:

$$|\boldsymbol{x}_n + \boldsymbol{y}_n - (\boldsymbol{x}+\boldsymbol{y})|_{\Phi} = |(\boldsymbol{x}_n - \boldsymbol{x}) + (\boldsymbol{y}_n - \boldsymbol{y})|_{\Phi} \leqslant M(|\boldsymbol{x}_n - \boldsymbol{x}|_{\Phi} + |\boldsymbol{y}_n - \boldsymbol{y}|_{\Phi}) \to 0.$$

For property (vii), we assume that $|\alpha_n| \leqslant 2^k$, for every n. Then

$$\begin{aligned} |\alpha_n \boldsymbol{x}_n - \alpha\boldsymbol{x}| &= |\alpha_n(\boldsymbol{x}_n - \boldsymbol{x}) + (\alpha_n - \alpha)\boldsymbol{x}|_{\Phi} \leqslant \\ &\leqslant M|\alpha_n(\boldsymbol{x}_n - \boldsymbol{x})|_{\Phi} + M|(\alpha_n - \alpha)\boldsymbol{x}|_{\Phi} \leqslant \\ &\leqslant M^k|(\boldsymbol{x}_n - \boldsymbol{x})|_{\Phi} + M|(\alpha_n - \alpha)\boldsymbol{x}|_{\Phi} \end{aligned}$$

and the last member tends to 0, by property (v).

Property (viii) is proved in the following way:

$$|\Phi(|x_n(t)|)-\Phi(|x(t)|)|\leqslant M\Phi(|x_n(t)|-|x(t)|)\leqslant$$
$$\leqslant M\Phi(|x_n(t)-x(t)|).$$

whence

$$||x_n|_\Phi-|x|_\Phi|=|\int[\Phi(|x_n(t)|)-\Phi(|x(t)|)]\,\mathrm{d}\mu(t)|\leqslant$$
$$\leqslant\int|\Phi(|x_n(t)|)-\Phi(|x(t)|)|\,\mathrm{d}\mu(t)\leqslant$$
$$\leqslant M\int\Phi(|x_n(t)-x(t)|)\,\mathrm{d}\mu(t)=M|x_n-x|_\Phi\to 0.$$

In this way property (ix) has been proved too.

29.6 PROPOSITION. *Let* (x_n) *be a sequence of vector fields of* $\mathcal{O}^\Phi_{\mathscr{A}}$, *such that* $\Sigma_{1\leqslant n<\infty}M^n|x_n|_\Phi<+\infty$. Then *the series* $\Sigma_{1\leqslant n<\infty}x_n(t)$ *is absolutely convergent,* μ*-almost everywhere. If we put* $x(t)=\Sigma_{1\leqslant n<\infty}x_n(t)$ *at the points of convergence and* $x(t)=0$ *at the other points, then* $x\in\mathcal{O}^\Phi_{\mathscr{A}}$ *and for every n we have*

$$|x-\Sigma_{k\leqslant n}x_k|_\Phi\leqslant\Sigma_{k>n}M^k|x_k|_\Phi.$$

Consider the (finite or infinite) function

$$f(t)=\Sigma_{1\leqslant n<\infty}|x_n(t)|.$$

Then

$$|f|_\Phi=|\Sigma_{1\leqslant n<\infty}x_n|\leqslant\Sigma_{1\leqslant n<\infty}M^n|x_n|_\Phi<+\infty$$

therefore $\Phi(f(t))<+\infty$, μ-almost everywhere.

It follows that $f(t)<+\infty$, μ-almost everywhere, therefore the series $\Sigma\,x_n(t)$ is absolutely convergent, consequently convergent, μ-almost everywhere. The vector field x, defined in the statement is μ-measurable and we have

$$|x(t)|\leqslant\Sigma_{1\leqslant n<\infty}|x_n(t)|=f(t).$$

Then

$$\int\Phi(|x(t)|)\,\mathrm{d}\mu(t)\leqslant\int\Phi(\Sigma|x_n(t)|)\,\mathrm{d}\mu(t)\leqslant$$
$$\leqslant\Sigma M^n\int\Phi(|x_n(t)|)\,\mathrm{d}\mu(t)<+\infty$$

hence $x\in\mathcal{O}^\Phi_{\mathscr{A}}$. For every n we have

$$|\boldsymbol{x}(t)-\Sigma_{k\leqslant n}\boldsymbol{x}_k(t)|\leqslant\Sigma_{k>n}|\boldsymbol{x}(t)|\,,$$

μ-almost everywhere, consequently

$$|\boldsymbol{x}-\Sigma_{k\leqslant n}\boldsymbol{x}_k|_\Phi\leqslant|\Sigma_{k>n}|\boldsymbol{x}_k|_\Phi\leqslant\Sigma_{k>n}M^k|\boldsymbol{x}_k|_\Phi$$

and the proposition is proved.

A sequence $(\boldsymbol{x}_n)$ of vector fields of $\mathcal{O}^\Phi_{\mathscr{A}}$ is called a Φ-Cauchy sequence if $|\boldsymbol{x}_n-\boldsymbol{x}_m|_\Phi\to 0$ when $n, m\to\infty$.

29.7 Proposition. *For every Φ-Cauchy sequence $(\boldsymbol{x}_n)$ of elements of $\mathcal{O}^\Phi_{\mathscr{A}}$ there exists a field $\boldsymbol{x}\in\mathcal{O}^\Phi_{\mathscr{A}}$ such that $\boldsymbol{x}_n\to\boldsymbol{x}$ with respect to Φ.*

Let $(\boldsymbol{x}_n)$ be a Φ-Cauchy sequence. For every $\varepsilon>0$ there exists an integer N_ε such that for $n, m>N_\varepsilon$ we have

$$|\boldsymbol{x}_n-\boldsymbol{x}_m|<\varepsilon/2M\,.$$

We can find then a strictly increasing sequence of natural numbers (n_k) such that

$$|\boldsymbol{x}_{n_{k+1}}-\boldsymbol{x}_{n_k}|\leqslant 1/2^k M^k\,.$$

If we put $\boldsymbol{y}_k=\boldsymbol{x}_{n_{k+1}}-\boldsymbol{x}_{n_k}$, then $M^k|\boldsymbol{y}_k|_\Phi\leqslant 1/2^k$ and $\Sigma_{1\leqslant k<\infty}M^k|\boldsymbol{y}_k|_\Phi<+\infty$. By proposition 29.6 the series $\Sigma_{1\leqslant n<\infty}\boldsymbol{x}_n(t)$ is absolutely convergent μ-almost everywhere to a vector field $\boldsymbol{y}\in\mathcal{O}^\Phi_{\mathscr{A}}$. Then the vector field $\boldsymbol{x}=\boldsymbol{y}+\boldsymbol{x}_{n_1}$ is the limit of the sequence $(\boldsymbol{x}_{n_k})$ and, by proposition 29.5, we have

$$|\boldsymbol{x}-\boldsymbol{x}_{n_k}|_\Phi<\varepsilon/2M,\ \text{ for } n>N'_\varepsilon\,.$$

Then

$$|\boldsymbol{x}-\boldsymbol{x}_n|_\Phi=|\boldsymbol{x}-\boldsymbol{x}_{n_k}+\boldsymbol{x}_{n_k}-\boldsymbol{x}_n|_\Phi\leqslant$$
$$\leqslant M(|\boldsymbol{x}-\boldsymbol{x}_{n_k}|_\Phi+|\boldsymbol{x}_{n_k}-\boldsymbol{x}_n|_\Phi)\leqslant\varepsilon/2+\varepsilon/2=\varepsilon$$

if $n>\max(N_\varepsilon, N'_\varepsilon)$, therefore $\boldsymbol{x}_n\to\boldsymbol{x}$ with respect to Φ.

29.8 Proposition. *The set $\mathscr{K}_{\mathscr{A}}(T)$ of continuous vector fields with compact support and the set $\mathscr{E}_{\mathscr{A}}(\mathscr{B})$ of the Borel step vector fields are vector subspaces of $\mathcal{O}^\Phi_{\mathscr{A}}$. For every $\boldsymbol{x}\in\mathcal{O}^\Phi_{\mathscr{A}}$, there exists a sequence $(\boldsymbol{x}_n)$ of bounded vector fields of $\mathcal{O}^\Phi_{\mathscr{A}}$ such that $\boldsymbol{x}_n\to\boldsymbol{x}$ with respect to Φ.*

Let $\boldsymbol{x}\in\mathscr{K}^\Phi_{\mathscr{A}}(T)$. Then $\Phi(|\boldsymbol{x}(t)|)$ is a bounded scalar function with compact support. It is also easy to see that this function is a Borel function, therefore it is μ-integrable, consequently $\boldsymbol{x}\in\mathcal{O}_{\mathscr{A}}^{\Phi}$.

In the same way one can show that $\mathscr{E}_{\mathscr{A}}(\mathscr{B})\subset\mathcal{O}_{\mathscr{A}}^{\Phi}$.

Let now $\boldsymbol{x}\in\mathcal{O}_{\mathscr{A}}^{\Phi}$. For each natural number n put $\boldsymbol{x}_n(t)=\boldsymbol{x}(t)/|\boldsymbol{x}(t)|$ *if* $|\boldsymbol{x}(t)|>n$, and $\boldsymbol{x}_n(t)=\boldsymbol{x}(t)$, if $|\boldsymbol{x}(t)|\leqslant n$.

We have $|\boldsymbol{x}_n(t)|\leqslant n$. The field $\boldsymbol{x}_n$ is measurable, bounded and Φ-integrable, since $|\boldsymbol{x}_n(t)|\leqslant|\boldsymbol{x}(t)|$.

For every $t\in T$ we have $\boldsymbol{x}_n(t)\to\boldsymbol{x}(t)$ and

$$|\boldsymbol{x}_n(t)-\boldsymbol{x}(t)|\leqslant|\boldsymbol{x}_n(t)|+|\boldsymbol{x}(t)|\leqslant 2|\boldsymbol{x}(t)|\,.$$

We deduce that for every $t\in T$ we have

$$\Phi(|\boldsymbol{x}_n(t)-\boldsymbol{x}(t)|)\to 0,$$

$$\Phi(|\boldsymbol{x}_n(t)-\boldsymbol{x}(t)|)\leqslant\Phi(2|\boldsymbol{x}(t)|)\leqslant M\Phi(|\boldsymbol{x}(t)|)$$

and, applying the Lebesgue theorem, it follows that

$$\lim\int\Phi(|\boldsymbol{x}_n(t)-\boldsymbol{x}(t)|)\mathrm{d}\mu(t)=0$$

that is, $\boldsymbol{x}_n\to\boldsymbol{x}$ with respect to Φ.

The function $\boldsymbol{x}\to|\boldsymbol{x}|_{\Phi}$ has properties similar to a seminorm; the Φ-convergence defines on $\mathcal{O}_{\mathscr{A}}^{\Phi}$ a topology consistent with the structure of a vector space of $\mathcal{O}_{\mathscr{A}}^{\Phi}$ (proposition 29.5, properties (v) and (viii)). With this topology, $\mathcal{O}_{\mathscr{A}}^{\Phi}$ is not locally convex, in general

29.9 Proposition. *The family of subsets of $\mathcal{O}_{\mathscr{A}}^{\Phi}\times\mathcal{O}_{\mathscr{A}}^{\Phi}$ of the form*

$$U_n=\{(\boldsymbol{x},\boldsymbol{y})\,|\,\boldsymbol{x},\boldsymbol{y}\in\mathcal{O}_{\mathscr{A}}^{\Phi},\ |\boldsymbol{x}-\boldsymbol{y}|_{\Phi}\leqslant n^{-1}\}$$

is a fundamental system of vecinities of a uniform structure $\mathscr{U}$ on $\mathcal{O}_{\mathscr{A}}^{\Phi}$.

It is easy to see that $(\boldsymbol{x},\boldsymbol{x})\in U_n$ for every $\boldsymbol{x}\in\mathcal{O}$, that $U_m\subset U_n$ if $n\leqslant m$ and that U_n are symmetric.

Let us show that if $2Mm\leqslant n$, then $U_n^2\subset U_m$.

In fact, if $(\boldsymbol{x},\boldsymbol{z})\in U_n$ and $(\boldsymbol{z},\boldsymbol{y})\in U_n$ we have

$$|\boldsymbol{x}-\boldsymbol{z}|_{\Phi}\leqslant n^{-1}$$

and

$$|\boldsymbol{z}-\boldsymbol{y}|_{\Phi}\leqslant n^{-1}\,,$$

therefore

$$|x-y|_\Phi = |x-z+-y+z|_\Phi \leqslant M(|x-z|_\Phi + |z-y|_\Phi) \leqslant 2M/n \leqslant 1/m\,,$$

consequently $(x, y) \in U_m$.

Thus the sequence (U_n) fulfills all the axioms of a fundamental system of vecinities.

Denote by τ the topology deduced on $\mathcal{O}^\Phi_{\mathscr{A}}$ from the uniform structure $\mathscr{U}$. Since the system of neighbourhoods of the origin in $\mathcal{O}^\Phi_{\mathscr{A}}$ is countable, there exists an semi-distance d on $\mathcal{O}^\Phi_{\mathscr{A}}$ which defines a uniform structure on $\mathcal{O}^\Phi_{\mathscr{A}}$, identical with $\mathscr{U}$.

29.10 Proposition. *A sequence (x_n) of $\mathcal{O}^\Phi_{\mathscr{A}}$ is convergent to $x \in \mathcal{O}^\Phi_{\mathscr{A}}$ for the topology τ, if and only if $x_n \to x$, with respect to Φ.*

In fact, assume first that $|x_n - x|_\Phi \to 0$. For every integer $k > 0$, there exists a number N_k such that $|x_n - x|_\Phi \leqslant k^{-1}$ for $n \geqslant N_k$. It follows that the section $S_k = \{x_n | n \geqslant N_k\}$ is contained in the neighbourhood $U_k(x)$ of x, therefore the elementary filter associated to the sequence (x_n) converges to x in the topology τ, that is, $x_n \to x$ in τ.

Conversely, if $x_n \to x$ in τ, the elementary filter associated to the sequence (x_n) is finer than the filter of the neighbourhoods of x in τ. For every integer $k > 0$ there exists a section $S_k \subset U_k(x)$, that is,

$$|x_n - x|_\Phi \leqslant k^{-1}, \text{ if } n \in S_k$$

which means that $x_n \to x$ with respect to Φ.

From this proposition we deduce that the topology τ is compatible with the structure of the vector space of $\mathcal{O}^\Phi_{\mathscr{A}}$.

29.11 Proposition. *The separated space $O^\Phi_{\mathscr{A}}$ associated to $\mathcal{O}^\Phi_{\mathscr{A}}$ is a complete metrizable vector space.*

The set $\mathscr{N}_{\mathscr{A}}$ of the negligible fields is a closed vector subspace in $\mathcal{O}^\Phi_{\mathscr{A}}$ and we have $O^\Phi_{\mathscr{A}} = \mathcal{O}^\Phi_{\mathscr{A}} / \mathscr{N}_{\mathscr{A}}$. Since $\mathcal{O}^\Phi_{\mathscr{A}}$ is complete, $O^\Phi_{\mathscr{A}}$ is complete. Since $O^\Phi_{\mathscr{A}}$ has a countable fundamental system of neighbourhoods of the origin, it is metrizable, the distance being deduced from the semi-distance d by factorisation.

Example. If $\Phi(u) = u^p$, $(1 \leqslant p < +\infty)$, condition (Δ) is satisfied with $M = 2^p$. In this case the spaces $\mathscr{L}^p_{\mathscr{A}}$ and $\mathcal{O}^\Phi_{\mathscr{A}}$ contain the same elements and their topologies are equivalent.

2 *Complementary Young functions.*

Let φ be a *positive* (with finite or $+\infty$ values) *increasing* and *left continuous* function, defined on $[0, +\infty]$ with $\varphi(0)=0$.

Consider the function ψ, "inverse" to the function φ, defined on $[0, +\infty]$, by:

$$\psi(0)=0 \text{ and } \psi(s)=\sup\{t;\, \varphi(t)<s\} \text{ for } 0<s\leqslant +\infty\,.$$

It is easy to see that ψ is also *positive* (with finite or $+\infty$ values), *increasing and left continuous.*

The function ψ can also be described in the following way: if φ has a jump at a point t, then $\psi(s)=t$ for $\varphi(t-0)<s\leqslant\varphi(t+0)$; if $\varphi(t)=c$ for $t\in(a, b]$ but $\varphi(t)<c$ for $t<a$, then $\psi(c)=a$; if φ is continuous at t and $\varphi(t)=s$, but φ is not constant on any interval of the form $(t, b]$ then $\psi(s)=t$ and ψ is also continuous at s.

Since φ and ψ are increasing, the set of their points of discontinuity is countable.

The function χ defined on $[0, +\infty]$ by the equalities

$$\chi(0)=0 \text{ and } \chi(t)=\sup\{s;\, \psi(s)<t\}, \text{ for } 0<t\leqslant +\infty\,,$$

is also positive, increasing and left continuous and coincides with φ in the common points of continuity. Since the functions φ and χ are left continuous and the set of points of discontinuity is countable, it follows that φ and χ coincides everywhere. Thus we have

$$\varphi(t)=\sup\{s;\, \psi(s)<t\}, \text{ for } 0<t\leqslant +\infty\,,$$

therefore φ is, in turn, "inverse" to the function ψ.

By the definition of the function ψ it follows immediately that: if $s=\varphi(t)$, then $\psi(s)\leqslant t$ and if $t=\psi(s)$ then $\varphi(t)\leqslant s$.

We remark that at least one of the functions φ and ψ has finite values on $[0, +\infty)$. In fact, if, for example, we had $\psi(s)<+\infty$ for $s<s_0<+\infty$ and $\psi(s)=+\infty$ for $s>s_0$, then $\varphi(+\infty)=s_0$, therefore, φ being increasing, $\varphi(t)\leqslant s_0$ for every $t\in[0, +\infty]$.

Also, at least one of the functions φ and ψ vanishes only at the origin. In fact, if, for example, we had $\psi(s)=0$ for $s\leqslant\delta\neq 0$, then $\varphi(t)\geqslant\delta$ for $t>0$.

That is why, we shall assume, in the sequel, that

$$0<\varphi(t)<+\infty \text{ for } 0<t<+\infty\,.$$

If $\lim_{t\to\infty} \varphi(t) = l < +\infty$, then

$$\psi(s) < +\infty \text{ for } s < l \text{ and } \psi(s) = +\infty \text{ for } s > l\,.$$

If $\lim_{t\to\infty} \varphi(t) = +\infty$, then we have

$$\psi(s) < +\infty \text{ for } s < +\infty \text{ and } \lim_{s\to\infty} \psi(s) = \infty\,.$$

If $\varphi(0+0) = \delta > 0$, then $\psi(s) = 0$ for $s \leqslant \delta$.
If φ is continuous at 0, then ψ is also continuous at 0.

29.12 Lemma. *If $\varphi(t) < s$ then $t \leqslant \psi(s)$. If $s < \varphi(t)$ then $\psi(s) < t$.*

In fact, the first assertion follows directly from the definition of the function ψ. If $s < \varphi(t)$, there exists a $p \geqslant 0$ with $s + p = \varphi(t)$, therefore $\psi(s) \geqslant \psi(s+p) \leqslant t$. But we cannot have $\psi(s) = t$, since in this case we would have $s \geqslant \varphi(t)$, which contradicts the hypothesis.

Thus, $\psi(s) < t$ and the lemma is proved.

Let us divide the first quadrant $Q = \{(t, s)\,;\, t \geqslant 0, s \geqslant 0\}$ of the plane into four sets

$$E_1 = \{(t, s) | s < \varphi(t)\}\,;$$
$$E_2' = \{(t, s) | s = \varphi(t)\}\,;$$
$$E_2'' = \{(t, s) | s > \varphi(t) \text{ and } t = \psi(s)\}\,;$$
$$E_3' = \{(t, s) | s > \varphi(t) \text{ and } \psi(s) > t\}\,.$$

The set $E_2 = E_2' \cup E_2''$ consists of points (t, s) satisfying at least one of the equalities $\varphi(t) = s$ or $t = \psi(s)$.

29.13 Lemma. *The sets E_1 and E_3 are open (relative to the first quadrant) and E_2 is closed and negligible for the Lebesgue measure in the plane.*

Let $(t_0, s_0) \in E_1$; therefore $s_0 < \varphi(t_0)$, hence $s_0 = \varphi(t_0) - 2p$ with $p \geqslant 0$. Since $\varphi(t_0) = \varphi(t_0 - 0)$, there exists $\delta > 0$ such that $\varphi(t) > \varphi(t_0) - p$ if $t \geqslant t_0 - \delta$. In the rectangle $|t - t_0| \leqslant \delta$, $|s - s_0| \leqslant \delta$, we have $s \leqslant s_0 + p = \varphi(t_0) - p < \varphi(t)$, therefore the intersection of this rectangle with the first quadrant, which is a neighbourhood of (t_0, s_0) in the first quadrant, is contained in the set E_1. Thus (t_0, s_0) is an interior point of E_1. As (t_0, s_0) was chosen arbitrarily, it follows that E_1 is open.

The fact that E_3 is open is proved similarly.

Since $E_2 = Q \setminus (E_1 \cup E_3)$ it follows that E_2 is closed. The set E_2 is negligible as union of the graphs of the functions φ and ψ.

The functions Φ and Ψ defined on $[0, +\infty]$ by the equalities

$$\Phi(u) = \int_0^u \varphi(t)\,dt \text{ and } \Psi(v) = \int_0^v \psi(s)\,ds$$

are said to be complementary in Young's sense.

It is easy to see that Φ is continuous and increasing and

$$\Phi(0) = 0, \ \lim_{u\to\infty} \Phi(u) = +\infty .$$

If $\varphi(t) = 0$ for $t > u_0$, then Φ is strictly increasing for $u > u_0$.

In particular, if $0 < \varphi(t) < +\infty$ for $0 < t < \infty$, then Φ is strictly increasing and

$$0 < \Phi(u) < +\infty \text{ for } 0 < u < +\infty .$$

The function Ψ is also increasing and continuous at all points, with the following exception:

$$\text{if } \lim_{t\to\infty} \varphi(t) = l < +\infty, \text{ then } \Psi(v) < +\infty \text{ for } v \leqslant l$$

and $\Psi(v) = +\infty$ for $v > l$; at the point l the function Ψ is left continuous.

If $\lim_{t\to\infty} \varphi(t) = +\infty$, then $\Psi(v) < +\infty$ for $v < +\infty$ and $\lim_{v\to\infty} \Psi(v) = +\infty$.

We have also $\Psi(0) = 0$, and if $\Psi(v_1) \neq 0$, then Ψ is strictly increasing for $v \geqslant v_1$.

Example. If $\varphi(t) = t^{p-1}$, $(1 < p < +\infty)$, then $\psi(s) = s^{q-1}$, $(p^{-1} + q^{-1} = 1)$. In this case $\Phi(u) = p^{-1} u^p$ and $\Psi(v) = q^{-1} v^q$. Both functions Φ and Ψ satisfy condition (Δ).

If $\varphi(t) = 1$ for $t \neq 0$, then $\psi(s) = 0$ for $s \leqslant 1$ and $\psi(s) = +\infty$ for $s > 1$. In this case $\Phi(u) = u$ and $\Psi(v) = 0$ for $v \leqslant 1$ and $\Psi(v) = +\infty$ for $v > 1$. The function Φ fulfills condition (Δ) but the function Ψ does not.

29.14 Proposition (*Young's inequality*). *If Φ and Ψ are complementary in Young's sense, we have*

$$uv \leqslant \Phi(u) + \Psi(v)$$

for every $u \geqslant 0$ and $v \geqslant 0$. The equality is obtained if and only if $v = \varphi(u)$ or $u = \psi(u)$.

Let $u_0 \geqslant 0$ and $v_0 \geqslant 0$ and denote $D = [0, u_0] \times [0, v_0]$. Denoting by m the Lebesgue measure in the plane we have

$$u_0 v_0 = m(D) = m(D \cap E_1) + m(D \cap E_2) + m(D \cap E_3) =$$
$$= m(D \cap E_1) + m(D \cap E_3) = \int_D \varphi_{E_1} \mathrm{d}m + \int_D \varphi_{E_3} \mathrm{d}m .$$

Using the Fubini theorem we obtain

$$\int_D \varphi_{E_1} \mathrm{d}m = \int \int_D \varphi_{E_1}(u, v) \mathrm{d}u \mathrm{d}v = \int_0^{u_0} \mathrm{d}u \int_0^{v_0} \varphi_{E_1}(u, v) \mathrm{d}v =$$
$$= \int_0^{u_0} \mathrm{d}u \int_0^{m(u)} \varphi_{E_1}(u, v) \mathrm{d}v ,$$

where $m(u) = \min(v_0, \varphi(u))$,therefore

$$\int_D \varphi_{E_1} \mathrm{d}m \leqslant \int_0^{u_0} \mathrm{d}u \int_0^{\varphi(u)} \mathrm{d}v = \int_0^{u_0} \varphi(u) \mathrm{d}u = \Phi(u_0)$$

with equality if and only if $v_0 \geqslant \varphi(u_0)$.

In the same way we deduce

$$\int_D \varphi_{E_3} \mathrm{d}m \leqslant \Psi(v_0)$$

with equality if and only if $u_0 \geqslant \varphi(v_0)$. Thus

$$u_0 v_0 \leqslant \Phi(u_0) + \Psi(v_0)$$

with equality if and only if we have at the same time $v_0 \geqslant \varphi(u_0)$ and $u_0 \geqslant \psi(v_0)$. But, by lemma 29.12, we cannot have at the same time $v_0 > \varphi(u_0)$ and $u_0 > \psi(v_0)$, hence the equality is obtained if and only if at least one of the equalities $v_0 = \varphi(u_0)$ or $u_0 = \psi(v_0)$ is satisfied.

Remark. If $\varphi(t) = t^{p-1}$ with $1 < p < +\infty$, the Young inequality has the form

$$uv \leqslant u^p/p + v^q/q, \quad 1/p + 1/q = 1 ,$$

with equality if and only if $v = u^{p-1}$.

3 *The $\|x\|_\Phi$ seminorms*

In the rest of this chapter we shall consider a fixed increasing and left continuous function φ, defined on $[0, +\infty]$, with $\varphi(0) = 0$,

$$0 < \varphi(t) < +\infty, \text{ for } 0 < t < +\infty ,$$

and the corresponding functions Φ and Ψ complementary in Young's sense.

The function Φ is positive, continuous and strictly increasing on $[0, +\infty]$ and we have

$$\Phi(0)=0,\ \lim_{u\to\infty}\Phi(u)=+\infty$$

and

$$0<\Phi(u)<+\infty \text{ for } 0<u<+\infty .$$

If Φ does not fulfill condition (Δ), the set $\mathcal{O}^{\Phi}_{\mathscr{A}}$ is not a vector space. But even if Φ fulfills this condition, the mapping $i \to |x|_{\Phi}$ is not a seminorm on this space.

For every vector field $x \in \mathscr{C}(\mathscr{E})$ put

$$||x||_{\Phi}=\sup_{|f|_{\Psi}\leqslant 1}\int^{*}|x(t)|\,|f(t)|\,d\mu(t)$$

$$||x||_{\Psi}=\sup_{|f|_{\Phi}\leqslant 1}\int^{*}|x(t)|\,|f(t)|\,d\mu(t),$$

the supremum being considered for all *measurable real* functions f defined on T.

In the sequel, we shall study only the function $||x||_{\Phi}$.

The properties of the function $||x||_{\Phi}$ remain valid also for $||x||_{\Psi}$, except when the contrary is mentioned explicitly.

From the properties of the upper integral, the following properties are immediately seen:

(1) $0\leqslant ||x||_{\Phi}\leqslant +\infty$;

(2) $||\alpha x||_{\Phi}=|\alpha|\,||x||_{\Phi}$;

(3) $||x+y||_{\Phi}\leqslant ||x||_{\Phi}+||y||_{\Phi}$.

29.15 PROPOSITION. *For every field* $x \in \mathscr{C}(\mathscr{E})$ *we have*

$$||x||_{\Phi}=\sup_{K}||x\varphi_{K}||_{\Phi},\ K \text{ compact.}$$

In fact

$$||x||_{\Phi}=\sup_{h}\int^{*}|x|\,|h|\,d\mu=\sup_{h}\sup_{K}\int^{*}|x\varphi_{K}|\,|h|\,d\mu=$$

$$=\sup_{K}\sup_{h}\int^{*}|x\varphi_{K}|\,|h|\,d\mu=\sup||x\varphi_{K}||_{\Phi}\,.$$

29.16 PROPOSITION. *Assume that there exists a continuous fundamental family* $\mathscr{A}'\subset\mathscr{C}(\mathscr{E}')$ *of functional fields, such that the scalar functions* $t\to\langle x(t),x'(t)\rangle$ *is continuous for every* $x\in\mathscr{A}$ *and* $x'\in\mathscr{A}'$. *In this case, if* x *is measurable (with respect to* $\mathscr{A}$ *and* μ*), we have*

$$\|\boldsymbol{x}\|_\Phi = \sup_{|\boldsymbol{x}'|_\Psi \leqslant 1} \int^* |\langle \boldsymbol{x}(t), \boldsymbol{x}'(t)\rangle| \, d\mu(t) = \\ = \sup_{|\boldsymbol{x}'|_\Psi \leqslant 1} \int^* |\boldsymbol{x}(t)| \, |\boldsymbol{x}'(t)| \, d\mu(t);$$

and if $\boldsymbol{x}$ *is measurable and* $\|\boldsymbol{x}\|_\Phi < +\infty$, *then*

$$\|\boldsymbol{x}\|_\Phi = \sup_{|\boldsymbol{x}'|_\Psi \leqslant 1} \left| \int \langle \boldsymbol{x}(t), \boldsymbol{x}'(t)\rangle \, d\mu(t) \right| .$$

We have first

$$\sup_{|\boldsymbol{x}'|_\Psi \leqslant 1} \int^* |\langle \boldsymbol{x}(t), \boldsymbol{x}'(t)\rangle| \, d\mu(t) \leqslant \sup_{|\boldsymbol{x}'|_\Psi \leqslant 1} \int^* |\boldsymbol{x}| \, |\boldsymbol{x}'| \, d\mu \leqslant \|\boldsymbol{x}\|_\Phi$$

even if $\boldsymbol{x}$ is not measurable. Assume that $\boldsymbol{x}$ is measurable; we prove the inequality

$$\|\boldsymbol{x}\|_\Phi \leqslant \sup_{|\boldsymbol{x}'|_\Psi \leqslant 1} \int^* |\langle \boldsymbol{x}(t), \boldsymbol{x}'(t)\rangle| \, d\mu(t) .$$

Let $f \in \mathcal{O}^\Psi$ with $|f|_\Psi \leqslant 1$. Let $0 < a < 1$. Since $\boldsymbol{x}$ is measurable, there exists a measurable field $\boldsymbol{y}' \in \mathscr{C}(\mathscr{E}')$ such that

$$a|\boldsymbol{x}(t)| \leqslant |\langle \boldsymbol{x}(t), \boldsymbol{y}'(t)\rangle| \text{ and } a \leqslant |\boldsymbol{y}'(t)| \leqslant 2-a$$

almost everywhere on T. The functional field

$$\boldsymbol{x}'(t) = \boldsymbol{y}'(t) |f(t)| / (2-a)$$

is measurable and $|\boldsymbol{x}'(t)| \leqslant |\boldsymbol{y}'(t)|$ almost everywhere, hence $|\boldsymbol{x}'|_\Psi \leqslant 1$.

We have then

$$|\langle \boldsymbol{x}(t), \boldsymbol{x}'(t)\rangle| = |\langle \boldsymbol{x}(t), \boldsymbol{y}'(t) |f(t)| / (2-a)\rangle| \geqslant |\boldsymbol{x}(t)| \, |f(t)| a/(2-a)$$

almost everywhere, consequently

$$\|\boldsymbol{x}\|_\Phi = \sup_{|f|_\Psi \leqslant 1} \int^* |\boldsymbol{x}| \, |f| \, d\mu \leqslant \\ \leqslant (2-a) a^{-1} \sup_{|\boldsymbol{x}'|_\Psi \leqslant 1} \int^* |\langle \boldsymbol{x}(t), \boldsymbol{x}'(t)\rangle| d\mu(t) .$$

Letting $a \to 1$ we obtain the desired inequality and the first series of equalities is proved.

Assume now that $\boldsymbol{x}$ is measurable and $\|\boldsymbol{x}\|_\Phi < +\infty$. Then the function $t \to \langle \boldsymbol{x}(t), \boldsymbol{x}'(t)\rangle$ is integrable, for every $\boldsymbol{x}' \in \mathcal{O}^\Psi_{\mathscr{A}'}$ with $|\boldsymbol{x}'| \leqslant 1$. If for each $\boldsymbol{y}' \in \mathcal{O}^\Psi_{\mathscr{A}'}$ with $|\boldsymbol{y}'|_\Psi \leqslant 1$ we put

$$\langle \boldsymbol{x}(t), \boldsymbol{y}'(t)\rangle = \rho(t) e^{i\theta(t)},$$

where we take $e^{i\theta(t)} = 1$ if $\rho(t) = 0$, then the functions $\rho(t)$ and $e^{i\theta(t)}$ are measurable and the field $\boldsymbol{x}'(t) = \boldsymbol{y}'(t)/e^{i\theta(t)}$ is measurable and $|\boldsymbol{x}'(t)| = |\boldsymbol{y}'(t)|$, therefore $|\boldsymbol{x}'|_\Psi \leqslant 1$; it follows that

$$\langle \boldsymbol{x}(t), \boldsymbol{x}'(t)\rangle = \rho(t) = |\langle \boldsymbol{x}(t), \boldsymbol{y}'(t)\rangle|,$$

whence it follows also the second inequality.

29.17 PROPOSITION. *A field* $\boldsymbol{x} \in \mathscr{C}(\mathscr{E})$ *is negligible if and only if* $\|\boldsymbol{x}\|_\Phi = 0$.

If $\boldsymbol{x}$ is negligible, then the function $|\boldsymbol{x}|\,|f|$ is negligible for every real function f, therefore $\|\boldsymbol{x}\|_\Phi = 0$. Conversely, assume that $\|\boldsymbol{x}\|_\Phi = 0$. This means that

$$\int^* |\boldsymbol{x}|\,|f|\,\mathrm{d}\mu = 0$$

for every function $f \in \mathcal{O}^\Psi$ with $|f|_\Psi \leqslant 1$.

Let $K \subset T$ be a compact set. Since $\lim_{v\to 0} \Psi(p) = 0$, there exists a number $p > 0$ with $u(K)\Psi(p) \leqslant 1$.

Let $f \in \mathcal{O}^\Psi$. Define the function h on T by $h(t) = p$ for $t \in K$ and $h(t) = 0$ for $t \notin K$.

The function h is measurable and $\int \Psi(|h(t)|)\mathrm{d}\mu(t) = \Psi(p)\mu(K) \leqslant 1$, consequently $\int^* |\boldsymbol{x}|\,|h|\,\mathrm{d}\mu = 0$. But

$$\int^* |\boldsymbol{x}|\,|h|\,\mathrm{d}\mu = p \int_K^* |\boldsymbol{x}|\,\mathrm{d}\mu$$

therefore $\boldsymbol{x}\varphi_K$ is a negligible field. As K is an arbitrary compact set, it follows that $\boldsymbol{x}$ is negligible.

4 *The* $\mathscr{L}^\Phi_\mathscr{A}$ *spaces*

We shall denote by $\mathscr{L}^\Phi_\mathscr{A}$ the set of all *measurable* vector fields $\boldsymbol{x} \in \mathscr{C}(\mathscr{E})$ with $\|\boldsymbol{x}\|_\Phi < +\infty$, and by $L^\Phi_\mathscr{A}$ the quotient space of $\mathscr{L}^\Phi_\mathscr{A}$ by the subspace of the negligible fields.

29.18 PROPOSITION. *The set* $\mathscr{L}^\Phi_\mathscr{A}$ *is a vector space and* $\|\boldsymbol{x}\|_\Phi$ *is a seminorm on this space.*

The assertion follows immediately from the properties of the function $\|\boldsymbol{x}\|_\Phi$.

The spaces $\mathscr{L}^\Phi_\mathscr{A}$ are called *Orlicz spaces*, since Orlicz has introduced these spaces, for numerical functions, defined on an interval.

29.19 PROPOSITION. *We have* $\mathcal{O}^\Phi_\mathscr{A} \subset \mathscr{L}^\Phi_\mathscr{A}$ *and*

$$\|\boldsymbol{x}\|_\Phi \leqslant |\boldsymbol{x}|_\Phi + 1, \text{ for every } \boldsymbol{x} \in \mathcal{O}^\Phi_\mathscr{A}.$$

For $\boldsymbol{x} \in \mathscr{C}(\mathscr{E})$ and $f: T \to R$, the Young inequality is written

$$|\boldsymbol{x}(t)|\,|f(t)| \leqslant \Phi(|\boldsymbol{x}(t)|) + \Psi(|f(t)|) \;;$$

and if $\boldsymbol{x} \in \mathcal{O}^{\Phi}_{\mathscr{A}}$ and $f \in \mathcal{O}^{\Psi}$, then

$$\int^* |\boldsymbol{x}(t)|\,|f(t)|\,\mathrm{d}\mu(t) \leqslant |\boldsymbol{x}|_{\Phi} + |f|_{\Psi} \,.$$

Taking in the left-hand side the supremum for $|f|_{\Psi} \leqslant 1$, we obtain

$$\|\boldsymbol{x}\|_{\Phi} \leqslant |\boldsymbol{x}|_{\Phi} + 1 \,,$$

therefore $\boldsymbol{y} \in \mathscr{L}^{\Phi}_{\mathscr{A}}$, consequently $\mathcal{O}^{\Phi}_{\mathscr{A}} \subset \mathscr{L}^{\Phi}_{\mathscr{A}}$.

Remark. In general, we have $\mathcal{O}^{\Phi}_{\mathscr{A}} \neq \mathscr{L}^{\Phi}_{\mathscr{A}}$. Nevertheless, from every field $\boldsymbol{x} \in \mathscr{L}^{\Phi}_{\mathscr{A}}$ we obtain a field of $\mathcal{O}^{\Phi}_{\mathscr{A}}$, by multiplication by a convenient constant, as theorem 29.21 will show. We shall prove first the following lemma:

29.20 LEMMA. *If $q \geqslant 1$ and $v \geqslant 0$ then $q\Psi(v) \leqslant \Psi(qv)$ and $q\Phi(v) \leqslant \Phi(qv)$.*

In fact

$$\Psi(qv) - \Psi(v) = \int_v^{qv} \psi(s)\,\mathrm{d}s \geqslant (q-1)v\,\psi(v) = (q-1)\int_0^v \psi(s)\,\mathrm{d}s \geqslant$$
$$\geqslant (q-1)\int_0^v \psi(s)\,\mathrm{d}s = (q-1)\psi(v) \,,$$

whence it follows that $q\Psi(v) \leqslant \Psi(qv)$. In the same way one can prove the inequality concerning Φ.

29.21 THEOREM. *If $\boldsymbol{x} \in \mathscr{L}^{\Phi}_{\mathscr{A}}$ and if $\|\boldsymbol{x}\|_{\Phi} \neq 0$, then*

$$\big|\,|\boldsymbol{x}|/\|\boldsymbol{x}\|_{\Phi}\big|_{\Phi} \leqslant 1 \,.$$

Let $\boldsymbol{x} \in \mathscr{L}^{\Phi}_{\mathscr{A}}$ with $\|\boldsymbol{x}\| \neq 0$ and $h \in \mathcal{O}^{\Psi}$. If $|h|_{\Psi} \leqslant 1$, we have

$$\int |\boldsymbol{x}|\,|h|\,\mathrm{d}\mu \leqslant \|\boldsymbol{x}\|_{\Phi} \,.$$

If $|h|_{\Psi} > 1$, we have, by lemma 29.20,

$$\Psi(|h(t)|\,|h|_{\Psi}^{-1}) \leqslant |h|_{\Psi}^{-1}\,\Psi(|h(t)|)$$

consequently $|h|h|^{-1}| < 1$. In this case, it follows

$$\int |\boldsymbol{x}|\,|h|\,|h|_{\Psi}^{-1}\,\mathrm{d}\mu \leqslant \|\boldsymbol{x}\|_{\Phi}$$

that is

$$\int |\boldsymbol{x}|\,|h|\,\mathrm{d}\mu \leqslant \|\boldsymbol{x}\|_\Phi\,|h|_\Psi\,.$$

If we put $\rho(h)=\max(1, |h|_\Psi)$, we have, for every $h\in \mathcal{O}^\Psi$,

$$\int |\boldsymbol{x}|\,|h|\,\mathrm{d}\mu \leqslant \|\boldsymbol{x}\|_\Phi\,\rho(h)\,.$$

Assume first that $\boldsymbol{x}$ is bounded and vanishes outside a compact set $K\subset T$.

The bounded functions $\Phi(|\boldsymbol{x}(t)|/\|\boldsymbol{x}\|_\Phi)$ and $\Psi(\varphi|\boldsymbol{x}(t)|/\|\boldsymbol{x}\|_\Phi))$ are integrable.

Consider the measurable function h defined on T by the equality $h(t)=\varphi(|\boldsymbol{x}(t)|/\|\boldsymbol{x}\|_\Phi)$. Taking in the Young inequality $u=|\boldsymbol{x}(t)|/\|\boldsymbol{x}\|_\Phi$ and $v=h(t)$, and remarking that in this case the Young inequality is satisfied with equality, we obtain

$$\rho(h)\geqslant \int |\boldsymbol{x}|/\|\boldsymbol{x}\|_\Phi\, h\,\mathrm{d}\mu = \int \Phi(|\boldsymbol{x}(t)|/\|\boldsymbol{x}\|_\Phi)\,\mathrm{d}\mu + \int \Psi(h(t))\,\mathrm{d}\mu = \\ = |\,|\boldsymbol{x}|/\|\boldsymbol{x}\|_\Phi|_\Phi + |h|_\Psi$$

consequently

$$|\boldsymbol{x}/\|\boldsymbol{x}\|_\Phi|_\Phi \leqslant \rho(h)-|h|_\Psi\,.$$

If $|h|_\Psi\geqslant 1$, then $\rho(h)=|h|_\Psi$, therefore $|\boldsymbol{x}/\|\boldsymbol{x}\|_\Phi|_\Phi=0$. If $|h|_\Psi<1$, then $\rho(h)=1$, hence $|\boldsymbol{x}/\|\boldsymbol{x}\|_\Phi|_\Phi\leqslant 1-|h|_\Psi\leqslant 1$.

If $\boldsymbol{x}$ is unbounded (but vanishes outside a compact set $K\subset T$), for each n we define the field $\boldsymbol{x}_n$ by $\boldsymbol{x}_n(t)=\boldsymbol{x}(t)$ if $|\boldsymbol{x}(t)|\leqslant n$, and $\boldsymbol{x}_n(t)=n\boldsymbol{x}(t)/|\boldsymbol{x}(t)|$ if $|\boldsymbol{x}(t)|>n$.

The sequence $(|\boldsymbol{x}_n|)$ is increasing and converges to $|\boldsymbol{x}|$ and $\|\boldsymbol{x}_n\|_\Phi\leqslant\|\boldsymbol{x}\|_\Phi$, consequently, if $\|\boldsymbol{x}_n\|_\Phi\neq 0$, then

$$|\boldsymbol{x}_n|/\|\boldsymbol{x}\|_\Phi\leqslant|\boldsymbol{x}_n|/\|\boldsymbol{x}_n\|_\Phi\,,$$

whence

$$|\boldsymbol{x}_n/\|\boldsymbol{x}\|_\Phi|_\Phi\leqslant|\boldsymbol{x}_n/\|\boldsymbol{x}_n\|_\Phi|_\Phi\leqslant 1\,.$$

Since the function Φ is continuous and increasing, we deduce

$$\Phi(|\boldsymbol{x}(t)|/\|\boldsymbol{x}\|_\Phi)=\lim\Phi(|\boldsymbol{x}_n(t)|/\|\boldsymbol{x}\|_\Phi)=\sup\Phi(|\boldsymbol{x}_n(t)|/\|\boldsymbol{x}\|\ )$$

for each $t\in T$, whence, the sequence $\Phi(|\boldsymbol{x}_n(t)|/\|\boldsymbol{x}\|_\Phi)$ being increasing for each $t\in T$, we deduce

$$\int\Phi(|\boldsymbol{x}(t)|/\|\boldsymbol{x}\|_\Phi)\,\mathrm{d}\mu=\sup\int\Phi(|\boldsymbol{x}_n(t)|/\|\boldsymbol{x}\|_\Phi)\,\mathrm{d}\mu$$

consequently,

$$|\boldsymbol{x}/||\boldsymbol{x}||_\Phi|_\Phi \leqslant 1\,.$$

Assume now that $\boldsymbol{x}$ is arbitrary, with $||\boldsymbol{x}||_\Phi \neq 0$. For each compact set $K \subset T$ we have

$$|\boldsymbol{x}\varphi_K/||\boldsymbol{x}||_\Phi|_\Phi \leqslant 1\,.$$

In fact, if $||\boldsymbol{x}\varphi_K||_\Phi = 0$, $\boldsymbol{x}\varphi_K$ is negligible and the inequality is evident. If $||\boldsymbol{x}\varphi_K||_\Phi \neq 0$, the desired inequality follows from the inequality

$$|\boldsymbol{x}\varphi_K|/||\boldsymbol{x}\varphi_K||_\Phi \leqslant |\boldsymbol{x}\varphi_K|/||\boldsymbol{x}||_\Phi\,.$$

As we have

$$|\boldsymbol{x}/||\boldsymbol{x}||_\Phi|_\Phi = \sup_K |\boldsymbol{x}\varphi_K/||\boldsymbol{x}||_\Phi|_\Phi,\ K \text{ compact}\,,$$

we deduce that $|\boldsymbol{x}/||\boldsymbol{x}||_\Phi|_\Phi \leqslant 1$ and the theorem is proved.

Remark. The theorem remains valid also for $\mathscr{L}^\Psi_{\mathscr{A}}$ but in case $\lim_{t\to\infty}\varphi(t) = l < +\infty$, we have $\Psi(v) = +\infty$ for $v > l$, and so the first part of the theorem needs a different proof.

We remark that if $\boldsymbol{x} \in \mathscr{L}^\Psi_{\mathscr{A}}$, then $|\boldsymbol{x}(t)|/||\boldsymbol{x}||_\Psi \leqslant l$ almost everywhere. In fact, if there existed an integrable set $A \subset T$ with $\mu(A) > 0$, such that $|\boldsymbol{x}(t)| > l||\boldsymbol{x}||_\Psi$ for every $t \in A$, then setting $h = l^{-1}\mu(A)^{-1}\varphi_A$ and remarking that $\Phi(u) \leqslant lu$, we would have

$$|h|_\Phi = \int \Phi(h(t))\,\mathrm{d}\mu(t) = \Phi(1/l\mu(A))\mu(A) \leqslant 1$$

consequently

$$\int |\boldsymbol{x}|\,h\,\mathrm{d}\mu = l^{-1}\mu(A)^{-1}\int |\boldsymbol{x}|\varphi_A\,\mathrm{d}\mu \geqslant ||\boldsymbol{x}||_\Psi$$

which is impossible. Thus, we have $|\boldsymbol{x}(t)|/||\boldsymbol{x}||_\Psi \leqslant l$ almost everywhere.

Assume first that $\boldsymbol{x} \in \mathscr{L}^\Psi_{\mathscr{A}}$ vanishes outside an integrable set $A \subset T$ and that $||\boldsymbol{x}||_\Psi \neq 0$ and choose $0 < \delta < 1$. Then $\Psi(\delta|\boldsymbol{x}(t)|/||\boldsymbol{x}||_\Psi)$ and $\Phi(\psi(|\boldsymbol{x}(t)|/||\boldsymbol{x}||_\Psi))$ are bounded and vanish outside the set A, therefore they are integrable. Put $h(t) = \psi(\delta|\boldsymbol{x}(t)|/||\boldsymbol{x}||_\Psi)\varphi_A(t)$. One can show as at the beginning of theorem 29.21 that

$$\int |\boldsymbol{x}|\,h\,\mathrm{d}\mu < ||\boldsymbol{x}||_\Psi\,\pi(h)\,,$$

where $\pi(h) = \max(1, |h|_\Phi)$ hence, we can write

$$\delta\pi(h) \geqslant \int \delta|\boldsymbol{x}(t)|/||\boldsymbol{x}||_\Psi\,h(t)\,\mathrm{d}\mu(t) = \int \Psi(\delta|\boldsymbol{x}(t)|/||\boldsymbol{x}||_\Psi)\,\mathrm{d}\mu(t) + |h|_\Phi$$

since the Young inequality becomes an equality. Then

$|h|_\Phi \leqslant \delta\pi(h) < \pi(h)$ therefore $\pi(h)=1$ and $\delta|x|/||x||_\Psi \leqslant \delta < 1$.

Taking $\delta \to 1$ and remarking that Ψ is left continuous at the point $v=l$, we obtain $|x/||x||_\Psi|_\Psi \leqslant 1$. For the case when x is arbitrary, we remark that

$$|x/||x||_\Psi|_\Psi = \sup_K |x\varphi_K/||x||_\Psi|_\Psi, \; K \text{ compact},$$

and we reason as in the proof of the theorem.

29.22 PROPOSITION. *If $x \in \mathscr{L}^\Phi_\mathscr{A}$, then x vanishes almost everywhere on the complement of the union of a sequence of compact sets.*

If $||x||_\Phi = 0$, x is negligible. Assume, hence, that $||x||_\Phi \neq 0$. Then

$$\int \Phi(|x(t)|/||x||_\Phi)\,\mathrm{d}\mu(t) \leqslant 1 < +\infty.$$

It follows that $\Phi(|x(t)|/||x||_\Phi)=0$ therefore $|x(t)|=0$ on the complement of the union of a sequence of compact sets and of a negligible set.

Remark. If $\Psi(v)>0$ for $v>0$, the proposition remains true also for $x \in \mathscr{L}^\Psi_\mathscr{A}$.

But if $\Psi(v_0)=0$ for some $v_0>0$, we may have $||x||_\Psi < +\infty$ for a field x not vanishing in any point, for example if $|x(t)| < v_0$ for every $t \in T$.

29.23 THEOREM. *The space $\mathscr{L}^\Phi_\mathscr{A}$ is complete.*

Let (x_n) be a Cauchy sequence of fields of $\mathscr{L}^\Phi_\mathscr{A}$. For every $\varepsilon>0$ there exists a number $N(\varepsilon)$ such that

$$\int |x_n(t) - x_m(t)|\,|h(t)|\,\mathrm{d}\mu(t) < \varepsilon$$

for every $n, m \geqslant N(\varepsilon)$ and $h \in \mathcal{O}^\Psi$ with $|h|_\Psi \leqslant 1$.

Let $K \subset T$ be a compact set. There exists a number $p>0$ such that $\mu(K)\Psi(p) \leqslant 1$. If we take $h = p\varphi_K$ we have $|h|_\Psi \leqslant 1$, therefore

$$\int |x_n(t) - x_m(t)|\,|h(t)|\,\mathrm{d}\mu(t) = p\int_K |x_n(t) - x_m(t)|\,\mathrm{d}\mu(t) < \varepsilon$$

for $n, m \geqslant N(\varepsilon)$, consequently

$$\lim_{n,m} \int |x_n\varphi_K - x_m\varphi_K|\,\mathrm{d}\mu = 0 .$$

Thus, $(x_n\varphi_K)$ is a Cauchy sequence in the Banach space $\mathscr{L}^1_\mathscr{A}(K)$ of integrable fields vanishing outside K. It follows that there exists a field $x_K \in \mathscr{L}^1_\mathscr{A}(K)$, uniquely determined almost everywhere on K, to which the sequence $(x_n\varphi_K)$ tends in the topology of the space $\mathscr{L}^1_\mathscr{A}(K)$. There exists also a subsequence (x_{n_p}) of the sequence (x_n), convergent to x_K, almost everywhere on K. If in the inequality

$$\int_K |x_n(t) - x_m(t)|\,|h(t)|\,d\mu(t) < \varepsilon ,$$

valid for $n, m > N(\varepsilon)$ and $|h|_\Psi \leqslant 1$, m runs over the sequence (n_p) of indices we obtain

$$\int |x_n \varphi_K - x_K|\,|h|\,d\mu < \varepsilon, \text{ for } n \geqslant N(\varepsilon) \text{ and } |h|_\Psi \leqslant 1,$$

whence $||x_n \varphi_K - x_K||_\Phi \leqslant \varepsilon$ for $n \geqslant N(\varepsilon)$.

It follows first that $x_n \varphi_K - x_K \in \mathscr{L}^\Phi_{\mathscr{A}}$, therefore $x_K \in \mathscr{L}^\Phi_{\mathscr{A}}$, since $x_n \varphi_K \in \mathscr{L}^\Phi_{\mathscr{A}}$. We deduce then that $||x_n \varphi_K - x_K||_\Phi \to 0$. We remark that for two compact sets K_1 and K_2 which are put into correspondence, as above, with the fields x_{K_1} and x_{K_2}, we have $x_{K_1}(t) = x_{K_2}(t)$, almost everywhere on $K_1 \cap K_2$. In fact, denoting $K = K_1 \cap K_2$, we have $x_n \varphi_{K_1} \to x_{K_1}$ therefore $x_n \varphi_K \to x_{K_1} \varphi_K$ in the topology of the space $\mathscr{L}^1_{\mathscr{A}}(K_1)$; on the other hand, $x_n \varphi_K \to x_K$ in the topology of the space $\mathscr{L}^1_{\mathscr{A}}(K) \subset \mathscr{L}^1_{\mathscr{A}}(K_1)$. It follows that $x_{K_1} \varphi_K = x_K$ almost everywhere. In the same way one can show that $x_{K_2} \varphi_K = x_K$ almost everywhere. It follows then that $x_{K_1} \varphi_K = x_{K_2} \varphi_K$ almost everywhere.

By proposition 11.23 there exists a measurable field $x \in \mathscr{C}(\mathscr{E})$ such that for each compact set $K \subset T$, we have $x(t) = x_K(t)$, almost everywhere on K. Thus $||x_n \varphi_K - x \varphi_K||_\Phi \to 0$ for every compact set $K \subset T$. As the sequence $(||x_n||_\Phi)$ of real numbers is a Cauchy sequence, it is bounded by a number M. For each compact set $K \subset T$ we have $||x \varphi_K||_\Phi = \lim ||x_n \varphi_K||_\Phi \leqslant M$. It follows that

$$||x||_\Phi = \sup_K ||x \varphi_K||_\Phi \leqslant M < +\infty ,$$

therefore $x \in \mathscr{L}^\Phi_{\mathscr{A}}$. On the other hand, if in the inequality

$$||x_n \varphi_K - x_m \varphi_K||_\Phi \leqslant \varepsilon$$

valid for $n, m \leqslant N(\varepsilon)$ and every compact set $K \subset T$ we take $m \to \infty$, we deduce $||x_n \varphi_K - x \varphi_K||_\Phi \leqslant \varepsilon$ for $n \geqslant N(\varepsilon)$ and every compact set $K \subset T$, hence

$$||x_n - x||_\Phi = \sup_K ||x_n \varphi_K - x \varphi_K||_\Phi \leqslant \varepsilon \text{ for } n \geqslant N(\varepsilon) .$$

It follows that $||x_n - x||_\Phi \to 0$, thus the theorem is completely proved.

29.24 PROPOSITION. *A measurable field $x \in \mathscr{C}(\mathscr{E})$ belongs to the space $\mathscr{L}^\Phi_{\mathscr{A}}$, if and only if $xh \in \mathscr{L}^1_{\mathscr{A}}$, for every function $h \in \mathscr{L}^\Psi_{\mathscr{A}}$. We have*

$$\int |x|\,|h|\,d\mu \leqslant ||x||_\Phi ||h||_\Psi .$$

Assume first that $x \in \mathscr{L}^{\Phi}_{\mathscr{A}}$. Let $h \in \mathscr{L}^{\Psi}$. If $\|x\|_{\Phi}=0$, the field x is negligible, therefore the field xh is negligible, consequently $xh \in \mathscr{L}^1_{\mathscr{A}}$ and the stated inequality is satisfied. If $\|x\|_{\Psi} \neq 0$, we have $|x/\|x\|_{\Phi}|_{\Phi} \leqslant 1$.

Since

$$\|h\|_{\Psi} = \sup_{|f|_{\Phi} \leqslant 1} \int^* |f|\,|h|\,\mathrm{d}\mu < +\infty\,,$$

taking, in particular, $f=|x|/\|x\|_{\Phi}$, we obtain

$$\int^* |x|/\|x\|_{\Phi}\,|h|\,\mathrm{d}\mu \leqslant \|h\|_{\Psi}\,,$$

whence

$$\int^* |x|\,|h|\,\mathrm{d}\mu \leqslant \|x\|_{\Phi}\|h\|_{\Psi} < +\infty\,.$$

It follows that $xh \in \mathscr{L}^1_{\mathscr{A}}$.

Conversely, assume that x is measurable and $xh \in \mathscr{L}^1_{\mathscr{A}}$ for every $h \in \mathscr{L}^{\Psi}$. In order to prove that $x \in \mathscr{L}^{\Phi}_{\mathscr{A}}$ it is sufficient to prove that $\|x\|_{\Phi} < \infty$.

Assume the contrary, that $\|x\|_{\Phi} = +\infty$. Since

$$\|x\|_{\Phi} = \sup_{|h|_{\Phi} \leqslant 1} \int^* |x|\,|h|\,\mathrm{d}\mu\,,$$

there exists a sequence (h_n) of measurable positive functions, with $|h_n|_{\Psi} \leqslant 1$, such that, for each n, we have

$$\int |x|\,h_n\,\mathrm{d}\mu > n^3\,.$$

Put $h=\Sigma_{1 \leqslant n<\infty} n^{-2}h_n$. Since $\|h_n\|_{\Psi} \leqslant |h_n|_{\Psi}+1 \leqslant 2$ for each n, we deduce

$$\|h\|_{\Psi} \leqslant \Sigma_{1 \leqslant n<\infty} n^{-2}\,\|h_n\|_{\Psi} < +\infty\,,$$

therefore $h \in \mathscr{L}^{\Psi}$. Since $h \geqslant n^{-2}h_n$ for each n, we have

$$\int^* |x|\,h\,\mathrm{d}\mu \geqslant n^{-2}\int^* |x|\,h\,\mathrm{d}\mu > n$$

for each n, consequently

$$\int^* |x|\,h\,\mathrm{d}\mu = +\infty\,,$$

that is xh is not integrable, and thus we obtain a contradiction.

It follows therefore that $\|x\|_{\Phi} < +\infty$, consequently $x \in \mathscr{L}^{\Phi}_{\mathscr{A}}$ and the proposition is proved.

29.25 PROPOSITION. *Assume that there exists a fundamental family $\mathscr{A}' \subset \mathscr{C}(\mathscr{E}')$ such that the scalar function $t \to \langle x(t), x'(t)\rangle$ is continuous for every $x \in \mathscr{A}$ and $x' \in \mathscr{A}'$. A measurable field $x \in \mathscr{C}(\mathscr{E})$ belongs to the space*

$\mathscr{L}^{\Phi}_{\mathscr{A}}$, *if and only if the scalar function* $t \to \langle x(t), x'(t) \rangle$ *is summable for every* $x' \in \mathscr{L}^{\Psi}_{\mathscr{A}'}$. *We have*

$$\int |\langle x(t), x'(t) \rangle| \, d\mu(t) \leqslant \int |x(t)| \, |x'(t)| \, d\mu(t) \leqslant \|x\|_{\Phi} \|x'\|_{\Psi} .$$

Let, first, $x \in \mathscr{L}^{\Phi}_{\mathscr{A}}$ and $x' \in \mathscr{L}^{\Psi}_{\mathscr{A}'}$. If $\|x\|_{\Phi} = 0$, x is negligible, therefore $\langle x(t), x'(t) \rangle$ is negligible for every $x' \in \mathscr{L}^{\Psi}_{\mathscr{A}'}$, and the stated inequality is satisfied. If $\|x\|_{\Phi} \neq 0$ we have

$$\int^* |\langle x(t), x'(t) \rangle| \, d\mu(t) \leqslant \int^* |x(t)| \, |x'(t)| \, d\mu(t) =$$
$$= \|x\|_{\Phi} \int^* |x(t)| \, \|x\|_{\Phi}^{-1} |x'(t)| \, d\mu(t) \leqslant \|x\|_{\Phi} \|x'\|_{\Psi} < +\infty,$$

since $|x \|x\|_{\Phi}^{-1}|_{\Phi} \leqslant 1$. It follows that $\langle x(t), x'(t) \rangle$ is integrable.

Conversely, assume that x is measurable and that the function $\langle x(t), x'(t) \rangle$ is integrable for every $x' \in \mathscr{L}^{\Psi}_{\mathscr{A}'}$. We must show that $\|x\|_{\Phi} < +\infty$.

Assume, by absurd, that $\|x\|_{\Phi} = +\infty$. By proposition 29.24, there exists a positive function $h \in \mathscr{L}^{\Psi}$ such that

$$\int^* |x| \, h \, d\mu = +\infty .$$

For every $0 < a < 1$, there exists a measurable functional field $x' \in \mathscr{C}(\mathscr{E}')$ such that

$$a|x(t)| \leqslant |\langle x(t), x'(t) \rangle \quad \text{and} \quad a \leqslant |x'(t)| \leqslant 2 - a ,$$

almost everywhere. Put $y' = x(2-a)^{-1} h$. Then $|y'(t)| = |x'(t)|(2-a)^{-1} h(t) \leqslant h(t)$, therefore $\|y'\|_{\Psi} \leqslant \|h\|_{\Psi} < +\infty$ that is $y' \in \mathscr{L}^{\Psi}_{\mathscr{A}'}$. But

$$|\langle x(t), y'(t) \rangle| = h(t)(2-a)^{-1} |\langle x(t), x'(t) \rangle| \geqslant a(2-a)^{-1} |x(t)| \, h(t),$$

consequently,

$$\int^* |\langle x(t), y'(t) \rangle| \, d\mu(t) \geqslant a(2-a)^{-1} \int^* |x(t)| \, h(t) \, d\mu(t) = +\infty$$

and we obtained a contradiction. It follows hence that $x \in \mathscr{L}^{\Phi}_{\mathscr{A}}$ and the proposition is proved.

In certain cases we have $\mathscr{L}^{\Phi}_{\mathscr{A}} = \mathscr{O}^{\Phi}_{\mathscr{A}}$.

29.26 PROPOSITION. *If there exists a number M such that* $\Phi(2u) \leqslant M\Phi(u)$ *for* $u \geqslant 0$, *then the spaces* $\mathscr{O}^{\Phi}_{\mathscr{A}}$ *and* $\mathscr{L}^{\Phi}_{\mathscr{A}}$ *contain the same elements.*

Since $\mathscr{O}^{\Phi}_{\mathscr{A}} \subset \mathscr{L}^{\Phi}_{\mathscr{A}}$ it is sufficient, in this case, to prove the converse inclusion. Let $x \in \mathscr{L}^{\Phi}_{\mathscr{A}}$. If $\|x\|_{\Phi} = 0$, then x is negligible, therefore $|x|_{\Phi} = 0$, hence $x \in \mathscr{O}^{\Phi}_{\mathscr{A}}$. If $\|x\|_{\Phi} \neq 0$ then (theorem 29.21) we have $|x / \|x\|_{\Phi}|_{\Phi} \leqslant 1$, consequently

$x/||x||_{\Phi} \in \mathcal{O}^{\Phi}_{\mathcal{A}}$. Let $p > 0$ be such that $||x||_{\Phi} \leqslant 2^p$. Then

$$\Phi(|x(t)| \leqslant \Phi(2^p|x(t)|\,||x||_{\Phi}^{-1}) \leqslant M^p \Phi(|x(t)|\,||x||_{\Phi})$$

for each $t \in T$, therefore $|x|_{\Phi} \leqslant M^p |x/||x||_{\Phi}|_{\Phi} \leqslant M^p < +\infty$, consequently $x \in \mathcal{O}^{\Phi}_{\mathcal{A}}$. Thus the proposition is proved.

29.27 Proposition. *If there exists a number M and a number $u_0 > 0$ such that $\Phi(2u) \leqslant M\Phi(u)$ for $u \geqslant u_0$, and if the measure μ is bounded, then $\mathcal{L}^{\Phi}_{\mathcal{A}} = \mathcal{O}^{\Phi}_{\mathcal{A}}$.*

Let $x \in \mathcal{L}^{\Phi}_{\mathcal{A}}$. Define the fields x_1 and x_2 by $x_1(t) = x(t)$ if $|x(t)| < ||x||_{\Phi} u_0$, and $x_1(t) = 0$ if $|x(t)| \geqslant ||x||_{\Phi} u_0$, and $x_2 = x - x_1$. The field x_1 is bounded and measurable, therefore the function $\Phi(|x_1(t)|)$ is bounded and measurable. Since μ is bounded, the function $\Phi(|x_1(t)|)$ is integrable.

The field x_2 is in general unbounded, but $x_2/||x_2||_{\Phi} \in \mathcal{O}^{\Phi}_{\mathcal{A}}$. If $p > 0$ is such that $||x_2||_{\Phi} \leqslant 2^p$, then

$$\Phi(|x_2(t)|) \leqslant \Phi(2^p|x_2(t)|\,||x_2||_{\Phi}^{-1}) \leqslant M^p \Phi(|x_2(t)|\,||x_2||_{\Phi}^{-1}) < +\infty$$

therefore $\Phi(|x_2(t)|)$ is integrable. Since x_1 and x_2 vanish on complementary sets, we have

$$\Phi(|x(t)|) = \Phi(|x_1(t)|) + \Phi(|x_2(t)|)$$

consequently $\Phi(|x(t)|$ is integrable, that is $x \in \mathcal{O}^{\Phi}_{\mathcal{A}}$ and the proposition is proved.

29.28 Proposition. *If there exists a number M such that $\Phi(2u) \leqslant M\Phi(u)$ for $u \geqslant 0$ then $|x|_{\Phi} \leqslant 1/M^p$ implies $||x||_{\Phi} \leqslant 2/2^p$. In particular if $|x_n - x|_{\Phi} \to 0$ then $||x_n - x||_{\Phi} \to 0$.*

We have

$$\int \Phi(2^p|x(t)|)\,\mathrm{d}\mu(t) \leqslant M^p \int \Phi(|x(t)|)\,\mathrm{d}\mu(t) = M^p|x|_{\Phi} \leqslant 1.$$

For every $h \in \mathcal{O}^{\Psi}$ with $|h|_{\Psi} \leqslant 1$, we have, by Young's inequality,

$$\int 2^p|x(t)|\,|h(t)|\,\mathrm{d}\mu(t) \leqslant \int \Phi(2^p|x(t)|)\,\mathrm{d}\mu(t) + \int \Psi(|h(t)|)\,\mathrm{d}\mu(t) \leqslant 2$$

consequently $||x||_{\Phi} \leqslant 2/2p$

Remark. Denote $L^{\Phi}_{\mathcal{A}} = \mathcal{L}^{\Phi}_{\mathcal{A}}/\mathcal{N}_{\mathcal{A}}$, where $\mathcal{N}_{\mathcal{A}}$ is the set of negligible fields. If $\Phi(2u) \leqslant M\Phi(u)$ for $u \geqslant 0$, the normed space $L^{\Phi}_{\mathcal{A}}$ contains the same elements as $O^{\Phi}_{\mathscr{A}} = \mathcal{O}^{\Phi}_{\mathscr{A}}/\mathscr{N}_{\mathscr{A}}$. By proposition 29.28, the topology of $O^{\Phi}_{\mathscr{A}}$ is finer than the topology of $L^{\Phi}_{\mathscr{A}}$. As both spaces are metrisable and complete, the two topologies are equivalent. It follows that the set of the bounded fields $\mathscr{L}^{\Phi}_{\mathscr{A}}$ is dense in $\mathscr{L}^{\Phi}_{\mathscr{A}}$, since this set is dense in $\mathcal{O}^{\Phi}_{\mathscr{A}}$.

29.29 PROPOSITION. *Assume that* $\lim_{t\to\infty} \varphi(t) = l < +\infty$.
If $x \in \mathcal{O}^{\Phi}_{\mathscr{A}}$ *then* $|x|_{\Phi} \leqslant lN_1(x)$, *and if* $x \in \mathcal{O}^{\Psi}_{\mathscr{A}}$ *then* $N_{\infty}(x) \leqslant l$.
If $x \in \mathscr{L}^{\Phi}_{\mathscr{A}}$ *then* $\|x\|_{\Phi} \leqslant lN_1(x)$, *and if* $x \in \mathscr{L}^{\infty}_{\mathscr{A}}$ *then* $N_{\infty}(x) \leqslant l\|x\|_{\Psi}$.

In fact, if $\lim_{t\to\infty} \varphi(t) = l < +\infty$, then $\Phi(u) \leqslant lu$ for $u \geqslant 0$, and $\Psi(v) = +\infty$ for $v > l$.

It follows that if $x \in \mathcal{O}^{\Phi}_{\mathscr{A}}$, then

$$|x|_{\Phi} = \int \Phi(|x(t)|)\,\mathrm{d}\mu(t) \leqslant l \int^* |x(t)|\,\mathrm{d}\mu(t) = lN_1(x)$$

and if $x \in \mathcal{O}^{\Psi}_{\mathscr{A}}$ then $\int^* \Psi(|x(t)|)\mathrm{d}\mu(t) < +\infty$, therefore $\Psi(|x(t)|) < +\infty$, almost everywhere, consequently $|x(t)| \leqslant l$ almost everywhere, whence $N_{\infty}(x) \leqslant l$.

Now let $x \in \mathscr{L}^{\Phi}_{\mathscr{A}}$. We have

$$\|x\|_{\Phi} = \sup_{|h|_{\Psi} \leqslant 1} \int^* |x|\,|h|\,\mathrm{d}\mu \leqslant \sup_{N_{\infty}(h/l) \leqslant 1} l \int^* |x|\,|h|\,l^{-1}\,\mathrm{d}\mu \leqslant lN_1(x)$$

since $|h|_{\Psi} \leqslant 1$ implies $N_{\infty}(h) \leqslant l$, therefore $N_{\infty}(h/l) \leqslant 1$.

If $x \in \mathscr{L}^{\Psi}_{\mathscr{A}}$, we have

$$N_{\infty}(x) = \sup_{N_1(h) \leqslant 1} \int^* |x|\,|h|\,\mathrm{d}\mu \leqslant \sup_{|x/l|_{\Phi} \leqslant 1} l \int^* |x|\,l^{-1}\,|h|\,\mathrm{d}\mu \leqslant l\|h\|_{\Psi}$$

since $N_1(h) \leqslant 1$ implies $|x|_{\Phi} \leqslant l$, therefore $|x/l|_{\Phi} \leqslant 1$.

29.30 PROPOSITION. *If* $\lim_{t\to\infty} \varphi(t) = l < +\infty$, *then* $\mathscr{L}^{1}_{\mathscr{A}} \subset \mathscr{L}^{\Phi}_{\mathscr{A}}$ *and* $\mathscr{L}^{\Psi}_{\mathscr{A}} \subset \mathscr{L}^{\infty}_{\mathscr{A}}$. *The topology of* $\mathscr{L}^{1}_{\mathscr{A}}$ *is finer than the topology induced on* $\mathscr{L}^{1}_{\mathscr{A}}$ *by the topology of* $\mathscr{L}^{\Phi}_{\mathscr{A}}$, *and the topology of* $\mathscr{L}^{\Psi}_{\mathscr{A}}$ *is finer than the topology induced on* $\mathscr{L}^{\Psi}_{\mathscr{A}}$ *by the topology of* $\mathscr{L}^{\infty}_{\mathscr{A}}$.

29.31 PROPOSITION. *Assume that* $\lim_{t\to\infty} \varphi(t) = l < +\infty$. *If the measure* μ *is bounded, or if* $\lim_{t\to 0} \varphi(t) = \delta > 0$, *then* $\mathscr{L}^{1}_{\mathscr{A}} = \mathscr{L}^{\Phi}_{\mathscr{A}}$ *and* $\mathscr{L}^{\Psi}_{\mathscr{A}} = \mathscr{L}^{\infty}_{\mathscr{A}}$.

Assume first that μ is bounded and let $x \in \mathscr{L}^{\infty}_{\mathscr{A}}$. We can find a number $p > 0$ such that $p|x(t)| \leqslant a < 1$ almost everywhere, therefore $\int \Psi(p|x(t)|)\mathrm{d}\mu(t) \leqslant \Psi(a)\|\mu\| < +\infty$, consequently $px \in \mathcal{O}^{\Psi}_{\mathscr{A}} \subset \mathscr{L}^{\Psi}_{\mathscr{A}}$ whence $x \in \mathscr{L}^{\Psi}_{\mathscr{A}}$. Thus $\mathscr{L}^{\infty}_{\mathscr{A}} \subset \mathscr{L}^{\Psi}_{\mathscr{A}}$, therefore by proposition 29.29, $\mathscr{L}^{\infty}_{\mathscr{A}} = \mathscr{L}^{\Psi}_{\mathscr{A}}$.

If $x \in \mathscr{L}^{\Phi}_{\mathscr{A}}$, the function $|x|\,|h|$ is integrable, for every $h \in \mathscr{L}^{\Psi}$; if we take $h(t) \equiv 1$, we have $h \in \mathscr{L}^{\infty} = \mathscr{L}^{\Psi}$, consequently $|x|$ is integrable, that is $x \in \mathscr{L}^1_{\mathscr{A}}$. Thus $\mathscr{L}^{\Phi}_{\mathscr{A}} \subset \mathscr{L}^1_{\mathscr{A}}$ whence $\mathscr{L}^{\Phi}_{\mathscr{A}} = \mathscr{L}^1_{\mathscr{A}}$.

Assume now that $\lim_{t \to 0} \varphi(t) = \delta > 0$. Then $\Psi(v) = 0$ for $v \leqslant \delta$. If $x \in \mathscr{L}^{\infty}_{\mathscr{A}}$, there exists a number $p > 0$ such that $p|x(t)| \leqslant \delta$ almost everywhere, therefore $|px|_{\Psi} = 0$, consequently $px \in \mathscr{O}^{\Psi}_{\mathscr{A}} \subset \mathscr{L}^{\Psi}_{\mathscr{A}}$, whence $x \in \mathscr{L}^{\Psi}_{\mathscr{A}}$. Thus $\mathscr{L}^{\infty}_{\mathscr{A}} \subset \mathscr{L}^{\Psi}_{\mathscr{A}}$, therefore $\mathscr{L}^{\infty}_{\mathscr{A}} = \mathscr{L}^{\Psi}_{\mathscr{A}}$. The equality $\mathscr{L}^{\Phi}_{\mathscr{A}} = \mathscr{L}^1_{\mathscr{A}}$ is proved as above.

29.32 PROPOSITION. *If there exists a number M such that $\Phi(2u) \leqslant M\Phi(u)$ for $u \geqslant 0$, then the set $\mathscr{E}_{\mathscr{A}}(\mathscr{B})$ of the Borel step fields with compact support is dense in the space $\mathscr{L}^{\Phi}_{\mathscr{A}}$, for the seminorm $\|x\|_{\Phi}$.*

Let $x \in \mathscr{L}^{\Phi}_{\mathscr{A}} = \mathscr{O}^{\Phi}_{\mathscr{A}}$. There exists a sequence of compact sets (C_n) outside which x vanishes almost everywhere. We can find a sequence (A_n) of mutually disjoint Borel sets, with $A_n \subset C_n$ for each n and $\cup C_n = \cup A_n$. Since x is measurable, each set A_n is the union of a negligible set and of a sequence of mutually disjoint compact sets $(C_{nm})_m$ such that the restriction of x to each C_{nm} is continuous. Thus, changing the indices of the sets C_{nm}, we can obtain a sequence (D_n) of mutually disjoint compact sets such that the restriction of x to each D_n is continuous and x vanishes almost everywhere outside the union of the sets D_n. The sets $K_n = \cup_{1 \leqslant i \leqslant n} D_i$ are compact, from an increasing sequence, the restriction of x to each K_n is continuous and denoting $x_n = x\varphi_K$, the sequence (x_n) converges almost everywhere to x. Then

$$\Phi(|x_n(t) - x(t)|) \to 0$$

almost everywhere. For each n we have

$$|x_n(t) - x(t)| \leqslant |x_n(t)| + |x(t)| \leqslant 2|x(t)|,$$

consequently $\Phi(|x_n(t) - x(t)|) \leqslant M\Phi(|x(t)|)$.

By Lebesgue's theorem we deduce

$$\lim \int \Phi(|x_n(t) - x(t)|)\, \mathrm{d}\mu(t) = 0$$

that is $|x_n - x|_{\Phi} \to 0$, therefore (proposition 29.38) $\|x_n - x\|_{\Phi} \to 0$.

We have proved thus that the set $\mathscr{C}$ of the fields with compact support, having continuous restriction to this support, is dense in $\mathscr{L}^{\Phi}_{\mathscr{A}}$ for the seminorm $\|x\|_{\Phi}$.

Let now $x \in \mathscr{C}$ and let K be the support of x. There exists a field $y \in \mathscr{K}_{\mathscr{A}}(T)$ such that $x\varphi_K = y\varphi_K$. There exists a sequence (x_n) of Borel step fields which converges uniformly to $\boldsymbol{y}$. The fields $\boldsymbol{x}_n = \boldsymbol{y}_n \varphi_K$ are also Borel fields, and converge uniformly on K to $\boldsymbol{y}\varphi_K$ consequently they converge uniformly on T to $\boldsymbol{x}$. It follows that $(\boldsymbol{x}_n)$ converge also to $\boldsymbol{x}$ in the topology of the space $\mathscr{L}_{\mathscr{A}}^{\Phi}$. Thus, the set $\mathscr{E}_{\mathscr{A}}(\mathscr{B})$ is dense in $\mathscr{C}$ for the seminorm $||\boldsymbol{x}||_{\Phi}$, therefore $\mathscr{E}_{\mathscr{A}}(\mathscr{B})$ is dense in $\mathscr{L}_{\mathscr{A}}^{\Phi}$ for $||\boldsymbol{x}||_{\Phi}$.

29.33 COROLLARY. *If* $\lim_{t\to 0}\varphi(t) = \delta > 0$ *and* $\lim_{t\to 0}\varphi(t) = l < +\infty$, *the set of Borel step field with compact support is dense in* $\mathscr{L}_{\mathscr{A}}^{\Phi}$ *for the seminorm* $||\boldsymbol{x}||_{\Phi}$.

In fact, we have $0 < \delta \leqslant \varphi(u) \leqslant l$ therefore $\delta u \leqslant \Phi(u) \leqslant lu$. Then

$$\Phi(2u) \leqslant 2lu = (2l/\delta)\delta u \leqslant (2/\delta)\Phi(u)$$

consequently, taking $M = 2l/\delta$, we have $\Phi(2u) \leqslant M\Phi(u)$ for $u \geqslant 0$ and we are in the conditions of the preceding proposition.

Example. If $\varphi(t) = t^{p-1}$, $(1 < p < +\infty)$, then $\psi(s) = s^{q-1}$, $(p^{-1} + q^{-1} = 1)$, $\Phi(u) = p^{-1}u^p$ and $\Psi(v) = q^{-1}v^q$. In this case we have $\Phi(2u) = 2^p\Phi(u)$ for $u \geqslant 0$ and $\Psi(2v) = 2^q\Psi(v)$ for $v \geqslant 0$. It follows that $\mathscr{O}_{\mathscr{A}}^{\Phi} = \mathscr{L}_{\mathscr{A}}^{\Phi}$ and $\mathscr{O}_{\mathscr{A}}^{\Psi} = \mathscr{L}_{\mathscr{A}}^{\Psi}$. We have then $||\boldsymbol{x}||_{\Phi} = q^{1/q}||\boldsymbol{x}||_p$ and $||\boldsymbol{x}||_{\Psi} = p^{1/p}\,||\boldsymbol{x}||_q$.

If $\varphi(0) = 0$ and $\varphi(t) \equiv 1$ for $t > 0$, then $\psi(s) = 0$ for $s \leqslant 1$ and $\psi(s) = +\infty$ for $s > 1$; also $\Phi(u) = u$ for $u \geqslant 0$, $\Psi(v) = 0$ for $v \leqslant 1$ and $\Psi(v) = +\infty$ for $v > 1$.

In this case $||\boldsymbol{x}||_{\Phi} = ||\boldsymbol{x}||_1$ and $||\boldsymbol{x}||_{\Psi} = ||\boldsymbol{x}||_{\infty}$ therefore $\mathscr{L}_{\mathscr{A}}^{\Phi} = \mathscr{L}_{\mathscr{A}}^{1}$ and $\mathscr{L}_{\mathscr{A}}^{\Psi} = \mathscr{L}_{\mathscr{A}}^{\infty}$.

5 *Linear operations on* $\mathscr{L}_{\mathscr{A}}^{\Phi}$

Let F be a Banach space and $U : \mathscr{L}_{\mathscr{A}}^{\Phi} \to E$ a linear mapping. Denote

$$|||U||| = \sup \Sigma |U(\varphi_{A_i}\boldsymbol{x}_i)| \text{ and } ||U|| = |\Sigma\, U(\varphi_{A_i}\boldsymbol{x}_i)|\,,$$

where the supremum is taken for all the Borel step fields $\boldsymbol{x} = \Sigma\,\varphi_{A_i}\boldsymbol{x}_i$, with disjoint sets A_i and $||\boldsymbol{x}||_{\Phi} \leqslant 1$. We have

$$||U|| < |||U||| \leqslant +\infty\,.$$

If $\Phi(2u) < M\,\Phi(u)$ for $u \geqslant 0$, the set $\mathscr{E}_{\mathscr{A}}(\mathscr{B})$ of the Borel step fields is dense in $\mathscr{L}_{\mathscr{A}}^{\Phi}$, therefore $||U||$ represents the usual norm of the linear operation U, if it is continuous.

29.34 PROPOSITION. *For every functional* $U: \mathscr{L}^{\Phi}_{\mathscr{A}} \to C$ *we have* $\|U\| = |||U||| \leqslant +\infty$.

In fact, let $\boldsymbol{x} = \Sigma\, \varphi_{A_i} \boldsymbol{x}_i$ be a Borel step field with A_i disjoint and $\|\boldsymbol{x}\|_{\Phi} \leqslant 1$. For each i there exists a complex number θ_i with $|\theta_i| = 1$ such that

$$|U(\varphi_{A_i} \boldsymbol{x}_i)| = \theta_i U(\varphi_{A_i} \boldsymbol{x}_i) = U(\varphi_{A_i} \theta_i \boldsymbol{x}_i)\,.$$

Then

$$\Sigma |U(\varphi_{A_i} \boldsymbol{x}_i)| = U(\Sigma\, \varphi_{A_i} \theta_i \boldsymbol{x}_i) \leqslant \|U\|\, \|\Sigma \varphi_{A_i} \theta_i \boldsymbol{x}_i\|_{\Phi} \leqslant \|U\|,$$

whence $|||U||| \leqslant \|U\|$ and then $\|U\| = |||U|||$.

Remark. If $F \neq C$, it is possible that $\|U\| < |||U|||$.

For each $t \in T$ denote $G(t) = \mathscr{L}(\mathscr{E}(t), F)$ and $\mathscr{G} = (G(t))_{t \in T}$.

29.35 THEOREM. *Assume that there exists* $M > 0$ *with* $\Phi(2u) \leqslant M\Phi(u)$ *for* $u \geqslant 0$. *For every continuous linear mapping* $U: \mathscr{L}^{\Phi}_{\mathscr{A}}(\mu) \to F$ *with* $|||U||| < \infty$, *its restriction* $\boldsymbol{m}$ *to* $\mathscr{K}_{\mathscr{A}}(T)$ *is a bounded measure, absolutely continuous with respect to* μ, *such that*

$$\int |\varphi|\, \mathrm{d}|\boldsymbol{m}| \leqslant |||U|||\, \|\varphi\|_{\Phi} \text{ for } \varphi \in \mathscr{K}(T)\,.$$

Conversely, every dominated measure $\boldsymbol{m}: \mathscr{K}_{\mathscr{A}}(T) \to F$, *absolutely continuous with respect to* μ, *for which there exists* $\alpha > 0$ *such that*

$$\int |\varphi|\, \mathrm{d}|\boldsymbol{m}| \leqslant \alpha \|\varphi\|_{\Phi} \text{ for } \varphi \in \mathscr{K}(T)\,,$$

can be extended uniquely to a linear operation $U: \mathscr{L}^{\Phi}_{\mathscr{A}}(\mu) \to F$ *with* $|||U||| < \infty$. *We have then* $|\boldsymbol{m}| = g\mu$ *with*

$$\tfrac{1}{2}\|g\|_{\Psi} \leqslant |||U||| \leqslant \|g\|_{\Phi}\,,$$

$\mathscr{L}^{\Phi}_{\mathscr{A}}\mu) \subset \mathscr{L}^{\Phi}_{\mathscr{A}}(\boldsymbol{m})$ *and*

$$U(\boldsymbol{x}) = \int \boldsymbol{x}\, \mathrm{d}\boldsymbol{m} \text{ and } \int |\boldsymbol{x}|\, \mathrm{d}|\boldsymbol{m}| \leqslant |||U|||\, \|\boldsymbol{x}\|_{\Phi}$$

for $\boldsymbol{x} \in \mathscr{L}^{\Phi}_{\mathscr{A}}(\mu)$.

(a) Let $U: \mathscr{L}^{\Phi}_{\mathscr{A}}(\mu) \to F$ be a linear mapping with $|||U||| < \infty$. The restriction of U to $\mathscr{K}_{\mathscr{A}}(T)$ is linear. Show that it is dominated:

For every $\varphi \in \mathscr{L}^{\Phi}$, every $\boldsymbol{x} \in \mathscr{A}$ and every relatively compact Borel set

$B \in \mathscr{B}$ we have $\varphi \boldsymbol{x} \varphi_B \in \mathscr{L}^{\Phi}_{\mathscr{A}}$, since $|\varphi \boldsymbol{x} \varphi_B| \leqslant \|\boldsymbol{x}\|_B |\varphi|$, where $\|\boldsymbol{x}\|_B = \sup_{t \in B} |\boldsymbol{x}(t)|$. Put $M_B(\varphi)\boldsymbol{x} = U(\varphi \boldsymbol{x} \varphi_B)$.

The mapping $M_B(\varphi): \boldsymbol{x} \to M_B(\varphi)\boldsymbol{x}$ of $\mathscr{A}$ into F is linear and continuous for the seminorm $\|\boldsymbol{x}\|_B$ on $\mathscr{A}$, since

$$|M_B(\varphi)\boldsymbol{x}| = |U(\varphi \boldsymbol{x} \varphi_B)| \leqslant \|U\| \, \|\boldsymbol{x}\|_B \, \|\varphi\|_\Phi ,$$

therefore $M_B(\varphi) \in \mathscr{L}(\mathscr{A}, F)$, $\mathscr{A}$ being endowed with the topology defined by the family of seminorms $(\|\boldsymbol{x}\|_B)_{B \in \mathscr{B}}$, and

$$\|M_B(\varphi)\|_B \leqslant \|U\| \, \|\varphi\|_\Phi$$

where, for $S \in \mathscr{L}(\mathscr{A}, F)$, we denoted

$$\|S\|_A = \sup_{\|\boldsymbol{x}\|_A \leqslant 1} |S\boldsymbol{x}| .$$

For each $A \in \mathscr{B}$ put $\boldsymbol{m}(A) = M_A(\varphi_A)$. For every $B \in \mathscr{B}$, such that $A \subset B$. we have $\boldsymbol{m}(A) = M_B(\varphi_A)$.

The set function $\boldsymbol{m}: \mathscr{B} \to \mathscr{L}(\mathscr{A}, F)$ has the following properties:

(1) For every $A \in \mathscr{B}$ and $\boldsymbol{x}, \boldsymbol{y} \in \mathscr{A}$ such that $\varphi_A \boldsymbol{x} = \varphi_A \boldsymbol{y}$ we have

$$\boldsymbol{m}(A)\boldsymbol{x} = \boldsymbol{m}(A)\boldsymbol{y} .$$

(2) $\boldsymbol{m}$ is addtive

$$\boldsymbol{m}(\cup_{1 \leqslant i \leqslant n} A_i) = \Sigma_{1 \leqslant i \leqslant n} \boldsymbol{m}(A_i)$$

If A_i are mutually disjoint.

(3) $\boldsymbol{m}$ has finite variation. Let $A \in \mathscr{B}$ and $(A_i)_{1 \leqslant i \leqslant n}$ be a partition of A consisting of Borel sets, $(c_i)_{1 \leqslant i \leqslant n}$ a family of numbers and $\varepsilon > 0$. For each i there exists a field $\boldsymbol{x}_i \in \mathscr{A}$ with $\|\boldsymbol{x}_i\|_A \leqslant 1$ and

$$\|\boldsymbol{m}(A_i)\|_A \leqslant |\boldsymbol{m}(A_i)\boldsymbol{x}_i| + \varepsilon/n|c_i|, \text{ if } c_i \neq 0,$$
$$\|\boldsymbol{m}(A_i)\|_A \leqslant |\boldsymbol{m}(A_i)\boldsymbol{x}_i| + \varepsilon/n, \text{ if } c_i = 0 .$$

Then for every i we have

$$\|\boldsymbol{m}(A_i)\|_A |c_i| \leqslant |\boldsymbol{m}(A_i) c_i \boldsymbol{x}_i| + \varepsilon/n ,$$

therefore

$$\Sigma_{1 \leqslant i \leqslant n} \|\boldsymbol{m}(A_i)\|_A |c_i| \leqslant \Sigma_{1 \leqslant i \leqslant n} |\boldsymbol{m}(A_i) c_i \boldsymbol{x}_i| + \varepsilon = \Sigma |U(\varphi_{A_i} \boldsymbol{x}_i c_i)| + \varepsilon \leqslant$$
$$\leqslant |||U||| \, \|\Sigma \varphi_{A_i} c_i \boldsymbol{x}_i\|_\Phi + \varepsilon \leqslant |||U||| \, \|\Sigma \varphi_{A_i} c_i\|_\Phi + \varepsilon .$$

ε being arbitrary, we deduce

$$\Sigma \|\boldsymbol{m}(A_i)\|_{A_i} |c_i| \leqslant |||U||| \, \|\Sigma \varphi_{A_i} c_i\|_\Phi .$$

Taking $c_i = 1$ for each i, we obtain

$$\Sigma \|\boldsymbol{m}(A_i)\|_A \leqslant |||U||| \, \|\varphi_A\|_\Phi$$

and then

$$v(A) = \sup \Sigma \|\boldsymbol{m}(A_i)\|_A \leqslant |||U||| \, \|\varphi_A\|_\Phi < \infty ,$$

the supremum being taken for all the partitions (A_i) of A. Thus $\boldsymbol{m}$ has finite variation and the variation v is additive on $\mathscr{B}$.

(4) We also have the inequality

$$\Sigma_{1 \leqslant i \leqslant n} v(A_i) |c_i| \leqslant |||U||| \, \|\Sigma_{1 \leqslant i \leqslant n} \varphi_{A_i} c_i\|_\Phi .$$

In fact, let $\varepsilon > 0$. For each i, there exists a partition (B_{ij}) of A_i, consisting of Borel sets, such that

$$v(A_i) \leqslant \Sigma_j \|\boldsymbol{m}(B_{ij})\|_A + \varepsilon/n|c_i|, \text{ if } c_i \neq 0,$$
$$v(A_i) \leqslant \Sigma_j \|\boldsymbol{m}(B_{ij})\|_A + \varepsilon/n, \text{ if } c = 0 ,$$

therefore for every i we have

$$v(A_i)|c_i| \leqslant \Sigma_j \|\boldsymbol{m}(B_{ij})\|_A |c_i| + \varepsilon/n,$$

whence

$$\Sigma_{1 \leqslant i \leqslant n} v(A_i)|c_i| \leqslant \Sigma_{ij} \|\boldsymbol{m}(B_{ij})\|_A |c_i| + \varepsilon \leqslant$$
$$\leqslant |||U||| \, \|\Sigma_{i,j} \varphi_{B_{ij}} c_i\|_\Phi + \varepsilon \leqslant |||U||| \, \|\Sigma \varphi_{A_i} c_i\|_\Phi + \varepsilon .$$

ε being arbitrary, we deduce

$$\Sigma v(A_i)|c_i| \leqslant |||U||| \, \|\Sigma \varphi_{A_i} c_i\|_\Phi ,$$

whence

$$|\Sigma v(A_i) c_i| \leqslant |||U||| \, \|\Sigma_{1 \leqslant i \leqslant n} \varphi_{A_i} c_i\|_\Phi .$$

(5) For every real Borel step function $\varphi = \Sigma \varphi_{A_i} c_i$ with A_i disjoint, put

$$\upsilon(\varphi) = \Sigma v(A_i) c_i .$$

It is easy to see that $\upsilon(\varphi)$ does not depend on the particular form in which φ is written as a step function, that υ is linear and

$$|\upsilon(\varphi)| \leqslant |||U||| \, \|\varphi\|_\Phi .$$

Since $\Phi(2u) \leqslant M\Phi(u)$ for $u \geqslant 0$, the Borel step functions are dense in $\mathscr{L}^{\Phi}$, therefore v can be extended by continuity to a linear continuous functional on $\mathscr{L}^{\Phi}$, denoted still by v, and we have

$$|v(\varphi)| \leqslant |||U|||\ ||\varphi||_{\Phi} \text{ for } \varphi \in \mathscr{L}^{\Phi}.$$

For every Borel step field $\boldsymbol{x} = \Sigma \varphi_{A_i} \boldsymbol{x}_i$, with A_i disjoint, we have

$$|U(\boldsymbol{x})| = |\Sigma\, \boldsymbol{m}(A_i)\, \boldsymbol{x}_i| \leqslant v(|\boldsymbol{x}|)\,.$$

Since the mappings $\boldsymbol{x} \to U(\boldsymbol{x})$ and $\boldsymbol{x} \to v(|\boldsymbol{x}|)$ are continuous on $\mathscr{L}^{\Phi}_{\mathscr{A}}$, the inequality

$$|U(\boldsymbol{x})| \leqslant v(|\boldsymbol{x}|)$$

remains valid for every $\boldsymbol{x} \in \mathscr{L}^{\Phi}_{\mathscr{A}}$. As the restriction of v to $\mathscr{K}(T)$ is a positive measure, we deduce that the restriction $\boldsymbol{m}$ on U to $\mathscr{K}_{\mathscr{A}}(T)$ is a dominated measure. It is, evidently, absolutely continuous with respect to μ.

From the inequality $|\boldsymbol{m}| \leqslant v$ we deduce

$$\int |\varphi|\, \mathrm{d}|\boldsymbol{m}| \leqslant v(|\varphi|) \leqslant |||U|||\ ||\varphi||_{\Phi} \text{ for } \varphi \in \mathscr{K}(T)\,.$$

(b) Conversely, let $\boldsymbol{m} : \mathscr{K}_{\mathscr{A}}(T) \to F$ be a dominated measure, absolutely continuous with respect to μ, such that there exists $\alpha > 0$ with

$$\int |\varphi|\, \mathrm{d}|\boldsymbol{m}| \leqslant \alpha\, ||\varphi||_{\Phi} \text{ for } \varphi \in \mathscr{K}(T)\,.$$

Since $\boldsymbol{m}$ is absolutely continuous with respect to μ, there exists a locally μ-integrable function $g \geqslant 0$ with $|\boldsymbol{m}| = g\mu$. For every $\varphi \in \mathscr{K}(T)$ we have

$$\int |\varphi|\, g\, \mathrm{d}\mu = \int |\varphi|\, \mathrm{d}|\boldsymbol{m}| \leqslant \alpha\, ||\varphi||_{\Phi} \leqslant \alpha(|\varphi|_{\Phi} + 1)\,,$$

therefore

$$||g||_{\Psi} \leqslant 2\alpha < \infty\,.$$

It follows that if $\varphi \in \mathscr{L}^{\Phi}(\mu)$, then $\varphi g \in \mathscr{L}^1(\mu)$, hence $\varphi \in \mathscr{L}^1(|\boldsymbol{m}|)$, consequently $\mathscr{L}^{\Phi}(\mu) \subset \mathscr{L}^1(|\boldsymbol{m}|)$ and $\mathscr{L}^{\Phi}_{\mathscr{A}}(\mu) \subset \mathscr{L}^1_{\mathscr{A}}(\boldsymbol{m})$. We have then

$$\int |\boldsymbol{x}|\, \mathrm{d}|\boldsymbol{m}| \leqslant \alpha\, ||\boldsymbol{x}||_{\Phi} \text{ for every } \boldsymbol{x} \in \mathscr{L}^{\Phi}_{\mathscr{A}}\,.$$

Put

$$U(\boldsymbol{x}) = \int \boldsymbol{x}\, \mathrm{d}\boldsymbol{m}, \text{ for } \boldsymbol{x} \in \mathscr{L}^{\Phi}_{\mathscr{A}}(\mu)\,.$$

The mapping $U : \mathscr{L}^{\Phi}_{\mathscr{A}}(\mu) \to F$ is linear and we shall show that $|||U||| < \infty$. Let $\boldsymbol{x} = \Sigma \varphi_{A_i} \boldsymbol{x}_i$ be a step field with A_i disjoint relatively compact Borel sets.

Then

$$\Sigma|U(\varphi_{A_i}\boldsymbol{x}_i)| = \Sigma|\int\varphi_{A_i}\boldsymbol{x}_i \mathrm{d}\boldsymbol{m}| \leqslant \Sigma\int\varphi_{A_i}|\boldsymbol{x}_i|\mathrm{d}|\boldsymbol{m}| = \\ = \int|\Sigma\varphi_{A_i}\boldsymbol{x}_i|\mathrm{d}|\boldsymbol{m}| \leqslant \alpha\|\boldsymbol{x}\|_{\Phi},$$

therefore $|||U||| \leqslant \alpha < \infty$.

(c) Let now $U : \mathscr{L}^{\Phi}_{\mathscr{A}}(\mu) \to F$ be a linear mapping with $|||U||| < \infty$ and $\boldsymbol{m}$ the corresponding dominated measure. From the first part we deduce that

$$\int|\boldsymbol{x}|\mathrm{d}|\boldsymbol{m}| \leqslant |||U|||\ \|\boldsymbol{x}\|_{\Phi} \text{ for } \boldsymbol{x}\in\mathscr{L}^{\Phi}_{\mathscr{A}}(\mu).$$

From the second part of the proof, with $\alpha = |||U|||$, we deduce that there exists a locally μ-integrable function $g \geqslant 0$ with $|\boldsymbol{m}| = g\mu$ and

$$\tfrac{1}{2}\|g\|_{\Psi} \leqslant |||U|||.$$

Let $\boldsymbol{x} = \Sigma\varphi_{A_i}\boldsymbol{x}_i$ be a step field with A_i disjoint and with $\|\boldsymbol{x}\|_{\Phi} \leqslant 1$. Then, as above, we have

$$\Sigma|U(\varphi_{A_i}\boldsymbol{x}_i)| \leqslant \int|\boldsymbol{x}|\mathrm{d}|\boldsymbol{m}| = \int|\boldsymbol{x}|g\,\mathrm{d}\mu \leqslant \|g\|_{\Psi},$$

whence $|||U||| \leqslant \|g\|_{\Psi}$.

Thus the theorem is proved.

29.36 THEOREM. *Let $Z \subset F'$ be a norming subspace for F, and $U : \mathscr{L}^{\Phi}_{\mathscr{A}}(\mu) \to F$ a linear mapping with $|||U||| < \infty$. Assume that there exists $M > 0$ with $\Phi(2u) \leqslant M\Phi(u)$ for $u \geqslant 0$ and that the family $\mathscr{A}$ satisfies axiom* (G). *There exists then an operation field $\boldsymbol{u}\in\mathscr{C}(\mathscr{G}_Z)$ with the following properties:*

(1) *The function $\boldsymbol{u}$ belongs to the space $\mathscr{L}^{\Psi}(\mu)$ and we have*

$$\tfrac{1}{2}\|\boldsymbol{u}\|_{\Psi} \leqslant |||U||| \leqslant \|\boldsymbol{u}\|_{\Psi}.$$

(2) *The field $\boldsymbol{u}$ is Z-weakly locally μ-integrable and we have*

$$\langle U(\boldsymbol{x}), z\rangle = \int\langle \boldsymbol{u}(t)\boldsymbol{x}(t), z\rangle\,\mathrm{d}\mu, \textit{ for } \boldsymbol{x}\in\mathscr{L}^{\Phi}_{\mathscr{A}} \textit{ and } z\in Z.$$

(3) *If ρ is a lifting of $\mathscr{L}^{\infty}(\mu)$ and $\mathscr{B} \subset \mathscr{A}$ is a countable set such that, for each $t\in T$, the set $\{\boldsymbol{u}(t); \boldsymbol{x}\in\mathscr{B}\}$ is dense in $E(t)$, then we can choose $\boldsymbol{u}$ such that*

$$\rho[\boldsymbol{u}\boldsymbol{x}] = \boldsymbol{u}\boldsymbol{x}, \text{ for } \boldsymbol{x}\in\mathscr{B}.$$

(4) *We have $\boldsymbol{u}\in\mathscr{C}(\mathscr{G}_F)$ in each of the following cases:*

(*a*) *$F = Z'$,*

(*b*) *for each $\boldsymbol{x}\in\mathscr{A}$, the convex balanced hull of the set $\{U(\varphi\boldsymbol{x};$*

$\varphi \in \mathscr{K}(T)$, $\int |\varphi| \mathrm{d}|\mu| \leqslant 1\}$ *is relatively compact in F for the topology* $\sigma(F, Z)$.

The proof is the same as that of theorem 18.42, using theorem 28.32, and theorem 29.35, instead of theorems 18.16, and 18.40.

Remarks.

1. If F is of countable type, the function $t \to \boldsymbol{u}(t)\boldsymbol{x}(t)$ is measurable for every $\boldsymbol{x} \in \mathscr{L}^{\Phi}_{\mathscr{A}}(\mu)$.

In this case the representation formula is written

$$U(\boldsymbol{x}) = \int \boldsymbol{u}(t)\boldsymbol{x}(t)\,\mathrm{d}\mu, \text{ for } \boldsymbol{x} \in \mathscr{L}^{\Phi}_{\mathscr{A}}.$$

2. Assume that there exists a fundamental family $\mathscr{D} \subset \mathscr{C}(\mathscr{G}_{Z'})$ such that the function $t \to \boldsymbol{v}(t)\boldsymbol{x}(t)$ is continuous for every $\boldsymbol{v} \in \mathscr{D}$ and $\boldsymbol{x} \in \mathscr{A}$.

In this case the field $\boldsymbol{u}$ is measurable with respect to $\boldsymbol{u}$ and $\mathscr{D}$, consequently $\boldsymbol{u} \in \mathscr{L}^{\Psi}_{\mathscr{D}}$.

29.37 Corollary. *Assume that there exists M with* $\Phi(2u) \leqslant M\Phi(u)$ *for* $u \geqslant 0$, *and that* $\mathscr{A}$ *satisfies axiom* (G). *Then, for every linear continuous functional U on* $\mathscr{L}^{\Phi}_{\mathscr{A}}$ *there exists a functional field* $\boldsymbol{u} \in \mathscr{C}(\mathscr{E}')$ *with*

$$\langle \boldsymbol{x}, U \rangle = \int \langle \boldsymbol{x}(t), \boldsymbol{u}(t) \rangle \,\mathrm{d}\mu, \textit{ for } \boldsymbol{x} \in \mathscr{L}^{\Phi}_{\mathscr{A}},$$

and

$$\tfrac{1}{2}\|\boldsymbol{u}\|_{\Psi} \leqslant \|U\| \leqslant \|\boldsymbol{u}\|_{\Psi}.$$

In fact, in this case $|||U||| = \|U\| < \infty$.

If there exists a fundamental family $\mathscr{A}' \subset \mathscr{C}(\mathscr{E}')$ such that the function $t \to \langle \boldsymbol{x}(t), \boldsymbol{x}'(t) \rangle$ is continuous for every $\boldsymbol{x} \in \mathscr{A}$ $\boldsymbol{x}' \in \mathscr{A}'$ and if $\mathscr{A}'$ satisfies axiom (G), then $\boldsymbol{u}$ is measurable, therefore $\boldsymbol{u} \in \mathscr{L}^{\Psi}_{\mathscr{A}'}$.

29.38 Corollary. *Assume that*: $\Phi(2u) \leqslant M\Phi(u)$ *and* $\Psi(2u) \leqslant M\Psi(v)$; *the spaces E(t) are reflexive; there exists a fundamental family* $\mathscr{A}' \subset \mathscr{C}(\mathscr{E}')$ *such that the function* $t \to \langle \boldsymbol{x}(t), \boldsymbol{x}'(t) \rangle$ *is continuous for* $\boldsymbol{x} \in \mathscr{A}$ *and* $\boldsymbol{x}' \in \mathscr{A}'$; $\mathscr{A}$ *and* $\mathscr{A}'$ *satisfy axiom* (G). *Then the spaces* $\mathscr{L}^{\Phi}_{\mathscr{A}}$ *and* $\mathscr{L}^{\Psi}_{\mathscr{A}}$ *are reflexive.*

Without any other restriction on the family $\mathscr{A}$ we have the following converse theorem.

29.39 THEOREM. *Let $Z \subset F'$ be a norming subspace for F and $\boldsymbol{u} \in \mathscr{C}(\mathscr{G}_F)$ a Z-weakly locally μ-integrable operation field. If*

(i) $||\boldsymbol{u}||_{\Psi} < \infty$,

there exists a linear mapping $U : \mathscr{L}^{\Phi}_{\mathscr{A}}(\mu) \to Z'$ such that

$$\langle U(x), z\rangle = \int \langle \boldsymbol{u}(t)\boldsymbol{x}(t), z\rangle \mathrm{d}\mu, \text{ for } \boldsymbol{x} \in \mathscr{L}^{\Phi}_{\mathscr{A}} \text{ and } z \in Z.$$

In this case we have

$$|||U||| \leqslant ||\boldsymbol{u}||_{\Psi}$$

(ii) *We have $U(\boldsymbol{x}) \in F$ for every $\boldsymbol{x} \in \mathscr{L}^{\Phi}_{\mathscr{A}}(\mu)$ in each of the following cases:*

(*a*) $F = Z'$,

(*b*) *$\boldsymbol{u}$ is simply μ-measurable; in particular F is of countable type;*

(*c*) *for each $x \in \mathscr{A}$ there exists a locally countable family $(K_j)_{j \in J}$ of disjoint compact subsets with $T \setminus \cup K_j$ μ-negligible, such that for each $j \in J$, the convex balanced hull of the set $\{\boldsymbol{u}(t)\boldsymbol{x}(t); t \in K_j\}$ is relatively compact in F for the topology $\sigma(F, Z)$.*

(iii) *If there exists $M > 0$ with $\Phi(2u) \leqslant M\Phi(u)$ for $u \geqslant 0$, then the inequalities*

$$\tfrac{1}{2}||\boldsymbol{u}||_{\Psi} \leqslant |||U||| \leqslant ||\boldsymbol{u}||_{\Psi}$$

are satisfied in each of the following cases:

(1) *$\mathscr{A}$ satisfies axiom (G) and there exists a lifting ρ of $\mathscr{L}^{\infty}(\mu)$ and a countable subset $\mathscr{B} \subset \mathscr{A}$, such that for every $t \in T$ the set $\{\boldsymbol{x}(t); \boldsymbol{x} \in \mathscr{B}\}$ is dense in $E(t)$ and*

$$\rho[\boldsymbol{u}\boldsymbol{x}] = \boldsymbol{u}\boldsymbol{x}, \text{ for } \boldsymbol{x} \in \mathscr{B}.$$

(2) *$\mathscr{A}$ satisfies axiom (G) and there exists a countable set $S \subset Z$ norming for F.*

(3) *$\mathscr{A}$ satisfies axiom (G) and $\boldsymbol{u}$ is simply μ-measurable.*

(4) *There exists a fundamental family $\mathscr{A}' \subset \mathscr{C}(\mathscr{E}')$ of continuous functional fields, such that the function $t \to \langle \boldsymbol{x}(t), \boldsymbol{x}'(t)\rangle$ is continuous for every $\boldsymbol{x} \in \mathscr{A}$ and every $\boldsymbol{x}' \in \mathscr{A}'$ and $\boldsymbol{u}$ is a functional field of $\mathscr{C}(\mathscr{E}')$ measurable with respect to $\mathscr{A}'$ and μ.*

BIBLIOGRAPHY

Alexiewicz, A.,

Linear operations among bounded measurable functions, I, II. *Ann. Soc. Polon. Math.*, 19, 140–161, 161–164 (1946).

Banach, S.,

Théories des opérations linéaires, Warsaw, 1932.

Bartle, R. G.,

A general bilinear vector integral, *Studia Math.*, 15, 337–352 (1956).

Bartle, R. G., Dunford, N., Schwartz, J.,

Weak compactness and vector measures, *Canadian J. Math.*, 7, 289–305 (1955).

Birkhoff, G.,

Integration of functions with values in a Banach space, *Trans. Amer. Math. Soc.*, 38, 357–378 (1935).

Bochner, S.,

Integration von Funktionen, deren Werte die Elemente eines Vektorraumes sind, *Fund. Math.*, 20, 262–276 (1933).

Bochner, S., Taylor, A. E.,

Linear functionals on certain spaces of abstractly-valued functions, *Ann. of Math.* (2), 39, 913–944 (1938).

Bourbaki, N.,

Topologie générale, chap. I–V, Hermann, Paris, 1940–1949,

Espaces vectoriels topologiques, chap. I–V, Hermann, Paris, 1953–1955,

Intégration, chap. I–VI, Hermann, Paris, 1952, 1956, 1959.

Cartan, H., Godement, R.,

Théorie de la dualité et analyse harmonique dans les groupes abéliens localement compacts, *Ann. Ecole Norm-Sup.*, 64, 79–99 (1947).

Colojoară, I., Dinculeanu N., Marinescu, G.,

Measures with values in locally convex spaces, *Bull. Sci. Math. Phys. R.P.R.*, 5 (53), 167–180 (1961).

Cristescu, R.,

Spaţii liniare ordonate, Bucureşti, 1959.

Noţiunea de integrală în spaţii semiordonate, *Com. Acad. R.P.R.*, 2, 205–208 (1952).

Integrarea în spaţii semiordonate. *Bull. ştiinţ. Acad. R.P.R.*, 4, 291–310 (1952).

O teoremă asupra reprezentării operaţiilor liniare, *Com. Acad. R.P.R.*, 5, 655–659 (1955).
Integrali vettoriali di Stieltjes ed operatori lineari. *Rendiconti Accad. Naz. Lincei*, 27, 31–34 (1959).
Intégrales vectorielles et représentation de certains opérateurs linéaires, *Bull. Math. Soc. Sci. Math. Phys. R.P.R.*, 3 (51), 7–15 (1959).
Stieltjes Integral in *K*-spaces (Russian), *Doklady Akad. Nauk, SSSR*, 133, 519–522 (1960).
Asupra unor integrale vectoriale, *Com. Acad. RPR*, 11, 1169–1173 (1961).

Cuculescu, I.,
Généralisations aux groupes quelconques d'un théorème de E. Hille concernant les fonctions facteurs, *Rendiconti Accad. Naz. Lincei*, 24, 15–19 (1958).

Daniell, P. I.,
A general form of integral, *Ann. of Math.*, (2), 19, 279–294 (1917–1918).

Day, M. M.,
The space L^p with $0 < p < 1$, *Bull. Amer. Math. Soc.*, 46, 816–823 (1940).
Normed linear spaces. *Ergebnisse der Math.*, Berlin–Göttingen–Heidelberg, 1958.

Dieudonné, J.,
Sur le théorème de Lebesgue–Nikodym, *Ann. of Math.*, (2) 42, 547–555 (1941).
Sur le théorème de Lebesgue–Nikodym, II, *Bull. Soc. Math. France*, 72, 193–239 (1944);
Sur le théorème de Lebesgue–Nikodym, III, *Ann. Inst. Fourier*, Grenoble, 23, 25–53 (1947–1948).
Sur le théorème de Lebesgue–Nikodym, IV, *J. Indian Math. Soc.*, 15, 77–86 (1951).
Sur le théorème de Lebesgue–Nikodym, V, *Canadian J. Math.*, 3, 129–139 (1951).
Sur les espaces de Köthe, *J. Analyse Math.*, 1, 81–115 (1951).
Sur la convergence des suites de mesures de Radon, *Anais Acad. Brasil. Ci.*, 23, 21–38, 277–282 (1951).
Sur le produit de composition, *Compositio Math.*, 12, 17–34 (1964).
Anâlise Harmônica, Rio de Janeiro, 1952.

Dinculeanu, N.,
Espaces d'Orlicz de champs de vecteurs, I, *Rendiconti Accad. Naz.*

Lincei, 22, 135–139 (1957);
Espaces d'Orlicz de champs de vecteurs, II, *Rend. Accad. Naz. Lincei*, 22, 269–275 (1957).
Spatii Orlicz de cîmpuri de vectori, *St. cerc. mat*, 8, 343–412 (1957).
Espaces d'Orlicz de champs de vecteurs, III, Opérations linéaires, *Studia Math.*, 17, 285–293 (1958).
Espaces d'Orlicz de champs de vecteurs, IV, (Opérations linéaires), *Studia Math.*, 19, 322–331 (1960).
On the integral representation of linear operations on Orlicz spaces (Russian), *Doklady Akad. Nauk, SSSR*, 146, 1255–1258 (1962).
Sur la représentation intégrale de certaines opérations linéaires, *C.R. Acad. Sci.*, Paris, 245, 1203–1205 (1957).
Mesures vectorielles et opèrations linéaires. *C.R. Acad. Sci.*, Paris, 246, 2328–2331 (1958).
Sur la répresentation intégrale de certaines opérations linéaires, II, *Compositio Math.*, 14, 1–22 (1959).
Sur la représentation intégrale de certaines opérations linéaires, III, *Proc. Amer. Math. Soc.*, 10, 59–68 (1959).
Mesures vectorielles sur les espaces localement compacts, *Bull. Math. Soc. Sci. Math. Phys. R.P.R.*, 2 (50), 137–164 (1958).
Remarks on the integral representation of vector measures and linear operations on L^p_A, *Revue math. pures et appl.*, 7, 287–300 (1962).
Despre seminormele N_p. *Analele Universităţii Bucureşti*, 11, 155–163 (1962).
On the regular vector measures, *Acta Sci. Math. Szeged*, 24, 236–243 (1963).
Regularity of vector measures. *Revue math. pures et appl.*, 9, 81–90 (1964).
Integral representation of vector measures and linear operations, *Studia Math.*, 25, 181–205 (1965).

Dinculeanu, N., Foias C.,
Mesures vectorielles et opérations linéaires sur L^p_E, *C.R. Acad. Sci.*, Paris, 248, 1759–1762 (1959).
Sur la représentation intégrale de certaines opérations linéaires, IV, *Canad. J. Math.*, 13, 529–556 (1961).

Dixmier, J.,
Mesures de Haar et trace d'un opèrateur, *C.R. Acad. Sci.*, Paris, 228, 152–154 (1949).

Les algèbres d'opérateurs dans l'espace hilbertien (*algèbres de von Neumann*), Paris, 1957.

Doob, J. L.,

Probability methods applied to the first boundary value problem, *Proc. of the third Berkeley Symposium on Math. Statistics and Probability*, 1954–1955, 49–80.

Stochastic Processes, New York, London (1953).

Dowker, Y. N.,

Finite and σ-finite invariant measures, *Ann. of Math.* (2), 54, 595–608 (1951).

On measurable transformations in finite measure spaces, *Ann. of Math.* 62, 504–516(1955).

Dunford, N.,

Uniformity in linear spaces. *Trans. Amer. Math. Soc.*, 44, 305–356 (1938).

Integration in general analysis, *Trans. Amer. Math. Soc.*, 37, 441–453 (1935).

Integration and linear operations. *Trans. Amer. Math. Soc.*, 40, 474–494 (1936).

Dunford, N., Pettis, B. J.,

Linear operations on summable functions, *Trans. Amer. Math. Soc.*, 47, 323–392 (1950).

Dunford, N., Schwartz, J.,

Linear operators, Part I, New York (1958).

Edwards, R. E.,

A theory of Radon measures on locally compact spaces, *Acta Math.* 89, 133–164 (1953).

Vector-valued measures and bounded variation in Hilbert space, *Mathematica Scandinavica*, 3, 90–96 (1955).

On certain algebras of measures, *Pacif. J. Math.*, 5, 379–389 (1955).

Ellis, H. W., Halperin, I.,

Functions spaces determined by a levelling length function, *Canad. J. Math.*, 5, 576–592 (1953).

Fichtenholz, G.,

Sur les opérations linéaires dans l'espace des fonctions continues, *Bull. Acad. Roy. Belg.*, 22, 26–33 (1936).

Fichtenholz, G., Kantorovitch, L. V.,

Sur les opérations linéaires dans l'espace des fonctions bornées, *Studia Math.*, 5, 69–98 (1934).

Foiaş, C.,

On a commutative extension of a commutative Banach algebra, *Pacific. J. Math.*, 8, 407–410 (1958).

Décompositions intégrales des familles spectrales et semi-spectrales en opérateurs qui sortent de l'espace hilbertien, *Acta. Sci. Math. Szeged*, 20, 117–154 (1959).

On some semi-groups of contractions, connected to the representation of involution algebras, I, (Russian) *Revue math. pures et appl.*, 7, 319–325 (1962).

Foiaş, C., Singer, I.,

Some remarks on the representation of linear operators in spaces of vector valued continuous functions, *Revue math. pures et appl.*, 5, 729–752 (1960).

Fortet, R., Mourier, E.,

Résultats complémentaires sur les éléments aléatoires dans un espace de Banach, *Bull. Sci. Math.* 78, 14–30 (1954).

Fullerton, R. E.,

The representation of linear operators from L^p to L, *Proc. Amer. Math. Soc.*, 5, 689–696 (1954).

Găină, S.,

Extension of vector measures, *Revue math. pures et appl.* 8, 151–154 (1963).

Gelfand, I.,

Abstrakte Funktionen und lineare Operatoren, *Matem. Sbornik*, 4 (46), 235–284 (1938).

On the theory of characters of topological abelian groups (Russian), *Matem. Sbornik*, 9 (51), 49–50 (1941).

Gelfand, I., Naimark, M. A.,

Rings with involution and their representation (Russian), *Izv. Akad. Nauk SSSR*, 12, 445–480 (1948).

Gelfand, I., Raikov D. A.,

On the theory of characters of topological commutative groups (Russian), *Doklady Akad. Nauk SSSR*, 28, 195–197 (1940).

Gelfand, I., Silov, G.,

Über verschiedene Methoden der Einführung der Topologie in die Menge der maximalen Ideale eines normierten Ringe, *Mat. Sbornik*, 9 (51) 25–40 (1941).

Godement, R.,

Théorie générale des sommes continues d'espaces de Banach, *C.R. Acad.*

Sci., Paris, 288, 1321–1323 (1949).

Sur la théorie des représentation unitaires. *Annals of Math.*, (2), 53, 68–124 (1951).

Gowurin, M.,

Über die Stieltjessche Integration abstrakter Funktionen. *Fund. Math.*, 27, 255–268 (1936).

Grothendiek, A.,

Sur les applications linéaires faiblement compactes d'espaces de type $C(K)$, *Canad. J. Math.*, 5, 129–173 (1953).

Produits tensoriels topologiques et espaces nucléaires, Chap. I–II, *Memoirs Amer. Math. Soc. Providence*, (1955).

Haar, A.,

Der Massbegriff in der Theorie der kontinuierlichen Gruppen, *Annals of Math.*, (2), 34, 147–169 (1933).

Halmos, P. R.,

Measure Theory, New York, (1950).

Halperin, I.,

Function spaces. *Canad. J. Math.*, 5, 273–288 (1953).

Uniform convexity in function spaces, *Duke Math. J.*, 21, 195–204 (1954).

Reflexivity in the L^λ function spaces, *Duke Math. J.*, 21, 205–208 (1954).

Hewitt, E.,

Linear functionals on spaces of continuous functions, *Fund. Math.*, 37, 161–189 (1950).

Integral representation of certain linear functionals, *Ark. för Mat.*, 2, 269–282 (1952).

Integration on locally compact spaces, I, *Univ. of Washington Publ. in Math.*, 3, 71–75 (1952).

Hildebrandt, T. H.,

Integration in abstract spaces, *Bull. Amer. Math. Soc.*, 59, 111–139 (1953).

Hille, E., Phillips, R.,

Functional analysis and semi-groups, *Amer. Math. Soc. Coll. Publ.*, 1957.

Ionescu Tulcea, A., Ionescu Tulcea, C.,

On the decomposition and integral representation of continuous linear operators, *Annali di Mat. pura et appl.*, 53, 63–87 (1961).

On the lifting property, I, *J. Math. Analysis and Appl*, 3, 573–546 (1961).

On the lifting property, II, *Representation of linear operators* on L^r_E, $1 \leqslant r < \infty$, *J. Math. Anal.*, 70, 193–197 (1964).

On the lifting property, III, *Bull. Amer. Math. Soc.*, 70, 193–197 (1964).
Ionescu Tulcea, C.,
Spaţii Hilbert, Bucureşti (1956).
Deux théorèmes concernant certains espaces de champs de vecteurs, *Bull. Sci. Math.*, 79, 106–111 (1955).
Sur certaines classes de fonctions de type positif, *Ann. Ec. Norm. Sup.*, 74, 231–248 (1957).
Kakutani S.,
Concrete representation of abstract (L)-spaces and the mean ergodic theorem, *Annals of Math.*, (2), 42, 523–537 (1941).
Concrete representation of abstract (M)-spaces (A characterisation of the space of continuous functions) *Annals of Math.*, (2), 42, 994–1024 (1941).
Kolmogoroff, A.,
Grundbegriffe der Wahrscheinlichkeitsrechnung, Berlin, 1933.
Krasnoselski, M. A., Rutitki, Ia., B.,
On the theory of Orlicz spaces. (Russian), *Doklady Akad. Nauk SSSR* 81, 497–500 (1951).
Integral linear operators in Orlicz spaces (Russian), *Doklady Akad. Nauk. SSSR*, 85, 33–38 (1952).
Linear functional in Orlicz spaces (Russian), *Doklady Akad. Nauk. SSSR*, 4, 68–124 (1954)
Convex functions and Orlicz spaces (Russian), Moskow, 1958.
Lebesgue, H.,
Leçons sur l'intégration et la recherche des fonctions primitives, Paris (1904).
Loomis, L. H.,
An introduction to abstract harmonic analysis, New York, (1953).
Lorentz, G. G.,
On the theory of spaces Λ, *Pacific J. Math.*, 1, 411–429 (1951).
Some new functional spaces, *Annals of Math.*, (2), 51, 37–55 (1950).
Lorentz, G. G., Wertheim, D. G.,
Representation of linear functionals on Köthe spaces, *Canadian J. Math.*, 5, 568–575 (1953).
Luxemburg, W. A. J.,
Banach function spaces, Thesis, Delft, (1955).
Luxemburg, W. A. J., Zaanen, A. C.,
Conjugate spaces of Orlicz spaces, *Indagationes Math.*, 18, 217–228 (1956).

Some remarks on Banach function spaces, *Indagationes Math.*, 18, 110–119 (1956).

Maharam, D.,

On two theorems of Jessen, *Proc. Amer. Math. Soc.*, 9, 995–999 (1958).

On a theorem of von Neumann, *Proc. Amer. Math. Soc.*, 9, 987–994 (1958).

Marcus, S.,

Atomic measures and Darboux property, *Revue math. pures et appl.*, 7, 327–332 (1962).

La mesure de Jordan et l'intégrale de Riemann dans un espace mesuré topologique, *Acta. Sci. Math. Szeged*, 20, 156–163 (1959).

Marczewski E.,

On compact measures, *Fund. Math.*, 40, 113–124 (1953).

Marinescu, G.,

Spaţii vectoriale normate, Buruceşti (1956).

Spaţii vectoriale topologice si pseudotopolotice, Bucureşti (1959).

Espaces vectoriels pseudotopologiques et théorie des distributions, Berlin (1963).

Morse, M.,

Bilinear functionals over $C \times C$, *Acta Sci. Math. Szeged*, 12, 41–48 (1950).

Bimeasures and their integral extensions, *Annali di Mat. pura ed. appl.*, 39, 345–356 (1955).

Morse, M., Transue, W.,

The representation of a C-bimeasure on a general rectangle, *Proc. Nat. Acad. Sci.*, 42, 89–95 (1956).

C-bimeasures and their integral extensions, *Annals of Math.*, 64, 480–504 (1956).

Semi-normed vector spaces with duals of integral type *J.d'Analyse Math.*, 4, 149–186 (1954–1955).

C-bimeasures Λ and their superior integrals Λ^*, *Rendiconti Circolo Mat. Palermo*, 4, 270–300 (1955).

Products of a C-measure and a locally integrable mapping, *Canad. J. Math.*, 9, 475–486 (1957).

Nagy, B. von Sz.,

Spektraldarstellung linearer Transformationen des Hilbertschen Raumes, *Ergebnisse der Math.*, Berlin (1942).

Naimark, M. A.,

Normed rings (Russian), Moskow, (1956).

von Neumann, J.,
Zum Haarschen Mass in topologischen Gruppen, *Comp. Math.*, 1, 106–114 (1934).
Zur Operatorenmethode in der klassischen Mechanik, *Annals of Math.* (2) 33, 587–642 (1932).
Algebraische Repräsentanten der Funktionen bis auf eine Menge von Masse Null, *J. Crelle*, 165, 109–115 (1931).
Nicolescu, M.,
Analiza matematică, I, II, III, Bucureşti, 1957, 1958, 1960.
Functii reale şi analiză functională, Bucureşti, (1962).
Sur quelques propriétés élémentaires de la mesure de Jordan, *Bul. Fac. şt. Cernăuti*, 6, 222–224 (1932).
Sur les fonctions mesurables (*J*), *Bull. Sci. Math.*, Paris 57, 276–281 (1933).
Nikodym, O.,
Sur une gènèralisation des intégrales de M. J. Radon, *Fund. Math.* 15, 131–179 (1930).
Onicescu, O.,
Asupra unei integrale şi aplicaţiile ei, *Analele Univ. Bucuresti*, 13, 9–10 (1957).
Fonctions somme sur une *b*-algèbre, *Bull. Math. Soc. Sci. Math. Phys. R.P.R.*, 3 (51), 77–91 (1959).
Note asupra *b*-algebrelor, *Analele Univ. Bucureşti*, 22, 17–22 (1959).
Onicescu, O., Mihoc, G., Ionescu Tulcea, C.,
Calculul probabilităţilor şi aplicaţii, Bucureşti, (1956).
Orlicz, W.,
Über eine gewisse Klasse von Räumen von Typus B, *Bull. Int. Acad. Sci. Polon. Sci. Ser. A.*, 207–220 (1932).
Pettis, B. J.,
On integration in vector spaces, *Trans. Amer. Math. Soc.*, 44, 277–304 (1938).
Linear functionals and completely additive set functions, *Duke Math. J.*, 4, 552–565 (1938).
Phillips, R. S.,
Integration in a convex linear topological space, *Trans. Amer. Math. Soc.*, 47, 114–145 (1940).
On linear transformations, *Trans. Amer. Math. Soc.*, 48, 516–541 (1940).
Plancherel, M.,

Integraldarstellungen willkürlicher Funktionen, *Math. Ann.*, 67, 519–534 (1909)

Pontriagin, L. S.,
Continuous groups (Russian), Moskow (1956).

Price, G. B.,
The theory of integration, *Trans. Amer. Math. Soc.*, 47, 1–50 (1940).

Radon, J.,
Theorie und Anwendungen der absolut additiven Mengenfunktionen, *S-B, Akad. Wiss.* Wien, 122, 1295–1438 (1913).

Radu, E.,
Măsuri Stieltjes vectoriale, *Bull. Soc. Sci. Math. Phys. RPR*, 6 (54), 79–86 (1962).

Raikov, D. A.,
On the functions of positive type (Russian), *Doklady Akad. Nauk, SSSR*, 26, 857–862 (1940).
Harmonic analysis of commutative groups with Haar measure; theory of characters (Russian), *Trudy Mat. Inst. Steklova*, 14, 1–86 (1945).

Rickart, C. E.,
Integration in a convex linear space, *Trans. Amer. Math. Soc.*, 52, 498–521 (1942).
An abstract Radon–Nikodym theorem, *Trans. Amer. Math. Soc.* 56, 50–66 (1944).
Decomposition of additive set functions, *Duke Math. J.*, 10, 653–665 (1943).

Riesz, F.,
Sur les opérations fonctionnelles linéaires, *C.R. Acad. Sci.*, Paris, 149, 974–977 (1909).
Sur la représentation des opérations fonctionnelles linéaires par des intégrales de Stieltjes, *Proc. Roy. Physiog. Soc. Lund*, 21, 16, 145–151 (1952).
Sur la convergence en moyenne, I, II, *Acth Sci. Math. Szeged*, 4, 58–64, 182–185 (1928–1929).

Riesz, F., Sz.-Nagy, B.,
Leçons d'analyse fonctionnelle, Budapest, (1952).

Ryll-Nardzewski, C.,
On quasi-compact measures, *Fund. Math.*, 40, 125–130 (1953).

Saks, S.,
Theory of integral, Warsaw, (1937).

Schwartz, L.,
Théorie des distributions, I, II, Paris (1951).

Segal, I. E.,
Invariant measures on locally compact spaces, *J. Indian Math. Soc.*, 13, 105–130 (1949).
A non-commutative extension of abstract integration, *Annals. of Math.*, 57, 401–457 (1953).

Sîmboan, G.,
Vectorial measures, *Revue math. pures et appl.*, 7, 383–415 (1962).
Asupra măsurilor compacte, *Com. Acad. R.P.R.*, 9, 105–110 (1959).
Măsuri în spaţii topologice ordonate, *Com. Acad. R.P.R.* 9, 237–243 (1959).

Singer, I.,
Linear functionals on the space of continuous mappings of a compact Hausdorff space into a Banach space, *Revue math. pures et appl.*, 2, 301–305 (1957).
Les duals de certains espaces de Banach de champs de vecteurs, I, II, *Bull. Sci. Math.* 82, 29–40 (1958); 83, 93–96 (1959).
Sur les applications linéaires intégrales de espaces de fonctions continus, I, *Revue math. pures et appl.*, 4, 391–401 (1959).
Sur les applications linéaires majorées de espaces de fonctions continues, *Rendiconti Accad. Naz. Lincei*, 27, 35–41 (1959).
Sur la représentation intégrale des applications linéaires continues des espaces $L_F^p (1 \leqslant p < +\infty)$, *Rendiconti Accad. Naz. Lincei*, 29, 28–32 (1960).
Sur une classe d'applications linéaires continues des espaces L_F^p $(1 \leqslant p < +\infty)$,*Ann. Ecole. Norm. Sup.*, 77, 235–256 (1960).
Sur les applications linéaires intégrales des espaces de fonctions continues à valeurs vectorielles, *Acta Sci. Math. Acad. Sci. Hung.*, 11, 3–13 (1960).

Stieltjes,
Recherches sur les fractions continues, *Ann. Fac. Sci*, Toulouse, (1), 8, 1–22 (1894).

Stone, M. H.,
Applications of the theory of Boolean rings to general topology, *Trans. Amer. Math. Soc.*, 41, 375–481 (1937).
The generalized Weierstrass approximation theorem, *Math. Mag.*, 21, 167–184, 237–254 (1947–1948).

Notes on integration, I–IV, *Proc. Nat. Acad. Sci. U.S.A.*, 34, 336–342, 447–455, 483–490 (1948); 35, 50–58 (1949).

The theory of representation for Boolean algebras, *Trans. Amer. Mat. Soc.*, 40, 37–111 (1936).

On the foundations of harmonic analysis, *Proc. Roy. Physiog. Soc. Lund.*, 21, 17, 152–172 (1952).

Takahashi, T.,

On the compactness of the function-set by the convergence in mean of general type, *Studia Math.*, 5, 141–150 (1934).

Topping D. M.,

Lebesgue spaces of summable functions, *Proc. Amer. Nath. Soc.*, 12, 773–777 (1961).

Vasiliu C.,

Asupra teoremei de descompunere a lui Lebesgue, *Com. Acad. R.P.R.*, 13, 863–869 (1936).

Vulih B. Z.,

Introduction in the theory of the ordered spaces, (Russian) Moscow, 1961.

Weil, A.,

L'intégration dans les groupes topologiques et ses applications, Paris (1953).

Sur les groupes topologiques et les groupes de mesures, *C.R. Acad. Sci.*, Paris, 202, 1147–1149 (1936).

Zaanen, A. C.,

On a certain class of Banach spaces, *Annals of math.* (2), 47, 654–666 (1946).

Integral transformations and their resolvent in Orlicz and Lebesgue spaces, *Compositio Math.*, 10, 56–94 (1952).

Note on a certain class of Banach spaces, *Nederl. Akad. Wetensch. Proc.*, 52, 448–499 (1949).

Linear Analysis, New York–Amsterdam (1953).

SUBJECT INDEX

INDEX OF SYMBOLS